Biofuel and Bioenergy Technology

Biofuel and Bioenergy Technology

Special Issue Editors

Wei-Hsin Chen
Keat Teong Lee
Hwai Chyuan Ong

MDPI • Basel • Beijing • Wuhan • Barcelona • Belgrade

MDPI

Special Issue Editors

Wei-Hsin Chen
National Cheng Kung University
Taiwan

Keat Teong Lee
Universiti Sains Malaysia
Malaysia

Hwai Chyuan Ong
Universiti Sains Malaysia
Malaysia

Editorial Office
MDPI
St. Alban-Anlage 66
4052 Basel, Switzerland

This is a reprint of articles from the Special Issue published online in the open access journal *Energies* (ISSN 1996-1073) from 2017 to 2019 (available at: https://www.mdpi.com/journal/energies/special_issues/biofuel_bioenergy)

For citation purposes, cite each article independently as indicated on the article page online and as indicated below:

LastName, A.A.; LastName, B.B.; LastName, C.C. Article Title. *Journal Name* **Year**, *Article Number*, Page Range.

ISBN 978-3-03897-596-0 (Pbk)
ISBN 978-3-03897-597-7 (PDF)

Contents

About the Special Issue Editors

Wei-Hsin Chen (Distinguished Professor) received a B.S. from the Department of Chemical Engineering, Tunghai University in 1988, completed his Ph.D. at the Institute of Aeronautics and Astronautics, National Cheng Kung University in 1993. After receiving his Ph.D., Dr. Chen worked in an iron and steel corporation as a process engineer for one and a half years (1994–1995). He joined the Department of Environmental Engineering and Science, Fooyin University in 1995 and was promoted to a full professor in 2001. In 2005, he moved to the Department of Marine Engineering, National Taiwan Ocean University. Two years later (2007), he moved to the Department of Greenergy, National University of Tainan. Currently, he is a faculty member and distinguished professor at the Department of Aeronautics and Astronautics, National Cheng Kung University. Professor Chen visited Princeton University, USA, from 2004 to 2005, the University of New South Wales, Australia, in 2007, the University of Edinburgh, UK, in 2009, the University of British Columbia, Canada, from 2012 to 2013, and the University of Lorraine, France, in 2017 and 2019 as a visiting professor and invited lecturer. His teaching courses at the National Cheng Kung University include Bioenergy, Materials Engineering and Science, Energy Experiments, and Engineering Mathematics. His research topics include bioenergy, hydrogen energy, clean energy, carbon capture, and aerosol physics. He has published over 500 papers in international and domestic journals and conferences. He is the editor, associate editor, guest editor, and editorial member of a number of international journals, including *Applied Energy, Energy Conversion and Management, International Journal of Hydrogen Energy, International Journal of Energy Research, and Energies*. He is also the author of several books concerning energy science and air pollution. His important awards include the 2015 Outstanding Research Award (Ministry of Science and Technology, Taiwan), the 2015 Highly Cited Paper Award (*Applied Energy*, Elsevier), the 2017 Outstanding Engineering Professor Award (Chinese Institute of Engineers, Taiwan) as well as the 2016, 2017, and 2018 Highly Cited Researcher Award (Web of Science).

Keat Teong Lee (Professor. Dr.) obtained his Ph.D in Chemical Engineering from Universiti Sains Malaysia (USM) in 2004. He currently holds the position of Professor at the School of Chemical Engineering, and he is also the Director of the Research Creativity & Management Office, Universiti Sains Malaysia. Dr. Lee has co-authored two books, 10 book chapters, 30 review papers and more than 100 research papers in peer reviewed international journals. Dr. Lee is currently the Co-Editor for *Energy Conversion and Management* (Elsevier) and an Editorial Board Member for *Bioresource Technology* (Elsevier) and *Energy Science & Engineering* (Wiley). He has also won numerous awards including the Young Scientist Award 2011 from The International Forum on Industrial Bioprocess and the 2012 Top Research Scientists Malaysia Award from the Academy Sciences of Malaysia. He is one of the four Malaysian researchers to be recognized as one of the most cited researchers in the latest Shanghai Academic Ranking of World Universities 2016 by Subjects (Energy Science & Engineering). Dr. Lee is now working on the production of biofuels (biodiesel and bioethanol) from biomass (including macro and microalgae) using various technologies. Apart from that, he also has a special interest in the social and sustainability aspects of biofuels.

Hwai Chyuan Ong (Dr.) obtained his B.Eng. (Hons.) in Mechanical Engineering from the Faculty of Engineering, University of Malaya, with distinction. Then, he obtained his Ph.D. in Mechanical Engineering from the same university in December 2012. His research interests are wide-ranging under the general umbrella of renewable energy. In particular, these include biofuel and bioenergy, solar thermal, and green technology and environmental science. He is currently appointed as a Senior Lecturer at the Department of Mechanical Engineering, University of Malaya. He is also a Chartered Engineer of Engineering Council (CEng) under the Institution of Mechanical Engineers (IMechE), United Kingdom. He has published more than 100 high impact SCI journal papers with an H-index of 25 (WOS). In 2017 and 2018, he received the Malaysia's Research Star Award (frontier researcher) and in 2016, he received the Malaysia's Rising Star Award (young researcher) from the Ministry of Higher Education and Clarivate Analytics. In 2018, he also received the Outstanding Research Award and the most Highly Cited Paper Award at the University of Malaya Excellence Awards. Currently, he is an Associate Editor of the *Journal of Renewable and Sustainable Energy* (IF: 1.135) and was a Guest Editor in *Energies* (SCI, IF = 2.262) in the Special Issues "Biofuel and Bioenergy Technology" and "Biomass Processing for Biofuels, Bioenergy and Chemicals".

energies

MDPI

Editorial

Biofuel and Bioenergy Technology

Wei-Hsin Chen [1],*, Keat Teong Lee [2],* and Hwai Chyuan Ong [3],*

[1] Department of Aeronautics and Astronautics, National Cheng Kung University, Tainan 701, Taiwan
[2] School of Chemical Engineering, Universiti Sains Malaysia, Nibong Tebal 14300, Pulau Pinang, Malaysia
[3] Department of Mechanical Engineering, University of Malaya, Kuala Lumpur 50603, Malaysia
* Correspondence: chenwh@mail.ncku.edu.tw or weihsinchen@gmail.com (W.-H.C.); ktlee@usm.my (K.T.L.);
   onghc@um.edu.my (H.C.O.)

Received: 24 December 2018; Accepted: 10 January 2019; Published: 18 January 2019

1. Introduction

Biomass is considered as a renewable resource because of its short life cycle, and biomass-derived biofuels are potential substitutes to fossil fuels. When biomass grows, all carbon in biomass comes from the atmosphere and is liberated into the environment when it is burned. Therefore, biomass is thought of as a carbon-neutral fuel. For these reasons, the development of bioenergy is an effective countermeasure to elongate fossil fuel reserves, lessen greenhouse gas (GHG) emissions, and mitigate global warming and climate change. Biomass can be converted into biofuels through a variety of routes such as physical, thermochemical, chemical, and biological methods. The common and important biofuels for bioenergy include charcoal, biochar, biodiesel, bioethanol, biobutanol, pyrolysis and liquefaction bio-oils, synthesis gas (syngas), biogas, and biohydrogen, etc. On account of the merit of bioenergy for environmental sustainability, biofuel and bioenergy technology plays a crucial role for renewable energy development. This Special Issue aims to publish high-quality review and research papers, addressing recent advances in biofuel and bioenergy. State-of-the-art studies of advanced techniques of biorefinery for biofuel production are also included. Research involving experimental studies, recent developments, and novel and emerging technologies in this field are covered. The particular topics of interest in the original call for papers included, but were not limited to:

- Novel and unexploited biomass resources for biofuel and bioenergy production
- New emerging technologies for biofuel and bioenergy production
- Development of thermochemical conversion routes for biofuel and bioenergy produciton
- Advanced biorefinery processes for biofuel and biochemicals production
- Bioreactors or microbial fuel cell for bioenergy and power production
- State-of-the-art review in the progress of biofuel and bioenergy technology

This Special Issue of *Energies* on the subject of "Biofuel and Bioenergy Technology" contains the successful invited submissions [1–27]. A total of twenty-seven technical papers which cover diversified biofuel and bioenergy technology related researches have shown critical results and contributed significant findings in biomass processing [1,2], bio-oil and biodiesel [3–11], syngas [12–14], biogas/methane [15–19], bioethanol and alcohol-based fuels [20–22], solid fuel [23–25] and also microbial fuel cell [13,26,27] developments.

2. Statistics of the Special Issue

The response to our call had the following statistics:

- Submissions (46);
- Publications (27);
- Rejections (19);

- Article types: research articles (25); review articles (2).

  The authors' geographical distribution (published papers) is:

- Taiwan (8);
- Korea (4);
- Czech Republic (3)
- Australia (3);
- USA (2);
- China (1);
- Malaysia (1);
- Mexico (1);
- Pakistan (1);
- Poland (1);
- Spain (1);
- The Netherlands (1).

Published submissions are related to the most important techniques and analysis applied to the biofuel and bioenergy technology. In summary, the edition and selections of papers for this special issue are very inspiring and rewarding. We thank the editorial staff and reviewers for their efforts and help during the process.

3. Brief Overview of the Contributions to This Special Issue

Table 1 provides some of the key information, including the research type, field of study, final product as well as the key findings. As observed, a majority of the publications (twenty-three papers) focus on experimental work to improve or explore novel technologies for energy-products synthesis, while three papers focus on modelling studies and two papers focus on literature review studies. The following discussion highlights and groups the research findings in accordance to the corresponding research field or work.

As the initial step in most synthesis routes, biomass processing can enhance the substrate's quality for other synthesis processes. Thus, commonly, these are treated as pretreatment to enhance the characteristics of the biomass. In two research works [1,2], the combination of physical treatment (ball milling) and chemical treatment (ethanol organosolv) showed improved glucan digestibility. Three different biomasses such as giant miscanthus, corn stover and wheat straw were pretreated with ball milling and ethanol organosolv and the overall biomass size was reduced as a result of the prolonged pre-treatment [1]. Due to the improved physicochemical characteristics resulting from the pre-treatment, a maximum of 91% glucan digestibility could be achieved. A parametric study on combined ball milling and organosolv was performed as well to optimize the glucan digestibility [2]. It was determined that at 170 °C, with reaction time of 90 min and ethanol concentration of 40% and liquid/solid ratio of 10, the pretreatment process achieved the best results. Thus, the biomass processing method could be beneficial in generating desired products.

Table 1. Key Information of the Publications Submitted to Special Issue.

Research Work	Research Type			Technology/ Field of Work	Product	Key Findings
	Experimental	Modelling	Review			
Anwar et al., 2018 [5]	×			Blending	Biodiesel blend of papaya seed oil	• Reduction in brake power, torque and brake thermal efficiency. • Significant effect on brake specific fuel consumption.
Anwar et al., 2018 [6]	×			Alkali-catalysed transesterification	Australian native stone fruit biodiesel	• Optimisation with response surface methodology. • Maximum biodiesel yield of 95.8%. • Met ASTM D6751 and EN14214 standards. • Potential second-generation biodiesel.
Bidabadi et al., 2018 [25]		×		Mathematic asymptotic technique	–	• Oxidizer and fuel Lewis number were between 0.4 and 1, the maximum flame temperature was ~1860 K. • Per unit of fuel Lewis number, the minimum thermophoretic force was -1.48×10^{-8} N. • Per unit of oxidizer Lewis number, the minimum thermophoretic force was -1.53×10^{-8} N. • Per unit of porosity factor, the minimum thermophoretic force was -1.28×10^{-8} N.
Brunerová et al., 2018 [24]	×			High-Pressure Densification	Bio-Briquette Fuel	• Low ash content for bamboo fibre (1.16%) and sugarcane skin (8.62%). • Satisfactory mechanical durability for bamboo fibre (97.80%) and sugarcane skin (97.70%). • These products can be used for bio-briquette fuel production.

Table 1. *Cont.*

Research Work	Research Type			Technology/ Field of Work	Product	Key Findings
	Experimental	Modelling	Review			
Černý et al., 2018 [15]	x			Biogas study with DNA analysis	Biogas (Hydrogen)	• Occurrence of potentially harmful microorganisms such as *Clostridium novyi* was detected at higher ratio (65.63%) in the population of the bioreactor.
Chein et al., 2018 [12]	x			Tri-Reforming Process	Syngas (hydrogen)	• First-Law Efficiency increased with increased reaction temperature for higher hydrogen and carbon monoxide yields. • Second-Law Efficiency decreased with increased reaction temperature due to more complete chemical reaction.
Chen et al., 2018 [3]	x			Pyrolysis	Pyrolytic Oil	• Optimisation with Taguchi Method. • Maximum pyrolytic oil yield of 10.19%. • Synthesis conditions: 450 °C, 60 min, 10 °C/min and nitrogen flow of 700 mL/min.
Chen et al., 2018 [26]	x			Microbial Fuel Cell	-	• Hydrodynamic boundary layer of 1.6 cm (thin layer) showed maximum voltage of 22 mV and charged transfer resistance of 39 Ω.
David et al., 2018 [18]	x			Thermophilic anaerobic digestion	Methane	• Food wastes (corn stover, prairie cordgrass and unbleached paper) undergone thermophilic anaerobic digestion. • Highest methane yield of 305.45 L/kg was achieved after 30 days of incubation at 60 °C at 100 rpm.

Table 1. *Cont.*

Research Work	Research Type			Technology/ Field of Work	Product	Key Findings
	Experimental	Modelling	Review			
Dziekońska-Kubczak et al., 2018 [20]	×			Nitric acid pretreatment for enzymatic hydrolysis and fermentation	Bioethanol	• Jerusalem artichoke stalks were converted into bioethanol with nitric acid as catalyst. • Nitric acid pretreated hydrolysates led to 30% improvement in ethanol yield (77–82% of theoretical yield).
Encinar et al., 2018 [7]	×			Transesterification with base-catalysed reactions	Biodiesel	• Ultrasonic accelerated rate of biodiesel transesterification reactions. • Reaction followed a pseudo-first order kinetic model.
Eri et al., 2018 [14]		×		Equilibrium constants modelling	-	• Simulations were performed with two different models (with and without tar). • The simulations were validated by experimental data.
Fernedas et al., 2018 [13]	×	×		Gasifier-Specific Solid Oxide Fuel Cell System	-	• Validation data showed good agreement between experimental and simulation data. • System efficiencies were estimated to be 33.7–34.5%.
Kim et al., 2018 [10]	×			Photobioreactor with coal-fired flue-gas	Microalgal biodiesel	• M082 strain showed maximum lipid content (397 mg fatty acid methyl ester (FAME)/g cell) with good tolerance to high temperature. • FAME produced met the international standards.

Table 1. *Cont.*

Research Work	Research Type		Technology/ Field of Work	Product	Key Findings
	Experimental	Modelling	Review		
Kim et al., 2018 [1]	x			Ball milling and ethanol organosolv	-	• Combined pretreatment on giant miscanthus, corn stover and wheat straw show varied results (increased of glucan content for giant miscanthus, removal of cellulose for corn stover). • Enzymatic digestibility was improved with 91% glucan digestibility.
Kim et al., 2018 [2]	x			Ball milling and ethanol organosolv	-	• Pretreatment was performed using a 30 L bench-scale ball mill reactor. • Pretreatment conditions were varied: room temperature to 170 °C, time from 30 to 120 min, ethanol concentration from 30% to 60%, liquid/solid ratio from 10 to 20. • Highest glucan digestibility was performed at 170 °C, reaction time to 90 min, 40% of ethanol concentration and L/S = 10.
Kuan et al., 2018 [8]	x			Transesterification	Biodiesel	• Acid-catalysed synthesis by 0.6 M sulphuric acid at 70 °C for 20 h yielded 111% of FAME. • Base-catalysed synthesis by 1.0 g/L of sodium hydroxide at 70 °C for 10 h yielded 102% of FAME. • Direct transesterification shortened the reaction time and improved FAME yield.

Table 1. *Cont.*

Research Work	Research Type			Technology/ Field of Work	Product	Key Findings
	Experimental	Modelling	Review			
Längauer et al., 2018 [4]	x			Simultaneous Extraction and Emulsification	Emulsified bio-oil	• Emulsified ratio (bio-oil to emulsifier, B/E ratio) at 1 showed higher solubility of 66.48 wt %. • At higher temperature, higher solubility was also observed. • Methanol as co-surfactant also improved better solubility from 58.83 to 70.96 wt %.
Li et al., 2018 [21]	x			Electrochemical Hydrogenation using polymer electrolyte membrane reactor	Isopropanol	• Polymer electrolyte membrane fuel cell was used to produce isopropanol as main product and diisopropyl ether as byproduct. • High selectivity and (>90%) and high current efficiency (59.7%) were observed at mild conditions of 65 °C and at atmospheric pressure.
Musa et al., 2018 [19]			x	Anaerobic Membrane Bioreactors (AnMBRs)	-	• Anaerobic digestion technologies were critically reviewed. • Factors on membrane fouling, microbial environment conditions as well as parameters on the operations of AnMBRs were discussed. • Microfiltration as the mean to reduce energy and water usage in the AnMBRs was suggested.
Nguyen et al., 2018 [11]	x			Liquid Lipase Catalyzed Esterification	Biodiesel	• Optimisation with Response Surface Methodology • Superadsorbent polymer (SAP), as water removal agent, was used in esterification. • The polymer improved the conversion to 96.73% at 35.25 °C, methanol to oleic acid molar ratio of 3.44:1, SAP loading of 10.55% and enzyme loading of 11.98%.

Table 1. *Cont.*

Research Work	Research Type		Technology/ Field of Work	Product	Key Findings
	Experimental	Modelling	Review		

Research Work	Experimental	Modelling	Review	Technology/ Field of Work	Product	Key Findings
Poudel et al., 2018 [23]	x			Torrefaction	Torrefied Biomass	• Wood waste was torrefied at 200–400 °C and 0–50 min. • 300 °C as the optimal temperature for torrefaction based on Van Krevelen diagram.
Rahman et al., 2018 [22]			x	Bio-hydrocarbon Production in Bacteria	-	• Bioenergy products (alcohols and *n*-alkene hydrocarbons (C_2 to C_{18}) as produced by engineered microorganisms showed promising energy potential. • The review discussed the complexity of metabolic networks to obtain these bio-hydrocarbon products.
Roubík, et al., 2018 [16]	x			Biogas Plant Study	Biogas (methane)	• Biogas composition was measured for 107 small-scale biogas plants, respectively. • Mean compositions as follows: For plants younger than 5 years, CH_4 was 65.44% and CO_2 was 29.31%; for plants older than 5 years, CH_4 was 64.57% and CO_2 was 29.93%.
Su et al., 2018 [9]	x			Two-step acid-catalysed transesterification	Biodiesel	• Soursop (*Annona muricata*) seeds were used to produce bio-oil (29.6% (*w/w*)). • Bio-diesel with highest of 97.02% was produced under acid-catalysed conditions of 65 °C, 1% sulphuric acid, reaction time of 90 min and methanol: oil ratio of 10:1 and under base-catalysed conditions of 65 °C, 0.6% NaOH, reaction time of 30 min and methanol: oil ratio of 8:1. • Produced biodiesel met the EN14214 and D6751 requirements.

Table 1. *Cont.*

Research Work	Research Type		Technology/ Field of Work	Product	Key Findings
	Experimental	Modelling	Review		
Valero et al., 2018 [17]	x			-	Biomethane	• Biochemical Methane Potential (BMP) showed that the addition of granular activated carbon (GAC) improved the methane yield by 34% for instance testing and 54% for 10 days of GAC biofilm development. • Addition of GAC can improve digester's anaerobic digestion performance.
Wu et al., 2018 [27]	x			Microbial Fuel Cell	-	• Different calcination temperatures (500–900 °C) of iron oxide (Fe_2O_3) were tested to investigate their photocatalytic properties within the cathodic chambers. • Calcinated Fe_2O_3 improved the bio-electro-Fenton microbial fuel cell (Bio-E-Fenton-MFC) on degrading oily wastewater. • Within one hour, oily water was best-degraded up to 99.3% with electrode material synthesised at 500 °C with maximum power density of 52.5 mW/m^2.

Bio-oils can be synthesized from sewage sludge by using pyrolysis techniques [3]. Taguchi optimization suggested the best pyrolysis was performed at 450 °C, 60 min and 10 °C/min, which also showed consistency with other research work. Nonetheless, under most conditions, pyrolytic oil/bio-oil requires further processing or upgrading for use as biodiesel. To maintain the stability of bio-oil, blending with emulsifier resulted in high solubility (58.83–70.96 wt %) [4]. These findings suggest that simple blending could improve the properties of biodiesel or bio-oil tremendously, which is worthy of further investigation. Aside from using bio-oil as a precursor for biodiesel [3], biodiesel could be directly synthesized using other oil materials such as Australian native stone fruit oil [6], rapeseed oil [7], *Rhodotorula glutinis* [8] and soursop seed oil [9] via transesterification techniques. Transesterification of Australian native stone biodiesel showed a high yield of 95.8% with the response surface methodology optimization and its quality fulfilled the ASTM D6751 and EN14214 requirements [6]. Kuan et al. [8] investigate both direct acid and base-transesterification on *Rhodotorula glutinis* biomass which gave 111% yield of FAME and 102% yield of FAME, respectively, which were regarded as of good biodiesel quality. Another research work by Su et al. [9] used soursop seed to produce bio-oil which was eventually upgraded to biodiesel using a two-step acid catalyzed transesterification. The biodiesel produced met both EN14214 and D6751 standards. Encinar et al. [7] performed rapeseed transesterification with KOH catalyst as well as with the aid of ultrasound whereby the kinetic behavior obeyed a pseudo-first order trend. Liquid lipase-based esterification was attempted and optimized using RSM to enhance the usage of water removal agent in the system [11]. In Anwar et al's, [5] work, it was found that by blending papaya oil biodiesel with varying contents (5–20%) with diesel could improve the engine testing properties. Microalgal biodiesel was generated using photobioreactor with coal-fired flue gas from three strains (M082, M134 and KR-1) [10]. Among the strains, M082 generated high lipid value of 397 mg/g which was regarded to be a suitable feedstock for biodiesel production.

In recent years, gasification also garners high interest due to its rapid processing step and high yield of syngas which could be directly used for combustion. One of the main constitutes of syngas is hydrogen which usually provides high calorific value. Based on Chein and Hsu's [12] work, the tri-reforming process could produce good quality syngas. In addition, it was also found that at higher reaction temperatures, more hydrogen and carbon monoxide were produced. In pilot plant study, the gasifier which was embedded with specific solid oxide fuel cell system in an industrial scale was investigated in details [13]. To further understanding the gasification process, a thermodynamic equilibrium constants derivation and modelling was performed for two cases, with and without tar. The simulated data were validated with experimental data [14]. These findings could serve as good guides for future development of gasification process.

Bio-digester or bioreactor could also be used for biogas production. Černý et al. [15] discovered that microorganisms such as *Clostridium novyi* were detected at higher ratio (65.63%) in the population of the bioreactor for the biogas production. Such detection could serve as an important reminder to seek ways to inhibit these harmful microorganisms in the system. In an investigation and survey of 107 biogas plants, it was also found that the younger plant (<5 years) produced higher CH_4 (65.44%) and CO_2 (29.31%) [16]. Addition of granular activated carbon (GAC) in the digestion system could also directly improve the methane production by more than 34% [17]. Thermophilic anaerobic digestion is another interesting field of research. David et al. [18] found that by co-digestion under such conditions, high yield of methane (up to 305.45 kg/L) could be achieved. Musa's work critically reviewed some of the more critical findings on anaerobic membrane reactors for biogas recovery especially on membrane fouling and parameters of operation [19].

The studies on alcohol-based biofuels are increasing due to its high energy-content and suitability as fuel products. Dziekońska-Kubczak et al. [20] report that nitric acid as a form of chemical pretreatment could enhance the bioethanol production up to 30%. A Polymer Electrolyte Membrane Fuel Cell was used to produce isopropanol from acetone for use as a biofuel [21]. The hydrogenation process consumes less energy and less chemical wastes compared to other techniques [21]. Bio-hydrocarbons (alcohol and alkenes) produced from bacteria and their synthesis mechanisms are reviewed by Rahman et al. [22], as well as future challenges and complexity.

High pressure densification and torrefaction are currently attracting attention among the research community as these methods can produce potential solid-based fuels which require no further upgrade and can be use directly. Both methods are usually applied in mild synthesis conditions which differ from common thermochemical conversion techniques like slow pyrolysis and fast pyrolysis. For example, biomass with low ash content (1.16–8.62%) and good mechanical durability (97%) such as bamboo fibre and sugarcane skin could be directly densified as bio-briquette fuel without any energy processing [24]. As for torrefaction, a form of mild pyrolysis, wood wastes could be converted into torrefied biomass as low as 300 °C [23]. As observed, both methods consume relatively lower energy requirement and are simpler in term of synthesis process. In a modelling study, *Lycopodium* particles were also simulated and modelled as biofuel and burned in air environment [25]. It was discovered that the particles of *Lycopodium* were greatly influenced by thermophoretic force. Microbial fuel cell studies were also being investigated thoroughly. The effect of the hydrodynamic layer thickness was found to be significant on the voltage and charged transfer resistance [26]. In another study, the calcination temperature on the cathodic chambers was studied and it was found that the electrode synthesized at 500 °C could degrade oily wastewater up to 99.3% [27]. Thus, the microbial fuel cell shows tremendous potential to be developed for other applications.

Conflicts of Interest: The authors declare no conflict of interest.

References

1.   Kim, S.; Um, B.; Im, D.; Lee, J.; Oh, K. Combined Ball Milling and Ethanol Organosolv Pretreatment to Improve the Enzymatic Digestibility of Three Types of Herbaceous Biomass. *Energies* **2018**, *11*, 2457. [CrossRef]
2.   Kim, T.; Im, D.; Oh, K.; Kim, T. Effects of Organosolv Pretreatment Using Temperature-Controlled Bench-Scale Ball Milling on Enzymatic Saccharification of Miscanthus × giganteus. *Energies* **2018**, *11*, 2657. [CrossRef]
3.   Chen, G.-B.; Li, J.-W.; Lin, H.-T.; Wu, F.-H.; Chao, Y.-C. A Study of the Production and Combustion Characteristics of Pyrolytic Oil from Sewage Sludge Using the Taguchi Method. *Energies* **2018**, *11*, 2260. [CrossRef]
4.   Längauer, D.; Lin, Y.-Y.; Chen, W.-H.; Wang, C.-W.; Šafář, M.; Čablík, V. Simultaneous Extraction and Emulsification of Food Waste Liquefaction Bio-Oil. *Energies* **2018**, *11*, 3031. [CrossRef]
5.   Anwar, M.; Rasul, M.; Ashwath, N. A Systematic Multivariate Analysis of Carica papaya Biodiesel Blends and Their Interactive Effect on Performance. *Energies* **2018**, *11*, 2931. [CrossRef]
6.   Anwar, M.; Rasul, M.; Ashwath, N.; Rahman, M. Optimisation of Second-Generation Biodiesel Production from Australian Native Stone Fruit Oil Using Response Surface Method. *Energies* **2018**, *11*, 2566. [CrossRef]
7.   Encinar, J.; Pardal, A.; Sánchez, N.; Nogales, S. Biodiesel by Transesterification of Rapeseed Oil Using Ultrasound: A Kinetic Study of Base-Catalysed Reactions. *Energies* **2018**, *11*, 2229. [CrossRef]
8.   Kuan, I.-C.; Kao, W.-C.; Chen, C.-L.; Yu, C.-Y. Microbial Biodiesel Production by Direct Transesterification of Rhodotorula glutinis Biomass. *Energies* **2018**, *11*, 1036. [CrossRef]
9.   Su, C.-H.; Nguyen, H.; Pham, U.; Nguyen, M.; Juan, H.-Y. Biodiesel Production from a Novel Nonedible Feedstock, Soursop (*Annona muricata* L.) Seed Oil. *Energies* **2018**, *11*, 2562. [CrossRef]
10. Kim, B.; Praveenkumar, R.; Choi, E.; Lee, K.; Jeon, S.; Oh, Y.-K. Prospecting for Oleaginous and Robust Chlorella spp. for Coal-Fired Flue-Gas-Mediated Biodiesel Production. *Energies* **2018**, *11*, 2026. [CrossRef]
11. Nguyen, H.; Huong, D.; Juan, H.-Y.; Su, C.-H.; Chien, C.-C. Liquid Lipase-Catalyzed Esterification of Oleic Acid with Methanol for Biodiesel Production in the Presence of Superabsorbent Polymer: Optimization by Using Response Surface Methodology. *Energies* **2018**, *11*, 1085. [CrossRef]
12. Chein, R.-Y.; Hsu, W.-H. Analysis of Syngas Production from Biogas via the Tri-Reforming Process. *Energies* **2018**, *11*, 1075. [CrossRef]
13. Fernandes, A.; Brabandt, J.; Posdziech, O.; Saadabadi, A.; Recalde, M.; Fan, L.; Promes, E.; Liu, M.; Woudstra, T.; Aravind, P. Design, Construction, and Testing of a Gasifier-Specific Solid Oxide Fuel Cell System. *Energies* **2018**, *11*, 1985. [CrossRef]

14. Eri, Q.; Wu, W.; Zhao, X. Numerical Investigation of the Air-Steam Biomass Gasification Process Based on Thermodynamic Equilibrium Model. *Energies* **2017**, *10*, 2163. [CrossRef]

15. Černý, M.; Vítězová, M.; Vítěz, T.; Bartoš, M.; Kushkevych, I. Variation in the Distribution of Hydrogen Producers from the Clostridiales Order in Biogas Reactors Depending on Different Input Substrates. *Energies* **2018**, *11*, 3270. [CrossRef]

16. Roubík, H.; Mazancová, J.; Le Dinh, P.; Dinh Van, D.; Banout, J. Biogas Quality across Small-Scale Biogas Plants: A Case of Central Vietnam. *Energies* **2018**, *11*, 1794. [CrossRef]

17. Valero, D.; Rico, C.; Canto-Canché, B.; Domínguez-Maldonado, J.; Tapia-Tussell, R.; Cortes-Velazquez, A.; Alzate-Gaviria, L. Enhancing Biochemical Methane Potential and Enrichment of Specific Electroactive Communities from Nixtamalization Wastewater using Granular Activated Carbon as a Conductive Material. *Energies* **2018**, *11*, 2101. [CrossRef]

18. David, A.; Govil, T.; Tripathi, A.; McGeary, J.; Farrar, K.; Sani, R. Thermophilic Anaerobic Digestion: Enhanced and Sustainable Methane Production from Co-Digestion of Food and Lignocellulosic Wastes. *Energies* **2018**, *11*, 2058. [CrossRef]

19. Musa, M.; Idrus, S.; Che Man, H.; Nik Daud, N. Wastewater Treatment and Biogas Recovery Using Anaerobic Membrane Bioreactors (AnMBRs): Strategies and Achievements. *Energies* **2018**, *11*, 1675. [CrossRef]

20. Dziekońska-Kubczak, U.; Berłowska, J.; Dziugan, P.; Patelski, P.; Pielech-Przybylska, K.; Balcerek, M. Nitric Acid Pretreatment of Jerusalem Artichoke Stalks for Enzymatic Saccharification and Bioethanol Production. *Energies* **2018**, *11*, 2153. [CrossRef]

21. Li, C.; Sallee, A.; Zhang, X.; Kumar, S. Electrochemical Hydrogenation of Acetone to Produce Isopropanol Using a Polymer Electrolyte Membrane Reactor. *Energies* **2018**, *11*, 2691. [CrossRef]

22. Rahman, Z.; Nawab, J.; Sung, B.; Kim, S. A Critical Analysis of Bio-Hydrocarbon Production in Bacteria: Current Challenges and Future Directions. *Energies* **2018**, *11*, 2663. [CrossRef]

23. Poudel, J.; Karki, S.; Oh, S. Valorization of Waste Wood as a Solid Fuel by Torrefaction. *Energies* **2018**, *11*, 1641. [CrossRef]

24. Brunerová, A.; Roubík, H.; Brožek, M. Bamboo Fiber and Sugarcane Skin as a Bio-Briquette Fuel. *Energies* **2018**, *11*, 2186. [CrossRef]

25. Bidabadi, M.; Ghashghaei Nejad, P.; Rasam, H.; Sadeghi, S.; Shabani, B. Mathematical Modeling of Non-Premixed Laminar Flow Flames Fed with Biofuel in Counter-Flow Arrangement Considering Porosity and Thermophoresis Effects: An Asymptotic Approach. *Energies* **2018**, *11*, 2945. [CrossRef]

26. Chen, Y.-M.; Wang, C.-T.; Yang, Y.-C. Effect of Wall Boundary Layer Thickness on Power Performance of a Recirculation Microbial Fuel Cell. *Energies* **2018**, *11*, 1003. [CrossRef]

27. Wu, J.-C.; Yan, W.-M.; Wang, C.-T.; Wang, C.-H.; Pai, Y.-H.; Wang, K.-C.; Chen, Y.-M.; Lan, T.-H.; Thangavel, S. Treatment of Oily Wastewater by the Optimization of Fe2O3 Calcination Temperatures in Innovative Bio-Electron-Fenton Microbial Fuel Cells. *Energies* **2018**, *11*, 565. [CrossRef]

energies

MDPI

Article

Numerical Investigation of the Air-Steam Biomass Gasification Process Based on Thermodynamic Equilibrium Model

Qitai Eri [1], Wenzhen Wu [1] and Xinjun Zhao [1,2,*]

[1] School of Energy and Power Engineering, Beihang University, Beijing 100191, China; eriqitai@buaa.edu.cn (Q.E.); wenzhenwu@yeah.net (W.W.)

[2] The 41st Research Institute, the Sixth Academy of CASIC, Hohhot 010010, China

[*] Correspondence: kunpengzhao@buaa.edu.cn; Tel.: +86-010-8233-9866

Received: 20 November 2017; Accepted: 14 December 2017; Published: 18 December 2017

Abstract: In the present work, the air-steam biomass gasification model with tar has been developed based on the equilibrium constants. The simulation results based on two different models (with and without tar) have been validated by the experimental data. The model with tar can well predict the tar content in gasification; meanwhile, the predicted gas yield (GY), based on the model with tar, is much closer to the experimental data. The energy exchange between the gasifier and the surrounding has been studied based on the dimensionless heat transfer ratio (DHTR), and the relationship between DHTR and the process parameters is given by a formula. The influence of process parameters on the syngas composition, tar content, GY, lower heating value (LHV), and exergy efficiency have been researched.

Keywords: air-steam gasification; equilibrium model; tar; energy exchange; exergy efficiency

1. Introduction

Biomass gasification is a thermochemical conversion method [1,2], which converts biomass into combustible gases through partial oxidation. Decomposition, partial oxidation, and various other chemical reactions occur during gasification, involving heat and mass transfer processes. Biomass gasification has gained attention worldwide due to its capacity to handle a wide range of biomass feedstocks and its high efficiency of energy conversion [3–5].

Due to its inherent complexity, simulating the gasification process and predicting the process performance is still at a preliminary stage [6]. In the simulation of gasification, thermodynamic equilibrium modeling does not consider the mechanism of transformation; meanwhile, it is independent of gasifier geometry and not limited to a specified range of operating conditions. Equilibrium model represents the gasification process as a single reaction, so it is useful to predict the exit gas composition and estimate the trend of the gasifier output with variations in process parameters [7].

Equilibrium models can be categorized into stoichiometric models based on equilibrium constants of the most important reactions [8] and non-stoichiometric models based on minimization of the Gibbs free energy [9–13]. Some researchers [14–16] simulated the gasification process based on non-stoichiometric models, whereas other researchers [17–19] paid more attention to stoichiometric models. However, both the approaches are essentially equivalent [20]. Air, steam, and air-steam are the generally used gasifying agents in gasification. Air is comprehensively used [21,22], since it is cheap. Steam gasification was found to produce gas with a high heating value [23,24] compared to air gasification due to the high content of H_2 in the syngas. The pure steam gasification process requires an indirect or external heat supply for endothermic gasification reaction, which increases the cost of gasification. Alternatively, air-steam gasification [25,26] decreases the cost, since the partial combustion reaction provides the heat. Meanwhile, the heating value of the syngas is still acceptable, so this process has received a lot of attention.

In the air-steam gasification process, when the energy balance equation is introduced to the equilibrium model, the value of energy exchange between the gasifier and the surrounding should be given. If this value is inaccurate, the simulation results are very likely to be misleading. However, this value is usually determined through experiments, which causes the data to be insufficient. If the relationship between the energy exchange and the process parameters can be determined by a formula, the value will be obtained accurately and conveniently. Tar is generally neglected in the simulation study based on the stoichiometric models [18,27]. In fact, tar is invariably present in the gasification process, especially at low gasification temperatures [28,29]. The components of tar are very complex, and the primary tars are generally obtained at temperatures below 973 K. As temperature increases, primary tars continue to decompose with the formation of secondary tars, comprising phenolics and olefins [30]. In some simulations of gasification, phenol was used as the model compound for tars [31].

Hence, in this study, the air-steam biomass gasification model with tar has been developed based on the equilibrium constants, and phenol is used as the model compound for tars. The effects of process parameters on the syngas composition, tar content, gas yield, the lower heating value of the syngas, and the exergy efficiency of gasification have been studied. Moreover, the relationship between the energy exchange and the process parameters has also been researched.

2. Thermodynamic Equilibrium Model

For the present air-steam thermodynamic equilibrium model, some assumptions have been made for the biomass gasification process: (1) temperature of the products is equal to the gasification temperature; (2) oxygen is consumed completely; (3) the gas mixture is in equilibrium and is homogeneous; (4) the residence time is long enough, and all the chemical reactions can reach an equilibrium state; (5) nitrogen does not take part in the reactions; (6) the biomass is composed of C, H, and O; (7) the solid products include only the char, which is composed of entirely of carbon; (8) ash is not considered in the simulation; (9) phenol is regarded as tar; (10) atmospheric pressure is assumed in the gasifier.

The overall gasification reaction with air and steam can be written as follows:

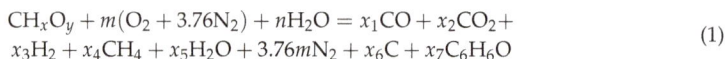

$$CH_xO_y + m(O_2 + 3.76N_2) + nH_2O = x_1CO + x_2CO_2 + \\ x_3H_2 + x_4CH_4 + x_5H_2O + 3.76mN_2 + x_6C + x_7C_6H_6O \tag{1}$$

where x and y are the number of atoms of hydrogen and oxygen for each atom of carbon in the biomass, respectively. m and n are the molar amounts of air and steam, respectively, fed per mole of biomass. $x_1, x_2, x_3, x_4, x_5, x_6,$ and x_7 are the number of moles of the corresponding products, respectively.

The element balances of carbon, hydrogen, and oxygen elements are given below:

$$1 = x_1 + x_2 + x_4 + x_6 + 6x_7 \tag{2}$$

$$x + 2n = 2x_3 + 4x_4 + 2x_5 + 6x_7 \tag{3}$$

$$y + 2m + n = x_1 + 2x_2 + x_5 + x_7 \tag{4}$$

2.1. Equilibrium Constants and Non-Equilibrium Factor

When biomass is added to the gasifier, the pyrolysis reactions occur instantaneously. This is followed by a number of competing intermediate reactions, including both heterogeneous reactions and homogeneous gas phase reactions. Char, tar, and other gases (CO, CO_2, H_2O, H_2, and CH_4) are produced during the gasification process. The three independent equilibrium reactions used for equilibrium calculation are considered to be:

$$C + H_2O \leftrightarrow CO + H_2 - 131 \text{ kJ/mol} \tag{5}$$

$$C + 2H_2 \leftrightarrow CH_4 + 75 \text{ kJ/mol} \tag{6}$$

$$CO + H_2O \leftrightarrow CO_2 + H_2 + 41 \text{ kJ/mol} \tag{7}$$

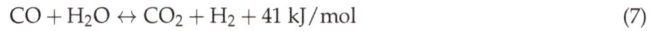

The above three reactions are water gas reaction, methane generation reaction, and water gas shift reaction, respectively. The equilibrium constants for these three reactions can be shown as follows:

$$K_1 = \frac{(P_{CO}/P_{total})(P_{H_2}/P_{total})}{(P_{H_2O}/P_{total})} = \frac{x_1 x_3}{x_5} \cdot \frac{1}{x_{total}} \tag{8}$$

$$K_2 = \frac{(P_{CH_4}/P_{total})}{(P_{H_2}/P_{total})^2} = \frac{x_4}{(x_3)^2} \cdot x_{total} \tag{9}$$

$$K_3 = \frac{(P_{CO_2}/P_{total})(P_{H_2}/P_{total})}{(P_{CO}/P_{total})(P_{H_2O}/P_{total})} = \frac{x_2 \cdot x_3}{x_1 \cdot x_5} \tag{10}$$

where K and P are equilibrium constant and pressure, respectively. x_{total} is the total number of moles of syngas.

The equilibrium constant can be calculated based on the Gibbs free energy; meanwhile, the standard Gibbs function of the reaction is dependent on the standard enthalpy change. Through mathematical conversion, the equilibrium constant is associated with the standard enthalpy change and a constant of integration, which can be found from the related reference [32]. From the data of standard enthalpy of formation and standard Gibbs function of formation, also shown in that paper, this constant of integration can be calculated. Thus, the correlations between the equilibrium constants and the temperatures for the Reactions (5)–(7) are obtained shown below:

$$\ln K_1 = \frac{-15702.01}{T} + 1.384 \ln T - 0.000621 T + \frac{39900}{T^2} + 7.642 \tag{11}$$

$$\ln K_2 = \frac{7082.848}{T} - 6.567 \ln T + 0.003733 T - 3.60667 \times 10^{-7} T^2 + \frac{35050}{T^2} + 32.541 \tag{12}$$

$$\ln K_3 = \frac{5872.461}{T} + 1.86 \ln T - 0.000269 T - \frac{58200}{T^2} - 18.014 \tag{13}$$

In the actual gasification system, it is a fact that methane generation reaction is not at equilibrium state. Some researchers [17,27] found that the methane yield was underestimated in the simulation based on the conventional thermodynamic equilibrium model. Hence, it is necessary to consider the non-equilibrium behavior. According to the linear regression analysis of many experimental data sets, the dependency of CH_4 with respect to the process parameters was obtained [17]. It was observed that the methane yield had a significant relationship with ER, and that was consistent with the results in the literature [33,34]. Meanwhile, the tar is formed in pyrolysis stage accompanied with H_2, CO, CO_2, H_2O, and CH_4, and it can also convert to these gases according to the decomposition reactions. So, it is reasonable to believe that the equilibrium relationship between these gases will not change, and the effect of the tar on these gases is just revealed from Equation (9). Hence, the non-equilibrium modification is needed to apply to K_2, which is shown below:

$$K_2^* = (38.75 - 30.7 \cdot ER) \cdot K_2 \tag{14}$$

where ER is equivalence ratio, and it is defined as the ratio of the actual amount of air fed to the gasifier to the stoichiometric amount of air necessary for the combustion of biomass.

Meanwhile, the Equations (8) and (9) can be combined to give the Equation (15) as shown below:

$$K_1 \cdot K_2 = \frac{x_1 \cdot x_4}{x_3 \cdot x_5} \tag{15}$$

In the gasification process, the carbon in biomass gets partially converted into gas and the rest is converted into char. Hence, the carbon conversion ratio is defined as the ratio of total number of moles of carbon in syngas to the number of moles of carbon in the biomass feed. Many studies [34–36] have found that carbon conversion ratio cannot be calculated directly, and the empirical formula obtained based on experimental data sets will be suitable. Generally, the carbon conversion ratio can be fitted as a function of process parameters, such as ER [34], or both ER and temperature [17,37]. It has been found that the temperature has a small effect on carbon conversion ratio. An empirical formula was proposed by Lim et al. [17] based on many experimental data points under different experimental conditions, and the temperature influence had been considered. So, this formula can be considered as a representative one, and it is shown as follows:

$$f = x_1 + x_2 + x_4 + 6x_7 = 0.901 + 0.439(1 - e^{(-ER+0.0003T)}) \tag{16}$$

2.2. Energy Balance in the Gasification Process

Based on the first law of thermodynamics, the equation for energy conservation of the reactants and products of gasification can be written as follows:

$$\begin{aligned}
Q = \sum (w_{out} \cdot H_{out}^0) + \int_{298.15}^{T_2} (\sum (w_{out} \cdot C_{p,out}(T)))dT \\
- \sum (w_{in} \cdot H_{in}^0) - \int_{298.15}^{T_1} (\sum (w_{in} \cdot C_{p,in}(T)))dT
\end{aligned} \tag{17}$$

where C_p is the specific heat at constant pressure, and T_1 and T_2 are the inlet temperature of reactants and the outlet temperature of products, respectively. H_0 is the standard enthalpy of formation, and the value for the gas can be found in the literature [8,32]. Q is the energy exchange between the gasifier and the environment, which is positive for the endothermic reaction and negative for the exothermic reaction. w is the molar amount of the corresponding material. The subscript "in" corresponds to the reactants and the subscript "out" corresponds to the products.

The equation for the complete combustion reaction of biomass is shown below:

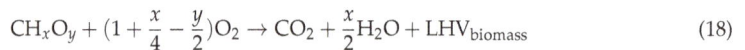

$$CH_xO_y + (1 + \frac{x}{4} - \frac{y}{2})O_2 \rightarrow CO_2 + \frac{x}{2}H_2O + LHV_{biomass} \tag{18}$$

The standard enthalpy of formation for biomass can be calculated as follows:

$$H_{f,CH_xO_y}^0 = H_{f,CO_2}^0 + \frac{x}{2}H_{f,H_2O}^0 + LHV_{biomass} \tag{19}$$

where $LHV_{biomass}$ is the lower heating value of the biomass in MJ/kg.

2.3. Exergy Efficiency of Gasification

Exergy is the maximum amount of theoretically available energy, which is used for evaluating the quality of energy. The conversion of exergy before and after gasification is controlled by the second law of thermodynamics, which considers the loss due to increase in entropy. The quality of energy could not be considered in the first law of thermodynamics during the gasification process. Hence, it is highly necessary to analyze the exergy efficiency.

Exergy includes two components [8]: chemical exergy and physical exergy.

The chemical exergy of the compound is calculated as follows [38]:

$$E_{chem,CHO} = w_C E_{chem,C} + 0.5w_{H_2} E_{chem,H_2} + 0.5w_{O_2} E_{chem,O_2} \tag{20}$$

The chemical exergies for pure components are given in the literature [39,40].

The physical exergy of a component is calculated as follows [41]:

$$E_{phy} = \int_{T_0}^{T} C_p dT - T_0 \int_{T_0}^{T} \frac{C_p}{T} dT = (h - h_0) - T_0(s - s_0) \tag{21}$$

where h and s are the enthalpy and entropy, respectively.

The exergy of biomass is calculated from its lower heating value as follows:

$$E_{biomass} = \beta \cdot LHV_{biomass} \tag{22}$$

where β is a multiplication factor [8], which is expressed in terms of oxygen-carbon and hydrogen-carbon ratios.

The exergy balance of the reaction process is as follows:

$$\sum E_{reactants} + W = \sum E_{products} + E_{loss} \tag{23}$$

where the exergy of reactants, $E_{reactants}$, comprises biomass exergy and the exergy of the inlet gas mixture. The exergy of products, $E_{products}$, comprises syngas exergy and unconverted carbon exergy. E_{loss} is the exergy loss during the gasification process. W is the exergy accompanying the energy exchange, which is positive for the endothermic reaction and negative for the exothermic reaction.

Exergy efficiency (η) corresponding to the endothermic reaction is defined as follows:

$$\eta = \frac{\sum E_{products}}{\sum E_{reactants} + W} \tag{24}$$

Exergy efficiency corresponding to the exothermic reaction is defined as follows:

$$\eta = \frac{\sum E_{products} - W}{\sum E_{reactants}} \tag{25}$$

2.4. Other Parameters

Steam to biomass ratio (SBR) is defined as the ratio of the steam mass flow rate to the biomass mass flow rate on dry basis.

Gas yield (GY) is calculated by dividing the rate of production of dry syngas, excluding nitrogen, by the biomass feeding rate, and its unit is Nm^3/kg.

When the combustible components of the syngas are CO, H_2, and CH_4, the lower heating value (LHV) of the syngas is calculated as follows [42]:

$$LHV = 12.64\phi_{CO} + 10.8\phi_{H_2} + 35.8\phi_{CH_4} \tag{26}$$

where LHV is expressed in MJ/Nm^3. ϕ is the corresponding molar fraction of combustible gas species in the dry syngas, which includes nitrogen.

3. Model Validation

In order to validate the equilibrium model, the experimental results reported by some researchers have been used as comparison, and the operating conditions of the samples chosen to validate the model are presented in Table 1. The validation results are shown through Figures 1–6.

Table 1. Operating conditions for the experimental runs of gasification.

Expt. Sample No.	Author(s)	Gasification Agent	T (°C)	SBR	ER
1	Campoy et al. [43]	Steam + Air	752	0.18	0.23
2	Campoy et al.	Steam + Air	727	0.28	0.19
3	Campoy et al.	Steam + Air	786	0.23	0.27
4	Campoy et al.	Steam + Air	755	0.43	0.27
5	Campoy et al.	Steam + Air	804	0.22	0.33
6	Campoy et al.	Steam + Air	789	0.45	0.33
7	Jarungthammachote et al. [44]	Steam + Air	909	0.17	0.34
8	Jarungthammachote et al.	Steam + Air	903	0.21	0.34
9	Jarungthammachote et al.	Steam + Air	887	0.28	0.34
10	Jarungthammachote et al.	Steam + Air	937	0.12	0.37
11	Jarungthammachote et al.	Steam + Air	911	0.26	0.37
12	Jarungthammachote et al.	Steam + Air	878	0.33	0.37
13	Salami et al. [45]	Steam + Air	770	0.23	0.29
14	Salami et al.	Steam + Air	770	0.4	0.29
15	Salami et al.	Steam + Air	770	0.51	0.29
16	Salami et al.	Steam + Air	770	0.6	0.29
17	Salami et al.	Steam + Air	770	0.67	0.29
18	Salami et al.	Steam + Air	770	0.7	0.29
19	Salami et al.	Steam + Air	770	0.75	0.29
20	Salami et al.	Steam + Air	861	0.23	0.29
21	Salami et al.	Steam + Air	861	0.4	0.29
22	Salami et al.	Steam + Air	861	0.51	0.29
23	Salami et al.	Steam + Air	861	0.6	0.29
24	Salami et al.	Steam + Air	861	0.67	0.29
25	Salami et al.	Steam + Air	861	0.7	0.29
26	Salami et al.	Steam + Air	861	0.75	0.29
27	Cheng et al. [46]	Steam	714	0.15	0
28	Cheng et al.	Steam	788	0.15	0
29	Cheng et al.	Steam	863	0.15	0
30	Cerone et al. [47]	Steam + Air	702	0.233	0.19
31	Cerone et al.	Steam + Oxygen	742	0.3	0.18
32	Ruoppolo et al. [48]	Steam + Oxygen + Nitrogen [1]	732	0.23	0.22

[1] The proportion of nitrogen is not presented in the experiment, so nitrogen is not considered for simulation.

The comparisons between the experimental results and numerical results for syngas composition, LHV, and GY are shown in Figures 1 and 2. The molar fraction of individual component of syngas is defined as the number of moles of individual component divided by the total number of moles of dry syngas, which includes nitrogen.

The error between experimental results and predicted results is estimated using the statistical parameter of root mean square (RMS) error, and it is defined as follows:

$$\text{RMS} = \sqrt{\frac{\sum_{i}^{N}(\text{Exp}_i - \text{Mod}_i)^2}{N}} \tag{27}$$

where Exp and Mod are the experimental molar fractions and the predicted molar fractions of syngas species, respectively. N is the number of data.

The RMS errors are calculated for each set of data point, and the maximum value of RMS error is 0.041 for sample No. 19, as shown in Figure 3.

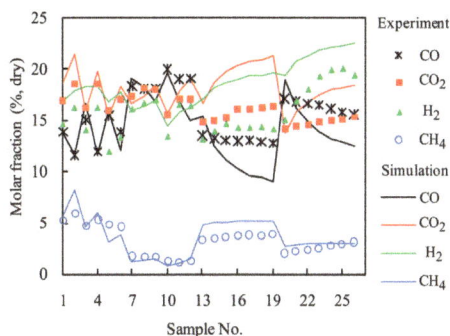

Figure 1. Comparison between experimental data and predicted data for syngas composition.

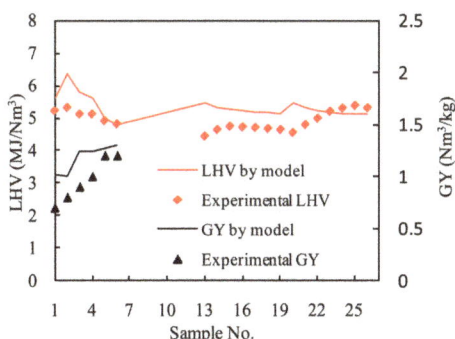

Figure 2. Comparisons between experimental data and predicted data for lower heating value (LHV) and gas yield (GY).

For the first 6 data points, the simulation values for syngas composition well fit the experimental data. Only the molar fraction of H_2 is slightly above the experimental data, so the predicted LHV and GY are just above the experimental data, as shown in Figure 2. For the next 6 data points, the predicted syngas composition also very well fits the experimental data. From sample No. 13 to 19, it shows that there is a difference in composition between the model results and the experimental data, as shown in Figures 1 and 3. However, at high gasification temperature, from sample No. 20 to 26, the RMS error increases less obviously with increasing SBR, as shown in Figure 3. Meanwhile, it is also observed that the predicted LHV is closer to the experimental data at high gasification temperature, as shown in Figure 2. The effect of SBR on the accuracy of calculation depends on the gasification temperature. When the same value of SBR is choosen, the difference between the simualtion result and experimental result at low gasification temperature is greater than the value at high gasification temperature. Through the present results, it is evident that the predicted molar fraction of CH_4 well fits with the experimental data at all the data set points.

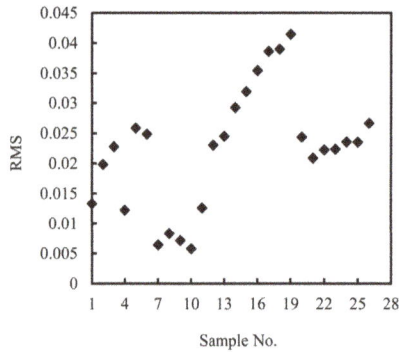

Figure 3. RMS errors from comparisons between experimental data and predicted results.

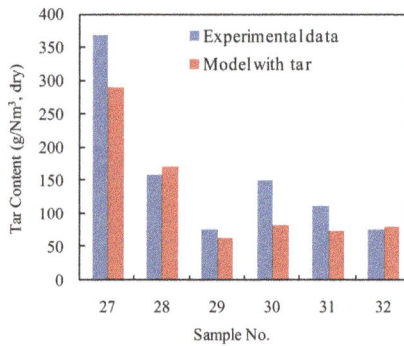

Figure 4. Comparison between experimental data and predicted date for tar content.

The comparisons between the experimental results and the numerical results for tar content and GY are shown in Figures 4 and 5. Effects of the models, with or without tar, on the simulation results are also researched. The tar content in g/Nm^3 is the weight of the tar divided by the volume of dry syngas, which includes nitrogen. From Figure 4, the tar content can be predicted very well. Moreover, the predicted GY based on the model with tar are much closer to the experimental data. For the model without tar, all the tar gets converted into syngas; hence, the predicted GY is larger than the value calculated from the model with tar.

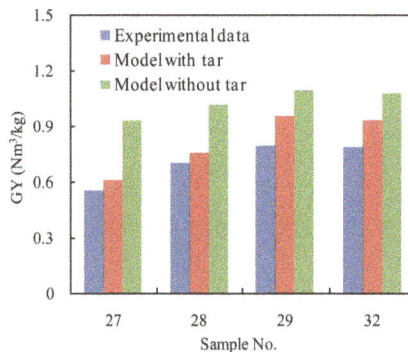

Figure 5. Comparison between experimental data and predicted data based on different models for GY.

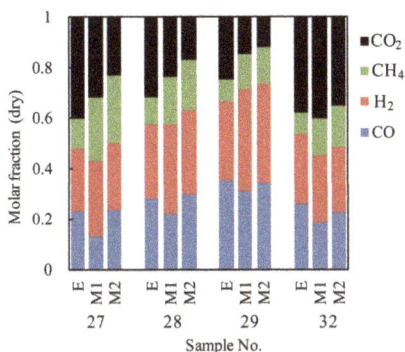

Figure 6. Comparison between experimental data and predicted data based on different models for syngas composition (E-experimental data, M1-model with tar, M2-model without tar).

From Figure 6, the predicted total molar fraction of the combustible gases based on the model without tar is larger than the predicted value based on the model with tar. From Figures 4–6, the predicted values based on the model with tar are closer to the experimental data. This shows that the gasification model with tar is reasonable and effective. At low gasification temperature with small ER and SBR, the tar content is very high, as shown in Table 1 and Figure 4. In this case the simulation result will be inaccurate if the tar is not considered in the gasification model.

Considering the effects of the process parameters on syngas composition, tar content, LHV, and GY, the proposed equilibrium model with tar is suitable for the simulation of gasification at relatively low gasification temperature with small ER and SBR.

4. Results and Discussion

In this study, gasification of woody biomass is studied with ER varying from 0.1 to 0.5, SBR from 0.15 to 0.6, and the gasification temperature from 900 K to 1100 K. The molecular formula for woody biomass is $CH_{1.4}O_{0.64}$ on the basis of an earlier reference [43]. The temperature of the inlet biomass is 298.15 K, and the temperatures of inlet air and steam are 673.15 K. The LHV of woody biomass from the same reference work is 17.1 MJ/kg, and the biomass is completely dry in the simulation study.

4.1. The Effect of Gasification Temperature

The effects of gasification temperature on syngas composition and tar content are shown in Figure 7. The tar content in g/kg is the weight of the tar divided by the weight of biomass. The result from Figure 7 reveals the expected strong effect of temperature on the gasification products. The molar fractions of CO and H_2 increase significantly as gasification temperature increases, whereas the molar fractions of CO_2 and CH_4 show an opposite trend, as mentioned in the literature [35,42]. Meanwhile, with increase in temperature, the tar content decreases with the value changing from 137 g/kg at 900 K to 0 g/kg at 1050 K. The value of 0 shows that the tar cannot be predicted based on the proposed model in some calculation conditions, which does not agree with the fact. Hence, from this perspective, the model has a certain application scope, which is consistent with the conclusion of previous section of this article. As the tar content decreases and the syngas volume increases with increase in temperature, GY increases with increase in temperature, as shown in Figure 8. Meanwhile, LHV also increases with increase in temperature due to the formation of large molar fractions of CO and H_2.

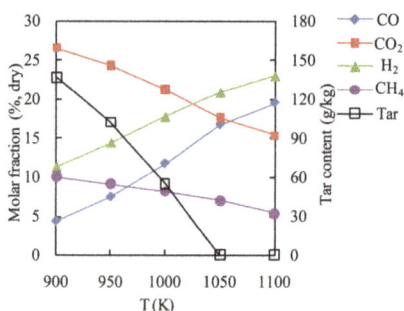

Figure 7. Effects of gasification temperature on syngas composition and tar content at ER = 0.2 and SBR = 0.3.

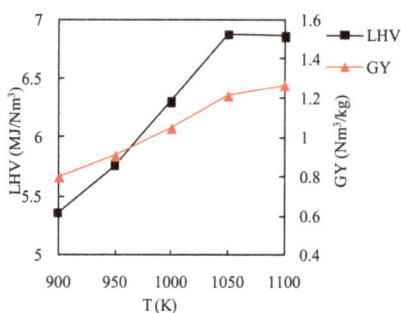

Figure 8. Effects of gasification temperature on LHV and GY at ER = 0.2 and SBR = 0.3.

At high ER, the influence of the gasification temperature on the syngas composition is shown in Figure 9. The varying trend of syngas composition shown in Figure 9 is similar to the trend shown in Figure 7, with increase in gasification temperature. However, the total molar fraction of the combustible gases decreases, and this results in the small value of LHV when comparing 4.04 shown in Figure 10 with 6.86 shown in Figure 8 at the same temperature of 1100 K. Meanwhile, tar is not formed throughout the range of temperature, so it is advisable to decrease the tar content by increasing ER.

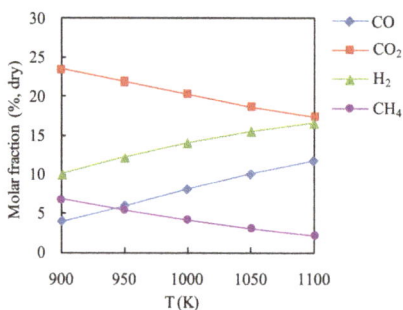

Figure 9. Effect of gasification temperature on syngas composition at ER = 0.4 and SBR = 0.5.

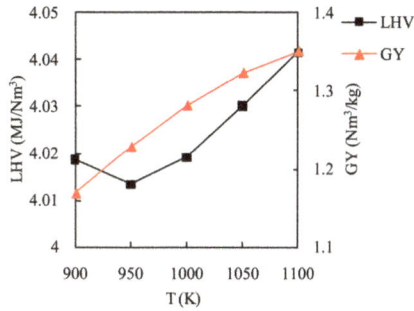

Figure 10. Effects of gasification temperature on LHV and GY at ER = 0.4 and SBR = 0.5.

4.2. The Effect of ER

At the gasification temperature of 950 K, the effects of ER on syngas composition and tar content are shown in Figure 11. The molar fractions of the combustible gases decrease with increase in ER due to increase in N_2 content, so LHV decreases, as shown in Figure 12. With the conversion of tar and participation of oxygen in the gasifier, GY increases with increase in ER.

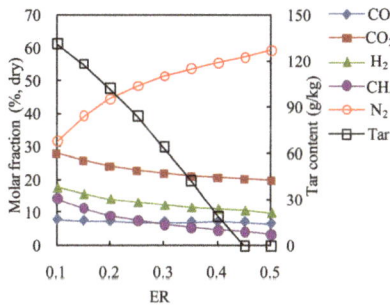

Figure 11. Effects of ER on syngas composition and tar content at *T* = 950 K and SBR = 0.3.

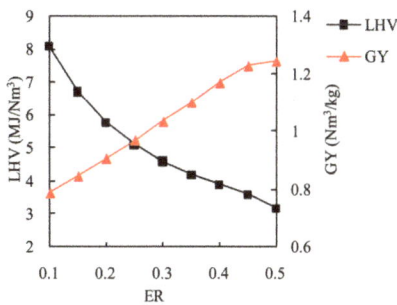

Figure 12. Effects of ER on LHV and GY at *T* = 950 K and SBR = 0.3.

At the gasification temperature of 1100 K, the influence of ER on syngas composition is shown in Figure 13. The percentage of CO_2 is defined as the ratio of the molar quantity of CO_2 to the total molar quantity of CO, CO_2, H_2, and CH_4 products. The varying trend of the syngas composition shown in Figure 13 is similar to the trend shown in Figure 11. The percentage of the CO_2 increases sharply with increase in ER due to combustion of more carbon into CO_2, and this is consistent with previous reports [17,34]. The molar fraction of CO_2 does not differ much with increasing ER, as seen in Figure 13, and in some studies this value increases slightly [9]. At high temperatures tar is not formed in the entire range of ER. This shows that temperature has an important effect on tar conversion in the gasification process. The effects of ER on LHV and GY shown in Figure 14 are similar to that shown in Figure 12. The GY increases as ER increases, whereas LHV decreases with increase in ER.

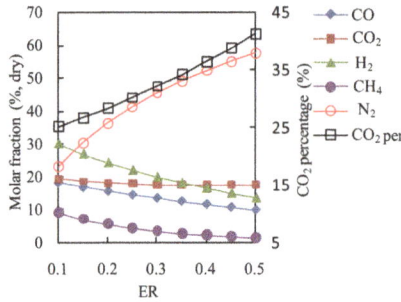

Figure 13. Effect of ER on syngas composition at $T = 1100$ K and SBR = 0.5.

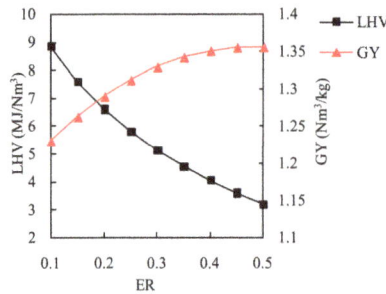

Figure 14. Effects of ER on LHV and GY at $T = 1100$ K and SBR = 0.5.

4.3. The Effect of SBR

Due to the significant influence of Reaction (7), the molar fractions of H_2 and CO_2 increase as SBR increases, whereas the value of CO decreases at the same time, as shown in Figure 15. The tar content decreases as SBR increases, which shows the same trend as that reported earlier [45]. Since the trend for the decrease in molar fraction of CO is not evident when SBR is below 0.4, the LHV increases as SBR increases over a maximum range of SBR, followed by a slight decrease, as shown in Figure 16. The GY increases as SBR increases, since Reaction (7) is favored as also mentioned earlier [17].

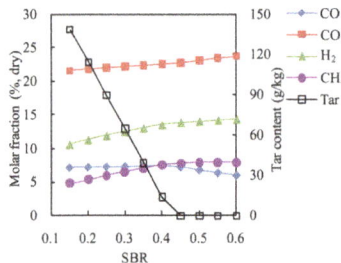

Figure 15. Effects of SBR on syngas composition and tar content at $T = 950$ K and ER = 0.3.

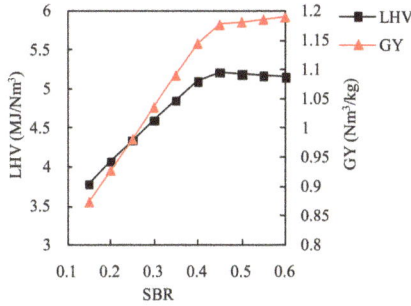

Figure 16. Effects of SBR on LHV and GY at T = 950 K and ER = 0.3.

From Figure 17, the molar fraction of CH_4 varies very less with increasing SBR, and the increase in the molar fraction of H_2 is not significant. However, at high gasification temperature the molar fraction of CO decreases apparently as SBR increases, which is very different to the trend shown in Figure 15. The result is consistent with earlier studies [9,43]. Meanwhile, in the formula for calculating LHV (Equation (26)), the coefficient corresponding to the molar fraction of CO is larger than that of H_2. Therefore, the LHV decreases as SBR increases, as shown in Figure 18 for wide range of SBR. This is different from the trend at low temperature, as shown in Figure 16. It can also be seen that increase in H_2 is always accompanied with a decrease in CO and an increase of CO_2. Hence, the value of SBR should not be too big, especially at high gasification temperature.

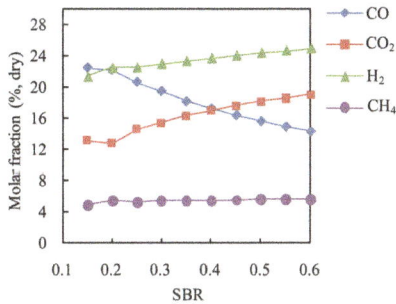

Figure 17. Effect of SBR on syngas composition at T = 1100 K and ER = 0.2.

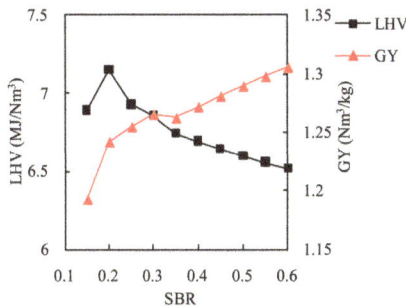

Figure 18. Effects of SBR on LHV and GY at T = 1100 K and ER = 0.2.

4.4. The Effect of Process Parameters on Energy Exchange

In the air-steam gasification process, the dimensionless heat transfer ratio (DHTR), which can be affected by the process parameters, is adopted to represent the value and direction of energy exchange between the gasifier and the environment, and it is defined as follows:

$$\text{DHTR} = \frac{Q}{\left| \sum \left(w_{\text{in}} \cdot \Delta H_{f,\text{in}}^0 \right) + \int_{298.15}^{T_1} \left(\sum \left(w_{\text{in}} \cdot C_{p,\text{in}}(T) \right) \right) dT \right|} \tag{28}$$

A DHTR value of zero corresponds to an auto-thermal reaction, which means there is no energy exchange between the gasifier and the environment. The value of more than zero corresponds to an endothermic reaction, whereas the value below zero corresponds to an exothermic reaction.

The distributions of DHTR are shown in Figures 19–22, and the air-steam biomass gasification is not an auto-thermal reaction process for most of the process parameters that are shown in the figures. At the same temperature and ER, the effect of SBR on DHTR is not evident, except when the temperature is low and ER is high, as shown in Figure 19. The influence of ER on DHTR is obvious, and the system changes from the endothermic state to the exothermic state as ER increases. This is attributed to an increase in the released heat due to a more complete combustion reaction. With increase in gasification temperature, the auto-thermal reaction zone moves from low ER area to high ER area. ER corresponding to the auto-thermal reaction zone changes from about 0.15 at 900 K temperature to about 0.3 at 1100 K gasification temperature.

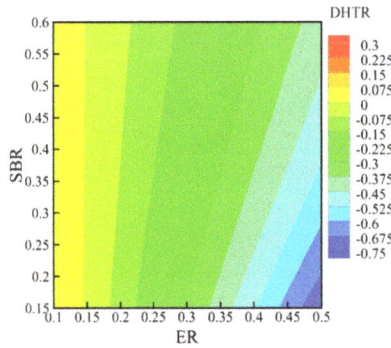

Figure 19. Dimensionless heat transfer ratio (DHTR) distribution at *T* = 900 K.

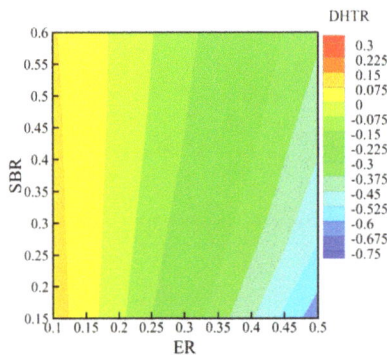

Figure 20. DHTR distribution at *T* = 950 K.

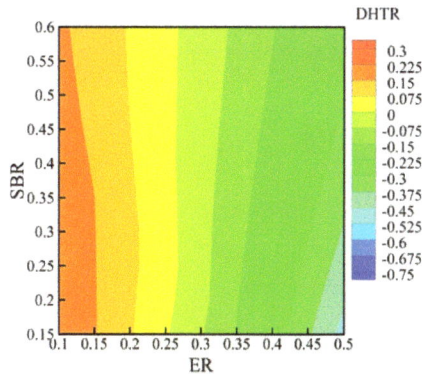

Figure 21. DHTR distribution at T = 1050 K.

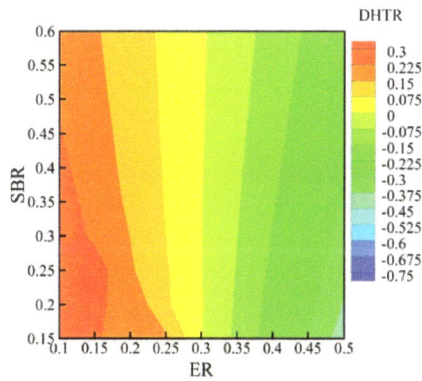

Figure 22. DHTR distribution at T = 1100 K.

Based on 450 numerical data of DHTR, the relationship between DHTR and the process parameters, which is evaluated in terms of the correlation coefficient, is determined as follows:

$$\text{DHTR} = -0.904090 - 1.426036 \times ER + 0.179806 \times SBR + 0.001144 \times T \tag{29}$$

The correlation coefficient is 0.983, which is a very high value, so it clearly shows that the formula is accurate. According to this formula, the value of energy exchange can be determined conveniently. In some cases, the energy balance equation is introduced to the equilibrium model. Based on Equation (29), the energy exchange can change with the changes of the process parameters, so the predicted results obtained by solving the simultaneous equations will be more reasonable.

4.5. The Effect of Process Parameters on Exergy Efficiency

The distributions of exergy efficiency are shown in Figures 23–25. The exergy efficiency decreases slightly as SBR increases when both gasification temperature and ER are small. From the figures, the exergy efficiency is between 0.8 and 0.9 under all simulation conditions. The effect of ER on exergy efficiency is obvious, and the exergy efficiency decreases as ER increases.

In exothermic state, based on Equation (25), the exergy efficiency (η) can also be expressed as follows:

$$\eta = \frac{\sum E_{\text{products}} - W}{\sum E_{\text{reactants}}} = \frac{\sum E_{\text{products}}}{\sum E_{\text{reactants}}} + \frac{-W}{\sum E_{\text{reactants}}} = \eta_{\text{p}} + \eta_{\text{e}} \tag{30}$$

where η_P is the exergy efficiency of products, which is termed as the overall exergy efficiency in some studies [9], and η_e is the exergy efficiency of energy exchange.

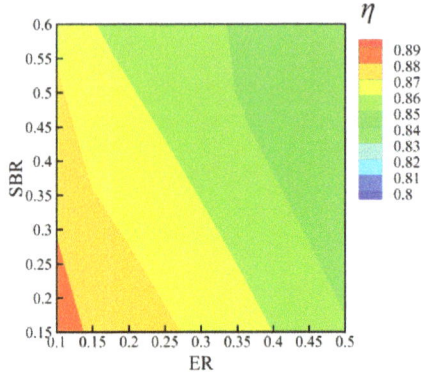

Figure 23. Exergy efficiency distribution at $T = 900$ K.

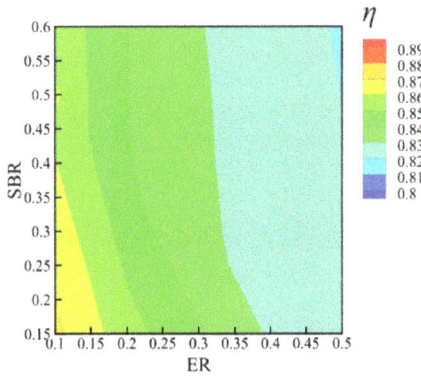

Figure 24. Exergy efficiency distribution at $T = 1000$ K.

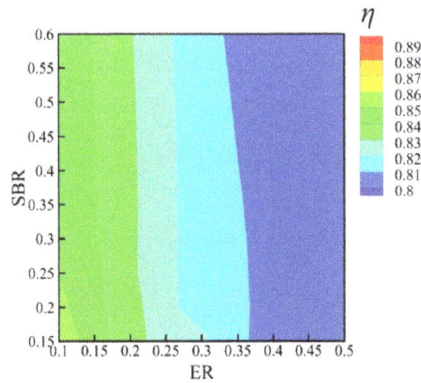

Figure 25. Exergy efficiency distribution at $T = 1100$ K.

For the exothermic reaction, the DHTR value decreases with increase in ER, as shown in Figures 19–22. By comparing the formulas of these two parameters, DHTR and η_e, the trends are just inverse as ER increases. So, the exergy efficiency of energy exchange increases with increase in ER. However, the total exergy efficiency decreases, as shown in Figures 23–25. So the exergy efficiency of products decreases rapidly, and this is the reason for the obvious decrease of exergy efficiency as ER increases in some studies [9].

5. Conclusions

In the present work, the air-steam biomass gasification model with tar has been developed based on the equilibrium constants, and phenol is used as the model compound for the tar. The simulation results based on two different models (with and without tar) have been validated by the experimental data reported in the references, and the model with tar can well predict the tar content in the gasification. Meanwhile, the predicted GY and total molar fraction of the combustible gases, based on the model with tar, agree better with the experimental data. So, the simulation results are much less accurate if the tar is not considered in the gasification model. Considering the effects of the process parameters on syngas composition, tar content, LHV, and GY, the proposed equilibrium model with tar is suitable for the simulation of the gasification process at a relatively low gasification temperature with a small ER and SBR.

The energy exchange between the gasifier and its surroundings has been studied based on DHTR. The influence of SBR on DHTR is not obvious, whereas the effects of ER and temperature on DHTR are evident. Moreover, the relationship between DHTR and the process parameters is given by a formula.

The influence of process parameters on syngas composition, tar content, GY, LHV, and exergy efficiency has also been parametrically researched. With increasing gasification temperature, the total molar fraction of combustible gases, GY, and LHV all increase, whereas the tar content and exergy efficiency all decrease. GY increases as ER increases, whereas tar content, total molar fraction of combustible gases, LHV, and exergy efficiency all decrease. With increasing SBR, the total molar fraction of combustible gases and GY increase, whereas tar content and exergy efficiency all decrease.

Author Contributions: Xinjun Zhao did the simulation and wrote the paper. Qitai Eri and Wenzhen Wu analyzed the data.

Conflicts of Interest: The authors declare no conflict of interest.

References

1. Pereira, E.G.; da Silva, J.N.; de Oliveira, J.L.; Machado, C.S. Sustainable energy: A review of gasification technologies. *Renew. Sustain. Energy Rev.* **2012**, *16*, 4753–4762. [CrossRef]
2. Jayathilake, R.; Rudra, S. Numerical and Experimental Investigation of Equivalence Ratio (ER) and Feedstock Particle Size on Birchwood Gasification. *Energies* **2017**, *10*, 1232. [CrossRef]
3. Baruah, D.; Baruah, D.C. Modeling of biomass gasification: A review. *Renew. Sustain. Energy Rev.* **2014**, *39*, 806–815. [CrossRef]
4. Purohit, P. Economic potential of biomass gasification projects under clean development mechanism in India. *J. Clean. Prod.* **2009**, *17*, 181–193. [CrossRef]
5. González-Vázquez, M.; García, R.; Pevida, C.; Rubiera, F. Optimization of a Bubbling Fluidized Bed Plant for Low-Temperature Gasification of Biomass. *Energies* **2017**, *10*, 306. [CrossRef]
6. Ahmed, T.Y.; Ahmad, M.M.; Yusup, S.; Inayat, A.; Khan, Z. Mathematical and computational approaches for design of biomass gasification for hydrogen production: A review. *Renew. Sustain. Energy Rev.* **2012**, *16*, 2304–2315. [CrossRef]
7. Sharma, S.; Sheth, P.N. Air–steam biomass gasification: Experiments, modeling and simulation. *Energy Convers. Manag.* **2016**, *110*, 307–318. [CrossRef]
8. Loha, C.; Chattopadhyay, H.; Chatterjee, P.K. Thermodynamic analysis of hydrogen rich synthetic gas generation from fluidized bed gasification of rice husk. *Energy* **2011**, *36*, 4063–4071. [CrossRef]

9. Tapasvi, D.; Kempegowda, R.S.; Tran, K.-Q.; Skreiberg, Ø.; Grønli, M. A simulation study on the torrefied biomass gasification. *Energy Convers. Manag.* **2015**, *90*, 446–457. [CrossRef]

10. Jangsawang, W.; Klimanek, A.; Gupta, A.K. Enhanced yield of hydrogen from wastes using high temperature steam gasification. *J. Energy Resour. Technol.* **2006**, *128*, 179–185. [CrossRef]

11. Rodríguez-Olalde, N.; Mendoza-Chávez, E.; Castro-Montoya, A.; Saucedo-Luna, J.; Maya-Yescas, R.; Rutiaga-Quiñones, J.; Ortega, J. Simulation of Syngas Production from Lignin Using Guaiacol as a Model Compound. *Energies* **2015**, *8*, 6705–6714. [CrossRef]

12. Suwatthikul, A.; Limprachaya, S.; Kittisupakorn, P.; Mujtaba, I. Simulation of Steam Gasification in a Fluidized Bed Reactor with Energy Self-Sufficient Condition. *Energies* **2017**, *10*, 314. [CrossRef]

13. He, J.; Göransson, K.; Söderlind, U.; Zhang, W. Simulation of biomass gasification in a dual fluidized bed gasifier. *Biomass Convers. Biorefin.* **2012**, *2*, 1–10. [CrossRef]

14. Rofouie, P.; Moshkelani, M.; Perrier, M.; Paris, J. A modified thermodynamic equilibrium model for woody biomass gasification. *Can. J. Chem. Eng.* **2014**, *92*, 593–602. [CrossRef]

15. Meng, W.X.; Banerjee, S.; Zhang, X.; Agarwal, R.K. Process simulation of multi-stage chemical-looping combustion using Aspen Plus. *Energy* **2015**, *90*, 1869–1877. [CrossRef]

16. Chutichai, B.; Patcharavorachot, Y.; Assabumrungrat, S.; Arpornwichanop, A. Parametric analysis of a circulating fluidized bed biomass gasifier for hydrogen production. *Energy* **2015**, *82*, 406–413. [CrossRef]

17. Lim, Y.-i.; Lee, U.-D. Quasi-equilibrium thermodynamic model with empirical equations for air–steam biomass gasification in fluidized-beds. *Fuel Process. Technol.* **2014**, *128*, 199–210. [CrossRef]

18. Karmakar, M.K.; Datta, A.B. Generation of hydrogen rich gas through fluidized bed gasification of biomass. *Bioresour. Technol.* **2011**, *102*, 1907–1913. [CrossRef] [PubMed]

19. Mendiburu, A.Z.; Carvalho, J.A.; Coronado, C.J.R. Thermochemical equilibrium modeling of biomass downdraft gasifier: Stoichiometric models. *Energy* **2014**, *66*, 189–201. [CrossRef]

20. Jarungthammachote, S.; Dutta, A. Thermodynamic equilibrium model and second law analysis of a downdraft waste gasifier. *Energy* **2007**, *32*, 1660–1669. [CrossRef]

21. Kim, Y.D.; Yang, C.W.; Kim, B.J.; Kim, K.S.; Lee, J.W.; Moon, J.H.; Yang, W.; Yu, T.U.; Lee, U.D. Air-blown gasification of woody biomass in a bubbling fluidized bed gasifier. *Appl. Energy* **2013**, *112*, 414–420. [CrossRef]

22. Sahoo, A.; Ram, D.K. Gasifier performance and energy analysis for fluidized bed gasification of sugarcane bagasse. *Energy* **2015**, *90*, 1420–1425. [CrossRef]

23. Umeki, K.; Yamamoto, K.; Namioka, T.; Yoshikawa, K. High temperature steam-only gasification of woody biomass. *Appl. Energy* **2010**, *87*, 791–798. [CrossRef]

24. Fremaux, S.; Beheshti, S.-M.; Ghassemi, H.; Shahsavan-Markadeh, R. An experimental study on hydrogen-rich gas production via steam gasification of biomass in a research-scale fluidized bed. *Energy Convers. Manag.* **2015**, *91*, 427–432. [CrossRef]

25. Gil-Lalaguna, N.; Sánchez, J.L.; Murillo, M.B.; Rodríguez, E.; Gea, G. Air–steam gasification of sewage sludge in a fluidized bed. Influence of some operating conditions. *Chem. Eng. J.* **2014**, *248*, 373–382. [CrossRef]

26. Ruoppolo, G.; Ammendola, P.; Chirone, R.; Miccio, F. H$_2$-rich syngas production by fluidized bed gasification of biomass and plastic fuel. *Waste Manag.* **2012**, *32*, 724–732. [CrossRef] [PubMed]

27. Huang, H.J.; Ramaswamy, S. Modeling biomass gasification using thermodynamic equilibrium approach. *Appl. Biochem. Biotechnol.* **2009**, *154*, 14–25. [CrossRef] [PubMed]

28. Mayerhofer, M.; Fendt, S.; Spliethoff, H.; Gaderer, M. Fluidized bed gasification of biomass—In bed investigation of gas and tar formation. *Fuel* **2014**, *117*, 1248–1255. [CrossRef]

29. Martínez-Lera, S.; Pallarés Ranz, J. On the development of a wood gasification modelling approach with special emphasis on primary devolatilization and tar formation and destruction phenomena. *Energy* **2016**, *113*, 643–652. [CrossRef]

30. Morf, P.; Hasler, P.; Nussbaumer, T. Mechanisms and kinetics of homogeneous secondary reactions of tar from continuous pyrolysis of wood chips. *Fuel* **2002**, *81*, 843–853. [CrossRef]

31. Ji, P.; Feng, W.; Chen, B. Production of ultrapure hydrogen from biomass gasification with air. *Chem. Eng. Sci.* **2009**, *64*, 582–592. [CrossRef]

32. Zainal, Z.A.; Ali, R.; Lean, C.H.; Seetharamu, K.N. Prediction of performance of a downdraft gasifier using equilibrium modeling for different biomass materials. *Energy Convers. Manag.* **2001**, *42*, 1499–1515. [CrossRef]

33. Gómez-Barea, A.; Leckner, B. Estimation of gas composition and char conversion in a fluidized bed biomass gasifier. *Fuel* **2013**, *107*, 419–431. [CrossRef]

34. Li, X.T.; Grace, J.R.; Lim, C.J.; Watkinson, A.P.; Chen, H.P.; Kim, J.R. Biomass gasification in a circulating fluidized bed. *Biomass Bioenergy* **2004**, *26*, 171–193. [CrossRef]

35. Meng, X.; de Jong, W.; Fu, N.; Verkooijen, A.H. M. Biomass gasification in a 100 kWth steam-oxygen blown circulating fluidized bed gasifier: Effects of operational conditions on product gas distribution and tar formation. *Biomass Bioenergy* **2011**, *35*, 2910–2924. [CrossRef]

36. Lv, P.M.; Xiong, Z.H.; Chang, J.; Wu, C.Z.; Chen, Y.; Zhu, J.X. An experimental study on biomass air-steam gasification in a fluidized bed. *Bioresour. Technol.* **2004**, *95*, 95–101. [CrossRef] [PubMed]

37. Li, X.; Grace, J.R.; Watkinson, A.P.; Lim, C.J.; Ergudenler, A. Equilibrium modeling of gasification: A free energy minimization approach and its application to a circulating fluidized bed coal gasifier. *Fuel* **2001**, *80*, 195–207. [CrossRef]

38. Srinivas, T.; Gupta, A.V.S.S.K.S.; Reddy, B.V. Thermodynamic equilibrium model and exergy analysis of a biomass gasifier. *J. Energy Resour. Technol.* **2009**, *131*, 98–105. [CrossRef]

39. Demirel, Y. Thermodynamic analysis. *Arab. J. Sci. Eng.* **2013**, *38*, 221–249. [CrossRef]

40. Pellegrini, L.; Deoliveirajr, S. Exergy analysis of sugarcane bagasse gasification. *Energy* **2007**, *32*, 314–327. [CrossRef]

41. Datta, A.; Ganguly, R.; Sarkar, L. Energy and exergy analyses of an externally fired gas turbine (EFGT) cycle integrated with biomass gasifier for distributed power generation. *Energy* **2010**, *35*, 341–350. [CrossRef]

42. Wang, R.; Huang, Q.; Lu, P.; Li, W.; Wang, S.; Chi, Y.; Yan, J. Experimental study on air/steam gasification of leather scraps using U-type catalytic gasification for producing hydrogen-enriched syngas. *Int. J. Hydrogen Energy* **2015**, *40*, 8322–8329. [CrossRef]

43. Campoy, M.; Gómez-Barea, A.; Vidal, F.B.; Ollero, P. Air-steam gasification of biomass in a fluidized bed under simulated autothermal and adiabatic conditions. *Ind. Eng. Chem. Res.* **2008**, *47*, 5957–5965. [CrossRef]

44. Jarungthammachote, S.; Dutta, A. Experimental investigation of a multi-stage air-steam gasification process for hydrogen enriched gas production. *Int. J. Energy Res.* **2012**, *36*, 335–345. [CrossRef]

45. Salami, N.; Skála, Z. Use of the steam as gasifying agent in fluidized bed gasifier. *Chem. Biochem. Eng. Q. J.* **2015**, *29*, 13–18. [CrossRef]

46. Cheng, G.; Li, Q.; Qi, F.; Xiao, B.; Liu, S.; Hu, Z.; He, P. Allothermal gasification of biomass using micron size biomass as external heat source. *Bioresour. Technol.* **2012**, *107*, 471–475. [CrossRef] [PubMed]

47. Cerone, N.; Zimbardi, F.; Villone, A.; Strjiugas, N.; Kiyikci, E.G. Gasification of wood and torrefied wood with air, oxygen, and steam in a fixed-bed pilot plant. *Energy Fuels* **2016**, *30*, 4034–4043. [CrossRef]

48. Ruoppolo, G.; Miccio, F.; Brachi, P.; Picarelli, A.; Chirone, R. Fluidized bed gasification of biomass and biomass/coal pellets in oxygen and steam atmosphere. *Chem. Eng. Trans.* **2013**, *32*, 595–600.

energies

MDPI

Article

Treatment of Oily Wastewater by the Optimization of Fe$_2$O$_3$ Calcination Temperatures in Innovative Bio-Electron-Fenton Microbial Fuel Cells

Jung-Chen Wu [1], Wei-Mon Yan [2], Chin-Tsan Wang [3,*], Chen-Hao Wang [1,*], Yi-Hao Pai [4], Kai-Chin Wang [1], Yan-Ming Chen [5], Tzu-Hsuan Lan [5] and Sangeetha Thangavel [2]

[1] Department of Materials Science and Engineering, National Taiwan University of Science and Technology, No. 43, Keelung Rd., Sec. 4, Da'an Dist., Taipei 10607, Taiwan; shlayex@yahoo.com.tw (J.-C.W.); v1234eszxcv@gmail.com (K.-C.W.)

[2] Department of Energy and Refrigerating Air-Conditioning Engineering, National Taipei University of Technology, 1, Sec. 3, Zhongxiao E. Rd., Taipei 10608, Taiwan; wmyan1234@gmail.com (W.-M.Y.); geetha.vishnu@gmail.com (S.T.)

[3] Department of Mechanical and Electro-Mechanical Engineering, National Ilan University, No. 1, Sec. 1, Shennong Rd., Yilan 26047, Taiwan

[4] Department of Opto-Electronic Engineering, National Dong Hwa University, No. 1, Sec. 2, Da Hsueh Rd., Shoufeng, Hualien 97401, Taiwan; paiyihao@mail.ndhu.edu.tw

[5] Department of Materials and Mineral Resources Engineering & Institute of Materials Science and Engineering National Taipei University of Technology, No. 1, Sec. 3, Zhong-Xiao E. Rd., Taipei 10608, Taiwan; ktotheo@hotmail.com (Y.-M.C.); jh2111900@hotmail.com (T.-H.L.)

* Correspondence: ctwang@niu.edu.tw (C.-T.W.); chwang@mail.ntust.edu.tw (C.-H.W.); Tel.: +886-3-935-7400 (ext. 7459) (C.-T.W.); +886-2-2730-3715 (C.-H.W.); Fax: +886-3-931-1326 (C.-T.W.); +886-2-2737-6544 (C.-H.W.)

Received: 20 January 2018; Accepted: 2 March 2018; Published: 6 March 2018

Abstract: Due to the fact that Iron oxide (Fe$_2$O$_3$) is known to have a good effect on the photochemical reaction of catalysts, an investigation in this study into the enhancement of the degradation performance of bio-electro-Fenton microbial fuel cells (Bio-E-Fenton MFCs) was carried out using three photocatalytic cathodes. These cathodes were produced at different calcination temperatures of Fe$_2$O$_3$ ranging from 500 °C to 900 °C for realizing their performance as photo catalysts within the cathodic chamber of an MFC, and they were compared for their ability to degrade oily wastewater. Results show that a suitable temperature for the calcination of iron oxide would have a significantly positive effect on the performance of Bio-E-Fenton MFCs. An optimal calcination temperature of 500 °C for Fe$_2$O$_3$ in the electrode material of the cathode was observed to produce a maximum power density of 52.5 mW/m^2 and a chemical oxygen demand (COD) degradation rate of oily wastewater (catholyte) of 99.3% within one hour of operation. These novel findings will be useful for the improvement of the performance and applications of Bio-E-Fenton MFCs and their future applications in the field of wastewater treatment.

Keywords: bio-electro-Fenton microbial fuel cells (Bio-E-Fenton MFCs); wastewater; photo catalyst; degradation; calcination; chemical oxygen demand (COD)

1. Introduction

The Bio-Electro-Fenton Microbial Fuel Cells (Bio-E-Fenton MFCs) is a new framework and has been operated extensively because of its simultaneous wastewater treatment and power generation capability. Significantly, the electro-Fenton reaction consists of iron ions (Fe^{2+}) and hydrogen peroxide, and has become a useful technology for treating organic pollutants in wastewater [1–10] as hydroxyl radicals (•OH) are generated in abundance within the electro-Fenton reaction [11]. Considering the working principle of

Bio-E-Fenton MFCs, electrons and protons are produced at the anode as a result of the microbial oxidation of substrates. Then, electrons move towards the cathode through an external circuit pathway, whereas the protons directly penetrate through a membrane and arrive at the cathode simultaneously. Finally, a complete reaction of redox is attained with the aid of oxygen in the cathodic reduction reaction of Bio-E-Fenton MFCs [12]. Although the electro-Fenton reaction is very feasible and effective in treating organic pollutants in wastewater, the COD removal is incomplete [13,14]. A new electro-Fenton system with an Fe@Fe$_2$O$_3$/CNT oxygen-fed gas diffusion cathode was able to degrade rhodamine B (RhB) reaching 91.5% within 120 min at neutral pH [11]. The anode was a Pt sheet with a surface area of 2.0 cm^2 and it also served as a pseudo reference electrode. Cyclic voltammetry experiments were performed and a couple of well-defined peaks were observed at ~0.62 V, which were ascribed to the overall redox potentials of the Fe^{3+}/Fe^{2+} couple. However, 74.1% and 60.1% of RhB was degraded on the Fe@Fe$_2$O$_3$/ACF and Fe@Fe$_2$O$_3$/graphite cathodes, respectively. Polyaniline (PANI)-Fe$_2$O$_3$ nanocomposite was used to successfully enhance the dye adsorption efficiency of acid violet 19 dye, as the dye concentration was increased from 20–80 mg/lt, and the percentage removal of dye decreased from 98.5–90% for the adsorbent at 10 gm/lt [15]. For another kind of photo catalytic material like MoS$_2$, Liu et al. [16] have indicated that a system with MoS$_2$-RGO composites has a better photo catalytic performance than using pure MoS$_2$. Here, the MoS$_2$-RGO composite with 0.5 wt % RGO achieved the highest methylene blue (MB) degradation rate of 99% within 60 min [16]. Nan Xu et al. [17] reported that the anode was carbon felt assisted with granular graphite to enhance the anodic power density. Anaerobic sludge collected from the Mangrove Wetland in Shenzhen was used as a biocatalyst in the anodic chamber. The system with E2 (17β-estrodial) in the catholyte enhanced the voltage output to 0.29 V and the E2 removal efficiency leveled off within 6 h, reaching 85.5%, 96.4%, 88.6%, and 75.6% with an external load of 50, 150, 500, and 1000 Ω, respectively, while the corresponding power density values were 3.47, 4.35, 3.01, and 2.41 W/m^3 [17].

Nevertheless, facing the issue of the weak performance of MFCs, many kinds of electrode materials such as carbon nanotube/polyaniline carbon paper, CNT/PANI carbon paper, have been developed [18]. Results showed that a lower ohm loss and a better power performance with a maximum power density of 1574 mW/m^2 were executed by using the CNT/PANI electrode. A voltage of 1.18 V and current of 12.8 mA in MFCs were achieved by using CNT/PANI carbon paper [18]. But some research works regarding the development of non-precious metal catalysts can be implemented for an effective oxygen reduction reaction (ORR) and the enhancement of MFC performance. Vitamin B12 (py-B12/C) pyrolyzed and supported by carbon black can be utilized as the cathode catalyst in the solid phase MFCs [19]. Therefore, an innovative endeavor of employing calcination iron oxide (Fe$_2$O$_3$) as a photo catalyst combined with the Bio-E-Fenton system, while simultaneously generating power, has not been previously reported for the treatment of organic wastewater. Fenton's reagent could function well under acidic conditions and during the anodic oxidation and cathode aeration resulted in a pH increase, inhibiting the generation of •OH and decreasing the COD removal rate [20–22]. In the experiment, 10% diluted sulfuric acid was added to the cathode solution to obtain a pH value of 3.0 [23]. The experimental mechanism of Fe$_2$O$_3$/carbon felt (CF) photo catalyst bio-electro-Fenton degradation of oily wastewater is illustrated in Figure 1. The first step ① in the cathode reaction: 2H$^+$ + 2e$^-$ + O$_2$ → H$_2$O$_2$ (1–1), resulted in H$_2$O$_2$ accumulation in the cathode chamber via the two-electron reduction of dissolved O$_2$ in the Bio-E-Fenton MFCs and then FeSO$_4$ powder was also added as a source of Fe^{2+}: 2H$^+$ + Fe → Fe^{2+} + H$_2$ (1–2); the second step ② Fe$_2$O$_3$ photo catalyst was added in the cathode chamber and a 300 W halogen lamp was used for the photo excitation of Fe$_2$O$_3$ in order to produce more hydroxyl radicals; the third step ③ finally showed the bio-electro-Fenton reaction at the cathode, where the hydroxyl radicals played a vital role in the degradation of products according to the reaction: Fe^{2+} + H$_2$O$_2$ → Fe^{3+} + •OH + OH$^-$ (1–3). This was followed by the reduction of Fe^{3+} to Fe^{2+}. This was a chain reaction where Fe^{3+} got reduced to Fe^{2+} and then Fe^{2+} was oxidized into Fe^{3+}. This was a completely physical reaction which took place at the cathode and microbes had no role to play. In the conclusion, the Bio-E-Fenton MFCs was reported to be an effective system to treat organic wastewater, as well as bio-refractory pollutants [13].

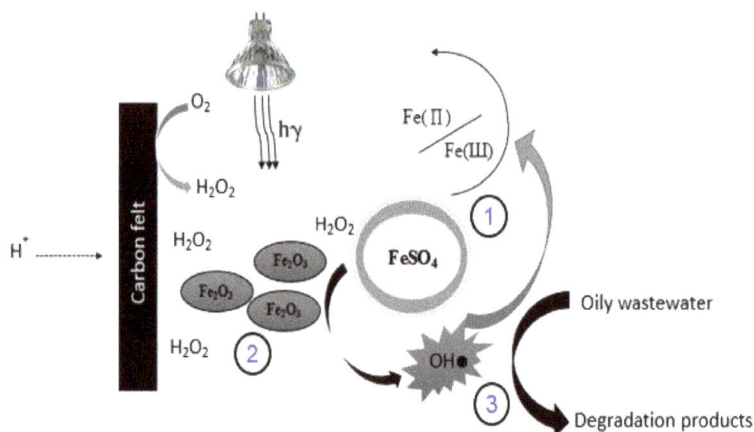

Figure 1. Proposed mechanism of Fe_2O_3/CF catalytic Bio-E-Fenton degradation of oily wastewater.

This research study has been carried out for gaining an in-depth understanding of the effects of calcination temperatures (500 °C to 900 °C) on the performance of Bio-E-Fenton MFCs. Based on the calcination temperature of Fe_2O_3, the photo catalyst structure will change and can be applied in this study for investigating the performance of MFCs. The results showed that the intensity of Fe_2O_3 will not vary greatly due to high calcination temperatures, but the crystalline particle size and surface morphology will be changed immensely [24]. This feature of Fe_2O_3 will enhance the COD removal efficiency of the MFC significantly [25–28]. Fe_2O_3/carbon felt (CF) has been selected to act as part of the cathode in Bio-E-Fenton MFCs because the photo catalysis of Fe_2O_3 can also produce hydroxyl radicals [11]. It can further combine with $FeSO_4$ to act as an iron source with H_2O_2 for producing more hydroxyl radicals (•OH), resulting in a higher degradation rate within a short time. In addition, a pH of 3 was maintained at the cathode chamber to treat refractory organic wastewater and this condition will not affect the pH of the anode chamber.

2. Experimental Design

2.1. Reactor Construction and Operation

An acrylic dual chambered tank with dimensions of 85 mm × 70 mm × 55 mm was selected for the construction of the Bio-E-Fenton MFC, as shown in Figure 2. Each chamber had a total volume of 200 mL with a proton exchange membrane Nafion-117 (80 mm × 70 mm) and carbon felt (CF, 60 mm × 60 mm × 5 mm) as the anode and cathode electrodes, respectively. The preparation method for the carbon felt is as follows: It was initially washed in hot H_2O_2 (10%, 90 °C) solution for 3 h to develop local Quinone sites on the carbon surface for improvement of the anode biocompatibility and increased quantity of anthraquinone [29]. In order to prevent the thermal heat transfer from the cathode to the anode, a Bakelite plate with dimensions of 100 mm × 50 mm × 20 mm (Figure 2) was employed in this study. Therefore, the power density would be influenced by the effect of the calcination temperature of Fe_2O_3 in the cathode chamber, but not by the temperature of the anode. In addition, a control reactor without Fe_2O_3 was not constructed and it was considered from the similar work of [30].

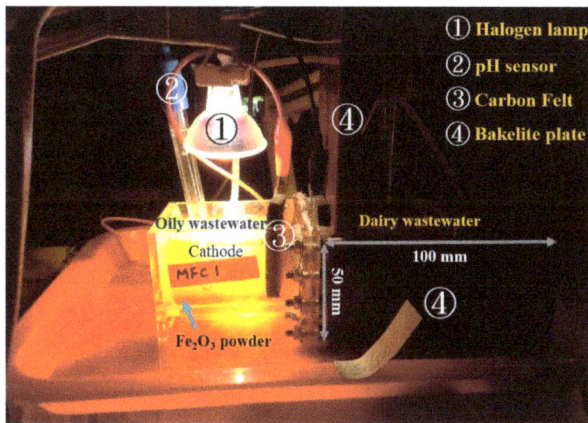

Figure 2. The experimental set-up of the proposed reactor.

2.2. Anolyte/Catholyte and Fe₂O₃ Catalyst Preparation

Kim et al. [31] proved that lactic acid wastewater can be employed to sustain the production of electrical energy in MFCs. Therefore, dairy wastewater was used in this study as the anolyte. During the early stage of fermentation of the dairy wastewater in the chamber, precipitate particulates were formed. Three layers were naturally formed when the wastewater was kept stagnant and were classified as top-level supernatant (clear fluid), a mid-level interface, and a bottom-level precipitate. Prior to the study, a conductivity experiment confirmed that the top-level supernatant in the chamber had the best conductivity of all, with an open-circuit voltage of 0.70 V, limiting current of 0.547 mA/m², and an achievable maximum power density of 101.4 mW/m². In addition, considering the fact that viscosity would influence microbial activity and also affect the power generation of MFCs [32], samples from the top-level supernatant were selected for further studies.

The water bodies [33] have long been contaminated by oil pollution [34–36], so artificially prepared oily wastewater was used as the catholyte in this study. A total of 1 mL of diesel was added to 1 L of water and heated in a magnetic heater and blended with a blender for 1 day at 50 °C. As diesel was immiscible in water, 10 g of emulsifying (Tween 80) agent was added to aid the dissolution of the diesel in water. Molognoni et al. [37] assessed the bio-electrochemical treatability of industrial (dairy) wastewater by MFCs, and an MFC was built and continuously operated for 72 days, during which time the anodic chamber was fed with dairy wastewater and the cathodic chamber with an aerated mineral solution. The study demonstrated that industrial effluents from agrifood facilities can be treated by bio-electrochemical systems (BESs) with >85% (average) organic matter removal. The study by Marashi and Kariminia [38] proved that the wastewater concentration had an influence on the MFC performance and using the raw wastewater with the concentration of 8000 mg COD L^{-1} resulted in the highest power density (65.6 mW m^{-2}) production. Kim et al. [39] reported that the MFC–Anaerobic fluidized membrane bioreactors (AFMBRs) achieved 89 ± 3% removal of the COD with an effluent of 36 ± 6 mg-COD/L over 112 days of operation.

A total of 270 mg of FeCl₃ (98%, Acros, Taipei, Taiwan) was dissolved in 30 mL of deionized water, and then 1 M of NaOH (98%, Fisher, Hampton, VA, USA) 1 mL and 0.5 M oxalic acid (98%, Acros) 750 μL were added to the Teflon tank, and the microwave was set up at 160 °C for 30 min. After that, the Fe₂O₃ sample solution was filtered through a membrane filter with a pore size of 0.2 μm (Advantec, Toyo, Japan) and mixed with the 80 °C deionized water and Fe₂O₃ powder (by freeze-drying at −50 °C) under a vacuum for 5 h. Calcination of Fe₂O₃ was achieved by loading it into a fused aluminum oxide boat and the temperatures were set to 500 °C, 700 °C, and 900 °C for 30 min per/sample in a high temperature furnace. Following the calcination, the furnace was cooled to room temperature by natural convection.

2.3. Experimental Analysis

Electrochemical analysis was performed by the workstation (Jiehan ECW-5600, Taipei, Taiwan) to measure the polarization performance of the Bio-E-Fenton MFCs. For COD analysis, an instrument (SUNTEX-V2000 photometer, Taipei, Taiwan) was utilized with solutions diluted till 100× with deionized water. The temperature for the calcination of Fe_2O_3 (500 °C–900 °C) was applied using high temperature furnaces (Riki JH-2, Taipei, Taiwan). The pH was measured in the anode/cathode chambers by using a pH meter (SUNTEX SP-2300, Taipei, Taiwan) and dissolved oxygen (DO) by using the DO analyzer (CLEAN DO-200, Taipei, Taiwan). For material analysis, instruments like an X-ray diffractometer (Bruker D2 Phaser), field emission scanning electron microscope (FE-SEM) (JSM-6500F), and Field Emission Gun Transmission Electron Microscopy, FEG-TEM (FEI Tecnai™ G2 F-20 S-TWIN), were used.

3. Results and Discussions

3.1. Performance of the Fe_2O_3-C500 °C/CF in Bio-E-Fenton MFCs Power Generation

Polarization curves for the Bio-E-Fenton MFCs using Fe_2O_3 calcination from 500 °C–900 °C are portrayed in Figure 3. The conditions adopted for the polarization test were listed as follows; two-electrode measurement, anode electrode connected (by copper wire) to the working electrode (WE); reference electrode 2 (RE2) was in the dairy wastewater solution; the cathode electrode was connected (by copper wire) to the counter electrode (CE) and reference electrode 1 (RE1) in oily wastewater solution. Here, the carbon felt (60 mm × 60 mm × 5 mm) was selected as the electrode and can be used to calculate the power density (power/working area of electrode) and current density (current/working area of electrode). It was clear that Fe_2O_3-500 °C /CF produced an open circuit voltage of 0.55 V, a maximum current density of 349 mA/m², and a maximum power density of 52.5 mW/m², which was 2.6 fold higher compared to Fe_2O_3-900 °C. A comparison between previous studies of Bio-E-Fenton MFCs and this study has been made and the results are shown in Table 1. In addition, an original report related to TiO_2 has been addressed for showing its optoelectronic ability to generate hydrogen in water [40]. Table 2 shows the list of related cases of photo catalysts.

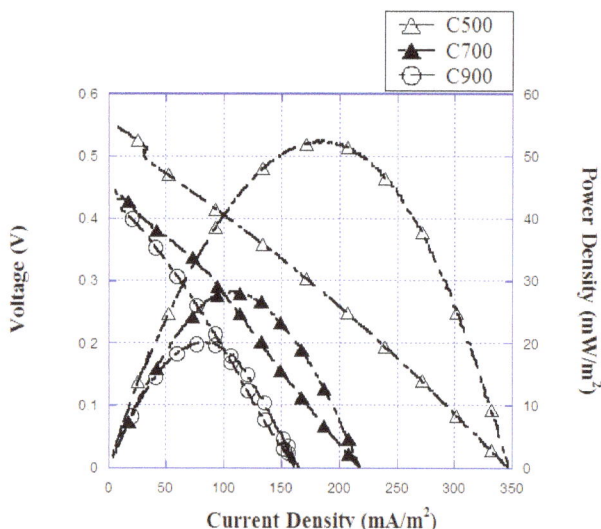

Figure 3. Polarization and power density curves of BEF-MFC at different calcination temperatures of Fe_2O_3.

Table 1. Research studies using Bio-E-Fenton systems.

Power Density (mW/m^2)	COD Removal (%/h)	Ref.
430	78/1080	[13]
625	88/48	[14]
52.5	99.3/1	This study

Table 2. Different kinds of photo catalyst systems employed.

Material	Wavelength (nm)	COD Removal (%/min)	pH	Ref.
Fe@Fe$_2$O$_3$/CNT	555	91.5/120	8	[11]
PANI-Fe$_2$O$_3$	545	98.5/90	8	[15]
MoS$_2$-RGO	450	99.0/60	7	[16]
Fe$_2$O$_3$-C500 °C/CF	410	99.3/60	3	This study

In consideration of the COD degradation rate, the Fe$_2$O$_3$-500 °C/CF performed better with 99.3% removal in 1 h and with an effluent concentration of 152 ± 5 mg-COD/mL/1 day. Generally speaking, a higher COD degradation rate was observed in this study compared to the other studies involving general bio-electro-Fenton systems [41] for the treatment of oily wastewater, with an efficiency that was 1.4 times higher (Table 3). The pH and DO at 700 °C were lower than those at 500 °C and 900 °C because under the constant resistance (1 KΩ) discharge at 700 °C, the DO and pH values affected the voltage output recorded from the long-term measurement. Nevertheless, the results of Table 3 indicated that a better power performance of the system and COD removal rate in the cathode chamber were observed at a calcination temperature of Fe$_2$O$_3$-500 °C, as it had a suitable pH and DO for befitting the biocompatibility [42–44] and electrical conductivity [45,46] in the system. The results showed that the maximum power density was 52.5 mW/m^2 and it was influenced by the calcination temperature of Fe$_2$O$_3$, which obviously showed that the power density decreased with increasing calcination temperature. In addition, the average temperature of the anode chamber was controlled at about 35~45 °C (by Bakelite plate) and the cathode temperature was higher than 70 °C (Shown in Table 3), which would be the effect of evaporation. Therefore, the measurement time was one hour and the time interval was based on the experimental reliability and neglecting the effect of evaporation in the cathode chamber.

Table 3. Performance of system @ calcination conditions of Fe$_2$O$_3$.

Parameter	500 °C	700 °C	900 °C	BF [30]
COD removal (%) @cathode	99.3	83.2	75.2	74.0
DO (ppm) @cathode	4.13	2.82	3.59	2.44
pH$_{average}$	2.1	1.8	2.4	3.0
PD$_{max}$ (mW/m^2)	52.5	28.1	20.1	26.0
Temp. $_{average}$/°C anode	34.6	43.7	42.7	22.6
Temp. $_{average}$/°C cathode	71.6	74.9	72.7	24.1

BF: Bio-electro-Fenton system [30].

To further understand the impact of Fe$_2$O$_3$-C500-/CF on the degradation rate, the following three reasons were formulated: (1) the in situ reaction between Fe^{2+} leached from Fe$_2$O$_3$/FeSO$_4$ and H$_2$O$_2$ generated from the CF was kinetically more favorable than the reaction between H$_2$O$_2$ from the CF and Fe^{2+} in the bulk solution; (2) the calcination of Fe$_2$O$_3$ resulted in the formation of different crystal structures that affected the COD removal efficiency [47]; and (3) the photo catalyst accelerated chemical reactions [48] so that the strong oxidation of the hydroxyl radicals could quickly break the covalent bonds of diesel and enhance the degradation of wastewater [49].

3.2. Characterizations of the Calcinated Fe₂O₃

Figure 4 displays XRD patterns of the calcinated Fe_2O_3. The characteristic peaks of 500 °C Fe_2O_3 were at 24.2°, 33.3°, 35.6°, 40.9°, 49.6°, 54.2°, 62.6°, and 64.2°, corresponding to facet indexes of (012), (104), (110), (113), (024), (116), (214), and (300), respectively. The diffraction peaks at a 2θ value of 35.6° were ascribed to the (110) reflection of Fe_2O_3 [11]. Fe_2O_3 crystallographic structures can be controlled by varying their calcination temperatures. At a high calcination temperature, the intensity did not vary greatly, but the crystalline particle size and surface morphology changed immensely [24].

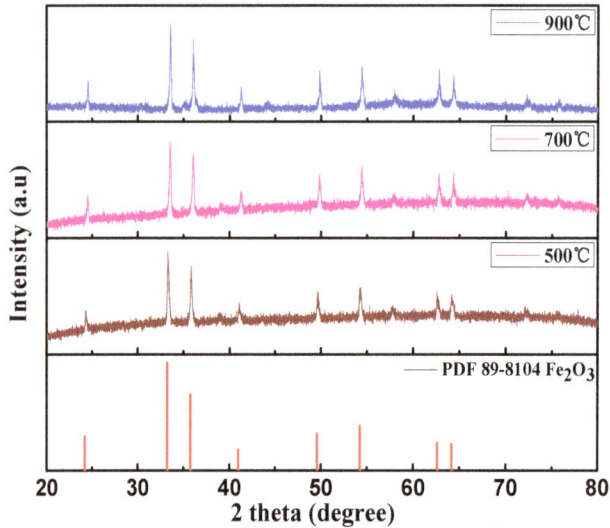

Figure 4. XRD patterns of Fe_2O_3 at different temperatures.

3.3. The Morphologies of the Fe₂O₃ at Different Calcination Temperatures

For analyzing the surface morphology of Fe_2O_3 at different temperatures, field emission scanning electron microscopy was used (Figure 5a–c). The Fe_2O_3 had an average particle size of 581 nm at 500 °C, 984 nm at 700 °C, and 1255 nm at 900 °C. Fe_2O_3 at 500 °C showed a layered stacked morphology and the particles were uniformly distributed with a large surface area. This showed that Fe_2O_3-C at 500 °C was better in terms of surface modification [50–52], which resulted in a maximum power density of 52.5 mW/m². In addition, the surface modification of Fe_2O_3 was indeed the main reason behind its best performance compared to the other systems.

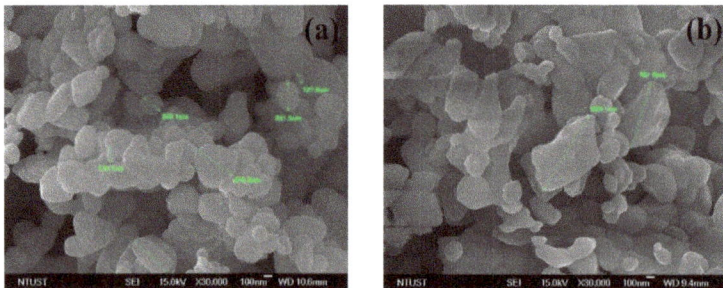

Figure 5. *Cont.*

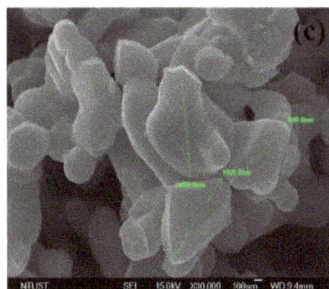

Figure 5. FE-SEM images of Fe_2O_3 at different calcination temperatures (**a**) 500 °C (**b**) 700 °C (**c**) 900 °C.

The TEM images depicted that highly-agglomerated Fe_2O_3 with a uniform size of ~28 nm in diameter affected the material structure of Fe_2O_3 after the calcination processes, as shown in Figure 6a, b. It should be noted that the structures after calcination at 500 °C to 700 °C led to a greater particle size and the lattice plane spacing of 0.21 to 0.26 nm. As shown in Figure 6c, calcination at 900 °C resulted in sheet-like structures and lattice plane spacing of 0.27 nm. The selected area diffraction pattern (SADP) showed excellent crystalline structure formations at 500 °C, which improved the COD degradation efficiency [25–28]. In addition, Fe_2O_3-C500 °C/CF had a better degradation rate of 99.3% in 1 h. This was a novel attempt where the Fe_2O_3-C500 °C/CF was combined with the bio-electro-Fenton reagent in MFCs.

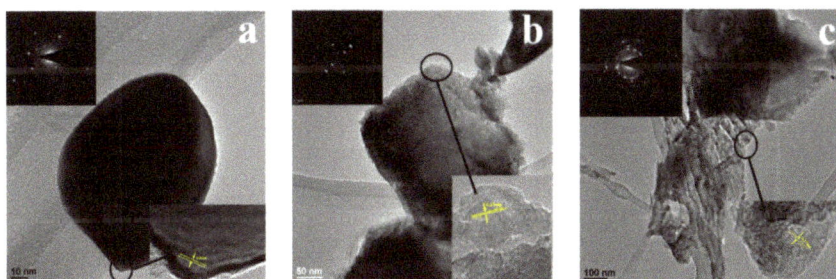

Figure 6. TEM images of Fe_2O_3 at different calcination temperatures (**a**) 500 °C (**b**) 700 °C (**c**) 900 °C.

4. Conclusions

In this study, the photo catalytic ability of iron oxide (Fe_2O_3) was investigated at different calcination temperatures ranging from 500 °C to 900 °C in a Bio-E-Fenton MFC for evaluating its effect on the degradation of dairy and oily wastewaters and power generation. A series of studies were executed and useful findings were addressed as follows: For the study of Fe_2O_3 in Bio-E-Fenton MFCs, firstly, the calcination at 500 °C brought about a better output voltage than in the case of the calcination at 900 °C, and its power density was also 2.6 times higher than at 900 °C. In addition, the COD degradation efficiency showed that Fe_2O_3-C at 500 °C/CF had a better performance of 99.3% removal in 1 h and with an effluent of (152 ± 5 mg-COD/mL/1 day) operation, which was 1.3 times higher than the Fe_2O_3-C at 900 °C/CF. Further, these evidences proved that calcination temperatures of 500 °C with 52.5 mW/m^2 would be optimal for Fe_2O_3 in Bio-E-Fenton MFCs because of its morphology with a uniform particle size and a larger surface area. In contrast, the degradation performance of 80.1% in 1 h was obtained by combining Fe_2O_3-C500 °C/CF with a bio-electro-Fenton reagent under non-illuminated conditions. Last but not the least, this innovative process was very promising for the treatment of organic wastewater treatment and the improved performance of related bio-electrochemical systems in the future.

Acknowledgments: The authors would like to acknowledge the generous funding support from MOST Taiwan under contract #MOST-104-2622-E-197-003-CC3, MOST-103-2221-E-197-022-MY3 and MOST-104-2628-E-011-003-MY3.

Author Contributions: Jung-Chen Wu: Operation of experiments, analysis and writing of article; Wei-Mon Yan: Support of experiments and integrated system; Chin-Tsan Wang: Guidance of experiments and modification of manuscript and reply of comments; Chen-Hao Wang: Guidance of chemical experiments; Yi-Hao Pai: Support and assistance of photo-catalyst; Kai-Chin Wang: Assistance of analysis in chemical part; Yan-Ming Chen: Assistance of analysis in material part; Tzu-Hsuan Lan: Assistance of analysis in material part; Sangeetha Thangavel: English modification of manuscript.

Conflicts of Interest: The authors declare no conflict of interest.

References

1. Khoufi, S.; Aloui, F.; Sayadi, S. Treatment of olive oil mill wastewater by combined process electro-Fenton reaction and anaerobic digestion. *Water Res.* **2006**, *40*, 2007–2016. [CrossRef] [PubMed]
2. Brillas, E.; Casado, J. Aniline degradation by Electro-Fenton® and peroxi-coagulation processes using a flow reactor for wastewater treatment. *Chemosphere* **2002**, *47*, 241–248. [CrossRef]
3. Liu, H.; Wang, C.; Li, X.; Xuan, X.; Jiang, C.; Cui, H.N. A novel electro-Fenton process for water treatment: Reaction-controlled pH adjustment and performance assessment. *Environ. Sci. Technol.* **2007**, *41*, 2937–2942. [CrossRef] [PubMed]
4. Panizza, M.; Oturan, M.A. Degradation of Alizarin Red by electro-Fenton process using a graphite-felt cathode. *Electrochim. Acta* **2011**, *56*, 7084–7087. [CrossRef]
5. Wang, A.; Qu, J.; Ru, J.; Liu, H.; Ge, J. Mineralization of an azo dye Acid Red 14 by electro-Fenton's reagent using an activated carbon fiber cathode. *Dyes Pigment.* **2005**, *65*, 227–233. [CrossRef]
6. Luo, M.; Yuan, S.; Tong, M.; Liao, P.; Xie, W.; Xu, X. An integrated catalyst of Pd supported on magnetic Fe_3O_4 nanoparticles: Simultaneous production of H_2O_2 and Fe^{2+} for efficient electro-Fenton degradation of organic contaminants. *Water Res.* **2014**, *48*, 190–199. [CrossRef] [PubMed]
7. Nidheesh, P.; Gandhimathi, R. Trends in electro-Fenton process for water and wastewater treatment: An overview. *Desalination* **2012**, *299*, 1–15. [CrossRef]
8. Nidheesh, P.; Gandhimathi, R.; Velmathi, S.; Sanjini, N. Magnetite as a heterogeneous electro Fenton catalyst for the removal of Rhodamine B from aqueous solution. *RSC Adv.* **2014**, *4*, 5698–5708. [CrossRef]
9. Ghoneim, M.M.; El-Desoky, H.S.; Zidan, N.M. Electro-Fenton oxidation of Sunset Yellow FCF azo-dye in aqueous solutions. *Desalination* **2011**, *274*, 22–30. [CrossRef]
10. Babuponnusami, A.; Muthukumar, K. A review on Fenton and improvements to the Fenton process for wastewater treatment. *J. Environ. Chem. Eng.* **2014**, *2*, 557–572. [CrossRef]
11. Ai, Z.; Mei, T.; Liu, J.; Li, J.; Jia, F.; Zhang, L.; Qiu, J. Fe@Fe_2O_3 core–shell nanowires as an iron reagent. 3. Their combination with CNTs as an effective oxygen-fed gas diffusion electrode in a neutral electro-Fenton system. *J. Phys. Chem. C* **2007**, *111*, 14799–14803. [CrossRef]
12. Zhuang, L.; Zhou, S.; Li, Y.; Liu, T.; Huang, D. In situ Fenton-enhanced cathodic reaction for sustainable increased electricity generation in microbial fuel cells. *J. Power Sources* **2010**, *195*, 1379–1382. [CrossRef]
13. Li, Y.; Lu, A.; Ding, H.; Wang, X.; Wang, C.; Zeng, C.; Yan, Y. Microbial fuel cells using natural pyrrhotite as the cathodic heterogeneous Fenton catalyst towards the degradation of biorefractory organics in landfill leachate. *Electrochem. Commun.* **2010**, *12*, 944–947. [CrossRef]
14. Wang, Y.; Feng, C.; Li, Y.; Gao, J.; Yu, C.-P. Enhancement of emerging contaminants removal using Fenton reaction driven by H_2O_2-producing microbial fuel cells. *Chem. Eng. J.* **2017**, *307*, 679–686. [CrossRef]
15. Patil, M.R.; Shrivastava, V.S. Adsorption removal of carcinogenic acid violet19 dye from aqueous solution by polyaniline-Fe_2O_3 magnetic nano-composite. *J. Mater. Environ. Sci* **2015**, *6*, 11–21.
16. Li, J.; Liu, X.; Pan, L.; Qin, W.; Chen, T.; Sun, Z. MoS_2–reduced graphene oxide composites synthesized via a microwave-assisted method for visible-light photocatalytic degradation of methylene blue. *RSC Adv.* **2014**, *4*, 9647–9651. [CrossRef]
17. Xu, N.; Zeng, Y.; Li, J.; Zhang, Y.; Sun, W. Removal of 17β-estrodial in a bio-electro-Fenton system: Contribution of oxidation and generation of hydroxyl radicals with the Fenton reaction and carbon felt cathode. *RSC Adv.* **2015**, *5*, 56832–56840. [CrossRef]
18. Wang, C.-T.; Huang, R.-Y.; Lee, Y.-C.; Zhang, C.-D. Electrode material of carbon nanotube/polyaniline carbon paper applied in microbial fuel cells. *J. Clean Energy Technol.* **2013**, *1*, 206–210. [CrossRef]

19. Wang, C.-H.; Wang, C.-T.; Huang, H.-C.; Chang, S.-T.; Liao, F.-Y. High stability pyrolyzed vitamin B12 as a non-precious metal catalyst of oxygen reduction reaction in microbial fuel cells. *RSC Adv.* **2013**, *3*, 15375–15381. [CrossRef]

20. Xu, N.; Zhou, S.; Yuan, Y.; Qin, H.; Zheng, Y.; Shu, C. Coupling of anodic biooxidation and cathodic bioelectro-Fenton for enhanced swine wastewater treatment. *Bioresour. Technol.* **2011**, *102*, 7777–7783. [CrossRef] [PubMed]

21. Fenton, H.J.H. LXXIII.—Oxidation of tartaric acid in presence of iron. *J. Chem. Soc. Trans.* **1894**, *65*, 899–910. [CrossRef]

22. Anotai, J.; Chen, C.-M.; Bellotindos, L.M.; Lu, M.-C. Treatment of TFT-LCD wastewater containing ethanolamine by fluidized-bed Fenton technology. *Bioresour. Technol.* **2012**, *113*, 272–275. [CrossRef] [PubMed]

23. Zhu, X.; Ni, J. Simultaneous processes of electricity generation and *p*-nitrophenol degradation in a microbial fuel cell. *Electrochem. Commun.* **2009**, *11*, 274–277. [CrossRef]

24. Fay, S.; Kroll, U.; Bucher, C.; Vallat-Sauvain, E.; Shah, A. Low pressure chemical vapour deposition of ZnO layers for thin-film solar cells: Temperature-induced morphological changes. *Sol. Energy Mater. Sol. Cells* **2005**, *86*, 385–397. [CrossRef]

25. Tian, C.; Zhang, Q.; Wu, A.; Jiang, M.; Liang, Z.; Jiang, B.; Fu, H. Cost-effective large-scale synthesis of ZnO photocatalyst with excellent performance for dye photodegradation. *Chem. Commun.* **2012**, *48*, 2858–2860. [CrossRef] [PubMed]

26. Yu, C.; Cao, F.; Li, X.; Li, G.; Xie, Y.; Jimmy, C.Y.; Shu, Q.; Fan, Q.; Chen, J. Hydrothermal synthesis and characterization of novel PbWO$_4$ microspheres with hierarchical nanostructures and enhanced photocatalytic performance in dye degradation. *Chem. Eng. J.* **2013**, *219*, 86–95. [CrossRef]

27. Zhou, W.; Sun, F.; Pan, K.; Tian, G.; Jiang, B.; Ren, Z.; Tian, C.; Fu, H. Well-Ordered Large-Pore Mesoporous Anatase TiO$_2$ with Remarkably High Thermal Stability and Improved Crystallinity: Preparation, Characterization, and Photocatalytic Performance. *Adv. Funct. Mater.* **2011**, *21*, 1922–1930. [CrossRef]

28. Xiong, J.; Cheng, G.; Li, G.; Qin, F.; Chen, R. Well-crystallized square-like 2D BiOCl nanoplates: Mannitol-assisted hydrothermal synthesis and improved visible-light-driven photocatalytic performance. *RSC Adv.* **2011**, *1*, 1542–1553. [CrossRef]

29. Feng, C.-H.; Li, F.-B.; Mai, H.-J.; Li, X.-Z. Bio-electro-Fenton process driven by microbial fuel cell for wastewater treatment. *Environ. Sci. Technol.* **2010**, *44*, 1875–1880. [CrossRef] [PubMed]

30. Oonnittan, A.; Sillanpaa, M.E.T. Water treatment by electro-Fenton process. *Curr. Org. Chem.* **2012**, *16*, 2060–2072. [CrossRef]

31. Kim, H.J.; Hyun, M.S.; Chang, I.S.; Kim, B.H. A microbial fuel cell type lactate biosensor using a metal-reducing bacterium, *Shewanella putrefaciens. J. Microbiol. Biotechnol.* **1999**, *9*, 365–367.

32. Diels, A.M.; Callewaert, L.; Wuytack, E.Y.; Masschalck, B.; Michiels, C.W. Inactivation of *Escherichia coli* by high-pressure homogenisation is influenced by fluid viscosity but not by water activity and product composition. *Int. J. Food Microbiol.* **2005**, *101*, 281–291. [CrossRef] [PubMed]

33. Chen, H.S.; Zou, X.G. Treatment in Water Bodies Pollution by Ecological Floating Bed Technology. *China Water Kesources* **2005**, *5*, 022.

34. Toyoda, M.; Inagaki, M. Heavy oil sorption using exfoliated graphite: New application of exfoliated graphite to protect heavy oil pollution. *Carbon* **2000**, *38*, 199–210. [CrossRef]

35. Blumer, M. Oil pollution of the ocean. In *Oil on the Sea*; Springer: Berlin, Germany, 1969; pp. 5–13.

36. Nelson-Smith, A. The problem of oil pollution of the sea. In *Advances in Marine Biology*; Elsevier: Amsterdam, The Netherlands, 1971; Volume 8, pp. 215–306.

37. Molognoni, D.; Chiarolla, S.; Cecconet, D.; Callegari, A.; Capodaglio, A.G. Industrial wastewater treatment with a bioelectrochemical process: Assessment of depuration efficiency and energy production. *Water Sci. Technol.* **2018**, *77*, 134–144. [CrossRef] [PubMed]

38. Marashi, S.; Kariminia, H. Performance of a single chamber microbial fuel cell at different organic loads and pH values using purified terephthalic acid wastewater. *J. Environ. Health Sci. Eng.* **2015**, *13*, 27. [CrossRef] [PubMed]

39. Kim, K.-Y.; Yang, W.; Ye, Y.; LaBarge, N.; Logan, B.E. Performance of anaerobic fluidized membrane bioreactors using effluents of microbial fuel cells treating domestic wastewater. *Bioresour. Technol.* **2016**, *208*, 58–63. [CrossRef] [PubMed]

40. Fujishima, A.; Honda, K. Electrochemical photolysis of water at a semiconductor electrode. *Nature* **1972**, *238*, 37–38. [CrossRef] [PubMed]

41. Xu, N.; Zhang, Y.; Tao, H.; Zhou, S.; Zeng, Y. Bio-electro-Fenton system for enhanced estrogens degradation. *Bioresour. Technol.* **2013**, *138*, 136–140. [CrossRef] [PubMed]

42. Jadhav, G.; Ghangrekar, M. Performance of microbial fuel cell subjected to variation in pH, temperature, external load and substrate concentration. *Bioresour. Technol.* **2009**, *100*, 717–723. [CrossRef] [PubMed]

43. Adani, F.; Baido, D.; Calcaterra, E.; Genevini, P. The influence of biomass temperature on biostabilization–biodrying of municipal solid waste. *Bioresour. Technol.* **2002**, *83*, 173–179. [CrossRef]

44. Liang, C.; Das, K.; McClendon, R. The influence of temperature and moisture contents regimes on the aerobic microbial activity of a biosolids composting blend. *Bioresour. Technol.* **2003**, *86*, 131–137. [CrossRef]

45. Wang, Y.; Li, B.; Zeng, L.; Cui, D.; Xiang, X.; Li, W. Polyaniline/mesoporous tungsten trioxide composite as anode electrocatalyst for high-performance microbial fuel cells. *Biosens. Bioelectron.* **2013**, *41*, 582–588. [CrossRef] [PubMed]

46. Wang, Z.; Zheng, Y.; Xiao, Y.; Wu, S.; Wu, Y.; Yang, Z.; Zhao, F. Analysis of oxygen reduction and microbial community of air-diffusion biocathode in microbial fuel cells. *Bioresour. Technol.* **2013**, *144*, 74–79. [CrossRef] [PubMed]

47. Chakrabarti, S.; Dutta, B.K. Photocatalytic degradation of model textile dyes in wastewater using ZnO as semiconductor catalyst. *J. Hazard. Mater.* **2004**, *112*, 269–278. [CrossRef] [PubMed]

48. Smejkal, Q.; Linke, D.; Bentrup, U.; Pohl, M.-M.; Berndt, H.; Baerns, M.; Brückner, A. Combining accelerated activity tests and catalyst characterization: A time-saving way to study the deactivation of vinylacetate catalysts. *Appl. Catal. A Gen.* **2004**, *268*, 67–76. [CrossRef]

49. Pignatello, J.J.; Oliveros, E.; MacKay, A. Advanced oxidation processes for organic contaminant destruction based on the Fenton reaction and related chemistry. *Crit. Rev. Environ. Sci. Technol.* **2006**, *36*, 1–84. [CrossRef]

50. Rezaei, F.; Richard, T.L.; Logan, B.E. Analysis of chitin particle size on maximum power generation, power longevity, and Coulombic efficiency in solid–substrate microbial fuel cells. *J. Power Sources* **2009**, *192*, 304–309. [CrossRef]

51. Xiao, L.; Damien, J.; Luo, J.; Jang, H.D.; Huang, J.; He, Z. Crumpled graphene particles for microbial fuel cell electrodes. *J. Power Sources* **2012**, *208*, 187–192. [CrossRef]

52. Fan, Y.; Xu, S.; Schaller, R.; Jiao, J.; Chaplen, F.; Liu, H. Nanoparticle decorated anodes for enhanced current generation in microbial electrochemical cells. *Biosens. Bioelectron.* **2011**, *26*, 1908–1912. [CrossRef] [PubMed]

energies

MDPI

Article

Effect of Wall Boundary Layer Thickness on Power Performance of a Recirculation Microbial Fuel Cell

Yan-Ming Chen [1], Chin-Tsan Wang [2,*] and Yung-Chin Yang [1,*]

[1] Institute of Materials Science and Engineering, National Taipei University of Technology, No.1, Sec. 3, Zhongxiao E. Rd., 106 Taipei, Taiwan; ktotheo@hotmail.com
[2] Department of Mechanical and Electro-Mechanical Engineering, National I-Lan University, No.1, Sec. 1, Shennong Rd., 26047 I Lan, Taiwan
* Correspondence: ctwang@niu.edu.tw (C.-T.W.); ycyang@ntut.edu.tw (Y.-C.Y.);
   Tel.: +886-3-935-7400 (ext. 7459) (C.-T.W.); +886-2-2771-2171 (ext. 2762) (Y.-C.Y.)

Received: 23 March 2018; Accepted: 17 April 2018; Published: 20 April 2018

Abstract: Hydrodynamic boundary layer is a significant phenomenon occurring in a flow through a bluff body, and this includes the flow motion and mass transfer. Thus, it could affect the biofilm formation and the mass transfer of substrates in microbial fuel cells (MFCs). Therefore, understanding the role of hydrodynamic boundary layer thicknesses in MFCs is truly important. In this study, three hydrodynamic boundary layers of thickness 1.6, 4.1, and 5 cm were applied to the recirculation mode membrane-less MFC to investigate the electricity production performance. The results showed that the thin hydrodynamic boundary could enhance the voltage output of MFC due to the strong shear rate effect. Thus, a maximum voltage of 21 mV was obtained in the MFC with a hydrodynamic boundary layer thickness of 1.6 cm, and this voltage output obtained was 15 times higher than that of MFC with 5 cm hydrodynamic boundary layer thickness. Moreover, the charge transfer resistance of anode decreased with decreasing hydrodynamic boundary layer thickness. The charge transfer resistance of MFC with hydrodynamic boundary layer of thickness 1.6 cm was 39 Ω, which was 0.79 times lesser than that of MFC with 5 cm thickness. These observations would be useful for enhancing the performance of recirculation mode MFCs.

Keywords: MFC; hydrodynamic boundary layer; recirculation mode; shear rate; voltage; charge transfer resistance

1. Introduction

In recent years, the development of renewable energy (solar energy, wind power, bio-energy, and fuel cell) and electric transportation/storage technologies (supercapacitors) has begun to flourish due to the consumption and pollution that is caused by fossil fuels [1–3]. Microbial fuel cells (MFCs) are renewable energy technology transducers [4], which has the potential for application in wastewater treatment plants [5,6], electrical devices [7,8], and biosensor [9,10], because it can use bacteria for generating electricity from waste water. However, the power densities of MFCs are still low due to their high internal resistances [11–13]. Thus, at present, there are many research works that are focused on MFCs. In MFCs, microorganisms function as a biocatalyst, which can convert organic matter into electricity by using three extracellular electron transfer (EET) pathways: (1) direct electron transfer (c-type cytochromes); (2) electron mediators; and, (3) solid conductive matrix (electrical conductive pili or nanowires) [14]. Nevertheless, the microbes using electron mediators are not suitable for commercial application as they are planktonic and could be easily washed out along with the effluent [15], and also has higher energy losses than other EET pathways [14]. Hence, it obvious that anode biofilm plays a key role in the electricity production of MFCs [16,17], as the bacteria that use direct electron transfer and solid conductive matrix pathway will be attached on the anode electrode.

Recently, there are different methods to enhance the anode biofilm formation. Li et al. demonstrated that the gravity effect being applied to the MFC not only facilitated easy attachment of microbes on the electrode, but also increased the biofilm thickness. Thus, the startup time was shorter than the control by 12%, and the maximum current density reached 8.41 ± 0.13 A/m^2, which was 29% higher than without the gravitational effect [18]. He et al. indicated that if the fiber diameter of anode fiber was lesser than the bacteria size, the thickness of biofilm will be less, which may result in less current density due to the fact that the fibers will form small pores and hinder the bacterial penetration [19]. Ren et al. found that the CNT-based electrode materials: Spin-Spray Layer-by-Layer (SSLbL) CNT can guarantee a thicker biofilm than other CNT materials. Therefore, it can result in a high current density of 2.59 A/m^2 [20]. Although, these studies stated that enhanced biofilm thickness lead to the enhanced electricity production of MFC, but most of the biofilm formation is dependent on natural adsorption [21]. In addition, very thick biofilms will not only block the electron transfer of MFCs [21], but also will become a diffusion barrier on the physical and chemical exchanges between the bacteria and substrate during the mass transfer processes [22].

In MFCs, the hydrodynamic effect is very significant because the major composition of anolyte is water and the physical effect of fluid motion can influence the MFC performance and behavior of biofilm microbial communities [23]. Nowadays, flow rate and recirculation rate are vital flow parameters in continuous and recirculation mode MFCs due to two reasons, first, the flow rate and recirculation rate can be easily set by peristaltic pump; second, the effect of these parameters is equally important as that of the effects of shear stress [23–26]. Increasing the flow/recirculation rate could enhance the shear rate [27,28]. Thus, resulting in a denser and thicker biofilm. Moreover, high shear rates can facilitate faster bacterial attachment on the electrode and shorten the startup time of MFC. Pham et al. indicated that the current and power output of MFC under the shear rate of about 120/s which was higher than that of low shear rate ($\cong 0.3/s$) by 2 to 3 times [27]. On the other hand, increasing the shear stress also can enhance the mass transfer rate between the substrate and the anode biofilm [28,29], and this will efficiently limit the mass transfer losses of MFCs [30,31]. Wang et al. reported that a high recirculation rate of 40 mL/min increased the mass transfer rate of MFC due to high shear stress. Therefore, warburg diffusion resistance of MFC was only 265.4 Ω, which is 73% less than the recirculation rate of 4 mL/min, than, the maximum current density of 50.19 mA/m^2 [31]. Therefore, the shear stress effect is worthy for investigation. Furthermore, the employment of honeycomb and avoiding the membrane in MFC was also important in this study. Because many researches indicated that the use of membranes would not only limit the MFC application [32], but also decrease system performance [33] and enhance the cost [32,34]. Adding a honeycomb pattern could maintain the liquid flow inside a reactor in a homogeneous, symmetrical, and uniform manner, which can enhance the mass transfer of substrates [31].

However, when anolyte with a value of flow/recirculation rate flows through the MFC anode chamber, viscous shearing forces occur at the chamber wall and the hydrodynamic boundary layer is formed on the chamber wall. The thickness of the hydrodynamic boundary layer will increase with an increase in distance from the leading edges of the chamber wall. The flow velocity gradient will result in a strong shear stress within the hydrodynamic boundary layer [35]. Therefore, the effect of the hydrodynamic boundary layer would not only affect the biofilm structure, but it would also influence the mass transfer [36,37]. So, an in-depth understanding of the effects of hydrodynamic boundary layer thickness is necessary. But, the effects of hydrodynamic boundary layer on MFC electricity production has not been well discussed yet. Thus, in this study, three different hydrodynamic boundary layer thicknesses (1.6, 4.1, and 5 cm) were applied to recirculation mode membrane-less MFCs with a honeycomb pattern for investigating the hydrodynamic boundary layer effect on internal resistance.

2. Results and Discussion

2.1. Voltage Output of MFCs

In order to investigate the effect of hydrodynamic boundary layer on MFCs, three different hydrodynamic boundary thickness (1.6, 4.1, and 5 cm) were applied to MFCs, and the voltage output was shown in Figure 1. All three MFCs obtained the similar output voltage at the beginning of the operation. However, it could be observed that the decrease in thickness of the hydrodynamic boundary layer shortened the startup time of MFCs. The voltage of MFC-1 started to rise up by the day 3.5, but MFC-2 and MFC-3 started voltage production 1.5 days later when compared to MFC-1. At day 5, the voltage of MFC-1 increased rapidly, and reached 11 mV at day 6. Although, the voltage of MFC-2 had the similar trend with MFC-1 at day 5, but it did not increase more than MFC-1. The voltage of MFC-2 at day 6 reached ≅4 mV, which was just 0.36 times that of MFC-1. Nevertheless, the voltage of MFC-2 was still higher than MFC-3 and the voltage of MFC-3 was still very similar to the voltage at the beginning of the operation. The reason why less hydrodynamic boundary thickness had faster and higher voltage production was due to the shear rate effect [27]. In a pipe flow, the shear rate is high at the inlet, but it will decrease gradually to the fully developed value [38], which means that a thin hydrodynamic boundary layer could result in a higher shear rate than that of thick boundary layer. The high shear rate could enhance the bacteria attachment to the electrode [27]. In this study, the MFC-1 had the highest shear rate of 0.047/s and MFC-3 has the lowest shear rate of 0.016/s (Table 1). Therefore, the MFC-1 had the shortest startup time, than MFC-2 and MFC-3. In addition, the shear rate of three MFCs were less than 120/s. Therefore, the bacteria were not detached from the electrode [27]. After day 6, the voltage of three MFCs started to decrease. Hence, the anolyte of three MFCs were replaced by fresh medium. However, from day 6 to 10, the voltage of three MFCs failed to rise up to the same level as day 6, and the voltage of MFC-1 even started decreasing. This might be due to the fact that a high concentration of glucose may have a toxic effect on bacteria [39]. Nonetheless, the voltage of MFC-1, MFC-2, and MFC-3 started to increase after day 11.7 and the MFC-1 reached a maximum voltage of 21 mV by day 16, which was 15 times that of MFC-3.

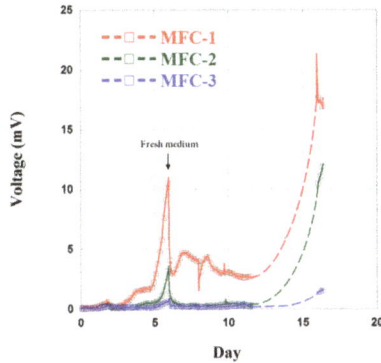

Figure 1. The voltage output versus time of microbial fuel cells (MFCs) with different hydrodynamic boundary layer thicknesses.

Table 1. Internal resistance of MFCs with respect to hydrodynamic boundary layer thickness.

Kinds	δ (cm)	G (1/S)	R_1 (Ω)	$R_{a,g}$ (Ω)	$C_{a,g}$ (C)	$R_{a,t}$ (Ω)	$C_{a,t}$ (C)	R_c (Ω)	C_c (C)	W_c (C)
MFC-1	1.6	0.047	47.4	39	0.0015	12.6	1.5×10^{-6}	15.8	1.6×10^{-6}	0.00086
MFC-2	4.1	0.02	45	45	0.0013	12	0.0025	3.85×10^{-6}	0.00098	0.00016
MFC-3	5	0.016	40.5	49.1	0.0018	11	1.7×10^{-6}	13.5	0.0008	0.00085

2.2. Internal Resistance of MFCs

In order to investigate the hydrodynamic boundary layer thickness effect on the internal resistance of the reactor, the equivalent circuit and internal resistance of MFCs were analyzed, as shown in Figure 2 and Table 1. According to Figure 2, it could be observed that the internal resistance of the MFCs was composed of ohmic losses, charge transfer losses of anode/cathode, and concentration losses. R_1 represented the ohmic losses in the MFCs [40] and the value of ohmic losses can be related to the conductivity of anolyte or the conductivity of electrode material. R_a, which represented the charge transfer losses of anode, which can indicate the activity of the biofilm [41]. Thus, two kind of charge transfer resistances ($R_{a,g}$: graphite felt sheet, $R_{a,t}$: titanium wire) occurred at anode. Capacitance $C_{a,g}$ and $C_{a,t}$ indicated the electrical double layer phenomenon [42] of graphite felt sheet and titanium wire, and this phenomenon occurred at the interface between the electrode and anolyte. R_c and C_c denoted the charge transfer losses and the electrical double layer of cathode. In addition, the W_c represented the concentration losses of MFCs [11], and the concentration losses could describe the diffusion of substrate, oxygen, or other reactants in MFCs.

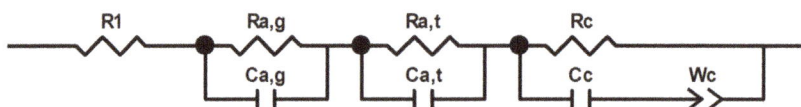

Figure 2. The equivalent circuit of MFC.

In Table 1, the ohmic losses were 47.4, 45, and 40.5 Ω for hydrodynamic boundary layers with thicknesses of 1.6, 4, and 5 cm, respectively. These results showed that the ohmic losses were similar due to the same conductivity of the Phosphate buffered saline (PBS) [31]. On the other hand, it can be clearly seen that the value of $R_{a,g}$ decreased when the hydrodynamic boundary layer thickness decreased. The charge transfer loss was 39 Ω with a hydrodynamic boundary layer thickness of 1.6 cm and increased to 49.1 Ω when the thickness was 5 cm, and these results were in accordance with the voltage output of MFCs (Figure 1). Due to the strong shear rate effect within the thin hydrodynamic boundary layer thickness [35,38], the physical properties of biofilms, such as density and strength, were influenced by fluid shear. Also, the easy attachment of bacteria on the electrode can result in a thicker and denser biofilm structure [27,28,43]. This biofilm can produce higher biomass than that of biofilm in lower shear rate environment [23]. Hence, the electron transfer rate could be enhanced and a lower charge transfer resistance could be obtained. Whereas, the effect of boundary layer thickness did not significantly influence the charge transfer losses of titanium wire and the values of $R_{a,t}$ was very similar. This may be due to the smooth surface of titanium wire, which cannot facilitate easy bacterial attachment [40]. The capacitance of MFC was related to the electron or ion accumulation phenomena in the biofilm or the substrate [36]. In this study, the capacitances ($C_{a,g}$) were 0.0015, 0.0013, 0.0018 C for the hydrodynamic boundary layers of thickness 1.6, 4.1, and 5 cm, respectively. These values were very similar to each other and it can be observed that the thickness of hydrodynamic boundary layer had no effect on the values of $C_{a,g}$ and these results were similar with the research of Ter Heijne et al. [36]. Moreover, the capacitances of $C_{a,t}$ also reported the similar values (around 1.5×10^{-6}), except for the hydrodynamic boundary layer thickness of 4.1 cm. However, this study suggests that electrochemical impedance spectroscopy (EIS) should be analyzed using three-electrode mode for the half-potential for data accuracy. Regarding the internal resistance of the cathode, the cathode of all three MFCs was installed at the top of the reactor, and, the oxygen was directed to the cathode and it did not pass through the hydrodynamic boundary layer. Thus, the thickness of the hydrodynamic boundary layer did not influence the diffusion of oxygen. Hence, the concentration losses (W_c) of three MFCs were similar.

2.3. The Mechanism of Hydrodynamic Boundary Layer Effects in Recirculation Mode MFCs

In recirculation mode MFCs, the hydrodynamics is significantly important as to the microbe, electrode biofilm, and the diffusion of reactant being affected by the flow field [44,45]. Thus, the effect of the flow field in the MFC chamber is necessary to be described. According to the results of this study and other references, the working mechanism model of the hydrodynamic boundary layer effect was illustrated in Figure 3. At first, when the anolyte flowed through the surge tank and passed through the honeycomb into the MFC chamber, the fluid particles of the anolyte in the layer, which were in contact with the chamber wall of MFC had completely stopped due to the no-slip condition. Particles of anolyte also slowed down gradually because of the friction in adjacent layers. Then, in order to compensate the decrease of mass flow rate near the MFC chamber wall, the velocity of the fluid at the midsection of the MFC chamber needed to be increased. Due to the velocity gradient (shear rate) that was developed in the MFC chamber, the velocity (hydrodynamic) boundary layer was formed on the MFC chamber wall [38]. The velocity boundary layer thickness increased along with the flow direction due to the viscosity of liquid [35,46]. Thus, the MFC-1 had the much thinner velocity boundary layer, and the MFC-3 had the thickest velocity boundary layer (see Figure 3).

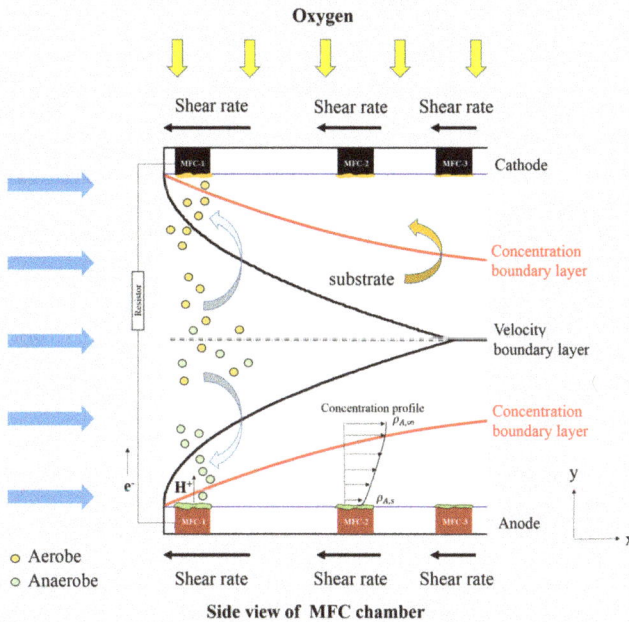

Figure 3. The mechanism of hydrodynamic boundary layer.

However, due to the shear rate effect, the efficiency of bacteria that were attached to the electrode was enhanced [27]. The anaerobic biofilm was formed at the anode electrode because the environment was more anaerobic than cathode, and the aerobic microbes moved to the cathode due to the presence of oxygen. Thin velocity boundary layer could result in a much thicker and denser biofilm than that of the thick velocity boundary layer due to the strong shear rate effect. This could enhance the current and power output of MFC [27]. The biofilm thickness reached maximum within the transition zone between laminar and turbulent because of two reasons, first, the diffusion of substrate was limited in laminar zone, second, the bacteria was detached from the biofilm in turbulent flow [47]. In this study, the inlet Reynolds number was Re = 89.6, which meant that the inertial forces were dominant in MFC, and that the velocity of boundary layer was laminar. Thus, the mass transfer of MFCs needed to be enhanced by increasing the Reynolds number or the flow convective [45]. Although the thickness of the biofilm would become a diffusion barrier [22] the thickness of the biofilm (about 40 μm) was smaller than the hydrodynamic boundary layer [37]. Therefore, the mass transfer of recirculation mode MFC was mainly determined by the thickness of hydrodynamic boundary layer [35,36]. Inside the velocity boundary layer, the concentration boundary layer occurred due to the concentration gradient between the bulk liquid and the chamber wall. The thickness of the concentration boundary layer increased with the flow direction until it reached the velocity boundary layer [48]. Therefore, the thickest concentration boundary layer occurred in MFC-3. In the concentration boundary layer, the mass transfer of reactant relied on diffusion, which meant that a higher mass transfer could be obtained in a thin concentration boundary layer due to a high concentration gradient. In recirculation mode MFCs, due to the moving fluid, the mass transfer not only depended on diffusion, but also on bulk fluid motion. The fluid motion enhanced the mass transfer as the reactant in the fluid can move from high a concentration area to a low concentration area [48]. Hence, the strong shear rate effect in thin velocity boundary layer not only reduced the charge transfer losses, but also the concentration losses of MFCs [31,35,36]. It is worth mentioning that the boundary layer had no effect on the oxygen diffusion as the oxygen was in direct contact with the cathode in this study. Moreover, research works had indicated that the viscous sublayer was also important [49–51] due to the fact that it could affect the biofilm accumulation and this could be suggested as an investigation for prominent future research.

2.4. Applications

Microbial fuel cells can be used in wastewater treatment plants or microfluidic devices in the future. However, the wastewater treatment plants and microfluidic devices are the continuous/recirculation flow mode systems. Thus, the effect of hydrodynamic boundary layer can not be ignored, and it must be considered. The thin hydrodynamic boundary layer could significantly enhance the biofilm formation and substrate mass transfer, and the maintainence of a thin hydrodynamic boundary layer is a challenge in MFC performance and commercialization. There are several methods to maintain the anode/cathode electrode in a thin hydrodynamic boundary layer. Firstly, the electrode placement should be place near the flow entrance, this suggestion could be indirect proof by other research. Ye et al. showed that the biofilm were thick at the beginning of the microchannel, and thin at the end of the microchannel [52]; Second, if the anolyte flows past the porous electrode (graphite/carbon felt) a thin porous electrode of 0.1 mm porosity should be chosen as the substrate transfer and the diffusion of protons from the biofilm will be limited if the flow passes through the thick felt [53,54]; Third, using the flow control method to inhibit the hydrodynamic boundary layer developing, such as increasing the number of anolyte inlet along the MFC flow channel. This method could let the new hydrodynamic boundary layer replace the original hydrodynamic boundary layer, and keep the hydrodynamic boundary layer thickness thin [35]. The results of this research significantly have great contribution on the design of any continuous/recirculation flow mode MFC systems in the future.

3. Materials and Methods

3.1. Reactor Construction

In order to investigate the effects of hydrodynamic boundary layer on the electricity production of MFC, three single chambered reactors were fabricated with Polymethylmethacrylate (PMMA) sheets (Figure 4A) [30]. The reactors were of symmetrical structure, and consisted of one MFC chamber, two honeycombs, and two surge tanks. No membranes were used, as it was a membrane-less reactor. The inner space dimension of MFC chamber was 10 cm × 10 cm × 10 cm. Honeycomb structure was made of plastic straws, and each one had an inner space dimension of 10 cm × 10 cm × 10 cm. The surge tanks had an inner space dimension of 10 cm × 17 cm × 17.5 cm. Two graphite felt sheets (each with L × W × D: 0.5 cm × 2.8 cm × 0.5 cm) were used as the anode and cathode electrode, and they were embedded at the bottom and top of the MFC chambers. The anode electrode was connected using a titanium wire to an external resistance (1000 Ω), which was further connected to the cathode electrode. In order to generate three different hydrodynamic boundary layer thicknesses in the MFC chamber, the anode and cathode electrodes were placed at the distances of 1 cm (MFC-1), 6 cm (MFC-2), and 9 cm (MFC-3), which were measured from the leading edges of the chamber wall.

3.2. Inoculation and Operational Conditions

The MFCs (MFC-1, MFC-2, MFC-3) were inoculated with 1:1 ratio of wastewater from the Department of Biotechnology and Animal Science, National Ilan University, Yilan, Taiwan [11] and 50 mM PBS [31]. 50 mM PBS contained Na_2HPO_4, 4.58 g/L; $NaH_2PHO_4·H_2O$, 2.45 g/L; NH_4Cl, 0.31 g/L; KCl, 0.13 g/L; and, 0.2 g/L glucose was used as the substrate. All of the MFCs were operated at the same flow rate of 480 mL/min from the beginning till the end of the experiment and the experimental setup was shown in Figure 4B.

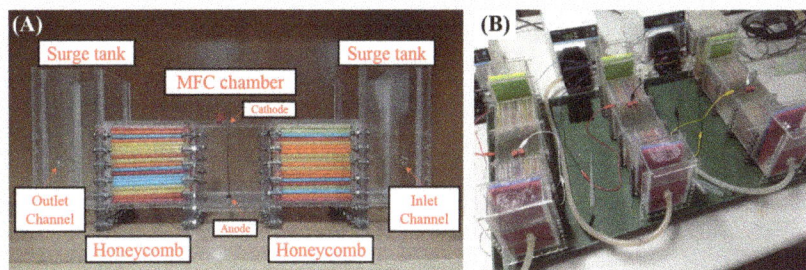

Figure 4. The (**A**) clear side view and (**B**) experimental setup of recirculation mode MFCs.

3.3. Electrochemical Analysis

Three MFCs (MFC-1, MFC-2, and MFC-3) were connected with an external resistance of 1000 Ω. The voltage of MFCs was measured by an automatic data acquisition system (Jiehan 5020, Jiehan Technology Corporation, Taiwan, Republic of China) with a sample rate of 1 point/min [55]. The EIS analysis of MFCs was measured by an impedance analyzer (HIOKY 3522-50 LCR HiTESTER, Hioki E.E. Corporation, Nagano, Japan), and the scan frequency was set as 100 KHz to 0.1 Hz with an AC amplitude of 10 mV [56]. The value of equivalent circuit of MFCs was analyzed by Zware.

3.4. Reynolds Number (Re)

Reynolds number is a dimensionless quantity that can be used to investigate the different fluid flow situations in fluid mechanics. Reynolds number (Re) is a ratio of inertia force to viscous forces, and is defined as shown in Equation (1) [57]:

$$Re = \rho V D_T / \mu \tag{1}$$

where ρ represents the density of water (997 kg/m^3); V indicate the inlet flow velocity (80×10^{-5} m/s, corresponding to the flow rate of 480 mL/min); D_T represents the hydraulic diameter (10 cm); and, μ is viscosity of water (8.9×10^{-4} kg/ms).

3.5. Flat Plate Boundary Layer Thickness (δ) and Shear Rate

The hydrodynamic boundary layer was formed by liquid flow through a flat plate due to the viscosity of liquid particle, and the thickness of flat plate boundary layer thickness, which can be defined as Equation (2) [46]:

$$\delta = 5\sqrt{\mu x/\rho V} \qquad (2)$$

where x is the electrode position of 1 cm (MFC-1), 6 cm (MFC-2), and 9 cm (MFC-3). (Corresponding to the hydrodynamic boundary layer thickness of 1.6, 4.1, and 5 cm [58]), and the schematic diagram of hydrodynamic boundary layer that was applied to MFCs were showed in Figure 5.

The hydrodynamic boundary layer thicknesses were chosen based on different flow field. hydrodynamic boundary of thickness 5 cm occurred at the hydrodynamically fully developed region. In this region, the velocity profile was fully developed and remained unchanged. The hydrodynamic boundary of thickness 1.6 cm occurred near the entrance of the MFC chamber. The velocity profile on this region was similar to the inlet flow.

Shear rate can influence the mass transfer and biofilm formation, and the shear rate is defined as shown in Equation (3) [27]:

$$G = V/d \qquad (3)$$

where V is the flow velocity at hydrodynamic boundary layer (0.99 times less than inlet flow velocity) (m/s), and d represents the thickness of the hydrodynamic boundary layer (m).

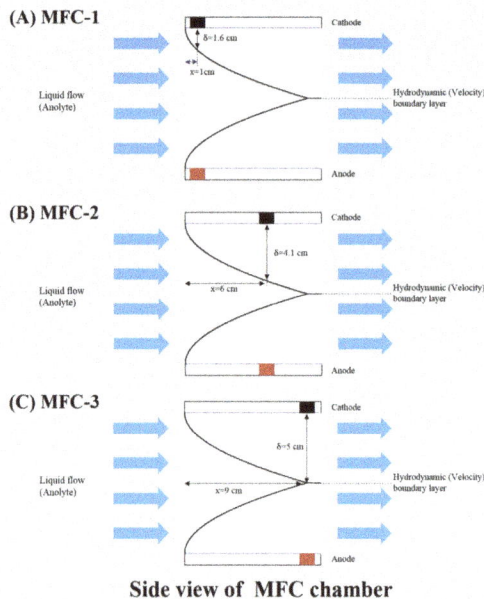

Figure 5. The schematic diagram of hydrodynamic boundary layer in the MFCs.

4. Conclusions

Hydrodynamic boundary layers of thickness 1.6, 4.1, and 5 cm were applied to the recirculation mode MFCs to investigate the electricity production in this study. It was observed that the voltage of MFCs increased with decreasing hydrodynamic boundary layer thicknesses. Due to the high shear rate of 0.047/s that occurred in the hydrodynamic boundary layer of 1.6 cm, the maximum voltage of 21 mV was observed in MFC-1, which was 15 times higher than that of MFC-3. These results were further confirmed by EIS analysis. The charge transfer losses of anode decreased with the decreasing hydrodynamic boundary layer thicknesses. The charge transfer resistance ($R_{a,g}$) of MFC-1 was only 39 Ω, which was only 0.79 times lesser than that of MFC-3. These findings were useful for the improvement in the performance of the continuous/recirculation flow mode MFCs.

Acknowledgments: The authors would like to acknowledge the generous funding support from MOST Taiwan under contract No. MOST 103-2221-E-197-022-MY3, 106-2923-E-197-001-MY3, 106-2622-E-197-007-CC3 and 106-2221-E-197-019, and thank Sangeetha Thangavel (Department of Energy and Refrigerating Air-Conditioning Engineering, National Taipei University of Technology, Taipei, Taiwan) for correcting the English grammar mistakes.

Author Contributions: Yan-Ming Chen: Operation of experiments, analysis and writing of article; Yung-Chin Yang: Guidance of material experiments; Chin-Tsan Wang: Guidance of experiments and modification of manuscript and reply of comments.

Conflicts of Interest: The authors declare no conflict of interest.

References

1. Repp, S.; Harputlu, E.; Gurgen, S.; Castellano, M.; Kremer, N.; Pompe, N.; Wörner, J.; Hoffmann, A.; Thomann, R.; Emen, F.M. Synergetic effects of Fe^{3+} doped spinel $Li_4Ti_5O_{12}$ nanoparticles on reduced graphene oxide for high surface electrode hybrid supercapacitors. *Nanoscale* **2018**, *10*, 1877–1884. [CrossRef] [PubMed]

2. Genc, R.; Alas, M.O.; Harputlu, E.; Repp, S.; Kremer, N.; Castellano, M.; Colak, S.G.; Ocakoglu, K.; Erdem, E. High-capacitance hybrid supercapacitor based on multi-colored fluorescent carbon-dots. *Sci. Rep.* **2017**, *7*, 11222. [CrossRef] [PubMed]

3. Wang, C.-T.; Chen, W.-J.; Huang, R.-Y. Influence of growth curve phase on electricity performance of microbial fuel cell by escherichia coli. *Int. J. Hydrogen Energy* **2010**, *35*, 7217–7223. [CrossRef]

4. Khan, M.D.; Khan, N.; Sultana, S.; Joshi, R.; Ahmed, S.; Yu, E.; Scott, K.; Ahmad, A.; Khan, M.Z. Bioelectrochemical conversion of waste to energy using microbial fuel cell technology. *Process Biochem.* **2017**, *57*, 141–158. [CrossRef]

5. Singh, H.M.; Pathak, A.K.; Chopra, K.; Tyagi, V.; Anand, S.; Kothari, R. Microbial fuel cells: A sustainable solution for bioelectricity generation and wastewater treatment. *Biofuels* **2018**, 1–21. [CrossRef]

6. Wu, Y.-C.; Wang, Z.-J.; Zheng, Y.; Xiao, Y.; Yang, Z.-H.; Zhao, F. Light intensity affects the performance of photo microbial fuel cells with *Desmodesmus* sp. A8 as cathodic microorganism. *Appl. Energy* **2014**, *116*, 86–90. [CrossRef]

7. Wang, C.-T.; Chen, Y.-M.; Qi, Z.-Q.; Wang, Y.-T.; Yang, Y.-C. Types of simplified flow channels without flow obstacles in microbial fuel cells. *Int. J. Hydrogen Energy* **2014**, *39*, 14306–14311. [CrossRef]

8. Yong, X.-Y.; Yan, Z.-Y.; Shen, H.-B.; Zhou, J.; Wu, X.-Y.; Zhang, L.-J.; Zheng, T.; Jiang, M.; Wei, P.; Jia, H.-H. An integrated aerobic-anaerobic strategy for performance enhancement of pseudomonas aeruginosa-inoculated microbial fuel cell. *Bioresour. Technol.* **2017**, *241*, 1191–1196. [CrossRef] [PubMed]

9. Chouler, J.; Cruz-Izquierdo, Á.; Rengaraj, S.; Scott, J.L.; Di Lorenzo, M. A screen-printed paper microbial fuel cell biosensor for detection of toxic compounds in water. *Biosens. Bioelectron.* **2018**, *102*, 49–56. [CrossRef] [PubMed]

10. Sun, J.-Z.; Kingori, G.P.; Si, R.-W.; Zhai, D.-D.; Liao, Z.-H.; Sun, D.-Z.; Zheng, T.; Yong, Y.-C. Microbial fuel cell-based biosensors for environmental monitoring: A review. *Water Sci. Technol.* **2015**, *71*, 801–809. [CrossRef] [PubMed]

11. Wang, C.-T.; Lee, Y.-C.; Ou, Y.-T.; Yang, Y.-C.; Chong, W.-T.; Sangeetha, T.; Yan, W.-M. Exposing effect of comb-type cathode electrode on the performance of sediment microbial fuel cells. *Appl. Energy* **2017**, *204*, 620–625. [CrossRef]

12. Liu, X.-W.; Huang, Y.-X.; Sun, X.-F.; Sheng, G.-P.; Zhao, F.; Wang, S.-G.; Yu, H.-Q. Conductive carbon nanotube hydrogel as a bioanode for enhanced microbial electrocatalysis. *ACS Appl. Mater. Interfaces* **2014**, *6*, 8158–8164. [CrossRef] [PubMed]

13. Wang, Q.-Q.; Wu, X.-Y.; Yu, Y.-Y.; Sun, D.-Z.; Jia, H.-H.; Yong, Y.-C. Facile in-situ fabrication of graphene/riboflavin electrode for microbial fuel cells. *Electrochim. Acta* **2017**, *232*, 439–444. [CrossRef]

14. Torres, C.I.; Marcus, A.K.; Lee, H.-S.; Parameswaran, P.; Krajmalnik-Brown, R.; Rittmann, B.E. A kinetic perspective on extracellular electron transfer by anode-respiring bacteria. *FEMS Microbiol. Rev.* **2009**, *34*, 3–17. [CrossRef] [PubMed]

15. Du, Z.; Li, Q.; Tong, M.; Li, S.; Li, H. Electricity generation using membrane-less microbial fuel cell during wastewater treatment. *Chin. J. Chem. Eng.* **2008**, *16*, 772–777. [CrossRef]

16. Feng, C.; Liu, Y.; Li, Q.; Che, Y.; Li, N.; Wang, X. Quaternary ammonium compound in anolyte without functionalization accelerates the startup of bioelectrochemical systems using real wastewater. *Electrochim. Acta* **2016**, *188*, 801–808. [CrossRef]

17. Wang, Z.; Mahadevan, G.D.; Wu, Y.; Zhao, F. Progress of air-breathing cathode in microbial fuel cells. *J. Power Sources* **2017**, *356*, 245–255. [CrossRef]

18. Li, T.; Zhou, L.; Qian, Y.; Wan, L.; Du, Q.; Li, N.; Wang, X. Gravity settling of planktonic bacteria to anodes enhances current production of microbial fuel cells. *Appl. Energy* **2017**, *198*, 261–266. [CrossRef]

19. He, G.; Gu, Y.; He, S.; Schröder, U.; Chen, S.; Hou, H. Effect of fiber diameter on the behavior of biofilm and anodic performance of fiber electrodes in microbial fuel cells. *Bioresour. Technol.* **2011**, *102*, 10763–10766. [CrossRef] [PubMed]

20. Ren, H.; Pyo, S.; Lee, J.-I.; Park, T.-J.; Gittleson, F.S.; Leung, F.C.; Kim, J.; Taylor, A.D.; Lee, H.-S.; Chae, J. A high power density miniaturized microbial fuel cell having carbon nanotube anodes. *J. Power Sources* **2015**, *273*, 823–830. [CrossRef]

21. Wang, X.; Lin, H.; Wang, J.; Xie, B.; Huang, W. Influence of the biofilm formation process on the properties of biofilm electrode material. *Mater. Lett.* **2012**, *78*, 174–176. [CrossRef]

22. Cabije, A.H.; Agapay, R.C.; Tampus, M.V. Carbon-nitrogen-phosphorus removal and biofilm growth characteristics in an integrated wastewater treatment system involving a rotating biological contactor. *Asia-Pac. J. Chem. Eng.* **2009**, *4*, 735–743. [CrossRef]

23. Oliveira, V.; Simões, M.; Melo, L.; Pinto, A. Overview on the developments of microbial fuel cells. *Biochem. Eng. J.* **2013**, *73*, 53–64. [CrossRef]

24. Ieropoulos, I.; Winfield, J.; Greenman, J. Effects of flow-rate, inoculum and time on the internal resistance of microbial fuel cells. *Bioresour. Technol.* **2010**, *101*, 3520–3525. [CrossRef] [PubMed]

25. Zhang, F.; Jacobson, K.S.; Torres, P.; He, Z. Effects of anolyte recirculation rates and catholytes on electricity generation in a litre-scale upflow microbial fuel cell. *Energy Environ. Sci.* **2010**, *3*, 1347–1352. [CrossRef]

26. Zhang, L.; Zhu, X.; Li, J.; Kashima, H.; Liao, Q.; Regan, J.M. Step-feed strategy enhances performance of unbuffered air-cathode microbial fuel cells. *RSC Adv.* **2017**, *7*, 33961–33966. [CrossRef]

27. Pham, H.T.; Boon, N.; Aelterman, P.; Clauwaert, P.; De Schamphelaire, L.; Van Oostveldt, P.; Verbeken, K.; Rabaey, K.; Verstraete, W. High shear enrichment improves the performance of the anodophilic microbial consortium in a microbial fuel cell. *Microb. Biotechnol.* **2008**, *1*, 487–496. [CrossRef] [PubMed]

28. Shen, Y.; Wang, M.; Chang, I.S.; Ng, H.Y. Effect of shear rate on the response of microbial fuel cell toxicity sensor to Cu (II). *Bioresour. Technol.* **2013**, *136*, 707–710. [CrossRef] [PubMed]

29. Celmer, D.; Oleszkiewicz, J.; Cicek, N. Impact of shear force on the biofilm structure and performance of a membrane biofilm reactor for tertiary hydrogen-driven denitrification of municipal wastewater. *Water Res.* **2008**, *42*, 3057–3065. [CrossRef] [PubMed]

30. Wang, C.-T.; Huang, Y.-S.; Sangeetha, T.; Chen, Y.-M.; Chong, W.-T.; Ong, H.-C.; Zhao, F.; Yan, W.-M. Novel bufferless photosynthetic microbial fuel cell (PMFCs) for enhanced electrochemical performance. *Bioresour. Technol.* **2018**, *255*, 83–87. [CrossRef] [PubMed]

31. Wang, C.-T.; Huang, Y.-S.; Sangeetha, T.; Yan, W.-M. Assessment of recirculation batch mode operation in bufferless bio-cathode microbial fuel cells (MFCs). *Appl. Energy* **2018**, *209*, 120–126. [CrossRef]

32. Jang, J.K.; Chang, I.S.; Kang, K.H.; Moon, H.; Cho, K.S.; Kim, B.H. Construction and operation of a novel mediator-and membrane-less microbial fuel cell. *Process Biochem.* **2004**, *39*, 1007–1012. [CrossRef]

33. Logan, B.E. *Microbial Fuel Cells*; John Wiley & Sons: Hoboken, NJ, USA, 2008.

34. Ghangrekar, M.; Shinde, V. Performance of membrane-less microbial fuel cell treating wastewater and effect of electrode distance and area on electricity production. *Bioresour. Technol.* **2007**, *98*, 2879–2885. [CrossRef] [PubMed]

35. Yang, Y.; Ye, D.; Liao, Q.; Zhang, P.; Zhu, X.; Li, J.; Fu, Q. Enhanced biofilm distribution and cell performance of microfluidic microbial fuel cells with multiple anolyte inlets. *Biosens. Bioelectron.* **2016**, *79*, 406–410. [CrossRef] [PubMed]

36. Ter Heijne, A.; Schaetzle, O.; Gimenez, S.; Fabregat-Santiago, F.; Bisquert, J.; Strik, D.P.; Barriere, F.; Buisman, C.J.; Hamelers, H.V. Identifying charge and mass transfer resistances of an oxygen reducing biocathode. *Energy Environ. Sci.* **2011**, *4*, 5035–5043. [CrossRef]

37. Ter Heijne, A.; Hamelers, H.V.; Saakes, M.; Buisman, C.J. Performance of non-porous graphite and titanium-based anodes in microbial fuel cells. *Electrochim. Acta* **2008**, *53*, 5697–5703. [CrossRef]

38. Yunus, A.C.; Cimbala, J.M. *Fluid Mechanics Fundamentals and Applications, International ed.*; McGraw Hill Publication: New York, NY, USA, 2006; pp. 325–326.

39. D'souza Rohan, V.D.; Rohan, G.; Satish, B. Bioelectricity production from microbial fuel using escherichia coli (glucose and brewery waste). *Int. Res. J. Biol. Sci.* **2013**, *2*, 50–54.

40. Chen, Y.-M.; Wang, C.-T.; Yang, Y.-C.; Chen, W.-J. Application of aluminum-alloy mesh composite carbon cloth for the design of anode/cathode electrodes in escherichia coli microbial fuel cell. *Int. J. Hydrogen Energy* **2013**, *38*, 11131–11137. [CrossRef]

41. Liu, P.; Liang, P.; Jiang, Y.; Hao, W.; Miao, B.; Wang, D.; Huang, X. Stimulated electron transfer inside electroactive biofilm by magnetite for increased performance microbial fuel cell. *Appl. Energy* **2018**, *216*, 382–388. [CrossRef]

42. Zhang, F.; Merrill, M.D.; Tokash, J.C.; Saito, T.; Cheng, S.; Hickner, M.A.; Logan, B.E. Mesh optimization for microbial fuel cell cathodes constructed around stainless steel mesh current collectors. *J. Power Sources* **2011**, *196*, 1097–1102. [CrossRef]

43. Stoodley, P.; Sauer, K.; Davies, D.G.; Costerton, J.W. Biofilms as complex differentiated communities. *Annu. Rev. Microbiol.* **2002**, *56*, 187–209. [CrossRef] [PubMed]

44. Wang, C.-T. Flow control in microbial fuel cells. In *Technology and Application of Microbial Fuel Cells*; InTech: Rijeka, Croatia, 2014.

45. Costerton, J.W.; Lewandowski, Z.; Caldwell, D.E.; Korber, D.R.; Lappin-Scott, H.M. Microbial biofilms. *Annu. Rev. Microbiol.* **1995**, *49*, 711–745. [CrossRef] [PubMed]

46. Schlichting, H.; Gersten, K. *Boundary-Layer Theory*, 7th ed.; McGraw-Hill Book Company: Columbus, OH, USA, 1989; p. 140.

47. Lewandowski, Z.; Walser, G. *Influence of Hydrodynamics on Biofilm Accumulation*; Environmental Engineering; ASCE: Reston, VA, USA, 1991; pp. 619–624.

48. Cengel, A. *Heat and Mass Transfer A Practical Approach*, 3rd ed.; McGraw-Hill Book Company: Columbus, OH, USA, 2006; p. 810.

49. Bryers, J.D.; Characklis, W.G. Processes governing primary biofilm formation. *Biotechnol. Bioeng.* **1982**, *24*, 2451–2476. [CrossRef] [PubMed]

50. Kim, T.W.; Micheli, F. Decreased solar radiation and increased temperature combine to facilitate fouling by marine non-indigenous species. *Biofouling* **2013**, *29*, 501–512. [CrossRef] [PubMed]

51. De Beer, D.; Stoodley, P.; Lewandowski, Z. Liquid flow in heterogeneous biofilms. *Biotechnol. Bioeng.* **1994**, *44*, 636–641. [CrossRef] [PubMed]

52. Ye, D.; Yang, Y.; Li, J.; Zhu, X.; Liao, Q.; Deng, B.; Chen, R. Performance of a microfluidic microbial fuel cell based on graphite electrodes. *Int. J. Hydrogen Energy* **2013**, *38*, 15710–15715. [CrossRef]

53. Sleutels, T.H.; Lodder, R.; Hamelers, H.V.; Buisman, C.J. Improved performance of porous bio-anodes in microbial electrolysis cells by enhancing mass and charge transport. *Int. J. Hydrogen Energy* **2009**, *34*, 9655–9661. [CrossRef]

54. Sleutels, T.H.; Hamelers, H.V.; Buisman, C.J. Effect of mass and charge transport speed and direction in porous anodes on microbial electrolysis cell performance. *Bioresour. Technol.* **2011**, *102*, 399–403. [CrossRef] [PubMed]

55. Wang, C.-T.; Chen, Y.-M.; Hu, Z.-Y.; Chong, W.-T. Dynamic power response of microbial fuel cells under external electrical exciting. *Int. J. Hydrogen Energy* **2017**, *42*, 22208–22213. [CrossRef]

56. Hutchinson, A.J.; Tokash, J.C.; Logan, B.E. Analysis of carbon fiber brush loading in anodes on startup and performance of microbial fuel cells. *J. Power Sources* **2011**, *196*, 9213–9219. [CrossRef]

57. Van Dyke, M.; Van Dyke, M. *An Album of Fluid Motion*; Parabolic Press: Stanford, CA, USA, 1982.

58. Chen, Y.-M.; Wang, C.-T.; Yang, Y.-C.; Wang, Y.-T. Effect of Boudary Layer Thickness on the Performance of Microbial Fuel Cell. In Proceedings of the EMChIE 2015, Tarragona, Spain, 10–12 June 2015.

energies

MDPI

Article

Microbial Biodiesel Production by Direct Transesterification of *Rhodotorula glutinis* Biomass

I-Ching Kuan, Wei-Chen Kao, Chun-Ling Chen and Chi-Yang Yu *

Department of Bioengineering, Tatung University, Taipei 10452, Taiwan; iching@ttu.edu.tw (I.-C.K.);
weijen0219@gmail.com (W.-C.K.); jojogigi38@gmail.com (C.-L.C.)
* Correspondence: chrisyu@ttu.edu.tw; Tel.: +886-2-2182-2928 (ext. 6330)

Received: 8 March 2018; Accepted: 19 April 2018; Published: 24 April 2018

Abstract: (1) Background: Lipids derived from oleaginous microbes have become promising alternative feedstocks for biodiesel. This is mainly because the lipid production rate from microbes is one to two orders of magnitude higher than those of energy crops. However, the conventional process for converting these lipids to biodiesel still requires a large amount of energy and organic solvents; (2) Methods: In this study, an oleaginous yeast, *Rhodotorula glutinis*, was used for direct transesterification without lipid pre-extraction to produce biodiesel, using sulfuric acid or sodium hydroxide as a catalyst. Such processes decreased the amount of energy and organic solvents required simultaneously; (3) Results: When 1 g of dry *R. glutinis* biomass was subject to direct transesterification in 20 mL of methanol catalyzed by 0.6 M H_2SO_4 at 70 °C for 20 h, the fatty acid methyl ester (FAME) yield reached 111%. Using the same amount of biomass and methanol loading but catalyzed by 1 g/L NaOH at 70 °C for 10 h, the FAME yield reached 102%. The acid-catalyzed process showed a superior moisture tolerance; when the biomass contained 70% moisture, the FAME yield was 43% as opposed to 34% of the base-catalyzed counterpart; (4) Conclusions: Compared to conventional transesterification, which requires lipid pre-extraction, direct transesterification not only simplifies the process and shortens the reaction time, but also improves the FAME yield.

Keywords: biodiesel; direct transesterification; *Rhodotorula glutinis*; single cell oil

1. Introduction

Biodiesel is one of the most promising renewable fuels in transportation. It can be used as a drop-in replacement fuel for existing diesel vehicles and boiler engines without major modifications. It is also compatible with current fuel infrastructure [1]. Biodiesel is defined as the fatty acid, alkyl monoesters, derived from renewable feedstocks such as vegetable oils, animal fats, and waste cooking oil. Compared to conventional petrodiesel, biodiesel is highly degradable, non-toxic, and cleaner in exhaust emissions, with the exception of NO_x [2]. The combustion properties of biodiesel are similar to those of petroleum diesel.

Currently, commercial biodiesel is produced by the transesterification of plant oils with short-chain alcohols, using alkaline catalysts such as NaOH and KOH; most of these plant oils are edible, including rapeseed, sunflower, palm, and soybean oil [3]. The biodiesel derived from these oils (first generation biodiesel) has two major drawbacks: (1) the high cost of feedstocks renders biodiesel unable to compete commercially with petrodiesel [4]; (2) the utilization of these edible plants for fuel production may endanger the world's food supply [5]. These problems led to the development of the second generation which mainly utilizes non-edible plant oils such as jatropha, jojoba, and waste cooking oil [6]. However, the supply of these non-edible oils is not likely to meet the global demand for biodiesel. Recently, the use of oil-accumulating microbes as feedstocks for biodiesel production has drawn a lot of attention [7]; the biodiesel derived from these oleaginous microalgae, bacteria, yeasts, and fungi is commonly referred to as the third-generation.

These microbial lipids, also known as single cell oils (SCO), have certain advantages over oils derived from plants. Firstly, these oleaginous microbes can be cultivated all year round with suitable substrates on non-arable lands; secondly, the oil production rate is one to two orders of magnitude higher than those of conventional energy crops [8]. In addition, the lipid compositions of many oleaginous microbes are not too different from those of vegetable oils [9]. This suggests that the fuel properties of biodiesel derived from microbial origins should be similar to those derived from vegetables oils. Among the oleaginous microbes, microalgae have been extensively studied for their potential as alternative feedstocks for biodiesel. Microalgae are mostly phototrophic and capable of accumulating a large amount of biomass and lipids rapidly [10]. Although microalgae are very effective in producing SCO, if grown photo synthetically, they require a large area of land to be cultivated, and are subject to daily and seasonal variations [9,10]. Compared to microalgae, oleaginous yeasts could be a better alternative for the production of SCO. The growth rates of these yeasts are faster than those of microalgae; the duplication time could be less than 1 h in certain cases [11]. Unlike phototrophic microalgae, the growth of oleaginous yeasts is not affected by the variation in weather and sunlight and is less prone to contamination by other microorganisms. Several yeasts are known for their ability to accumulate SCO, such as *Cryptococcus curvatus*, *Lipomyces starkeyi*, *Rhodosporidium toruloides*, and *Rhodotorula glutinis* [12]. *R. glutinis* caught our attention as an ideal SCO-producing strain because it could be cultivated with a variety of low cost carbon sources such as crude glycerol [13] and lignocellulosic biomass hydrolysate [14]. *R. glutinis* is also very effective in accumulating SCO with oil content of up to 72% [9]. *R. glutinis* is also capable of synthesizing carotenoids including β-carotene, torulene, and torularhodin [15]; the composition of these carotenoids depends on the cultivation conditions.

The conventional process for converting yeast SCO to biodiesel includes the following steps in this order: cell disruption, oil extraction, separation and transesterification. In order to decrease the energy expenditure and the amount of solvents used in these steps, many researchers combine the aforementioned steps into one, which is usually described as direct transesterification [16]. The simplified process reduces the overall high cost of biodiesel derived from microbial origins. A fatty acid methyl esters (FAME) yield of up to 98% was reached by direct transesterification of dry *R. toruloides* biomass catalyzed by either H_2SO_4 or HCl [17]. Using the same dry biomass, a similar yield was also obtained by direct transesterification using NaOH [18].

In this work, dry *R. glutinis* biomass was converted to FAME by direct transesterification with either acidic or basic catalysts. The effects of catalyst amount, reaction temperature, incubation time, and methanol loading on FAME yield were studied. In order to evaluate the feasibility of using wet biomass directly, the influences of moisture content on the FAME yield were also examined. Finally, under optimized reaction conditions, the yields and FAME compositions from direct transesterification were compared with those from the conventional processes.

2. Results and Discussion

2.1. Effect of Catalyst Concentration on Transesterification

The average biomass concentration of *R. glutinis* from fermentation was 16 ± 5 g/L, with a lipid content of $39 \pm 6\%$. Simple calculation revealed that 1 g of FAME is expected for 1 g of microbial lipid under complete transesterification. Thus the highest theoretical yield is 100%. When 0.1 and 0.2 g/L NaOH were used as catalysts, no FAME product was detected (Figure 1a). The added NaOH could be consumed by the saponification side reaction because there was an excess of yeast lipids. The FAME yield increased to 18% when the catalyst concentration was 0.5 g/L; the highest FAME yield of 94% was observed with 1 g/L NaOH. However, the FAME yield started to decrease rapidly as the catalyst concentration increased further. No FAME was observed with 4 g/L NaOH; a high alkaline catalyst concentration may facilitate saponification, thus leading to this low FAME yield [19]. Our results indicated that only a narrow range of NaOH concentration was suitable for FAME production. Similar results were observed by others using dry *R. toruloides* biomass as a feedstock [18]. The NaOH concentration of 1 g/L was selected for later experiments.

When 0.05 M H_2SO_4 was used as a catalyst, the FAME yield was only 46% (Figure 1b). The FAME yield increased with the concentration of H_2SO_4, and the highest FAME yield of 103% was observed with 0.6 M H_2SO_4. However, a further increase in the H_2SO_4 concentration to 0.8 M led to a decrease in FAME yield; a similar observation was made by others using a dry *R. toruloides* biomass as a feedstock, and the decrease in FAME yield was explained as a result of side reactions such as polymerization under harsh conditions [17]. The transesterification reaction mixture appeared orange when H_2SO_4 concentration was below 0.1 M, but the color changed to light yellow as the catalyst concentration increased further; the color change may be due to the oxidation of the caroteinoids present in the biomass [20]. The H_2SO_4 concentration of 0.6 M was selected for later experiments.

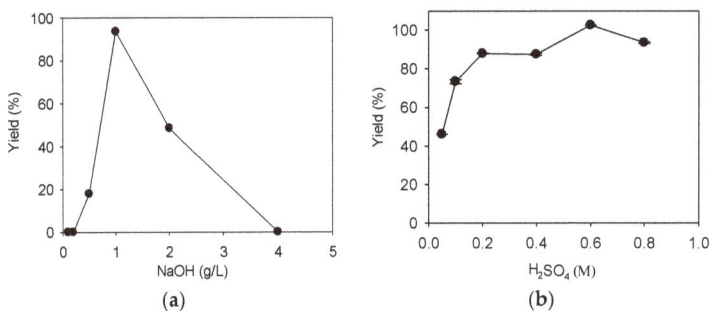

Figure 1. The effect of catalyst concentration on fatty acid methyl ester (FAME) yield. (a) NaOH; (b) H_2SO_4. For both catalysts, direct transesterification on 1 g of dry *R. glutinis* biomass was carried out with a methanol loading of 1:20 (biomass: methanol, w/v) at 70 °C for 15 h. Data were obtained by averaging three individual measurements and the standard deviations were indicated with error bars.

2.2. Effect of Reaction Time and Temperature on Transesterification

The FAME yield obtained by NaOH catalysis was only 49% after 2 h of incubation, while increasing dramatically up to a yield of 102% after 10 h of incubation (Figure 2). Further extension of the reaction time did not improve the yield, indicating that the reaction reached completion after 10 h. When H_2SO_4 was used as a catalyst, the FAME yield increased with the reaction time of up to 20 h with a yield of 111%. More FAME was formed than the available lipid which could be explained by the following reasons: (1) The lipid content within the biomass was underestimated; (2) some phopholipids were also converted to FAME [21,22]. Further increase in the reaction time beyond 20 h did not improve the FAME yield. For later experiments, the reaction times of 10 and 20 h were used for NaOH and H_2SO_4 catalysts, respectively.

With a 10 h reaction time and NaOH as a catalyst; the FAME yields at 50 and 60 °C were only 61% and 94%, respectively; even increasing the reaction duration up to 20 h did not enhance the FAME yield. Thus, 70 °C was optimal for FAME production catalyzed by NaOH. A similar observation was also obtained with catalysis by H_2SO_4. A previous study reported an optimal temperature of 50 °C for converting the dry biomass of oleaginous *R. toruloides* directly to biodiesel within a lidded test tube [18]. The difference in the optimal reaction temperature might be attributed to a different reaction setup. In later experiments, for both NaOH and H_2SO_4, a reaction temperature of 70 °C was used.

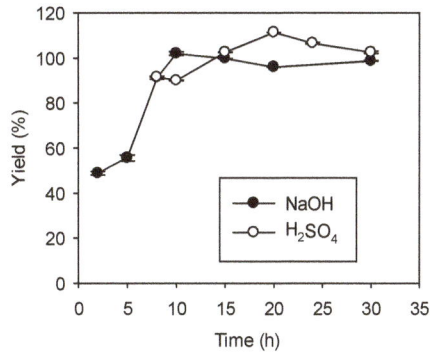

Figure 2. The effect of reaction time on FAME yield. Direct transesterification on 1 g of dry *R. glutinis* biomass was carried out with a methanol loading of 1:20 (biomass: methanol, *w/v*) at 70°C using 1 g/L NaOH or 0.6 M H_2SO_4 as a catalyst. Data were obtained by averaging three individual measurements and the standard deviations were indicated with error bars.

2.3. Effect of Methanol Loading on Transesterification

Methanol serves as one of the reactants and also acts as a solvent to weaken and disrupt the cell walls [23]. The effect of methanol loading on transesterification is shown in Figure 3. When NaOH was used as a catalyst, almost no FAME was formed with a biomass to methanol ratio (*w/v*) of 1:10; this could be because themethanol loading is insufficient to disrupt the cell walls effectively. When the biomass to methanol ratio was increased to 1:20, the highest yield of 102% was obtained. Nevertheless, the yield started to decrease as the ratio was further increased; similar results were also observed by others [24]. One possible explanation for the decreased yield at high methanol loading is that excess methanol dilutes the biomass and catalyst concentrations, thus giving rise to a lower conversion [18,24]. On the other hand, when H_2SO_4 was employed as a catalyst, the methanol loading had little influence on the FAME yield within the range we studied. For both catalysts, a biomass to methanol ratio (*w/v*) of 1:20 was used for later experiments.

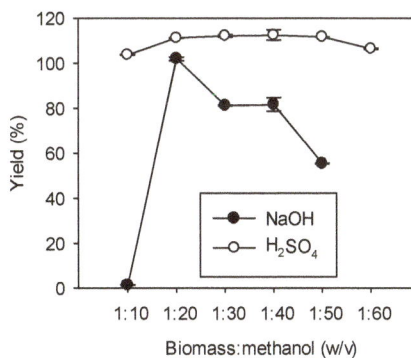

Figure 3. The effect of methanol loading on transesterification. Direct transesterification on 1 g of dry *R. glutinis* biomass was carried out at 70 °C with 1 g/L NaOH and 0.6 M H_2SO_4 for 10 and 20 h, respectively. Data were obtained by averaging three individual measurements and the standard deviations were indicated with error bars.

2.4. Effect of Moisture on Transesterification

In an esterification reaction, the moisture content should be minimized because the presence of water favors the generation of free fatty acids from the hydrolysis of triglycerides, thus enhancing the degree of saponification and lowering the FAME yield. Previous studies on the transesterification of vegetable oils have shown that the presence of water even at very low concentrations has a detrimental effect on the FAME yield [25]. To determine the effect of moisture content on the FAME yield, various amounts of water were added back to 1 g of dry biomass; the moisture content was defined by the following equation:

$$\text{moisture content (\%)} = \text{amount of added water (g)}/(\text{amount of added water (g)} + 1 \text{ g of biomass})$$ (1)

Using NaOH as a catalyst, the FAME yield was decreased with an increase in the moisture content (Figure 4); the FAME yield was 34% when the biomass contained 70% moisture. Our results were quite different from those reported for *R. toruloides* (also using NaOH as a catalyst), which showed that there was almost no formation of FAME with moisture content above 10% [18]. In the case of H_2SO_4, the process tolerated moisture fairly well up to 50%. However, the FAME yield decreased rapidly to 85% and 43% when the biomass contained 60% and 70% moisture, respectively. Similar results were obtained by others using dry microalgae biomass, *Chaetoceros gracilis*, as a feed stock with a higher methanol loading of 1:40 [21]. We have also examined the feasibility of using a wet *R. glutinis* biomass collected from centrifugation without drying (the moisture content was 72%) for direct transesterification. However, this included a heating step at 80°C for 10 min before centrifugation. Consequently, the FAME yield could reach 73%. Yellapu et al. reported that the yield could be further improved by pre-treating the wet biomass with surfactants such as *N*-lauroyl sarcosine, and the reaction time may also be shortened by carrying out the reaction under sonication [26].

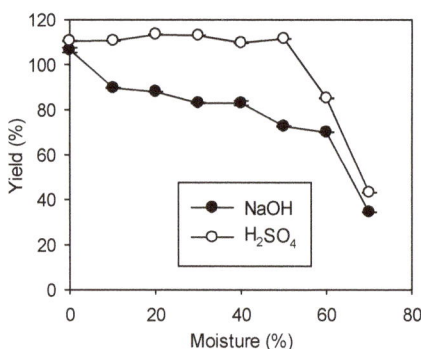

Figure 4. The effect of moisture on FAME yield. Direct transesterification on 1 g of dry *R. glutinis* biomass was carried out at 70 °C with 1 g/L NaOH and 0.6 M H_2SO_4 for 10 and 20 h, respectively. A methanol loading of 1:20 (biomass: methanol, *w/v*) was used for both catalysts. Data were obtained by averaging three individual measurements and the standard deviations were indicated with error bars.

2.5. Comparison of Yields and Compositions of FAME Derived from Different Transesterification Methods

The yields and compositions of FAME produced by different transesterification methods are listed in Table 1. The conventional methods resulted in lower FAME yields of 78% and 85% using NaOH and H_2SO_4 as catalysts, respectively; the acidic method had a yield similar to that of the previous report (81%), using the same biomass [27]. Direct transesterification, regardless of the catalyst used, resulted in significantly higher yields in a shorter time (approximately 4 h less) than conventional methods, which involved cell disruption, lipid extraction, and transesterification in that order. For direct transesterification, the compositions of FAME obtained with either catalyst were almost the same; the oleic acid methyl ester was the dominant FAME, followed by linoleic, and then palmitic acid methyl ester. However, a previous report showed that the FAME composition was catalyst dependent, especially for oleic and linolenic acid methyl esters [18]. The content of linoleic acid methyl ester from direct transesterification was about 20% higher than the linoleic acid content of *R. glutinis* cultured with glycerol as a carbon source [13,14]. The higher amount of linoleic acid methyl ester was explained by the fact that some FAME was derived from the phospholipids in the cell membrane [22,23]. When conventional methods were used, the linoleic acid methyl ester was the dominant FAME, followed by oleic acid methyl ester instead.

Table 1. Yields and compositions of FAME derived from *R. glutinis* using different transesterification methods.

Method	FAME Yield (%) [a]	FAME Composition (%)					
		C16:0	C16:1	C18:0	C18:1	C18:2	C18:3
Direct acid-catalyzed	111.0 ± 0.1	16.8	0.7	4.1	42.2	33.8	2.5
Direct base-catalyzed	102.0 ± 0.8	18.8	0.7	4	41.1	32.9	2.6
Conventional acid-catalyzed [b]	85.2 ± 0.9	18.1	1.6	3	33.3	41.1	4.1
Conventional base-catalyzed [c]	77.9 ± 0.7	21	1	2.5	30.5	41.2	3.7

[a] Data were obtained by averaging three individual measurements. [b] The lipid was first extracted as described in Methods (Total lipid analysis) except that 1 g of dry biomass was used and reagents required were scaled-up accordingly. Transesterification of extracted lipid was carried out with 0.6 M H_2SO_4 for 20 h at 70 °C, using a methanol loading of 1:20 (biomass: methanol, w/v). [c] The lipid was first extracted as described in Methods (Total lipid analysis) except that 1 g of dry biomass was used and the reagents required were scaled-up accordingly. The transesterification of extracted lipid was carried out with 1 g/L NaOH for 10 h at 70 °C, using a methanol loading of 1:20 (biomass: methanol, w/v).

3. Materials and Methods

3.1. Microorganism and Medium

The seed medium for culturing *R. glutinis* BCRC (Bioresource Collection and Research Center, Hsinchu, Taiwan) 22,360 contained 3 g of yeast extract, 3 g of malt extract, 5 g of peptone, and 10 g of dextrose per liter. The fermentation medium contained 80 g of crude glycerol (Yu-Hwa Biodiesel Company, Changhua, Taiwan), 2 g of yeast extract, 2 g of $(NH_4)_2SO_4$, 1 g of KH_2PO_4, 0.5 g of $MgSO_4 \cdot 7H_2O$, 0.1 g of $CaCl_2$, and 0.1 g of NaCl per liter. The crude glycerol, dark brown in color and slightly viscous, was used directly without any pretreatment or purification.

3.2. Production of Biomass

Under aerobic conditions, 300 mL of seed culture was prepared in a seed medium after a 24-h incubation at 24 °C with orbital shaking at 150 rpm. The seed culture was inoculated into a 5-L stirred-tank fermentor (BIOSTAT® A plus, Sartorius, Gottingen, Germany) containing 2.7 L of fermentation medium (3 L working volume). The pH was maintained at 5.5 by automatically feeding 1.0 N NaOH or 1.0 N HCl solution into the medium. The fermentor was operated at 24 °C with dissolved oxygen controlled at 30, ±2% of the saturation level. The agitation during the process was limited to a range from 200 to 400 rpm to avoid potential damage resulting from high shear force. The biomass was harvested after 72 h of incubation by centrifugation at $12,021 \times g$, and then the pellets were washed with deionized water to remove any residual fermentation medium. The biomass was stored at −80 °C for 24 h before lyophilized in a Freezemobile 12 XL freeze dryer (Virtis). The dried biomass was stored in a refrigerator for later use.

3.3. Total Lipid Analysis

The extraction of lipids from a biomass was modified from the procedure reported by Bligh and Dyer [28]. Fifty milliliters of chloroform/methanol mixture (2:1, v/v) was mixed with 0.5 g of finely ground powder of dry biomass. The suspension was then subject to ultrasonication with a Misonix XL2020 sonicator for 6 min on an ice bath; the ultrasonic horn was directly immersed in the suspension and the on/off cycle was set to 60/60 s in order to minimize the heat generated. The suspension was incubated at room temperature for 1 h on a rocking mixer at 100 rpm. The solvent phase was collected by centrifugation at $12,100 \times g$ for 10 min; the biomass was extracted with an additional 30 mL of chloroform/methanol mixture for 1 h. The combined solvent phase was evaporated at 65 °C under vacuum (337 mbar), and then the remaining lipid was weighted. The lipid content was defined as below:

$$\text{lipid content (\%)} = \text{weight of lipid (g)} / \text{weight of biomass (g)} \times 100 \tag{2}$$

3.4. Direct Transesterification

One gram of dry biomass and various amounts of methanol (biomass to methanol ratio in the range of 1:10 to 1:60, w/v) were added to a 100 mL round-bottom flask fitted with a condenser. NaOH (from 0.1 to 4 g/L) or H_2SO_4 (from 0.05 to 0.8 M) dissolved in methanol was used as a catalyst. The suspension was heated at 40 to 70 °C in a water bath under atmospheric pressure with vigorous mixing for 2 to 30 h, followed by a centrifugation at $12,100 \times g$ for 10 min and a collection of the supernatant. The residual biomass was then washed with 10 mL of *n*-hexane and the washing fraction was collected after centrifugation as described above. The supernatant and the washing fraction were combined and extracted with equal volume of *n*-hexane. The *n*-hexane layer was collected and the solvent was evaporated at 65 °C under vacuum (337 mbar). The biodiesel product was collected and weighted. The experiments were performed in triplicate; the average from triplicate was plotted in the figures and the standard deviation was indicated with an error bar.

3.5. Analysis of Fatty Acid Methyl Esters

The content of FAME in the transesterification product was measured according to the standard of Taiwan, CNS-15051. The biodiesel product was first treated with sodium sulfate followed by centrifugation at $2040 \times g$ for 5 min. Fifty microliters of the treated sample was mixed with 1 mL of 10 mg/mL methyl heptadecanoate in hexane as an internal standard. One microliter of the sample was injected into a gas chromatograph (GC-2014, Shimadzu, Kyoto, Japan) equipped with a flame-ionization detector (FID). A BPX70 capillary column (30 m × 0.25 mm i.d.; SGE Analytical Science, Ringwood, Australia) with nitrogen as carrier gas was used. The injector and FID temperatures were both set at 250 °C. The oven temperature was initially held at 150 °C for 30 s and then increased to 180 °C at 10 °C/min, finally to 198 °C at 1.5 °C/min. The content of FAME was calculated using the following equation:

$$\text{FAME content (\%)} = \frac{\sum A - A_{EI}}{A_{EI}} \times \frac{C_{EI} \times V_{EI}}{m} \times 100 \tag{3}$$

where $\sum A$ is the summation of peak area of all the FAME peaks (from C14:0 to C24:1); A_{EI} is the peak area of internal standard, methyl heptadecanoate; C_{EI} is the concentration of methyl heptadecanoate, V_{EI} is the volume of methyl heptadecanoate, and m is the mass of the biodiesel sample. The FAME yield was calculated using the following equation:

$$\text{FAME yield (\%)} = \frac{\text{FAME content (\%)} \times \text{weight of biodiesel (g)}}{\text{lipid content (\%)} \times \text{weight of biomass (g)}} \times 100 \tag{4}$$

4. Conclusions

In this study, the oleaginousyeast, *R. glutinis*, was used for the direct transesterification without lipid pre-extraction to produce FAME. The use of chloroform, a highly toxic organic solvent often applied in lipid extraction, was avoided. The highest FAME yield of 111% was obtained with 0.6 M H_2SO_4 as a catalyst and a methanol loading of 1:20 (biomass to methanol, w/v), the reaction was carried out at 70 °C for 20 h. Using the same methanol loading and temperature, a slightly lower yield of 102% was obtained with 1 g/L NaOH after 10 h of incubation. Although basic catalysts are often avoided because of the potential saponification caused by the free fatty acids present in SCO, the NaOH catalyst was still able to achieve similar FAME yield in half the reaction time required by the H_2SO_4 catalyst. However, the acid-catalyzed direct transesterification tolerates moisture content much better than the base-catalyzed process; thus, the H_2SO_4 catalyst is more suitable for wet biomass feedstocks. Compared to conventional transesterification, the direct process is not only simpler, but also improves the FAME yield by 24–29% with a shorter reaction time and a lower energy consumption. The reaction processes reported in this work simplify for the production of SCO-derived biodiesel and reduce its cost. Our preliminary work also indicates that wet biomass can be used directly as a feedstock with some minor adjustments in reaction conditions, suggesting that further process simplification and cost reduction are possible.

Author Contributions: Chi-Yang Yu and I-Ching Kuan designed the experiments and analyzed the data. Wei-Chen Kao and Chun-Ling Chen performed the experiments using the basic and acidic catalyst, respectively. Chi-Yang Yu and I-Ching Kuan prepared the manuscript. All authors read and approved the final manuscript. All authors contributed equally to this work.

Acknowledgments: We thank Hong-Wei Yen (Department of Chemical and Materials Engineering, Tunghai University, Taiwan) and Yaw-Nan Chang (Department of Biotechnology, National Formosa University, Taiwan) for providing *R. glutinis* BCRC 22360.This work was supported by the Ministry of Science and Technology (MOST 105-2621-M-036-001-MY2) in Taiwan.

Conflicts of Interest: The authors declare no conflict of interest.

References

1. Sitepu, I.R.; Garay, L.A.; Sestric, R.; Levin, D.; Block, D.E.; German, J.B.; Boundy-Mills, K.L. Oleaginous yeasts for biodiesel: Current and future trends in biology and production. *Biotechnol. Adv.* **2014**, *32*, 1336–1360. [CrossRef] [PubMed]

2. Canakci, M.; Sanli, H. Biodiesel production from various feedstocks and their effects on the fuel properties. *J. Ind. Microbiol. Biotechnol.* **2008**, *35*, 431–441. [CrossRef] [PubMed]

3. Fukuda, H.; Kondo, A.; Noda, H. Biodiesel fuel production by transesterification of oils. *J. Biosci. Bioeng.* **2001**, *92*, 405–416. [CrossRef]

4. Demirbas, A. Importance of biodiesel as transportation fuel. *Energy Policy* **2007**, *35*, 4661–4670. [CrossRef]

5. Escobar, J.C.; Lora, E.S.; Venturini, O.J.; Yáñez, E.E.; Castillo, E.F.; Almazan, O. Biofuels: Environment, technology and food security. *Renew. Sustain. Energy Rev.* **2009**, *13*, 1275–1287. [CrossRef]

6. Kulkarni, M.G.; Dalai, A.K. Waste cooking oil-an economical source for biodiesel: A review. *Ind. Eng. Chem. Res.* **2006**, *45*, 2901–2913. [CrossRef]

7. Huang, C.; Chen, X.-F.; Xiong, L.; Chen, X.-D.; Ma, L.-L.; Chen, Y. Single cell oil production from low-cost substrates: The possibility and potential of its industrialization. *Biotechnol. Adv.* **2013**, *31*, 129–139. [CrossRef] [PubMed]

8. Chen, C.-Y.; Yeh, K.-L.; Aisyah, R.; Lee, D.-J.; Chang, J.-S. Cultivation, photobioreactor design and harvesting of microalgae for biodiesel production: A critical review. *Bioresour. Technol.* **2011**, *102*, 71–81. [CrossRef] [PubMed]

9. Meng, X.; Yang, J.; Xu, X.; Zhang, L.; Nie, Q.; Xian, M. Biodiesel production from oleaginous microorganisms. *Renew. Energy* **2009**, *34*, 1–5. [CrossRef]

10. Chisti, Y. Biodiesel from microalgae. *Biotechnol. Adv.* **2007**, *25*, 294–306. [CrossRef] [PubMed]

11. Yousuf, A.; Khan, M.R.; Islam, M.A.; Wahid, Z.A.; Pirozzi, D. Technical difficulties and solutions of direct transesterification process of microbial oil for biodiesel synthesis. *Biotechnol. Lett.* **2017**, *39*, 13–23. [CrossRef] [PubMed]

12. Ratledge, C.; Cohen, Z. Microbial and algal oils: Do they have a future for biodiesel or as commodity oils? *Lipid Technol.* **2008**, *20*, 155–160. [CrossRef]

13. Yen, H.-W.; Zhang, Z. Effects of dissolved oxygen level on cell growth and total lipid accumulation in the cultivation of *Rhodotorula glutinis*. *J. Biosci. Bioeng.* **2011**, *112*, 71–74. [CrossRef] [PubMed]

14. Yen, H.-W.; Chang, J.-T. Growth of oleaginous *Rhodotorula glutinis* in an internal-loop airlift bioreactor by using lignocellulosic biomass hydrolysate as the carbon source. *J. Biosci. Bioeng.* **2015**, *119*, 580–584. [CrossRef] [PubMed]

15. Kot, A.M.; Błażejak, S.; Kurcz, A.; Gientka, I.; Kieliszek, M. *Rhodotorula glutinis*—Potential source of lipids, carotenoids, and enzymes for use in industries. *Appl. Microbiol. Biotechnol.* **2016**, *100*, 6103–6117. [CrossRef] [PubMed]

16. Cheirsilp, B.; Louhasakul, Y. Industrial wastes as a promising renewable source for production of microbial lipid and direct transesterification of the lipid into biodiesel. *Bioresour. Technol.* **2013**, *142*, 329–337. [CrossRef] [PubMed]

17. Liu, B.; Zhao, Z. Biodiesel production by direct methanolysis of oleaginous microbial biomass. *J. Chem. Technol. Biotechnol.* **2007**, *82*, 775–780. [CrossRef]

18. Thliveros, P.; Uçkun Kiran, E.; Webb, C. Microbial biodiesel production by direct methanolysis of oleaginous biomass. *Bioresour. Technol.* **2014**, *157*, 181–187. [CrossRef] [PubMed]

19. Leung, D.Y.C.; Guo, Y. Transesterification of neat and used frying oil: Optimization for biodiesel production. *Fuel Process. Technol.* **2006**, *87*, 883–890. [CrossRef]

20. Boon, C.S.; McClements, D.J.; Weiss, J.; Decker, E.A. Factors influencing the chemical stability of carotenoids in foods. *Crit. Rev. Food Sci. Nutr.* **2010**, *50*, 515–532. [CrossRef] [PubMed]

21. Wahlen, B.D.; Willis, R.M.; Seefeldt, L.C. Biodiesel production by simultaneous extraction and conversion of total lipids from microalgae, cyanobacteria, and wild mixed-cultures. *Bioresour. Technol.* **2011**, *102*, 2724–2730. [CrossRef] [PubMed]

22. Vicente, G.; Bautista, L.F.; Rodríguez, R.; Gutiérrez, F.J.; Sádaba, I.; Ruiz-Vázquez, R.M.; Torres-Martínez, S.; Garre, V. Biodiesel production from biomass of an oleaginous fungus. *Biochem. Eng. J.* **2009**, *48*, 22–27. [CrossRef]

23. Zhang, X.; Yan, S.; Tyagi, R.D.; Surampalli, R.Y.; Valéro, J.R. Ultrasonication aided *in-situ* transesterification of microbial lipids to biodiesel. *Bioresour. Technol.* **2014**, *169*, 175–180. [CrossRef] [PubMed]

24. Patil, P.D.; Gude, V.G.; Mannarswamy, A.; Cooke, P.; Munson-McGee, S.; Nirmalakhandan, N.; Lammers, P.; Deng, S. Optimization of microwave-assisted transesterification of dry algal biomass using response surface methodology. *Bioresour. Technol.* **2011**, *102*, 1399–1405. [CrossRef] [PubMed]

25. Canakci, M.; Van Gerpen, J. Biodiesel production via acid catalysis. *Trans. ASAE* **1999**, *42*, 1203–1210. [CrossRef]

26. Yellapu, S.K.; Kaur, R.; Tyagi, R.D. Detergent assisted ultrasonication aided in situ transesterification for biodiesel production from oleaginous yeast wet biomass. *Bioresour. Technol.* **2017**, *224*, 365–372. [CrossRef] [PubMed]

27. Dai, C.-C.; Tao, J.; Xie, F.; Dai, Y.-J.; Zhao, M. Biodiesel generation from oleaginous yeast *Rhodotorula glutinis* with xylose assimilating capacity. *Afr. J. Biotechnol.* **2007**, *6*, 2130–2134.

28. Bligh, E.G.; Dyer, W.J. A rapid method of total lipid extraction and purification. *Can. J. Biochem. Physiol.* **1959**, *37*, 911–917. [CrossRef] [PubMed]

energies

MDPI

Article

Analysis of Syngas Production from Biogas via the Tri-Reforming Process

Rei-Yu Chein * and Wen-Hwai Hsu

Department of Mechanical Engineering, National Chung Hsing University, Taichung City 40227, Taiwan; dawnxcirno@gmail.com
* Correspondence: rychein@dragon.nchu.edu.tw; Tel.: +886-4-2284-0433

Received: 25 March 2018; Accepted: 25 April 2018; Published: 27 April 2018

Abstract: The tri-reforming process was employed for syngas production from biogas at elevated pressures in this study. In the tri-reforming process, air and water were added simultaneously as reactants in addition to the main biogas components. The effects of various operating parameters such as pressure, temperature and reactant composition on the reaction performance were studied numerically. From the simulated results, it was found that methane and carbon dioxide conversions can be enhanced and a higher hydrogen/carbon monoxide ratio can be obtained by increasing the amount of air. However, a decreased hydrogen yield could result due to the reverse water–gas shift reaction. A higher level of methane conversion and hydrogen/carbon monoxide ratio can be obtained with increased water addition. However, negative carbon dioxide conversion could result due to the water–gas shift and reverse carbon dioxide methanation reactions. The dry reforming reaction resulting in positive carbon dioxide conversion can only be found at a high reaction temperature. For all cases studied, low or negative carbon dioxide conversion was found because of carbon dioxide production from methane oxidation, water–gas shift, and reverse carbon dioxide methanation reactions. It was found that carbon dioxide conversion can be enhanced in the tri-reforming process by a small amount of added water. It was also found that first-law efficiency increased with increased reaction temperature because of higher hydrogen and carbon monoxide yields. Second-law efficiency was found to decrease with increased temperature because of higher exergy destruction due to a more complete chemical reaction at high temperatures.

Keywords: biogas; tri-reforming process; syngas; methane and carbon dioxide conversion; hydrogen/carbon monoxide ratio; first-law/second-law efficiency

---

1. Introduction

The efficient production of syngas (a mixture of hydrogen and carbon monoxide) is gaining significant attention worldwide as it is a versatile feedstock that can be used to produce a variety of fuels and chemicals, such as methanol, Fischer–Tropsch fuels, H_2, and dimethyl ether (DME) [1]. Using CH_4 as the primary material, syngas can be produced from steam reforming (SR), partial oxidation (POX), autothermal reforming (ATR), and dry reforming (DR). The tri-reforming (TR) process for syngas production from CH_4 has received growing attention because of its technical simplicity and flexible operation [2–5]. In the TR process, the syngas is produced by combining SR, DR, and POX in a single step. The TR process was proposed originally for syngas production from power plant flue gas [6,7]. There are several advantages for syngas production from the TR process. As CO_2 is one of the reactants, there is no need for CO_2 separation from the flue gas [6,7]. The H_2/CO ratio in syngas can be altered by adjusting the relative amounts of the reactants. In addition, the presence of H_2O and O_2 in the feedstock helps to mitigate carbon deposition, and catalyst deactivation can be prevented [8,9].

As fossil energy resources reduce sharply and environmental pollution becomes more serious, searching for new materials for syngas production plays an important role in future energy development [10–12]. Biogas is receiving much attention because of its considerable economic and environmental benefits [13]. The biogas composition is related to the starting substrate, but is basically composed of CH_4 and CO_2 with the volume ratio of 2 [14–16]. Both CH_4 and CO_2 are regarded as major greenhouse gases (GHGs), which pose a serious threat to the global climate and environment. Using biogas for syngas production, both CH_4 and CO_2 emissions into the atmosphere can be reduced. Because of its potential for reducing global warming, further understanding of syngas production from biogas is essential. Moreover, syngas can also be used for H_2 production. In this case, H_2 can be enriched via the water–gas shift reaction using syngas and H_2O as feedstock. Among the various alternative energy forms, hydrogen is considered an important energy carrier in the future [17]. It is also an important raw material in the chemical industry and can be used as a fuel in fuel cells to produce electrical energy. For reasons of sustainability, the use of renewable fuel sources such as biogas or biomass for hydrogen production has received considerable attention [18–20].

Several studies have reported on syngas production from biogas via the TR process experimentally. In the study of Vita et al. [21], tri-reforming simulated biogas over a Ni/ceria based catalysts was carried out and the H_2O/CH_4 and O_2/CH_4 molar ratios, reaction temperature, and nickel content effects on the catalyst's performance were studied. They found that the H_2/CO ratio could be flexibly adjusted using added amounts of oxygen and steam in order to meet the requirements of downstream processes. In the study of Lau et al. [22], biogas was used as the fuel source in dry reforming and combined dry/oxidative reforming reactions. The gas stream temperature and reactor space velocity effects were examined experimentally. Their results indicated that an increase in the O_2/CH_4 ratio at low temperature promotes hydrogen production. In dry/oxidative reforming, they found that biogas dry reforming is dominant and the overall reaction is net endothermic when the reaction temperature is higher than 600 °C. In the study of Zhu et al. [23], biogas reforming with added O_2 through a spark-shade plasma was conducted under an O_2/CH_4 ratio of 0.60 and CO_2/CH_4 ratios ranging from 0.17 to 1.00. Their results indicated that O_2 and CH_4 conversions decreased when the CO_2/CH_4 ratio was increased. They also reported that the partial oxidation of methane contributed mostly to CH_4 conversion and the reverse water–gas shift (WGS) reaction dominated in CO_2 conversion.

In addition to experimental work, several numerical TR process models using biogas as the feedstock have also been reported in the literature. In the study of Corigliano and Fragiacomo [24], biogas dry reforming analysis under various operating conditions was carried out using a numerical model. The CO_2/CH_4 ratio, pressure and temperature effects on reaction performance were reported. In the study of Hernández and Martín [25], a process based on mass and energy balances, chemical and phase equilibria, and rules of thumb was developed to optimize the production of methanol using biogas as the raw material. Based on the production cost and carbon footprint, the optimized CH_4/CO_2 ratio contained in the biogas was found. In the study of Hajjaji et al. [26], a H_2 production system via biogas reforming was investigated using life-cycle assessment (LCA). They found that the total GHG emissions from the system were about half of the life-cycle GHG of conventional H_2 production systems via steam methane reforming. In the study of Zhang et al. [27], the effects of various factors including reaction temperature, reactor pressure and CH_4 flow rate on the syngas compositions obtained from the TR process were investigated numerically. An optimum operating condition for syngas production with a target ratio and maximized CO_2 conversion were obtained.

In this work, the TR process is employed for syngas production using biogas as the feedstock. The effects of various operating conditions such as pressure, temperature, biogas composition, air addition, and H_2O additions are investigated. The novelty of this paper is the focus on CO_2 conversion in the TR process, which is seldom reported in the literature. Air is used as the added reactant in this study instead of pure oxygen in the conventional TR process.

2. Modeling

2.1. Chemical Reaction

The following reactions are coupled and carried out in a single reactor in the TR porcess:

Steam reforming (SR):

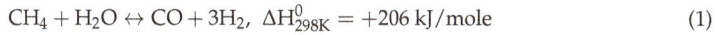

$$CH_4 + H_2O \leftrightarrow CO + 3H_2, \ \Delta H^0_{298K} = +206 \ kJ/mole \tag{1}$$

Dry reforming (DR):

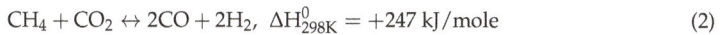

$$CH_4 + CO_2 \leftrightarrow 2CO + 2H_2, \ \Delta H^0_{298K} = +247 \ kJ/mole \tag{2}$$

Partial oxidation (POX):

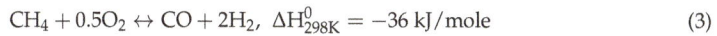

$$CH_4 + 0.5O_2 \leftrightarrow CO + 2H_2, \ \Delta H^0_{298K} = -36 \ kJ/mole \tag{3}$$

As shown in Equations (1)–(3), the TR process combines the endothermic SR and DR reactions and the exothermic POX reaction. The heat released from POX is used as the heat supply for SR and DR and makes the TR process energy efficient [28]. As noted by Cho et al. [29], the chemical reactions involved in the TR process can be alternatively described using Equation (1) along with the following reactions:

Reverse CO_2 methanation (RCM):

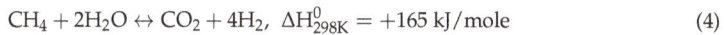

$$CH_4 + 2H_2O \leftrightarrow CO_2 + 4H_2, \ \Delta H^0_{298K} = +165 \ kJ/mole \tag{4}$$

Water-gas shift (WGS):

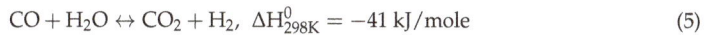

$$CO + H_2O \leftrightarrow CO_2 + H_2, \ \Delta H^0_{298K} = -41 \ kJ/mole \tag{5}$$

Complete oxidation of methane (COM):

$$CH_4 + 2O_2 \leftrightarrow CO_2 + 2H_2O, \ \Delta H^0_{298K} = -803 \ kJ/mole \tag{6}$$

Note that with the chemical reactions described in Equations (1) and (4)–(6), the TR process becomes the well-known catalytic partial oxidation of methane (CPOM). In the literature, there are many studies devoted to the analysis of kinetic mechanisms for CPOM [30,31]. Similar to other reforming process of CH_4, many reactions are likely to occur in the TR process. In addition to the study of Cho et al. [29], studies of De Groote and Froment [32], Scognamiglio et al. [33], Chan and Wang [34], and Izquierdo et al. [35] also reported that the reaction mechanism of CPOM is indirect in which the process can be described by combining reactions of methane oxidation, methane–steam reforming, and water–gas shift. According to these studies, reactions such as CO oxidation, H_2 oxidation, and the Boudouard reaction were not included.

Equations (1), (4) and (5) are the reactions involved in the conventional SR reaction. In this study, syngas under high pressure is of interest for further fuel synthesis. The kinetic model for the SR reaction over a nickel catalyst given by Xu and Froment [36] is adopted,

SR:

$$r_1 = \frac{k_1}{P_{H_2}^{2.5}} \left[P_{CH_4} P_{H_2O} - \frac{P_{H_2}^3 P_{CO}}{K_{eq,1}} \right] / DEN^2 \tag{7}$$

WGS:

$$r_2 = \frac{k_2}{P_{H_2}} \left[P_{CO} P_{H_2O} - \frac{P_{H_2} P_{CO_2}}{K_{eq,2}} \right] / DEN^2 \tag{8}$$

RCM:

$$r_3 = \frac{k_3}{P_{H_2}^{3.5}} \left[P_{CH_4} P_{H_2O}^2 - \frac{P_{H_2}^4 P_{CO_2}}{K_{eq,3}} \right] / DEN^2 \qquad (9)$$

$$DEN = 1 + K_{CH_4} P_{CH_4} + K_{CO} P_{CO} + K_{H_2} P_{H_2} + \frac{K_{H_2O} P_{H_2O}}{P_{H_2}} \qquad (10)$$

For COM, the kinetic model of Trimm and Lam [37] is adopted in this study,

COM:

$$r_4 = \frac{k_{4a} P_{CH_4} P_{O_2}}{\left(1 + K_{CH_4}^C P_{CH_4} + K_{O_2}^C P_{O_2}\right)^2} + \frac{k_{4b} P_{CH_4} P_{O_2}}{\left(1 + K_{CH_4}^C P_{CH_4} + K_{O_2}^C P_{O_2}\right)} \qquad (11)$$

Equation (11) was derived over Pt-based catalyst support, while the model adsorption parameters are adjusted for a Ni-based catalyst [38]. In Equations (7)–(11), r_i is the reaction rate for SR (i = 1), WGS (i = 2), RCM (i = 3), and COM (i = 4); $K_{eq,i}$ and k_i are the chemical equilibrium constant and rate constant for reaction i (i = 1,2,3,4); p_j (j = CH_4, CO_2, H_2O, H_2, and CO) is the partial pressure of species j; and K_j and K_j^C are the adsorption constants of species j. All of these kinetic parameters are given in the Arrhenius function type and are functions of temperature, and can be found in the literature [36,37]. It is noted that catalyst deactivation due to the thermal effect and carbon deposition is neglected in this study [39]. For a reforming reaction involving CH_4, carbon formation is inevitable. The carbon deposition on the catalyst surface is one of the reasons that causes catalyst deactivation. In the tri-reforming process, the appearances of O_2 and H_2O may suppress carbon formation [40,41]. Therefore, catalyst deactivation due to carbon deposition on the catalyst surface is neglected in this study.

2.2. Process Simulation

In this study, Aspen Plus (v.10) is employed to carry out the TR process using biogas as the feedstock. The flow process is depicted in Figure 1. The simulation is performed for a steady state. The biogas stream is assumed to be purely composed of CH_4 and CO_2 with the designated molar ratio. The air stream is composed of 21% O_2, 78% N_2, and 1% H_2. The purpose of H_2 addition is to avoid the singularity in chemical reaction rate computation. A 1% H_2 addition is determined through sensitivity analysis [42,43]. In order to produce high-pressure syngas for future use in fuel synthesis, two compressors (COM-1 and COM-2) are used to increase the biogas and air pressures. In the H_2O stream, a pump is used to increase the water pressure and it is then superheated in a boiler with heat supplied from the high-temperature product stream. After mixing in a mixer, the reactant mixture (TRI) is heated to a certain temperature before entering the insulated Rplug reactor (TR). The gas mixture from the reactor (TRO) is sent to the boiler where the heat is recovered for superheating the water. The TR process performance is characterized using the following dimensionless groups,

CH_4 and CO_2 conversions:

$$X_i = \frac{n_{i,in} - n_{i,out}}{n_{i,in}} \times 100\%, \ i = CH_4, CO_2 \qquad (12)$$

H_2 yield:

$$Y_{H_2} = \frac{n_{H_2,out} - n_{H_2,in}}{n_{CH_4,in}} \qquad (13)$$

CO yield:

$$Y_{CO} = \frac{n_{CO,out} - n_{CO,in}}{n_{CH_4, in}} \qquad (14)$$

H_2/CO ratio:

$$H_2/CO = \frac{Y_{H_2}}{Y_{CO}} \qquad (15)$$

where $n_{i,in}$ is the molar flow rate of the i-th species supplied to the process; and $n_{i,out}$ is the molar flow rate of the i-th species at reactor outlet. Based on these definitions, CH_4 conversion is the ratio of the CH_4 consumption rate to the fed CH_4 flow rate at the reactor inlet. Similarly, CO_2 conversion is the ratio of the CO_2 consumption rate to the fed CO_2 flow rate at the reactor inlet. The H_2 and CO yields are defined as the net increased amounts of H_2 and CO from the reaction per fed CH_4 flow rate. The H_2/CO ratio is defined as the ratio of H_2 yield to CO yield. Note that all these variables are dimensionless.

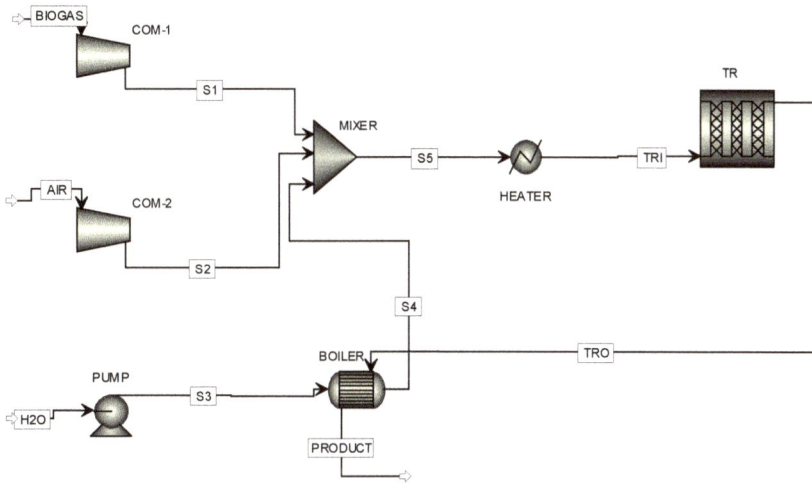

Figure 1. Tri-reform process for syngas production using biogas as feedstock.

In addition to the reactant conversion and product yield, energy and exergy analyses based on the resulting product stream flowing out of the TR reactor are also carried out. For the chemical reaction, there are several ways to define the first- and second-law efficiencies [44,45]. Since the objective of the TR process is to convert biogas into syngas, and noting that CO_2 has zero low heating value (LHV), the first-law efficiency is then defined as,

$$\eta_I = \frac{n_{CO,out}LHV_{CO} + n_{H_2,out}LHV_{H_2}}{n_{CH_4,in}LHV_{CH_4} + W_{comp} + W_{pump} + Q_{heat}} \times 100\% \tag{16}$$

where W_i (i = comp, pump) and Q_{heat} are the input work and heat input, respectively. The main heat input occurs at the heater at which the mixed reactant is heated to a certain inlet temperature. The exergetic analysis is carried out by considering three exergy transfers:

Exergy due to work:

$$Ex_W = W_{comp} + W_{pump} \tag{17}$$

Exergy due to heat transfer:

$$Ex_Q = Q_{heat}\left(1 - \frac{T_0}{T}\right) \tag{18}$$

Exergy due to mass flow:

$$Ex_{f,i} = N_i\{[(h - h_0) - T_0(s - s_0)] + \sum x_k e_k^{CH} + RT_0 \sum x_k \ln x_k\} \tag{19}$$

In these equations, the subscript 0 denotes the reference state (25 °C and 1 atm). The exergy due to mass flow is contributed by physical exergy, chemical exergy and mixing exergy as shown on the right-hand side of Equation (19). For second-law efficiency, this is generally defined as the ratio of exergy recovered to the exergy supplied,

$$\eta_{II} = \frac{Ex_{out}}{Ex_{in}} \times 100\% \tag{20}$$

where Ex_{in} and Ex_{out} are the exergies supplied to and recovered from the system, respectively. Based on Figure 1, Ex_{in} and Ex_{out} are expressed as,

$$Ex_{in} = Ex_{biogas} + Ex_{air} + Ex_{H_2O} + Ex_Q + Ex_W, \; Ex_{out} = Ex_{product} \tag{21}$$

3. Results and Discussion

The TR process using biogas as the feedstock is similar to the tri-reforming of methane (TRM). The only difference is the CH_4 and CO_2 composition. We developed this work from our previous study [46] and focused on using biogas as the feedstock. To verify the correctness of the model built in Aspen Plus, the TRM using the reactor geometry and reactant composition reported in the studies of Chein et al. [46] and Arab Aboosadi et al. [47] was carried out using the built model in Aspen Plus. Figure 2 shows the comparison between the temperature and gas species distributions predicted from a two-dimensional model [46] and from a model built in Aspen Plus. As shown in Figure 2, the agreements for both temperature distribution shown in Figure 2a and species mole fractions shown in Figure 2b are quite good at the reactor downstream. The discrepancies in the region near the reactor inlet zone is believed due to the difference between one- and two-dimensional modeling. Since the TR process performance is evaluated using results at the reactor outlet, good agreement between one- and two-dimensional results is expected. In addition to the comparisons between numerical models, experimental verification of the numerical model was given in our previous study [46]. Also note that O_2 is consumed rapidly as it enters the reactor shown in Figure 2b. That is, there will be no O_2 available for oxidation of CO or H_2 in the downstream of the reactor.

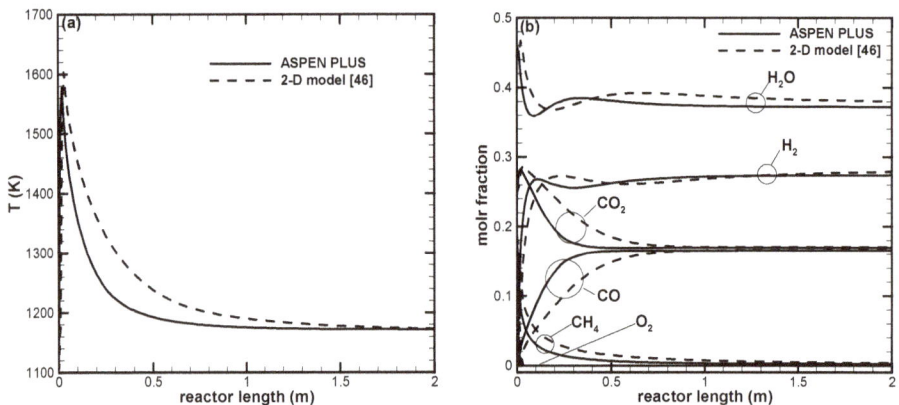

Figure 2. Comparison of results predicted from Aspen Plus and two-dimensional model [46] using the optimized reactant composition reported by Arab Aboosadi et al. [47]. T_{in} = 1100 K, p = 20 atm, and reactant composition $CH_4/CO_2/H_2O/O_2$ = 1/1.3/2.46/0.47. (**a**) Temperature and (**b**) species mole fraction variations along the reactor center line.

Based on the comparisons discussed above, the model built in Aspen Plus can be correctly extended to the TR process using biogas as the feedstock. The base operating conditions are listed in Table 1. The parameters listed in Table 1 are adopted from our previous study except for the feedstock composition and total volumetric flow rate [46]. The molar ratio of the reactants are chosen as CH_4:CO_2:Air:H_2O = 1:0.5:2:1 and the volume flow rate is fixed as 0.0723 L/min. As compared with the previous study [46], a higher volume flow is used in this study because of the presence of N_2 in the air. For economy, air is added instead of pure oxygen. The advantage of this is to avoid the cost of oxygen separation from air, but this obviously results in increased reactor volume. In the following, the TR process performance is examined using reactant inlet temperature T_{in} as the primary parameter. The effects of various pressures, catalyst weight/volume flow rate (W/F) ratios, CO_2/CH_4 ratio in biogas, amounts of air and H_2O on TR process performance are discussed.

Table 1. Reactor geometry and base operation conditions [46,47].

Parameter	Value
Reactor length, L	2 m
Reactor diameter, d(=2R_b)	10 mm
Inlet pressure, p_{in}	20 atm
Inlet temperature, T_{in}	300~1000 °C
Reactant flow rate, F	0.0723 L min^{-1}
Molar ratio of biogas CH_4:CO_2:Air:H_2O	1:0.5:2:1
Catalyst	Ni/Al_2O_3
Catalyst size, d_p	0.42 mm
Catalyst weight, W	0.25 g
W/F ratio	0.0576 ghL^{-1}
Heat-transfer condition	Adiabatic

Figure 3 shows the TR process performance using the base operations listed in Table 1. In Figure 3a the temperature variation along the reactor length for T_{in} = 900 °C is shown. Due to the methane oxidation reaction, the maximum temperature occurs in the near entrance region. The energy produced from methane oxidation is used for steam reforming and dry reforming in the reactor downstream. This causes the temperature to decrease along the reactor length. The CH_4 and CO_2 conversions are shown in Figure 3b. The abrupt increase in CH_4 conversion occurs at T_{in} = 550 °C. This indicates that T_{in} should be higher than 550 °C in order to activate the catalyst. With temperature higher than 550 °C, CH_4 conversion increases gradually with increased T_{in}. In Figure 3b, negative CO_2 conversion results for the low T_{in} regime. From the TR process chemical reactions, CO_2 is produced by the methane oxidation and WGS reactions and consumed by the dry reforming reaction. For low T_{in}, the WGS reaction is dominated and CO_2 consumption by DR is low. This results in negative CO_2 conversion. However, positive CO_2 conversion can result when T_{in} becomes higher than 700 °C, indicating that DR is active. DR contributes to increase the H_2 and CO yield in the high T_{in} regime. Figure 3b also indicates that a complex interaction between CO_2 consumption and production reactions occurs for T_{in} in the 500~600 °C range. The conversions of CH_4 and CO_2 from an equilibrium TR process obtained from an Aspen Plus simulation are also shown in Figure 3b using the parameters listed in Table 1. Since the results from the equilibrium process can be regarded as the theoretical limit of the reaction, it can be seen that CH_4 conversion from the catalytic reaction is lower than that from the equilibrium reaction. Due to more CO_2 production, lower CO_2 conversion results from the equilibrium reaction. For T_{in} higher than 800 °C, CO_2 conversion from the equilibrium reaction is higher than that from the catalytic reaction. From Figure 3c, the H_2 yield, CO yield and H_2/CO ratio are shown. It can be seen that when T_{in} is lower than 500 °C, the H_2 and CO yields are very low due to inactive catalytic reactions at low temperatures. In this low T_{in} regime, CO yield is much lower than H_2 yield and results in a high H_2/CO ratio. As T_{in} is higher than 550 °C, the H_2/CO ratio decreases with T_{in} slowly with a value close to 2. The decrease in H_2/CO with T_{in} is due to increased CO production from the DR reaction while H_2 decreases due to the reverse WGS reaction. In Figure 3d, the H_2 yield, CO yield, and H_2/CO ratio from the equilibrium TR process are also shown. As with conversions of CH_4 and CO_2 shown in Figure 3b, both yields of H_2 and CO from the catalytic reaction are lower than the equilibrium reaction. At high temperature, the H_2/CO ratio from both equilibrium and catalytic reactions is about the same. In Figure 3d, the first- and second-law efficiencies are shown. Based on Equation (16), the first-law efficiency depends on the H_2 and CO yields. Because of higher H_2 and CO yields at higher T_{in}, η_I increases with increased T_{in}. However, the variation in η_{II} is opposite that of η_I. Increased T_{in} implies that the chemical reaction is more complete towards the product side. Since the chemical reaction is a highly irreversible process, high exergy destruction due to the chemical reaction is expected. This results in decreased η_{II} as T_{in} increases. For the low T_{in} regime, exergy destruction due to the chemical reaction is low because of low catalytic activity. Moreover, the contributions of exergy destruction from compressors, pump, heaters and mixers are small. This leads to high η_{II} in the low T_{in} regime.

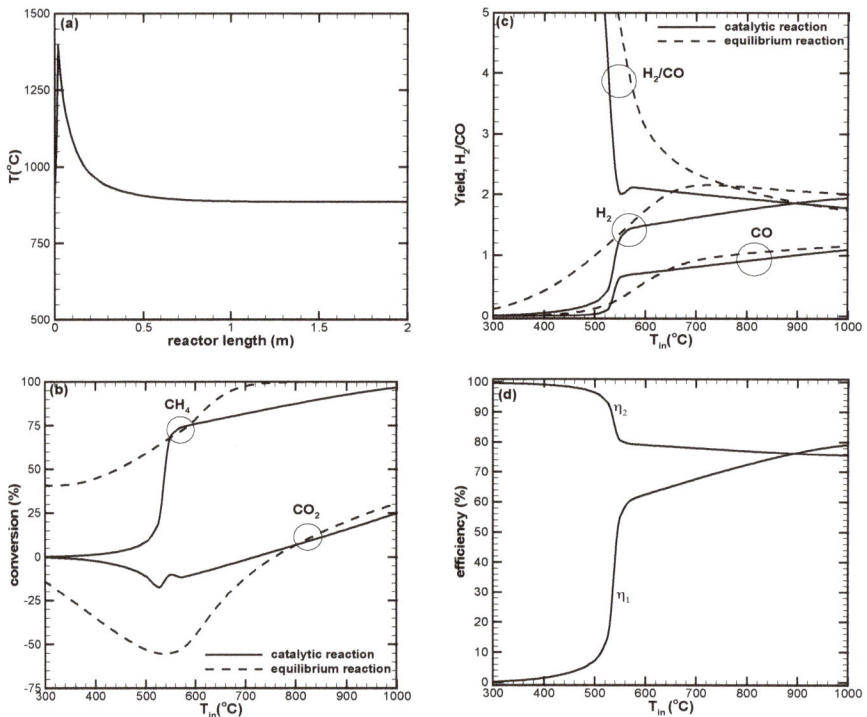

Figure 3. Performance of tri-reforming (TR) process obtained using the base operation conditions listed in Table 1. (**a**) Temperature variation along reactor with T_{in} = 900 °C; (**b**) CH_4 and CO_2 conversions; (**c**) H_2 yield, CO yield, and H_2/CO ratio; and (**d**) First- and second-law efficiencies.

In Figure 4, the variation of species mole fraction of the TR process using the base operation conditions listed in Table 1 is shown. It can be seen that significant mole fraction variation can be found when T_{in} is higher than 500 °C. The mole fractions of reactants (CH_4, CO_2, H_2O, O_2, and N_2) decrease while the mole fractions of products (CO and H_2) increase as T_{in} increases. Due to a highly active methane oxidation reaction, O_2 is consumed completely when T_{in} is greater than 550 °C. Also note that the variation trend of mole fractions of CO and H_2 are similar to the yields of CO and H_2 presented in Figure 3c. The yields of H_2 and CO are used to characterize the TR process performance in this study.

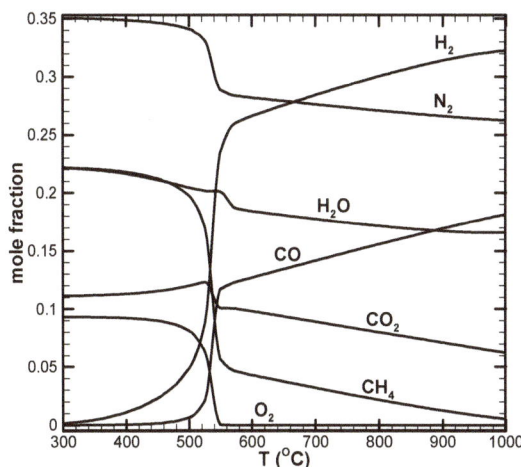

Figure 4. Variations of species mole fraction of the TR process using the base operation conditions listed in Table 1.

In the following, parametric studies based on the base operation conditions listed in Table 1 are carried out. As listed in Table 1, the inlet temperature of the reactant is the primary parameter and the amount of CH_4 fed is used as the reference for the species contained in the reactant and product. The catalyst weight (W) is varied from 0.025 g to 2.5 g; the operation pressure (P) is varied from 10 to 30 atm; the air/CH_4 ratio is varied from 1 to 3; H_2O/CH_4 is varied from 1 to 3; and CO_2/CH_4 is varied from 0.25 to 0.75.

In Figure 5 the effect of W/F ratios on the TR process is examined. The results shown in Figure 4 were obtained by varying the catalyst weight, while other parameters listed in Table 1 were kept fixed. That is, higher W/F ratio results when the catalyst weight is increased. As shown in Figure 5a, a higher temperature along the reactor length is obtained for the W/F = 0.00576 ghL^{-1} case. This indicates that a smaller amount of energy released from methane oxidation reaction is used for endothermic SR and DR reactions. For W/F = 0.0576 and 0.576 ghL^{-1} cases, temperature variations are identical at the reactor downstream. That is, there is a limiting W/F ratio for the reaction. Increasing the W/F ratio (either increasing catalyst weight or decreasing reactant volumetric flow rate) may not lead to further improved reaction performance. In Figure 5b, CH_4 and CO_2 conversions are shown. Due to low catalyst activity, CH_4 conversion is low when T_{in} is low. It can be seen that the T_{in} at which CH_4 conversion abruptly increases can be decreased by increasing the W/F ratio. That is, the catalyst activation temperature can be lowered with increased W/F ratio. As shown in Figure 5b, the T_{in} at which CH_4 conversion increases abruptly are 700 °C, 500 °C and 400 °C for W/F = 0.00576, 0.0576, and 0.576 gLh^{-1}, respectively. CH_4 conversions for the W/F = 0.0576 and 0.576 gLh^{-1} cases become identical when T_{in} is higher than 550 °C. As discussed above, limited performance results when the W/F ratio is increased. From Figure 5b, CO_2 conversion has a negative value except in the high T_{in} regime. This is due to CO_2 formation in the methane oxidation and WGS reactions while DR is less active. At high temperatures, CO_2 is consumed via the dry reforming reaction, leading to positive CO_2 conversion. In Figure 5c, the H_2/CO ratios for various W/F ratios are shown. It can be seen that H_2/CO ratio is about the same for the three W/F ratios studied when T_{in} is high. The H_2/CO ratio close to a value of 2 can be obtained for the W/F range studied. In Figure 5d, variations in η_I and η_{II} are shown. It can be seen that η_I can be enhanced by increasing the W/F ratio. However, η_{II} decreases when the W/F ratio is increased because of a more complete chemical reaction.

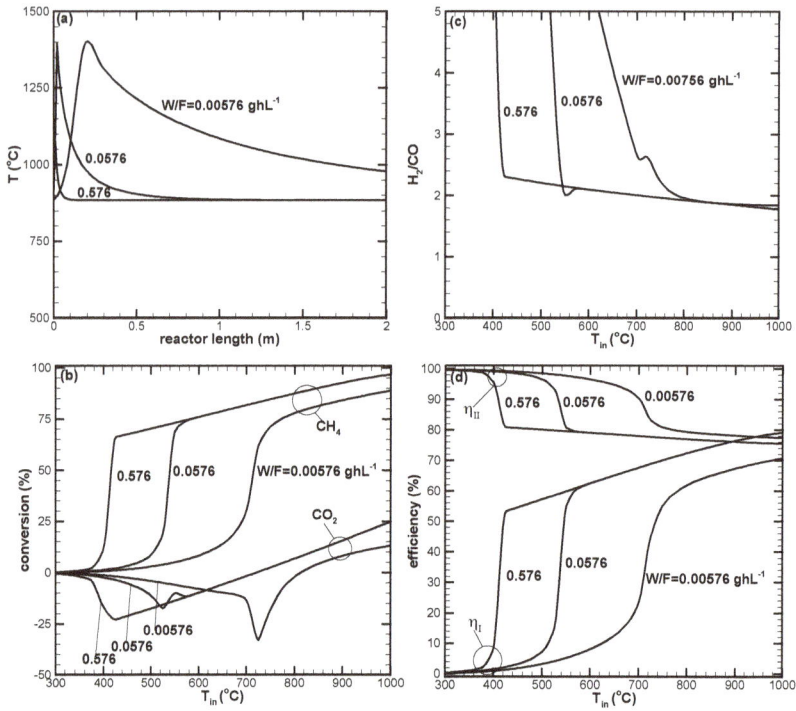

Figure 5. Effect of catalyst weight/volume flow rate (W/F) ratios on the TR process. Catalyst weight is varied from 0.025 to 2.5 g while the other parameters were kept unchanged, as listed in Table 1. (**a**) Temperature variation along reactor with T_{in} = 900 °C; (**b**) CH_4 and CO_2 conversions; (**c**) H_2/CO ratio; and (**d**) first- and second-law efficiencies.

In Figure 6, the reactor operating pressure effect on TR process performance is examined. From Figure 6a the highest temperature increases with increased operating pressure. As the high temperature in the near-entrance region of the reactor is due to the methane oxidation reaction, this implies that methane oxidation can be enhanced by increasing the operating pressure. Due to the enhanced methane oxidation reaction, the T_{in} at which an abrupt increase in CH_4 conversion occurs can be decreased by increasing the pressure, as shown in Figure 6b. Figure 6b also shows that CH_4 conversion can be increased in the low T_{in} regime when the pressure is increased. That is, increased operating pressure can enhance catalyst activity at lower temperatures. In the high T_{in} regime, CH_4 conversion is slightly decreased as the pressure is increased. Although higher CH_4 conversion can be obtained from lower pressure operations, the resulting syngas may not be suitable for further use because most applications involve high-pressure synthetic processes. A H_2/CO ratio with a value close to 2 is obtained for all the pressures studied when T_{in} is high, as shown in Figure 6c. Because of the reduced CH_4 conversion at a high T_{in} regime, it can be seen that η_I decreases with increased T_{in} and pressure, as shown in Figure 6d. However, Figure 6d shows that η_{II} increases with decreasing pressure because of less exergy destruction by the chemical reaction.

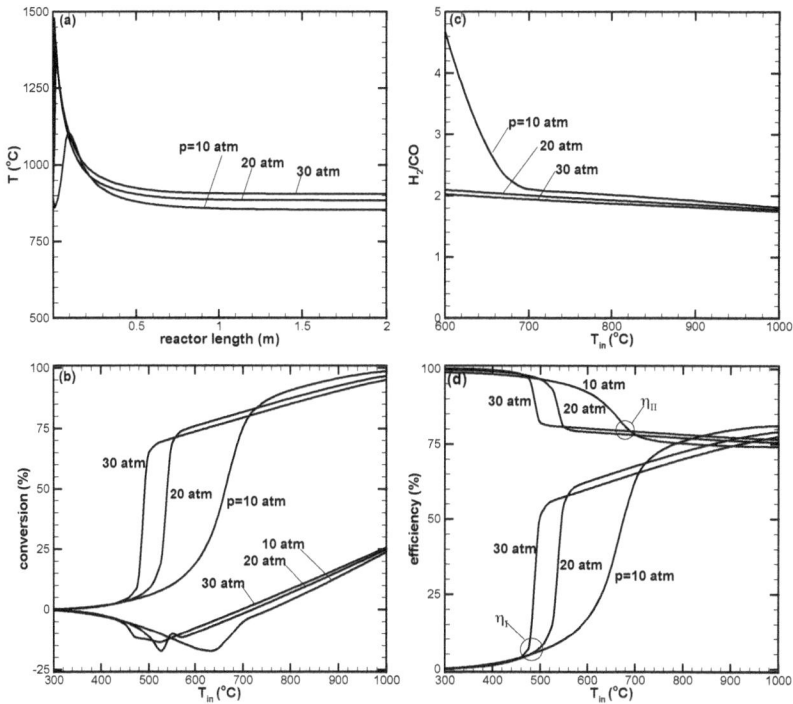

Figure 6. Effect of pressure on the TR process. Pressure is varied from 10 to 30 atm while the other parameters were kept unchanged, as listed in Table 1. (**a**) Temperature variation along reactor with T_{in} = 900 °C; (**b**) CH_4 and CO_2 conversions; (**c**) H_2/CO ratio; and (**d**) first- and second-law efficiencies.

The variation in reactant composition effect on TR process performance is examined in the following. Figure 7 shows the air amount effect. Figure 7a shows that temperature can be increased using more air as the reactant. That is, a more complete methane oxidation reaction is achieved when the air supply is increased. With the increase in air amount, both CH_4 and CO_2 conversions can be enhanced, as shown in Figure 7b. For the Air/CH_4 = 3 case, 100% CH_4 conversion can be reached for T_{in} higher than 550 °C. Due to the increased energy supply, dry reforming can occur in the lower T_{in} regime resulting in increased CO_2 conversion. However, negative CO_2 conversion is still found when T_{in} is low. Although more N_2 is also introduced, increasing the volumetric flow rate of the entire reactant, it does not affect CH_4 and CO_2 conversions. As shown in Figure 7c, a H_2/CO ratio with a value higher than 2 can be obtained for the Air/CH_4 = 3 case because DR is more active when the temperature is high. For the Air/CH_4 = 1 case, the H_2/CO value is lower than 2. This is due to the reverse WGS reaction at high temperatures, reducing the H_2 amount. Because of decreased H_2 yield, lower η_I in the higher T_{in} regime is obtained, as shown in Figure 7d. The reverse WGS reaction also causes η_{II} to increase with T_{in} in the high T_{in} regime.

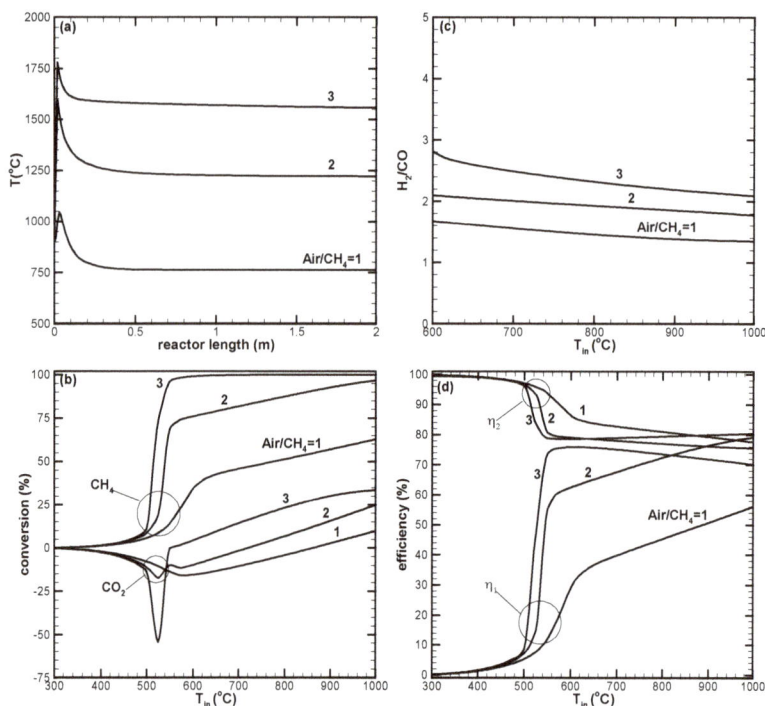

Figure 7. Air/CH$_4$ ratio effect on the TR process. The Air/CH$_4$ ratio is varied from 1 to 3 while the other parameters are kept unchanged, as listed in Table 1. (**a**) Temperature variation along the reactor with T$_{in}$ = 900 °C; (**b**) CH$_4$ and CO$_2$ conversions; (**c**) H$_2$/CO ratio; and (**d**) first- and second-law efficiencies.

Figure 8 shows the H$_2$O amount effect on TR process performance. With increased H$_2$O in the reaction, lower temperature results at the reactor entrance region, as shown in Figure 8a, because of an increased reactant volumetric flow rate and endothermic SR reaction. The increased H$_2$O amount does not affect CH$_4$ conversion, as shown in Figure 8b. However, more negative CO$_2$ conversion results. In addition to CO$_2$ produced from methane oxidation, CO$_2$ may also be produced from WGS and RCM reactions, as indicated in Equations (4) and (5) when H$_2$O is increased. As shown in Figure 8c, a higher H$_2$/CO ratio is obtained when H$_2$O is increased because of increased H$_2$ yield. Figure 8d shows lower η_I results when the H$_2$O amount is increased. This is because higher heating to the reactant is required when the H$_2$O amount is increased. η_{II} increases with increased H$_2$O amount, indicating that less exergy destruction results as H$_2$O is increased.

Figure 9 shows the amount of CO$_2$ contained in the biogas effect on the TR process. As shown in Figure 9a, the amount of CO$_2$ does not affect the reaction temperature to a large extent. The temperature increases slightly as the CO$_2$ amount is decreased. As shown in Figure 9b, CH$_4$ conversion is affected insignificantly by the CO$_2$ amount. However, CO$_2$ conversion is always negative for the CO$_2$/CH$_4$ = 0.25 case. That is, more CO$_2$ is produced as a result of SR and WGS reactions than that consumed by DR and reverse WGS reactions. In Figure 9c, higher H$_2$/CO results when CO$_2$ is decreased. This may be due to less CO formed from CO$_2$ conversion. As with CH$_4$ conversion, the CO$_2$ amount effect on first- and second-law efficiencies is not significant, as shown in Figure 9d.

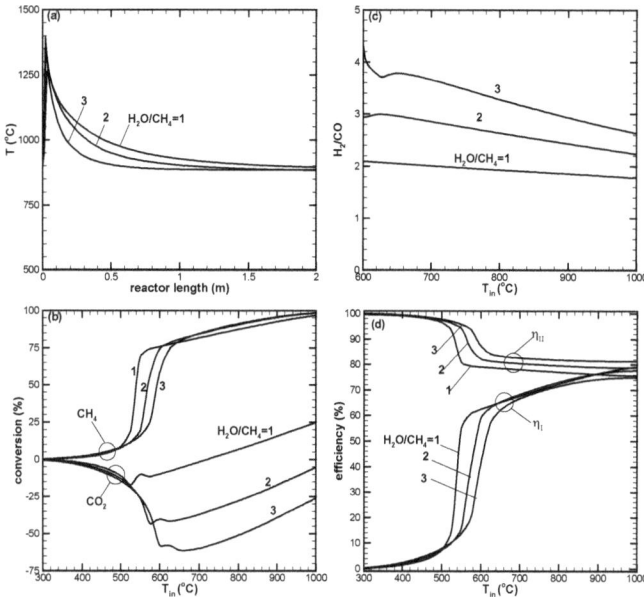

Figure 8. Effect of H_2O/CH_4 ratios on the TR process. H_2O/CH_4 is varied from 1 to 3 while the other parameters were kept unchanged, as listed in Table 1. (**a**) Temperature variation along reactor with T_{in} = 900 °C; (**b**) CH_4 and CO_2 conversions; (**c**) H_2/CO ratio; and (**d**) first- and second-law efficiencies.

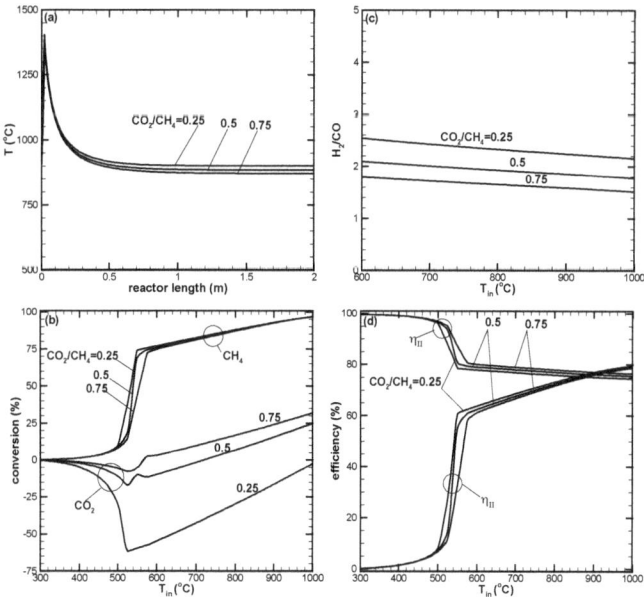

Figure 9. Effect of CO_2/CH_4 ratios on the TR process. CO_2/CH_4 is varied from 0.25 to 0.75 while the other parameters were kept unchanged, as listed in Table 1. (**a**) Temperature variation along reactor with T_{in} = 900 °C; (**b**) CH_4 and CO_2 conversions; (**c**) H_2/CO ratio; and (**d**) first- and second-law efficiencies.

Typical H_2 and CO yield results are shown in Figure 10 for various air and H_2O amounts. In Figure 10a, the H_2 and CO yields increase with the increased air added in the reactant. As a high temperature results in the Air/CH_4 = 3 case, the H_2 yield decreases with increased T_{in} due to the reverse WGS reaction. Figure 10b shows that H_2 yield can be enhanced by increased H_2O addition. CO yield also decreases with H_2O addition because of inactive DR and reverse WGS reactions. As a result, a higher H_2/CO ratio is obtained, as shown in Figure 8b.

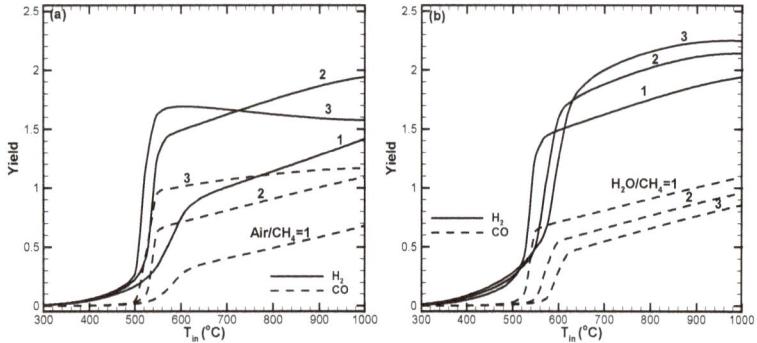

Figure 10. H_2 and CO yields for various (**a**) air and (**b**) H_2O amounts added.

From the results shown above, the CO_2 conversion is low or negative (CO_2 production). Positive CO_2 conversion only occurs in the high T_{in} regime. It is, then, desirable to determine the way to enhance CO_2 conversion in the TR process. After several sets of numerical experiments, it was found that high CO_2 conversion can be obtained when H_2O is low. In this case, the TR process approaches the dry reforming of methane (DRM). Figure 11 shows TR process performance with H_2O/CH_4 = 0.001. As shown in Figure 11a, the temperature drop occurs in the region very near the entrance because DRM is a highly endothermic reaction. Large amounts of required heat leads to this temperature drop. When the methane oxidation becomes active, energy release causes a temperature increase in the reactor downstream. In Figure 11b, CH_4 and CO_2 conversions are shown for the H_2O/CH_4 = 0.001 case. CO_2 conversion is always positive and increases with increased T_{in}. The CO_2 conversion is lower than that of CH_4 because of a low CO_2/CH_4 ratio in the biogas. Because of small amounts of H_2O, the H_2/CO ratio is close to unity, which is the stoichiometric H_2/CO ratio of DRM, as shown in Figure 11c. Figure 11d shows that η_I increases with increased T_{in} because of higher H_2 and CO yield. The second-law efficiency states that η_{II} decreases with increased T_{in} because of higher exergy destruction when the chemical reaction is more complete.

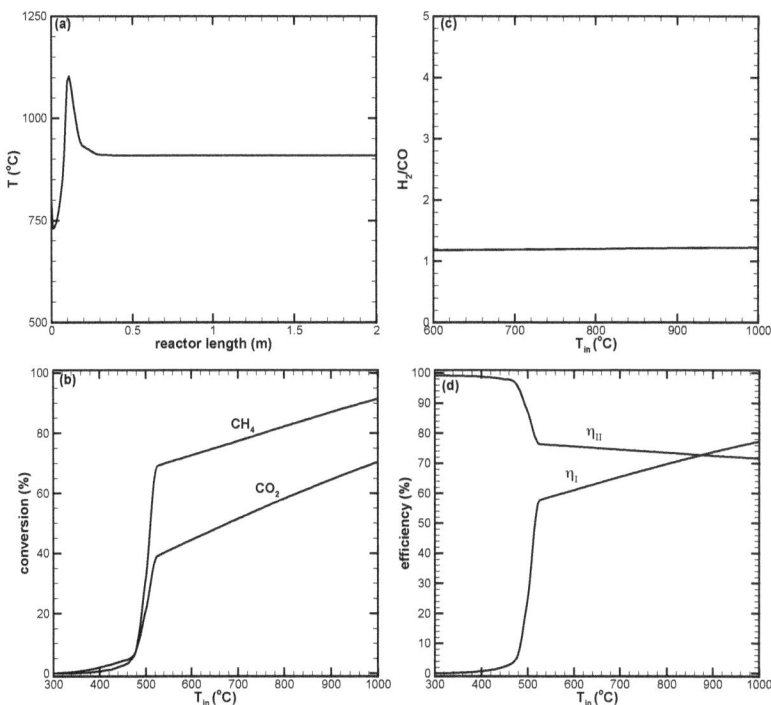

Figure 11. TR process with $H_2O/CH_4 = 0.001$ while the other parameters were kept unchanged, as listed in Table 1. (**a**) Temperature variation along reactor with $T_{in} = 900$ °C; (**b**) CH_4 and CO_2 conversions; (**c**) H_2/CO ratio; and (**d**) first- and second-law efficiencies.

4. Conclusions

The tri-reforming process was used in this study for syngas production from biogas. The effects of various operating parameters such as pressure, temperature and reactant composition were studied based on a model built in Aspen Plus. Based on the results obtained, the following conclusions can be drawn:

(1) There appears to be a limiting space velocity for the reaction. Beyond this limiting value, the reaction approaches the same performance. Lowering the reaction pressure could lead to higher CH_4 conversion, but the syngas produced may not be suitable for further applications.

(2) CH_4 and CO_2 conversions can be enhanced by increasing the amount of air in the reactant. Higher amounts of air could result in decreased H_2 yield due to the reverse water–gas shift reaction, which is favorable at high reaction temperatures.

(3) A higher H_2/CO ratio can be obtained by increasing H_2O addition. However, the dry reforming reaction is suppressed, leading to low CO_2 or negative conversion.

(4) Dry reforming of CO_2 can only be found when the reaction temperature is high. This results in positive CO_2 conversion and contributes to increased H_2 and CO yields.

(5) Higher CO_2 conversion can be obtained for the low H_2O addition case. However, low H_2/CO with a value close to unity results.

(6) The first-law efficiency increases with the increased reaction temperature because of higher H_2 and CO yields. The second-law efficiency decreases with the increased temperature because of higher exergy destruction due to a more complete chemical reaction at high temperatures.

Author Contributions: This paper is a result of the collaboration between all co-authors. Rei-Yu Chein conceived the idea, designed the study, and wrote the paper. Wen-Hwai Hsu established the simulation model and data analysis.

Acknowledgments: Financial support from the Ministry of Science and Technology of Taiwan (MOST 106-2221-E-005-073-MY3) is acknowledged.

Conflicts of Interest: The authors declare no conflict of interest.

Abbreviation

Nomenclature

d_p	catalyst particle diameter, m
Ex	exergy, kJ
e^{CH}	chemical exergy, kJ mol^{-1}
F	reactant volumetric flow rate, m^3 s^{-1}
K_j	surface adsorption equilibrium constant of species j, Pa^{-1}
K_j^C	surface adsorption equilibrium constant of species j in combustion reaction, Pa^{-1}
$K_{eq,i}$	equilibrium constant of reaction i
k_i	rate constant of reaction i, mol Pa$^{0.5}$kg$_{cat}$ s^{-1}, or mol Pa kg$_{cat}$ s^{-1}
L	length of reactor, m
LHV	lower heating value, kJ mol^{-1}
N_i	total molar flow rate of a stream i, mole s^{-1}
n_j	molar flow rate of species j, mole s^{-1}
p	pressure, Pa
Q	heat transfer, W
R	universal gas constant, 8.314 J mol^{-1} K^{-1}
R_b	reactor radius, m
r_i	kinetic rate of reaction i, mol kg$_{cat}$ s^{-1}
s	entropy, kJ mol^{-1} K^{-1}
T	temperature, K
W	catalyst weight, g
W_{comp}	compressor work, W
W_{pump}	pump work, W
X	species conversion
x	mole fraction
Y	species yield

Subscript

in	inlet
out	outlet
0	reference state

Greek symbols

ΔH	heat of reaction, kJ/mol
η	efficiency

References

1. Raju, A.S.K.; Park, C.S.; Norbeck, J.M. Synthesis gas production using steam hydrogasification and steam reforming. *Fuel Process. Technol.* **2009**, *90*, 330–336. [CrossRef]
2. Rathod, V.P.; Shete, J.; Bhale, P.V. Experimental investigation on biogas reforming to hydrogen rich syngas production using solar energy. *Int. J. Hydrog. Energy* **2016**, *41*, 132–138. [CrossRef]
3. Su, B.; Han, W.; Jin, H. Proposal and assessment of a novel integrated CCHP system with biogas steam reforming using solar energy. *Appl. Energy* **2017**, *206*, 1–11. [CrossRef]
4. Damanabi, A.T.; Bahadori, F. Improving GTL process by CO$_2$ utilization in tri-reforming reactor and application of membranes in Fischer-Tropsch reactor. *J. CO$_2$ Util.* **2017**, *21*, 227–237. [CrossRef]
5. Dwivedi, A.; Gudi, R.; Biswas, P. An improved tri-reforming based methanol production process for enhanced CO$_2$ valorization. *Int. J. Hydrog. Energy* **2017**, *42*, 23227–23241. [CrossRef]

6. Song, C.; Pan, W. Tri-reforming of methane: A novel concept for catalytic production of industrially useful synthesis gas with desired H_2/CO ratios. *Catal. Today* **2004**, *98*, 463–484. [CrossRef]

7. Majewski, A.J.; Wood, J. Tri-reforming of methane over Ni@SiO$_2$ catalyst. *Int. J. Hydrog. Energy* **2014**, *39*, 12578–12585. [CrossRef]

8. Kumar, N.; Shojaee, M.; Spivey, J.J. Catalytic bi-reforming of methane: From greenhouse gases to syngas. *Curr. Opin. Chem. Eng.* **2015**, *9*, 8–15. [CrossRef]

9. Choudhary, V.R.; Mondal, K.C.; Choudhary, T.V. Oxy-CO$_2$ reforming of methane to syngas over CoOx/MgO/SA-5205 catalyst. *Fuel* **2006**, *85*, 2484–2488. [CrossRef]

10. Weiland, P. Biogas production: Current state and perspectives. *Appl. Microbiol. Biotechnol.* **2010**, *85*, 849–860. [CrossRef] [PubMed]

11. Nicoletti, G.; Arcuri, N.; Nicoletti, G.; Bruno, R. A technical and environmental comparison between hydrogen and some fossil fuels. *Energy Convers. Manag.* **2015**, *89*, 205–213. [CrossRef]

12. Zhao, X.; Zhou, H.; Sikarwar, V.S.; Zhao, M.; Park, A.A.; Fennell, P.S.; Shen, L.; Fan, L. Biomass-based chemical looping technologies: The good, the bad and the future. *Energy Environ. Sci.* **2017**, *10*, 1885–1910. [CrossRef]

13. Samson, R.; LeDuy, A. Biogas production from anaerobic digestion of spirulina maxima algal biomass. *Biotechnol. Bioeng.* **2012**, *24*, 1919–1924. [CrossRef] [PubMed]

14. Rasi, S.; Veijanen, A.; Rintala, J. Trace compounds of biogas from different biogas production plants. *Energy* **2017**, *32*, 1370–1380. [CrossRef]

15. Hagman, L.; Blumenthal, A.; Eklund, M.; Svensson, N. The role of biogas solutions in sustainable biorefineries. *J. Clean. Prod.* **2018**, *172*, 3982–3989. [CrossRef]

16. Meyer, A.K.P.; Ehimen, E.A.; Holm-Nielsen, J.B. Future European biogas: Animal manure, straw and grass potentials for a sustainable European biogas production. *Biomass Bioenergy* **2018**, *111*, 154–164. [CrossRef]

17. Molino, A.; Larocca, V.; Chianese, S.; Musmarra, D. Biofuels production by biomass gasification: A review. *Energies* **2018**, *11*, 811. [CrossRef]

18. Chianese, S.; Loipersböck, J.; Malits, M.; Rauch, R.; Hofbauer, H.; Molino, A.; Musmarra, D. Hydrogen from the high temperature water gas shift reaction with an industrial Fe/Cr catalyst using biomass gasification tar rich synthesis gas. *Fuel Process. Technol.* **2015**, *132*, 39–48. [CrossRef]

19. Molino, A.; Migliori, M.; Blasi, A.; Davoli, M.; Marino, T.; Chianese, S.; Catizzone, E.; Giordano, G. Municipal waste leachate conversion via catalytic supercritical water gasification process. *Fuel* **2017**, *206*, 155–161. [CrossRef]

20. Chianese, S.; Fail, S.; Binder, M.; Rauch, R.; Hofbauer, H.; Molino, A.; Blasi, A.; Musmarra, D. Experimental investigations of hydrogen production from CO catalytic conversion of tar rich syngas by biomass gasification. *Catal. Today* **2016**, *277*, 182–191. [CrossRef]

21. Vita, A.; Pino, L.; Cipitì, F.; Laganà, M.; Recupero, V. Biogas as renewable rawmaterial for syngas production by tri-reforming process over NiCeO$_2$ catalysts: Optimal operative condition and effect of nickel content. *Fuel Process. Technol.* **2014**, *127*, 47–58. [CrossRef]

22. Lau, C.S.; Tsolakis, A.; Wyszynski, M.L. Biogas upgrade to syn-gas (H_2-CO) via dry and oxidative reforming. *Int. J. Hydrog. Energy* **2011**, *36*, 397–404. [CrossRef]

23. Zhu, X.; Li, K.; Liu, J.; Li, X.; Zhu, A. Effect of CO_2/CH_4 ratio on biogas reforming with added O$_2$ through an unique spark-shade plasma. *Int. J. Hydrog. Energy* **2014**, *39*, 13902–13908. [CrossRef]

24. Corigliano, O.; Fragiacomo, P. Technical analysis of hydrogen-rich stream generation through CO_2 reforming of biogas by using numerical modeling. *Fuel* **2015**, *158*, 538–548. [CrossRef]

25. Hernández, B.; Martín, M. Optimal process operation for biogas reforming to methanol: Effects of dry reforming and biogas composition. *Ind. Eng. Chem. Res.* **2016**, *55*, 6677–6685. [CrossRef]

26. Hajjaji, N.; Martinez, S.; Trably, E.; Steyer, J.; Helias, A. Life cycle assessment of hydrogen production from biogas reforming. *Int. J. Hydrog. Energy* **2016**, *41*, 6064–6075. [CrossRef]

27. Zhang, Y.; Zhang, S.; Benson, T. A conceptual design by integrating dimethyl ether (DME) production with tri-reforming process for CO_2 emission reduction. *Fuel Process. Technol.* **2015**, *131*, 7–13. [CrossRef]

28. Solovev, S.A.; Kurilets, Y.; Orlik, S.N. Tri-reforming of methane on structured Ni-containing catalysts. *Theor. Exp. Chem.* **2012**, *48*, 199–205. [CrossRef]

29. Cho, W.; Song, T.; Mitso, A.; McKinnon, T.J.; Ko, G.H.; Tolsma, J.E. Optimal design and operation of a natural gas tri-reforming reactor for DME synthesis. *Catal. Today* **2009**, *139*, 261–267. [CrossRef]

30. York, A.P.E.; Xiao, T.; Green, M.L.H. Brief overview of the partial oxidation of methane to synthesis gas. *Top. Catal.* **2003**, *22*, 345–358. [CrossRef]

31. Horn, R.; Williams, K.A.; Degenstein, N.J.; Bitsch-Larsen, A.; Dalle Nogare, D.; Tupy, S.A.; Schmidt, L.D. Methane catalytic partial oxidation on autothermal Rh and Pt foam catalysts: Oxidation and reforming zones, transport effects, and approach to thermodynamic equilibrium. *J. Catal.* **2007**, *249*, 380–393. [CrossRef]

32. De Groote, A.M.; Froment, D. Simulation of the catalytic partial oxidation of methane to syngas. *Appl. Catal. A* **1996**, *138*, 245–264. [CrossRef]

33. Scognamiglio, D.; Russo, L.; Maffettone, P.L.; Salemme, L.; Simeone, M.; Crescitelli, S. Modeling temperature profiles of a catalytic autothermal methane reformer with nickel catalyst. *Ind. Eng. Chem. Res.* **2009**, *48*, 1804–1815. [CrossRef]

34. Chan, S.H.; Wang, M.H. Thermodynamic analysis of natural gas fuel processing for fuel cell applications. *Int. J. Hydrog. Energy* **2000**, *25*, 441–449. [CrossRef]

35. Izquierdo, U.; García-García, I.; Gutierrez, Á.M.; Arraibi, J.R.; Barrio, V.L.; Cambra, J.F.; Arias, P.L. Catalyst deactivation and regeneration processes in biogas tri-reforming process. The effect of hydrogen sulfide addition. *Catalysts* **2018**, *8*, 12. [CrossRef]

36. Xu, J.; Froment, G.F. Methane steam reforming, methanation and water-gas shift: I. Intrinsic kinetics. *AIChE J.* **1989**, *35*, 88–96. [CrossRef]

37. Trimm, D.L.; Lam, C.W. The combustion of methane on platinum-alumina fibre catalysts. I. Kinetics and mechanism. *Chem. Eng. Sci.* **1908**, *35*, 1405–1413. [CrossRef]

38. De Smet, C.R.H.; de Croon, M.H.J.M.; Berger, R.J.; Marin, G.B.; Schouten, J.C. Design of adiabatic fixed-bed reactors for the partial oxidation of methane to synthesis gas. *Chem. Eng. Sci.* **2001**, *56*, 4849–4861. [CrossRef]

39. Bartholomew, C.H. Mechanisms of catalyst deactivation. *Appl. Catal. A Gen.* **2001**, *212*, 17–60. [CrossRef]

40. Velasco, J.A.; Fernandez, C.; Lopez, L.; Cabrera, S.; Boutonnet, M.; Jaras, S. Catalytic partial oxidation of methane over nickel and ruthenium based catalysts under low O_2/CH_4 ratios and with addition of steam. *Fuel* **2015**, *153*, 192–201. [CrossRef]

41. Ahmed, S.; Lee, S.H.E.; Ferrandon, M.S. Catalytic steam reforming of biogas e Effects of feed composition and operating conditions. *Int. J. Hydrog. Energy* **2015**, *40*, 1005–1015. [CrossRef]

42. Barbieri, G.; Di Maio, F.P. Simulation of the methane steam re-forming process in a catalytic Pd-membrane reactor. *Ind. Eng. Chem. Res.* **1997**, *36*, 2121–2127. [CrossRef]

43. Kim, J.H.; Choi, B.H.; Yi, J. Modified simulation of methane steam reforming in Pd-membrane/pack-bed type reactor. *J. Chem. Eng. Jpn.* **1999**, *32*, 760–769. [CrossRef]

44. Lutz, A.E.; Bradshaw, R.W.; Keller, J.O.; Witmer, D.E. Thermodynamic analysis of hydrogen production by steam reforming. *Int. J. Hydrog. Energy* **2003**, *28*, 159–167. [CrossRef]

45. Simpson, A.P.; Lutz, A.E. Exergy analysis of hydrogen production via steam methane reforming. *Int. J. Hydrog. Energy* **2007**, *32*, 4811–4820. [CrossRef]

46. Chein, R.; Wang, C.; Yu, C. Parametric study on catalytic tri-reforming of methane for syngas production. *Energy* **2017**, *118*, 1–17. [CrossRef]

47. Arab Aboosadi, Z.; Jahanmiri, A.H.; Rahimpour, M.R. Optimization of tri-reformer reactor to produce synthesis gas for methanol production using differential evolution (DE) method. *Appl. Energy* **2011**, *88*, 2691–2701. [CrossRef]

energies

MDPI

Article

Liquid Lipase-Catalyzed Esterification of Oleic Acid with Methanol for Biodiesel Production in the Presence of Superabsorbent Polymer: Optimization by Using Response Surface Methodology

Hoang Chinh Nguyen [1,†], Dinh Thi My Huong [2,3,†], Horng-Yi Juan [3,†], Chia-Hung Su [3,*] and Chien-Chung Chien [3]

[1] Faculty of Applied Sciences, Ton Duc Thang University, Ho Chi Minh City 700000, Vietnam; nguyenhoangchinh@tdt.edu.vn
[2] Faculty of Chemical Engineering, University of Technology and Education—The University of Danang, Danang City 550000, Vietnam; myhuongdinh@gmail.com
[3] Department of Chemical Engineering, Ming Chi University of Technology, New Taipei City 24301, Taiwan; hyjuan@mail.mcut.edu.tw (H.-Y.J.); m06138117@o365.mcut.edu.tw (C.-C.C.)
* Correspondence: chsu@mail.mcut.edu.tw; Tel.: +88-622-908-9899 (ext. 4665)
† These authors contributed equally to this work.

Received: 23 March 2018; Accepted: 26 April 2018; Published: 28 April 2018

Abstract: Liquid lipase-catalyzed esterification of fatty acids with methanol is a promising process for biodiesel production. However, water by-product from this process favors the reverse reaction, thus reducing the reaction yield. To address this, superabsorbent polymer (SAP) was used as a water-removal agent in the esterification in this study. SAP significantly enhanced the conversion yield compared with the reaction without SAP. The lipase-catalyzed esterification in the presence of SAP was then optimized by response surface methodology to maximize the reaction conversion. A maximum conversion of 96.73% was obtained at a temperature of 35.25 °C, methanol to oleic acid molar ratio of 3.44:1, SAP loading of 10.55%, and enzyme loading of 11.98%. Under these conditions, the Eversa Transform lipase could only be reused once. This study suggests that the liquid lipase-catalyzed esterification of fatty acids using SAP as a water-removal agent is an efficient process for producing biodiesel.

Keywords: biodiesel; esterification; liquid lipase; superabsorbent polymer; response surface methodology

---

1. Introduction

Extensive energy consumption and environmental pollution have stimulated the development of renewable energy sources. Biodiesel, a renewable fuel derived from vegetable oil, is increasingly considered a promising alternative to petrodiesel because of its superior combustion properties, compatibility with diesel engines, and environmental benefits [1–4]. Therefore, biodiesel is being produced globally to reduce the consumption of petrodiesel.

Biodiesel is commonly produced from edible feedstocks such as soybean, sunflower, and rapeseed oils [5–8]; however, the use of these feedstocks for biodiesel production is restricted because of their high cost (which accounts for 75% of the production cost) and competition with demand for the food supply [9–11]. Therefore, inedible and waste oils have been developed as potential feedstocks for biodiesel production [12–15]. These inedible and waste oils usually contain a high level of free fatty acids, which must be esterified into biodiesel before the transesterification [16,17]. In recent years, the esterification of fatty acids for biodiesel production has been widely investigated [18–20]. The common method is acid-catalyzed esterification [16,17,21]. Although biodiesel is successfully produced from fatty acids through acid catalysis, this process retains several drawbacks such as equipment corrosion and negative environmental effects [10,22,23]. Esterification using lipase as a biocatalyst is considered a promising alternative for biodiesel production, because this method is ecofriendly and proceeds at mild reaction conditions, thus reducing the energy consumption and adverse environmental effects [10,24,25]. To improve the stability and reusability of lipase, immobilized lipases have been developed and used for the reaction [10,14,24]. However, the rate of the reaction catalyzed by immobilized lipase is relatively low because of the mass transfer limitation between the enzyme and substrate [26,27]. Moreover, the high cost of immobilized lipase is the main drawback that limits its industrial application [28].

Liquid lipase formulations have increasingly attracted attention as a promising alternative to immobilized lipase for industrial applications because of their low cost (30- to 50-fold lower than that of immobilized lipase) and high catalytic activity [29–31]. Recent studies have shown that liquid lipases can be used for biodiesel production with high yield [26,29,32,33]. However, high water content (from the feedstock and produced from the esterification of fatty acid and methanol) favors the reverse reaction, thus lowering the reaction rate and production yield [32,34]. Efforts have been made to remove the water from the reaction mixture, including the use of a molecular sieve, alumina, or silica gel as adsorbents [35–37]. Although these adsorbents efficiently remove water from the reaction solution, they cannot prevent the inactivation of lipase caused by water [37]. Superabsorbent polymer (SAP) has been widely used for soil water conservation, sewage treatment, mineral dewatering, and drug drying [37,38]. SAP demonstrates rapid water absorption and high water retention capacity [37] and has been employed to remove water formed during the transesterification of corn oil and dimethyl carbonate [37]. However, no report has mentioned the use of SAP as a water-removal agent for the esterification process; this is an attractive research direction.

This study examined the potential use of SAP as a water-removal agent in the esterification of fatty acid with methanol when using liquid lipase for biodiesel production. Oleic acid was used as a model substrate, because it is one of the most common fatty acids in plant oils and animal fats [34]. Response surface methodology (RSM) was employed to analyze the effects of reaction conditions (temperature, reaction time, SAP loading, and enzyme loading) on the reaction conversion. Liquid lipase was also studied for its reusability.

2. Materials and Methods

2.1. Materials

Eversa Transform lipase (liquid lipase produced by *Thermomyces lanuginosus*) with activity of 100,000 PLU/g was obtained from Novozymes A/S (Bagsvaerd, Denmark). The SAP was provided by Formosa Plastic Corp. (Kaohsiung, Taiwan). The SAP used in this study is mainly produced from sodium polyacrylate and its properties are absorption capacity (0.9% NaCl) of 60 g/g, centrifuge retention capacity (0.9% NaCl) of 38 g/g, and particle size distribution of 470 μm. Oleic acid (99%) was provided by Showa Chemical Industry Co., Ltd. (Tokyo, Japan). Methanol, ethanol, and other reagents were analytical grade and obtained from Echo Chemical Co. Ltd. (Miaoli, Taiwan).

2.2. Effect of SAP on the Esterification

A comparative study conducted lipase-catalyzed esterification of oleic acid with methanol with and without the presence of SAP (5%, *w/w*) to investigate the effects of SAP on reaction conversion. The reaction was initiated by adding 10% Eversa Transform lipase into reaction mixtures containing methanol and oleic acid at a molar ratio of 3:1 and various amounts of water (0–30%, *w/w*). The reaction was subsequently kept at 35 °C with stirring for 150 min. The sample was regularly withdrawn for determination of the reaction conversion.

The amount of oleic acid during the esterification was determined using a previously reported procedure [39]. A sample was withdrawn from the reaction mixture, weighed, and dissolved in a

20-mL ethanol–diethyl ether solution (1:1, *v/v*). The sample was subsequently titrated against 0.1 M KOH using phenolphthalein as the indicator to determine the acid value (AV). The reaction conversion was then calculated as follows [40]:

$$\text{Reaction conversion } (\%) = \frac{AV_1 - AV_2}{AV_1} \times 100 \tag{1}$$

where AV_1 is the initial acid value, and AV_2 is the acid value after esterification.

2.3. Optimization of Esterification Using RSM

A four-level and four-factorial central composite design was used to study the effects of reaction factors on the reaction conversion. Esterifications with different reaction temperatures (30–50 °C), methanol:oleic acid molar ratios (1:1–9:1), SAP loadings (5–15%), and enzyme loadings (5–15%) were performed in 100-mL screw-cap glass bottles with stirring for 150 min. After the reaction, a sample was withdrawn from the reaction mixture to determine the reaction conversion. A quadratic equation was then used to establish the relationship between the determined reaction conversion and reaction factors:

$$\begin{aligned} Y = {} & \beta_0 + \beta_1 X_1 + \beta_2 X_2 + \beta_3 X_3 + \beta_4 X_4 + \beta_{11} X_1^2 + \beta_{22} X_2^2 + \beta_{33} X_3^2 + \beta_{44} X_4^2 + \beta_{12} X_1 X_2 \\ & + \beta_{13} X_1 X_3 + \beta_{14} X_1 X_4 + \beta_{23} X_2 X_3 + \beta_{24} X_2 X_4 \\ & + \beta_{34} X_3 X_4 \end{aligned} \tag{2}$$

where Y is the reaction conversion; X_1 is the reaction temperature; X_2 is the molar ratio of methanol to oleic acid; X_3 is the SAP loading; X_4 is the enzyme loading; β_0 is the regression coefficient for the intercept term; β_1–β_4 are linear parameters; β_{12}, β_{13}, β_{14}, β_{23}, β_{24}, and β_{34} are interaction parameters; and β_{11}, β_{22}, β_{33}, and β_{44} are quadratic parameters. These parameters were determined using the least-squares method [41], and an empirical model was subsequently employed to determine the optimal reaction conditions for maximizing reaction conversion [41]. Minitab 16 (Minitab Inc., State College, PA, USA) was employed to develop the empirical model, perform analysis of variance (ANOVA), and determine the optimal reaction conditions.

2.4. Enzyme Reuse

Eversa Transform lipase was reused in the esterification in the presence of SAP. The reaction was conducted under the optimal conditions determined through RSM. After the reaction was completed, the reaction mixture was centrifuged for phase separation. The oil phase was collected for the reaction conversion determination, and the water phase containing liquid lipase was subsequently remixed with fresh reactants and SAP to start a new reaction.

3. Results and Discussion

3.1. Effect of SAP on the Reaction Conversion

This study compared liquid lipase-catalyzed esterification of oleic acid with methanol in various water contents with and without the presence of SAP. As shown in Figure 1, the conversion of the reaction without SAP increased when the amount of water increased. This indicated that a certain amount of water is required for the activity of liquid lipase. Nevertheless, a further increase in water content caused a significant decrease in the conversion of the reaction without SAP. This is attributed to high water content driving the reaction equilibrium to the reverse reaction [35,36,42] and decreasing the activity of liquid lipase [32,35,37]. This result agrees with those of other studies [29,32,37]. Studies have reported that a minimal amount of water is required for optimal enzyme activity, but excess water adversely affects enzyme activity and stability, thus reducing the reaction conversion [32,35,37]. To overcome this obstacle, SAP was used as a water-removal agent for the reaction. The results showed that the conversion of the reaction with SAP increased and reached the highest conversion when increasing the water content from 0% to 5%. Remarkably, higher water contents resulted in no loss in reaction conversion, indicating that excess water had no negative effect on the conversion of the reaction with SAP. This is attributed to the efficient water absorption of the SAP that reduced the negative effects caused by water [37]. In biodiesel production, the feedstock always contains various amount of water; the presence of water in feedstock is the main concern, because it can reduce the conversion yield [32,43,44]. The feedstock is thus treated to remove water before being used for the reaction [44–46]. However, based on the results of this study, the water-removal can be eliminated by adding SAP directly into the reaction solution. Therefore, the liquid lipase-catalyzed esterification using SAP as a water-removal agent is a promising process for biodiesel synthesis from feedstock containing high water content.

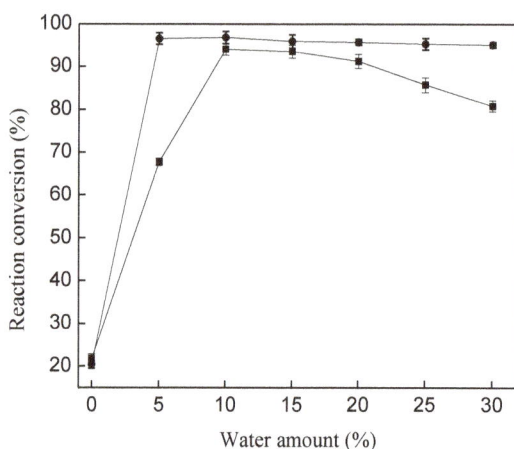

Figure 1. Effects of SAP on the esterification with varied water content. The reaction was conducted under the following conditions: molar ratio of methanol to oleic acid of 3:1, a temperature of 35 °C, enzyme loading of 10%, in the presence of 5% SAP (●), or without SAP (■).

3.2. The RSM Model Development

A central composite RSM model was employed to establish the relationships between reaction conversion (measured response) and reaction factors (input variables)—temperature, molar ratio of methanol to oleic acid, SAP loading, and enzyme loading. Table 1 shows the input variables with their coded and un-coded values. Table 2 illustrates the experimental design for obtaining the optimal reaction conditions. Based on the results shown in Table 2, the measured responses and input variables in term of code values were input into an empirical model as the following quadratic polynomial equation:

$$Y = 89.5 - 6.6X_1 - 6.39X_2 + 1.26X_3 + 5.9X_4 - 3.77X_1^2 - 4.53X_2^2 - 3.45X_3^2 - 3.75X_4^2$$
$$+ 0.86X_1X_2 + 1.21X_1X_3 - 0.26X_1X_4 - 1.99X_2X_3 + 0.15X_2X_4 \qquad (3)$$
$$- 0.21X_3X_4$$

where X_3, X_4, X_1X_2, X_1X_3, and X_2X_4, have positive effects on the response, and the other parameters have adverse effects.

Table 1. Coded values of the input variables for the central composite RSM design.

Variables	Symbols	Variable Levels				
		−2	−1	0	1	2
Temperature (°C)	X_1	30	35	40	45	50
Methanol:oleic acid molar ratio	X_2	1	3	5	7	9
SAP loading (%)	X_3	5	7.5	10	12.5	15
Enzyme loading (%)	X_4	5	7.5	10	12.5	15

Table 2. Experimental design for the influences of the four independent variables on the reaction conversion in coded values and experimental results.

Run	Variable				Response, Y
	X_1	X_2	X_3	X_4	
1	1	1	1	1	66.87
2	1	−1	1	1	88.13
3	−2	0	0	0	85.83
4	1	1	−1	1	68.91
5	0	0	2	0	77.68
6	0	2	0	0	60.76
7	−1	−1	1	1	96.20
8	1	−1	−1	1	74.42
9	−1	−1	−1	1	95.88
10	0	0	0	2	81.21
11	1	1	1	−1	55.43
12	1	−1	1	−1	70.65
13	−1	1	−1	−1	65.83
14	−1	−1	−1	−1	80.98
15	−1	−1	1	−1	85.09
16	1	1	−1	−1	55.23
17	1	−1	−1	−1	64.56
18	0	−2	0	0	79.16
19	2	0	0	0	60.15
20	−1	1	1	1	78.20
21	0	0	0	−2	64.98
22	0	0	−2	0	70.88
23	−1	1	−1	1	83.59
24	−1	1	1	−1	65.38
25	0	0	0	0	90.92
26	0	0	0	0	89.90
27	0	0	0	0	88.79
28	0	0	0	0	90.92
29	0	0	0	0	87.92
30	0	0	0	0	88.36
31	0	0	0	0	89.72

Repeated experiments based on the central runs (25–31) showed a low coefficient of variance (1.33%), indicating the high reproducibility and precision of the experiments. The model was evaluated for statistical significance using the F test for ANOVA (Table 3). Results showed a very low p value (<0.0001) of the model in the F test, confirming that the regression was statistically significant at the 95% confidence level. The coefficient of determination (R^2) was determined to evaluate the quality of the developed model. Result showed that a high R^2 value (0.97) was achieved, signifying high reliability of the model for predicting reaction conversion. As shown in Figure 2, the model predictions were in good agreement with experimental values, indicating that the established model provided satisfactory and accurate results. Table 4 presents the overall effects of the input variables on the reaction conversion, which were examined using t tests. Low p values (<0.05) of the intercept term, three linear terms (X_1, X_2, and X_4), all quadratic terms, and an interaction term (X_2X_3) indicated their significant effects on the reaction. The developed model can therefore be used to forecast the optimal reaction conditions for obtaining maximal responses.

Table 3. Analysis of variance for the empirical model.

Source	DF [b]	SS [b]	MS [b]	F Value	Probability (P) > F
Model [a]	14	4319.58	308.54	32.91	<0.0001
Residual (error)	16	150.01	9.38	-	-
Total	30	4469.59	-	-	-

[a] Coefficient of determination (R^2) = 0.97; adjusted R^2 = 0.94. [b] DF, degree of freedom; SS, sum of squares; MS, mean square.

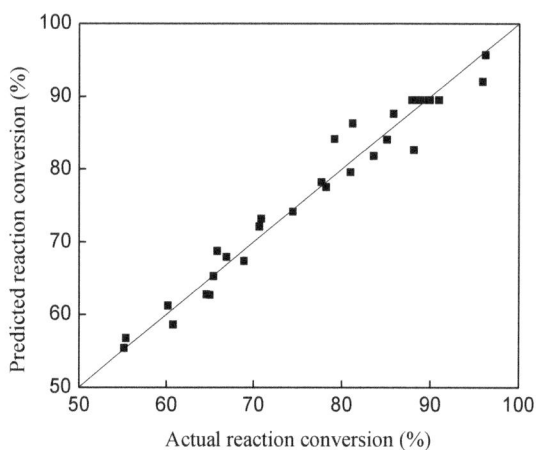

Figure 2. Correlation between experimental and fitted conversions of the reaction.

Table 4. Significance of the coefficients in the empirical model.

Model Term	Parameter Estimate	Standard Error	t Value [a]	p Value
β_0	89.50	1.16	77.34	0.000 [b]
β_1	−6.60	0.63	−10.55	0.000 [b]
β_2	−6.39	0.63	−10.22	0.000 [b]
β_3	1.26	0.63	2.01	0.062
β_4	5.90	0.63	9.43	0.000 [b]
β_{11}	−3.77	0.57	−6.59	0.000 [b]
β_{22}	−4.53	0.57	−7.91	0.000 [b]
β_{33}	−3.45	0.57	−6.03	0.000 [b]
β_{44}	−3.75	0.57	−6.54	0.000 [b]
β_{12}	0.86	0.77	1.13	0.275
β_{13}	1.21	0.77	1.58	0.133
β_{14}	−0.26	0.77	−0.34	0.740
β_{23}	−1.99	0.77	−2.61	0.019 [b]
β_{24}	0.15	0.77	0.19	0.850
β_{34}	−0.21	0.77	−0.27	0.788

[a] $t_{\alpha/2,n-p} = t_{0.025,19} = 2.093$. [b] $p < 0.05$ indicates that the model terms are significant.

3.3. Effect of Reaction Factors on Reaction Conversion

Figure 3 shows the effects of the methanol:oleic acid molar ratio and SAP loading on the reaction conversion while maintaining temperature and enzyme loading at their central levels. Results showed a significant interaction between the methanol:oleic acid molar ratio and SAP loading. At high methanol:oleic acid molar ratios, SAP loading slightly affected the reaction conversion. However, at a low methanol:oleic acid molar ratio, increasing the SAP loading significantly enhanced the reaction conversion. This is attributed to the SAP absorbing water from the reaction mixture and enhancing the enzyme activity to create a suitable microenvironment for an efficient reaction [37]. However, a higher SAP loading resulted in a significant decrease in reaction conversion. Because liquid lipase requires a minimal amount of water for optimal activity, high SAP loading caused less water content to be present the reaction solution [35,37]. Consequently, the enzyme activity was reduced, leading to a reduction in reaction conversion.

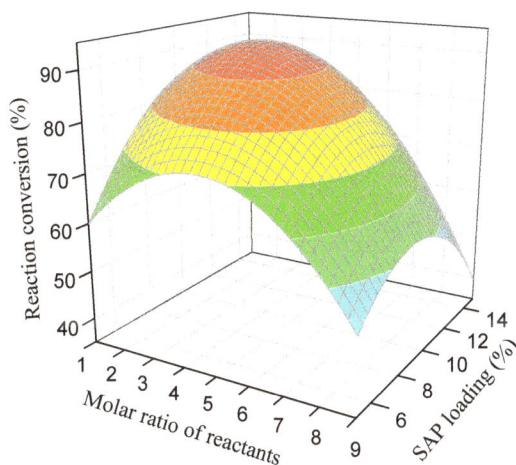

Figure 3. Response surface plot of combined effects of the reactant molar ratio and SAP loading on the conversion of the reaction at a constant temperature (40 °C) and enzyme loading (10%).

Figure 4 presents a response surface curve obtained by plotting the reaction conversion against temperature and molar ratio of methanol to oleic acid while maintaining the other factors at their central levels. At any temperature, the reaction conversion significantly increased when increasing the methanol:oleic acid molar ratio. This was because a high amount of methanol is required for esterification [20,47,48]. However, the reaction conversion decreased with a further increase in the methanol:oleic acid molar ratio. This result is similar to that of other studies [10,14]. Studies have reported that high methanol content in the reaction mixture can inhibit the activity of lipase [10,14]. In this study, the maximal reaction conversion was obtained at a methanol:oleic acid molar ratio of 3.34:1.

Figure 5 presents the effects of temperature and enzyme loading on the reaction conversion while maintaining methanol:oleic acid molar ratio and SAP loading at their central levels. At any enzyme loading level, the reaction conversion increased when increasing the temperature. Nevertheless, a high temperature resulted in a decrease in the reaction conversion because the enzyme becomes inactive at high temperatures [14,32,49]. This result agrees with that of other studies [32,33]. Studies have demonstrated that lipases are sensitive to temperature, and therefore low or elevated temperatures caused a dramatic decrease in their activity [14,33,49]. Similar to temperature, the enzyme loading also significantly affected the reaction conversion. At a given temperature, increasing enzyme loading led to a significant increase in reaction conversion. Studies have reported that an increase in the amount of the enzyme increased contact between the enzyme active surface area and the reactants, thus enhancing the reaction [14,50]. However, a further increase in enzyme loading resulted in a slight decrease in reaction conversion. Excess enzyme possibly caused enzyme aggregation, which limited the enzyme flexibility to react with the oleic acid–methanol interface, thus lowering the conversion of the reaction [14,50].

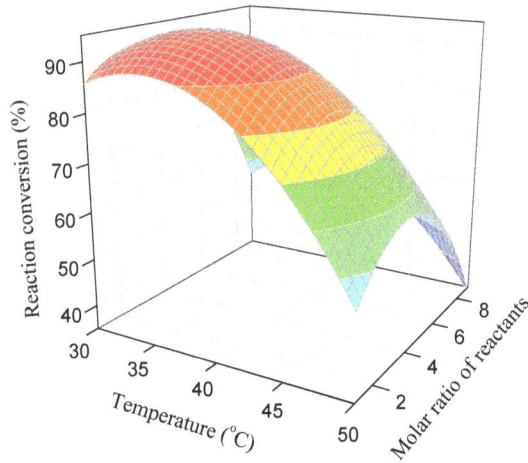

Figure 4. Response surface plot of the combined effects of temperature and reactant molar ratio on the conversion of the reaction at a constant SAP loading (10%) and enzyme loading (10%).

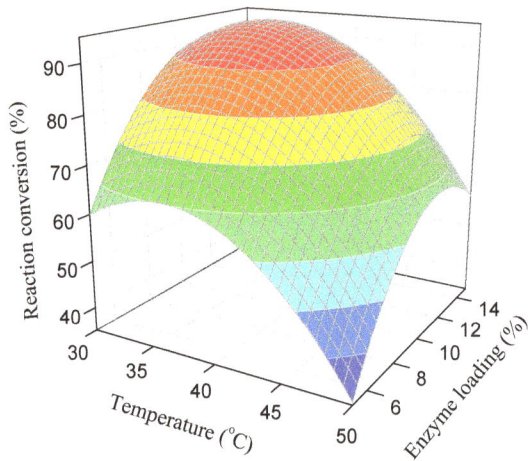

Figure 5. Response surface plot of the combined effects of temperature and enzyme loading on the conversion of the reaction at a constant reactant molar ratio (5:1) and SAP loading (10%).

3.4. Obtaining Optimal Reaction Conditions

Based on the empirical model [Equation (3)], the canonical method was used to forecast the optimal reaction conditions for maximizing reaction conversion. The highest conversion was predicted to be 97.83% at 35.25 °C, methanol:oleic acid molar ratio of 3.34:1, SAP loading of 10.55%, and enzyme loading of 11.98%. An experiment was then carried out under optimal conditions to verify the prediction. A conversion of 96.73% ± 0.15% was obtained, indicating consistent with the empirical model prediction. The developed RSM model can be therefore used to describe the relationships between the response and the variables in the liquid lipase-catalyzed esterification of oleic acid with methanol. Furthermore, the reaction conversion was comparable with other process but exhibited a shorter reaction time [25,29]. In this study, high reaction conversion (96.73%) was obtained under the reaction time of 2.5 h whereas 24 h was required to yield conversion of 90.8% in the esterification of oleic acid with methanol using lipase without SAP reported by Rosset et al. [25]. This suggests that the liquid lipase-catalyzed esterification of fatty acids with methanol using SAP as a water-removal agent is a promising process for producing biodiesel.

3.5. Reusability of Liquid Lipase

Although immobilized lipase is being used for its stability and reusability, its high cost restricts its industrial application. To solve this concern, liquid lipase was developed as an alternative for the reaction. Liquid lipase can be reused several times without significant loss in activity [29]. This study investigated the reusability of Eversa Transform lipase through the esterification of oleic acid and methanol under the optimal reaction conditions. As indicated in Figure 6, the Eversa Transform lipase could only be reused once to drive the reaction to high conversion. After one cycle, the reaction conversion decreased sharply. This was attributed to the inactivation effect of methanol on the enzyme [10,14] and the low stability of liquid enzyme [51]. Further investigation is required to address this limitation. Although liquid lipase demonstrated low reusability in this study, the enzyme remains a promising alternative to immobilized lipase for industrial application because of its low cost [33,52]. Studies have demonstrated that the cost of liquid lipase is 30- to 50-fold lower than that of immobilized lipase [30]. Additionally, the preparation of the liquid lipase is much simpler. Because of these merits, liquid lipase is suggested as a potential alternative for the reaction to improve economic viability.

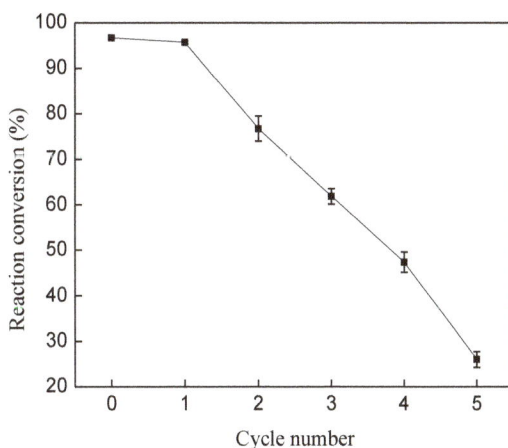

Figure 6. Reusability of liquid lipase for esterification of oleic acid and methanol.

4. Conclusions

This paper reports the liquid lipase-catalyzed esterification of oleic acid with methanol using SAP as a water-removal agent for biodiesel production. The use of SAP significantly enhanced the reaction conversion by suppressing the reverse reaction. The reaction conditions were then optimized to obtain maximal conversion yield using RSM. A maximal conversion of 96.73% was achieved and verified the optimization calculations. Although Eversa Transform lipase was ineffectively reused due to the deactivation of enzyme caused by methanol, liquid lipase-catalyzed esterification in the presence of SAP is promising for biodiesel production from feedstocks containing high water content.

Author Contributions: Hoang Chinh Nguyen and Chia-Hung Su conceived and designed the experiments; Dinh Thi My Huong and Horng-Yi Juan performed the experiments; Hoang Chinh Nguyen and Chien-Chung Chien analyzed the data; Chia-Hung Su contributed reagents and materials; Hoang Chinh Nguyen and Chia-Hung Su wrote the paper.

Acknowledgments: The authors thank for the financial support from the Ministry of Science and Technology (MOST) of Taiwan, R.O.C, under grant number 106-2221-E-131-028.

Conflicts of Interest: The authors declare no conflict of interest.

References

1. Da Silva Araújo, F.D.; do Nascimento Cavalcante, A.; Sousa, M.D.D.B.; de Moura, C.V.R.; Chaves, M.H.; Aued-Pimentel, S.; Fernandes Caruso, M.S.; Tozetto, L.J.; Kaline Morais Chaves, S. Biodiesel production from *Bombacopsis glabra* oil by methyl transesterification method. *Energies* **2017**, *10*, 1360. [CrossRef]

2. Kakati, J.; Gogoi, T. Biodiesel production from Kutkura (*Meyna spinosa* Roxb. Ex.) fruit seed oil: Its characterization and engine performance evaluation with 10% and 20% blends. *Energy Convers. Manag.* **2016**, *121*, 152–161. [CrossRef]

3. Anwar, M.; Rasul, M.G.; Ashwath, N. Production optimization and quality assessment of papaya (*Carica papaya*) biodiesel with response surface methodology. *Energy Convers. Manag.* **2018**, *156*, 103–112. [CrossRef]

4. Mosarof, M.; Kalam, M.; Masjuki, H.; Alabdulkarem, A.; Ashraful, A.; Arslan, A.; Rashedul, H.; Monirul, I. Optimization of performance, emission, friction and wear characteristics of palm and *Calophyllum inophyllum* biodiesel blends. *Energy Convers. Manag.* **2016**, *118*, 119–134. [CrossRef]

5. Kim, K.H.; Lee, E.Y. Simultaneous production of transformer insulating oil and value-added glycerol carbonates from soybean oil by lipase-catalyzed transesterification in dimethyl carbonate. *Energies* **2017**, *11*, 82. [CrossRef]

6. Jin, H.; Kolar, P.; Peretti, S.W.; Osborne, J.A.; Cheng, J.J. Kinetics and mechanism of NaOH-impregnated calcined oyster shell-catalyzed transesterification of soybean oil. *Energies* **2017**, *10*, 1920. [CrossRef]

7. Vahid, B.R.; Haghighi, M. Biodiesel production from sunflower oil over $MgO/MgAl_2O_4$ nanocatalyst: Effect of fuel type on catalyst nanostructure and performance. *Energy Convers. Manag.* **2017**, *134*, 290–300. [CrossRef]

8. Saka, S.; Kusdiana, D. Biodiesel fuel from rapeseed oil as prepared in supercritical methanol. *Fuel* **2001**, *80*, 225–231. [CrossRef]

9. Poudel, J.; Shah, M.; Karki, S.; Oh, S.C. Qualitative analysis of transesterification of waste pig fat in supercritical alcohols. *Energies* **2017**, *10*, 265. [CrossRef]

10. Nguyen, H.C.; Liang, S.H.; Doan, T.T.; Su, C.H.; Yang, P.C. Lipase-catalyzed synthesis of biodiesel from black soldier fly (*Hermetica illucens*): Optimization by using response surface methodology. *Energy Convers. Manag.* **2017**, *145*, 335–342. [CrossRef]

11. Martindale, W.; Trewavas, A. Fuelling the 9 billion. *Nat. Biotechnol.* **2008**, *26*, 1068–1070. [CrossRef] [PubMed]

12. Branco-Vieira, M.; San Martin, S.; Agurto, C.; Santos, M.A.D.; Freitas, M.A.; Mata, T.M.; Martins, A.A.; Caetano, N.S. Potential of *Phaeodactylum tricornutum* for biodiesel production under natural conditions in Chile. *Energies* **2017**, *11*, 54. [CrossRef]

13. Poudel, J.; Karki, S.; Sanjel, N.; Shah, M.; Oh, S.C. Comparison of biodiesel obtained from virgin cooking oil and waste cooking oil using supercritical and catalytic transesterification. *Energies* **2017**, *10*, 546. [CrossRef]

14. Nguyen, H.C.; Liang, S.H.; Chen, S.S.; Su, C.H.; Lin, J.H.; Chien, C.C. Enzymatic production of biodiesel from insect fat using methyl acetate as an acyl acceptor: Optimization by using response surface methodology. *Energy Convers. Manag.* **2018**, *158*, 168–175. [CrossRef]

15. Nguyen, H.C.; Liang, S.H.; Li, S.Y.; Su, C.H.; Chien, C.C.; Chen, Y.J.; Huong, D.T.M. Direct transesterification of black soldier fly larvae (*Hermetia illucens*) for biodiesel production. *J. Taiwan Inst. Chem. Eng.* **2018**, *85*, 165–169. [CrossRef]

16. Yang, S.; Li, Q.; Gao, Y.; Zheng, L.; Liu, Z. Biodiesel production from swine manure via housefly larvae (*Musca domestica* L.). *Renew. Energy* **2014**, *66*, 222–227. [CrossRef]

17. Berchmans, H.J.; Hirata, S. Biodiesel production from crude *Jatropha curcas* L. seed oil with a high content of free fatty acids. *Bioresour. Technol.* **2008**, *99*, 1716–1721. [CrossRef] [PubMed]

18. Anguebes-Franseschi, F.; Abatal, M.; Bassam, A.; Escalante Soberanis, M.A.; May Tzuc, O.; Bucio-Galindo, L.; Cordova Quiroz, A.V.; Aguilar Ucan, C.A.; Ramirez-Elias, M.A. Esterification optimization of crude African palm olein using response surface methodology and heterogeneous acid catalysis. *Energies* **2018**, *11*, 157. [CrossRef]

19. Bhuyan, M.S.U.S.; Alam, A.H.M.A.; Chu, Y.; Seo, Y.C. Biodiesel production potential from littered edible oil fraction using directly synthesized S-TiO$_2$/MCM-41 catalyst in esterification process via non-catalytic subcritical hydrolysis. *Energies* **2017**, *10*, 1290. [CrossRef]

20. Chongkhong, S.; Tongurai, C.; Chetpattananondh, P.; Bunyakan, C. Biodiesel production by esterification of palm fatty acid distillate. *Biomass Bioenergy* **2007**, *31*, 563–568. [CrossRef]

21. Hayyan, A.; Alam, M.Z.; Mirghani, M.E.; Kabbashi, N.A.; Hakimi, N.I.N.M.; Siran, Y.M.; Tahiruddin, S. Reduction of high content of free fatty acid in sludge palm oil via acid catalyst for biodiesel production. *Fuel Process. Technol.* **2011**, *92*, 920–924. [CrossRef]

22. Campanelli, P.; Banchero, M.; Manna, L. Synthesis of biodiesel from edible, non-edible and waste cooking oils via supercritical methyl acetate transesterification. *Fuel* **2010**, *89*, 3675–3682. [CrossRef]

23. Minami, E.; Saka, S. Kinetics of hydrolysis and methyl esterification for biodiesel production in two-step supercritical methanol process. *Fuel* **2006**, *85*, 2479–2483. [CrossRef]

24. Voulgaris, S.; Papadopoulou, A.A.; Alevizou, E.; Stamatis, H.; Voutsas, E. Measurement and prediction of solvent effect on enzymatic esterification reactions. *Fluid Phase Equilib.* **2015**, *398*, 51–62. [CrossRef]

25. Rosset, I.G.; Cavalheiro, M.C.H.; Assaf, E.M.; Porto, A.L. Enzymatic esterification of oleic acid with aliphatic alcohols for the biodiesel production by *Candida antarctica* lipase. *Catal. Lett.* **2013**, *143*, 863–872. [CrossRef]

26. Andrade, T.A.; Errico, M.; Christensen, K.V. Evaluation of reaction mechanisms and kinetic parameters for the transesterification of castor oil by liquid enzymes. *Ind. Eng. Chem. Res.* **2017**, *56*, 9478–9488. [CrossRef]

27. He, Y.; Li, J.; Kodali, S.; Balle, T.; Chen, B.; Guo, Z. Liquid lipases for enzymatic concentration of n-3 polyunsaturated fatty acids in monoacylglycerols via ethanolysis: Catalytic specificity and parameterization. *Bioresour. Technol.* **2017**, *224*, 445–456. [CrossRef] [PubMed]

28. Alves, J.S.; Vieira, N.S.; Cunha, A.S.; Silva, A.M.; Ayub, M.A.Z.; Fernandez-Lafuente, R.; Rodrigues, R.C. Combi-lipase for heterogeneous substrates: A new approach for hydrolysis of soybean oil using mixtures of biocatalysts. *RSC Adv.* **2014**, *4*, 6863–6868. [CrossRef]

29. Ren, H.; Li, Y.; Du, W.; Liu, D. Free lipase-catalyzed esterification of oleic acid for fatty acid ethyl ester preparation with response surface optimization. *J. Am. Oil Chem. Soc.* **2013**, *90*, 73–79. [CrossRef]

30. Cesarini, S.; Diaz, P.; Nielsen, P.M. Exploring a new, soluble lipase for FAMEs production in water-containing systems using crude soybean oil as a feedstock. *Process Biochem.* **2013**, *48*, 484–487. [CrossRef]

31. Lv, L.; Dai, L.; Du, W.; Liu, D. Effect of water on lipase NS81006-catalyzed alcoholysis for biodiesel production. *Process Biochem.* **2017**, *58*, 239–244. [CrossRef]

32. Andrade, T.A.; Errico, M.; Christensen, K.V. Influence of the reaction conditions on the enzyme catalyzed transesterification of castor oil: A possible step in biodiesel production. *Bioresour. Technol.* **2017**, *243*, 366–374. [CrossRef] [PubMed]

33. Adewale, P.; Vithanage, L.N.; Christopher, L. Optimization of enzyme-catalyzed biodiesel production from crude tall oil using Taguchi method. *Energy Convers. Manag.* **2017**, *154*, 81–91. [CrossRef]

34. Pan, Y.; Alam, M.A.; Wang, Z.; Wu, J.; Zhang, Y.; Yuan, Z. Enhanced esterification of oleic acid and methanol by deep eutectic solvent assisted Amberlyst heterogeneous catalyst. *Bioresour. Technol.* **2016**, *220*, 543–548. [CrossRef] [PubMed]

35. Giacometti, J.; Giacometti, F.; Milin, Č.; Vasić-Rački, Đ.A. Kinetic characterisation of enzymatic esterification in a solvent system: Adsorptive control of water with molecular sieves. *J. Mol. Catal. B Enzym.* **2001**, *11*, 921–928. [CrossRef]

36. Duan, Y.; Du, Z.; Yao, Y.; Li, R.; Wu, D. Effect of molecular sieves on lipase-catalyzed esterification of rutin with stearic acid. *J. Agric. Food Chem.* **2006**, *54*, 6219–6225. [CrossRef] [PubMed]

37. Gu, J.; Xin, Z.; Meng, X.; Sun, S.; Qiao, Q.; Deng, H. Studies on biodiesel production from DDGS-extracted corn oil at the catalysis of Novozym 435/super absorbent polymer. *Fuel* **2015**, *146*, 33–40. [CrossRef]

38. Ma, Z.; Li, Q.; Yue, Q.; Gao, B.; Xu, X.; Zhong, Q. Synthesis and characterization of a novel super-absorbent based on wheat straw. *Bioresour. Technol.* **2011**, *102*, 2853–2858. [CrossRef] [PubMed]

39. Su, C.H. Recoverable and reusable hydrochloric acid used as a homogeneous catalyst for biodiesel production. *Appl. Energy* **2013**, *104*, 503–509. [CrossRef]

40. Li, Q.; Zheng, L.; Cai, H.; Garza, E.; Yu, Z.; Zhou, S. From organic waste to biodiesel: Black soldier fly, *Hermetia illucens*, makes it feasible. *Fuel* **2011**, *90*, 1545–1548. [CrossRef]

41. Arteaga, G.; Li-Chan, E.; Vazquez-Arteaga, M.; Nakai, S. Systematic experimental designs for product formula optimization. *Trends Food Sci. Technol.* **1994**, *5*, 243–254. [CrossRef]

42. Li, S.F.; Wu, W.T. Lipase-immobilized electrospun PAN nanofibrous membranes for soybean oil hydrolysis. *Biochem. Eng. J.* **2009**, *45*, 48–53. [CrossRef]

43. Cheng, J.; Yu, T.; Li, T.; Zhou, J.; Cen, K. Using wet microalgae for direct biodiesel production via microwave irradiation. *Bioresour. Technol.* **2013**, *131*, 531–535. [CrossRef] [PubMed]

44. Kusdiana, D.; Saka, S. Effects of water on biodiesel fuel production by supercritical methanol treatment. *Bioresour. Technol.* **2004**, *91*, 289–295. [CrossRef]

45. Wahidin, S.; Idris, A.; Shaleh, S.R.M. Rapid biodiesel production using wet microalgae via microwave irradiation. *Energy Convers. Manag.* **2014**, *84*, 227–233. [CrossRef]

46. Halim, R.; Danquah, M.K.; Webley, P.A. Extraction of oil from microalgae for biodiesel production: A review. *Biotechnol. Adv.* **2012**, *30*, 709–732. [CrossRef] [PubMed]

47. Veillette, M.; Giroir-Fendler, A.; Faucheux, N.; Heitz, M. Esterification of free fatty acids with methanol to biodiesel using heterogeneous catalysts: From model acid oil to microalgae lipids. *Chem. Eng. J.* **2017**, *308*, 101–109. [CrossRef]

48. Berrios, M.; Siles, J.; Martin, M.; Martin, A. A kinetic study of the esterification of free fatty acids (FFA) in sunflower oil. *Fuel* **2007**, *86*, 2383–2388. [CrossRef]

49. Köse, Ö.; Tüter, M.; Aksoy, H.A. Immobilized *Candida antarctica* lipase-catalyzed alcoholysis of cotton seed oil in a solvent-free medium. *Bioresour. Technol.* **2002**, *83*, 125–129. [CrossRef]

50. Waghmare, G.V.; Rathod, V.K. Ultrasound assisted enzyme catalyzed hydrolysis of waste cooking oil under solvent free condition. *Ultrason. Sonochem.* **2016**, *32*, 60–67. [CrossRef] [PubMed]

51. Rotková, J.; Šuláková, R.; Korecká, L.; Zdražilová, P.; Jandová, M.; Lenfeld, J.; Horák, D.; Bílková, Z. Laccase immobilized on magnetic carriers for biotechnology applications. *J. Magn. Magn. Mater.* **2009**, *321*, 1335–1340. [CrossRef]

52. Nielsen, P.; Rancke-Madsen, A.; Holm, H.; Burton, R. Production of biodiesel using liquid lipase formulations. *J. Am. Oil Chem. Soc.* **2016**, *93*, 905–910. [CrossRef]

energies

MDPI

Article

Valorization of Waste Wood as a Solid Fuel by Torrefaction

Jeeban Poudel [1], Sujeeta Karki [2] and Sea Cheon Oh [2],*

[1] Waste & Biomass Energy Technology Center, Kongju National University, 1223-24 Cheonan-Daero, Seobuk, Chungnam 330-717, Korea; jeeban1985@kongju.ac.kr
[2] Department of Environmental Engineering, Kongju National University, 1223-24 Cheonan-Daero, Seobuk, Chungnam 330-717, Korea; karkisujeeta17@gmail.com
* Correspondence: ohsec@kongju.ac.kr; Tel.: +82-41-521-9423

Received: 8 May 2018; Accepted: 15 June 2018; Published: 23 June 2018

Abstract: The aim of this study was to investigate the optimal temperature range for waste wood and the effect torrefaction residence time had on torrefied biomass feedstock. Temperature range of 200–400 °C and residence time of 0–50 min were considered. In order to investigate the effect of temperature and residence time, torrefaction parameters, such as mass yield, energy yield, volatile matter, ash content and calorific value were calculated. The Van Krevelen diagram was also used for clarification, along with the CHO index based on molecular C, H, and O data. Torrefaction parameters, such as net/gross calorific value and CHO increased with an increase in torrefaction temperature, while a reduction in energy yield, mass yield, and volatile content were observed. Likewise, elevated ash content was observed with higher torrefaction temperature. From the Van Krevelen diagram, it was observed that at 300 °C the torrefied feedstock came in the range of lignite. With better gross calorific value and CHO index, less ash content and nominal mass loss, 300 °C was found to be the optimal torrefaction temperature for waste wood.

Keywords: waste wood; torrefaction; energy yield; mass yield; CHO index; gross calorific value; Van Krevelen diagram

---

1. Introduction

In recent years, biomass has obtained remarkable attention because of the potential it holds to replace the energy derived from fossil fuel. Biomass is considered to be an important renewable fuel and the most widespread technology, and can be grouped into thermochemical (torrefaction, pyrolysis, combustion, etc.), chemical (alkaline hydrolysis, etc.) and biochemical (fermentation, anaerobic digestion, etc.) categories [1]. Biomass can be considered as a flexible source of energy as it can be transformed into numerous energy products, for instance, bio-oil, syngas, and so forth. However, biomass has numerous challenges, such as but not limited to, high moisture content, poor grindability, hydroscopicity, low heating value, fibrous in nature, and so forth. [2,3]. These challenges confine the combustion performance and escalate the handling and transportation cost of biomass. Therefore, in order to overcome these challenges, several pretreatment approaches have been suggested. Thermal degradation of woody biomass is a complex topic in itself and comprises of numerous fractions with various thermal behaviors [4].

One of the promising approaches is to pretreat the biomass in the temperature range from 200–300 °C in an inert environment, described as torrefaction. Torrefaction improves the dimensional stability and durability of the wood [5]. This advancement can be curbed by the wood's moisture; therefore, the reduction of the wood's moisture content is a must. In addition, the grindability of the wood increases with torrefaction and reduces the energy required [6]. This reduction in wood commination serves as major advantages for numerous energy derived applications, such as co-firing, cement kilns, steel industry, coke, and so forth. [7]. Moreover, torrefied biomasses are more suitable for further energy conversion processes, such as fast pyrolysis [8], gasification [9], and in the steel industry, as a substitute for coke [10]. However, Koppenjan et al. [11] highlights the need to address numerous issues, such as energy integration, volatile gases handling and the applicability of feedstock for the advancement of torrefaction technologies. In addition, in spite of these advantages, some of the detrimental effects are related to torrefaction degradation of strength and toughness [12,13]. The reduction in the cohesive nature of wood is associated with the disintegration of the hemicellulose matrix and depolymerization of cellulose during torrefaction [14]. Although not spread in a large commercial scale, there has been an enormous amount of studies for torrefaction of biomass in the last decade. For instance, a comprehensive study by Rodriques et al. [15] evaluates the potential torrefaction enhancement for production of bioenergy in Portugal. Similarly, Li et al. [16] highlights the advantages of carbon dioxide torrefaction, such as elevated thermal stability and combustion reactivity. In addition, Benavente and Fullana [17] utilized two-phase olive mill waste (TPOMW) to determine the optimum torrefaction temperature range, and concluded that torrefaction of TPOMW around 200 °C min^{-1} is best for energy utilization. The assessment of the overall product yield, right from the production of the raw materials to the utilization of the final energy, can be carried out by conducting Life Cycle Assessment scenarios [18].

Decisive parameters like temperature, residence time, or properties of torrefaction products, have crucial impact on torrefaction behavior, properties of the torrefied product, and comprehensive mass and energy conversion efficacy [19]. However, the characteristic feature of the biomass material used for torrefaction cannot be overlooked. The main objective of the present work is to study the effect torrefaction has on the thermochemical properties of wood. The torrefied wood properties are based on the Gross Calorific Value (GCV) (MJ/kg), elemental composition, proximate and gas analysis, and energy yield (Y_{energy} (%)) and mass yield (Y_{mass} (%)). The result of this work is to define the optimal torrefaction temperature range for waste wood suitable for other energy conversion processes or energy production use.

2. Materials and Methods

The following sections elaborate the reactor, equations and experimental setup of the process for this study.

2.1. Materials

In this study, waste wood was investigated. The mixed waste wood collected from a waste wood collection center in Cheonan, South Korea was first screened, and all the non-woody and coarse parts were removed. The waste wood was homogeneously mixed and dried at 105 °C for 24 h for the torrefaction experiment. In Table 1, the properties of waste wood indicates about 13.5% moisture content, 1.5% ash content and 20.6 MJ/kg GCV. The inorganic components of the raw waste wood are referred as "other" in the table.

Table 1. Properties of waste wood (raw).

	C	46.96
	H	5.90
	N	3.07
Elemental Analysis (wt %) [a]	O	42.30
	S	0.07
	Cl	0.57
	Other	1.13
	Cd	20.02
	Pb	16.02
Heavy Metal Analysis (mg/kg)	Cr	4.61
	Hg	ND
	As	ND
Moisture (%) [b]		13.5
Volatile fraction (%) [b]		85.0
Ash (%) [b]		1.5
GCV (MJ/kg) [a]		20.6
LHV (MJ/kg) [b]		17.81

[a] Dry Basis; [b] Wet Basis; LHV (Lower Heating Value).

2.2. Torrefaction Experiment Setup and Procedure

The schematic diagram of the horizontal tubular reactor used for this study is illustrated in Figure 1. The internal diameter and length of the reactor were 150 mm and 60 mm, respectively. For each experimental run, 20 g of sample were weighed and flushed with nitrogen (21 min^{-1}) until the level of oxygen was below 1%. Torrefaction was then carried out for the temperature range of 200–400 °C for a residence time of 0–50 min. After completion of each experiment, the sample was removed from the reactor and was weighed. The initial and final weights of the samples were measured to determine mass yield (Y_{mass} (%)), which was calculated with Equation (1):

$$\text{Mass Yield } (Y_{mass}(\%)) = \frac{\text{mass after torrefaction}}{\text{mass of raw sample}} \times 100\% \tag{1}$$

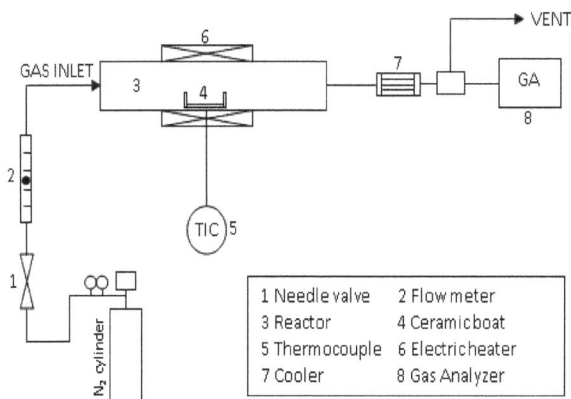

Figure 1. Schematic diagram of the horizontal tubular reactor used in this study.

In addition, the GCV of the raw and torrefied biomass was measured using bomb calorimeter (Parr Instrument Co., Model 1672, Moline, IL, USA) and the energy yield (Y_{energy} (%)) was calculated using Equation (2):

$$\text{Energy Yield } \left(Y_{energy}(\%)\right) = Y_{mass} \times \frac{\text{HHV (torrefied sample)}}{\text{HHV(raw sample)}} \times 100\% \tag{2}$$

Thermogravimetric analysis (TGA) was carried out using a thermogravimetric analyzer (TA Instruments, Q50, New Castle, DE, USA), 5 mg of sample was placed in a crucible and the experiment was conducted in an inert environment with a nitrogen flow rate of 60 mL·min^{-1}. The temperature range was allowed to rise up to 600 °C during various heating rates, ranging from 10–30 °C·min^{-1}. The changes in the atomic composition of waste wood were investigated by using the Van Krevelen diagram. It is constructed using the molar ratio of hydrogen:carbon (hydrogen index) as the ordinate to the molar ratio of oxygen:carbon (oxygen index) in the abscissa. The raw and torrefied waste wood were plotted, and the specific location of these plots helped to identify the alterations in the atomic composition of torrefied waste wood.

3. Results and Discussions

In this study, parameters such as thermogravimetric analysis, effect of torrefaction temperature and residence time on torrefied product, and heavy metal analysis were used for the determination of optimal temperature range for torrefaction of waste wood.

3.1. Thermogravimetric Analysis

The weight loss characteristics of waste wood as a function of torrefaction temperature are illustrated in Figure 2. It can be seen from the figure, that until 250 °C, there was no significant weight loss for all three heating rates. However, the bulk weight loss of the waste wood is in the temperature range of 250–370 °C. There was an insignificant difference in the weight loss with respect to heating rate, i.e., although the weight loss for the heating rate of 30 °C·min^{-1} was highest followed by 20 °C·min^{-1} and the lowest was for 10 °C and the differences were minimal. After the main decomposition, i.e., after 370 °C, weight loss was relatively slow.

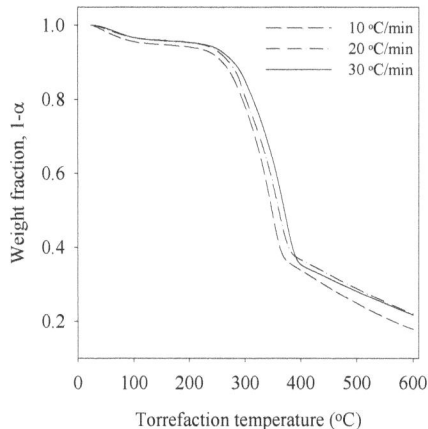

Figure 2. TGA analysis of waste wood with respect to torrefaction temperature.

3.2. Effect of Temperature on Torrefaction Parameters

The effect of temperature on torrefaction parameters, such as Y_{mass} %, Y_{energy} (%), volatile matter (VM_c (%)), GCV (MJ/kg), and ash content (A_c (%)), are illustrated in Figure 3. It can be seen from the graph that overall, there is a decrease in Y_{energy} (%), Y_{mass} (%), and VM_c (%), with an increase in torrefaction temperature. Conversely, there is an increase in GCV (MJ/kg), and A_c (%), with increase in torrefaction temperature.

As from the Figure 3, it can be observed that until 300 °C, the mass loss was not significant after which there was a loss of 42.77% at 350 °C. This drastic alteration of mass loss can be attributed to the drying process and thermal degradation of stable wood constituents [7,20,21]. Similar changes can be observed for Y_{energy} (%), where a significant reduction of 27.80% was observed at 350 °C. In contrast, an increment in the GCV (MJ/kg) is observed with increase in torrefaction temperature. At elevated torrefaction temperature, VM_c (%) decreases while the A_c (%) increases, the results obtained in this study is well supported by the results obtained by Pelaez-Samaniego et al. [22] for ponderosa pine wood species.

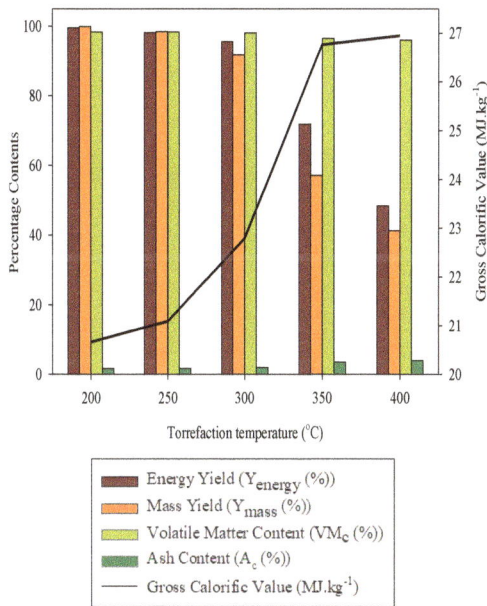

Figure 3. Effect of torrefaction temperature on Energy Yield (Y_{energy} (%)), Mass Yield (Y_{mass} (%)), Volatile Matter (VM_c (%)), Ash Content (A_c (%)) and Gross Calorific Value (GCV (MJ/kg)).

The GCV (MJ/kg) is one of the most important characteristics of a solid fuel. Although with increasing torrefaction temperature, the GCV (MJ/kg) of the biomass feedstock keeps increasing; it must be noted that Y_{mass} (%) decreases with elevated torrefaction temperature. The loss of mass can reach a point where the torrefaction process is regarded as incomplete. Therefore, this study aims to provide the optimal range of temperatures that totally benefit torrefaction. Therefore, the temperature range of 250–350 °C was taken into account to further investigating the effect of torrefaction residence time on torrefied waste wood for this study.

Coal contains higher percentage of carbon than biomass due to which the GCV (MJ/kg) of coal is much higher than that of biomass. The importance of the hydrogen and oxygen index on the GCV (MJ/kg) of waste wood is illustrated in Figure 4, presenting the Van Krevelen diagram, a graphical illustration which portrays the decomposition of waste wood during torrefaction. During the pretreatment, waste wood undergoes physiochemical changes that involve the loss or gain of integral amounts of elemental composition, such as C, H, N, and O, leading to specific changes in the Van Krevelen diagram. From the Van Krevelen diagram, it can be seen that as the torrefaction temperature increases the molar ratio, i.e., H/C and O/C ratio decreases, bringing the waste wood closer towards lignite and coal. Figure 4 also clearly indicates, that as the temperature increases to 300 °C, the waste wood appears closer to lignite. Torrefaction temperatures above 250 °C, the Van Krevelen plot demonstrates that the elemental composition of waste wood is shifted towards that of coal and lignite. During torrefaction, carbon dioxide and water is released making the torrefied product more valuable for thermal treatment processes, such as gasification and combustion. Similar observations have been made by Granados et al. [23], and have reasoned it due to the large decomposition of cellulose and hemicellulose leaving a lignin-rich material. In addition, not many changes are observed around 200–250 °C, and the product is still in the range of biomass. Thus, from Figures 3 and 4, the optimal torrefaction temperature for waste wood is 300 °C, with elevated GCV (MJ/kg), nominal mass loss, and less Ac (%).

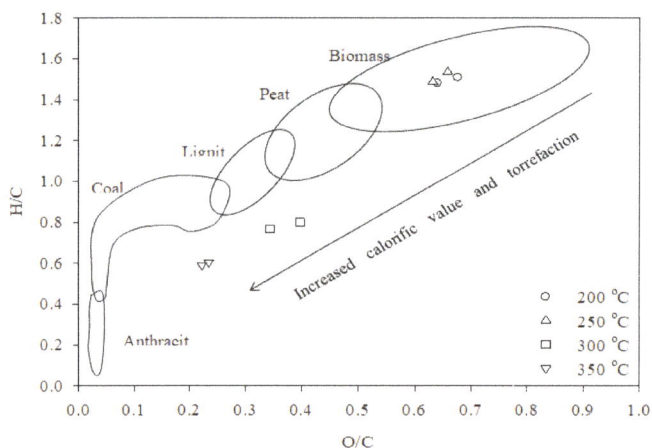

Figure 4. Van Krevelen diagram of waste wood with respect to torrefaction temperature.

To describe the oxidation state of organic compound present in organic materials, Mann et al. [24] suggested the use of the CHO index, defined as:

$$CHO = \frac{2[O] - [H]}{[C]} \tag{3}$$

where, [O], [C] and [H] are mole fraction of oxygen, carbon, and hydrogen, respectively. Figure 5a shows the CHO index values of waste wood (current study) and for different lignocellulosic biomass obtained from [25], and Figure 5b denotes the CHO index obtained in this study with respect to torrefaction temperature. The higher the CHO index, the greater the number of oxygenated compounds. Whereas, a lower CHO index refers to the lowered amount of oxygenated compounds, denoting a relatively lower amount of oxygen and higher amount of relative hydrogen content. The increase in torrefaction temperature leads to a decrease the amount of oxygen and hydrogen, which in turn increases the relative amount of carbon content, which provides a good CHO index. This value can be correlated to the Van Krevelen diagram and an increase in GCV (MJ/kg) as well.

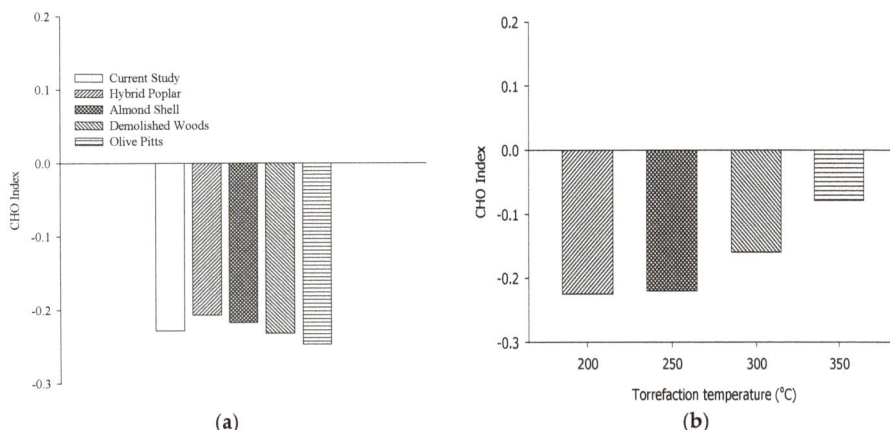

Figure 5. CHO index values of (**a**) biomass feedstocks (data adapted from Reference [21]) (**b**) waste wood with respect to torrefaction temperature.

3.3. Effect of Residence Time on Torrefaction Parameters

The effect of torrefaction residence time on Y_{energy} (%), Y_{mass} (%), VM_c (%), A_c (%) and GCV (MJ/kg) is shown in Table 2. Holding time or residence time can be described as the total amount of time the biomass feedstock is inside the torrefaction reactor [26]. The residence time had a significant effect on all of the parameters except for VM_c (%), where the overall decrease at 300 °C was 1.6% at 50 min residence time. Additionally, at 50 min residence time, the Y_{mass} (%) decreased by 59% at 350 °C for 50 min and A_c (%) increased by 82% at 300 °C for 50 min residence time. In addition, the GCV (MJ/kg) showed an increment of 24.75% at 250 °C for 50 min residence time.

Table 2. Effect of torrefaction residence time on Energy Yield (Y_{energy} (%)), Mass Yield (Y_{mass} (%)), Volatile Matter (VM_c (%)), Ash Content (A_c (%)) and Gross Calorific Value (GCV (MJ/kg)).

	0 min Residence Time		
	250 °C	**300 °C**	**350 °C**
Y_{energy} (%)	99.58	98.10	71.85
Y_{mass} (%)	99.88	91.70	57.16
VM_c (%)	98.34	98.03	96.52
A_c (%)	1.65	1.97	3.47
GCV (MJ/kg)	21.08	22.77	26.76
	10 min Residence Time		
	250 °C	**300 °C**	**350 °C**
Y_{energy} (%)	87.10	59.15	49.87
Y_{mass} (%)	79.84	53.06	39.68
VM_c (%)	97.89	97.38	96.21
A_c (%)	2.11	2.61	3.79
GCV (MJ/kg)	23.09	23.73	26.75
	20 min Residence Time		
	250 °C	**300 °C**	**350 °C**
Y_{energy} (%)	70.27	56.44	43.41
Y_{mass} (%)	62.55	47.96	33.11
VM_c (%)	97.55	96.33	95.72
A_c (%)	2.45	3.67	4.27
GCV (MJ/kg)	23.77	25.06	27.92

Table 2. *Cont.*

	30 min Residence Time		
	250 °C	**300 °C**	**350 °C**
Y_{energy} (%)	74.23	53.45	52.66
Y_{mass} (%)	65.45	43.87	38.33
VM_c (%)	96.39	95.87	95.61
A_c (%)	3.60	4.13	4.39
GCV (MJ/kg)	24.00	25.94	29.25
	40 min Residence Time		
	250 °C	**300 °C**	**350 °C**
Y_{energy} (%)	82.81	57.31	54.74
Y_{mass} (%)	74.26	45.12	39.61
VM_c (%)	96.65	96.25	95.31
A_c (%)	3.35	3.75	4.68
GCV (MJ/kg)	23.59	27.04	29.42
	50 min Residence Time		
	250 °C	**300 °C**	**350 °C**
Y_{energy} (%)	89.24	59.00	53.00
Y_{mass} (%)	71.79	44.42	39.57
VM_c (%)	97.41	96.39	95.74
A_c (%)	2.58	3.60	4.26
GCV (MJ/kg)	26.31	28.28	28.52

The Y_{energy} (%) and Y_{mass} (%) demonstrates significant alterations at elevated residence time. However, this conversion rate may vary depending on the type of biomass used, for instance, agricultural residue demonstrates higher conversion rate compared to woody biomass because of the higher hemicellulose content [27]. In addition, a study by Chen et al. [20] reflects on how the polymeric structure of feedstock influences the reactivity of the torrefaction reaction.

Various works have shown that the effect of torrefaction temperature has a more pronounced effect than residence time, for instance, Chin et al. [28] concluded that although the GCV (MJ/kg) increased with both residence time and temperature, the value was more influenced by torrefaction temperature. In contrast, the present study has found that the overall increase in the GCV (MJ/kg) is comparable for both temperature and residence time.

3.4. Heavy Metal Analysis of Waste Wood

The heavy metal analysis of the waste wood and the torrefied products for temperature ranging from 250 °C to 400 °C (Table 3). The presence of heavy metals has a potentially damaging effect on human physiology and other biological systems, when the tolerance levels are exceeded. It is important to state that the cadmium problem in biomass utilization is caused by the technology itself. It is imported via deposition onto the forest. The source of this deposition is the dissipation of cadmium by anthropogenic processes, most notably fossil energy use [29]. The Cd, Pb and Cr content of the raw WW were 20.02, 16.02 and 4.61 mg/kg, respectively (Table 2). Hg and As were not detected. From Table 2, it was found that the concentrations of heavy metals increased following torrefaction. This is due to the fact that the heavy metals were relatively concentrated by reducing of volatile components in the biomass.

Table 3. Heavy metal concentration of WW as a function of the torrefaction temperature.

		Raw	250 °C	300 °C	350 °C
	Cd	20.02	24.51	94.25	103.66
Heavy Metal (mg/kg)	Pb	16.02	22.17	62.14	88.52
	Cr	4.61	7.34	25.65	29.44
	Hg	ND	ND	ND	ND
	As	ND	ND	ND	ND

ND = Not Detected.

4. Conclusions

In the present work, waste wood under the influence of torrefaction residence time and temperature was investigated. The results showed that there was an increase in the GCV (MJ/kg) of waste wood with the increase in torrefaction temperature and residence time. However, after 350 °C, there was a negligible increase in the GCV (MJ/kg), but the mass loss was maximum; therefore, for 0–50 min residence time, the temperature range of 250–350 °C was taken into consideration. Comparing the present data with the Van Krevelen diagram, we found that with an increase in torrefaction severity the torrefied biomass behaved more like lignite and coal. Also, the CHO index showed a positive influence with an increase in torrefaction temperature, i.e., the amount of oxygenated compound decreased. Unlike other works, the effect of both residence time and temperature were significant on Y_{energy} (%), Y_{mass} (%), VM_c (%), A_c (%) and GCV (MJ/kg) of the waste wood. However, a decisive conclusion cannot be made and more investigation considering other torrefaction parameters, such as grindability and heat-induced variations must be studied in detail. Calculating all of the parameters, 300 °C was considered as an optimal torrefaction temperature for waste wood, i.e., higher GCV (MJ/kg), lesser mass loss, comparatively low A_c (%) and the properties resembling more towards lignite.

Author Contributions: J.P. and S.K. were responsible for the overall experiment, data analysis and arrangement along with preparation of the manuscript. S.C.O. was the supervisor of the project and was mainly responsible for the data and manuscript confirmation. All the authors were equally responsible for finalizing the manuscript and submission.

Funding: This work was supported by a Grant from the Human Resources Development Program (No. 20154030200940) of the Korea Institute of Energy Technology Evaluation and Planning (KETEP) funded by the Ministry of Trade, Industry and Energy of the Korean government.

Conflicts of Interest: The authors declare no conflict of interest.

References

1. *Biochemical Conversion of Biomass in Liquid Transportation Fuels from Coal and Biomass: Technological Status, Costs, and Environmental Impacts*; The National Academies Press: Washington, DC, USA, 2009; pp. 117–162.
2. Doddapaneni, T.R.K.C.; Konttinen, J.; Hukka, T.I.; Moilanen, A. Influence of torrefaction pretreatment on the pyrolysis of Eucalyptus clone: A study on kinetics, reaction mechanism and heat flow. *Ind. Crops Prod.* **2016**, *92*, 244–254. [CrossRef]
3. Van der Stelt, M.; Gerhauser, H.; Kiel, J.; Ptasinski, K. Biomass upgrading by torrefaction for the production of biofuels: A review. *Biomass Bioenergy* **2011**, *35*, 3748–3762. [CrossRef]
4. Peng, J.H.; Bi, H.T.; Lim, C.J.; Sokhansanj, S.A. Study on density, hardness, and moisture uptake of torrefied wood pellets. *Energy Fuels* **2013**, *27*, 967–974. [CrossRef]
5. Kamdem, D.; Pizzi, A.; Jermannaud, A. Durability of heat-treated wood. *Holz als Roh-und Werkstoff* **2002**, *60*, 1–6. [CrossRef]
6. Bridgeman, T.; Jones, J.; Williams, A.; Waldron, D. An investigation of the grindability of two torrefied energy crops. *Fuel* **2010**, *89*, 3911–3918. [CrossRef]
7. Phanphanich, M.; Mani, S. Impact of torrefaction on the grindability and fuel characteristics of forest biomass. *Bioresour. Technol.* **2011**, *102*, 1246–1253. [CrossRef] [PubMed]

8. Boateng, A.; Mullen, C. Fast pyrolysis of biomass thermally pretreated by torrefaction. *J. Anal. Appl. Pyrolysis* **2013**, *100*, 95–102. [CrossRef]

9. Svoboda, K.; Pohořelý, M.; Hartman, M.; Martinec, J. Pretreatment and feeding of biomass for pressurized entrained flow gasification. *Fuel Process. Technol.* **2009**, *90*, 629–635. [CrossRef]

10. Thrän, D.; Witt, J.; Schaubach, K.; Kiel, J.; Carbo, M.; Maier, J.; Ndibe, C.; Koppejan, J.; Alakangas, E.; Majer, S. Moving torrefaction towards market introduction–Technical improvements and economic-environmental assessment along the overall torrefaction supply chain through the SECTOR project. *Biomass Bioenergy* **2016**, *89*, 184–200. [CrossRef]

11. Koppejan, J.; Sokhansanj, S.; Melin, S.; Madrali, S. Status overview of torrefaction technologies. *IEA Bioenergy Task* **2012**, *32*, 1–54.

12. Stamm, A.J. Thermal degradation of wood and cellulose. *Ind. Eng. Chem.* **1956**, *48*, 413–417. [CrossRef]

13. Stanzl-Tschegg, S.; Beikircher, W.; Loidl, D. Comparison of mechanical properties of thermally modified wood at growth ring and cell wall level by means of instrumented indentation tests. *Holzforschung* **2009**, *63*, 443–448. [CrossRef]

14. Bergman, P.C.; Boersma, A.; Zwart, R.; Kiel, J. *Torrefaction for Biomass Co-Firing in Existing Coal-Fired Power Stations "Biocoal"*; Report No.ECN-C-05-013; Energy Centre of Netherlands: Petten, The Netherlands, 2005.

15. Rodrigues, A.; Loureiro, L.; Nunes, L.J.R. Torrefaction of woody biomasses from poplar SRC and Portuguese roundwood: Properties of torrefied products. *Biomass Bioenergy* **2018**, *107*, 55–65. [CrossRef]

16. Li, S.-X.; Zou, J.-Y.; Li, M.-F.; Wu, X.-F.; Bian, J.; Xue, Z.-M. Structural and thermal properties of *Populus tomentosa* during carbon dioxide torrefaction. *Energy* **2017**, *124*, 321–329. [CrossRef]

17. Benavente, V.; Fullana, A. Torrefaction of olive mill waste. *Biomass Bioenergy* **2015**, *73*, 186–194. [CrossRef]

18. Christoforou, E.A.; Fokaides, P.A. Life cycle assessment (LCA) of olive husk torrefaction. *Renew Energy* **2016**, *90*, 257–266. [CrossRef]

19. Prins Mark, J.; Ptasinski, K.; Janssen, F. Torrefaction of wood Part 2. *Analysis of products J. Anal. Appl. Pyrolysis* **2006**, *77*, 35–40. [CrossRef]

20. Chen, W.H.; Kuo, P.C. A study on torrefaction of various biomass materials and its impact on lignocellulosic structure simulated by a thermogravimetry. *Energy* **2010**, *35*, 2580–2586. [CrossRef]

21. Hill, S.J.; Grigsby, W.J.; Hall, P.W. Chemical and cellulose crystallite changes in Pinus radiata during torrefaction. *Biomass Bioenergy* **2013**, *56*, 92–98. [CrossRef]

22. Pelaez-Samaniego, M.R.; Yadama, V.; Garcia-Perez, M.; Lowell, E.; McDonald, A.G. Effect of temperature during wood torrefaction on the formation of lignin liquid intermediates. *J. Anal. Appl. Pyrolysis* **2014**, *109*, 222–233. [CrossRef]

23. Granados, D.; Basu, P.; Chejne, F.; Nhuchhen, D. Detailed investigation into torrefaction of wood in a two-stage inclined rotary torrefier. *Energy Fuels* **2016**, *31*, 647–658. [CrossRef]

24. Mann, B.F.; Chen, H.; Herndon, E.M.; Chu, R.K.; Tolic, N.; Portier, E.F.; Chowdhury, T.R.; Robinson, E.W.; Callister, S.J.; Wullschleger, S.D. Indexing permafrost soil organic matter degradation using high-resolution mass spectrometry. *PLoS ONE* **2015**, *10*. [CrossRef] [PubMed]

25. Jenkins, B.; Baxter, L.; Miles, T., Jr.; Miles, T. Combustion properties of biomass. *Fuel Process. Technol.* **1998**, *54*, 17–46. [CrossRef]

26. Strandberg, M.; Olofsson, I.; Pommer, L.; Wiklund-Lindström, S.; Åberg, K.; Nordin, A. Effects of temperature and residence time on continuous torrefaction of spruce wood. *Fuel Process. Technol.* **2015**, *134*, 387–398. [CrossRef]

27. Bridgeman, T.G.; Jones, J.M.; Shield, I.; Williams, P.T. Torrefaction of reed canary grass, wheat straw and willow to enhance solid fuel qualities and combustion properties. *Fuel* **2008**, *87*, 844–856. [CrossRef]

28. Chin, K.; H'ng, P.; Go, W.; Wong, W.; Lim, T.; Maminski, M.; Paridah, M.; Luqman, A. Optimization of torrefaction conditions for high energy density solid biofuel from oil palm biomass and fast growing species available in Malaysia. *Ind. Crops Prod.* **2013**, *49*, 768–774. [CrossRef]

29. Narodoslawsky, M.; Obernberger, I. From waste to raw material—The route from biomass to wood ash for cadmium and other heavy metals. *J. Hazard. Mater.* **1996**, *50*, 157–168. [CrossRef]

energies **MDPI**

Review

Wastewater Treatment and Biogas Recovery Using Anaerobic Membrane Bioreactors (AnMBRs): Strategies and Achievements

Mohammed Ali Musa [1,2], Syazwani Idrus [1,*], Hasfalina Che Man [1] and Nik Norsyahariati Nik Daud [1]

[1] Department of Civil Engineering, Faculty of Engineering, University Putra Malaysia, Serdang 43400, Selangor, Malaysia; alisulezee@gmail.com (M.A.M.); hasfalina@upm.edu.my (H.C.M.); niknor@upm.edu.my (N.N.N.D.)

[2] Department of Civil and Water Resources Engineering, University of Maiduguri, P.M.B. 1069, Maiduguri 600230, Borno State, Nigeria

* Correspondence: syazwani@upm.edu.my, Tel.:+60-13-692-2301

Received: 17 May 2018; Accepted: 19 June 2018; Published: 27 June 2018

Abstract: Anaerobic digestion is one of the most essential treatment technologies applied to industrial and municipal wastewater treatment. Membrane-coupled anaerobic bioreactors have been used as one alternative to the conventional anaerobic digestion process. They are presumed to offer the advantage of completely reducing or minimizing the volume of sludge and increasing biogas production. However, researchers have consistently reported different kinds of fouling that resulted in the reduction of membrane life span. Depending on the strength of the effluent, factors such as high suspended and dissolved solids, fats, oil and grease, transmembrane pressure (TMP) and flux were reported as major contributors to the membrane fouling. Moreover, extracellular polymeric substances (EPSs) are an important biological substance that defines the properties of sludge flocs, including adhesion, hydrophobicity and settling and have been found to accelerate membrane fouling as well. Extensive studies of AnMBR have been done at laboratory while little is reported at the pilot scale. The significance of factors such as organic loading rates (OLRs), hydraulic retention time (HRT), pH and temperature on the operations of AnMBRs have been discussed. Microbial environmental conditions also played the most important role in the production of biogas and the chemical oxygen demand (COD) removal, but adverse effects of volatile fatty acids formation were reported as the main inhibitory effect. Generally, evaluating the potential parameters and most cost effective technology involved in the production of biogas and its inhibitory effects as well as the effluent quality after treatment is technically challenging, thus future research perspectives relating to food to microorganism F/M ratio interaction, sufficient biofilm within the reactor for microbial attachment was recommended. For the purpose of energy savings and meeting water quality discharge limit, the use of micro filtration was also proposed.

Keywords: anaerobic digestion; biogas production; wastewater treatment; membrane bioreactors

1. Introduction

Over the past century, there has been quite a number of studies on the most economic, efficient and environmental friendly wastewater treatment technologies. Conventional aerobic methods have existed for over a century now, but they have major drawbacks that include sludge production, high energy use for aeration, large operating space and a higher maintenance cost. Moreover, the systems are characterized by uncontrolled release of potential atmospheric greenhouse gases such as methane (CH_4), carbon dioxide (CO_2), and nitrogen oxide (N_2O) that contribute immensely to deterioration of the environment [1], although in the aerobic process quality effluents are produced which comply with standards limits set by the different regulatory bodies, especially in developed countries. However, one obvious disadvantage is the lack of material and energy recovery [2,3]. Hence, interest continues to grow in finding the best alternative.

The advent of anaerobic digestion in the field of wastewater treatment marks the beginning of economic and efficient technology [4]. This is seen in the quality of the effluents discharged, material recovered and energy generated as well as the mode of sludge production, handling and processing [5]. One of the early constraints detected from the onset of this technology was the long hydraulic retention time coupled with the slow growing methanogenic bacteria [6]. However, towards attaining higher efficiency, research focus was moved towards coupling anaerobic bioreactors with membrane filtration units. These systems were seen to offer a more unique and prudent technique over conventional anaerobic method [7]. It simply combines the anaerobic process and membrane technology operating simultaneously. Table 1 presents a comparison of conventional anaerobic treatment and anaerobic membrane bioreactor (AnMBR).

Table 1. Comparison of conventional anaerobic treatment and anaerobic membrane bioreactor (AnMBR) [8].

Feature	Conventional Anaerobic Treatment	AnMBR
Discharge quality	Moderate-Poor	High
Sludge volume	low	low
substrate loading concentration	High	High
Removal efficiency (Effluent)	High	High
Biomass retention	Low	Complete
Footprint	High-Moderate	Low
Alkalinity Requirement	Depends on microbial activity	Depends on microbial activity
Nutrient requirement	Low	Low
Startup period	2–4 Months	Less than 2 Weeks
Temperature requirement	Low-Moderate	Low-Moderate
Energy input	Low	Low
Pre-treatment requirement	Not necessary	Mostly for high solid substrates
Biogas recovery	Yes	Yes

The combination of anaerobic bioreactors with membranes is presumed to reduce the overall energy demand and to facilitate the retention of microorganisms so as to operate with high biomass concentration [9]. Depending on the intended use of the treated effluent, membranes such as microfiltration, ultrafiltration, nano-filtration and reverse osmosis are usually coupled to anaerobic reactors in both industrial and municipal wastewater treatment systems [10]. Buntner et al. [11] studied the combination of an upflow anaerobic sludge blanket (UASB) reactor with ultrafiltration membranes for dairy wastewater treatment at ambient temperature. The intent of this combination was to decrease the COD of the dairy wastewater whilst producing biogas rich in methane and diminish the overall sludge production as well as obtaining high quality effluent. About 150 L/kg COD of biogas was achieved out of which 73% was methane gas. The organic loading rate was up to 4.85 kg COD/m^3·day and 95% COD removal was realized, reaching 99% during its stable operation period. However, this is lower compared to the work of Deowan et al. [12]. Both studies were reported at pilot scales, but one configured the membrane externally while the other was submerged within the bioreactor. The aim of this review paper is to gain insight knowledge on the strategies and achievements recorded in the recent past on the use of AnMBR technology, since the idea was presented as a possible solution or substitute that might overcome the disadvantages of conventional wastewater treatment in terms of space utilization, superior effluent quality, energy generation and the overall operation and maintenance cost.

2. Fundamentals of AnMBR

An anaerobic reactor coupled with a membrane was first commercialized by Dorr-Oliver in the early 1980s [13]. It was constructed to treat high strength wastewater (whey) and was named anaerobic membrane bioreactor system. Since then, research on the feasibility and treatment efficiencies of AnMBRs continue, most especially on the type of material and configurations for treating low, medium, and high strength wastewater [14–17]. The two forms of membrane configuration includes internal and external, as shown in Figure 1, but the external configuration is the most widely reported.

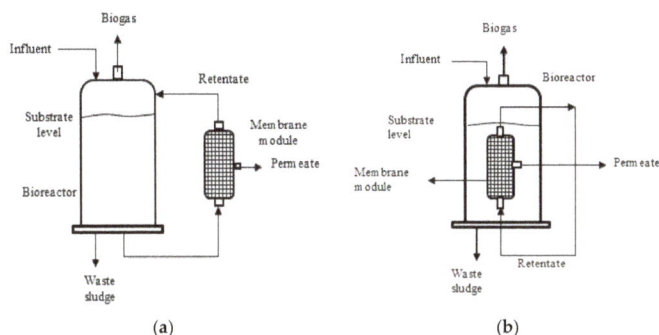

Figure 1. A schematic of AnMBR configurations—(**a**) Side or external membrane (**b**) submerged membrane.

Membrane Configuration Performance

A pilot scale study of municipal wastewater treatment with an external membrane configuration was reported by Huang et al. [14]. The system achieved COD removal efficiency close to 90%, but a slow and linear increase in the filtration resistance was observed under critical flux conditions and subsequently resulted in fouling due to solid accumulation on the surface of the membrane. However, it was observed that gas sparging and additional shear were needed to control membrane fouling. According to Chu et al. [10], high shear stress may interfere with the biological activities during anaerobic digestion of biomass. Furthermore, studies of Zhao et al. [15] showed that the external membrane configuration system is easier to maintain and monitor. However, these require more energy with high hydraulic shear force which might also disrupt anaerobic bio-solids favoring substrates with smaller particle sizes and this in turn causes membrane fouling. In another development, thorough investigations of submersible membrane module coupled to anaerobic bioreactors were reported in [16–19] and it is believed to overcome the numerous drawbacks of conventional methods of wastewater treatment. However, it also appears that the technology is more suitable for low strength organic loads, especially municipal wastewater as examined in the previous research of [20–22].

Martinez-Sosa et al. [3] also examined a submerged anaerobic membrane bioreactor and found 97% COD removal. The final effluent COD was less than 20 mg/L at volumetric organic loading rate of 0.5 to 12.5 kg/m^3·day. The system could be termed efficient even though, the temperature fluctuates between 12 °C and 26 °C. Likewise the membrane permeability was not outstanding due to intermittent suction mode and membrane flux that lead to frequent membrane cleaning. Furthermore, investigation of Deowan et al. [12] on a submerged anaerobic membrane bioreactor (SAMBR) treating textile wastewater revealed COD removal efficiency around 90% with negligible fluctuations in some phases. However, 60% color removal was reported in the first phase and subsequently dropped to between 20–50% after the system stabilized. This phenomenon was attributed to color adsorption on the surface of the membrane. In a similar configuration manner, the defects in COD removal were ascribed to the presence of high total nitrogen (TN) and total phosphorus (TP) [23]. In view of that, application of anoxic conditions or an aerobic process or a combination of the two could enhance removal of both TN and TP. Katayon et al. [24], experimented on the effect of vertical and horizontal membrane configurations, in which two procedures consisting of low and high mixed liquor suspended solids (MLSS) concentrations were considered. The results showed that horizontally placed membrane was able to removed 99.2% total solids and 99.73% turbidity at lower MLSS concentration at a mean flux value of 5.03 L/m²h. This is quite higher than those found in vertical modules with high MLSS concentrations and mean flux value of 2.27 L/m²h. Hence, for higher efficiency, maintaining high biomass concentration and sufficient microbial activity with minimal energy utilization would enhance the process of anaerobic membrane bioreactor operation processes.

3. Membrane Performance of Various Wastewater Treatments

3.1. Industrial Wastewater

Industrial regulatory issues towards meeting the stringent water quality discharge permissible limits motivate researchers to study more on finding lasting solutions to the problems associated with the effluent released to the environment. Most industrial wastewaters are regarded as high strength wastewater, because they contain large amounts of settleable, dissolved and suspended solids or other elements such as heavy metals in greater proportion [25,26]. However, the strength of the wastewater could differ from one industry to another owing to the different types of operations. Generally, industrial wastewater comes from streams such as production lines, cooling towers or boiler and cleaning processes that contain diverse substances. These may include organic and inorganic compounds, viruses, bacteria and toxic compounds. Therefore, applying less energy and achieving high COD removal with minimum sludge production is of utmost priority to the water and wastewater treatment industry. For instance, a pilot scale study of SAMBR treating high strength wastewater (raw tannery wastewater) achieved higher COD removal efficiency up to 90% at organic loading rate OLR of 6 g/L·day and biogas yield (0.160 L/g COD removed) [27]. The system performed efficiently, but was strongly characterized by a high hydraulic retention time (HRT) (40 h) and as such, high energy was expended, although the permeate flux remained at (6.8 LMH) which was well below the critical flux (17.5 LMH) as determine in the earlier work of Hu et al. [28]. Similarly, Fush et al. [29] reported treatment of three different high strength wastewater effluents (artificial wastewater, animal slaughterhouse, and sauerkraut brine) using continuous stirrer tank reactor CSTR coupled with membrane filtration units. The COD removal in all the reactors were >90% at OLR of 20 g COD/L·day, 8 g COD/L·day and 6–8 g COD/L·day respectively. The methane yields were in the range of 0.17–0.30, 0.20–0.34, and 0.12–0.32 Ln/g·COD fed. On the other hand, Saddoud et al. [30] showed that an anaerobic cross-flow ultrafiltration membrane bioreactor exhibited high efficiency removal of suspended solid (SS), biochemical oxygen demand BOD_5, COD and microorganisms. It reached >100%, 90%, 88%, and 100% respectively. Biogas production got to 30 liters per day with average of 0.27 L CH4/g·COD yield. Interestingly, this high yield of biogas was achieved at low organic loading rate (OLR) of 2 g COD/L·day.

Most recently, the filtration performance of an AnMBR treating high strength lipid-rich wastewater and corn-to-ethanol thin stillage was conducted by Dereli et al. [31]. The reactors delivered a high COD removal efficiency of up to 99% under stable operating conditions with an average OLR of 8.3, 7.8, and 6.1 kg COD/m³·day. However, the permeate quality turned out to be inferior in quality with increased in solid retention time (SRT). Table 2 present some examples of the biological and membrane performance of different AnMBR applications for treatment of industrial wastewater. Unfortunately, information regarding the membrane performance with respect to biogas production is quite limited in most of the studies.

3.2. Municipal Wastewater

Effluents characterized by low organic strength and high particulate organic matter are mostly reported as municipal wastewater [32]. The ability of AnMBRs in retaining biomass within the reactor has made it a subject of research for municipal wastewater treatment as a possible alternative to the conventional anaerobic treatment process. According to Ho et al. [33], Kocadagistan and Nazmi [34], bioreactors coupled with membranes used for the treatment of municipal wastewater (MWW) have shown excellent effluent quality that meets the stringent discharge standards in terms of COD, suspended solids and pathogen counts removals when compared to conventional anaerobic methods. Moreover, the use membrane bioreactor (MBR), for anaerobic treatment of municipal wastewater in high temperate climate is still a challenge. This is because, domestic wastewater is usually complex in nature and characterized by high fraction of particulate organic matter. It could be in the form of proteins, suspended solids of different origin, and fatty acids with large to moderate biodegradability portions [35]. Based on these characteristics, operation of MWW treatment reactor under psychrophilic temperature (<20 °C) might be some how difficult. Smith et al. [36] compared simulated and actual domestic wastewater (DWW) at the bench-scale using submerged flat-sheet microfiltration membranes. A psychrophilic temperature of 15 °C was set. An average removal efficiency of 92 ± 5% COD was achieved, which corresponds to an average permeate COD of 36 ± 21 mg/L in simulated DWW, while 69 ± 10% was realized during actual DWW treatment. However, it is obvious the membrane in this study utilizes a lot of energy with relatively high concentration of dissolved methane. Although previous references [37–41] have reported the performance of simulated and actual DWW treatments, only a few of them were performed at psychrophilic temperature (15 °C and below). Table 3 summarizes some studies on the use of AnMBRs for domestic wastewater treatment.

Another investigation of Smith et al. [52] showed a simulated domestic wastewater using AnMBR at psychrophilic temperatures of 15, 12, 9, 6, and 3 °C. Remarkably, a total reduction of COD > 95% was realized at a temperature of 6 °C. The success was attributed to viable microbial activity in the membrane biofilm but subsequently give rise to high dissolved methane oversaturation in permeate and consequently fell to 86% at 3 °C. However, decreasing temperature resulted in increased soluble COD in the bioreactor. Thus, this signifies a reduction in suspended biomass activity. Dolejs et al. [53] conducted a research on the effect of psychrophilic temperature shocks on a gas-lift anaerobic membrane bioreactor (Gl-AnMBR) used for treating synthetic domestic wastewater. The stability of the system was measured by transiting between mesophilic and psychrophilic stages having several psychrophilic shocks (12–48 h). The result showed an average COD removal of 94 ± 2% at mesophilic with an average methane yield of 0.19 L CH_4/g COD removed (including psychrophilic shocks). More than 80% of the influent COD accumulated in the reactor under psychrophilic in comparison to 39% under mesophilic conditions.

A pilot scale study of anaerobic urban wastewater treatment in a submerged hollow fiber membrane bioreactor was reported by Gimenez et al. [20]. They assessed the effect of a number of operational variables on both biological and physical separation process performance. Mesophilic temperature of 33 °C, at 70 days SRT, and HRT ranging from 20 h down to 6 h. COD removal stood almost at 90%, with no trace of irrevocable fouling observed but yet the methane yield was very low and this was mainly attributed to influent COD/SO_4–S ratio. The work of Gouveia et al. [54] on a pilot scale AnMBR coupled to an external ultrafiltration treating MWW also attained COD removal efficiency of 87 ± 1% with specific methane yield of 0.18 and 0.23 Nm^3 CH_4/kg COD removed at a lower temperature of 18 ± 2 °C. More recently, a toxicity reduction in wastewater (synthetic wastewater) aiming at reuse possibilities was performed at relatively psychrophilic temperature (25 °C) using submerged anaerobic membrane bioreactor with forward osmosis membrane (FO-AnMBR) [55]. The FO-AnMBR process exhibited >96% removal of organic carbon, nearly 100% of total phosphorus and 62% of ammonia-nitrogen respectively. This suggests better removal efficiency than the conventional AnMBR. The average COD removal was 96.7% corresponding to the influent COD concentration of 460 mg/L and a methane production of 0.21 L CH4/g COD. Thus, the system demonstrates high feasibility of energy recovery.

Table 2. References on anaerobic membrane bioreactors (AnMBRs) for industrial wastewater.

Type of System/Module Configuration/Membrane Configuration	Membrane Type/Material/Characteristic	Wastewater Treated	Operating Condition	Reactor Working Volume + Scale	Influent COD mg/L	Effluent COD Mg/L	Maximum COD Removal (%)	Biogas Production (L CH₄/g COD removed)	Reference
AnMBR/External cross flow membrane/Tubular	-/ceramic (ZrO$_2$-TiO$_2$)/ Pore size = 0.2 μm Area = 0.25 m²	Industrial wastewater	Flux = 8.4 L/m²·h, HRT = 1.7-5, 3.5, 3-3.5, SRT = 120-450 day, Temp = 35-37 °C, MLSS = OLR = 2.5 g, COD L/day	V = 50 L S = P	4300	830	78	0.5 and 0.6	[42]
SAMBR/Submerged/Flat sheet and hollow fiber	MF (Polipropilen-PP)/Chlorinated polyethylene/ Pore size = 0.05 μm Area = 0.66 m²	Synthetic industrial wastewater	Flux = 3-0.9 L/m²·h, HRT = 390, 167, 168, SRT = ∞, Temp = 35 °C, MLSS = OLR = 0.3-0.54 g, COD/L-day	V = 4 L S = L	20,000-23,000	152	85-90	-	[43]
Submerged anaerobic membrane bioreactor (SAnMBR)/Submerged/ Hollow fiber	MF/Curtain-type/ Pore size = 0.2 μm Area 5.4 m²	Synthetic industrial wastewater	Flux = 6 L/m²·h, HRT = 2.2 h, SRT = -, Temp = 35 °C, MLSS = 10.9 g/L, OLR = 3.0 kg COD/m³·day	V = 25 L S = P	223 ± 111	50 ± 22	87	0.12	[44]
SAnMBR/Submerged/ Hollow fiber	MF/Curtain-type/ Pore size = 0.4 μm Area = 0.040 m²	Paper mill wastewater	Flux = 7.2 L/m²·h, HRT = 35 h, SRT = 40 day, Temp = 21 °C, MLSS = 12.90 g/L, OLR = 7.0 kg COD/m³·day	V = 10 L S = L	11,415 ± 15	228.3 ± 5	98	-	[45]
AnMBR	UF/Hollow fiber/Polyvinylidene fluoride (PVDF) Pore size = - Area = 20 m²	Synthetic anti-biotic solvent	Flux = 20 L/m²·h, HRT = 48, 36, 24, and 18 h, Temp = 35 ± 1 °C, MLSS = 16.5-12.4 g/L, OLR = 3.9-12.7 kg, COD/m³·day	V = 4.4 m³ S = P	7892-21,986	8056.5 1218.2 1523.5 1828.3 2207.7	93.6 and 98.7	94 and 130.7	[46]
AnMBR	UF/Hollow fiber/ Polyvinylidene fluoride (PVDF) Pore size = - Area = 20 m²	Synthetic anti-biotic solvent	Flux = 20 L/m²·h, HRT = 48 h, Temp = 37 °C, MLSS = 52.2 g/L, OLR = 3.79 kg, COD/m³·day	V = 4.4 m³ S = P	15,000-25,000	4000	96.5	-	[47]
AnMBR	UF/Hollow fiber/ Polyvinylidene fluoride (PVDF) Pore size = - Area = 20 m²	Synthetic anti-biotic solvent	Flux = 20 L/m²·h, HRT = 48-24 h, Temp = (35 ± 3 °C, 25 ± 3 °C, 15 ± 3 °C, 25 ± 3 °C), MLSS = 16.5 g/L, OLR = 10.0 kg, COD/m³·day	V = 4.4 m³ S = P	1000-25,000	-	95	-	[48]
AnMBR	UF/Hollow fiber/ Pore size = 0.04 μm Area = 0.047 m²	Brewery wastewater	Flux = 8.64 ± 0.69, L/m²·h, HRT = 44 h, Temp = (35 °C), MLSS = 2.8 g/L, OLR = 3.5-11.5 g, COD/L.d	V = 15 L S = L	19,100	171	99	0.53 ± 0.015	[49]

Table 2. *Cont.*

Type of System/Module Configuration/Membrane Configuration	Membrane Type/Material/Characteristic	Wastewater Treated	Operating Condition	Reactor Working Volume + Scale	Influent COD mg/L	Effluent COD Mg/L	Maximum COD Removal (%)	Biogas Production (L CH₄/g COD removed)	Reference
AnMBR and B-AnMBR	UF/Hollow fiber / Polyvinylidene fluoride (PVDF) Pore size = 0.02 μm Area = 0.07 m	Bamboo wastewater	Flux = 33.4–16.2 L/m²·h, HRT = 3 d, Temp = 32 ± 2 °C, MLSS = 16 g/L, OLR = 6 kg, COD/m³·day	V = 15 L S = L	17,160 ± 814	278.9 ± 4.2 mg/L and 125.3 ± 3.2 mg/L	94.5 ± 2.9 and 89.1 ± 3.1	13.2 ± 1.2, 10.3 ± 0.8	[50]
C-AnMBR and B-AnMBR	UF/Hollow fiber / Polyvinylidene fluoride (PVDF) Pore size = 0.04 μm Area = 0.047 m²	Pharmaceutical wastewater	Flux = 6 L/m²·h, HRT = 30.6 h, Temp = 27 ± 1.0 °C, MLSS = 9500–10,200 and 10,000 mg/L, OLR = 13.0 ± 0.6 kg, COD/m³·day	V = 10 L S = L	16,249 ± 714	8723 ± 593 and 6432 ± 445	46.1 ± 2.9 and 60.3 ± 2.8	24.5 ± 2.1 And 17.6 ± 1.5	[51]

AnMBR = anaerobic membrane bioreactor, COD = chemical oxygen demand, HRT = hydraulic retention time, MLSS = Mixed liquor suspended solids, OLR = organic loading rate, P = pilot scale, - = not reported.

Table 3. References on anaerobic membrane bioreactors (AnMBRs) for municipal wastewater treatment.

Type of System/Module Configuration/Membrane Configuration	Membrane Type/Material/Characteristic	Wastewater Treated	Operating Condition	Reactor Working Volume + Scale	Influent COD mg/L	Effluent COD mg/L	Maximum COD Removal (%)	Biogas Production (L CH₄/g COD removed)	Reference
AFMBR /Submerged/ Hollow fiber	-/Non-woven Fibrous (chlorinated polyethylene)/ Pore size = 0.1 μm Area 0.022 m²	Raw municipal wastewater	Flux = 30 L/m²·h, HRT = 4 h, SRT = ∞, Temp = 23 ± 1 °C, MLSS = -, OLR = 3–6	V = 2.7 L S = L	-	23.5	97.07	2.11	[56]
SAMBRs/Submerged / Hollow fiber	MF/Polyvinylidene fluoride/ Pore size = 0.02 μm Area = -	Municipal (wastewater from the ethanol fermentation of food waste)	Flux = 9.72 L/m²·h, HRT 110 h, SRT = -, Temp = 23–25 °C, MLSS = 8–12 g/L, OLR = 2.3–3.6 g/L·day	V = 0.628 L S = L	15,000 ± 1000	202 ± 23	98.2	-	[57]
UASB/Submerged/Tubular	UF/Polyvinylidene fluoride/ Pore size = - Area = 0.2375 m²	Raw municipal wastewater	Flux = 2.5 L/m²·h, HRT = 8 h, Temp = 18–21 °C, MLSS = -, OLR = -	V = 0.7 m³ S = P	525 ± 174, 657 ± 235	222 ± 61, 130 ± 55	68.6	-	[58]
CG-AnMBR and SG-AnMBR/Submerged/ Hollow fiber	-/A polyvinylidence (PVDF) Pore size = 0.22 μm Area = 0.06 m²	Synthetic domestic wastewater	Flux = ±3 L/m²·h, HRT = 12 h, SRT = 25–30 day, Temp = 20 °C, MLSS = 20.50 ± 1.53 g/L, OLR = -	V = 3 L S = L	330–370	-	90	156.3 ± 5.8	[59]
MBR/Submerged /Flat sheet	MF/-/ Pore size = 0.4 μm Area = 16 m²	Real municipal wastewater	Flux = 7.8 L/m²·h, HRT = 35 h, SRT = 25–30 day, Temp = 6.5–21 °C/7.0–20 °C, MLSS = 5300–9800 mg/L, OLR = -	V = 3 m³ S = P	896 GC mL⁻¹	18.7 ± 2.9	93	-	[60]

Table 3. *Cont.*

Type of System/Module Configuration/Membrane Configuration	Membrane Type/Material/Characteristic	Wastewater Treated	Operating Condition	Reactor Working Volume + Scale	Influent COD mg/L	Effluent COD mg/L	Maximum COD Removal (%)	Biogas Production (L CH$_4$/g COD removed)	Reference
AnCMBRs/Submerged/ Flat sheet	-/Ceramic membrane/ Pore size = 80 nm Area = 0.08 m^2	Domestic wastewater	Flux = 8 L/m^2·h, HRT = 5.8 h, SRT = 60 day, Temp = 25 °C, MLSS = -, OLR =10.0 kg COD/m^3·day	V = 3.6 L S = L	417 ± 61	-	87	-	[61]
SAnMBR/Submerged/Flat-sheet	-/Polyethylene terephthalate/ Pore size = 0.2 μm Area = 0.116 m^2	Synthetic municipal wastewater (alcohol ethoxylates)	Flux = -, HRT = 42–12 h SRT = -, Temp = 25 ± 1 °C MLSS = -, OLR = 3.0–6.0 kg COD/m^3·day	V = 6 L S = L	-	17.1	95.5–98.8	2.30–4.25	[62]
SAnMBR/Submerged/Flat-sheet	UF/Polyvinylidene fluoride/ Pore size =0.2 μm Area 0.735 m^2	Domestic wastewater	Flux = 8.3–9.5 L/m^2·h and 6.0–6.7 L/m^2·h, HRT = 5.8–4.8 and 8.0 –7.1 h, SRT = 50 day, Temp = 35 °C and 25 °C, MLSS = -, OLR = 0.43–0.90 kg COD/m^3·day	V = 15 L S = L	400	-	90	276 ± 13	[63]
AnMBR/External/Tubular	UF/Polyethersulfone/ Pore size = 30 μm Area 0.11 m^2	Synthetic municipal wastewater	Flux 12.3 L/m^2·h, HRT = 6 h, SRT = 126 day, Temp = 25 °C and 15 °C, MLSS = -, OLR = 2 kg/m^3·day	V = 7 L S = L	530 ± 30	42 and 52	92 and 90	-	[64]
AnMBR/External/Tubular	UF/Polyethersulfone/ Pore size = 30 μm Area 0.0038 m^2	Synthetic municipal wastewater	Flux 12.3 L/m^2·h, HRT = 6 h, SRT = 126 day, Temp = 25–15 °C, MLSS = -, OLR = 2 kg/m^3·day	V = 7 L S = L	530 ± 30	149 ± 5.9 to 42 ± 4.4,	92	-	[65]

AnMBR = anaerobic membrane bioreactor, COD = chemical oxygen demand, HRT = hydraulic retention time, MLSS = Mixed liquor suspended solids, OLR = organic loading rate, P = pilot scale, - = not reported.

3.3. Synthetic Wastewater

Compounds such as starch, glucose, molasses, peptone, yeast, and cellulose are usually used as synthetic substrates to test new concept of AnMBR. The results of a number of studies are summarized in Table 4. The COD removal efficiencies of those investigations were generally >90%, with OLR less than 10 kg COD/m^3·day. Effect of HRT and SRT on treatment performance of submerged AnMBR for synthetic low-strength wastewater reveals a total COD removal efficiencies higher than 97% at all the operating conditions with a maximum biogas production in terms of mixed liquor volatile suspended solid (MLVSS) removals (0.056 L CH4/g MLVSS) at infinite SRT [66]. Though increasing in OLR with short hydraulic time HRT and long SRT boosted the methanogenic environment, but membrane fouling was worsened due to a decrease in HRT which heightened the growth of biomass and accumulation of soluble microbial products (SMP). The experiment performed by Jeison et al. [67] on synthetic wastewater using submerged anaerobic membrane bioreactor showed a reversible cake layer formation on short term basis. Additionally, cake consolidation was detected during a long-term operation at a flux close to the critical point. Remarkably, increasing OLR from 50–60 g COD/L·day towards the end of the operation presented high COD removal efficiency greater than 90%.

Fallah et al. [68] reported a significant COD removal of (<99%). In their studies, they considered two hydraulic retention times (24 h and 18 h) using a MBR to remove styrene from a synthetic wastewater having a chemical oxygen demand and styrene concentration of 1500 mg/L and 50 mg/L. nonetheless, reduction in HRT to 18 h caused a release of extracellular polymeric substance (EPS) from the bacterial cells that led to the rise in soluble microbial product (SMP) and sludge deflocculation. Moreover, the dramatic rise in transmembrane pressure TMP which was operating fairly low and constant for a number of days give rise to severe membrane fouling. This trend was attributed to the rise in SMP concentrations and the decrease in mean floc size. Similar research was reported in Ho et al. [69]. The HRT were varied during treatment of synthetic municipal wastewater. The permeate quality was outstanding, irrespective of HRT differences with over 90% COD removal at HRT of 6 h. Methane produce was 0.21 to 0.22 CH$_4$/g COD removed. Conversely, the fraction of methane recovered from the synthetic municipal wastewater declined from 48 to 35% with the reduction of HRT from 12 to 6 h. Subsequently, the result of the increased mixed-liquor soluble COD which was precluded and accumulated in the AnMBR drastically affects the performance.

Table 4. References on anaerobic membrane bioreactors for synthetic wastewater treatment.

Type of System/Module Configuration/Membrane Configuration	Membrane Type/Material/Characteristic	Wastewater Treated	Operating Condition	Reactor Working Volume + Scale	Influent COD mg/L	Effluent COD mg/L	Maximum COD Removal (%)	Biogas Production (L CH$_4$/g COD removed)	Reference
SAMBR/ Submerged/ Flat sheet	-/Non-woven fibrous (chlorinated polyethylene)/ Pore size = 0.2 μm Area = 0.116 m²	Synthetic municipal wastewater (linear alkyl benzene sulfonate concentration in sewage)	Flux = -, HRT = 24–12 h, SRT = Temp = 25 ± 1 °C MLSS = -, OLR = 3–6 kg COD/m³·day	V = 6 L S = L	-	23.5	97.07	2.11	[74]
SAMBRs/ Submerged/ Flat sheet	MF/Polymethyl methacrylate/ Pore size = 0.2 μm Area = 0.116 m²	Synthetic sewage	Flux = 15 L/m²·h, HRT = 12, 8, 6, 4 and 2 h, SRT = 200 day, Temp = 35 ± 1 °C, MLSS = 6000 mg/L and 7000 mg/L, OLR = -	V = 3 L S = L	544 ± 22	14 ± 2	>97	252 ± 27, 236 ± 27, 249 ± 29, 264 ± 46, 134 ± 23	[75]
SMBR/-/-	MF/Polymethyl methacrylate/ Pore size = 0.04 μm Area = 0.047 m²	Synthetic wastewater	Flux = 10.64L/m²·h, HRT = 8 h, SRT = 140 day, Temp = 18 ± 15 °C, MLSS = 6000 mg/L, OLR = 1.77 ± 0.03 g COD/L·day	V = 4 L S = L	-	-	91 and 98	-	[76]
AnOMBR/submerged/-	Cathode/Stainless steel mesh/ Pore size = - Area 1.5 m²	Synthetic wastewater	Flux = 3.532 LMH HRT = -, SRT = - Temp = 35 ± 1 °C MLSS = 16.31 g/L OLR = -	V = 6.8 L S = P	2000	-	71.1	0.254	[77]
SAMBRs/Submerged/Flat sheet	MF/Non-woven fibrous (chlorinated polyethylene)/ Pore size = 0.2 μm Area = 0.116 m²	Synthetic sewage	Flux = 5 L/m²·h, HRT = 12–6 h, SRT = ,Temp = 35 °C, MLSS = -,OLR = 3–6 kg COD/m³·day	V = 3.2 L S = L	-	29 ± 8	93.2 ± 6.6	1.7 ± 0.3	[78]
Integrated anaerobic fluidized bed membrane (IAFMBR)/-/-	-/Hollow fiber/ Pore size = 0.4 μm Area = 0.21 m²	Synthetic benzothiazole wastewater	Flux = 11.3 L/m²·h, HRT = 24 h, Temp = 35 °C, MLSS = -, OLR = -	V = 6.1 L S = L	-	230	96	0.31 ± 0.02, 0.32 ± 0.03 and 0.31 ± 0.01	[79]
SAMBRs/Submerged/Flat sheet	MF/Chlorinated polyethylene/ Pore size = 0.2 μm Area = 0.116 m²	Synthetic sewage	Flux = -, HRT = 48, 24, 12, 6 h, Temp = 25, 15, 10 °C, MLSS = -, OLR = -	V = 6 L S = L	-	134	94	-	[80]
AnMBRs/External /-	-/Polyvinylidene (PVDF) hollow fiber/ Pore size = 0.22 μm Area = 0.06 m²	Synthetic sewage	Flux = 20.2 ± 8.5 L/m²·h, HRT = 3.5 ± 1.1, 2.2 ± 0.5, 2.3 ± 0.3 day, Temp = 35 °C, MLSS = 4.78 ± 1.9 g L^{-1}, OLR = 3.4 kg COD/m³·day	V = 3.5 L S = L	6752 ± 663	0.095 ± 0.150.14 ± 0.230.28 ± 0.35	96.7 ± 2.7	0.50 ± 0.17	[81]
EG-AnMBR and SG-AnMBR	-/Polyvinylidene (PVDF) hollow fiber/ Pore size = 0.22 μm Area = 0.06 m²	Synthetic sewage	Flux = 7 L/m²·h, HRT = 12 h, Temp = 20 °C, MLSS = 22.34 ± 0.41 g/L, OLR = 0.53–0.59 kg COD/m³	V = 4 L S = L	-	-	<90%	160	[82]

AnMBR = anaerobic membrane bioreactor, COD = chemical oxygen demand, HRT = hydraulic retention time, MLSS = Mixed liquor suspended solids, OLR = organic loading rate, P = pilot scale, - = not reported.

4. Effect of Microbial Activity on Anaerobic Membrane Performance

During microbial activity, the concentration of volatile fatty acids (VFA) usually reflects the state of anaerobic digestion performance, specifically in the acetogenic and methanogenic phase. According to Wijekoon et al. [70], 85–96% of high strength molasses-based synthetic wastewater was removed as total chemical oxygen demand (COD) at optimum organic loading rate of 8 ± 0.3 kg COD/m^3·day. Though it was at higher temperature, but biogas production of 15, 20 and 35 L/day at OLR 5.1 ± 0.1, 8.1 ± 0.3 and 12.0 ± 0.2 kg COD/m^3·day respectively was achieved. It was seen from the results of the treatment, increasing loading rate amounts to increase in hydrolytic and methanogenic activities. However, it was also observed that the process performance reached its maximum level with continuing increase in loading rate which subsequently reduce the biological activities of the system.

In anaerobic digestion, rapid methane formation is mostly attributed to the presence of acetic acid and butyric acid. However, acetic acid contributes 70% of the total acidic portion. This might be due to the fact that, all the volatile acids are converted into acetate during metabolism. The presence of propionic acid tends to upset anaerobic digestion processes due to its toxicity among the VFAs, but this effect could be counteracted under thermophilic conditions. The experiments of Speece et al. [71] showed that propionic acid degradation rate to other intermediate was very poor, largely because of lower partial pressure (H_2) demand, but it also supports process start up and stabilization under strict anaerobic conditions.

Moreover, methanogenic activity might be inhibited with propionic acid concentrations greater than 1–2 g/L. It could also withstand acetic and butyric acid concentrations up to 10 g/L. Ahmed et al. [72] have found that, microbial respiratory quinones are essential parts of the bacterial respiratory chain that perform a vital role in electron transfer during the period of microbial respiration. However, an investigation of Hiraishi et al. [73] demonstrated that alterations in microbial community structure in a mixed culture of microbes could effectively be quantified using quinone profiles.

Recent research on the effects of temperature and temperature shock on the performance and microbial community structure of a submerged anaerobic membrane bioreactor (SAnMBR) was demonstrated by Gao et al. [21]. The result obtained indicates that not only the diversity, but also the species richness of microbial populations are affected by temperature variation. It further proves that submerged AnMBR performance under temperature shock conditions have no effect on the COD removal ability of the reactor. The biogas production rates were 0.21 ± 0.03, 0.20 ± 0.03 and 0.21 ± 0.02 L/g COD removed. Moreover, no major change is observed in the production and composition of biogas. Nevertheless, temporary production of biogas occurred at temperature shock which affected the abundance and diversity of microbial populations.

According to Iranpour et al. [83], the similarities that exist among the microbial community working under variable temperatures may be connected to the development of thermomesophiles rather than thermophiles. During microbial activity in anaerobic digestion, the release of soluble organic compounds called SMP in normal biomass metabolism is of great interest, not only in terms of achieving discharge standard limits, but also in setting the lower limit for treatments [84]. The significance of SMP in all kinds of wastewater treatment is at the moment objectively well recognized, but complications still come about in trying to measure SMP and draw conclusions when they are present in effluents from plants treating highly complex feeds. However, researchers like Sciener et al. [85] and Chidoba et al. [86] have studied this and conclude that large portions of the soluble organic matter in the effluent from the biological treatment processes are actually SMP.

The work of Judd [87] and Li et al. [88] reveals that microbial activity decreases under prolonged SRT in AnMBR. This is because microbes take a long time to degrade the inorganic matter and require a high concentration of biomass to ensure all the organics are totally degraded. Still, the application of high SRT is advantageous, since it favors biomass growth which is responsible for biodegradability of organic pollutants. It also provides room for higher MLSS in AnMBR that establish starvation conditions to achieve good quality effluent and create a low F/M ratio [89,90].

5. Operational and Performance Parameters

5.1. Temperature

In anaerobic digestion (AD) processes, temperature is a major factor that plays an important role in the stabilization and performance of the whole system [91,92]. Under strict anaerobic conditions, some bacteria thrive well under psychrophilic (<25 °C), mesophilic (25–40 °C) and thermophilic (>45 °C) conditions [93]. According to El-Mashad et al. [92] and Duran et al. [94], thermophilic conditions offer more advantage in terms of specific growth and metabolic rates with comparatively less ammonia inhibition than mesophiles. However, it could also cause higher microbial death rates, poor supernatant quality and reduced process stability due to chronically high propionate concentrations than mesophilic bacteria. Studies on biogas production and wastewater treatment at varying temperature such as psychrophilic (0–20 °C), mesophilic (20–42 °C), and Thermophilic (42–75 °C) were well investigated in [10,12,14,94].

The diversity and activities of microbial communities coupled with thermodynamic equilibrium of the biochemical reactions are adversely affected by temperature [95]. It usually shifts the abundance and activities of specific microbial populations and determines the roles of specific taxa in the AD food chain. It all begins with substrate hydrolysis, followed by acidogenesis, then acetogenesis, and ends with methanogenesis [96]. Study on temperature adaptation was examined by Chidoba, 1985 [86]. It was initially set at mesophilic condition (37 °C) and gradually switched to thermophilic (55 °C). A drastic reduction in biogas production by 15% from the original state of mesophilic was observed. This was attributed to the change in temperature along with increase in VFA to 3000 mg/L. More detailed studies were reported by Song et al. [97] on thermophilic and mesophilic temperature co-phase anaerobic digestions. Their examination exclusively focused on the sewage sludge using the exchange process of digesting sludge between spatially separated mesophilic and thermophilic digesters and compared with single-stage mesophilic and thermophilic anaerobic digestions. It was confirmed that the system stability, effluent quality, specific methane production during single-stage operation was greater in mesophilic than thermophilic., but volatile solids (VS) reduction and total coliform destruction were much higher in single-stage thermophilic than mesophilic digestion.

In the past, study on the effect of temperature range between 40–64 °C using cow manure as the main substrate was well studied by Angelidaki, 1994 [98]. Two different ammonia concentrations (2.5 and 6.0 g N/L) were continuously fed to the lab-scale reactor. It was observed at some stage, precisely (HRT 15 days); the high temperature and ammonia loading resulted in poor process performance. Consequently, a significant change in the amount of biogas production and process stability was seen when the ammonia load was high and temperature reduces to below 55 °C.

5.2. pH

Volatile fatty acid (VFA) concentration usually determines the pH of the effluent, which is one of the influential factor in anaerobic digestion (AD) processes [99]. Different range of pH is required for bacterial growth in AD, comprehensively between 4.0 to 8.5 [100]. However, a pH level of 6.8 to 7.2 suitably favors methanogenic bacteria [101,102]. On the other hand, hydrolysis and acidogenesis thrive well in the pH range between 5.5 and 6.5 [103,104]. It was also shown that, excessively alkaline pH may result in microbial granules disintegration and subsequent failure [105]. The pH adjustment studies in [101] favorably increased methane yield. The maximum cumulative biogas production reached 16,607 mL at pH 7.0 (0.4535 L methane/g VS). However, the yield decreased to 6916 and 9739 mL at pH 6.0 and 8.0, equivalent to 0.1889 L methane/g VS and 0.2659 L methane/g VS, respectively.

Recent research on novel biogas-pH automation control strategy using a combined gas-liquor phase monitoring was developed by Yu et al. [106] using an AnMBR treating high starch wastewater COD (27.53 g/L). The biogas pH progressed with threshold between biogas production rate >98 NmL·h^{-1} preventing overload and pH > 7.4 preventing under load. The OLR and the effluent COD was doubled to 11.81 kg COD/m^3·day and halved as 253.4 mg/L, respectively. In another development, a ternary contour was performed by Mao et al. [107] to picture the pH dissimilarities in ternary buffer system. The variation was controlled by a system composing of VFAs, ammonia and carbonate. However, the accurate simulation of pH variation in AnMBR during methanogenesis is extremely challenging due to enormous magnitude of electrolytes, and as such the ternary macro-quantity buffer salts were chosen for the visualization which denotes the critical pH buffer capacity to during methanogenic stage.

Kim et al. [104] demonstrated that one major problem faced with COD removal especially for starch wastewater is the provision of sufficient carbonate alkalinity. The effect of high pH shocks (pH 8.0, 9.1 and 10.0) on the performance of submerged AnMBR was well reported in [105]. It was found that; pH 8.0 had slight influence, while pH 9.1 and 10.0 shocks have put forth vigorous impact on biogas production, COD removal, and membrane filtration performance. In addition, colloids and solutes accumulation in the sludge suspension further accelerates deterioration of membrane performance. Interestingly, when neutral pH (7) was taken up again, it cost the reactor approximately 1, 6, and 30 days to recover from the previous shocks.

An experiment on the effect of mixed liquor (pH 5 and 9) in the removal of trace organics in both acidic and basic environment was also reported in [106]. A reduction in total organic carbon TOC and total nitrogen TN removal efficiencies was detected with ionisable trace organic contaminants. Hence, the removal efficiencies were extremely pH dependent. Conversely, the biological performance was favorable at near optimum pH (i.e., approximately pH 6–7) with respect to TOC and TN removal efficiencies. As result of the differences in attainment of optimum pH by acidogenic bacteria (5.5–6.5) and methanogens (6.5–8.2), several studies have been conducted on phase separation of acidogenesis and methanogenesis using AnMBRs. For example, Mao et al. [107] demonstrated a two-stage reactor configuration and optimizes each one phase as an entity. VFA accumulation was drastically reduced and the system stability improves to the extent of forbearing greater loading rate and toxicity. Therefore, maintenance and adherence to these features would positively escalate the chance of achieving a high methane yield.

5.3. Effect of OLR

The amount of volatile solids VS fed to a bioreactor at an interval of time under continuous operation is termed as the OLR. Under normal operating conditions, it is expected that an increase in organic loading rate would also increase the quantity of biogas. AnMBR processes have the advantage of tolerating changes in organic loading similar to tolerance to fluctuations in temperature. Organic loadings ranging from 0.5 to 12.5 kg/m^3·day was applied to AnMBR for the treatment of domestic wastewater. The system achieved 97% (COD) removal with less than 20 mg/L effluent COD [108]. In a similar experiment by Vincent et al. [109], less than 50 mg/L soluble COD in the effluent was obtained at organic loading rate 0.25 kg/m^3·day and 0.7 kg/m^3·day using AnMBR. This is not the case in the work of Qiao et al. [110]. The system exhibited very poor effluent quality after a long HRT. Study of the performance of AnMBR by Yingyu et al. [111] showed that biogas yield rose linearly with increasing organic loading. A similar trend was observed by Wijekoon et al. [70] using a two-stage thermophilic AnMBR with continuous increasing loading rate from 5 to 12 kg COD/m^3·day. Bornare et al. [112] achieved biogas production increased from 159 to 289 L/day, but they observed a decreased in yield from 0.48 to 0.42 L biogas/g CODremoved when OLR was increased from 0.62 to 1.32 kg COD/m^3·day. However, Dereli et al. [113] indicated that OLR could not be an independent parameter and therefore should be assessed along with SRT.

High performance integrated anaerobic–aerobic fixed-film pilot-scale reactor with an arranged media for treating slaughter house wastewater showed a removal efficiency of 93% at OLR of 0.77 kg COD with methane yield of 0.38 m^3 CH$_4$/kg COD [114]. High mixing as a result of system integration causes low extension of the anaerobic process and consequently affects the methanogenic activities. A fact known to most researchers in the field of anaerobic digestion is that an increase in OLR could result in excessive VFA formation which may inhibit microbial activities and deteriorate the system. For example, the research of Saddoud et al. [115] confirms that accumulation of VFA was the main reason for methanogenic inhibition, thus, resulting in a decrease in methane yield at OLR of 16.3 kg COD/m^3 in one-phase AnMBR. To counteract this effect, they proposed the use of a two-stage AnMBR coupled with an anaerobic filter as acidogenic reactor while a jet flow AnMBR was used as methanogenic reactor at high OLR and realized a substantial improvement in biogas production in the subsequent stage. Jeison et al. [116] presented removal efficiency below 50% using continuous stirrer tank reactor CSTR fed with acidified and partially acidified substrate. Even though the organic loading rates reached 10–17 kg COD/m^3·day, the low removal efficiency might be connected to soluble microbial product level within the AnMBR and these could cause irreversible pore fouling in the membranes.

5.4. Effect of HRT and SRT

Operational parameters such as hydraulic retention time (HRT) and solid retention time (SRT) are two factors that play a vital role in the treatment performance of AnMBRs. A study by Mei et al. [61] demonstrated using submerged anaerobic membrane bioreactors (SAnMBRs). SRTs of 30, 60 and infinite and HRT of 12, 10 and 8 h were used. A total COD removal efficiency >97% was observed at all operating conditions. Biogas production rate reached 0.056 L CH4/g MLVSS day at an infinite SRT. However membrane fouling occurred as a result of shorter HRT and infinite SRT. It was also seen that, longer SRT was the main cause of higher SMP production. The presence of SMP subsequently introduces more nutrients onto the membrane surface that caused the blockage of the pores and enhanced biocake formation. Dong et al. [117] reported the influence of SRT and HRT on bioprocess performance of pilot and bench scale AnMBRs using municipal wastewater as substrate. The SRT and HRT applied were 40–10 day and 2.5 to 8.5 h. A good permeate quality with COD concentration (40 mg/L) and BOD$_5$ (10 mg/L) was observed in all conditions. The range of values tested for SRT and HRT have not considerably interfered with COD and BOD$_5$ removal efficiencies. Moreover prolonged SRTs caused a reduced sludge production and increase methane yield.

Similarly, in the research of Ozgun et al. [65], reduction in permeate COD concentrations from 16.5 to 5 mg/L resulted in an increased methane yield from 0.12 to 0.25 L CH$_4$/g COD. The improvement in the quality of effluent and the methane was attributed to prolonged SRT from 30 days to infinite. Salazar et al. [39] and Hu et al. [28] revealed a permeate COD concentration increase from 10 to 50 mg/L with decreasing HRT. However, Liao et al. [35] strongly suggested that, lower HRTs give room for shorter contact time between microorganisms and substrate and therefore might pave way for a part of influent COD leaving the reactor without proper treatment. One very important aspect of AnMBRs is the enabling environment that allows SRT to be completely independent from HRT irrespective of the sludge properties. Conversely, experience and frequent practice shows that longer SRTs operation yield more quantities of biogas. This is because any reduction in the SRT may decrease the extent of reactions required for stable digestion. For instance Huang et al. [66] clearly reported a methane yield of (0.670 ± 0.203 L CH$_4$/day), (0.906 ± 0.357 L CH$_4$/day), (1.290 ± 0.267 L CH$_4$/day) at longer SRT of 30, 60, and infinite days respectively. Therefore, it is apparent that, longer SRT in AnMBRs operations give room for minimal sludge production, and hence cuts disposal cost significantly. Based on the studies on the effects of HRT and SRT conducted so far, it could be seen that prolonging HRT may result in inadequate utilization of AnMBRs' volume and its reduction may lead to rapid VFA accumulation which may hinder methanogenic activities. Prolonged SRT might also result to membrane fouling. It could also encourage the release of soluble microbial products (SMP) as well as rapid cake formation and excessive decline in flux.

6. Inhibitors

Inhibitory substances are the primary cause of disparity in anaerobic reactors. This might also lead to entire failure of the digestion processes when the concentration is beyond tolerable limits. Anaerobic digestion usually presents different variations in the level of inhibition and toxicity. Mechanisms such as synergism, acclimation, and complexity substrate could considerably upset the phenomenon of inhibition. Nevertheless, bioreactor failures are regularly reported as the result of high ammonia inhibition which directly affects the microbial activity [118].

Kayhanian et al. [119], reported ammonia as one of the inhibitory substances found in anaerobic digestion. It is mainly as a result of biological degradation of nitrogenous matter in the form of proteins and urea. Inorganic ammonia nitrogen like ammonium ion $NH_4{}^+$ and free ammonia (FA) are usually found in aqueous form. FA is considered as the primary cause of inhibition since it is freely membrane-permeable [120,121]. Calli et al. [122] reveal an apparent COD removal of 78–96% in a study of the effect of high free ammonia concentrations in synthetic wastewater using UASB reactor. The reactor was fed at OLR of 1.2 kg COD m^3/day with total ammonia nitrogen concentration total ammonia nitrogen (TAN) increasing from 1000 to 6000 mg/L. However, Tchobanoglous et al. [123] showed that, the accumulation of propionic acid along with reduction in eubacterial suggest a sensitivity of propionate degrading acetogenic bacteria to free ammonia than methanogenic archaea. A theoretical stoichiometric relationship of estimating the quantity of ammonia that could be generated from organic substrate in anaerobic biodegradation is shown in Equation (1):

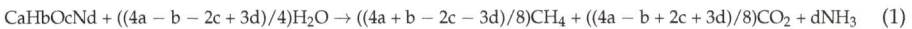

$$CaHbOcNd + ((4a - b - 2c + 3d)/4)H_2O \rightarrow ((4a + b - 2c - 3d)/8)CH_4 + ((4a - b + 2c + 3d)/8)CO_2 + dNH_3 \quad (1)$$

According to Kayhanian [124], methanogens are the most sensitive to ammonia inhibition among the four bacteria types that exist in anaerobic digestion processes. The effect might even cause the bacteria to cease to grow. Study on the effects of ammonia on propionate degradation and microbial community in bioreactors showed that, using propionate as a sole carbon source result to reactor failure after four hydraulic retention times [125]. A total ammonia nitrogen (TAN) concentration of 2.5 g NL^{-1} at OLR of 0.8 g propionic acid (HPr)/L·day were observed and 95% of the degraded HPr was converted to methane. On the average, the degradation rate of HPr is below 53%, likewise an average of 74% and 99% HPr degradation and methane recovery rates was also recorded during the last HRT. Thus, these behaviors demonstrate an alteration of the microbial community. Frequent maintenance, change in intracellular pH of methanogens and inhibition of a specific enzyme reactions are some of the numerous suggested pathways for overcoming ammonia inhibition. Thorough understanding of the ammonia toxicity occurrence which ammonia may affect methanogenic bacteria is not readily available but, is the few available studies with unadulterated cultures this was revealed to influence the treatment in two ways: (i) direct inhibition of methane- producing enzymes by ammonium ion and/or (ii) the hydrophobic nature of ammonia molecules which may diffuse passively into bacterial cells causing proton imbalance [126].

Another important anaerobic digestion inhibiting parameter is sulfate. Sulfate is a common constituent of many industrial wastewaters that is converted in to sulfide by the sulfate-reducing bacteria (SRB) [127]. Major SRB includes complete and incomplete oxidizers. Complete oxidizers convert acetate to CO_2 and $HCO_3{}^-$ completely. Compounds such as lactate are reduced to acetate and CO_2 by incomplete oxidizers. Furthermore, primary and secondary inhibitions are the two stages that exist during sulfate reduction [128]. Methane production is suppressed by primary inhibitors as a result of competition for common organic and inorganic substrate by the SBR, but secondary inhibition is caused by various bacterial groups due to the toxicity of sulfide [129,130].

7. Membrane Fouling

Membrane fouling is the result of gradual accumulation of suspended solids (SS) and dissolved solids (DS) mostly in the form of fats, oil and grease onto the surface of the membrane [131]. Mechanisms that propel fouling occur via: (1) deposition of sludge flocs onto the surface, (2) adsorption of solutes within the membrane, (3) surface cake layer formation, and (4) the shear force that exists between membrane surface, soluble microbial products (SMPs) and extracellular polymeric substances

(EPS) [132]. The surface blockage or clogging phenomenon is always attributed to the type of membrane itself, biomass and operating conditions. Fouling due to the membrane itself could be attributed to the nature of the material, configuration, hydrophobicity, porosity and pore size of the membrane [8]. Moreover, biomass characteristic such as MLSS, EPS or SMP, flock structure and dissolved matter contribute immensely to causing fouling [4]. Conditions at which membrane bioreactor operates might also increase fouling if not properly monitored. Other factors like cross flow velocity, HRT or SRT, aeration and transmembrane pressure could also contribute significantly to fouling [133].

Zhang et al. [134] further revealed that, the use of adsorbent/flocculants in (AnMBR) could rarely overcome the effect of fouling completely. However, their research revealed that, among the combination of eight additives (three powdered activated carbon PACs, two granular activated carbons, one cationic polymer, and two metal salts), 400 mg/L of PAC was able to reduce transmembrane pressure rise from 0.94 to 0.06 kPa/h. This outcome signified an outstanding fouling reduction technique. Still, the effect of increasing in OLR and reduction of HRT (24–18 h) is clearly confirmed as the main factor responsible for severe membrane fouling, but research by Jeison et al. [67] showed that, extracellular polymeric compounds (EPS) from microbial cells was responsible for the release of SMP and a drastic increase in TMP during the long term operation that led to the occurrence of fouling. Filtration performance of membrane bioreactors is largely dependent on the mixed liquor suspended solids. In the experiments performed by Meng et al. [135], filtration resistance including membrane resistance (12%), cake resistance (80%), blocking and irremovable fouling resistance (8%), depicts that the formation of cake layer is the main cause of membrane fouling. Lee et al. [136] pointed out that cake layers are not uniformly distributed on the entire surface of all of the membrane fibers. It could be wholly covered by a static sludge cake that could not be removed by the sheer force due to aeration, and partially by a thin sludge film that might frequently washed away due to aeration turbulence. A study of Jeison [67] concluded that cake layer formation could be removed despite longer operation period (>200 days). However, during these periods, cake consolidation was seen close to the critical flux. A normal backwashing cycle was unable to remove the consolidated cake and as such physical external treatment was applied. Similarly, Cho et al. [131] and Di-Bella et al. [137] showed that cake layer formations on AnMBRs coupled with external aeration zone could easily be removed. This study suggest that AnMBRs operating with aerobic zones external to the membrane module could achieve higher cake layer removal as compared to cake layer formed on membrane embedded within an anaerobic bioreactor.

8. Conclusions

They has been a quite significant development of technologies towards meeting the stringent environmental regulatory discharge requirements and biogas production. These can be seen in the manner, in which MBRs are configured, variations of operational parameters, and the different techniques adopted towards provision of the basic living conditions for microorganisms to strive. However, membrane lifespan is still the main concern of stakeholders in the water and wastewater treatment industries. The efficiency of physical, chemical and biological methods of reversing fouling on membrane surface is being exploited. Though physical and chemical methods have been sufficient, the disadvantages are huge. A lot of energy is consumed during aeration and considerable amount of chemicals are utilized which does not favor the players in this field with regards to cost and environmentally-wise. Research is still ongoing on the most economical biological methods of producing quality effluent and biogas production using AnMBRs, and future studies should take in to account the following factors:

1.   Food to microorganism F/M ratio interaction and HRT of bioreactors operating at thermophilic, mesophilic or psychrophilic temperature. These will offer more information on biogas production and effluent quality.

2. Provision of lining or rough surfaces within the bioreactor (biofilm) would facilitate the retention of large microbial populations preventing them from exiting the reactor along with the effluent if shorter HRT is to be used.

3. If the purpose of the treatment is to meet discharge standard limits, and subsequently discharge the effluent to water bodies, then the use of microfiltration should be encouraged rather than ultrafiltration. This is because; smaller size membrane surface area would require more energy for water to pass through and as such applying ultrafiltration for such treatment would be costly.

Author Contributions: M.A.M. and S.I. conceived the concept for the review; M.A.M. and S.I. analyzed the existing literature; H.C.M. and N.N.N.D. wrote and edited the paper. All authors have read and approved this review.

Acknowledgments: The work required for the preparation and writing of this review was funded by Universiti Putra Malaysia (UPM) baseline funding GP-IPS/9633400 awarded to Mohammed Ali Musa.

Conflicts of Interest: The authors declare no conflicts of interest.

References

1. Tock, J.Y.; Lai, C.L.; Lee, K.T.; Tan, K.T.; Bhatia, S. Banana Biomass as Potential Renewable Energy Resource: A Malaysian Case Study. *Renew. Sustain. Energy Rev.* **2010**, *14*, 798–805. [CrossRef]
2. Verstraete, W.; van de Caveye, P.; Diamantis, V. Maximum use of resources present in domestic used water. *Bioresour. Technol.* **2009**, *100*, 5537–5545. [CrossRef] [PubMed]
3. Martinez-Sosa, D.; Helmreich, B.; Netter, T.; Paris, S.; Bischof, F.; Horn, H. Anaerobic Submerged Membrane Bioreactor (AnSMBR) for Municipal Wastewater Treatment under Mesophilic and Psychrophilic Temperature Conditions. *Bioresour. Technol.* **2011**, *102*, 10377–10385. [CrossRef] [PubMed]
4. Yoon, Y.; Kim, S.; Oh, S.; Kim, C. Potential of Anaerobic Digestion for Material Recovery and Energy Production in Waste Biomass from a Poultry Slaughterhouse. *Waste Manag.* **2014**, *34*, 204–209. [CrossRef] [PubMed]
5. Kim, K.Y.; Yang, W.; Ye, Y.; LaBarge, N.; Logan, B.E. Performance of Anaerobic Fluidized Membrane Bioreactors Using Effluents of Microbial Fuel Cells Treating Domestic Wastewater. *Bioresour. Technol.* **2016**, *208*, 58–63. [CrossRef] [PubMed]
6. Kanai, M.; Ferre, V.; Wakahara, S.; Yamamoto, T.; Moro, M. A Novel Combination of Methane Fermentation and MBR—Kubota Submerged Anaerobic Membrane Bioreactor Process. *Desalination* **2010**, *250*, 964–967. [CrossRef]
7. Harb, M.; Hong, P. Anaerobic Membrane Bioreactor Effluent Reuse: A Review of Microbial Safety Concerns. *Fermentation* **2017**, *3*, 39. [CrossRef]
8. Lin, H.; Peng, W.; Zhang, M.; Chen, J.; Hong, H.; Zhang, Y. A Review on Anaerobic Membrane Bioreactors. Applications, Membrane Fouling and Future Perspectives. *Desalination* **2013**, *314*, 169–188. [CrossRef]
9. Padmasiri, S.I.; Zhang, J.; Fitch, M.; Norddahl, B.; Morgenroth, E.; Raskin, L. Methanogenic Population Dynamics and Performance of an Anaerobic Membrane Bioreactor (AnMBR) Treating Swine Manure under High Shear Conditions. *Water Res.* **2007**, *41*, 134–144. [CrossRef] [PubMed]
10. Chu, L.B.; Yang, F.L.; Zhang, X.W. Anaerobic Treatment of Domestic Wastewater in a Membrane-Coupled Expended Granular Sludge Bed (EGSB) Reactor under Moderate to Low Temperature. *Process Biochem.* **2005**, *40*, 1063–1070. [CrossRef]
11. Buntner, D.; Sanchez, A.; Garrido, J.M. Feasibility of Combined UASB and MBR System in Dairy Wastewater Treatment at Ambient Temperatures. *Chem. Eng. J.* **2013**, *230*, 475–481. [CrossRef]
12. Deowan, S.A.; Galiano, F.; Hoinkis, J.; Figoli, A.; Drioli, E. Submerged Membrane Bioreactor (SMBR) for Treatment of Textile Dye Wastewatertowards Developing Novel MBR Process. *APCBEE Procedia* **2013**, *5*, 259–264. [CrossRef]
13. Skouteris, G.; Hermosilla, D.; López, P.; Negro, C.; Blanco, Á. Anaerobic Membrane Bioreactors for Wastewater Treatment. A Review. *Chem. Eng.* **2012**, *198–199*, 138–148. [CrossRef]
14. Huang, Z.; Ong, S.L.; Ng, H.Y. Feasibility of Submerged Anaerobic Membrane Bioreactor (SAMBR) for Treatment of Low-Strength Wastewater. *Water Sci. Technol.* **2008**, *58*, 1925–1931. [CrossRef] [PubMed]

15. Zhao, L.; Liu, D.; Wang, X. Effect of Several Factors on Peracetic Acid Pretreatment of Sugarcane Bagasse for Enzymatic Hydrolysis. *J. Chem. Technol. Biotechnol.* **2007**, *82*, 115–121. [CrossRef]

16. Van Zyl, P.J.; Wentzel, M.C.; Ekama, G.A.; Riedel, K.J. Design and Start-up of a High Rate Anaerobic Membrane Bioreactor for the Treatment of a Low pH, High Strength, Dissolved Organic Waste Water. *Water Sci. Technol.* **2008**, *57*, 291–295. [CrossRef] [PubMed]

17. Akram, A.; Stuckey, D.C.; Kram, A.A.; Tuckey, D.C.S. Biomass Acclimatisation and Adaptation during Start-Up of a Submerged Anaerobic Membrane Bioreactor (SAMBR) Biomass Acclimatisation and Adaptation during Start-Up of a Submerged Anaerobic Membrane Bioreactor (Sambr). *Environ. Technol.* **2014**, *29*, 1053–1065. [CrossRef] [PubMed]

18. Vyrides, I.; Stuckey, D.C. Saline Sewage Treatment Using a Submerged Anaerobic Membrane Reactor (SAMBR): Effects of Activated Carbon Addition and Biogas-Sparging Time. *Water Res.* **2009**, *43*, 933–942. [CrossRef] [PubMed]

19. Lin, H.; Liao, B.; Chen, J.; Gao, W.; Wang, L.; Wang, F.; Lu, X. New Insights into Membrane Fouling in a Submerged Anaerobic Membrane Bioreactor Based on Characterization of Cake Sludge and Bulk Sludge. *Bioresour. Technol.* **2011**, *102*, 2373–2379. [CrossRef] [PubMed]

20. Giménez, J.B.; Robles, A.; Carretero, L.; Durán, F.; Ruano, M.V.; Gatti, M.N.; Ribes, J.; Ferrer, J.; Seco, A. Experimental Study of the Anaerobic Urban Wastewater Treatment in a Submerged Hollow-Fibre Membrane Bioreactor at Pilot Scale. *Bioresour. Technol.* **2011**, *102*, 8799–8806. [CrossRef] [PubMed]

21. Gao, W.J.; Leung, K.T.; Qin, W.S.; Liao, B.Q. Effects of temperature and temperature Shock on the Performance and Microbial Community Structure of a Submerged Anaerobic Membrane Bioreactor. *Bioresour. Technol.* **2011**, *102*, 8733–8740. [CrossRef] [PubMed]

22. Zhang, X.; Wang, Z.; Wu, Z.; Wei, T.; Lu, F.; Tong, J.; Mai, S. Membrane Fouling in an Anaerobic Dynamic Membrane Bioreactor (AnDMBR) for Municipal Wastewater Treatment: Characteristics of Membrane Foulants and Bulk Sludge. *Process Biochem.* **2011**, *46*, 1538–1544. [CrossRef]

23. Lin, H.; Chen, J.; Wang, F.; Ding, L.; Hong, H. Feasibility evaluation of submerged anaerobic membrane bioreactor for municipal secondary wastewater treatment. *Desalination* **2011**, *280*, 120–126. [CrossRef]

24. Katayon, S.; Noor, M.J.M.M.; Ahmad, J.; Ghani, L.A.A.; Nagaoka, H.; Aya, H. Effects of Mixed Liquor Suspended Solid Concentrations on Membrane Bioreactor Efficiency for Treatment of Food Industry Wastewater. *Desalination* **2004**, *167*, 153–158. [CrossRef]

25. Hogye, S.P. *National Small Flows Clearinghouse at West Virginia University, Morgantown*; National Environmental Services Center (NESC): Morgantown, WV, USA, 2003.

26. Robinson, A.H. *New Developments in the Application of Membrane Bioreactors (MBR) for Industrial Wastewater Treatment*; Wehrle Environmental: Witney, UK, 2001.

27. Umaiyakunjaram, R.; Shanmugam, P. Study on Submerged Anaerobic Membrane Bioreactor (SAMBR) Treating High Suspended Solids Raw Tannery Wastewater for Biogas Production. *Bioresour. Technol.* **2016**, *216*, 785–792. [CrossRef] [PubMed]

28. Hu, A.Y.; Stuckey, D.C. Treatment of dilute wastewaters using a Novel Submerged Anaerobic Membrane Bioreactor. *J. Environ. Eng.* **2006**, *132*, 190–198. [CrossRef]

29. Fuchs, W.; Binder, H.; Mavrias, G.; Braun, R. Anaerobic Treatment of Wastewater with High Organic Content Using a Stirred Tank Reactor Coupled with a Membrane Filtration Unit. *Water Res.* **2003**, *37*, 902–908. [CrossRef]

30. Saddoud, A.; Ellouze, M.; Dhouib, A.; Sayadi, S. Anaerobic Membrane Bioreactor Treatment of Domestic Wastewater in Tunisia. *Desalination* **2007**, *207*, 205–215. [CrossRef]

31. Dereli, R.K.; Heffernan, B.; Grelot, A.; van der Zee, F.P.; van Lier, J.B. Influence of High Lipid Containing Wastewater on Filtration Performance and Fouling in AnMBRs Operated at Different Solids Retention Times. *Sep. Purif. Technol.* **2015**, *139*, 43–52. [CrossRef]

32. Van Lier, J.B. High-Rate Anaerobic Wastewater Treatment: Diversifying from End-of-the-Pipe Treatment to Resource-Oriented Conversion Techniques. *Water Sci. Technol.* **2008**, *57*, 1137–1148. [CrossRef] [PubMed]

33. Ho, J.; Sung, S. Methanogenic Activities in Anaerobic Membrane Bioreactors (AnMBR) Treating Synthetic Municipal Wastewater. *Bioresour. Technol.* **2010**, *101*, 2191–2196. [CrossRef] [PubMed]

34. Kocadagistan, E.; Topcu, N. Treatment Investigation of the Erzurum City Municipal Wastewaters with Anaerobic Membrane Bioreactors. *Desalination* **2007**, *216*, 367–376. [CrossRef]

35. Liao, B.; Kraemer, J.T.; Bagley, D.M. Anaerobic Membrane Bioreactors. Applications and Research Directions Anaerobic Membrane Bioreactors. Applications and Research Directions. *Crit. Rev. Environ. Sci. Technol.* **2006**, *36*, 489–530. [CrossRef]

36. Smith, A.L.; Skerlos, S.J.; Raskin, L. Psychrophilic Anaerobic Membrane Bioreactor Treatment of Domestic Wastewater. *Water Res.* **2013**, *47*, 1655–1665. [CrossRef] [PubMed]

37. Lew, B.; Tarre, S.; Beliavski, M.; Dosoretz, C.; Green, M. Anaerobic Membrane Bioreactor (AnMBR) for Domestic Wastewater Treatment. *Desalination* **2009**, *243*, 251–257. [CrossRef]

38. Martinez-Sosa, D.; Helmreich, B.; Horn, H. Anaerobic Submerged Membrane Bioreactor (AnSMBR) Treating Low-Strength Wastewater under Psychrophilic Temperature Conditions. *Process Biochem.* **2012**, *47*, 792–798. [CrossRef]

39. Salazar-Peláez, M.L.; Morgan-Sagastume, J.M.; Noyola, A. Influence of Hydraulic Retention Time on Fouling in a UASB Coupled with an External Ultrafiltration Membrane Treating Synthetic Municipal Wastewater. *Desalination* **2011**, *277*, 164–170. [CrossRef]

40. Gao, D.W.; Zhang, T.; Tang, C.Y.Y.; Wu, W.M.; Wong, C.Y.; Lee, Y.H.; Yeh, D.H.; Criddle, C.S. Membrane Fouling in an Anaerobic Membrane Bioreactor: Differences in Relative Abundance of Bacterial Species in the Membrane Foulant Layer and in Suspension. *J. Membr. Sci.* **2010**, *364*, 331–338. [CrossRef]

41. Giménez, J.B.; Martí, N.; Ferrer, J.; Seco, A. Methane Recovery Efficiency in a Submerged Anaerobic Membrane Bioreactor (SAnMBR) Treating Sulphate-Rich Urban Wastewater: Evaluation of Methane Losses with the Effluent. *Bioresour. Technol.* **2012**, *118*, 67–72. [CrossRef] [PubMed]

42. Svojitka, J.; Dvorak, L.; Studer, M.; Straub, J.O.; Fromelt, H.; Wintgens, T. Performance of an Anaerobic Membrane Bioreactor for Pharmaceutical Wastewater Treatment. *Bioresour. Technol.* **2017**, *229*, 180–189. [CrossRef] [PubMed]

43. Kaya, Y.; Bacaksiz, A.M.; Bayrak, H.; Gnder, Z.B.; Vergili, I.; Hasar, H.; Yilmaz, G. Treatment of Chemical Synthesis-Based Pharmaceutical Wastewater in an Ozonation-Anaerobic Membrane Bioreactor (AnMBR) System. *Chem. Eng. J.* **2017**, *322*, 293–301. [CrossRef]

44. Mei, X.; Wang, Z.; Miao, Y.; Wu, Z. A Pilot-Scale Anaerobic Membrane Bioreactor under Short Hydraulic Retention Time for Municipal Wastewater Treatment: Performance and Microbial Community Identification. *J. Water Reuse Desalin.* **2017**. [CrossRef]

45. Erkan, H.S.; Engin, G.O. The Investigation of Paper Mill Industry Wastewater Treatment and Activated Sludge Properties in a Submerged Membrane Bioreactor. *Water Sci. Technol.* **2017**. [CrossRef] [PubMed]

46. Hu, D.; Su, H.; Chen, Z.; Cui, Y.; Ran, C.; Xu, J.; Xiao, T.; Li, X.; Wang, H.; Tian, Y.; et al. Performance Evaluation and Microbial Community Dynamics in a Novel AnMBR for Treating Antibiotic Solvent Wastewater. *Bioresour. Technol.* **2017**, *243*, 218–227. [CrossRef] [PubMed]

47. Hu, D.; Xiao, T.; Chen, Z.; Wang, H.; Xu, J.; Li, X.; Su, H.; Zhang, Y. Effect of the High Cross Flow Velocity on Performance of a Pilot-Scale Anaerobic Membrane Bioreactor for Treating Antibiotic Solvent Wastewater. *Bioresour. Technol.* **2017**, *243*, 47–56. [CrossRef] [PubMed]

48. Hu, D.; Xu, J.; Chen, Z.; Wu, P.; Wang, Z.; Wang, P.; Xiao, T.; Su, H.; Li, X.; Wang, H.; et al. Performance of a Pilot Split-Type Anaerobic Membrane Bioreactor (AnMBR) Treating Antibiotics Solvent Wastewater at Low Temperatures. *Chem. Eng. J.* **2017**, *325*, 502–512. [CrossRef]

49. Chen, H.; Chang, S.; Guo, Q.; Hong, Y.; Wu, P. Brewery Wastewater Treatment Using an Anaerobic Membrane Bioreactor. *Biochem. Eng. J.* **2016**, *105*, 321–331. [CrossRef]

50. Xia, T.; Gao, X.; Wang, C.; Xu, X.; Zhu, L. An Enhanced Anaerobic Membrane Bioreactor Treating Bamboo Industry Wastewater by Bamboo Charcoal Addition: Performance and Microbial Community Analysis. *Bioresour. Technol.* **2016**, *220*, 26–33. [CrossRef] [PubMed]

51. Ng, K.K.; Shi, X.; Ng, H.Y. Evaluation of System Performance and Microbial Communities of Abioaugmented Anaerobic Membrane Bioreactor Treating Pharmaceutical Wastewater. *Water Res.* **2015**, *81*, 311–324. [CrossRef] [PubMed]

52. Smith, A.L.; Skerlos, S.J.; Raskin, L. Anaerobic Membrane Bioreactor Treatment of Domestic Wastewater at Psychrophilic Temperatures Ranging from 15 °C to 3 °C. *Environ. Sci. Water Res. Technol.* **2015**, *1*, 56–64. [CrossRef]

53. Dolejs, P.; Ozcan, O.; Bair, R.; Ariunbaatar, J.; Bartacek, J.; Lens, P.N.L.; Yeh, D.H. Effect of Psychrophilic Temperature Shocks on a Gas-Lift Anaerobic Membrane Bioreactor (Gl-AnMBR) Treating Synthetic Domestic Wastewater. *J. Water Process Eng.* **2017**, *16*, 108–114. [CrossRef]

54. Gouveia, J.; Plaza, F.; Garralon, G.; Fdz-polanco, F.; Peña, M. Long-Term Operation of a Pilot Scale Anaerobic Membrane Bioreactor (AnMBR) for the Treatment of Municipal Wastewater under Psychrophilic Conditions. *Bioresour. Technol.* **2015**, *185*, 225–233. [CrossRef] [PubMed]

55. Chen, L.; Gu, Y.; Cao, C.; Zhang, J.; Ng, J.W.; Tang, C. Performance of a Submerged Anaerobic Membrane Bioreactor with Forward Osmosis Membrane for Low-Strength Wastewater Treatment. *Water Res.* **2014**, *50*, 114–123. [CrossRef] [PubMed]

56. Wu, B.; Li, Y.; Lim, W.; Lee, S.L.; Guo, Q.; Fane, A.G.; Liu, Y. Single-Stage versus Two-Stage Anaerobic Fluidized Bed Bioreactors in Treating Municipal Wastewater: Performance, Foulant Characteristics, and Microbial Community. *Chemosphere* **2017**, *171*, 158–167. [CrossRef] [PubMed]

57. Zhu, X.; Li, M.; Zheng, W.; Liu, R.; Chen, L. Performance and Microbial Community of a Membrane Bioreactor System-Treating Wastewater from Ethanol Fermentation of Food Waste. *J. Environ. Sci.* **2017**, *53*, 284–292. [CrossRef] [PubMed]

58. Cerón-Vivas, A.; Noyola, A. Fouling Membrane in an Anaerobic Membrane Bioreactor Treating Municipal Wastewater. *Water Pract. Technol.* **2017**, *12*, 314–321. [CrossRef]

59. Chen, C.; Guo, W.S.; Ngo, H.H.; Liu, Y.; Du, B.; Wei, Q.; Wei, D.; Nguyen, D.D.; Chang, W.S. Evaluation of a Sponge Assisted-Granular Anaerobic Membrane Bioreactor (SG-AnMBR) for Municipal Wastewater Treatment. *Renew. Energy* **2017**, *111*, 620–627. [CrossRef]

60. Gurung, K.; Ncibi, M.C.; Sillanpää, M. Assessing Membrane Fouling and the Performance of Pilot-Scale Membrane Bioreactor (MBR) to Treat Real Municipal Wastewater during Winter Season in Nordic Regions. *Sci. Total Environ.* **2017**, *579*, 1289–1297. [CrossRef] [PubMed]

61. Mei, X.; Quek, P.J.; Wang, Z.; Ng, H.Y. Alkali-Assisted Membrane Cleaning for Fouling Control of Anaerobic Ceramic Membrane Bioreactor. *Bioresour. Technol.* **2017**, *240*, 25–32. [CrossRef] [PubMed]

62. Nie, Y.; Tian, X.; Zhou, Z.; Li, Y. Impact of Food to Microorganism Ratio and Alcohol Ethoxylate Dosage on Methane Production in Treatment of Low-Strength Wastewater by a Submerged Anaerobic Membrane Bioreactor. *Front. Environ. Sci. Eng.* **2017**, *11*, 6. [CrossRef]

63. Mei, X.; Wang, Z.; Miao, Y.; Wu, Z. Recover Energy from Domestic Wastewater Using Anaerobic Membrane Bioreactor: Operating Parameters Optimization and Energy Balance Analysis. *Energy* **2016**, *98*, 146–154. [CrossRef]

64. Ozgun, H.; Tao, Y.; Ersahin, M.E.; Zhou, Z.; Gimenez, J.B.; Spanjers, H.; van Lier, J.B. Impact of Temperature on Feed-Flow Characteristics and Filtration Performance of an Upflow Anaerobic Sludge Blanket Coupled Ultrafiltration Membrane Treating Municipal Wastewater. *Water Res.* **2015**, *83*, 71–83. [CrossRef] [PubMed]

65. Ozgun, H.; Gimenez, J.B.; Ersahin, M.E.; Tao, Y.; Spanjers, H.; van Lier, J.B. Impact of Membrane Addition for Effluent Extraction on the Performance and Sludge Characteristics of Upflow Anaerobic Sludge Blanket Reactors Treating Municipal Wastewater. *J. Membr. Sci.* **2015**, *479*, 95–104. [CrossRef]

66. Huang, Z.; Ong, S.L.; Ng, H.Y. Submerged Anaerobic Membrane Bioreactor for Low-Strength Wastewater Treatment: Effect of HRT and SRT on Treatment Performance and Membrane Fouling. *Water Res.* **2011**, *45*, 705–713. [CrossRef] [PubMed]

67. Jeison, D.; van Lier, J.B. Cake Formation and Consolidation: Main Factors Governing the Applicable Flux in Anaerobic Submerged Membrane Bioreactors (AnSMBR) Treating Acidified Wastewaters. *Sep. Purif. Technol.* **2007**, *56*, 71–78. [CrossRef]

68. Fallah, N.; Bonakdarpour, B.; Nasernejad, B.; Moghadam, M.R.A. Long-Term Operation of Submerged Membrane Bioreactor (MBR) for the Treatment of Synthetic Wastewater Containing Styrene as Volatile Organic Compound (VOC): Effect of Hydraulic Retention Time (HRT). *J. Hazard. Mater.* **2010**, *178*, 718–724. [CrossRef] [PubMed]

69. Ho, J.; Sung, S. Anaerobic Membrane Bioreactor Treatment of Synthetic Municipal Wastewater at Ambient Anaerobic Membrane Bioreactor Treatment of Synthetic Municipal Wastewater at Ambient Temperature. *Water Environ. Res.* **2009**, *81*, 922–928. [CrossRef] [PubMed]

70. Wijekoon, K.C.; Visvanathan, C.; Abeynayaka, A. Effect of Organic Loading Rate on VFA Production, Organic Matter Removal and Microbial Activity of a Two-Stage Thermophilic Anaerobic Membrane Bioreactor. *Bioresour. Technol.* **2011**, *102*, 5353–5360. [CrossRef] [PubMed]

71. Speece, R.E.; Boonyakitsombut, S.; Kim, M.; Azbar, N.; Ursillo, P. Overview of Anaerobic Treatment: Thermophilic and Propionate Implications—Keynote Address, Association of Environmental Engineering and Science Professors, 78th Annual Water Environment Federation Technical Exposition and Conference, Washington, DC, USA, 27 October–2 November 2005. *Water Environ. Res.* **2006**, *78*, 460–473. [PubMed]

72. Ahmed, Z.; Cho, J.; Lim, B.-R.; Song, K.-G.; Ahn, K.-H. Effects of Sludge Retention Time on Membrane Fouling and Microbial Community Structure in a Membrane Bioreactor. *J. Membr. Sci.* **2007**, *287*, 211–218. [CrossRef]

73. Hiraishi, A.; Ueda, Y.; Ishihara, J. Quinone Profiling of Bacterial Removal Quinone Profiling of Bacterial Communities in Natural and Synthetic Sewage Activated Sludge for Enhanced Phosphate Removal. *Appl. Environ. Microbiol.* **1998**, *64*, 992–998. [PubMed]

74. Nie, Y.; Chen, R.; Tian, X.; Li, Y.-Y. Impact of Water Characteristics on the Bioenergy Recovery from Sewage Treatment by Anaerobic Membrane Bioreactor via a Comprehensive Study on the Response of Microbial Community and Methanogenic Activity. *Energy* **2017**, *139*, 459–467. [CrossRef]

75. Kunacheva, C.; Soh, Y.N.A.; Trzcinski, A.P.; Stuckey, D.C. Soluble Microbial Products (SMPs) in the Effluent from a Submerged Anaerobic Membrane Bioreactor (SAMBR) under Different HRTs and Transient Loading Conditions. *Chem. Eng. J.* **2017**, *311*, 72–81. [CrossRef]

76. Zolfaghari, M.; Drogui, P.; Seyhi, B.; Brar, S.K.; Buelna, G.; Dubé, R.; Klai, N. Investigation on Removal Pathways of Di 2-Ethyl Hexyl Phthalate from Synthetic Municipal Wastewater Using a Submerged Membrane Bioreactor. *J. Environ. Sci.* **2015**, *37*, 37–50. [CrossRef] [PubMed]

77. Zhang, H.; Jiang, W.; Cui, H. Performance of Anaerobic Forward Osmosis Membrane Bioreactor Coupled with Microbial Electrolysis Cell (AnOMEBR) for Energy Recovery and Membrane Fouling Alleviation. *Chem. Eng. J.* **2017**, *321*, 375–383. [CrossRef]

78. Xiao, Y.; Yaohari, H.; de Araujo, C.; Sze, C.C.; Stuckey, D.C. Removal of Selected Pharmaceuticals in an Anaerobic Membrane Bioreactor (AnMBR) with/without Powdered Activated Carbon (PAC). *Chem. Eng. J.* **2017**, *321*, 335–345. [CrossRef]

79. Li, Y.; Hu, Q.; Chen, C.H.; Wang, X.L.; Gao, D.W. Performance and Microbial Community Structure in an Integrated Anaerobic Fluidized-Bed Membrane Bioreactor Treating Synthetic Benzothiazole Contaminated Wastewater. *Bioresour. Technol.* **2017**, *236*, 1–10. [CrossRef] [PubMed]

80. Watanabe, R.; Nie, Y.; Wakahara, S.; Komori, D.; Li, Y.Y. Investigation on the Response of Anaerobic Membrane Bioreactor to Temperature Decrease from 25 °C to 10 °C in Sewage Treatment. *Bioresour. Technol.* **2017**, *243*, 747–754. [CrossRef] [PubMed]

81. Basset, N.; Santos, E.; Dosta, J.; Matalvarez, J. Start-up and Operation of an AnMBR for Winery Wastewater Treatment. *Ecol. Eng.* **2016**, *86*, 279–289. [CrossRef]

82. Chen, C.; Guo, W.; Ngo, H.H.; Chang, S.W.; Nguyen, D.D.; Nguyen, P.D.; Bui, X.T.; Wu, Y. Impact of Reactor Configurations on the Performance of a Granular Anaerobic Membrane Bioreactor for Municipal Wastewater Treatment. *Int. Biodeterior. Biodegrad.* **2017**, *121*, 131–138. [CrossRef]

83. Iranpour, R.; Cox, H.H.J.; Shao, Y.J.; Moghaddam, O.; Kearney, R.J.; Deshusses, M.A.; Stenstrom, M.K.; Ahring, B.K. Changing Mesophilic Wastewater Sludge Digestion into Thermophilic Operation at Terminal Island Treatment Plant. *Water Environ. Res.* **2002**, *74*, 494–507. [CrossRef] [PubMed]

84. Barker, D.J.; Stuckey, D.C. A Review of Soluble Microbial Products (SMP) in Wastewater Treatment Systems. *Water Res.* **1999**, *33*, 3063–3082. [CrossRef]

85. Sciener, P.; Nachaiyasit, S.; Stuckey, D.C. Production of Soluble Microbial Products (SMP) in an Anaerobic Baffled Reactor: Composition, Biodegradability and the Effect of Process Parameters. *Environ. Technol.* **1998**, *4*, 391–400. [CrossRef]

86. Chudoba, J.A.N. Inhibitory effect of refractory organic compounds produced by activated sludge micro-organisms on microbial activity and flocculation. *Water Res.* **1985**, *19*, 197–200. [CrossRef]

87. Judd, S. *Principles and Applications of Membrane Bioreactors in Water and Wastewater Treatment*; Butterworth-Heinemann: Oxford, UK, 2006.

88. Li, H.; Yang, M.; Zhang, Y.; Yu, T.; Kamagata, Y. Nitrification Performance and Microbial Community Dynamics in a Submerged Membrane Bioreactor with Complete Sludge Retention. *J. Biotechnol.* **2006**, *123*, 60–70. [CrossRef] [PubMed]

89. Van Den Broeck, R.; van Dierdonck, J.; Nijskens, P.; Dotremont, C.; Krzeminski, P. The Influence of Solids Retention Time on Activated Sludge Bioflocculation and Membrane Fouling in a Membrane Bioreactor (MBR). *J. Membr. Sci.* **2012**, *401–402*, 48–55. [CrossRef]

90. De Vrieze, J.; Saunders, A.M.; He, Y.; Fang, J.; Nielsen, P.H.; Verstraete, W.; Boon, N. Ammonia and Temperature Determine Potential Clustering in the Anaerobic Digestion Microbiome. *Water Res.* **2015**, *75*, 312–323. [CrossRef] [PubMed]

91. Pap, B.; Györkei, Á.; Boboescu, I.Z.; Nagy, I.K.; Bíró, T.; Kondorosi, É.; Maróti, G. Temperature-Dependent Transformation of Biogas-Producing Microbial Communities Points to the Increased Importance of Hydrogenotrophic Methanogenesis under Thermophilic Operation. *Bioresour. Technol.* **2015**, *177*, 375–380. [CrossRef] [PubMed]

92. El-Mashad, H.M.; Zeeman, G.; van Loon, W.K.P.; Bot, G.P.A.; Lettinga, G. Effect of Temperature and Temperature Fluctuation on Thermophilic Anaerobic Digestion of Cattle Manure. *Bioresour. Technol.* **2004**, *95*, 191–201. [CrossRef] [PubMed]

93. Van Lier, J.B. Thermophilic Anaerobic Wastewater Treatment; Temperature Aspects and Process Stability. Ph.D. Thesis, Wageningen University & Research, Wageningen, The Netherlands, 1995.

94. Duran, M.; Speece, R.E. Temperature-Staged Anaerobic Processes. *Environ. Technol.* **1997**, *18*, 747–753. [CrossRef]

95. Wilson, C.A.; Murthy, S.M.; Fang, Y.; Novak, J.T. The Effect of Temperature on the Performance and Stability of Thermophilic Anaerobic Digestion. *Water Sci. Technol.* **2008**, *57*, 297–304. [CrossRef] [PubMed]

96. Lin, Q.; de Vrieze, J.; Li, J.; Li, X. Temperature Affects Microbial Abundance, Activity and Interactions in Anaerobic Digestion. *Bioresour. Technol.* **2016**, *209*, 228–236. [CrossRef] [PubMed]

97. Song, Y.C.; Kwon, S.J.; Woo, J.H. Mesophilic and Thermophilic Temperature Co-Phase Anaerobic Digestion Compared with Single-Stage Mesophilic- and Thermophilic Digestion of Sewage Sludge. *Water Res.* **2004**, *38*, 1653–1662. [CrossRef] [PubMed]

98. Angelidaki, I. Anaerobic thermophilic digestion of manure at different ammonialoads: Effect of temperature. *Water Res.* **1994**, *28*, 727–731. [CrossRef]

99. Chandra, R.; Vijay, V.K.; Subbarao, P.M.V.; Khura, T.K. Production of Methane from Anaerobic Digestion of Jatropha and Pongamia Oil Cakes. *Appl. Energy* **2012**, *93*, 148–159. [CrossRef]

100. Appels, L.; van Assche, A.; Willems, K.; Degrève, J.; van Impe, J.; Dewil, R. Peracetic Acid Oxidation as an Alternative Pre-Treatment for the Anaerobic Digestion of Waste Activated Sludge. *Bioresour. Technol.* **2011**, *102*, 4124–4130. [CrossRef] [PubMed]

101. Zhou, J.; Zhang, R.; Liu, F.; Yong, X.; Wu, X.; Zheng, T.; Jiang, M.; Jia, H. Biogas Production and Microbial Community Shift through Neutral pH Control during the Anaerobic Digestion of Pig Manure. *Bioresour. Technol.* **2016**, *217*, 44–49. [CrossRef] [PubMed]

102. Wang, S.; Nomura, N.; Nakajima, T.; Uchiyama, H. Case Study of the Relationship between Fungi and Bacteria Associated with High-Molecular-Weight Polycyclic Aromatic Hydrocarbon Degradation. *J. Biosci. Bioeng.* **2012**, *113*, 624–630. [CrossRef] [PubMed]

103. Kim, J.; Park, C.; Kim, T.-H.; Lee, M.; Kim, S.; Kim, S.-W.; Lee, J. Effects of Various Pretreatments for Enhanced Anaerobic Digestion with Waste Activated Sludge. *J. Biosci. Bioeng.* **2003**, *95*, 271–275. [CrossRef]

104. Kim, M.; Gomec, C.Y.; Ahn, Y.; Speece, R.E. Hydrolysis and Acidogenesis of Particulate Organic Material in Mesophilic and Thermophilic Anaerobic Digestion. *Environ. Technol.* **2003**, *24*, 1183–1190. [CrossRef] [PubMed]

105. Sandberg, M.; Ahring, B.K. Anaerobic Treatment of Fish Meal Process Waste Water in a UASB Reactor at High pH. *Appl. Microbiol.* **1992**, *36*, 800–804. [CrossRef]

106. Yu, D.; Liu, J.; Sui, Q.; Wei, Y. Biogas-pH Automation Control Strategy for Optimizing Organic Loading Rate of Anaerobic Membrane Bioreactor Treating High COD Wastewater. *Bioresour. Technol.* **2016**, *203*, 62–70. [CrossRef] [PubMed]

107. Mao, C.; Feng, Y.; Wang, X.; Ren, G. Review on Research Achievements of Biogas from Anaerobic Digestion. *Renew. Sustain. Energy Rev.* **2015**, *45*, 540–555. [CrossRef]

108. Wen, C.; Huang, X.; Qian, Y. Domestic Wastewater Treatment Using an Anaerobic Bioreactor Coupled with Membrane Filtration. *Process Biochem.* **1999**, *35*, 335–340. [CrossRef]

109. Vincent, N.M.; Tong, J.; Yu, D.; Zhang, J.; Wei, Y. Membrane Fouling Characteristics of a Side-Stream Tubular Anaerobic Membrane Bioreactor (AnMBR) Treating Domestic Wastewater. *Processes* **2018**, *6*, 50. [CrossRef]

110. Qiao, W.; Takayanagi, K.; Niu, Q.; Shofie, M.; Li, Y.Y. Long-Term Stability of Thermophilic Co-Digestion Submerged Anaerobic Membrane Reactor Encountering High Organic Loading Rate, Persistent Propionate and Detectable Hydrogen in Biogas. *Bioresour. Technol.* **2013**, *149*, 92–102. [CrossRef] [PubMed]

111. An, Y.Y.; Yang, F.L.; Bucciali, B.; Wong, F.S. Municipal Wastewater Treatment Using a UASB Coupled with Cross-Flow Membrane Filtration. *J. Environ. Eng.* **2009**, *135*, 86–91. [CrossRef]

112. Bornare, J.; Kalyyanraman, V.; Sonde, R.R. Application of Anaerobic Membrane Bioreactor (AnMBR) for Low-Strength Wastewater Treatment and Energy Generation. In *Industrial Wastewater Treatment, Recycle and Reuse*; Butterworth-Heinemann: Oxford, UK, 2014; pp. 399–434.

113. Dereli, R.K.; Ersahin, M.E.; Ozgun, H.; Ozturk, I.; Jeison, D.; van der Zee, F.; van Lier, J.B. Potentials of Anaerobic Membrane Bioreactors to Overcome Treatment Limitations Induced by Industrial Wastewaters. *Bioresour. Technol.* **2012**, *122*, 160–170. [CrossRef] [PubMed]

114. Del Pozo, R.; Diez, V. Integrated Anaerobic-Aerobic Fixed-Film Reactor for Slaughterhouse Wastewater Treatment. *Water Res.* **2005**, *39*, 1114–1122. [CrossRef] [PubMed]

115. Saddoud, A.; Sayadi, S. Application of Acidogenic Fixed-Bed Reactor prior to Anaerobic Membrane Bioreactor for Sustainable Slaughterhouse Wastewater Treatment. *J. Hazard. Mater.* **2007**, *149*, 700–706. [CrossRef] [PubMed]

116. Jeison, D.; Plugge, C.M.; Pereira, A.; van Lier, J.B. Effects of the Acidogenic Biomass on the Performance of an Anaerobic Membrane Bioreactor for Wastewater Treatment. *Bioresour. Technol.* **2009**, *100*, 1951–1956. [CrossRef] [PubMed]

117. Dong, Q.; Parker, W.; Dagnew, M. Influence of SRT and HRT on Bioprocess Performance in Anaerobic Membrane Bioreactors Treating Municipal Wastewater. *Water Environ. Res.* **2016**, *88*, 158–167. [CrossRef] [PubMed]

118. Hejnfelt, A.; Angelidaki, I. Anaerobic Digestion of Slaughterhouse by-Products. *Biomass Bioenergy* **2009**, *33*, 1046–1054. [CrossRef]

119. Kayhanian, M. Ammonia Inhibition in High-Solids Biogasification: An Overview and Practical Solutions. *Environ. Technol.* **1999**, *20*, 355–365. [CrossRef]

120. De Baere, L.A.; Devocht, M.; van Assche, P.; Verstraete, W. Influence of High NaCl and NH₄Cl salt levels on methanogenic associations. *Water Res.* **1984**, *18*, 543–548. [CrossRef]

121. Kroeker, E.J.; Shulte, D.D.; Sparling, A.B.; Lapp, H.M. Anaerobic Process Treatment Stability. *J. WPCF* **1979**, *51*, 718–727.

122. Calli, B.; Mcrtoglu, B.; Inanc, B.; Yenigun, O. Effects of High Free Ammonia Concentrations on the Performances of Anaerobic Bioreactors. *Process Biochem.* **2005**, *40*, 1285–1292. [CrossRef]

123. Tchobanoglous, G.; Theisen, H.; Vigil, S. *Integrated Waste Management: Engineering Principles and Management Issues*; McGraw-Hill: New York, NY, USA, 1993.

124. Kayhanian, M. Performance of a High-Soilds Anaerobic Digestion Process under Various Ammonia Concentrations. *J. Chem. Technol. Biotechnol.* **1994**, *59*, 349–352. [CrossRef]

125. Li, Y.; Zhang, Y.; Kong, X.; Li, L.; Yuan, Z.; Dong, R.; Sun, Y. Effects of Ammonia on Propionate Degradation and Microbial Community in Digesters Using Propionate as a Sole Carbon Source. *J. Chem. Technol. Biotechnol.* **2017**. [CrossRef]

126. Gallert, C.; Bauer, S.; Winter, J. Effect of Ammonia on the Anaerobic Degradation of Protein by a Mesophilic and Thermophilic Biowaste Population. *Appl. Microbiol. Biotechnol.* **1998**, *50*, 495–501. [CrossRef] [PubMed]

127. Koster, I.W.; Rinzema, A.; de Vegt, A.L.; Lettinga, G. Sulfide Inhibition of the Methanogenic activity of granular sludge at various pH-levels. *Water Res.* **1986**, *20*, 1561–1567. [CrossRef]

128. Harada, H.; Uemura, S.; Momonoi, K. Interaction Between Sulfate Reducing Bacteria and Methane Producing Bacteria In Uasb Reactors Fed With Low Strength Wastes Containing Different Levels of Sulfate. *Water Res.* **1994**, *28*, 355–367. [CrossRef]

129. Stefanie, J.W.H.; Elferink, O.; Visser, A.; Pol, L.W.H.; Stams, A.J.M. Sulfate Reduction in Methanogenic Bioreactors. *FEMS Microbiol. Rev.* **1994**, *15*, 119–136. [CrossRef]

130. Colleran, E.; Pender, S.; Philpott, U.; O'Flaherty, V.; Leahy, B. Full-Scale and Laboratory-Scale Anaerobic Treatment of Citric Acid Production Wastewater. *Biodegradation* **1998**, *9*, 233–245. [CrossRef] [PubMed]

131. Cho, J.; Song, K.G.; Yun, H.; Ahn, K.H.; Kim, J.Y.; Chung, T.H. Quantitative Analysis of Biological Effect on Membrane Fouling in Submerged Membrane Bioreactor. *Water Sci. Technol.* **2005**, *51*, 9–18. [CrossRef] [PubMed]

132. Colleran, E.; Finnegan, S.; Lens, P. Anaerobic Treatment of Sulfate-Containing Waste Streams. *Antonie Van Leeuwenhoek Int. J. Gen. Mol. Microbiol.* **1995**, *67*, 29–46. [CrossRef]

133. Wang, Z.; Ma, J.; Tang, C.Y.; Kimura, K.; Wang, Q. Membrane Cleaning in Membrane Bioreactors: A Review. *J. Membr. Sci.* **2014**, *468*, 276–307. [CrossRef]

134. Zhang, Q.; Singh, S.; Stuckey, D.C. Fouling Reduction Using Adsorbents/Flocculants in a Submerged Anaerobic Membrane Bioreactor. *Bioresour. Technol.* **2017**, *239*, 226–235. [CrossRef] [PubMed]

135. Meng, F.; Chae, S.R.; Drews, A.; Kraume, M.; Shin, H.S.; Yang, F. Recent Advances in Membrane Bioreactors (MBRs): Membrane Fouling and Membrane Material. *Water Res.* **2009**, *43*, 1489–1512. [CrossRef] [PubMed]

136. Lee, J.; Ahn, W.; Lee, C. Comparison of the Filtration Characteristics Between Attached and Suspended Growth Microorganisms in Submerged Membrane Bioreactor. *Water Res.* **2001**, *35*, 2435–2445. [CrossRef]

137. Di Bella, G.; Durante, F.; Torregrossa, M.; Viviani, G.; Mercurio, P.; Cicala, A. The Role of Fouling Mechanisms in a Membrane Bioreactor. *Water Sci. Technol.* **2007**, *55*, 455–464. [CrossRef] [PubMed]

energies

MDPI

Article

Biogas Quality across Small-Scale Biogas Plants: A Case of Central Vietnam

Hynek Roubík [1], Jana Mazancová [1], Phung Le Dinh [2], Dung Dinh Van [2] and Jan Banout [1,*]

[1] Department of Sustainable Technologies, Faculty of Tropical AgriSciences, Czech University of Life Sciences Prague, Kamýcká 129, 165 00 Prague, Czech Republic; roubik@ftz.czu.cz (H.R.); mazan@ftz.czu.cz (J.M.)

[2] Department of Animal Sciences and Veterinary Medicine, Hue University of Agriculture and Forestry, Hue University, 102 Phung Hung Hue City Vietnam, Thua Thien Hue 53000, Vietnam; phung.ledinh@huaf.edu.vn (P.L.D.); dinhvandung@huaf.edu.vn (D.D.V.)

* Correspondence: banout@ftz.czu.cz; Tel.: +420-224-384-186

Received: 25 May 2018; Accepted: 26 June 2018; Published: 9 July 2018

Abstract: Production of bioenergy by the fermentation reaction is gaining attraction due to its easy operation and the wide feedstock selection. Anaerobic fermentation of organic waste materials is generally considered a cost-effective and proven technology, allowing simultaneous waste management and energy production. Small-scale biogas plants are widely and increasingly used to transform waste into gas through anaerobic fermentation of organic materials in the developing world. In this research, the quality of biogas produced in small-scale biogas plants was evaluated, as it has a direct effect on its use (as fuel for biogas cookers), as well as being able to influence a decision making process over purchasing such technology. Biogas composition was measured with a multifunctional portable gas analyser at 107 small-scale biogas plants. Complementary data at household level were collected via the questionnaire survey with the owners of biogas plants ($n = 107$). The average daily biogas production equals 0.499 m^3, not covering the demand of rural households which are using other sources of energy as well. Related to the biogas composition, the mean content of methane (CH_4) was 65.44% and carbon dioxide (CO_2) was 29.31% in the case of biogas plants younger than five years; and CH_4 was 64.57% and CO_2 was 29.93% for biogas plants older than five years. Focusing on the age of small-scale biogas plants there are no, or only minor, differences among tested values. In conclusion, the small-scale biogas plants are sustaining a stable level of biogas quality during their life-span.

Keywords: anaerobic digestion; methane; carbon dioxide; small-scale biogas plants; developing countries

1. Introduction

In developing countries, environmental pollution and access to energy sources still represent challenges, especially in relation to human and environmental health and economic development [1]. Energy influences the status and pace of development; hence, a current challenge for the developing world lies in the available supply of affordable and sustainable eco-friendly renewable energy (SDG 7) [2]. Energy poverty is exhibited by a lack of access to electricity and clean cooking facilities, which are two elements that are essential to meeting basic human needs [3]. Therefore, bioenergy production by fermentation reaction is gaining attraction due to its easy operation and a wide selection of organic wastes feedstock [4]. Anaerobic fermentation of organic materials is generally considered as a major cost-effective and matured technology [5] with its dual benefits as a waste management tool and simultaneous energy production [6]. Small-scale biogas plants are widely and increasingly used to transform waste into valuable gas in the developing world [7,8] and may represent an economically feasible technology [9,10], which is producing biogas as a main product and simultaneously producing digestate (which may be used as fertilizer) as a by-product through waste degradation [9,11].

There are many factors affecting household energy consumption (as well as CO_2 emissions), such as socio-economic factors, household characteristics and geographic factors [12].

The small-scale biogas plants also create a number of indirect environmental, economic, and societal benefits, such as a reduction in deforestation, fewer hours devoted to fuelwood collection or savings on fuelwood/fossil fuels purchasing, decreasing the need to purchase propane for cooking, the creation of jobs, the decrease of organic matter in effluent waters, the decrease of odour, the production of less indoor smoke than other fuels, the production of digestate as a fertilizer and a reduction in greenhouse gas emissions into the atmosphere, if used appropriately [6,7,13–15]. According to Zhang et al. [12] biogas households indicate having over 50% lower greenhouse gas emissions than non-biogas households. Due to above-mentioned benefits, small-scale biogas technology has been widely promoted and financially supported by governments and development aid donors in Asia, including Vietnam [13] Long-term, stable running and maintenance are key points to maximize the benefits of small-scale biogas plant [13,16]. However, if these key points are not met, the benefits of this technology may be compromised [5,13,17].

In comparison to other forms of renewable energy (solar energy, biodiesel, bioethanol, wind energy, etc.) biogas production through small-scale biogas plants is relatively simple, decentralised, and can operate under various conditions in tropical regions, such as Southeast Asia, particularly Vietnam. The most common feedstock material is animal dung or human faeces, as it is usually the most problematic waste material in terms of waste management for rural households [18]. An important advantage of small-scale biogas technology and one of the main reasons for the government support is that the technology is a cost-effective method of reducing greenhouse gas emissions and odours from animal manure, if used properly [13,19]. The biogas produced is mainly used for cooking, heating and lighting, therefore replacing energy sources such as fuelwood, dried dung, coal, or liquid petroleum gas (LPG), commonly used for these purposes in rural households [5,9,13]. It is always difficult to adopt new and unknown digester technology within households. Therefore, recommendations for various models implemented within the country are needed. The design of the biogas plants varies based on geographical locations, availability of feedstock and climatic conditions. The most common types of feedstock for chosen Asian countries can be seen in Table 1. In Asia, the fixed dome model is the most commonly used [5,16]. However, there are two exceptions—Indonesia, where various models were applied according to the regions and islands, and India with a prevailing number of the floating drum model followed by the fixed dome [20].

Table 1. Most common feedstock for small-scale biogas plants in selected Asian countries.

Country	Most Common Feedstock	Reference
Vietnam	Pig manure	[7,21]
Cambodia	Combination of pig manure and cow manure	[22]
Bangladesh	Cow and buffaloes manure	[5,23]
Laos	Cow manure	[24]
Nepal	Cow manure	[25]
India	Livestock manure	[26]
Indonesia	Cow manure	[27,28]

In Vietnam, anaerobic digestion of animal manure has already been practiced since the 1960s [7,10]. Since then its popularity has grown, mainly due to the promotion of this technology by the government and international organizations, e.g., SNV (Netherland Development Organization). The Ministry of Agriculture and Rural Development of Vietnam (MARD), together with SNV (using 10% government subsidy for support of capital costs of small-scale biogas technology), installed between 2003 and 2013 over 200,000 small-scale biogas plants [5,29]. The follow-up biogas programme by SNV and MARD aimed to build 140,000 biogas digesters between 2006 and 2011. The current number of small-scale biogas plants in Vietnam is more than 500,000 [30]. The target was reached, and digesters serve over 600,000 people with cooking fuel with CO_2 savings of around 260,000 tons per year [31]. However, in Vietnam, the biogas technology is still far below its potential for utilizing available livestock and agricultural wastes [30].

As the small-scale biogas technology is one of the fastest growing and highly promising renewable energy sources, mainly for rural households, the main objective of this paper is to evaluate the quality of biogas produced in the small-scale biogas plants installed in central Vietnam in terms of chemical and physical parameters in relation to the age of installed biogas plants. The quality of biogas has a direct effect on its use (as fuel for biogas cookers), which may, in return, influence an individual decision of purchasing such a technology. Furthermore, biogas quality evaluation is needed to provide sufficient information for authorities to have their future policy decisions well supported.

2. Materials and Methods

The research was carried out in two districts, Huong Tra and Phong Dien, Thua Thien Hue Province, in Central Vietnam (Figure 1). Huong Tra is a rural district in the northern part of the central coast of Vietnam with a population of over 115,000 inhabitants covering an area of 521 km². The district is located on the northern outskirts of Hue (provincial city of Thua Thien Hue province) and can, therefore, be considered as a peri-urban area. Phong Dien has a population over 105,000 inhabitants and covers an area of 954 km². The district has a varied topography with mountains, plains, and coastline.

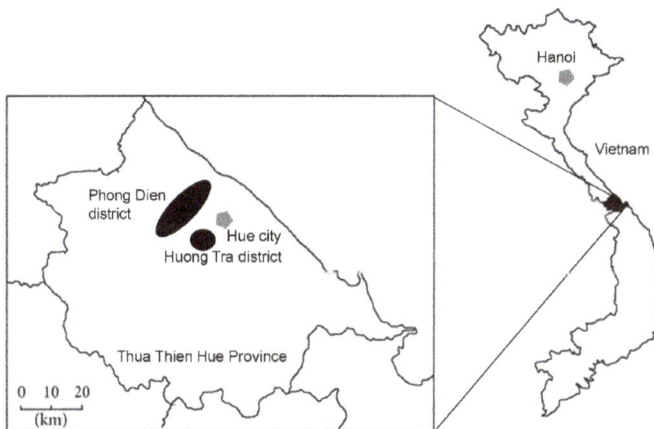

Figure 1. Thua Thien Hue province and the target area.

2.1. Description of Biogas Technology in Target Area

The study was done on two types of small-scale biogas plants, specifically KT1 (Figure 2a) and KT2 (Figure 2b). Both types are predominant in the target area. Both types are varieties of the Chinese fixed dome, where KT1 is an appropriate type for a good structure of soil to be easily excavated. KT2 is used in places where soil excavation is difficult or where high levels of groundwater or floods are reported. Both types are unheated and usually built underground, in order to minimize the temperature fluctuations and for space saving reasons. The digester is filled in through the inlet tank and the inlet pipe. The produced biogas is accumulated at the upper part of the digester and the difference between the slurry inside the digester and the digestate in the compensation tank creates a gas pressure. The slurry flows back into the digester from the compensation tank after the gas is released through the gas pipe. Both types and their potential problems are described in detail in the study by [21]. The Vietnamese small-scale biogas plants operate at the temperature of the surrounding soil as they are built underground. The time of the year significantly influences temperatures in the air, the slurry mixing tank, the soil, and the digesters. The average summer temperatures in Central Vietnam are around 34 °C (mesophilic conditions), creating a suitable environment for the bacterial fermentation; however, during winter time the temperature is in the range of 15–25 °C, which might cause lower biogas production [9,32].

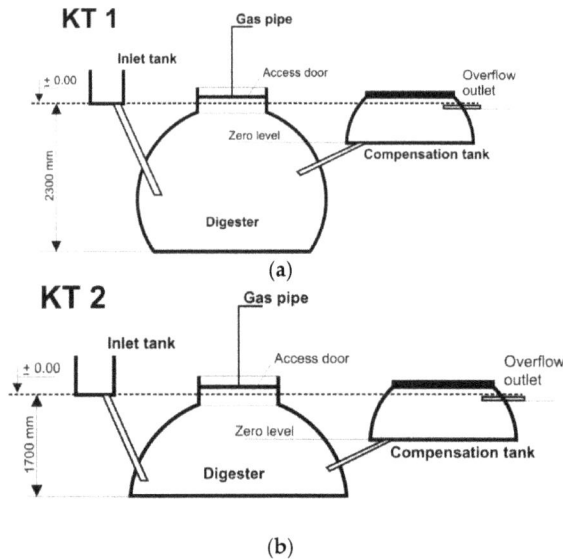

(a)

(b)

Figure 2. (**a**) Small-scale biogas plant—fixed dome model (KT1). Reprinted from [21]. (**b**) Small-scale biogas plant—fixed dome model (KT2). Reprinted from [21].

2.2. Data Collection—Questionnaire Survey

The questionnaire survey was carried out with the owners of small-scale biogas plants from June to July 2013. Biogas plants were randomly selected from recipients of government subsidies ($n = 107$; corresponding to 20% share of total subsidy recipients in the area) listed by the local unit of the Ministry of Agriculture and Rural Development. The recipients were, at the same time, beneficiaries of one of two running projects on building small-scale biogas plants—one supported by the SNV, the other one by the Czech Development Agency (CzDA). The questionnaire included nine questions (Table 2). Furthermore, the data results were cross-checked with the local facilitators during field trips in July and August 2016 in order to increase their validity and reliability.

Table 2. Overview of variables in the questionnaire.

Variable	Type of the Question	Value	Unit
Capacity of the digester	Close-ended question	No.	m^3
Investor of the construction	Close-ended question	SNV/CzDA	-
Digester type	Close-ended question	KT1/KT2	-
Digester connection to the toilet	Open-ended question	-	-
Animal stable	Open-ended question	-	-
Feedstock materials for the BGP	Open-ended question	-	-
No. of animals	Open-ended question	No.	Heads
No. of other applications powered by biogas	Open-ended question	No.	-
Digestate practices	Open-ended question	-	-

2.3. Data Collection—Biogas Composition

There are many analytical methods for biogas quality evaluation based on its final use [33] From the technical point of view, the most important parameter is the content of the potentially corrosive components (oxygen, hydrogen, water, carbon dioxide, hydrogen sulphide, and chlorine and fluorine compounds) [34,35]. Biogas composition was measured using a GA5000 multifunctional portable gas analyser (Geotech, Leamington Spa, UK), which is adapted to measurements of CH_4, CO_2, O_2, H_2, and H_2S with the following measurement accuracy of: CH_4 (0–70 vol % ± 0.5%), CH_4 (70–100 vol % ± 1.5%), CO_2 (0–60 vol % ± 0.5%), CO_2 (60–100 vol % ± 1.5%), O_2 (0–25 vol % ± 1.0%), H_2S (0–5000 ppm ± 2.0%), and H_2S (0–10,000 ppm ± 5.0%). For measurements of media (CH_4/CO_2) a dual-wavelength infrared sensor was used, for O_2/H_2S an internal electrochemical sensor was used. The measurements were taken upstream of the H_2S filter to eliminate measurement inaccuracies. Obtained values are the mean of three measurements in the interval of one hour at each biogas plant. Calorific values were set as the quantity of heat produced by complete combustion of a unit of a combustible compound. In total, the measurements were taken in 81 KT1 small-scale biogas plants and 26 KT2 small-scale biogas plants at districts Huong Tra (n = 49) and Phong Dien (n = 58), Thua Thien Hue Province.

2.4. Data Analysis

The collected data were categorized, coded, and analysed with descriptive and inferential statistics using the SPSS version 18. Effects of variables such as type of biogas plant, age of the biogas plant, and size of the biogas plant on biogas composition were analysed by the analysis of covariance model. Dummy variables included the type of biogas plant (KT1 and KT2), the age of the biogas plant (>5 years old and <5 years old), and a continuous variable of the size of the biogas plants (m^3).

$$Y_i = \alpha + \beta x_1 + \gamma_1 D_{i1} + \gamma_2 D_{i2} + \varepsilon_i \tag{1}$$

Y_i = Biogas composition;
α = Intercept;
x_1 = Biogas plant size (m^3);
D_{i1} = Biogas plant types (D_{i1} = 1 = KT2; D_{i1} = 0 = KT1);
D_{i2} = Biogas plant age (D_{i2} = 1 5 years; D_{i1} = 0 ≥ 5 years);
β = Regression coefficient of biogas plant size on biogas composition;
γ_1 = The difference of biogas composition between KT2 and KT1;
γ_2 = The difference of biogas composition between <5 year and ≥5 year old biogas plants; and
ε_i = Error term.

3. Results and Discussion

3.1. Feedstock Used for Biogas Production

The majority of respondents (90%) are farmers producing mainly rice. However, many of them are also involved in additional off-farm activities, such as trade, rice noodle production, and rice wine production. All questioned households use pig slurry as their main feedstock for biogas plants and, in all cases, pigs were housed in concrete pigpens (with a concrete floor). Feedstock manure from other animals (65%) is also used as an additive within the surveyed households. This included chicken manure (29%) and human excreta (36%) (Figure 3). These additives are added if they are available in sufficient quantities. In every household, the pigpen is connected directly to the biogas plant, and in 37% of cases toilet outflows are connected to the biogas plant as well. Only in one case was a chicken shed was connected to the biogas plant; in the rest of cases the chicken manure is put in the digester inlet manually. Generally, the feedstock input was unified as biogas owners were recipients of one of two running projects on building small-scale biogas plants and there were criteria on the necessary number of pigs. In addition, further details regarding manure management practices of small-scale farmers in Vietnam can be seen in our previous study [6].

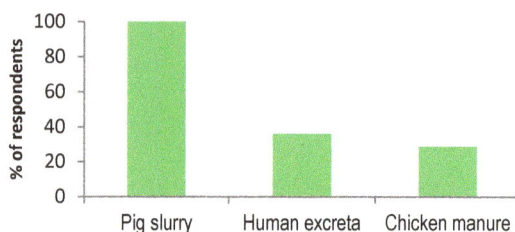

Figure 3. Feedstock used for biogas production (Multiple choices were possible) (*n* = 107).

3.2. Use of Products of Small-Scale Biogas Technology within the Rural Households in Central Vietnam

Small-scale biogas plants have been applied as an optimal livestock waste treatment in Vietnam since the 1960s and, although the history of this technology is rather old, the number of the constructed biogas plants is still limited. With the current number of units around 500,000 [5,23] it is still far below the real demand on livestock waste treatment [10,36] that has increased significantly in the last decade [7]. The primary use of biogas is for cooking [21]. However, it could also be used for lighting in remote areas where electrification is limited [29]. Furthermore, biogas plants produce residue from the process in the form of digestate, which can be applied as organic fertilizer to enhance agriculture production [10,17]. On a daily basis, a minimum of 20 kg of organic waste is required to operate the smallest biogas plants (4 m^3) in the area. Therefore, there is a required number (five growing pigs) of pigs (manure, respectively) to meet the feedstock needs of these plants. This quota is usually met, as the average number of animals in the area is 13–14 piglets and 2–3 sows [6,21].

3.2.1. Biogas for Cooking and Water Boiling

In developing countries, biogas produced from the household biogas plants is used mainly for cooking [29]. This is also applied to the present study, where 100% of households use biogas primarily for cooking, with an average time of 2.8 h/day. Biogas is usually used primarily for household cooking and boiling water and, afterwards, for cooking of feed for pigs. An average daily production of biogas is 0.499 m^3 (±0.086). Such an amount used for cooking purposes may represent about 8–10 m^3 and 96–120 m^3 of biogas per month and year, respectively. However, according to the study by [37], the biogas volume needed for a typical farming household of six people is 0.8 to 1 m^3 per day. This difference between average daily biogas productions could be explained by [38], as he states that the fixed dome biogas digesters can annually leak around 55% of CH$_4$ and the production of biogas is also dependent on the temperature of the feedstock. Therefore, the majority of respondents (60%) are still using additional energy sources in form of LPG and/or electricity (for cooking rice in rice cookers) and fuelwood (usually only for cooking of feed for animals).

3.2.2. Biogas for Lightning and Power Generation

The other major possible application of biogas may be for lighting and power generation. Biogas lamps are more efficient than kerosene-powered lamps, but their efficiency is quite low compared to electric-powered lamps [18]. In addition, electricity is now widely available in Vietnam; therefore, use of biogas lamps is very occasional. In the case of our respondents, less than 10% were using biogas lamps. Farmers usually prefer biogas for cooking instead of lighting; also, from the reason that 1 m^3 of biogas is equal to lighting of 60–100 watt bulbs for around 6 h, or cooking 2–3 meals a day for 5–6 persons. As observed during interviews with the farmers, power generation is favoured when farmers have an abundance of biogas. In that case, they purchase a combustion engine, which converts biogas into the mechanical energy in a heat engine and, consequently, the mechanical energy activates a generator to produce electrical power.

3.2.3. Digestate

The residue remaining after treatment (anaerobic digestion) in the biogas plant is called digestate. The use of digestate as a fertilizer is considered beneficial since it provides nutrients (N, P, K) which are easily accessible to plants. Digestate can be applied directly through the overflow outlet or manually. Another option is through pre-treatment (e.g., drying) before application. However, this possibility is used only sporadically. The most common practice is the usage of the digestate directly to the surrounding household home gardens and use of mainly solid parts of digestate as a crop fertilizer for rice. Another way of usage (especially of liquid manure and slurry) is partly limited by a long distance between the biogas plant location and the rice field. In 25% of cases, farmers use digestate as a fertilizer for vegetable and home-garden, which is a very popular way because of its simplicity and convenience for the farmers. Study also showed that usage of digestate for fish feeding is still not adopted in this area, as none of the respondents used digestate for such purpose even though its benefits were proven in the study by [39] focused on pig-biogas-fish systems.

3.3. Biogas Composition in Various Types of Small-Scale Biogas Plant (KT1 and KT2)

The performance of two models of biogas plants (KT1 and KT2) was observed in order to show the differences between these two types. It was revealed that the KT2 digester has demonstrated a slightly higher production of CH_4 (66.23%) and, at the same time, lower production of carbon dioxide (28.27%) (Table 3). Furthermore, slight differences might be seen among other variables, such as content of O_2, NH_3, or H_2S (Figure 4). Nitrogen, hydrogen, and water vapour were collectively marked as NHW, where slight differences were also recognized. However, factors, such as organic input, or maintenance might be the cause of these discrepancies. Furthermore, Table 4 shows a comprehensive review of biogas composition from small-scale biogas plants reported by other studies from developing countries. As is shown, CH_4 varies from 50% to 75%, and CO_2 from 25% to 50%. Other elements (N_2, CO, O_2, and H_2S) are below 2% in general and, respectively, 1% in case of H_2.

Table 3. Biogas composition according to the type of biogas plant, KT1 ($n = 83$) and KT2 ($n = 24$).

Variable	Type of Biogas Plant	Mean	95% Confidence Interval	
			Lower Bound	Upper Bound
CH_4 (vol %)	KT1	63.79	62.94	64.63
	KT2	66.23	64.70	67.76
CO_2 (vol %)	KT1	30.97	30.04	31.89
	KT2	28.27	26.59	29.94
NH_3 (vol %)	KT1	0.05	0.04	0.05
	KT2	0.04	0.03	0.05
H_2S (vol %)	KT1	0.10	0.07	0.14
	KT2	0.16	0.09	0.22
CH:CO_2 index	KT1	2.14	2.02	2.25
	KT2	2.24	2.03	2.44
Calorific value (MJ/m^3)	KT1	21.60	21.31	21.89
	KT2	22.36	21.84	22.88

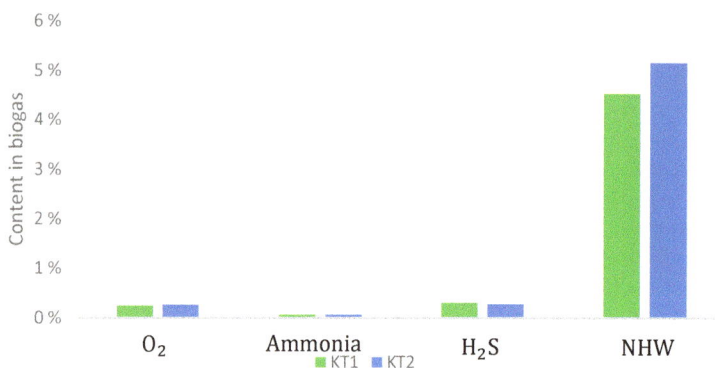

Figure 4. Performance of two models of biogas plants.

There is also the need to take into consideration the substances that can cause operation difficulties (dust, oils, and siloxanes) [33]. For the purposes of this study, another biogas quality indicator represented by methane and carbon dioxide index (CH_4:CO_2) was set up. This parameter characterises the relation between the content of CH_4 and CO_2 as two major substances influencing the final quality of biogas. From this point of view, a higher index means a higher quality of biogas.

Table 4. Biogas composition from small-scale biogas plants recognized in previous studies.

Feedstock	CH_4 (vol %)	CO_2 (vol %)	N_2 (vol %)	CO (vol %)	O_2 (vol %)	H_2 (vol %)	H_2S (vol %)	Type of BGP	Country	Reference
Animal wastewater	61–72	-	-	-	-	-	0.0043–0.0084	Tubular PVC	Costa Rica	[40]
Livestock manure	64	34	1.05	0.3	-	0.6	0.05	Floating dome	Pakistan	[8]
Livestock manure	56.2	39.51	-	1.91	-	-	1.84	Laboratory conditions	Nigeria	[41]
Livestock manure	50–75	25–45	<2	-	<2	<1	<1	Not specified	Developing countries	[42]
Livestock manure	60	35–40	-	-	-	-	-	Not specified	Malaysia	[28]
Organic waste	50–75	25–50	-	-	-	-	-	Not specified	Developing countries	[43]
Organic waste	60	-	-	-	-	-	-	Fixed dome	Sub-Saharan Africa	[44]
Generalized values	50–75	25–50	<2	<2	<2	<1	<2	-	-	-

All surveyed small-scale biogas plants (*n* = 107) showed in the average content of methane (CH_4) in biogas of 64.57% (±2.85) and the carbon dioxide (CO_2) of 30.20% (±3.10). The average presence of NH_3 was 0.05% (±0.02) and the presence of H_2S of 0.25% (±0.12). The average value of CH_4:CO_2 index was 2.20 (±0.35). The average calorific value of biogas produced by plants was 21.83 MJ/m^3 (±0.96), which corresponds with the typical value of 21–24 MJ/m^3 [42].

3.4. Biogas Composition According to Various Ages of Small-Scale Biogas Plants

The results presented in Table 5 is uncovering differences between small-scale biogas plants younger than five years and older than five years and its effect on various aspects of biogas quality. However, as shown in Table 5, there are no, or only minor, differences among tested values. This fact leads us to the conclusion that small-scale biogas plants using pig slurry as a main feedstock are sustaining a stable level of biogas quality during their life-span. Especially for the main indicators, which are volume of CH_4 in biogas, CH_4:CO_2 index and calorific value of biogas.

Table 5. Biogas composition from small-scale biogas plants younger than five years ($n = 82$) and older than five years ($n = 25$).

Variable	Small-Scale Biogas Plants Younger than Five Years			Small-Scale Biogas Plants Older than Five Years		
	Mean	95% Confidence Interval		Mean	95% Confidence Interval	
		Lower Bound	Upper Bound		Lower Bound	Upper Bound
CH_4 (vol %)	65.44	64.58	66.30	64.57	63.05	66.09
CO_2 (vol %)	29.31	28.37	32.25	29.93	28.27	31.58
NH_3 (vol %)	0.04	0.04	0.05	0.04	0.03	0.05
H_2S (vol %)	0.12	0.08	0.16	0.14	0.07	0.20
CH_4:CO_2 index	2.26	2.14	2.38	2.12	1.91	2.32
Calorific value (MJ/m^3)	22.08	21.79	22.39	21.87	21.36	22.39

3.5. Biogas Composition as Affected by Type, Age, and Capacity of the Biogas Plant

There was an effort to identify the factors that fundamentally influence the biogas quality (Table 6), therefore, the effects of variables of the type of biogas plant, the age of the biogas plants, and the size of the biogas plants on biogas composition were analysed. Firstly, the age of the biogas plant was tested as a relevant factor potentially influencing various aspects of the biogas quality. As demonstrated in Table 6, all biogas composition factors (including amounts of CH_4, CO_2, NH_3, H_2S, CH_4:CO_2 index, and calorific value) were not significantly affected by the age of biogas plant ($p > 0.05$). Secondly, the type of the biogas plant was tested as a relevant factor, using KT1 and KT2 models for comparison. As shown in Table 6, CH_4, CO_2, and calorific value were recognized as significantly influenced by the type of biogas plant ($p < 0.01$). The results show that KT2 model demonstrated a higher percentage of CH_4 and, consequently, a higher calorific value and a lower percentage of CO_2 of produced biogas. Another factor under the examination was the digester capacity (size). The results show different CH_4 contents, CH_4:CO_2 index values, and calorific values according to the digester capacity (size).

Table 6. Biogas composition as affected by type, age, and capacity of the biogas plant.

Dependent Variable	Parameter	Coefficient	Std. Error	t-Value	p
CH_4 (vol %)	Intercept	62.02	1.92	32.32	0.00
	Type of digester [a]	−2.44	0.82	−2.99	0.00
	Age [b]	0.87	0.81	1.08	0.28
	Digester capacity (m^3)	0.52	0.22	2.36	0.02
CO_2 (vol %)	Intercept	30.00	2.10	14.32	0.00
	Type of digester [a]	2.70	0.89	3.03	0.00
	Age [b]	−0.62	0.88	−0.70	0.49
	Digester capacity (m^3)	−0.20	0.24	−0.81	0.42
NH_3 (vol %)	Intercept	0.04	0.02	2.64	0.01
	Type of digester [a]	0.01	0.01	0.81	0.42
	Age [b]	0.00	0.01	0.67	0.51
	Digester capacity (m^3)	0.00	0.00	−0.25	0.80
H_2S (vol %)	Intercept	0.23	0.08	2.73	0.01
	Type of digester [a]	−0.05	0.04	−1.54	0.13
	Age [b]	−0.02	0.04	−0.51	0.62
	Digester capacity (m^3)	−0.01	0.01	−0.87	0.39
CH_4:CO_2 index	Intercept	1.56	0.26	5.96	0.00
	Type of digester [a]	−0.10	0.11	−0.87	0.39
	Age [b]	0.14	0.11	1.29	0.20
	Digester capacity (m^3)	0.08	0.03	2.78	0.01
Calorific value (MJ/m^3)	Intercept	20.96	0.65	32.14	0.00
	Type of digester [a]	−0.76	0.28	−2.73	0.01
	Age [b]	0.21	0.27	0.77	0.45
	Digester capacity (m^3)	0.18	0.07	2.38	0.02

[a] The base type is KT1; [b] the base age is >5 years.

Energies **2018**, *11*, 1794

4. Conclusions

Small-scale biogas plants can be a very useful tool for manure management and may help reduce global warming impacts if used appropriately. This technology offers a unique set of benefits, as it is a sustainable source of energy, it is benefiting the environment, and it provides a way to treat and reuse manure. However, if used inappropriately, its benefits may be compromised. In this study, the most common feedstock for a small-scale biogas plant was pig slurry, followed by a combination of pig slurry and human excreta. The majority of biogas plants were connected with the pig stable, or by latrine and stable. An average daily production of biogas equals to 0.499 m^3, which does not cover the demand of rural household with six members. Hence, 60% of surveyed households are still using other sources of energy as well. Biogas composition was measured with a multifunctional portable gas analyser. The mean content of methane (CH_4) was 65.44%, and for carbon dioxide (CO_2) was 29.31% in the case of biogas plants younger than five years and, respectively, CH_4 was 64.57% and CO_2 29.93% for biogas plants older than five years. The only dependent factor influencing the biogas quality was between biogas plant size and biogas composition, which was proven at CH_4, CH_4:CO_2 index, and the calorific value. Furthermore, type of the biogas plant affected CH_4, CO_2, and the calorific values of the biogas. Focusing on the influence of age of small-scale biogas plants there are no, or only minor, differences among tested qualitative biogas parameters. Concluding, that small-scale biogas plants sustaining a stable level of biogas quality during their life-span.

Author Contributions: Conceptualization, H.R., J.M. and J.B.; Methodology, H.R., J.M. and J.B.; Data collection, J.B., D.D.V.; Data analysis, H.R., J.B., P.L.D. and D.D.V.; Writing-Original Draft Preparation, H.R.; Writing-Review & Editing, H.R., J.M., P.L.D., D.D.V. and J.B.; Funding Acquisition, H.R., J.M. and J.B.

Funding: This research was conducted with the financial support of project number 20165003 of Internal Grant Agency of the Czech University of Life Sciences, Prague. Further support was provided by the Internal Grant Agency of the Faculty of Tropical AgriSciences, Czech University of Life Sciences, Prague, and project numbers 20185010 and 20185016.

Conflicts of Interest: The authors declare no conflict of interest.

References

1. Ahuja, D.; Tatsutani, M. Sustainable energy for developing countries. *Surv. Perspect. Integr. Environ. Soc.* **2009**, *2*, 1–16.
2. Mengistu, M.G.; Simane, B.; Eshete, G.; Workneh, T.S. Factors affecting households' decisions in biogas technology adoption, the case of Ofla and Mecha Districts, northern Ethiopia. *Renew. Energy* **2016**, *93*, 215–227. [CrossRef]
3. Rahman, M.M.; Hasan, M.M.; Paatero, J.V.; Lahdelma, R. Hybrid application of biogas and solar resources to fulfil household energy needs: A potentially viable option in rural areas of developing countries. *Renew. Energy* **2014**, *68*, 35–45. [CrossRef]
4. Thi, N.B.D.; Lin, C.-Y.; Kumar, G. Waste-to-wealth for valorisation of food waste to hydrogen and methane towards creating a sustainable ideal source of bioenergy. *J. Clean. Prod.* **2016**, *122*, 29–41. [CrossRef]
5. Kinyua, M.N.; Rowse, L.E.; Ergas, S.J. Review of small-scale tubular anaerobic digesters treating livestock waste in the developing world. *Renew. Sustain. Energy Rev.* **2016**, *58*, 896–910. [CrossRef]
6. Roubík, H.; Mazancová, J.; Phung, L.D.; Banout, J. Current approach to manure management for small-scale Southest Asian farmers—Using Vietnamese biogas and non-biogas farms as an example. *Renew. Energy* **2018**, *115*, 362–370. [CrossRef]
7. Huong, L.Q.; Madsen, H.; Anh, L.X.; Ngoc, P.T.; Dalsgaard, A. Hygienic aspects of livestock manure management and biogas systems operated by small-scale pig farmers in Vietnam. *Sci. Total Environ.* **2014**, *470*, 53–57. [CrossRef] [PubMed]
8. Mushtaq, K.; Zaida, A.A.; Askari, S.J. Design and performance analysis of floating dome type portable biogas plant for domestic use in Pakistan. *Sustain. Energy Technol. Assess.* **2016**, *14*, 21–25. [CrossRef]
9. Cu, T.T.T.; Cuong, P.H.; Hang, L.T.; Chao, N.V.; Anh, L.X.; Trach, N.X.; Sommer, S.G. Manure management practices on biogas and non-biogas pig farms in developing countries—Using livestock farms in Vietnam as an example. *J. Clean. Prod.* **2012**, *27*, 64–71. [CrossRef]

10. Rodolfo, S.; Anh, L.H.; Konrad, K. Feasibility assessment of anaerobic digestion technologies for household wastes in Vietnam. *J. Vietnam. Environ.* **2016**, *7*, 1–8. [CrossRef]
11. Chiumenti, A.; Chiumenti, R.; Chiumenti, A. Complete nitrification-denitrification of swine manure in a full-scale, non-conventional composting system. *Waste Manag.* **2015**, *46*. [CrossRef] [PubMed]
12. Zhang, X.; Luo, L.; Skitmore, M. Household carbon emission research: An analytical review of measurement, influencing factors and mitigation prospects. *J. Clean. Prod.* **2015**, *103*, 873–883. [CrossRef]
13. Bruun, S.; Jensen, L.S.; Vu, V.T.K.; Sommer, S. Small-scale household biogas digesters: An option for global warming mitigation or a potential climate bomb? *Renew. Sustain. Energy Rev.* **2014**, *33*, 736–741. [CrossRef]
14. Neupane, M.; Basnyat, B.; Fischer, R.; Froeschl, G.; Wolbers, M.; Rehfuess, E.A. Sustained use of biogas fuel and blood pressure among women in rural Nepal. *Environ. Res.* **2015**, *136*, 343–351. [CrossRef] [PubMed]
15. Mengistu, M.G.; Simane, B.; Eshete, G.; Workneh, T.S. The environmental benefits of domestic biogas technology in rural Ethiopia. *Biomass Bioenergy* **2016**, *90*, 131–138. [CrossRef]
16. Zhang, L.X.; Wang, C.B.; Song, B. Carbon emission reduction potential of a typical household biogas system in rural China. *J. Clean. Prod.* **2013**, *47*, 415–421. [CrossRef]
17. Truc, N.T.T.; Nam, T.S.; Ngan, N.V.C.; Bentzen, J. Factors Influencing the Adoption of Small-scale Biogas Digesters in Developing Countries—Empirical Evidence from Vietnam. *Int. Bus. Res.* **2017**, *10*. [CrossRef]
18. Rajendran, K.; Aslanzadeh, S.; Taherzadeh, J. Household Biogas Digesters—A Review. *Energies* **2012**, *5*, 2911–2942. [CrossRef]
19. Chiumenti, R.; Chiumenti, A.; da Borso, F.; Limina, S. Anaerobic Digestion of Swine Manure in Conventional and Hybrid Pilot Scale Plants: Performance and Gaseous Emissions Reduction. In Proceedings of the International Syposium ASABE 2009, Reno, NV, USA, 21–24 June 2009.
20. Bhattacharya, S.C.; Jana, C. Renewable energy in India: Historical developments and prospects. *Energy* **2009**, *34*, 981–991. [CrossRef]
21. Roubík, H.; Mazancová, J.; Banout, J.; Verner, V. Addressing problems at small-scale biogas plants: A case study from central Vietnam. *J. Clean. Prod.* **2016**, *112*, 2784–2792. [CrossRef]
22. Thy, S.; Preston, T.R.; Ly, J. Effect of retention time on gas production and fertilizer value of digester effluent. *Livest. Res. Rural Dev.* **2003**, *15*. Available online: http://www.lrrd.org/lrrd15/7/sant157.htm (accessed on 2 July 2018).
23. Khan, E.U.; Martin, A.R. Optimization of hybrid renewable energy polygeneration system with membrane distillation for rural households in Bangladesh. *FUEL* **2015**, *93*, 1116–1127. [CrossRef]
24. Phanthavongs, S.; Saikia, U. Biogas Digesters in Small Pig Farming Systems in Lao PDR: Evidence of Impact. *Livest. Res. Rural Dev.* **2013**, *25*. Available online: http://www.lrrd.org/lrrd25/12/phan25216.htm (accessed on 2 July 2018).
25. Katuwal, H.; Bohara, A.K. Biogas: A promising renewable technology and its impact on rural households in Nepal. *Renew. Sustain. Energy Rev.* **2009**, *13*, 2668–2674. [CrossRef]
26. Kaur, G.; Brar, Y.S.; Kothari, D.P. Potential of Livestock Generated Biomass: Untapped Energy Source in India. *Energies* **2017**, *10*, 847. [CrossRef]
27. Usack, J.G.; Wiratni, W.; Angenent, L.T. Improved design of anaerobic digesters for household biogas production in Indonesia: One cow, one digester, and one hour of cooking per day. *Sci. World J.* **2014**, 1–8. [CrossRef] [PubMed]
28. Abdeshahian, P.; Lim, J.S.; Ho, W.S.; Hashim, H.; Lee, C.T. Potential of biogas production from farm animal waste in Malaysia. *Renew. Sustain. Energy Rev.* **2016**, *60*, 714–723. [CrossRef]
29. Ghmire, P.C. SNV supported domestic biogas programmes in Asia and Africa. *Renew. Energy* **2013**, *49*, 90–94. [CrossRef]
30. Khan, E.U.; Martin, A.R. Review of biogas digester technology in rural Bangladesh. *Renew. Sustain. Energy Rev.* **2016**, *62*, 247–259. [CrossRef]
31. SNV. Biogas Programme in Vietnam, SNV Vietnam. Available online: http://www.snv.org/ (accessed on 1 February 2018).
32. Pham, C.H.; Vu, C.C.; Sommer, S.G.; Bruun, S. Factors Affecting Process Temperature and Biogas Production in Small-scale Biogas Digesters in Winter in Northern Vietnam. *Asian-Australas. J. Anim. Sci.* **2014**, *27*, 1050–1056. [CrossRef] [PubMed]
33. Zamorska-Wojdyla, D.; Gaj, K.; Holtra, A.; Sitarska, M. Quality evaluation of biogas and selected methods of its analysis. *Ecol. Chem. Eng.* **2012**, *19*, 77–87. [CrossRef]

34. Rasi, S. *Biogas Composition and Upgrading to Biomethane*; University of Jyväskylä: Jyväskylä, Finland, 2009; ISBN 978-951-39-3618-1.

35. Rasi, S.; Lantela, J.; Rintala, J. Trace compounds affecting biogas energy utilization—A review. *Energy Convers. Manag.* **2011**, *52*, 3369–3375. [CrossRef]

36. Nguyen, V.C.N. Small-scale anaerobic digesters in Vietnam—Development and challenges. *J. Vietnam. Environ.* **2011**, *1*, 12–18. [CrossRef]

37. Vu, Q.D.; Tran, T.M.; Nguyen, P.D.; Vu, C.C.; Vu, V.T.K.; Jensen, L.S. Effect of biogas technology on nutrient flows for small- and medium-scale pig farms in Vietnam. *Nutr. Cycl. Agroecosyst.* **2012**, *94*, 1–13. [CrossRef]

38. Sasse, L.; Kellner, C.; Kimaro, A. *Improved Biogas Unit for Developing Countries. Deutsche Gesellschaft für Technische Zusammenarbeit (GTZ) GmbH*; Vieweg & Sohn Verlagsgesellschaft Braunschweig: Eschborn, Germany, 1991.

39. Nhu, T.T.; Dewulf, J.; Serruys, P.; Huysveld, S.; Nguyen, C.V.; Sorgeloos, P.; Schaubroeck, T. Resource usage of integrated Pig-Biogas-Fish system: Partitioning and substitution within attributional life cycle assessment. *Resour. Conserv. Recycl.* **2015**, *102*, 27–38. [CrossRef]

40. Lansing, S.; Botero, R.B.; Martin, J.F. Waste treatment and biogas quality in small-scale agricultural digesters. *Bioresour. Technol.* **2008**, *99*, 5881–5890. [CrossRef] [PubMed]

41. Adegun, I.K.; Yaru, S.S. Cattle Dung Biogas as a *Renew. Energy* Source for Rural Laboratories. *J. Sustain. Technol.* **2013**, *4*, 1–8.

42. Bond, T.; Templeton, M.R. History and future of domestic biogas plants in the developing world. *Energy Sustain. Dev.* **2011**, *15*, 347–354. [CrossRef]

43. Surendra, K.C.; Takara, D.; Hashimoto, A.G.; Khanal, S.K. Biogas as sustainable energy source for developing countries: Opportunities and challenges. *Renew. Sustain. Energy Rev.* **2014**, *31*, 846–859. [CrossRef]

44. Tumwesige, V.; Fulford, D.; Davidson, G.C. Biogas appliances in Sub-Saharan Africa. *Biomass Bioenergy* **2014**, *70*, 40–50. [CrossRef]

energies

MDPI

Article

Design, Construction, and Testing of a Gasifier-Specific Solid Oxide Fuel Cell System

Alvaro Fernandes [1,*], Joerg Brabandt [2], Oliver Posdziech [2], Ali Saadabadi [1], Mayra Recalde [1], Liyuan Fan [1], Eva O. Promes [1], Ming Liu [1], Theo Woudstra [1] and Purushothaman Vellayan Aravind [1]

[1] Energy Technology Section, Department of Process and Energy, Delft University of Technology, Leeghwaterstraat 39, 2628 CB Delft, The Netherlands; S.A.Saadabadi@tudelft.nl (A.S.); mayra.recalde@tudelft.nl (M.R.); L.Fan@tudelft.nl (L.F.); e.j.o.promes@tudelft.nl (E.O.P.); m.Liu@tudelft.nl (M.L.); T.Woudstra@tudelft.nl (T.W.); p.v.aravind@tudelft.nl (P.V.A.)

[2] Sunfire GmbH, Gasanstaltstraße 2, 01237 Dresden, Germany; joerg.brabandt@sunfire.de (J.B.); oliver.posdziech@sunfire.de (O.P.)

* Correspondence: A.B.MonteiroFernandes@tudelft.nl; Tel.: +31-(0)15-278-36-88

Received: 26 May 2018; Accepted: 26 July 2018; Published: 31 July 2018

Abstract: This paper describes the steps involved in the design, construction, and testing of a gasifier-specific solid oxide fuel cell (SOFC) system. The design choices are based on reported thermodynamic simulation results for the entire gasifier- gas cleanup-SOFC system. The constructed SOFC system is tested and the measured parameters are compared with those given by a system simulation. Furthermore, a detailed exergy analysis is performed to determine the components responsible for poor efficiency. It is concluded that the SOFC system demonstrates reasonable agreement with the simulated results. Furthermore, based on the exergy results, the components causing major irreversible performance losses are identified.

Keywords: SOFC; validation; simulation; exergy; syngas

1. Introduction

The production of electricity, biofuels, and chemicals is increasingly using biomass sources. Indeed, by 2015, Europe had installed a net maximum capacity of 35.4 GW from energy sources including municipal waste, biogas, wood and wood residues, and other solid residues. [1].

Biomass is a storable feedstock that is being employed for power generation in biomass-fired plants. These plants are typically steam cycle or organic Rankine cycle (ORC) power systems capable of achieving electrical efficiencies from 15% (small plants) to 40% (large plants) [2,3].

Alternatively, biomass can be processed into gaseous fuels such as syngas and biogas for further use in power-producing steam engines, which have a modest efficiency of approximately 20%; or gas engines, gas turbines, and fuel cells, especially SOFC, which can achieve efficiencies up to 50% [4–7].

Nevertheless, there are major problems in the use of biomass for power generation, namely the logistics of collection and seasonal availability, which create inefficient biomass power chains. To overcome these issues, the installation of small-scale decentralized biomass power plants is an economically viable and efficient solution [5,8]. At this scale, gas engines and gas turbines suffer from lower efficiency (i.e., a reduction in power production capacity), compared with SOFCs. Moreover, SOFCs also have the advantage of operating at very high efficiencies in part-load windows. Furthermore, they are less susceptible to variations in fuel composition [9,10].

To accommodate the fluctuating electricity demands of both grid and off-grid installations, SOFC systems should be capable of operating within a wide part-load window. Consequently, it is fundamental that SOFC systems have an adequate system configuration and components. The selection of components must be based on, among other factors, the material limitations, variations in gas composition, thermal management, and carbon suppression.

System modelling is a rapid and cost-effective method for predicting system performance and off-design operation conditions. Many studies on the simulation of SOFC systems [11] focus on the performance under specific design conditions or the transient performance of SOFCs under a varying electric load [12], whereas others consider the transient performance of the entire SOFC system. These system models are usually validated by mathematical models, although some have been validated using experimental data for a number of the components.

Rokni [13] investigated the re-powering of a steam power plant with gas turbines and SOFCs. In his work, three system configurations were simulated: base case (steam plant), steam plant with gas turbines, and steam plant with SOFCs. Only the latter system was calibrated with experimental data for a planar SOFC, while for other components a similar modeling approach to other available studies was followed. It was reported that the plant with an SOFC system could achieve an optimized efficiency above 66% for an operating temperature of 1013 K, current density of 200 A/cm^2, and a fuel utilization of 80%.

Similarly, Ugartemendia et al. [14] validated his dynamic model of an SOFC-steam cycle with SOFC experimental data from the literature. Other components such as heat exchangers were assumed to have constant thermal effectiveness. The authors concluded that the operating temperature of 1173 K and a fuel utilization of 65% were the optimal conditions for achieving the higher power output.

Chung et al. [15] studied the influence of operating parameters on the plant efficiencies of a methane gas-fed SOFC power generation system. The results obtained from a mathematical model simulated under the design conditions revealed that the air-to-fuel ratio (A/F) was the most important parameter in terms of system efficiency. The pre-reforming rate of fuel was found to be relatively insignificant in terms of efficiency, but could be used as an auxiliary tuner for the operating temperature of the SOFCs, in addition to A/F.

Chitsaz et al. [16] conducted a thermodynamical evaluation of an integrated tri-generation system driven by an SOFC. Through steady-state system simulations based on mathematical correlations, and partially validated by experimental data obtained from an SOFC setup, a maximum system exergy efficiency of 46% was achieved. The main sources of irreversibility were observed to be the air heat exchanger, SOFC, and afterburner.

A similar analysis was performed by Stamatis et al. [17] for an SOFC and a hybrid SOFC-gas turbine system fuelled by ethanol. The models were partially validated by available experimental data from the literature for the SOFC component. A system efficiency of up to 60% was achieved under certain operating conditions. It was also disclosed that the SOFC and burner-reformer components were the major sources of irreversibilities within the systems.

Xu et al. [18] investigated the influence of various design parameters on the SOFC thermal behaviour and system performance of a natural-gas-fuelled 1 kW conceptual design for a residential combined heat and power system. This system was also modelled based on mathematical correlations and partially validated by experimental data obtained from an SOFC experimental setup. The results indicated that the cell output voltage, system inlet fuel flow rate, and SOFC stack inlet air temperature had a dramatic effect on the electrical efficiency and cogeneration efficiency.

Somekawa et al. [19] investigated the influence of various design parameters on a manufactured multi-stack SOFC system coupling an anode regenerator between stacks. The regenerator consisted of a CO_2 absorber and water vapour condenser to selectively remove these compounds from the anode off-gas mixture. The system models were validated by experimental data collected from an SOFC setup and a hot module designed especially for the study. A remarkable total fuel utilization of 92.0% and an electrical efficiency of 77.8% were achieved with this design.

A diesel-fed SOFC power system is being designed and developed for maritime applications under the SchIBZ-project [20]. In a first stage, a system model was prepared and validated for the reformer and SOFC components with experimental data available in the literature. The authors will further validate the whole system with experimental lab work. The system model will then be used for system performance analysis and the determination of optimal operation conditions through an exergetic analysis.

The aforementioned studies all use hydrogen, methane, or natural gas as the fuel. In a recent study, D'Andrea et al. [21] performed a dynamic simulation for the proof-of-concept of a biogas-fed SOFC polygeneration system. The SOFC model was primarily validated by reproducing similar test conditions as used by the manufacturer and comparing with data provided by the manufacturer. A second validation was performed by including the SOFC model in the system model for comparison with data collected from the proof-of-concept. This model was developed to investigate both the stack and balance-of-plant (BoP) thermal behaviour under abnormal operation conditions, namely fast load current ramps, fault cathode air, and different rates of internal reforming of the fuel in the SOFC. They concluded that in the event of the two first abnormal conditions, the SOFC may overheat and be damaged. To prevent this, the system control should shift the system to the open circuit condition. It was also concluded that the percentage of fuel that is internally reformed in the SOFC can be adjusted to control the temperature.

To the best of the authors' knowledge, this paper is the first to describe all of the steps involved in the design, construction, and testing of a gasifier-specific SOFC system. For the first time, system testing was conducted to compare the recorded performance with the predicted thermodynamic performance obtained by system simulation, aiming at validating the exergy flow model of the system. In addition to the design point analysis, we investigated part-load and off-design conditions to develop a deeper understanding of the variation in system performance.

2. Selection of the Design of SOFC System

The SOFC system was developed to be integrated with a plasma gasifier and a gas cleanup unit (Figure 1) at TU Delft. Prior to determining the system configuration, TU Delft performed a thermodynamic analysis of the power plant in which two suitable configurations for the SOFC system were evaluated [22]. The main difference in the configurations was the mechanism for suppressing carbon formation.

Both system configurations considered the SOFC materials, approaches for suppressing carbon formation, SOFC temperature control strategy, and thermal management.

2.1. SOFC Materials

Nickel–Gadolinium-doped Ceria (Ni/GDC) anode cells were selected for their anticipated advantages with hydrocarbon fuels [23,24]. It has been reported that Ni/GDC is more resistant to carbon deposition and poisoning by typical syngas contaminants than Nickel/Yttria-stabilized Zirconia (Ni/YSZ) anodes [25–27].

2.2. Carbon Suppression Technique

The major difference between the system configurations in the study of Liu et al. [22] is the technique employed for suppressing carbon formation. One system uses steam produced in a heat recovery steam generator unit that is mixed with the anode flow before it enters the SOFC module, whereas the other uses a catalytic partial oxidation unit (CPOx) to suppress carbon formation and pre-reform some of the hydrocarbons before it enters the SOFC module. In our system, the last technique is employed.

2.3. SOFC Temperature Control Strategy

In our system, the SOFC temperature is controlled by varying the cathode flow rate. An additional degree of control is offered by changing the fuel utilization.

2.4. Thermal Management

The syngas exits the gas cleanup unit at a relatively low temperature (approximately 673 K) and the cathode air enters the system at environment temperature. Therefore, both flows must be heated before they enter the SOFC module. This is accomplished by using heat exchangers and a CPOx unit to suppress carbon formation. Flue gas produced in an afterburner is used as heat source to heat the cathode air, and is also used to pre-heat/heat the syngas. For the system configuration using steam as the carbon suppression, the flue gas is also used as heat source to generate steam in a heat recovery steam generator. The flue gas subsequently moves to the dryer and is used to pre-dry the feedstock that is fed to the gasifier.

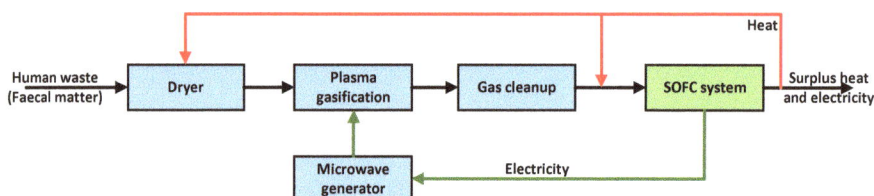

Figure 1. Flow of the integrated plasma gasifier-gas cleanup-solid oxide fuel cell (SOFC) system. Reproduced with permission from [22], Copyright Elsevier, 2018.

Previous results suggest that the system configuration with a CPOx unit could achieve higher electrical efficiency [22]. As this unit also increases the flexibility in terms of thermal integration, it was selected for construction.

3. Description of the SOFC System

In this section, the components in the SOFC system and the construction process are briefly described. The equipment used to record data for the test runs and the safety control strategies are also listed.

3.1. System Construction Design

The SOFC system (Figure 2) consists of two hot boxes: a BoP hot box and an SOFC hot box [28]; as well as a make-up gases panel. The hot boxes are connected by four insulated pipes (i.e., anode and cathode inflows/outflows). The BoP hot box is also connected to the supply gases, syngas, and intake air, and has an exhaust flue gas pipe. Each hot box is thermally insulated with low-thermal-conductivity panels to reduce heat losses to the surroundings. Moreover, the BoP hot box is filled with a granulated microporous insulation material that offers good resistance to heat transfer between BoP components.

The BoP hot box contains the air heater, fuel pre-heater, CPOx unit, and afterburner. Both the fuel pre-heater and air heater are counter-flow plate heat exchangers made of high-temperature alloy. Under the design conditions, the air heater increases the cathode air temperature to 923 K. The fuel pre-heater increases the fuel temperature to approximately 523–623 K, after which the fuel is supplied to the CPOx unit. This component increases the oxygen-to-carbon (O/C) ratio to prevent carbon formation, pre-reforms hydrocarbons, and increases the fuel temperature to 973–1073 K. Both the CPOx and afterburner are catalytic burners impregnated with adequate catalysts.

The SOFC hot box contains the SOFC module, consisting of two towers of SOFC stacks that are electrically connected in parallel. Each tower is composed of three 30-cell stacks that are electrically connected in series. This module was tested and achieved a nominal power of 4 kW in reference conditions (40% H_2/60% N_2, fuel utilization (U_f) = 0.75, cell voltage (V_{cell}) > 0.65 V, $T_{cathode\ outlet}$ < 1133 K). A reduction in the electrical output of approximately 10–15% is expected for syngas. The make-up gases panel contains the hydrogen and nitrogen rotameters, which provide the forming gas during system start-up, and the hydrogen, nitrogen, and air (CPOx air) mass flow controllers, which provide the gases for the test runs. Other gases such as carbon monoxide and carbon dioxide are measured by mass flow controllers located in the gas cleanup unit and supplied by the syngas inlet pipe.

The intake air is fed to the system using two blowers (i.e., a cathode blower and a CPOx blower). The cathode blower can provide 900 nL of air per minute under reference conditions. A filter and a flow measurement device are located on the discharge side. The CPOx blower provides 15 nL of air per minute under reference conditions and is coupled with a filter on the admission side. The mass flow controller is located on the discharge side of the CPOx blower.

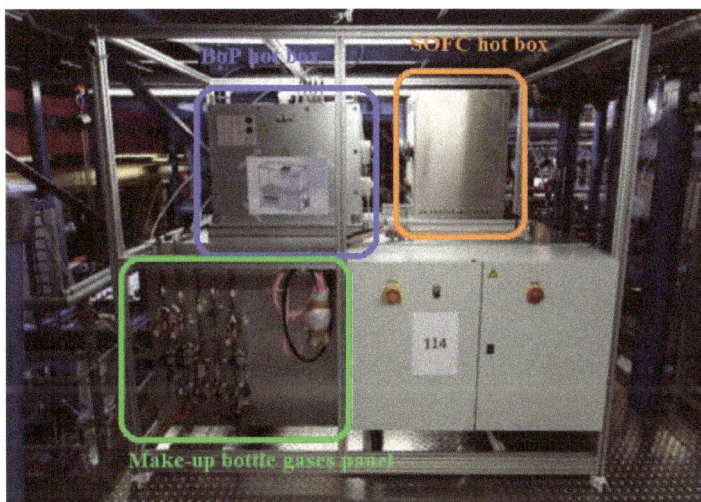

Figure 2. The SOFC 3.5 KW$_e$ system at TU Delft with description of the various parts. The air blowers are located behind the make-up bottle gases panel.

3.2. Safe Operation Parameters

The safe operating parameters are primarily intended to protect expensive components and users from risks associated with the operation of the SOFC system. This is accomplished by implementing safety functions in a programmable logic controller (PLC) that are capable of shifting the system to a safe operation mode when required. The safety functions process information acquired from installed measurement devices in the system and signals from external safeguards such as the gas detection and ventilation systems.

The parameters that can limit the extent of the system part-load operation are illustrated in Figure 3. The safe upper limit operating temperature, maximum flows of cathode air and CPOx air, and minimum cell operating voltage are the main parameters that influence the part-load operation window. Safe upper limit operating temperatures are primarily determined by material limitations, whereas the air flows limit the cooling capacity and the capability for suppressing carbon formation.

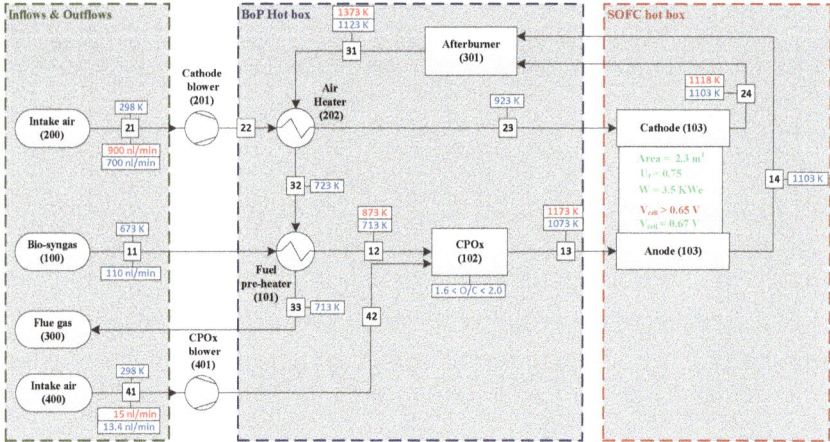

Figure 3. Limiting operation and preliminary design parameters of the SOFC system. Red parameters: maximum volume flows and temperatures; blue parameters: preliminary design parameters. BoP: balance-of-plant; CPOx: catalytic partial oxidation unit.

4. System Modelling

The system was simulated using Cycle-Tempo® software (version 5.0, Delft University of Technology, Delft, The Netherlands), a Fortran-based software package designed for analysing the first and second laws of thermodynamics of power plants [29].

Figure 4 illustrates the SOFC system model. Three inlet flows are defined: syngas (source 100), cathode air (source 200), and CPOx air (source 400). There is one outlet flow: flue gas (sink 300). The operating parameters of the SOFC are inserted to accurately determine the amount of chemical energy converted into electricity. Specifically, these parameters are the fuel utilization (U_f), equivalent resistance (R_{eq}), and electric power ($\dot{W}_{SOFC}$). Other parameters such as the CPOx O/C ratio and SOFC outflows temperature must also be specified.

The CPOx air mole flow (Φ_{41}) is calculated as

$$\Phi_{41} = \frac{\Phi_{11} \cdot \left[y_{11.CO} \cdot (O/C - 1) + y_{11.CO_2} \cdot (O/C - 2) \right]}{2 \cdot y_{41.O_2}}, \tag{1}$$

where $y_{pipe.species}$ is the mole fraction of the species (CO, CO_2, or O_2) in the mixture (pipe) and Φ_{11} is the inlet flow (mol/s) of the syngas in the system (pipe 11). Assuming a gas composition constant in the cross-section and that the process occurs at constant temperature and pressure, Φ_{11} is calculated in the SOFC component by

$$\Phi_{11} = \frac{I}{z \cdot F \cdot U_f \cdot (y_{13.H_2} + y_{13.CO})} - \Phi_{41} \cdot y_{41.N_2}, \tag{2}$$

$$U_f = \frac{\Phi_{13} \cdot (y_{13.H_2} + y_{13.CO}) - \Phi_{14} \cdot (y_{14.H_2} + y_{14.CO})}{\Phi_{13} \cdot (y_{13.H_2} + y_{13.CO})}, \tag{3}$$

$$I = \frac{\dot{W}_{SOFC}}{V_{cell}}, \tag{4}$$

$$V_{cell} = V_{Nernst,x} - \frac{I_x}{Area_x} \cdot R_{eq}, \tag{5}$$

$$V_{\text{Nernst},x} = -\frac{\triangle g^o}{z \cdot F} + \frac{\bar{R} \cdot T_{\text{cell}}}{z \cdot F} \cdot \ln\left(\frac{y_{H_2,x} \cdot y_{O_2,x}^{0.5}}{y_{H_2O,x}} \cdot p_{\text{cell}}^{0.5}\right), \tag{6}$$

where I and V_{cell} are the current produced by all cells and the cell voltage in amperes and volts, respectively, z is the number of electrons involved in a single reaction, F is Faraday's constant (96485 C/mol), $V_{\text{Nernst},x}$ is the local Nernst potential (V), and the ratio of I_x and $Area_x$ is the local current density (A/cm^2). More detailed information can be found in [29,30].

The cathode air mole flow (Φ_{21}) is calculated by an energy balance in the SOFC as

$$\Phi_{21} = \frac{\Phi_{24} \cdot \bar{h}_{24} + \Phi_{14} \cdot \bar{h}_{14} - \Phi_{13} \cdot \bar{h}_{13} + \dot{W}_{\text{SOFC}}}{\bar{h}_{23}}, \tag{7}$$

where $\bar{h}$ is the enthalpy in kJ/mol.

The modelling of the heat exchangers is based on the first law of thermodynamics for the energy balance in the heat exchanger,

$$\dot{Q}_{\text{Trans}} = \Phi_{m,h} \cdot c_{p,h} \cdot (T_{h,in} - T_{h,out}) = \Phi_{m,c} \cdot c_{p,c} \cdot (T_{c,out} - T_{c,in}), \tag{8}$$

where c_p is the specific heat (kJ/(kg· K)) of the medium, Φ_m is the flow of the fluids (kg/s), h is the hot medium, and c is the cold medium. The transmitted heat flow in kW ($\dot{Q}_{\text{Trans}}$) can be calculated as

$$\dot{Q}_{\text{Trans}} = U \cdot A \cdot LMTD, \tag{9}$$

where U is the overall heat transfer coefficient (W/m^2·K) and A is the heat transfer surface area (m^2). The logarithm mean temperature difference ($LMTD$) for a counter-flow heat exchanger is

$$LMTD = \frac{(T_{h,in} - T_{c,out}) - (T_{h,out} - T_{c,in})}{\ln\left(\frac{T_{h,in} - T_{c,out}}{T_{h,out} - T_{c,in}}\right)}. \tag{10}$$

The off-design calculations of Cycle-Tempo® use the approach developed by Miedema [31], in which the variation in the overall heat transfer coefficient multiplied by the heat transfer surface area is proportional to the variation of the mass flow:

$$(U.A)_i = (U.A)_D \cdot \left(\frac{\Phi_{m,i}}{\Phi_{m,D}}\right)^{\eta_{cf}}, \tag{11}$$

where i refers to an off-design operating point, D refers to the design operating point of the heat exchanger, and η_{cf} is an exponential correction factor. Knowing the inlet temperatures of both media in the heat exchanger, Cycle-Tempo® iteratively calculates the outlet temperature of both flows. The input parameters are listed in Table 1.

Table 1. Input parameters used in the system modelling.

Item	Id	Parameters
Syngas	(100)	15.6% H_2, 18.3% CO, 6.1% CO_2, 60% N_2 (vol. basis)
		$T_{11} = 298$ K
SOFC	(103)	$R_{eq} = 1.1$ $\Omega \cdot cm^2$ Area = 2300 cm^2
		$T_{14,24} = 1073$ K, $U_f = 0.75$
Intake air	(200; 400)	79% N_2, 21% O_2 (vol. basis)
		$p_{21,41} = 101.3$ kPa and $T_{21,41} = 298$ K
Air heater	(202)	$\Phi_{m,D} = 13.41$ kg/s; $(U.A)_D = 55.87$ W/K; $\eta = 1.45$
Fuel pre-heater	(101)	$\Phi_{m,D} = 1.4$ kg/s; $(U.A)_D = 3.44$ W/K; $\eta = 0.8$
Blowers	(201; 401)	$\eta_{is} = 0.6$; $\eta_{mec,el} = 0.6$
All components	-	$\Delta p = 5\% \cdot p_{stream}$

5. Test Runs with the SOFC System

The SOFC system was designed and developed to integrate a gasifier-gas cleaning unit-SOFC plant fed by faeces from human waste. The faeces, after being separated from urine in the sanitation system, were sent to a pre-drying unit to reduce the moisture content from approximately 80% to acceptable levels for the plasma gasification process (20–40%) [22]. The syngas composition used for the design of the SOFC system was based up on the model developed and presented in [22]. The model was based on equilibrium assumptions for the gasifier considering that very-high-temperature gasification processes approach equilibrium conditions.

The SOFC system was tested over the range allowed by the limited testing conditions in the laboratory. The syngas composition was modified from the initial design composition by increasing the nitrogen concentration by 20% to enhance the accuracy of the nitrogen flow measurements. The syngas composition and flow rate were controlled by mixing bottled gases that were individually measured in mass flow controllers before entering the system. As a consequence, the syngas was fed in at the environmental temperature on a dry basis, which also diverged from the design conditions. Thus, some parameters like the flow rate of the CPOx blower or cathode air flow rate did limit the window of operation points in the performed experiments.

Operating points were taken at constant intervals of 10 mA/cm^2 current density. Operating points were only considered for analysis if the system achieved the steady state without crossing any of the safe operating parameters.

During the test runs, the following operating parameters that are calculated or displayed in the interface of the software were maintained:

- Constant SOFC fuel utilization of 0.75 ± 0.01,
- Constant O/C ratio of 1.6 ± 0.002 in the CPOx unit,
- Constant outlet temperature of 1073 ± 10 K at the SOFC cathode.

Table 2 lists the measurement devices installed in the system and their accuracy.

Table 2. List of measurement devices and their accuracy.

Measurement Type	Measurement Device	Accuracy
Temperature	Thermocouples type N	± 1.1 K or 0.4% of Rd ± 0.3% terminal (PLC)
Cell voltage	In-situ wiring	± 0.3% signal analog input (PLC)
CPOx air flow	Mass flow controller	± 0.5% of Rd ± 0.1 of Fs (16 nL/min)
H_2 flow	Mass flow controller	± 0.5% of Rd ± 0.1 of Fs (20 nL/min)
N_2 flow	Mass flow controller	± 0.5% of Rd ± 0.1 of Fs (20 nL/min)
CO flow	Mass flow controller	± 0.8% of Rd ± 0.2 of Fs (5 nL/min)
CO_2 flow	Mass flow controller	± 1% of Rd ± 1% of Fs (1 nL/min)
Cathode air	Flow meter	± 2% of Rd ± 0.3 signal analog input (PLC)
Current	Electronic load	± 0.2% of Fs (160 A)

Rd: reading value; Fs: full scale/nominal; PLC: programmable logic controller.

6. Results and Discussion

The results of this study are presented in four main parts. Firstly, the system model results for different syngas supply conditions and SOFC temperatures are compared with those under the design conditions. Secondly, data recorded in the test runs are compared with results obtained by the system model. Thirdly, the system efficiency is analysed to determine the major sources of inefficiency. Finally, based on the results, some considerations for improved design strategies are given.

6.1. Thermodynamic Comparison for Different Syngas Supply Conditions and SOFC Temperatures

A thermodynamic model was prepared considering both the design conditions and the experimental conditions. The results are depicted in Figure 4 and Table 3.

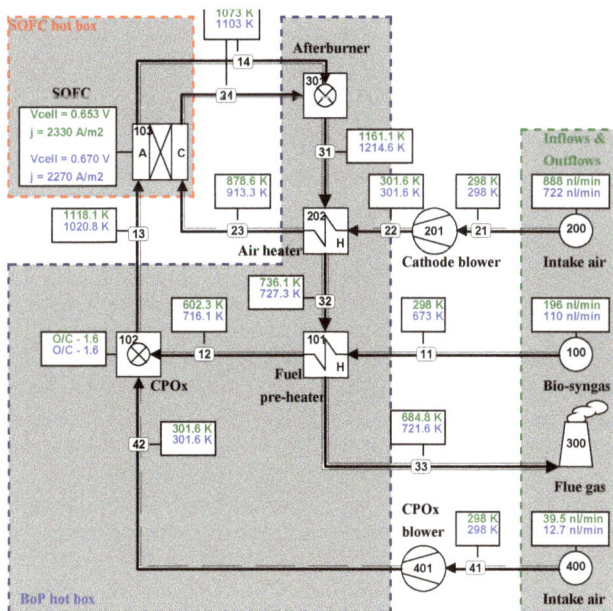

Figure 4. SOFC system model. Blue text refers to system simulation with design conditions, green text refers to experimental conditions for 3.5 kW production by the SOFC module.

Table 3. Comparison of the results of the main streams as well as system model efficiencies.

	Simulation Considering Design Conditions						
	Pipe number						
	11	13	14	23	24	31	41
mol % H_2	26.95	21.70	5.57				
mol % CO	22.00	18.55	4.54				
mol % CO_2	7.01	8.01	22.05			3.90	
mol % N_2	37.85	43.09	43.04	79.0	81.04	76.06	79.0
mol % O_2				21.0	18.96	15.57	21.0
mol % H_2O	6.19	8.60	24.81			4.46	
Mole flow (mol/s)	0.082	0.089	0.089	0.537	0.523	0.608	0.009
Vol. flow (nL/min)	110.0	119.9	119.9	721.8	703.3	817.2	12.7
Exergy (kW)	10,382	10,118	3984	4773	6958	10,287	~0

Auxiliaries consumption = 0.092 kW   System electric efficiency = 32.8%

	Simulation Considering Experimental Conditions						
	Pipe number						
	11	13	14	23	24	31	41
mol % H_2	15.60	8.71	2.23				
mol % CO	18.36	13.21	3.25				
mol % CO_2	6.10	7.84	17.80			4.38	
mol % N_2	60.0	65.51	65.50	79.0	80.70	77.99	79.0
mol % O_2				21.0	19.30	14.82	21.0
mol % H_2O		4.74	11.22			2.80	
Mole flow (mol/s)	0.146	0.169	0.169	0.661	0.647	0.811	0.029
Vol. flow (nL/min)	195.8	227.0	227.0	888.0	869.2	1090	39.5
Exergy (kW)	12,545	11,946	5166	5381	8129	12,606	~0

Auxiliaries consumption = 0.116 kW   System electric efficiency = 27.0%

The supply of a syngas composition with no water vapor and higher nitrogen content (experimental conditions) resulted in an increase of all other flows. Higher syngas flow was fed to compensate the loss of chemical energy in the CPOx unit by oxidation with air to achieve the specific O/C ratio. For the effect, a higher amount of CPOx air was also supplied (Figure 4). As a major consequence, a higher amount of heat was produced that resulted in higher outlet temperature of pipe 13 of the CPOx unit, which was also higher than the SOFC temperature. Therefore, the cooling requirements of the SOFCs were enhanced and, subsequently, a higher cathode air flow was needed.

The higher cathode flow resulted in a lower operating temperature in the afterburner (53 K) and lower temperature of the cathode inflow (pipe 23) of approximately 35 K.

The syngas fed at 298 K was heated in the fuel pre-heater to 100 K less than with the preliminary design conditions, even though higher heat flow was transmitted by the flue gas. As a consequence, the temperature of the flue gas leaving the system (pipe 33) was slightly reduced.

Finally, the system performance was substantially lower for the experimental conditions of 5%. A substantial amount of chemical energy converted in the CPOx and, consequently, higher syngas flow was required to produce equal electric power. Nonetheless, no significant deviations were found in terms of SOFC performance and temperatures and, therefore, the model showed to be reliable for comparison and analysis of the experimental work.

6.2. Comparison of System Simulation with Test Runs

In the test runs, only three steady-state operating points were achieved, at 70, 80 and 90 mA/cm^2, resulting in a part-load operating window of 37–47%. For lower current densities than 70 mA/cm^2, a steady state could not be reached because of the safety protection for the fuel pre-heater (T_{33} should not exceed 573 K), whereas for higher current densities than 90 mA/cm^2, no sufficient CPOx air could be provided to achieve the specified O/C ratio.

Table 4 compares the data measured in the test run at a current density of 90 mA/cm^2 with the values calculated in the system simulations. Good agreement between the results can be observed.

Figure 5 compares the SOFC performance (average cell voltage) and cathode air flow required to maintain an outlet cathode temperature of 1073 K for experimental and simulation results. A good match in the average cell voltage was seen, and thus equal power was produced. The cell voltage was also in accordance with values reported in many studies in syngas-fed-SOFCs [32,33]. Moreover, as equal fuel flow rate and composition were supplied, it can be concluded that the heat production in the subsystem was similar between test runs and the simulation model. The temperatures of the main streams in the BoP box are shown in Figures 6–8. There was reasonable agreement between the values measured in the real system during the test runs and those provided by the system simulations. In general, the maximum deviations were on the order of 10 K. The exception was the temperature of the syngas after the CPOx (T_{13}), where the deviation was approximately 20 K for a current density of 90 mA/cm^2.

These deviations were caused by various factors, such as simplifications and constraints imposed in the system simulation (non-heat-losses approach, equilibrium calculations for the CPOx component, and inaccuracies in the heat exchanger design), and the accuracy of measurement devices in the system, among others.

Table 4. Comparison of results for the 90 mA/cm^2 operating point.

Apparatus/Pipe	Parameter	Units	Simulation	Test Runs
11	Syngas flow	nL/min	75.6	75.7 ± 0.54
21	Cathode air flow	nL/min	631.8	633.2 ± 18
41	CPOx air flow	nL/min	15.3	15.3 ± 0.09
103	Cell voltage, V_{cell}	V	0.789	0.791 ± 0.0023
103	SOFC power, $\dot{W}_{SOFC}$	kW	1.632	1.631 ± 0.013
13	Temperature after CPOx, T_{13}	K	1049.9	1065.4 ± 5.5
23	Cathode inlet temp., T_{23}	K	996.7	996.5 ± 5.0
24	Temperature cathode outlet, T_{24}	K	1073	1073 ± 5.6
31	Temp. after afterburner, T_{31}	K	1126.1	1127.8 ± 6.0
33	Temp. exhaust flue gas, T_{33}	K	516.9	518.9 ± 1.8

6.3. System Efficiency

The system performance (electric efficiency) was evaluated at the three operating points obtained in the test runs. The thermodynamic calculations are similar to those described in [34]. Figure 9 shows the distribution (percentage) of the total input exergy (syngas flow) across the various types of exergy output by the system, namely electricity, exergy destruction, and exergy loss.

(a)

(b)

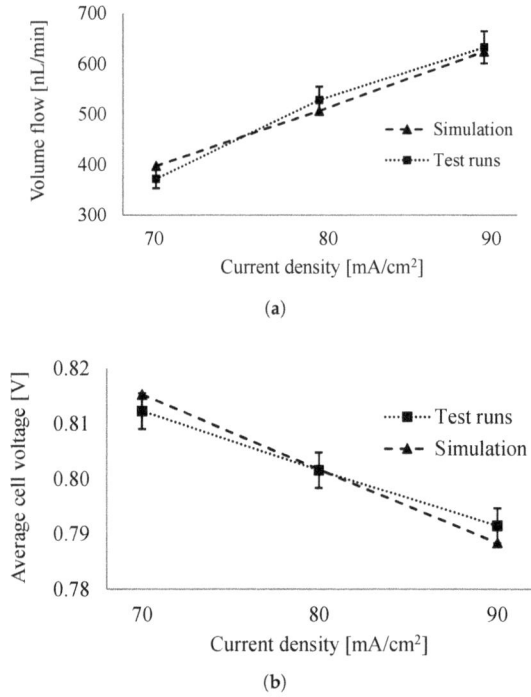

Figure 5. Comparison of parameters acquired in the test runs with simulated ones. (**a**) Cathode volume flow; (**b**) Average cell voltage (V_{cell}).

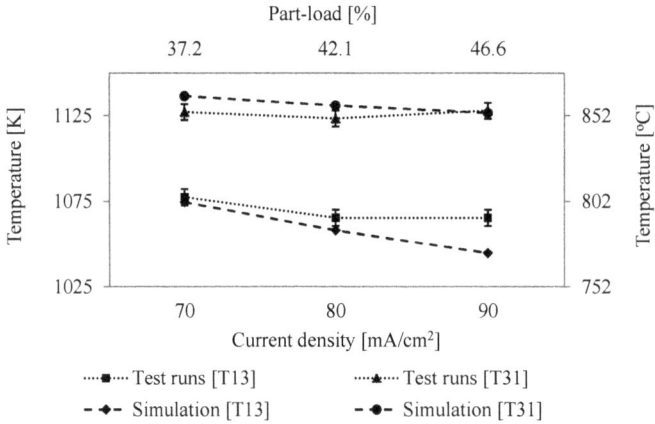

Figure 6. Comparison of simulated results with those acquired in the test runs. Syngas from the CPOx (T_{13}) Flue gas from afterburner (T_{31}) .

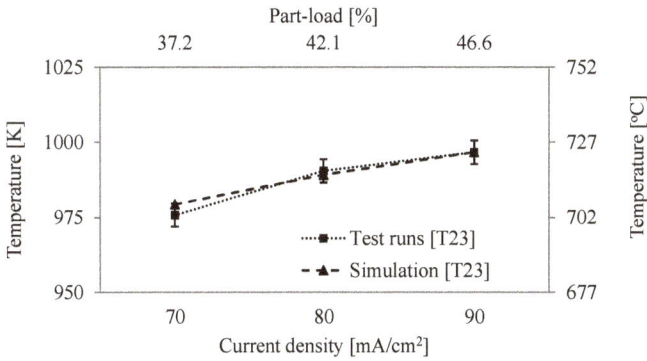

Figure 7. Comparison of simulated results with those acquired in the test runs. Cathode air from air heater (T_{23}).

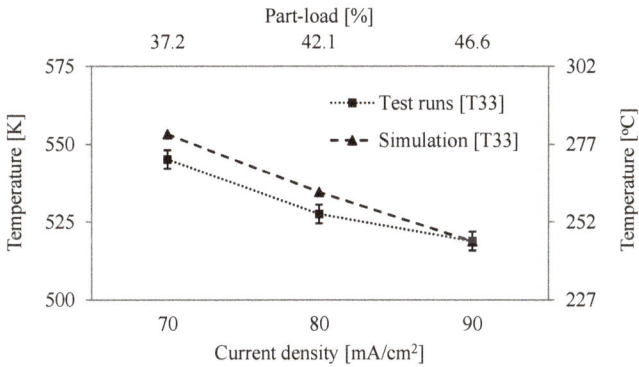

Figure 8. Comparison of simulated results with acquired ones in the test runs. Temperature of flue gas leaving the system (T_{33}).

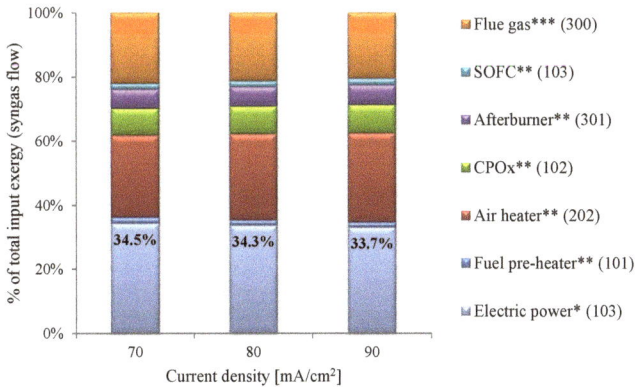

Figure 9. Exergy distribution into the various types of exergy produced in the system. * Net electric power; ** Exergy destruction per component; *** Exergy loss carried by the flue gas.

The air heater contributed to a 25–30% reduction in the efficiency of the system, making it the major source of irreversibility. The impact of this type of component on the system performance has been highlighted in many studies (e.g., [35,36]), and is caused by heat transfer at limited temperature as well as great temperature difference between hot and cold media. Flue gas losses appeared as the second contributor, carrying approximately 20% of the exergy. The high physical exergy carried by the stream due to high exhaust temperature and mass flow rate were the aspects contribution for such value. The afterburner and CPOx also caused a significant reduction in system performance of approximately 14% as a consequence of the irreversible nature of the oxidation process. The latter component also contributed to the great exergy contained in flue gas. Although it is not clearly demonstrated in Figure 9, a significant amount of the chemical exergy of the syngas was converted into heat exergy in the CPOx. The main consequence was that, for equal electric power production, a larger amount of syngas needed to be fed into the system, which subsequently reduced the system efficiency. Finally, the SOFC and fuel pre-heater were minor sources of irreversibility. Nevertheless, they collectively corresponded to approximately 6% in exergy destruction. This resulted in a modest system electrical efficiency of 33.7–34.5%.

Table 5. System performance of various stationary SOFC systems. Modified after [37].

System Manufacturer	Output Power (kW)	η (%)	Fuel Processing	Ref.
Bloom energy	250 AC	>53	-	[38]
Wärtsilä	24 DC	47	SR/AOGR	[39]
FZJ	20 DC	41	External SR	[37]
VTT(2010)	7 AC	43	SR/AOGR	[40]
VTT(2011)	8 AC	49	SR/AOGR	[41]
CFCL	1.5 AC	60	SR/WR	[42]
ENE-farm	0.7 AC	41–47	SR/WR	[43,44]
Hexis	1 AC	35	CPOx	[45]
SOFCpower	1 AC	32	CPOx	[46]
IKTS	1.26 AC	39	CPOx	[47]

SR: steam reforming; AOGR: anode off-gas recycling; WR: water recycling.

The system performance was comparable with other systems employing a CPOx unit, as can be seen in Table 5. It can also be observed that all SOFC systems employing steam reforming as the fuel processing had higher system efficiency. This aspect is associated with the steam reforming process, which increases the chemical energy of the fuel by using available heat exergy in the system.

Nevertheless, it is should also be highlighted that the system performance is affected by various aspects such as the fuel composition, selected load operation as well as the system design choices. In the next section, some considerations for improving the design are given.

7. Future Considerations for Improving the Design

As mentioned previously, high-temperature flue gas is exhausted as a consequence of the use of a CPOx unit for suppressing carbon formation. The replacement of this unit with a steam reformer would potentially enhance the electrical efficiency of the system. The produced steam would be mixed with the supply syngas. To enable this, a heat recovery steam generator should be installed and the heat exchangers should be rearranged. In this new arrangement, the flue gas from the afterburner would first be used as a heat source in the fuel pre-heater in order to increase the anode flow temperature to the specified value at inlet of the SOFC anode, then to the air heater, and finally to the heat recovery steam generator.

8. Summary

In this study, we investigated the design, construction, and testing of a gasifier-specific SOFC system. This paper makes the following contributions to the study of advanced SOFC technologies:

1. System development: this is the first study to describe the design, development, and testing of an SOFC system to be integrated with a gasifier. The gasifier considered in this study is a plasma reactor with the capacity to process 8.84 kW of human waste (before pre-drying) [22,48]. The SOFC system was designed based on discussions between TU Delft and Sunfire GmbH.
2. Calculated results exhibited good agreement with experimentally recorded values under different operating conditions. This clearly demonstrates the advantage of a rigorous thermodynamic model of new fuel cell power systems when they are being designed and built.
3. The validated model clearly indicates where the thermodynamic losses are occurring and provides indications on how to minimize these losses in such a system, resulting in improved designs in the future.
4. System efficiencies of 33.7–34.5% were estimated. The CPOx unit and heat exchangers, especially the air heater, were identified as the major contributors to reductions in efficiency.

Author Contributions: The experimental work was carried out by A.S., M.R., L.F. and E.O.P. The SOFC simulation was supported by M.L. O.P. and J.B. jointly with M.L., E.O.P. and P.V.A. decided the conceptual design SOFC system. J.B. and O.P. led and supervised the construction of the SOFC system. T.W., P.V.A. and J.B. contributed in the results discussion. A.F. led the experimental work, developed the SOFC models, and prepared the first draft and final version of the paper. The paper was corrected and reviewed by J.B. and P.V.A.

Funding: This research was founded by The Bill & Melinda Gates foundation to build the SOFC system.

Acknowledgments: The Bill & Melinda Gates Fundation is thanked for the financial support. Also, the Portuguese Fundação para a Ciência e Tecnologia (FCT), through the Grant—SFRH/BD/77042/2011, and the Ecuador Secretaría de Educación Superior, Ciencia, Tecnología e Innovación (SENESCY) are thanked for the partial financial support to the first authors.

Conflicts of Interest: The authors declare no conflict of interest.

References

1. Eurostat. The Key to European Statistics. 2017. Available online: http://ec.europa.eu/eurostat/web/energy/data/database (accessed on 22 April 2017).
2. Guercio, A.; Bini, R. 15—Biomass-fired Organic Rankine Cycle combined heat and power systems. In *Organic Rankine Cycle (ORC) Power Systems*; Macchi, E., Astolfi, M., Eds.; Woodhead Publishing: Sawston, UK, 2017; pp. 527–567. [CrossRef]
3. Hurskainen, M.; Vainikka, P. 7—Technology options for large-scale solid-fuel combustion. In *Fuel Flexible Energy Generation*; Oakey, J., Ed.; Woodhead Publishing: Boston, MA, USA, 2016; pp. 177–199. [CrossRef]
4. Chacartegui, R.; Sanchez, D.; de Escalona, J.M.; Monje, B.; Sanchez, T. On the effects of running existing combined cycle power plants on syngas fuel. *Fuel Process. Technol.* **2012**, *103*, 97–109. [CrossRef]
5. Asadullah, M. Barriers of commercial power generation using biomass gasification gas: A review. *Renew. Sustain. Energy Rev.* **2014**, *29*, 201–215. [CrossRef]
6. Santhanam, S.; Schilt, C.; Turker, B.; Woudstra, T.; Aravind, P. Thermodynamic modeling and evaluation of high efficiency heat pipe integrated biomass Gasifier-Solid Oxide Fuel Cells-Gas Turbine systems. *Energy* **2016**, *109*, 751–764. [CrossRef]
7. Bellomare, F.; Rokni, M. Integration of a municipal solid waste gasification plant with solid oxide fuel cell and gas turbine. *Renew. Energy* **2013**, *55*, 490–500. [CrossRef]
8. Singh, J. Management of the agricultural biomass on decentralized basis for producing sustainable power in India. *J. Clean. Prod.* **2017**, *142*, 3985–4000. [CrossRef]
9. Bae, C.; Kim, J. Alternative fuels for internal combustion engines. *Proc. Combust. Inst.* **2017**, *36*, 3389–3413. [CrossRef]
10. Hagos, F.Y.; Aziz, A.R.A.; Sulaiman, S.A. Study of syngas combustion parameters effect on internal combustion engine. *Asian J. Sci. Res.* **2013**, *6*, 187–196. [CrossRef]

11. Barelli, L.; Bidini, G.; Ottaviano, A. Part load operation of SOFC/GT hybrid systems: Stationary analysis. *Int. J. Hydrogen Energy* **2012**, *37*, 16140–16150. [CrossRef]

12. Andersson, D.; Aberg, E.; Yuan, J.; Sunden, B.; Eborn, J. Dynamic Modeling of a Solid Oxide Fuel Cell System in Modelica. In Proceedings of the ASME 2010 8th International Conference on Fuel Cell Science, Engineering and Technology, Brooklyn, NY, USA, 14–16 June 2010; American Society of Mechanical Engineers: New York, NY, USA, 2010; pp. 65–72.

13. Rokni, M. Performance comparison on repowering of a steam power plant with gas turbines and solid oxide fuel cells. *Energies* **2016**, *9*, 399. [CrossRef]

14. Ugartemendia, J.; Ostolaza, J.X.; Zubia, I. Operating point optimization of a hydrogen fueled hybrid solid oxide fuel cell-steam turbine (SOFC-ST) plant. *Energies* **2013**, *6*, 5046–5068. [CrossRef]

15. Chung, T.D.; Hong, W.T.; Chyou, Y.P.; Yu, D.D.; Lin, K.F.; Lee, C.H. Efficiency analyses of solid oxide fuel cell power plant systems. *Appl. Therm. Eng.* **2008**, *28*, 933–941. [CrossRef]

16. Chitsaz, A.; Mahmoudi, S.M.S.; Rosen, M.A. Greenhouse gas emission and exergy analyses of an integrated trigeneration system driven by a solid oxide fuel cell. *Appl. Therm. Eng.* **2015**, *86*, 81–90. [CrossRef]

17. Stamatis, A.; Vinni, C.; Bakalis, D.; Tzorbatzoglou, F.; Tsiakaras, P. Exergy analysis of an intermediate temperature solid oxide fuel cell-gas turbine hybrid system fed with ethanol. *Energies* **2012**, *5*, 4268–4287. [CrossRef]

18. Xu, H.; Dang, Z.; Bai, B.F. Analysis of a 1 kW residential combined heating and power system based on solid oxide fuel cell. *Appl. Therm. Eng.* **2013**, *50*, 1101–1110. [CrossRef]

19. Somekawa, T.; Nakamura, K.; Kushi, T.; Kume, T.; Fujita, K.; Yakabe, H. Examination of a high-efficiency solid oxide fuel cell system that reuses exhaust gas. *Appl. Therm. Eng.* **2017**, *114*, 1387–1392. [CrossRef]

20. Huerta, G.V.; Álvarez Jordán, J.; Dragon, M.; Leites, K.; Kabelac, S. Exergy analysis of the diesel pre-reforming solid oxide fuel cell system with anode off-gas recycling in the SchIBZ project. Part I: Modeling and validation. *Int. J. Hydrogen Energy* **2018**. [CrossRef]

21. D'Andrea, G.; Gandiglio, M.; Lanzini, A.; Santarelli, M. Dynamic model with experimental validation of a biogas-fed SOFC plant. *Energy Convers. Manag.* **2017**, *135*, 21–34. [CrossRef]

22. Liu, M.; Woudstra, T.; Promes, E.; Restrepo, S.; Aravind, P. System development and self-sustainability analysis for upgrading human waste to power. *Energy* **2014**, *68*, 377–384. [CrossRef]

23. Din, Z.U.; Zainal, Z. The fate of SOFC anodes under biomass producer gas contaminants. *Renew. Sustain. Energy Rev.* **2016**. [CrossRef]

24. Cassidy, M.; Connor, P.; Irvine, J.; Savaniu, C. 5—Anodes. In *High-Temperature Solid Oxide Fuel Cells for the 21st Century (Second Edition)*, 2nd ed.; Kendall, K., Kendall, M., Eds.; Academic Press: Boston, MA, USA, 2016; pp. 133–160. [CrossRef]

25. Kuramoto, K.; Hosokai, S.; Matsuoka, K.; Ishiyama, T.; Kishimoto, H.; Yamaji, K. Degradation behaviors of SOFC due to chemical interaction between Ni-YSZ anode and trace gaseous impurities in coal syngas. *Fuel Process. Technol.* **2017**, *160*, 8–18. [CrossRef]

26. Liu, M.; van der Kleij, A.; Verkooijen, A.; Aravind, P. An experimental study of the interaction between tar and SOFCs with Ni/GDC anodes. *Appl. Energy* **2013**, *108*, 149–157. [CrossRef]

27. Zhang, L.; Jiang, S.P.; He, H.Q.; Chen, X.; Ma, J.; Song, X.C. A comparative study of H₂S poisoning on electrode behavior of Ni/YSZ and Ni/GDC anodes of solid oxide fuel cells. *Int. J. Hydrogen Energy* **2010**, *35*, 12359–12368. [CrossRef]

28. GmbH, S. Power Core—Efficient Electricity Generator. 2017. Available online: http://www.sunfire.de/en/products-technology/power-core (accessed on 6 October 2017).

29. Delft, T. Cycle-Tempo 5.0. 2015. Available online: http://www.asimptote.nl/software/cycle-tempo/ (accessed on 22 September 2017).

30. De Groot, A. Advanced Energy Analysis of High Temperature Fuel Cell Systems. Ph.D. Thesis, Delft University of Technology, Delft, The Netherlands, 2004.

31. Miedema, J.A. CYCLE: A general computer code for thermodynamic cycle computations. Studies of cogeneration in district heating systems. In *NASA STI/Recon Technical Report N*; NASA: Washington, DC, USA, 1981; Volume 82.

32. Suwanwarangkul, R.; Croiset, E.; Entchev, E.; Charojrochkul, S.; Pritzker, M.; Fowler, M.; Douglas, P.; Chewathanakup, S.; Mahaudom, H. Experimental and modeling study of solid oxide fuel cell operating with syngas fuel. *J. Power Sources* **2006**, *161*, 308–322. [CrossRef]

33. Baldinelli, A.; Cinti, G.; Desideri, U.; Fantozzi, F. Biomass integrated gasifier-fuel cells: Experimental investigation on wood syngas tars impact on NiYSZ-anode Solid Oxide Fuel Cells. *Energy Convers. Manag.* **2016**, *128*, 361–370. [CrossRef]

34. Fernandes, A.; Woudstra, T.; van Wijk, A.; Verhoef, L.; Aravind, P. Fuel cell electric vehicle as a power plant and SOFC as a natural gas reformer: An exergy analysis of different system designs. *Appl. Energy* **2016**, *173*, 13–28. [CrossRef]

35. Lee, Y.D.; Ahn, K.Y.; Morosuk, T.; Tsatsaronis, G. Exergetic and exergoeconomic evaluation of an SOFC-Engine hybrid power generation system. *Energy* **2018**, *145*, 810–822. [CrossRef]

36. Hosseinpour, J.; Chitsaz, A.; Eisavi, B.; Yari, M. Investigation on performance of an integrated SOFC-Goswami system using wood gasification. *Energy* **2018**, *148*, 614–628. [CrossRef]

37. Peters, R.; Blum, L.; Deja, R.; Hoven, I.; Tiedemann, W.; Küpper, S.; Stolten, D. Operation experience with a 20 kW SOFC system. *Fuel Cells* **2014**, *14*, 489–499. [CrossRef]

38. Bloom Energy Corporation. Bloom Energy Server ES5—250 kW. Available online: https://www.bloomenergy.com/ (accessed on 8 July 2018).

39. Noponen, M.; Hottinen, T.; Sandstrom, C. WFC20 Biogas Unit Operation. In Proceedings of the 9th European Solid Oxide Fuel Cell Forum, Lucerne, Switzerland, 29 June–2 July 2010; pp. 2–90.

40. Halinen, M.; Rautanen, M.; Saarinen, J.; Pennanen, J.; Pohjoranta, A.; Kiviaho, J.; Pastula, M.; Nuttall, B.; Rankin, C.; Borglum, B. Performance of a 10 kW SOFC demonstration unit. *ECS Trans.* **2011**, *35*, 113–120.

41. Halinen, M.; Pennanen, J.; Himanen, O.; Kiviaho, J.; Silvennoinen, P.; Backman, J.; Salminen, P.; Pohjoisranta, A. Sofc Power 2007–2012 Project—The Finnish SOFC System Research Flagship. *J. Fuel Cell Technol.* **2013**, *13*, 32–39.

42. Payne, R.; Love, J.; Kah, M. Generating electricity at 60% electrical efficiency from 1–2 kWe SOFC products. *ECS Trans.* **2009**, *25*, 231–239.

43. Suzuki, M.; Takuwa, Y.; Inoue, S.; Higaki, K. Durability verification of residential SOFC CHP system. *ECS Trans.* **2013**, *57*, 309–314. [CrossRef]

44. Yoshida, H.; Seyama, T.; Sobue, T.; Yamashita, S. Development of residential SOFC CHP system with flatten tubular segmented-in-series cells stack. *ECS Trans.* **2011**, *35*, 97–103.

45. Mai, A.; Iwanschitz, J.; Schuler, A.; Denzler, R.; Nerlich, V.; Schuler, A. Hexis and the SOFC System Galileo 1000 N—Past, Present, Future. In Proceedings of the 11th European SOFC SOE Forum A, Lucerne, Switzerland, 1–4 July 2014; pp. 3–11.

46. Buchli, O.; Bertoldi, M.; Modena, S.; Ravagni, A. Development and manufacturing of SOFC based products at SOFCpower SpA. *ECS Trans.* **2013**, *57*, 81–88. [CrossRef]

47. Pfeifer, T.; Chakradeo, A.; Ahire, N.; Baade, J.; Barthel, M.; Dosch, C.; Näke, R.; Hartmann, M. Development of a SOFC Power Generator for the Indian Market. *Fuel Cells* **2017**, *17*, 550–561. [CrossRef]

48. Sturm, G.S.J.; Muñoz, A.N.; Aravind, P.V.; Stefanidis, G.D. Microwave-Driven Plasma Gasification for Biomass Waste Treatment at Miniature Scale. *IEEE Trans. Plasma Sci.* **2016**, *44*, 670–678. [CrossRef]

energies

MDPI

Article

Prospecting for Oleaginous and Robust *Chlorella* spp. for Coal-Fired Flue-Gas-Mediated Biodiesel Production

Bohwa Kim [1,†], Ramasamy Praveenkumar [2,†], Eunji Choi [3], Kyubock Lee [4], Sang Goo Jeon [3] and You-Kwan Oh [5,*]

[1] Advanced Biomass R&D Center, Korea Advanced Institute of Science and Technology (KAIST), Daejeon 34141, Korea; kbh87@kaist.ac.kr
[2] Aquatic Biology Laboratory, and Renewable Materials and Nanotechnology Group, KU Leuven, Campus Kulak, 8500 Kortrijk, Belgium; praveen.ramasamy@kuleuven.be
[3] Biomass and Waste Energy Laboratory, Korea Institute of Energy Research (KIER), Daejeon 34129, Korea; eunsie@hanmail.net (E.C.); sgjeon@kier.re.kr (S.G.J.)
[4] Graduate School of Energy Science and Technology, Chungnam National University (CNU), Daejeon 34134, Korea; kyubock.lee@cnu.ac.kr
[5] School of Chemical and Biomolecular Engineering, Pusan National University (PNU), Busan 46241, Korea
* Correspondence: youkwan@pusan.ac.kr; Tel.: +82-51-510-2395; Fax: +82-51-512-8563
† These authors contributed equally to this work.

Received: 2 July 2018; Accepted: 1 August 2018; Published: 3 August 2018

Abstract: Prospecting for robust and high-productivity strains is a strategically important step in the microalgal biodiesel process. In this study, 30 local strains of *Chlorella* were evaluated in photobioreactors for biodiesel production using coal-fired flue-gas. Three strains (M082, M134, and KR-1) were sequentially selected based on cell growth, lipid content, and fatty acid composition under autotrophic and mixotrophic conditions. Under autotrophic conditions, M082 and M134 showed comparable lipid contents (*ca.* 230 mg FAME [fatty acid methyl esters derived from microalgal lipids]/g cell) and productivities (*ca.* 40 mg FAME/L·d) versus a reference strain (KR-1) outdoors with actual flue-gas (CO_2, 13%). Interestingly, under mixotrophic conditions, M082 demonstrated, along with maximal lipid content (397 mg FAME/g cell), good tolerance to high temperature (40 °C). Furthermore, the fatty acid methyl esters met important international standards under all of the tested culture conditions. Thus, it was concluded that M082 can be a feedstock of choice for coal-fired, flue-gas-mediated biodiesel production.

Keywords: *Chlorella*; coal-fired flue-gas; screening; biodiesel property; mixotrophic cultivation

1. Introduction

Microalgae have attracted much global attention for their potential as biodiesel feedstock [1,2]. This interest reflects not only microalgae's higher photosynthetic efficiency and lipid yield compared with conventional oil crops but also their ability to mitigate atmospheric CO_2 and to grow in non-arable land and a variety of wastewaters [3–7]. Moreover, microalgae are easily isolated: they are ubiquitously present in nature and represent a vast diversity of (>50,000) species existing over a wide range of environment conditions [1,8].

Despite its several advantages, microalgae-based biodiesel production, when considered on a commercial scale, remains challenging. First, for competitive and sustainable biodiesel production, the cost of the culture system needs to be significantly reduced, and lipid productivity needs to be substantially improved [1]. Industrial exhaust flue-gases are an inexpensive and rich source of CO_2 (~14% *v/v*), the utilization of which for microalgal biomass production would be a fiscally and environmentally better option than compressed CO_2 [9,10]. However, high concentrations of CO_2 and the presence of inhibitory compounds such as NOx, Sox, and CO in flue-gases can significantly suppress the metabolic activities of microalgae [6,10]. Hence, for the successful application of flue-gases to microalgal biomass production, the selection of a flue-gas-tolerant and high-lipid-yield strain is warranted.

On the other hand, several microalgae, if not all, can grow mixotrophically by utilizing organic carbons along with light, thereby yielding high lipid productivity. Liang et al. reported that a mixotrophic cultivation of *Chlorella vulgaris* with 1–2% glucose yielded a 10-to-20-fold increase in the biomass (254 mg/L·d) and lipid (54 mg/L·d) productivities relative to an autotrophic cultivation [11]. Exploiting such a mixotrophic cultivation system with a suitable microalga would open up opportunities to utilize waste organics such as carbon sources and, at the same time, help to bring down cultivation costs.

Amongst the several oleaginous microalgae, *Chlorella* is one of the most widely studied genus for its potential to grow in outdoors at high cell densities and accumulate high levels of intracellular triacylglycerol (TAG), a desirable biodiesel feedstock [11–15]. However, the literature indicates that the growth rates and lipid contents of *Chlorella* are both species-specific and possibly significantly variant according to culture conditions. Recently, Sun et al. reported the differences in biomass productivity (0.41–0.58 g/L·d) and lipid content (36–49 wt. %) among 9 *Chlorella* spp. cultured with artificial CO_2 gas under autotrophic conditions [15].

What also must be noted is the fact that many highly productive microalgae selected under laboratory conditions fail to withstand the actual, fluctuating environmental conditions and contamination outdoors [7,16,17]. While accounting for such differential behavior among the *Chlorella* spp., it is imperative to select a native isolate that is appropriate not only in having a high lipid content but also in its capacity to adapt to environmental fluctuations under a suitable cultivation system.

The aim of the present work was to screen various native *Chlorella* spp. to identify a robust, fast-growing, and high-lipid-accumulating strain for high-quality biodiesel production under autotrophic/mixotrophic systems mediated by coal-fired flue-gas (CO_2, 13%). The employed screening strategies included comparisons of growth rate, lipid content, high-temperature tolerance, and fatty acids quality with respect to key biodiesel properties. The robustness of each of the selected high-productivity stains was finally evaluated under outdoor culture conditions mediated by coal-fired flue-gas.

2. Results and Discussion

2.1. Preliminary Screening Based on Growth Rate

In a preliminary screening, 30 local strains of *Chlorella* were evaluated for their growth rates in bubble-column photobioreactors (b-PBRs) with 10% CO_2 (v/v) under autotrophic and mixotrophic conditions. The mixotrophic cultures were supplemented with 5 g glucose/L. After 50 h incubation, and the growths (OD_{660}) of the strains were compared (Figure 1). Most of the strains showed relatively higher growths under the mixotrophic than under the autotrophic conditions. This phenomenon is consistent with other studies that have reported organic carbon's induction of growth promotion in *Chlorella* [11,18,19]. The 30 stains, as based on their growth characteristics, were divided into the following three groups.

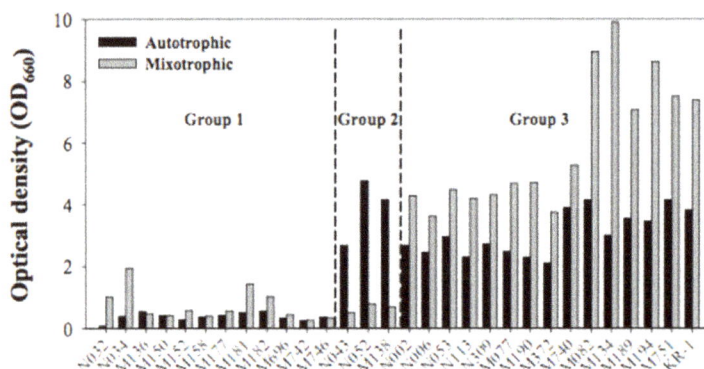

Figure 1. Growth of *Chlorella* spp. (30 strains) under autotrophic and mixotrophic conditions. The optical density was measured at 660 nm at the end of 50 h cultivation in b-PBRs. The mixotrophic cultures were supplemented with 5 g glucose/L.

Group 1: Twelve strains of the tested *Chlorella* showed a very low growth rate under both autotrophic and mixotrophic conditions. This delayed growth was expected, based on their demonstrated intolerance to the high concentrations of CO_2 (10% v/v in air) applied in this study. Many microalgae reportedly tolerate CO_2 concentrations up to 5% (v/v in air), whereas concentrations above this can be harmful, not to mention inhibitive of microalgal growth [9]. Hence, these strains might not be suitable for flue-gas-mediated cultivation, in which CO_2 concentrations often exceed 10%.

Group 2: Interestingly, this group of 3 strains (N043, N052, and M138) exhibited decreased growth under mixotrophic, compared with autotrophic, conditions. Lower growth rate under mixotrophy could possibly be due to the inhibitory effects of glucose on the microalgal cells. Such effects are species-dependent; high concentrations of glucose have been reported to substantially inhibit *Chlorella* and Nannochloropsis growths [11,18]. The three group 2 strains, seemingly sensitive to organic carbon supplementation, were not further studied, considering their poor future prospects for utilization in lipid-productivity-improvement using waste organics.

Group 3: For these 15 strains of *Chlorella*, the mixotrophic cultures resulted in higher growth rates than the autotrophic condition. That is to say, these 15 strains utilized glucose efficiently to achieve impressive growth. Among them, six (M082, M134, M189, M194, M751, and KR-1) exhibited the highest growth under mixotrophic conditions and also appeared to tolerate a high concentration of CO_2 (10% v/v in air). Hence, these six strains, thus identified as fast-growing, were held over for further investigation.

2.2. Secondary Screening Based on Lipid Content and Composition

Growth rate and lipid content are the most important factors in assessing prospective microalga for biodiesel production [1,20]. Additionally, the composition of the fatty acids produced by the microalga determines the quality of the biodiesel produced [21,22]. Hence, the six selected fast-growing strains (M082, M134, M189, M194, M751, and KR-1) were allowed to grow for a longer period of time (until 96 h), and then they were further evaluated for their lipid contents (in this study, expressed as FAME, i.e., amount of fatty acid methyl esters derived from microalgal lipids) and their compositions under autotrophic and mixotrophic conditions (Figure 2). After 96 h of cultivation, the microalgal cells would have already experienced nitrogen depletion [6]. Indeed, despite their fast-growing nature, strains M082 and M134 accumulated high levels of intracellular lipids: 283 and 298 mg FAME/g DCW (dry cell weight), respectively, under autotrophic conditions. Moreover, under mixotrophic conditions, these strains accumulated 397 and 310 mg FAME/g DCW, respectively, which is slightly less than the corresponding content for control strain KR-1 (411.3 mg FAME/g DCW). Notwithstanding, owing to their higher biomass productions, the small differences in the lipid contents of M082 and M134 could be very well compensated by their higher biomass productivities (Figure S1). As for *Chlorella* sp. KR-1, it has been well studied in the past for its fast-growth and high-lipid-accumulation potential under mixotrophic conditions [6,23]. Thus, M082 and M134 strains, based on their indoor performances, could be suggested to be potentially comparable with KR-1.

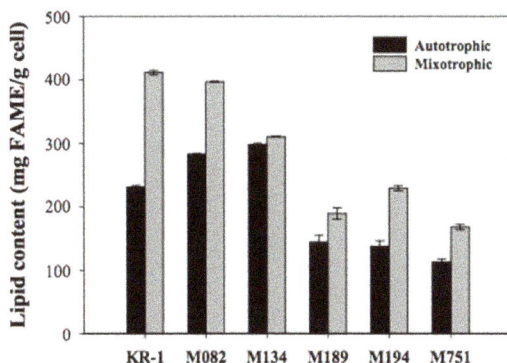

Figure 2. Lipid contents of six fast-growing *Chlorella* strains under autotrophic and mixotrophic conditions. The lipid contents of the cultures were estimated by FAME analysis after 96 h cultivation. The mixotrophic cultures were supplemented with 5 g glucose/L.

Further, the FAME compositions of the biodiesel produced from lipids by the six fast-growing strains also were examined (Table 1). The data were similar under both autotrophic and mixotrophic conditions: the C16 and C18 species, the common fatty acids in vegetable oils, were the main constituents [15,22,24]. Also, at different distribution (%) levels, major fatty acid methyl esters such as palmitate (C16:0), stearate (C18:0), oleate (C18:1n9C), linoleate (C18:2n6C), and linolenate (C18:3n3) were observed (Figure S2). Among these, palmitate, oleate, and linoleate accounted for more than 70% of the total fatty acids identified. The balance between these fatty acids determines the key qualities of biodiesel, such as oxidative stability, cold flow point, lubricity, viscosity, cetane number, iodine number, heat of combustion, and NOx emission [15,22,24]. The details are discussed in a later section below (see Section 2.4.—estimation of biodiesel properties from fatty acids profiles). In consideration of the high lipid contents and FAME compositions of the 3 *Chlorella* strains M082, M134, and KR-1, these were chosen for further investigation in an outdoor culture mediated by coal-fired flue-gas.

2.3. Outdoor Flue-Gas Cultivation and High-Temperature Tolerance Experimentation

The 3 high-productivity strains (M082, M134, and KR-1) were cultivated outdoors in 1 L b-PBRs for 160 h with a supply of actual flue-gas collected from a 2 MW demonstration-scale coal-burning power plant [6]. The changes in the light intensity and temperature were continuously monitored; the maximum sunlight was 1000 μmol/m^2·s with an average daylight of 500 μmol/m^2·s; meanwhile, the day temperature ranged between 30 and 38 °C and came down to 19 °C during the night (Figure 3a). M082 and M134 were tolerant to flue-gas and grew well in the outdoor conditions: M134, like the reference strain KR-1, readily adapted to the conditions, starting to grow in quick time. Irrespective of the sharp lag phase, M082, very similarly to the other two strains, grew rapidly to reach the final OD$_{660}$ (Figure 3b). At the end of 160 h of cultivation, the lipid contents of M082 and M134 were estimated to be 228 and 235 mg FAME/g cell, for FAME productivities of 39 and 42 mg/L·d, respectively (Figure 3c). These values were comparable with the lipid productivity of the reference strain KR-1 (41 mg FAME/L·d). Under the optimized culture conditions, the FAME productivity of KR-1 was shown to have increased as high as 155 mg/L·d [6], which highlights the likelihood of improving the productivities of the selected strains M082 and M134. Thus, despite growing outdoors, with a supply of actual flue-gas and under fluctuating environmental conditions, the productivities of these two strains did not deteriorate; moreover, the FAME compositions remained less altered when compared with the indoor cultivations (Figure S3).

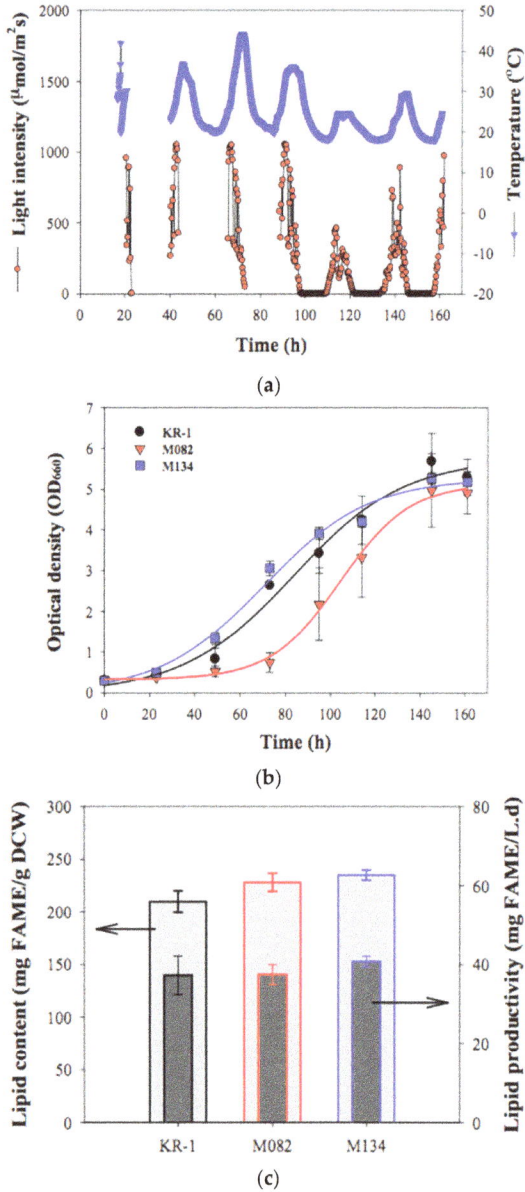

(a)

(b)

(c)

Figure 3. (a) Time-course changes in light intensity and temperature during cultivation; (b) growth; and (c) lipid content of three high-productivity *Chlorella* strains in outdoor b-PBR culture supplied with coal-fired flue-gas.

Table 1. FAME compositions of lipid extracted from six fast-growing *Chlorella* strains under autotrophic and mixotrophic conditions.

| Fatty Acid Methyl Ester | | Ratio (%) of Each FAME to Total FAME (Mean ± Standard Deviation) | | | | | | | | | | | |
| | | M082 | | M134 | | M189 | | M194 | | M751 | | KR-1 | |
		Auto	Mixo	Auto	Mixo	Auto	Mixo	Auto	Mixo	Auto	Mixo	Auto	Mixo
Caprate	C10:0	0.05 ± 0.07	0.08 ± 0.00	nd	0.1 ± 0.01	nd	nd	nd	nd	nd	nd	nd	0.19 ± 0.01
Tridecanoate	C13:0	0.26 ± 0.01	nd	0.26 ± 0.01	nd	0.34 ± 0.03	0.18 ± 0.00	0.37 ± 0.01	0.14 ± 0.01	0.49 ± 0.06	0.32 ± 0.00	0.23 ± 0.03	0.11 ± 0.01
Myristate	C14:0	nd	0.94 ± 0.01	nd	0.57 ± 0.00	0.61 ± 0.03	0.59 ± 0.00	0.64 ± 0.03	0.48 ± 0.00	0.36 ± 0.02	0.51 ± 0.51	0.23 ± 0.00	0.22 ± 0.00
Pentadecanoate	C15:0	0.12 ± 0.00	0.19 ± 0.01	0.1 ± 0.00	0.22 ± 0.00	nd	0.15 ± 0.00	0.21 ± 0.01	0.14 ± 0.00	nd	0.16 ± 0.00	nd	0.13 ± 0.00
Palmitate	C16:0	23 ± 0.08	26.78 ± 0.06	25.99 ± 0.05	26.79 ± 0.06	22.36 ± 0.99	24.55 ± 0.37	22.21 ± 1.34	24.6 ± 0.08	20.96 ± 0.16	23.91 ± 0.52	24.49 ± 0.30	28.55 ± 0.02
Palmitoleate	C16:1	0.49 ± 0.01	0.29 ± 0.00	0.55 ± 0.01	0.37 ± 0.00	0.35 ± 0.01	0.4 ± 0.01	0.31 ± 0.00	0.36 ± 0.00	0.14 ± 0.20	0.3 ± 0.02	0.36 ± 0.01	0.32 ± 0.00
Stearate	C18:0	3.72 ± 0.01	6.26 ± 0.01	4.32 ± 0.01	5.71 ± 0.02	0.78 ± 0.00	1.01 ± 0.00	0.72 ± 0.03	1.52 ± 0.35	0.5 ± 0.04	0.97 ± 0.00	4.57 ± 0.02	4 ± 0.02
Oleate	C18:1n9c	21.45 ± 0.02	21.8 ± 0.05	20.76 ± 0.03	19.25 ± 0.03	11.54 ± 0.03	16.06 ± 0.10	10.07 ± 0.22	22.02 ± 0.03	7.28 ± 0.01	15.26 ± 0.05	19.07 ± 0.10	18.7 ± 0.02
Linoleate	C18:2n6c	24.64 ± 0.04	25.42 ± 0.17	23.66 ± 0.07	27.47 ± 0.14	20.61 ± 0.55	21.43 ± 0.18	20.2 ± 0.69	21.6 ± 0.02	18.3 ± 0.21	19.43 ± 0.25	25.6 ± 0.30	24.08 ± 0.03
γ-Linolenate	C18:3n6	0.34 ± 0.00	0.31 ± 0.00	0.32 ± 0.00	0.35 ± 0.01	0.4 ± 0.03	0.35 ± 0.03	0.43 ± 0.01	0.33 ± 0.01	0.4 ± 0.02	0.38 ± 0.00	0.35 ± 0.01	0.29 ± 0.01
Linolenate	C18:3n3	6.69 ± 0.01	4.22 ± 0.06	6.22 ± 0.01	4.43 ± 0.01	12.87 ± 0.38	12.12 ± 0.16	13.09 ± 0.56	9.89 ± 0.01	14.52 ± 0.19	12.18 ± 0.21	6.5 ± 0.09	7.49 ± 0.02
Arachidate	C20:0	0.14 ± 0.19	0.35 ± 0.00	0.25 ± 0.01	0.15 ± 0.01	nd	nd	nd	nd	nd	nd	0.32 ± 0.03	0.26 ± 0.01
cis-11-icosenoate	C20:1	0.12 ± 0.01	0.05 ± 0.07	nd	nd	nd	nd	nd	nd	nd	nd	nd	nd
Behenate	C22:0	nd	0.19 ± 0.00	nd	0.05 ± 0.07	nd	nd	nd	nd	nd	nd	nd	0.1 ± 0.00
Lignocerate	C24:0	nd	0.15 ± 0.00	nd	0.15 ± 0.05	nd	nd	nd	nd	nd	nd	nd	nd
Saturated		27.41 ± 0.19	34.95 ± 0.07	30.91 ± 0.06	33.75 ± 0.16	24.1 ± 1.05	26.49 ± 0.36	24.14 ± 1.40	26.62 ± 0.05	22.31 ± 0.09	25.87 ± 0.53	29.85 ± 0.32	33.55 ± 0.04
Monounsaturated		22.08 ± 0.21	22.14 ± 0.02	21.31 ± 0.03	19.63 ± 0.03	11.89 ± 0.03	16.47 ± 0.11	10.38 ± 0.22	22.38 ± 0.04	7.42 ± 0.22	15.55 ± 0.07	19.43 ± 0.09	19.02 ± 0.02
Polyunsaturated		31.66 ± 0.05	29.95 ± 0.22	30.2 ± 0.07	32.25 ± 0.14	33.89 ± 0.90	33.9 ± 0.32	33.73 ± 1.26	31.82 ± 0.01	33.22 ± 0.39	31.99 ± 0.46	32.45 ± 0.40	31.85 ± 0.03
Others		18.85 ± 0.04	12.96 ± 0.13	17.58 ± 0.02	14.37 ± 0.01	30.12 ± 0.19	23.15 ± 0.15	31.75 ± 0.36	19.18 ± 0.09	37.05 ± 0.51	26.59 ± 0.13	18.26 ± 0.01	15.57 ± 0.09
Total		100	100	100	100	100	100	100	100	100	100	100	100

Auto: autotrophic condition; Mixo: mixotrophic condition; nd: not detected.

The three strains were further tested for their tolerances to high temperatures (35–40 °C) under indoor culture condition. Selection of a high-temperature-tolerant strain carries an added advantage for practical application of microalgal culture to higher-temperature flue-gas from emission sites, especially in summer [23]. Although *Chlorella* sp. KR-1 has been extensively studied due to its high biomass/lipid productivity and flue-gas tolerance [2,5,6,23], its lowered metabolic activities, especially at 40 °C, should be noticed. Interestingly, *Chlorella* sp. M082 demonstrated better tolerance to temperatures up to 40 °C than those of M134 and the reference strain KR-1 (Figure S4). The results indicated, thereby, that *Chlorella* sp. M082, because it is endowed with not only a high temperature/flue-gas tolerance but also high lipid productivity, might be a highly robust strain.

2.4. Theoretical Estimation of Biodiesel Properties from Fatty Acid Profiles

Based on fatty acid profiles (Table 1) and literature-based correlations [21,24], the biodiesel properties (especially viscosity, specific gravity, cloud point, cetane number, iodine number, and HHV-high heating value) of the six *Chlorella* strains (M082, M134, M189, M194, M751, and KR-1) grown under indoor and three strains under outdoor cultures were estimated (Table 2). The values of these key properties were then compared with the Europe (EN14214) and USA (ASTM D6751) standards and a variety of vegetable oil values.

Kinematic viscosity (mm/s^2) is one of the important properties of a biodiesel: the value must be high enough to provide sufficient lubrication for the engine parts, but also low enough not to affect the flow properties at the operational temperature. The viscosities of the *Chlorella* strains under the indoor and outdoor culture conditions were determined to be approximately 4.6 and 4.5 mm/s^2, respectively, which meet the specifications established by the international standards. These values, it should also be emphasized, are very similar to the viscosities of the vegetable oils. Sun et al. (2015), similarly, reported that the kinematic viscosities of 9 *Chlorella* strains all measured around 4.4 mm/s^2 [15]. Indeed, in the present study, no significant viscosity value changes were observed between the indoor and outdoor culture conditions, which suggests that these conditions have a negligible effect on such changes, especially in saturated fatty acids. The specific gravity of a biodiesel is the ratio of the density of the FAME to that of water at 15 °C. A denser biodiesel has a higher energy content and thus will deliver better engine performance. Irrespective of the culture conditions, the biodiesels from all of the tested strains tested in the present investigation measured a specific gravity of 0.88, which is similar to the values of the vegetable oils and petroleum diesel, and which, in any cases, falls within the limits of international standards.

The cloud point (°C) is defined as the temperature at which crystals appear in biodiesel as it becomes cooler. The cloud point increases with increased saturated fatty acids. The lower the cloud point, the better the fuel quality. Insignificant cloud point differences were observed among the strains under the different culture conditions; however, there were no common trends. In general, the cloud point values were within the 6.7–8.4 °C range indoors and the 5.4–6.5 °C range outdoors, whose values are slightly higher than those of the Jatropa oil that is commonly used as a biodiesel in many countries. Cloud point values ranging from 3.6 to 5.5 °C have been reported for different *Chlorella* strains tested under autotrophic conditions [15]. Nevertheless, the addition of anti-gel additives is widely recommended in order to lower the cloud point of biodiesel sufficiently to prevent filters from clogging under lower temperatures [24]. The cetane number (CN) represents the combustion quality of a biodiesel during compressor ignition. In a particular diesel engine, higher-CN fuels will have shorter ignition delay periods compared with lower-CN ones. The CN values of the tested strains were within a 56.26–57.10 range indoors and a 55.58–56.16 range in the flue-gas-mediated outdoor cultures, whose values were well above the standard specifications. Moreover, the microalgal biodiesels had higher CN values compared with Jatropa and most of the other vegetable oils.

The iodine value (IV) indicates the total unsaturation degree regardless of the relative proportions of mono to polyunsaturated compounds. A high IV value for a fuel represents a decreased oxidation stability, which causes the formation of various degradation products that can negatively affect engine operability. The IV values of the tested *Chlorella* strains ranged from 77.05 to 94.50 under the different culture conditions and were under the maximum limit of 120 established in the European biodiesel standard (no limits have been specified for IV in the U.S. standard). The higher heating value (HHV, MJ/kg) shows the energy content, otherwise known as the calorific value, of biodiesel. Hence, a higher HHV generally is recommended for diesel automobiles. The HHV of a given biodiesel increases with increasing fatty acid chain length and decreases with unsaturation [25]. The HHVs of the present tested strains were unaffected by the culture conditions and were around 40 MJ/kg (Table 2). Similar values of HHV (40.5 MJ/kg) had been reported among 9 *Chlorella* strains under indoor cultivation conditions [15]. Moreover, these values were almost the same as those of other vegetable oils. There are no specifications in the international standards. It should be noted that overall, irrespective of the culture conditions, the biodiesel properties of all of the tested *Chlorella* strains were within the ranges specified in the international standards.

Table 2. Biodiesel properties of *Chlorella* strains under autotrophic and mixotrophic conditions indoors and outdoors compared with literature and standards.

	Chlorella Strain	Viscosity [mm²/s²] Auto	Viscosity Mixo	*Specific Gravity Auto	Mixo	Cloud Point [°C] Auto	Mixo	Cetane Number Auto	Mixo	Iodine Value Auto	Mixo	HHV [MJ/kg] Auto	Mixo
Indoor culture	KR-1	4.63 ± 0.00	4.63 ± 0.00	0.88 ± 0.00	0.88 ± 0.00	7.81 ± 0.07	7.91 ± 0.01	56.79 ± 0.04	56.84 ± 0.00	80.53 ± 0.42	80.02 ± 0.05	40.14 ± 0.01	40.13 ± 0.00
	M082	4.62 ± 0.00	4.66 ± 0.00	0.88 ± 0.00	0.88 ± 0.00	7.67 ± 0.01	8.44 ± 0.05	56.72 ± 0.00	57.10 ± 0.03	81.35 ± 0.05	77.05 ± 0.29	40.16 ± 0.00	40.05 ± 0.01
	M134	4.65 ± 0.00	4.64 ± 0.01	0.88 ± 0.00	0.88 ± 0.00	8.21 ± 0.01	8.12 ± 0.03	56.99 ± 0.00	56.94 ± 0.02	78.34 ± 0.06	78.83 ± 0.16	40.09 ± 0.00	40.09 ± 0.00
	M189	4.62 ± 0.01	4.59 ± 0.00	0.88 ± 0.00	0.88 ± 0.00	7.58 ± 0.20	7.07 ± 0.06	56.68 ± 0.10	56.42 ± 0.04	81.83 ± 1.11	84.65 ± 0.35	40.17 ± 0.03	40.23 ± 0.01
	M194	4.63 ± 0.01	4.60 ± 0.00	0.88 ± 0.00	0.88 ± 0.00	7.79 ± 0.27	7.14 ± 0.00	56.78 ± 0.14	56.46 ± 0.00	80.66 ± 1.52	84.28 ± 0.02	40.14 ± 0.04	40.23 ± 0.00
	M751	4.64 ± 0.01	4.62 ± 0.00	0.88 ± 0.00	0.88 ± 0.00	8.14 ± 0.11	7.69 ± 0.11	56.95 ± 0.05	56.74 ± 0.05	78.74 ± 0.61	81.20 ± 0.56	40.09 ± 0.02	40.15 ± 0.02
¶Outdoor culture	KR-1	4.52 ± 0.00	na	0.88 ± 0.00	Na	5.37 ± 0.04	na	55.58 ± 0.02	na	94.12 ± 0.23	na	40.46 ± 0.01	na
	M082	4.55 ± 0.00	na	0.88 ± 0.00	Na	6.02 ± 0.00	na	55.90 ± 0.00	na	90.50 ± 0.01	na	40.37 ± 0.00	na
	M134	4.57 ± 0.00	na	0.88 ± 0.00	Na	6.54 ± 0.02	na	56.16 ± 0.01	na	87.64 ± 0.12	na	40.31 ± 0.00	na
Vegetable oil	†Jatropa	4.48		0.88		4.67		55.23		98.02		40.55	
	†Palm	4.61		0.87		14.00		61.90		54.00		40.60	
	†Rapeseed	4.50		0.88		-3.00		53.70		116.10		41.10	
	†Soy bean	4.26		0.88		0.00		51.30		125.50		39.70	
‡Petroleum diesel		2-3		0.85		Country specific		40-45		na		na	
International standard	EN14214	3.5-5.0		0.86-0.9		na		Minimum 51.0		Maximum 120		na	
	ASTM D6751	1.9-6.0		0.85-0.9		na		Minimum 47.0		na		na	

* Specific gravity = (1/1000) density at 15 °C; ¶ outdoor autotrophic are mediated by coal-fired flue-gas; † values from literature [21]; ‡ values from literature [24]; na: not available.

3. Materials and Methods

3.1. Microalgal Strains and Culture Medium

Twenty-nine strains of *Chlorella* were evaluated for biodiesel production in comparison with a reference strain *Chlorella* sp. KR-1 (Table 3). These strains were obtained from the Korean Collection for Type Cultures (KCTC; Daejeon, Korea). Meanwhile, *Chlorella* sp. KR-1, which has been extensively studied for its high lipid productivity and flue-gas tolerance, was maintained at the Korea Institute of Energy Research (KIER; Daejeon, Korea) [2,5,6,23,26,27]. The modified N8 medium used in this study was prepared and filter-sterilized through a 0.2 μm membrane, and its pH was adjusted to 6.5 [16].

Table 3. List of tested *Chlorella* strains and their origins.

No.	Strain	Species	Origin
1	N002	*Chlorella vulgaris*	Paddy field/Ssukgol, Gyeonggi
2	N006	*C. vulgaris*	Paddy field/Suwon, Gyeonggi
3	N032	*C. vulgaris*	Reservoir/Andongho, Gyeongbuk
4	N034	*C. vulgaris*	Reservoir/Andongho, Gyeongbuk
5	N043	*C. vulgaris*	Reservoir/Imhaho, Gyeongbuk
6	N052	*C. vulgaris*	Reservoir/Youngsanho, Chungnam
7	N053	*Chlorella* sp.	Reservoir/ Andong, Gyeongbuk
8	N113	*Chlorella* sp.	Youngpyung, Gangwon
9	N309	*Chlorella* sp.	Local fresh water body
10	M077	*Chlorella* sp.	Local fresh water body
11	M082	*Chlorella* sp.	Local fresh water body
12	M134	*Chlorella* sp.	Local fresh water body
13	M136	*Chlorella* sp.	Local fresh water body
14	M138	*Chlorella* sp.	Local fresh water body
15	M150	*Chlorella* sp.	Local fresh water body
16	M152	*Chlorella* sp.	Local fresh water body
17	M158	*Chlorella* sp.	Local fresh water body
18	M177	*Chlorella* sp.	Local fresh water body
19	M181	*Chlorella* sp.	Local fresh water body
20	M182	*Chlorella* sp.	Local fresh water body
21	M189	*Chlorella* sp.	Local fresh water body
22	M190	*Chlorella* sp.	Local fresh water body
23	M194	*Chlorella* sp.	Local fresh water body
24	M372	*Chlorella* sp.	Local fresh water body
25	M696	*C. vulgaris*	Reservoir/Yongsoho Jeju
26	M740	*C. sorokiniana*	Reservoir/Daedongho, Hampyung
27	M742	*C. vulgaris*	Reservoir/Daedongho, Hampyung
28	M746	*C. vulgaris*	Reservoir/Daedongho, Hampyung
29	M751	*C. vulgaris*	Local fresh water body
30	KR-1	*Chlorella* sp.	Yeongwol, Gangwon

3.2. Seed Flask Culture

All of the strains were initially cultured in 250 mL Erlenmeyer flasks (working vol., 100 mL). The flasks were incubated under continuous illumination (white fluorescent lamps, *ca.* 40 μmol photons/m^2·s) in a shaking incubator (IS-971RF, Lab Companion, Korea) at 150 rpm and 25 °C.

3.3. Indoor Culture Using Photobioreactor with 10% (v/v) CO_2

Pyrex glass bubble-column photobioreactor (b-PBRs; working vol., 500 mL) [23] was utilized to culture *Chlorella* strains. The biomass from the flask cultures was used as the b-PBRs inoculum, a concentration of which was fixed to an initial optical density (OD) of 0.1 at 660 nm. The strains were grown under either autotrophic or mixotrophic conditions for 96 h. For the latter, the modified N8 medium was supplemented with filter-sterilized glucose (5 g/L, Sigma, USA; 0.2 μm Minisart

High-Flow filter, 16532-K, Sartorius Stedium Biotech., Germany). The b-PBR was continuously supplied with 10% CO_2 in air (v/v) from the bottom of the reactor. The supplied gas was conveyed by a 0.2 μm PTFE syringe filter (Minisart SRP15, Satorius Stedium Biotech., Germany) and controlled by mass-flow controllers (MKP, Korea) and flow meters (RM Rate-Master, Dwyer instruments Inc., Michigan, IN, USA). The reactor was maintained under continuous illumination (white fluorescent lamps, ca. 170 μmol/m^2·s) in a temperature-controlled room (28–31 °C). Preliminary screening of 30 *Chlorella* spp. was done using single PBR, whereas the rest of the experiments were carried out with duplicate PBRs.

A high-temperature-tolerance experiment was carried out for the fast-growing three strains under autotrophic conditions including continuous illumination (white fluorescent lamps, ca. 69 μmol/m^2 s) in a temperature-controlled growth chamber (GC-300, Jeio Tech, Korea). The b-PBR was incubated at 35 °C for 45 h, and then the temperature was increased to 40 °C, where it was maintained for a further 25 h.

3.4. Outdoor Culture Using Photobioreactor with Actual Flue-gas

The three strains selected under indoor experimentation were grown outdoors in b-PBRs under autotrophic conditions for 160 h. The biomass from the flask cultures was used as an inoculum with a fixed initial OD of 0.3 at 660 nm. The cultures were mediated by a supply of coal-fired flue-gas (obtained from a 2 MW demonstration-scale coal-burning power plant located at KIER, Daejeon, Korea) at a flow rate of 0.6 vvm [23]. The typical composition of the flue-gas supplied was CO_2, 13.3%; O_2, 7.6%; CO, 39.1ppm; NOx, 6.9 ppm [6]. Details on the storage, pre-processing, and transfer of flue-gas are available elsewhere [6]. Throughout the experiment, the cultures were maintained at ambient temperature, and the average day time light intensity in this period was 500 μmol/m^2·s.

3.5. Analytical Methods

Growth of the microalgal strains was evaluated based on their OD values measured at 660 nm using a UV-Vis spectrophotometer (Optizen 2120UV, Mecasys Co., Daejeon, Korea). The dry-cell weight (DCW) was measured by GF/C filtration, washing with distilled water, and 105 °C drying. The pH and light intensity were determined using a pH meter (DKK-TOA Co., Japan) and a quantum meter (LI-250A, LI-COR Inc., Lincoln, NE, USA), respectively. For outdoor cultures, data on parameters such as temperature and light intensity were continuously recorded. The temperature probe (CIMON-SCADA V 2.10, KDT Systems, Seongnam, Korea) was immersed inside the b-PBR in the culture medium. Changes in the light intensity were monitored through a light sensor (LI-250A, LI-COR Inc., USA) that was fixed on the top of the b-PBR-setting panel, and the data was collected by a data logger (LI-1400, LI-COR Inc., USA). The compositions (CO, NO, NO_2, SO_2, CO_2, and O_2) of the flue gas supplied for the cultures were analyzed using a portable flue-gas analyzer (Vario Plus, MRU, Neckarsulm, Germany). The FAME (fatty acids methyl ester) contents of the dried cells were prepared by following a modified *in-situ* transesterification method [18] and analyzed by gas chromatography (GC; Agilent 6890, Agilent Technologies, Wilmington, DE, USA). The relevant detailed methodologies have previously been reported [3,23]. All the analyses were done in triplicates, and the results were represented as mean ± standard deviation.

3.6. Estimation of Biodiesel Properties

Important biodiesel properties including viscosity, specific gravity, cloud point, cetane number, iodine number, and higher heating value (HHV) were estimated from the fatty acid compositions based on the empirical formulae proposed by Tanimura et al. (2014) [21].

4. Conclusions

In this study, 30 local strains of *Chlorella* were evaluated for application to flue-gas-mediated biodiesel production. Among the 30 strains, 3 strains (M082, M134, and KR-1) could be selected on the

Energies **2018**, *11*, 2026

basis of their growth rates, lipid contents, and fatty acid profiles under autotrophic (10% CO_2) and mixotrophic (10% CO_2 plus 5 g glucose/L) nutrition modes. These strains were further tested under outdoor conditions using the actual coal-fired flue-gas from demonstration-scale plant. M082 showed good lipid production performances (228 mg FAME/g cell and 39 mg FAME/L·d), as well as a high, up-to-40 °C temperature tolerance. Thus, *Chlorella* sp. M082 can be considered a potential strain for simultaneous utilization of CO_2 from flue-gas and of waste organics for biodiesel production.

Supplementary Materials: The following are available online at http://www.mdpi.com/1996-1073/11/8/2026/s1, Figure S1: Growth curve of 6 fast-growing *Chlorella* strains under autotrophic and mixotrophic conditions for 96 h of cultivations, Figure S2: FAME distributions of 6 fast-growing *Chlorella* strains under autotrophic and mixotrophic indoor conditions for 96 h of cultivation, Figure S3: FAME distribution of 3 high-productivity *Chlorella* strains in outdoor culture mediated by coal-fired flue-gas, and Figure S4: Growth patterns of 3 high-productivity *Chlorella* strains under high-temperature conditions.

Author Contributions: Y.-K.O. and B.K. conceived and designed the experiments; B.K. and E.C. performed the experiments; Y.-K.O. and R.P. analyzed the data; Y.-K.O., K.L. and S.G.J. contributed reagents/materials/analysis tools; R.P., B.K. and Y.-K.O. wrote the paper.

Funding: This work was supported by a 2-Year Research Grant of Pusan National University.

Conflicts of Interest: The authors declare no conflict of interest.

References

1. Mata, T.M.; Martins, A.A.; Caetano, N.S. Microalgae for biodiesel production and other applications: A review. *Renew. Sustain. Energ. Rev.* **2010**, *14*, 217–232. [CrossRef]
2. Praveenkumar, R.; Kim, B.; Lee, J.; Vijayan, D.; Lee, K.; Nam, B.; Jeon, S.G.; Kim, D.M.; Oh, Y.K. Mild pressure induces rapid accumulation of neutral lipid (triacylglycerol) in *Chlorella* spp. *Bioresour. Technol.* **2016**, *220*, 661–665. [CrossRef] [PubMed]
3. Cho, S.; Luong, T.T.; Lee, D.; Oh, Y.K.; Lee, T. Reuse of effluent water from a municipal wastewater treatment plant in microalgae cultivation for biofuel production. *Bioresour. Technol.* **2011**, *102*, 8639–8645. [CrossRef] [PubMed]
4. Lee, K.; Lee, S.Y.; Na, J.G.; Jeon, S.G.; Praveenkumar, R.; Kim, D.M.; Chang, W.S.; Oh, Y.K. Magnetophoretic harvesting of oleaginous *Chlorella* sp. by using biocompatible chitosan/magnetic nanoparticle composites. *Bioresour. Technol.* **2013**, *149*, 575–578. [CrossRef] [PubMed]
5. Lee, K.; Lee, S.Y.; Praveenkumar, R.; Kim, B.; Seo, J.Y.; Jeon, S.G.; Na, J.G.; Park, J.Y.; Kim, D.M.; Oh, Y. Repeated use of stable magnetic flocculant for efficient harvest of oleaginous *Chlorella* sp. *Bioresour. Technol.* **2014**, *167*, 284–290. [CrossRef] [PubMed]
6. Praveenkumar, R.; Kim, B.; Choi, E.; Lee, K.; Park, J.Y.; Lee, J.S.; Lee, Y.C.; Oh, Y.K. Improved biomass and lipid production in a mixotrophic culture of *Chlorella* sp. KR-1 with addition of coal-fired flue-gas. *Bioresour. Technol.* **2014**, *171*, 500–505. [CrossRef] [PubMed]
7. Praveenkumar, R.; Shameera, K.; Mahalakshmi, G.; Akbarsha, M.A.; Thajuddin, N. Influence of nutrient deprivations on lipid accumulation in a dominant indigenous microalga *Chlorella* sp., BUM11008: Evaluation for biodiesel production. *Biomass Bioenergy* **2012**, *37*, 60–66. [CrossRef]
8. Remmers, I.M.; Wijffels, R.N.; Barbosa, M.; Lamers, P.P. Can We Approach Theoretical Lipid Yields in Microalgae? *Trend Biotechnol.* **2018**, *36*, 265–276. [CrossRef] [PubMed]
9. Chiu, S.Y.; Kao, C.Y.; Huang, T.-T.; Lin, C.J.; Ong, S.C.; Chen, C.D.; Chang, J.S.; Lin, C.S. Microalgal biomass production and on-site bioremediation of carbon dioxide, nitrogen oxide and sulfur dioxide from flue gas using *Chlorella* sp. cultures. *Bioresour. Technol.* **2011**, *102*, 9135–9142. [CrossRef] [PubMed]
10. Hende, S.V.D.; Vervaeren, H.; Boon, N. Flue gas compounds and microalgae:(Bio-) chemical interactions leading to biotechnological opportunities. *Biotechnol. Adv.* **2012**, *30*, 1405–1424. [CrossRef] [PubMed]
11. Liang, Y.; Sarkany, N.; Cui, Y. Biomass and lipid productivities of *Chlorella vulgaris* under autotrophic, heterotrophic and mixotrophic growth conditions. *Biotechnol. Lett.* **2009**, *31*, 1043–1049. [CrossRef] [PubMed]
12. Guccione, A.; Biondi, N.; Sampietro, G.; Rodolfi, L.; Bassi, N.; Tredici, M.R. *Chlorella* for protein and biofuels: From strain selection to outdoor cultivation in a green wall panel photobioreactor. *Biotechnol. Biofuel* **2014**, *7*, 84–95. [CrossRef] [PubMed]

13. Muthuraj, M.; Kumar, V.; Palabhanvi, B.; Das, D. Evaluation of indigenous microalgal isolate *Chlorella* sp. FC2 IITG as a cell factory for biodiesel production and scale up in outdoor conditions. *J. Ind. Microbiol. Biotechnol.* **2014**, *41*, 499–511. [CrossRef] [PubMed]

14. Nascimento, I.A.; Marques, S.S.I.; Cabanelas, I.T.D.; Pereira, S.A.; Druzian, J.I.; de Souza, C.O.; Vich, D.V.; de Carvalho, G.C.; Nascimento, M.A. Screening microalgae strains for biodiesel production: Lipid productivity and estimation of fuel quality based on fatty acids profiles as selective criteria. *Bioenergy Res.* **2013**, *6*, 1–13. [CrossRef]

15. Sun, Z.; Zhou, Z.G.; Gerken, H.; Chen, F.; Liu, J. Screening and characterization of oleaginous *Chlorella* strains and exploration of photoautotrophic *Chlorella protothecoides* for oil production. *Bioresour. Technol.* **2015**, *184*, 53–62. [CrossRef] [PubMed]

16. Li, L.; Cui, J.; Liu, Q.; Ding, Y.; Liu, J. Screening and phylogenetic analysis of lipid-rich microalgae. *Algal Res.* **2015**, *11*, 381–386. [CrossRef]

17. Xia, L.; Song, S.; He, Q.; Yang, H.; Hu, C. Selection of microalgae for biodiesel production in a scalable outdoor photobioreactor in north China. *Bioresour. Technol.* **2014**, *174*, 274–280. [CrossRef] [PubMed]

18. Cheirsilp, B.; Torpee, S. Enhanced growth and lipid production of microalgae under mixotrophic culture condition: Effect of light intensity, glucose concentration and fed-batch cultivation. *Bioresour. Technol.* **2012**, *110*, 510–516. [CrossRef] [PubMed]

19. Kim, B.; Praveenkumar, R.; Lee, J.; Nam, B.; Kim, D.M.; Lee, K.; Lee, Y.C.; Oh, Y.K. Magnesium aminoclay enhances lipid production of mixotrophic *Chlorella* sp. KR-1 while reducing bacterial populations. *Bioresour. Technol.* **2016**, *219*, 608–613. [CrossRef] [PubMed]

20. Abomohra, A.E.-F.; El-Sheekh, M.; Hanelt, D. Screening of marine microalgae isolated from the hypersaline Bardawil lagoon for biodiesel feedstock. *Renew. Energy* **2017**, *101*, 1266–1272. [CrossRef]

21. Tanimura, A.; Takashima, M.; Sugita, T.; Endoh, R.; Kikukawa, M.; Yamaguchi, S.; Sakuradani, E.; Ogawa, J.; Shima, J. Selection of oleaginous yeasts with high lipid productivity for practical biodiesel production. *Bioresour. Technol.* **2014**, *153*, 230–235. [CrossRef] [PubMed]

22. Vello, V.; Phang, S.M.; Chu, W.L.; Majid, N.A.; Lim, P.E.; Loh, S.K. Lipid productivity and fatty acid composition-guided selection of *Chlorella* strains isolated from Malaysia for biodiesel production. *J. Appl. Phycol.* **2014**, *26*, 1399–1413. [CrossRef]

23. Praveenkumar, R.; Kim, B.; Choi, E.; Lee, K.; Cho, S.; Hyun, J.; Park, J.; Lee, Y.; Lee, H.; Lee, J. Mixotrophic cultivation of oleaginous *Chlorella* sp. KR-1 mediated by actual coal-fired flue gas for biodiesel production. *Bioproc. Biosyst. Eng.* **2014**, *37*, 2083–2094. [CrossRef] [PubMed]

24. Hoekman, S.K.; Broch, A.; Robbins, C.; Ceniceros, E.; Natarajan, M. Review of biodiesel composition, properties, and specifications. *Renew. Sustain. Energy Rev.* **2012**, *16*, 143–169. [CrossRef]

25. Valdez-Ojeda, R.; González-Muñoz, M.; Us-Vázquez, R.; Narváez-Zapata, J.; Chavarria-Hernandez, J.C.; López-Adrián, S.; Barahona-Pérez, F.; Toledano-Thompson, T.; Garduño-Solórzano, G.; Medrano, R.M.E.-G. Characterization of five fresh water microalgae with potential for biodiesel production. *Algal Res.* **2015**, *7*, 33–44. [CrossRef]

26. Sung, K.D.; Lee, J.S.; Shin, C.S.; Park, S.C.; Choi, M.J. CO_2 fixation by *Chlorella* sp. KR-1 and its cultural characteristics. *Bioresour. Technol.* **1999**, *68*, 269–273. [CrossRef]

27. Yun, M.; Oh, Y.K.; Praveenkumar, R.; Seo, Y.S.; Cho, S. Contaminated bacterial effects and qPCR application to monitor a specific bacterium in *Chlorella* sp. KR-1 culture. *Biotechnol. Bioprocess. Eng.* **2017**, *22*, 150–160. [CrossRef]

energies

MDPI

Article

Thermophilic Anaerobic Digestion: Enhanced and Sustainable Methane Production from Co-Digestion of Food and Lignocellulosic Wastes

Aditi David [1], Tanvi Govil [1], Abhilash Kumar Tripathi [1], Julie McGeary [2], Kylie Farrar [3] and Rajesh Kumar Sani [1,4,*]

[1] Department of Chemical and Biological Engineering, South Dakota School of Mines and Technology, Rapid City, SD 57701, USA; aditi.david@mines.sdsmt.edu (A.D.); tanvi.govil@mines.sdsmt.edu (T.G.); abhilashkumar.tripathi@mines.sdsmt.edu (A.K.T.)

[2] Biology Teacher, Central High School, Rapid City, SD 57701, USA; julie.mcgeary@k12.sd.us

[3] Science Teacher, Russell Middle School, Colorado Springs, CO 80918, USA; kyliefarrar23@gmail.com

[4] BuG ReMeDEE Consortium, South Dakota School of Mines and Technology, Rapid City, SD 57701, USA

* Correspondence: rajesh.sani@sdsmt.edu; Tel.: +1-605-394-1240

Received: 6 July 2018; Accepted: 4 August 2018; Published: 8 August 2018

Abstract: This article aims to study the codigestion of food waste (FW) and three different lignocellulosic wastes (LW) (Corn stover (CS), Prairie cordgrass (PCG), and Unbleached paper (UBP)) for thermophilic anaerobic digestion to overcome the limitations of digesting food waste alone (volatile fatty acids accumulation and low C:N ratio). Using an enriched thermophilic methanogenic consortium, all the food and lignocellulosic waste mixtures showed positive synergistic effects of codigestion. After 30 days of incubation at 60 °C (100 rpm), the highest methane yield of 305.45 L·kg^{-1} volatile solids (VS) was achieved with a combination of FW-PCG-CS followed by 279.31 L·kg^{-1} VS with a mixture of FW-PCG. The corresponding volatile solids reduction for these two co-digestion mixtures was 68% and 58%, respectively. This study demonstrated a reduced hydraulic retention time for methane production using FW and LW.

Keywords: thermophilic anaerobic digestion; corn stover; prairie cord grass; unbleached paper; digester performance; process stability; synergistic effects; microbial community; *Methanothermobacter*

1. Introduction

Anaerobic digestion for the production of biogas is an environmentally friendly multi-step process employing complex consortia of microorganisms. These consortia comprise various facultative or obligate anaerobic microbial groups which work synergistically and convert complex organic substrates into biogas. Biogas with 60–70% of its component being methane is a combustible renewable energy that can be used as an alternative energy source to replace fossil fuels, either by direct combustion to generate heat and electricity, or through upgradation to be used as vehicle fuel and injection into the gas grid [1–4]. After the on-site demand of the produced biogas is met, the remaining biogas is usually stored as compressed natural gas (CNG) or liquefied biomethane (LBM) for future use. While the biogas industry in Europe is well established, with more than 10,000 biogas producing Anaerobic digestion (AD) plants in operation, in the United States, the biogas industry is still growing. According to the American Biogas Council, 1241 wastewater treatment plants and 236 farms have functional Biogas plants [5]. Currently, the feedstocks used for biogas production in these anaerobic digesters are limited, and therefore it is important to explore new substrates to meet the growing energy needs.

Food waste (FW) is an attractive substrate for AD because of its low total solids (TS) and high content of soluble organics which make it readily biodegradable. Additionally, FW offers low-cost alternatives for methane production and is abundantly available. The Food and Agriculture Organization of the United Nations reports a wastage of 1.3 billion tons of food per year. Therefore, using FW to produce methane appears to be an ideal way to strengthen the world's energy security while addressing waste management and nutrient recycling [6]. The AD of FWs, however, depends on the delicate balance between the acidification process and methanogenesis [7]. Many studies have attributed the low stability and efficiency of individual FW fermentation to the low C/N ratio, ammonia accumulation, and the readily biodegradable organic fraction that causes acidification [8]. If the acidification process is rapid, the accumulation of volatile fatty acids (VFAs) occurs, causing an abrupt fall in pH, which stresses and inhibits the methanogenic archaea [9,10]. The reactor acidification through reactor overload is one of the most common reasons for process deterioration in anaerobic digesters [11]. Therefore, the codigestion of FW with other organic wastes, such as municipal wastewater treatment plant sludge, animal manure, and agricultural biomass, has become popular, as it provides the ability to alter the carbon and nitrogen ratios which prevent accumulation of intermediate inhibitory products (NH_4^+ and VFAs) [12–14] and increase the methane yield, as well as improve the utilization efficiency [8]. The use of co-substrates for AD has shown to produce higher biogas yields due to positive synergisms established in the digestion medium and the supply of missing nutrients by the co-substrates [15,16]. Mixing organic substrates often results in the formation of a codigestion mixture with a C/N ratio included in the optimal range of 20–30% as reported in the literature [16–18]. The additional benefits of the codigestion process are (1) the dilution of the potential toxic compounds present in any of the co-substrates involved; (2) the adjustment of the moisture content and pH; (3) supply of the necessary buffer capacity to the mixture; (4) the increase of the biodegradable material content; and (5) the widening the range of bacterial strains taking part in the process [17].

Like food waste, lignocellulosic wastes (LW), which include agricultural residues (corn stover, rice straw), herbaceous crops (switchgrass, prairie cordgrass) and waste paper, are extensively available low-cost substrates. Lignocellulosic wastes are in fact the most abundant renewable organics, reaching, annually, over 150 billion tons in the form of plant biomass [19]. Due to their chemical composition based on sugars and other compounds of interest, they could be utilized to produce several value-added products, such as ethanol, biogas, food additives, organic acids, enzymes, and others [20–23]. Therefore, besides the environmental problems caused by their accumulation, the non-utilization of these materials constitutes a loss of potentially valuable sources [24]. Previous studies report that the methane generation potential is expected to be much higher if lignocellulosic biomass resources are used. It is estimated that 4.2 trillion cubic feet per year (about 4,318 trillion British thermal units [25]) LW are available, the biogasification of which can displace about 46% of current natural gas consumption in the electric power sector and the entire natural gas consumption in the transportation sector [26]. Moreover, being a second-generation feedstock for biofuel, the use of lignocellulosic residues does not compete for arable land [27,28]. Thus, utilizing LW in the AD process is a promising option. Still, the AD of LW alone has several limitations, such as the high C:N ratio creating nitrogen deficiency, the risk of producing inhibitors (e.g., furfural and hydroxymethylfurfural), a less digestible biomass, a high heat demand [29,30], and long digestion time due to their low cellulose, hemicellulose, and lignin conversion rates [8].

Performing anaerobic codigestion of FW with LW can overcome the limitations of their respective mono-digestions. Previous studies have also demonstrated the role of codigestion of FW and LW in the adjustment of C:N ratio [31], in the reduction of the start-up time and volatile fatty acid (VFA) accumulation [32], and in the enhancement of methane yield. Yong et al., 2015 examined the potential of co-digestion of food waste and straw at 35 °C. The study showed increased methane yield reaching 0.392 $m^3 \cdot kg^{-1}$ VS at an optimal mixing ratio of 5:1 (FW: straw), which marked a 39.5% and 149.7% increase in methane yield compared with individual digestion results of food waste and straw, respectively [8]. In another study, Jabeen et al., 2015, codigested FW with rice husk in a pilot-scale mesophilic (37 °C) anaerobic reactor. They obtained a daily biogas production of 196 $L \cdot d^{-1}$, at an organic loading rate of 6 kg $VS \cdot m^{-3} \cdot d^{-1}$, which decreased to 136 $L \cdot d^{-1}$ at a loading rate of

9 kg VS/m^3/d. They attributed the decrease in biogas production, reactor stability, and volatile solids removal efficiency to the increase in organic loading rate [33]. These effects of high organic loading rate are more profound in mesophilic anaerobic reactors as compared to thermophilic reactors.

In all the studies mentioned above, the anaerobic co-digestion of FW and LW improved process performance and increased biogas production. However, a major disadvantage of using LW is the recalcitrant nature of their cell wall. The intricate composition of lignocellulosic materials, where cellulose fiber is tightly linked to hemicellulose and lignin, hinders their biodegradability and thus limits their use as co-substrates for AD. The recalcitrant nature of the LW necessitates the inclusion of the pretreatment step (physical, chemical or biological) for their efficient conversion into biofuels [34]. The use of thermophilic digesters can eliminate the use of costly and sometimes hazardous pretreatment steps. Thermophilic anaerobic digestion (TAD) allows for the better degradation of LW by increasing the microbial hydrolysis rate which is considered the slowest as well as rate limiting step [35]. Besides aiding methanogenesis through increased hydrolysis rate, TAD also decreases the level of pathogens. This reduces the potential health hazards for the biogas plant operators as well as ensures safe disposal of the digestate after process completion [36].

Most of the studies to date on the codigestion of FW and LW have been conducted under mesophilic conditions [31,33,37–39] which require higher hydraulic retention times (HRTs) and lower organic loading rates. Here, however, we attempt to assess the methane production from codigestion of FW and LW under thermophilic conditions. The objectives of this study were to (1) evaluate the methane potential and biodegradability of FW, corn stover (CS), prairie cordgrass (PCG), and unbleached paper (UBP) at 60 °C and (2) describe the trends of methane yield, process stability, and digester performance, when these are codigested using a pre-acclimatized thermophilic methanogenic consortium. In South Dakota, PCG is abundant and can offer local alternatives to corn for the production of biofuels. This study is the first attempt at investigating the co-digestion of FW with PCG under thermophilic conditions (60 °C). The results of this study could provide baseline data for the adoption of Thermophilic Anaerobic Digestor (TAD) using readily available PCG as feedstocks for biogas generation.

2. Materials and Methods

2.1. Feedstock and Inoculum

The FW and three different types of LW-CS, PCG, and UBP—were used as feedstocks for TAD. CS and PCG were kindly provided by Dr. K. Muthukumarappan from South Dakota State University, Brookings, SD. The FW and UBP were collected from the cafeteria at the South Dakota School of Mines and Technology and stored at 4 °C. LW were reduced in size using a cutting mill and sieved through sieve between 15 and 10 mm pores (SM 200, Rectch GmbH, Haan, Germany). The particle size of the FW was also reduced by crushing it in an electrical kitchen blender, and the resultant FW slurry was sieved to remove coarse particles larger than 15 mm. The effect of shredding paper had been tested previously and did not influence the AD performance [40], so the UBP was hand cut. Inoculum used in this study was effluent procured from an anaerobic digester at the Waste Water Reclamation Plant, Rapid City, SD which was enriched in our lab. The enrichment details are included in our previous report [41]. This enriched thermophilic methanogenic consortia (called TMC) was used as the inoculum for this study and stored in an air-tight container at −20 °C. The characteristics of substrates and inocula are presented in Table 1. Each data point presented was the average of triplicate measurements of the same feedstock.

2.2. Experimental setup for Batch Tests

Batch tests were conducted in 500 mL serum bottles with a working volume of 200 mL containing 180 mL of anaerobic culture medium and 20 mL (10%, *v/v*) TMC. The composition of the anaerobic culture medium is given in Table 1. To analyze the feasibility of the individual FW and LW for anaerobic digestion, different organic loadings (1%, 2%, 5%, and 10% w/v VS) of these samples were first tested, which constituted 16 mono-digestion tests (Table 2). Based on this, seven FW samples (1 g VS each) were separately mixed with CS, PCG, or UBP for the anaerobic co-digestion study in mixing ratios as given in Table 3. The substrate-free serum bottles containing inoculum, and the media served as controls. The experiments were repeated three times under the same conditions using the same co-digestion mixing ratios. The average pH of digestion mixtures T1, T2, T3, T4, T5, T6, and T7 after mixing with inoculum were 7.1, 7.1, 7.2, 7.1, 7.2, 7.3, and 7.3, respectively.

The substrates (according to Tables 1 and 2), media and 10% (v/v) inoculum were added in the serum bottles to make a total working volume of 200 mL. These digester bottles were capped with butyl rubber stoppers, crimped with aluminum caps, and nitrogen gas was purged into them for 20 min to simulate anaerobic conditions. The bottles were then incubated at 60 °C (100 rpm) for up to 30 days. Methane production was monitored at different time intervals, and the methane yield was reported as mL CH_4 per gram of Volatile solids (mL CH_4/g VS).

Table 1. The composition of the anaerobic culture medium.

Components	Composition (g/L)
K_2HPO_4	0.30
KH_2PO_4	0.30
NaCl	0.10
$CaCl_2$	0.05
NH_4Cl	1.00
$MgCl_2 \cdot 6H_2O$	0.50
KCl	0.30
Cysteine.HCl	0.50
Yeast extract	0.05
$Na_2S \cdot 9H_2O$	0.003
$NaHCO_3$	20 mM
Nitsch trace element	2.5 mL

Table 2. The feedstock composition in the batch tests with individual wastes.

Individual Substrate Type	Food Waste (FW)				Corn Stover (CS)				Prairie Cordgrass (PCG)				Unbleached Paper (UBP)			
Organic Loading (%) s	1%	2.5%	5%	10%	1%	2.5%	5%	10%	1%	2.5%	5%	10%	1%	2.5%	5%	10%
Composition (g VS)	2	5	10	20	2	5	10	20	2	5	10	20	2	5	10	20

Table 3. The feedstock composition in batch tests (60 °C) with codigestion mixtures.

Test Bottles	T1	T2	T3	T4	T5	T6	T7
Composition	FW	FW + CS	FW + PCG	FW + UBP	FW + PCG + CS	FW + PCG + UBP	FW + UBP + CS
FW (g VS)	2.5	2.5	2.5	2.5	2.5	2.5	2.5
CS (g VS)	-	1.5	-	-	0.75	-	0.75
PCG (g VS)	-	-	1.5	-	0.75	0.75	-
UBP (g VS)	-	-	-	1.5	-	0.75	0.75
Mixing Ratio	-	5:3	5:3	5:3	5:1.5:1.5	5:1.5:1.5	5:1.5:1.5
C:N	19:1	23.9	25.6	28.2	24.4	26.8	25.6

(Note: FW = Food wastes; CS = Corn stover; PCG = Prairie cordgrass; UBP = Unbleached paper).

2.3. Analytical Methods

Total Solids (TS) and Volatile solids (VS) of the inoculum and feedstock were analyzed according to APHA Standard Methods [42]. The cellulose, hemicellulose, and lignin content were determined according to standard NREL analytical procedures [43]. The product gas was sampled using a 100 µL gas-tight glass syringe (Hamilton Company, Reno, NV, USA), and measured using gas chromatography (Agilent Technologies 7890A) equipped with a thermal conductivity detector at 200 °C and a Supelco

Porapak Q stainless steel packed column (6 ft × 1/8 in). Nitrogen was the carrier gas at a rate of 10 mL/min. The conditions for CH_4 analysis were as follows: injector temperature: 70 °C; detector temperature: 100 °C; and oven temperature: 1 min at 35 °C, followed by a 5 °C/min ramp to 50 °C with a hold time of 2 min (total run time 4.5 min). The VFAs (acetic acid, propionic, and butyric acid) were measured using an Aminex HPX-87 H column with 0.005 N sulfuric acid as the mobile phase. Liquid samples were sent to Atlantic Microlab, Norcross, GA for elemental analysis.

Methane yield (mL/g VS) was calculated as the volume of methane as produced per g of VS feedstock loaded into the digester bottles initially and corrected by subtracting the methane yield obtained from the control bottle [44]. VS represents the organic portion of the material solids that can be digested, while the remainder of the solids is considered as fixed. The 'fixed' solids are non-biodegradable [45]. The pH value of each mixture before and after the digestion process was measured with an Oakton pH 700 meter. Process stability was assessed in terms of final pH and VFA accumulation, and performance was measured in terms of methane production and VS removal efficiencies. The carbon to nitrogen ratio was calculated using the following equation [37]:

$$C:N = \frac{(VS * TOC)_{FW} + (VS * TOC)_{LW}}{(VS * TN)_{FW} + (VS * TN)_{LW}}$$

where TOC = the total organic carbon (%VS) and TN = total nitrogen (%VS).

2.4. Statistical Analysis

The batch tests were conducted in triplicates, and each data point was the average of triplicate readings for all the chemical analysis and measurements. The statistical significance of the results was analyzed by Analysis of Variance (Two-way ANOVA) with a 95% level of confidence ($p = 0.05$) [46]. Statistical analysis was done using GraphPad Prism version 7.04 (GraphPad Software, San Diego, CA, USA). The results of the analysis are given in the supplementary files.

3. Results and Discussion

3.1. Feedstock and Inoculum Characteristics

The results of the feedstock characterization are shown in Table 4. FW had a moisture content of 84.8% and a VS/TS ratio of 90.8%. These results comply with previous reports in the literature where the moisture content and VS/TS ratio of FW was 69–93% and 85–95%, respectively [47]. FW containing a high amount of digestible organic matter is suitable for anaerobic microbial growth and attaining high methane yield. However, elemental analysis showed that FW had a C:N ratio of 13:1, which was too low to maintain nutrient balance in the anaerobic digester. TS and VS contents of the three LW (CS, PCG, UBP) were higher than that of FW, and their VS/TS ratios were between 85% to 95% making them suitable feedstocks for TAD (Table 1). The C:N ratio was found to be the highest for UBP (124:1), followed by CS (55:1) and PCG (41:1), which was higher than the optimum range required (20–30:1) [48]. The cellulose content was the highest in UBP (84.5%) and lowest in PCG (30.3%). CS had the highest hemicellulose content among the tested feedstocks and PCG had the highest lignin content. UBP had the lowest hemicellulose and lignin content.

The inoculum (TMC) used in this study was enriched through sub-culturing techniques [41]. Briefly, 10% (v/v) effluent from the wastewater reclamation plant was inoculated in 200 mL of anaerobic medium (Table 1) in 500 mL serum bottles. One-gram VS of mixed wastes (containing equal amounts of FW, CS, PCG, and UBP) was used as the substrate and the bottles were incubated at 60 °C and 100 rpm. When methane production reached a stable level, 10% (v/v) of the actively growing anaerobic culture was transferred into fresh media (200 mL) containing another 1 g of mixed waste. After 10 serial transfers, a methanogenic consortium growing at 60 °C was obtained that produced methane using FW, CS, PCG, and UBP as a carbon and energy source. The enriched TMC had a VS content, TS content, and C:N ratio of 6.1%, 8.3%, and 2.7, respectively. During optimization, a pH of 7.5, 2–3% (w/v)

substrate loading, and 10% (v/v) inoculum density had a profound effect on consortium growth and subsequent methane production. The microbial community analysis of the inoculum highlighted the role of the bacterial orders—*Clostridiales, Bacillales, Bacteroidales*, and *Thermoanaerobacteriales*—in the anaerobic degradation of the complex organic polymers present in FW and LW. As expected, the hydrogenotrophic pathway was found to be the dominant pathway of methanogenesis where *Methanothermobacter* were the predominant archaea in TAD (Supplementary Figure S2A–D).

Table 4. The characteristics of inoculum and feedstock [a].

Parameters	Inoculum	FW	CS	PCG	UBP
TS (% w/w)	8.3	15.2	94.3	87.6	96.2
VS (% w/w) [b]	6.1	13.8	92.6	76.9	84.7
Ash (% w/w) [b]	93.9	86.2	7.4	23.1	15.3
VS/TS (%)	37.2	90.8	98.2	87.8	88.0
C:N [c]	2.7	13:1	55:1	41:1	124:1
VFA/alkalinity	1.3	ND	ND	ND	ND
Lignin (%) [b]	ND	ND	13.5	22.9	2.1
Cellulose (%) [b]	ND	ND	38.2	30.3	84.5
Hemicellulose (%) [b]	ND	ND	32.4	25.7	11.2

ND = Not determined; [a] Data shown are average values based on duplicate runs; [b] Based on the TS of the sample; [c] Based on the total weight of the sample.

3.2. Thermophilic Anaerobic Digestion of Individual Waste Substrates

Cumulative methane production for organic loadings of 1, 2.5, 5, and 10% (w/v VS) of FW is shown in Figure 1. The highest methane yield of 321.5 L·kg^{-1} VS was obtained with an organic loading of 1% FW. The further increase of organic loading resulted in a significant decrease in methane production ($p < 0.05$), indicating inhibition of methanogenesis. This is due to the rapid rise in the concentration of VFAs with increasing organic load, causing an abrupt fall in pH (<5.3) and the consequent acidification of the anaerobic system (Figure 2). Previous studies have also reported a similar trend, in which the accumulation of superfluous VFAs was found to be inhibitory towards methanogens, thus causing a drop in their activity [49,50]. The instability of the FW digesting system can also be attributed to the low C:N ratio of FW. A low C:N ratio causes ammonia accumulation, reduced substrate degradation, and even an inhibition of methanogenesis [8].

The codigestion of FW with lignocellulosic biomass has shown to solve the problem by bringing in an opportunity to balance the nutrients content [15,51]. However, the complex structures of lignin and other cell wall polysaccharides make lignocellulosic waste materials hard to biodegrade and to be used by anaerobic microorganisms, leading to a lower biogas production. Therefore, to assess the biodegradability of LW by the anaerobic consortia used in this study, CS, PCG, and UBP were digested individually at different organic loading rates (1, 2.5, 5, and 10%; w/v VS). Figure 1 presents the results of the mono-digestion study. The methane yield and solid content for all the substrates tested showed an inverse relationship where 1% w/v substrate gave the highest methane yield. Cumulative methane yields obtained with 1% LW were 236.14 L·kg^{-1} VS for UBP followed by CS (111.05 L·kg^{-1} VS) and PCG (94.87 L·kg^{-1} VS). The methane yields obtained in this study were comparable with values (80 to 530 L·kg^{-1} VS) reported by others for the same or similar agricultural and energy crops [39,52–54]. Substrate concentrations of more than 1% w/v showed a significant decrease in methane production ($p < 0.05$). Nevertheless, these results indicated that the TMC can efficiently degrade LW anaerobically.

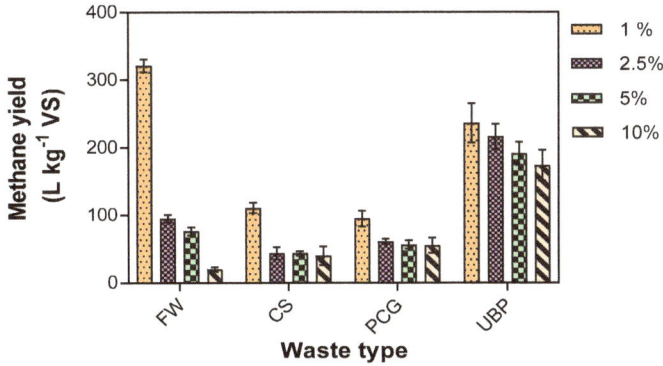

Figure 1. The methane yield from different organic loadings of food waste (FW), corn stover (CS), prairie cord grass (PCG), and unbleached paper (UBP). The data represented are the average values of the triplicate bottles. The error bars represent the standard deviation.

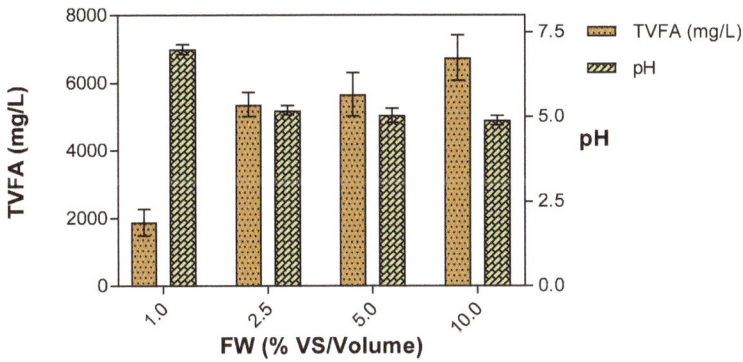

Figure 2. The Total Volatile Fatty Acid (TVFA) concentration and pH at the end of batch tests (the data represented are the average values of the triplicate bottles. The error bars represent the standard deviation).

3.3. Effect of Codigestion on Methane Yield and Digester Performance

High organic loading rates are desirable to reduce the digester volume and minimize capital losses, however, increased organic loading rates often result in VFA accumulation, nutrient imbalance, and sometimes the direct inhibition of methanogenesis, leading to complete digester failure [55–57]. Codigestion experiments were carried out with an organic loading of 2.5% FW to analyze the efficacy of codigestion in overcoming the limitations of high organic loadings observed in mono-digestion experiments. In the first set of tests, FWs (2.5% VS) were codigested with LW individually in three separate serum bottle digesters (T2, T3, and T4) where the concentration of CS, PCG, and UBP was 1.5% w/v (VS basis). As shown in Figure 3a, methane volume produced in these co-digestion studies with two waste mixtures was considerably higher than that obtained from the mono-digestion of FW ($p < 0.05$). This demonstrates positive synergistic effects among codigested wastes. The highest methane yield of 279.31 L·kg^{-1} was observed in T3 (FW + PCG), followed by 251.90 L·kg^{-1} in T2 (FW + CS) and 177.48 L·kg^{-1} in T4 (FW + UBP). Because a higher reduction in VS is indicative of better digester performance, a direct correlation between methane production and VS reduction was observed in this study ($p < 0.05$). Digester bottles with higher VS reduction yielded higher methane.

The highest VS reduction of 45.7% was achieved in T3 (FW + PCG) followed by 38.7% in T2 (FW + CS), and 38.3% in T4 (FW + UBP). The lowest biodegradability of the FW + UBP mixture is reflected in the mixture's lower VS removal and consequently in its lowest methane yield.

The second set of experiments where the co-digestion mixtures comprised of 2.5% w/v FW and two LW (0.75 % w/v each on VS basis) showed a similar pattern of results (Figure 3b). The highest methane yield of 305.45 L·kg^{-1} VS and a VS reduction of 68.5% were obtained with FW:CS:PCG (T5) mixture, while the lowest yield of 219.9 L·kg^{-1} VS and a VS reduction of 49.8% was observed with FW + UBP + CS (T7) mixture. The higher methane production for some codigestion mixtures over others can be attributed to the C/N ratio for these mixtures which were in the recommended range of 20:1 and 30:1 [18,58].

At a constant organic loading of 4 g VS, the highest methane yield among all the codigested mixtures was obtained with FW + PCG + CS (305.4 L·kg^{-1} VS), followed by FW + PCG (279.3 L·kg^{-1} VS). This marked a 94% and 74% increase in methane production compared to the mono-digestion of FW. Previous studies have also shown similar results but were performed at mesophilic temperatures and had higher hydraulic retention times (>30 days) in comparison to our study (18 days). Xu and Li, 2012 reported the highest methane yield of 304.4 L/kg VS feed with a codigestion mixture containing an equal amount of corn stover and dog food, 109% compared to methane yield from mono-digestion dog food, respectively [32]. In another study conducted by Brown and Li, 2013, the increased methane yields and volumetric productivities were observed with the co-digestion of food waste with yard waste. Food waste to yard waste mixture of 1:9 gave the highest volumetric productivity of 8.6 L methane L^{-1}, with a 43% VS reduction [31]. They attributed nutrient balance as one of the reasons for enhanced synergism during co-digestion.

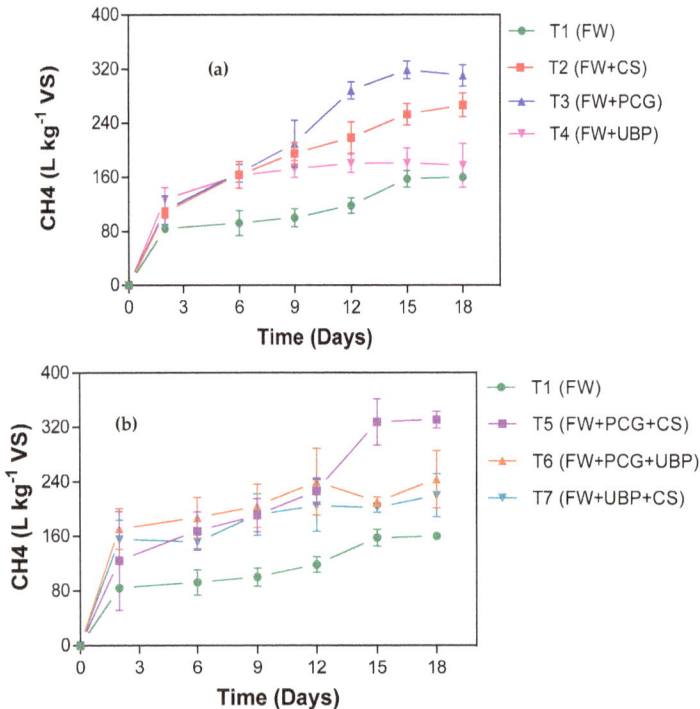

Figure 3. (**a**,**b**): methane production from the co-digestion of food and lignocellulosic waste (the data represented are the average values of triplicate bottles. The error bars represent the standard deviation).

3.4. Effect of Codigestion on Process Stability

Reactor acidification due to the buildup of VFAs is one of the most common reasons for the process deterioration and instability in anaerobic digesters [48]. VFAs are naturally produced by acidogenic and acetogenic bacteria and accumulate during conditions of substrate overload in the reactor. Additionally, a high VFA concentration is usually associated with a drop in pH and a breakdown of the buffering capacity of the reactor [31,59]. Hence, the final pH values in batch tests are another measure of system stability. Maintenance of the system pH in the proper range (6.5 and 7.6) is required for efficient anaerobic digestion [60]. Analysis of VFA concentrations and pH were done at the end of the batch experiments, and the results are shown in Figure 4. While acetic acid represented major portion of the total VFAs produced, propionic acid and butyric acid were produced in lower amounts (Supplementary Figure S1). The total VFA concentration was highest for T1 (FW) followed by T4 (FW + UBP) and T7 (FW + UBP + CS) which dropped the pH below 6 in all the three digesters creating instability. This instability directly affected the methane production which is evident from the lowest methane yield observed in T1, T4, and T7 (Table 5). The results indicated that these anaerobic reactors were operating under higher loading rates. Statistical analysis also showed that VFA concentration has a significant effect on methane yield with a p value of <0.05. The serum bottles having VFA concentrations of 1600–3300 mg/L did not show a drastic decrease in the pH of the system.

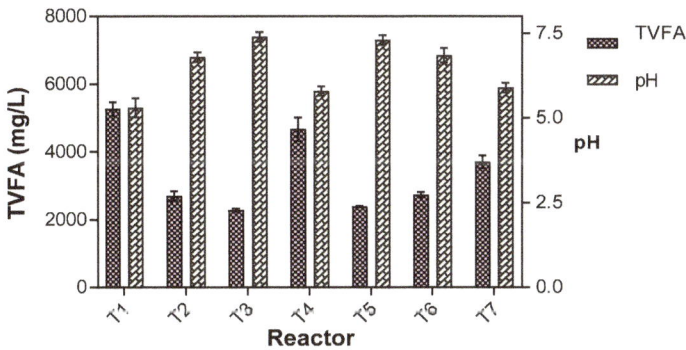

Figure 4. The Total Volatile Fatty acids (TVFA) and pH at the end of the codigestion tests (the data represented are the average values of triplicate bottles. The error bars represent the standard deviation).

Table 5. The effect of codigestion on digester performance.

Test Bottles	T1	T2	T3	T4	T5	T6	T7
Methane yield (L·kg^{-1} VS)	159.8 ± 2.4	251.9 ± 26.6	279.3 ± 7.2	177.5 ± 16.1	305.4 ± 23.8	243.6 ± 20.9	219.9 ± 10.9
VS reduction (%)	44 ± 1.4	57.4 ± 2.7	58.0 ± 0.9	47.4 ± 1.8	68.5 ± 2.2	51.1 ± 2.6	49.8 ± 1.5
Initial pH	7.1 ± 0.0	7.1 ± 0.0	7.2 ± 0.1	7.1 ± 0.1	7.2 ± 0.0	7.3 ± 0.3	7.3 ± 0.2
Final pH	5.3 ± 0.0	6.8 ± 0.1	7.4 ± 0.2	5.8 ± 0.1	7.3 ± 0.2	6.8 ± 0.1	5.9 ± 0.0

4. Conclusions

Biogas production using the abundant wastes as the substrate is a promising technology both in renewable energy and solid waste management sectors. FW and LW are attractive co-substrates for TAD as their complementary characteristics can overcome the limitations faced by their mono-digestion. The results suggested that codigesting FW with LW not only improved the system stability but also enhanced methane production, thus improving overall digester performance. At a constant organic loading of 4 g VS, the highest methane yield of 305.4 L·kg^{-1} VS was obtained with FW + PCG + CS followed by 279.3 L·kg^{-1} VS with FW + PCG. This marked a 94% and 74% increase in methane production compared to the mono-digestion of FW. Since the digester bottles with these codigestion mixtures showed increased system stability as well, they will be employed for scale-up studies.

The pretreatment of the LW increases the amount of available sugars for AD, but also increases the bioprocessing cost. As this study was conducted without employing any pretreatment, consortium development became a crucial step when the codigestion of LW was performed. The use of thermophilic temperature may increase the costs of the overall process but gives the added advantage of digesting higher organic loading of wastes at reduced hydraulic retention times. In addition, the surplus biogas produced can be used for heating purposes to maintain the reactor temperature in the thermophilic range. Nevertheless, the consortium developed in this study is significant for the co-digestion and can be used for scale-up studies. Codigestion mixtures showing the best performance and stability in this study will be used for scale-up studies to conduct the cost analysis. The present batch study can also direct further studies for setting up semi-continuous or continuous commercial scale plants for the codigestion of regionally abundant PCG with food waste and help in tackling the conditions limiting methane production from higher organic loadings.

Supplementary Materials: The following are available online at http://www.mdpi.com/1996-1073/11/8/2058/s1.

Author Contributions: A.D. conceived and designed the experiments. A.D., J.M., and K.F. performed the experiments. T.G. and A.K.T. analyzed the data, and A.D. and T.G. wrote the paper. R.K.S. contributed in the idea, experiment planning, writing, validity and interpretation of the results.

Funding: This research was supported by the US Air Force under the Biological Waste to Energy Project (Award #FA4819-14-C-0004) and Sustainable Development-Research Experience for Teachers (SD-RET) (Award #1711946). National Science Foundation also provided financial support in the form of the BuG ReMeDEE initiative (Award #1736255) The authors are grateful to the Department of Chemical and Biological Engineering at the South Dakota School of Mines and Technology for the research support.

Conflicts of Interest: The authors declare no conflict of interest.

Nomenclature

FW	Food wastes
LW	Lignocellulosic wastes
CS	Corn Stover
PCG	Prairie cordgrass
UBP	Unbleached paper
AD	Anaerobic digestion
VFA	Volatile Fatty acids
C/N	Carbon to nitrogen
TAD	Thermophilic anaerobic digestion
VS	Volatile solids
TS	Total solids
TMC	Thermophilic methanogenic consortia

References

1. Qian, Y.; Sun, S.; Ju, D.; Shan, X.; Lu, X. Review of the state-of-the-art of biogas combustion mechanisms and applications in internal combustion engines. *Renew. Sustain. Energy Rev.* **2017**, *69*, 50–58. [CrossRef]

2. Börjesson, P.; Mattiasson, B. Biogas as a resource-efficient vehicle fuel. *Trends Biotechnol.* **2008**, *26*, 7–13. [CrossRef] [PubMed]

3. Hengeveld, E.J.; van Gemert, W.J.T.; Bekkering, J.; Broekhuis, A.A. When does decentralized production of biogas and centralized upgrading and injection into the natural gas grid make sense? *Biomass Bioenergy* **2014**, *67*, 363–371. [CrossRef]

4. Ptak, M.; Koziołek, S.; Derlukiewicz, D.; Słupiński, M.; Mysior, M. Analysis of the Use of Biogas as Fuel for Internal Combustion Engines. In Proceedings of the 13th International Scientific Conference, Wrocław, Poland, 22–24 June 2016; Rusiński, E., Pietrusiak, D., Eds.; Springer International Publishing: Cham, Switzerland, 2017; pp. 441–450.

5. Dahl, R. A Second Life for Scraps: Making Biogas from Food Waste. *Environ. Health Perspect.* **2015**, *123*, A180–A183. [CrossRef] [PubMed]

6. Islam, M.; Park, K.-J.; Yoon, H.-S. Methane Production Potential of Food Waste and Food Waste Mixture with Swine Manure in Anaerobic Digestion. *J. Biosyst. Eng.* **2012**, *37*, 100–105. [CrossRef]

7. Braguglia, C.M.; Gallipoli, A.; Gianico, A.; Pagliaccia, P. Anaerobic bioconversion of food waste into energy: A critical review. *Bioresour. Technol.* **2018**, *248*, 37–56. [CrossRef] [PubMed]

8. Yong, Z.; Dong, Y.; Zhang, X.; Tan, T. Anaerobic co-digestion of food waste and straw for biogas production. *Renew. Energy* **2015**, *78*, 527–530. [CrossRef]

9. Staley, B.F.; de los Reyes, F.L.; Barlaz, M.A. Effect of Spatial Differences in Microbial Activity, pH, and Substrate Levels on Methanogenesis Initiation in Refuse. *Appl. Environ. Microbiol.* **2011**, *77*, 2381–2391. [CrossRef] [PubMed]

10. Manyi-Loh, C.E.; Mamphweli, S.N.; Meyer, E.L.; Okoh, A.I.; Makaka, G.; Simon, M. Microbial Anaerobic Digestion (Bio-Digesters) as an Approach to the Decontamination of Animal Wastes in Pollution Control and the Generation of Renewable Energy. *Int. J. Environ. Res. Public Health* **2013**, *10*, 4390–4417. [CrossRef] [PubMed]

11. Franke-Whittle, I.H.; Walter, A.; Ebner, C.; Insam, H. Investigation into the effect of high concentrations of volatile fatty acids in anaerobic digestion on methanogenic communities. *Waste Manag.* **2014**, *34*, 2080–2089. [CrossRef] [PubMed]

12. Lin, J.; Zuo, J.; Gan, L.; Li, P.; Liu, F.; Wang, K.; Chen, L.; Gan, H. Effects of mixture ratio on anaerobic co-digestion with fruit and vegetable waste and food waste of China. *J. Environ. Sci.* **2011**, *23*, 1403–1408. [CrossRef]

13. Zhang, C.; Xiao, G.; Peng, L.; Su, H.; Tan, T. The anaerobic co-digestion of food waste and cattle manure. *Bioresour. Technol.* **2013**, *129*, 170–176. [CrossRef] [PubMed]

14. Fitamo, T.; Boldrin, A.; Boe, K.; Angelidaki, I.; Scheutz, C. Co-digestion of food and garden waste with mixed sludge from wastewater treatment in continuously stirred tank reactors. *Bioresour. Technol.* **2016**, *206*, 245–254. [CrossRef] [PubMed]

15. Mata-Alvarez, J.; Dosta, J.; Macé, S.; Astals, S. Codigestion of solid wastes: A review of its uses and perspectives including modeling. *Crit. Rev. Biotechnol.* **2011**, *31*, 99–111. [CrossRef] [PubMed]

16. Mao, C.; Feng, Y.; Wang, X.; Ren, G. Review on research achievements of biogas from anaerobic digestion. *Renew. Sustain. Energy Rev.* **2015**, *45*, 540–555. [CrossRef]

17. Esposito, G.; Frunzo, L.; Panico, A.; d'Antonio, G. Mathematical modelling of disintegration-limited co-digestion of OFMSW and sewage sludge. *Water Sci. Technol.* **2008**, *58*, 1513–1519. [CrossRef] [PubMed]

18. Wang, X.; Lu, X.; Li, F.; Yang, G. Effects of temperature and carbon-nitrogen (C/N) ratio on the performance of anaerobic co-digestion of dairy manure, chicken manure and rice straw: Focusing on ammonia inhibition. *PLoS ONE* **2014**, *9*, e97265. [CrossRef] [PubMed]

19. Zhu, S.; Wu, Y.; Yu, Z.; Zhang, X.; Li, H.; Gao, M. The effect of microwave irradiation on enzymatic hydrolysis of rice straw. *Bioresour. Technol.* **2006**, *97*, 1964–1968. [CrossRef] [PubMed]

20. Zabed, H.; Sahu, J.N.; Boyce, A.N.; Faruq, G. Fuel ethanol production from lignocellulosic biomass: An overview on feedstocks and technological approaches. *Renew. Sustain. Energy Rev.* **2016**, *66*, 751–774. [CrossRef]

21. Zheng, Y.; Zhao, J.; Xu, F.; Li, Y. Pretreatment of lignocellulosic biomass for enhanced biogas production. *Prog. Energy Combust. Sci.* **2014**, *42*, 35–53. [CrossRef]

22. Ravindran, R.; Jaiswal, A.K. Microbial Enzyme Production Using Lignocellulosic Food Industry Wastes as Feedstock: A Review. *Bioengineering* **2016**, *3*, 30. [CrossRef] [PubMed]

23. Bellasio, M.; Mattanovich, D.; Sauer, M.; Marx, H. Organic acids from lignocellulose: Candida lignohabitans as a new microbial cell factory. *J. Ind. Microbiol. Biotechnol.* **2015**, *42*, 681–691. [CrossRef] [PubMed]

24. Kumar, M.; Revathi, K.; Khanna, S. Biodegradation of cellulosic and lignocellulosic waste by Pseudoxanthomonas sp R-28. *Carbohydr. Polym.* **2015**, *134* (Suppl. C), 761–766. [CrossRef] [PubMed]

25. The National Petroleum Council. *Renewable Natural Gas for Transportation: An Overview of the Feedstock Capacity, Economics, and GHG Emission Reduction Benefits of RNG as a Low-Carbon Fuel*; The National Petroleum Council: Washington, DC, USA, 2013; p. 44.

26. (EIA), U.S. Energy Information Administration. *Monthly Energy Review: Energy Consumption by Sector*; U.S. Energy Information Administration: Washington, DC, USA, 2013.

27. Valentine, J.; Clifton-Brown, J.; Hastings, A.; Robson, P.; Allison, G.; Smith, P. Food vs. fuel: The use of land for lignocellulosic 'next generation' energy crops that minimize competition with primary food production. *GCB Bioenergy* **2012**, *4*, 1–19. [CrossRef]

28. Awais, M.; Alvarado-Morales, M.; Tsapekos, P.; Gulfraz, M.; Angelidaki, I. Methane Production and Kinetic Modeling for Co-digestion of Manure with Lignocellulosic Residues. *Energy Fuels* **2016**, *30*, 10516–10523. [CrossRef]

29. Kabir, M.M.; del Pilar Castillo, M.; Taherzadeh, M.J.; Sárvári Horváth, I. Effect of the *N*-Methylmorpholine-*N*-Oxide (NMMO) Pretreatment on Anaerobic Digestion of Forest Residues. *BioResources* **2013**, *8*, 5409–5423. [CrossRef]

30. Achinas, S.; Achinas, V.; Euverink, G.J.W. A Technological Overview of Biogas Production from Biowaste. *Engineering* **2017**, *3*, 299–307. [CrossRef]

31. Brown, D.; Li, Y. Solid state anaerobic co-digestion of yard waste and food waste for biogas production. *Bioresour. Technol.* **2013**, *127*, 275–280. [CrossRef] [PubMed]

32. Xu, F.; Li, Y. Solid-state co-digestion of expired dog food and corn stover for methane production. *Bioresour. Technol.* **2012**, *118*, 219–226. [CrossRef] [PubMed]

33. Jabeen, M.; Zeshan; Yousaf, S.; Haider, M.R.; Malik, R.N. High-solids anaerobic co-digestion of food waste and rice husk at different organic loading rates. *Int. Biodeterior. Biodegr.* **2015**, *102*, 149–153. [CrossRef]

34. Mulat, D.G.; Huerta, S.G.; Kalyani, D.; Horn, S.J. Enhancing methane production from lignocellulosic biomass by combined steam-explosion pretreatment and bioaugmentation with cellulolytic bacterium Caldicellulosiruptor bescii. *Biotechnol. Biofuels* **2018**, *11*, 19. [CrossRef] [PubMed]

35. Gerardi, M.H. *The Microbiology of Anaerobic Digesters*; John Wiley & Sons, Inc.: Hoboken, NJ, USA, 2003; p. 177.

36. Lloret, E.; Pastor, L.; Pradas, P.; Pascual, J.A. Semi full-scale thermophilic anaerobic digestion (TAnD) for advanced treatment of sewage sludge: Stabilization process and pathogen reduction. *Chem. Eng. J.* **2013**, *232*, 42–50. [CrossRef]

37. Haider, M.R.; Zeshan; Yousaf, S.; Malik, R.N.; Visvanathan, C. Effect of mixing ratio of food waste and rice husk co-digestion and substrate to inoculum ratio on biogas production. *Bioresour. Technol.* **2015**, *190*, 451–457. [CrossRef] [PubMed]

38. Li, Y.; Zhang, R.; Liu, X.; Chen, C.; Xiao, X.; Feng, L.; He, Y.; Liu, G. Evaluating Methane Production from Anaerobic Mono- and Co-digestion of Kitchen Waste, Corn Stover, and Chicken Manure. *Energy Fuels* **2013**, *27*, 2085–2091. [CrossRef]

39. Li, L.; Yang, X.; Li, X.; Zheng, M.; Chen, J.; Zhang, Z. The Influence of Inoculum Sources on Anaerobic Biogasification of NaOH-treated Corn Stover. *Energy Sources Part A* **2010**, *33*, 138–144. [CrossRef]

40. Pommier, S.; Llamas, A.M.; Lefebvre, X. Analysis of the outcome of shredding pretreatment on the anaerobic biodegradability of paper and cardboard materials. *Bioresour. Technol.* **2010**, *101*, 463–468. [CrossRef] [PubMed]

41. Dhiman, S.S.; Shrestha, N.; David, A.; Basotra, N.; Johnson, G.R.; Chadha, B.S.; Gadhamshetty, V.; Sani, R.K. Producing methane, methanol and electricity from organic waste of fermentation reaction using novel microbes. *Bioresour. Technol.* **2018**, *258*, 270–278. [CrossRef] [PubMed]

42. Rice, E.W.; Bridgewater, L. *Standard Methods for the Examination of Water and Wastewater*; American Public Health Association: Washington, DC, USA, 2012.

43. Sluiter, A.; Hames, B.; Ruiz, R.; Scarlata, C.; Sluiter, J.; Templeton, D.; Crocker, D. *Determination of Structural Carbohydrates and Lignin in Biomass*; National Renewable Energy Laboratory: Golden, CO, USA, 2008.

44. Kaparaju, P.; Serrano, M.; Thomsen, A.B.; Kongjan, P.; Angelidaki, I. Bioethanol, biohydrogen and biogas production from wheat straw in a biorefinery concept. *Bioresour. Technol.* **2009**, *100*, 2562–2568. [CrossRef] [PubMed]

45. Dhar, H.; Kumar, P.; Kumar, S.; Mukherjee, S.; Vaidya, A.N. Effect of organic loading rate during anaerobic digestion of municipal solid waste. *Bioresour. Technol.* **2016**, *217*, 56–61. [CrossRef] [PubMed]

46. Bibra, M.; Kumar, S.; Wang, J.; Bhalla, A.; Salem, D.R.; Sani, R.K. Single pot bioconversion of prairie cordgrass into biohydrogen by thermophiles. *Bioresour. Technol.* **2018**, *266*, 232–241. [CrossRef] [PubMed]

47. Zhang, L.; Lee, Y.-W.; Jahng, D. Anaerobic co-digestion of food waste and piggery wastewater: Focusing on the role of trace elements. *Bioresour. Technol.* **2011**, *102*, 5048–5059. [CrossRef] [PubMed]

48. Tanimu, I.; Mohd Ghazi, T.; Razif Harun, M.; Idris, A. Effect of Carbon to Nitrogen Ratio of Food Waste on Biogas Methane Production in a Batch Mesophilic Anaerobic Digester. *Int. J. Innovation Technol.* **2014**, *5*, 116–119.

49. Zhai, N.; Zhang, T.; Yin, D.; Yang, G.; Wang, X.; Ren, G.; Feng, Y. Effect of initial pH on anaerobic co-digestion of kitchen waste and cow manure. *Waste Manag.* **2015**, *38*, 126–131. [CrossRef] [PubMed]

50. Zhang, T.; Mao, C.; Zhai, N.; Wang, X.; Yang, G. Influence of initial pH on thermophilic anaerobic co-digestion of swine manure and maize stalk. *Waste Manag.* **2015**, *35*, 119–126. [CrossRef] [PubMed]

51. Zainab Ziad Ismail, A.R.T. Assessment of anaerobic co-digestion of agro wastes for biogas recovery: A bench scale application to date palm wastes. *Int. J. Energy Environ.* **2014**, *5*, 591–600.

52. Tong, X.; Smith, L.H.; McCarty, P.L. Methane fermentation of selected lignocellulosic materials. *Biomass* **1990**, *21*, 239–255. [CrossRef]

53. Ge, X.; Xu, F.; Li, Y. Solid-state anaerobic digestion of lignocellulosic biomass: Recent progress and perspectives. *Bioresour. Technol.* **2016**, *205*, 239–249. [CrossRef] [PubMed]

54. Shiralipour, A.; Smith, P.H. Conversion of biomass into methane gas. *Biomass* **1984**, *6*, 85–92. [CrossRef]

55. Schmidt, T.; Pröter, J.; Scholwin, F.; Nelles, M. Anaerobic digestion of grain stillage at high organic loading rates in three different reactor systems. *Biomass Bioenergy* **2013**, *55*, 285–290. [CrossRef]

56. Ferguson, R.M.W.; Coulon, F.; Villa, R. Organic loading rate: A promising microbial management tool in anaerobic digestion. *Water Res.* **2016**, *100*, 348–356. [CrossRef] [PubMed]

57. Cantrell, K.B.; Ducey, T.; Ro, K.S.; Hunt, P.G. Livestock waste-to-bioenergy generation opportunities. *Bioresour. Technol.* **2008**, *99*, 7941–7953. [CrossRef] [PubMed]

58. Esposito, G.; Frunzo, L.; Giordano, A.; Liotta, F.; Panico, A.; Pirozzi, F. Anaerobic co-digestion of organic wastes. *Rev. Env. Sci. Biotechnol.* **2012**, *11*, 325–341. [CrossRef]

59. Ahring, B.K. Methanogenesis in thermophilic biogas reactors. *Antonie Leeuwenhoek* **1995**, *67*, 91–102. [CrossRef] [PubMed]

60. Labatut, R.A.; Gooch, C.A. *Monitoring of Anaerobic Digestion Process to Optimize Performance and Prevent System Failure*; Department of Biological and Environmental Engineering, Cornell University: Ithaca, NY, USA, 2014.

energies

MDPI

Article

Enhancing Biochemical Methane Potential and Enrichment of Specific Electroactive Communities from Nixtamalization Wastewater using Granular Activated Carbon as a Conductive Material

David Valero [1], Carlos Rico [2], Blondy Canto-Canché [3], Jorge Arturo Domínguez-Maldonado [1], Raul Tapia-Tussell [1], Alberto Cortes-Velazquez [4] and Liliana Alzate-Gaviria [1,*]

[1] Renewable Energy Unit, Yucatan Center for Scientist Research, Mérida CP 97302, Mexico; david.valero@cicy.mx (D.V.); joe2@cicy.mx (J.A.D.-M.); rtapia@cicy.mx (R.T.-T.)
[2] Department of Water and Environmental Science and Technologies, University of Cantabria, 39005 Santander, Spain; carlos.rico@unican.es
[3] Biotechnology Unit, Yucatan Center for Scientist Research, Mérida CP 97205, Mexico; cantocanche@cicy.mx
[4] GeMBio Laboratory, Yucatan Center for Scientist Research, Mérida CP 97205, Mexico; betocv@cicy.mx
* Correspondence: lag@cicy.com; Tel.: +52-999-930-0760

Received: 11 July 2018; Accepted: 1 August 2018; Published: 13 August 2018

Abstract: Nejayote (corn step liquor) production in Mexico is approximately 1.4×10^{10} m^3 per year and anaerobic digestion is an effective process to transform this waste into green energy. The biochemical methane potential (BMP) test is one of the most important tests for evaluating the biodegradability and methane production capacity of any organic waste. Previous research confirms that the addition of conductive materials significantly enhances the methane production yield. This study concludes that the addition of granular activated carbon (GAC) increases methane yield by 34% in the first instance. Furthermore, results show that methane production is increased by 54% when a GAC biofilm is developed 10 days before undertaking the BMP test. In addition, the electroactive population was 30% higher when attached to the GAC than in control reactors. Moreover, results show that electroactive communities attached to the GAC increased by 38% when a GAC biofilm is developed 10 days before undertaking the BMP test, additionally only in these reactors *Geobacter* was identified. GAC has two main effects in anaerobic digestion; it promotes direct interspecies electron transfer (DIET) by developing an electro-active biofilm and simultaneously it reduces redox potential from −223 mV to −470 mV. These results suggest that the addition of GAC to biodigesters, improves the anaerobic digestion performance in industrial processed food waste.

Keywords: biochemical methane potential; redox potential reduction; direct interspecies electron transfer; electroactive biofilm; Nejayote; granular activated carbon

---

1. Introduction

The Mexican corn tortilla industry produces between 2.2–3.5 m$^3 \cdot$t^{-1} of wastewater from processed corn [1,2]. The wastewater (Nejayote) presents an environmental problem for Mexican society due to its malodorous nature and its composition. It is imperative to have an explicit waste management strategy to deal with the unpleasant by-products, and it is also essential to ascertain the potential uses of the wastewater in order to harness and capitalise on the large quantities that are produced through nixtamalization. Nixtamalization is the ancient Aztec process of cooking corn with lime to produce corn masa [3]. Water, corn, and lime are cooked at 80 °C during three hours and is then steeped for 15 hours [4]. Nowadays, the process was carried out on an industrial scale but based on the process carried out by the Aztecs known as nixtamalization. This complex waste has a pH in the

9–14 range, a chemical oxygen demand (COD) between 3.5 and 40 g·L^{-1} and a phenol concentration of 4.2 g·L^{-1} [3–6]. This wide range is a result of the proportions of water corn and lime that vary in different nixtamal plants [4,5,7]. Its organic matter content makes nejayote a suitable waste for anaerobic digestion process rather than any aerobic process due to the huge amount of energy that is required to degrade its high organic load [8–10]. A biochemical methane potential (BMP) test is the most efficient means of identifying the methane production capacity, in anaerobic digestion, for any organic waste [11–13]. Additionally, it can be useful to identify future inhibition and adaptation problems before scaling the process [14]. Given the lack of previous studies in relation to BMP testing of nejayote, this article is the first step in determining the biochemical methane potential via direct interspecies electro transfer (DIET) of nejayote in anaerobic digestion and it provides an opportunity to develop a perspective on its possible performance on a larger scale.

Anaerobic digestion process is based in four main steps: hydrolysis, acidogenesis, acetogenesis, and methanogenesis. In methanogenesis, step electrons are transported from acetoclastic bacteria to hydrogenotrophic archaea to reduce carbon dioxide to methane. This transport can be done by shuttle moleculas as hydrogen or formate. Traditionally, interspecies hydrogen transfer (IHT) and interspecies formate transfer (IFT) have been intrinsically linked with interspecies electron transfer (IET). However, these can present problems in relation to H$_2$ partial pressure and formate concentration, respectively [15]. More recently, there has been a greater production of studies referring to DIET, which has become more widely accepted. DIET is another form of IET that can be produced through pili or employing conductive materials [16–19]. Exoelectrogenic bacteria and electrotrophic methanogens are the partners that participate in DIET [20–22]. Short chain volatile fatty acids (VFA) and alcohols are the biodegradable compounds in the syntrophic association between exoelectrogenic bacteria and electrotrophic methanogens [23,24]. For the first time in 2014, the DIET between exoelectrogenic bacteria and electrotrophic methanogens was evidenced through pili in a *Geobacter-Methanosaeta* co-culture [18]. Not all exoelectrogenic bacteria are able to generate pili; however, the addition of conductive materials increases the genera of bacteria that are able to supply pili the genera of bacteria that are able to participate in DIET. In cases where the bacteria are not able to generate pili, conductive materials fulfil the role of pili in terms of facilitating electron exchange [23,25,26]. Granular activated carbon has shown high performance in promoting DIET in different studies [26,27]. Moreover, its low cost [28,29] and its high conductivity (3000 µS·cm^{-1}) [21] when compared to other conductive materials such as biochar (4 µS·cm^{-1}) [30], magnetite (160 µS·cm^{-1}) [31], or stainless steel (667 µS·cm^{-1}) [32], make it a preferential material to promote DIET in order to improve methane production yields [17,33]. Additionally, metallic materials, such as stainless steel, have corrosion problems [34].

Within the past three years, through the microorganism identification of biofilms developed in conductive materials, new exoelectrogenic bacteria have been discovered, such as *Thauera*, *Sporanaerobacter*, *Enterococcus*, *Pseudomonas*, *Anaerolinaceae*, *Bacteroides*, *Streptococcus*, *Syntrophomonas*, *Sulfurospirillum*, *Caloramator*, *Tepidoanaerobacter*, *Coprothermobacter*, *Clostridium senso stricto*, *Peptococcaceae*, and *Bacillaceae*, also new electrotrophic methanogen genera have been found able to participate in DIET, such as *Methanosarcina*, *Methanobacterium*, *Methanolinea*, *Methanothrix*, *Methanoregula*, and *Methanospirillum* [17,27,35–44]. However, the number of bacteria that are capable of participating in DIET through conductive materials is still unknown [45].

Industrial processed food is one of the most important industries that employ anaerobic digestion to degrade water pollution due to its high COD [10]. Most previous studies have used simple substrates such as ethanol or short chain VFA, or commercial waste, such as dog food [17,23,27]. At present, there are no studies that use a substrate of industrial food waste using conductive materials to promote DIET.

This study is the first step towards developing an understanding of the behaviour and growth of electroactive communities in a real wastewater, and thus providing an indication of the improvements in methane production and VFA and COD degradation, which can be made by implementing DIET through conductive materials on a large scale.

2. Materials and Methods

2.1. Nejayote and Inoculum

Nejayote was collected from an industrial corn flour plant in Yucatan State (Mexico). Nejayote was preserved in a cold box at 4 °C from the corn flour plant to the laboratory, which is in accordance with the Standard Test Method 1060 for the collection and preservation of samples [46]. Prior to undertaking the BMP test, the nejayote was characterized according to the following parameters: pH, total suspended solids (TSS), total solids (TS) and volatile solids (VS), COD, total nitrogen (TN), phosphates (PO_4^{-3}), ammonia nitrogen (NH_3N), sulfate (SO_4^{-2}), and total alkalinity (TA).

The inoculum composition employed was made according to Poggi Varaldo et al. [47], while using 30 g·L^{-1} of deep soil, 300 g·L^{-1} of cattle manure, 150 g·L^{-1} of pig manure, 1.5 g·L^{-1} of commercial Na_2CO_3, and 1 L of tap water. The inoculum was characterized for TA, TS, and VS.

2.2. BMP Test

The trial was carried out in triplicate in 110 mL serum bottles that were capped with rubber septum sleeve stoppers. Three experimental conditions were assayed. In controls, the reactors (N) were filled with 60 g of inoculum and nejayote at a ratio of 2 based on VS [12]. In three reactors (N0) 3 g of GAC was added [33] at the beginning of the BMP test. In the other three reactors (N10), inoculum and 3 g of GAC were added to serum bottles ten days before undertaking the BMP test. After this time, nejayote was introduced in anoxic conditions. All of the tests were carried out for 30 days at 38 °C with an automatic agitation of 100 rpm [48]. All reactors were flushed with nitrogen in an anoxic chamber to avoid the presence of oxygen. Three blanks with 60 g of inoculum were tested to measure the inoculum methane potential. The methane production measurements are expressed at 0 °C and standard pressure of 760 mmHg (NCTP) in dry conditions.

Gas production measurement and numerical calculation was according to Valero et al. [49]. Headspace pressure was measured employing a digital pressure transducer (ifm Germany type PN78 up to 2 bars) with a syringe that was connected to pierce the septum. Statistical analysis was performed with the Analysis ToolPak in Microsoft Excel (Excel 2016 (v16.0)).

Once the BMP test was finished, COD, VFA, and redox potential were measured for all of the assays.

2.3. GAC Conductivity

GAC was introduced in PVC tubes, of specific lengths and known surface areas, to calculate its conductivity. A source meter (Keithley 2400, Cleveland, OH, USA) was used to measure the resistance of GAC. Once the resistance was determined, conductivity was calculated employing the following equations:

$$R = \rho \frac{A}{L} \tag{1}$$

$$K = \frac{1}{\rho} \tag{2}$$

where R is the resistance measured with Keithley 2400, A is the area of PVC tube, L is the tube length, ρ is the GAC resistivity, and K is the GAC conductivity. The process was undertaken in triplicate and the result was 3690 µS·cm^{-1} [50].

2.4. Analytical Methods

Biogas characterization was determined on a Molesieve column (30 m long, 0.53 mm internal diameter, and 0.25 µm film thickness) in a gas chromatograph (Clarus 500-Perkin Elmer, Waltham, MA, USA) with the thermal conductivity detector (TCD). Nitrogen was used as gas carrier and temperatures of 75, 30 and 200 °C were used for the injector, oven and detector, respectively. VFA were determined by gas chromatography using (Clarus 500-Perkin Elmer, Waltham, MA, USA). The column employed was Agilent J&W (30 m long, 0.53 mm internal diameter) with a flame ionization detector (FID). Before determining VFA by chromatography, samples were filtered, acidified with phosphoric acid, and then centrifuged. Conductivity, temperature, pH, TSS, TS, and VS were analyzed, following standard methods [46]. Colorimetric methods (Hach Company DR-890, Loveland, CO, USA) were used to determine COD, TN, PO_4^{-3}, NH_3-N, and SO_4^{-2}. Redox potential was determined employing a redox potential sensor (Extech RE300, Nashua, NH, USA).

2.5. Microbial Community Analysis

After the BMP test, sludge from N, N0, and N10 and GAC from N0 and N10 was kept at $-80\ °C$ [27]. DNA was extracted from the five samples N sludge (S), N0 sludge (S0), N10 sludge (S10), GAC biofilm from N0 (C0), and GAC biofilm from N10 (C10). Three DNA extraction of each sample were undertook, and then the DNA from each of samples were mixed before perform the sequencing. DNA from each sample was isolated with a DNA extraction kit ZymoBIOMICSTM DNA miniprep Kit, (Zymo Research, Irvine, CA, USA), in accordance with manufacturer´s protocol. The DNA pellet was resuspended in 50 µL of TE buffer (10mM Tris-HCl (pH 8) and 1mM EDTA (pH 8)). DNA purity and DNA concentration were checked by employing a NanoDrop ND-1000 spectrophotometer (NanoDrop Technologies, Seattle, WA, USA). The A260/280 ratios were over 1.6 and DNA concentration in the range of 50–100 ng·µL^{-1}. The primers employed for amplifying 16S rRNA genes (bacteria and archaea) were 341F and 805R. DNA was sent to Macrogen Inc. (Seoul, Korea) who performed sequencing on an Illumina MiSeq platform. MG-RAST software was used to analyze bacteria and archaeal communities through GREENGENES and RDP II databases [51].

2.6. Scanning Electron Microscopy (SEM)

GAC was analyzed by SEM (SEM, model JSM-6360LV, JEOL, Tokyo, Japan). In order to establish the existence of microbial communities attached to GAC, three samples of N10 were mounted on a metallic stub using double-sided adhesive tape, each being coated with a 15 nm gold layer and observed at 20 kV.

3. Results

3.1. Nejayote and Inoculum

Nejayote collected from industrial corn flour plant showed basic pH values of 10.2. The inoculum, developed in the laboratory, had a basic pH but with values that were lower than nejayote, in this case, the pH reached 8.1. The VS percentage obtained for nejayote was $0.85 \pm 0.01\%$, whereas inoculum values reached $2.69 \pm 0.01\%$. TS results were $1.22 \pm 0.01\%$ and $4.62 \pm 0.01\%$ for nejayote and inoculum, respectively. Nejayote and inoculum VS/TS ratios that were obtained were 0.69 and 0.55, respectively. TA concentration results were similar for nejayote and inoculum, reaching values of 1799 ± 116 and 1652 ± 147 mg $CaCO_3$ L^{-1}, respectively. COD concentration determined for nejayote was $15,433 \pm 827$ mg·L^{-1}. TSS levels for nejayote of 2676 ± 512 mg·L^{-1} were recorded. Other parameters that were characterized for nejayote were NH_3-N 4.6 ± 0.1 mg·L^{-1}, TN 95 ± 4 mg·L^{-1}, SO_4^{-2} 22 ± 2 mg·L^{-1}, and PO_4^{-3} 59 ± 1.2 mg·L^{-1}.

3.2. BMP Test

Methane production did not peak for the three assays N, N0, and N10 until the third day of BMP testing. The rate at which methane was produced spiked during the third day and it resulted in 39, 46, and 45 L CH_4 kg^{-1}VS day^{-1} for N, N0, and N10, respectively. As shown in Figure 1A, the first significant difference between N10 reactors and N and N0 reactors, was on the fifth day (t-student $p < 0.05$). N10 reactors resulted in higher methane production levels of 26 L CH_4 kg^{-1}VS day^{-1}, whilst N and N0 reactors produced 17 and 10 L CH_4 kg^{-1}VS day^{-1}, respectively.

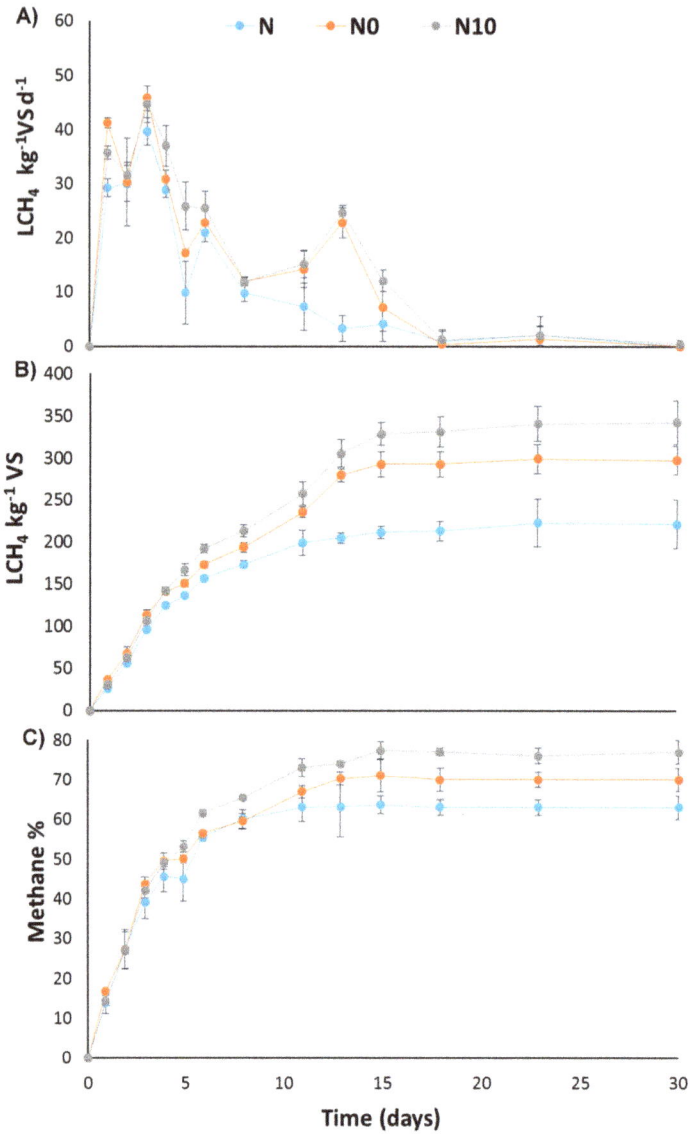

Figure 1. Biochemical Methane Potential (BMP) curves for control reactor (N), reactors with granular activated carbon (GAC) (N0) and reactors with biofilm GAC developed before undertaking the BMP test (N10). (**A**) Daily methane production (**B**) Cumulative methane production (**C**) methane biogas percentage. Mean values ± SD from triplicate assays.

DIET or IHT occurs in the last step of the anaerobic digestion process, however adding GAC promotes DIET instead of IHT in N0 and N10 reactors [45]. The only compounds that can be oxidized by exoelectrogenic bacteria are alcohols and VFA [24,52]. When these compounds are available to stimulate the production of methane through DIET, a significant difference can be seen between DIET reactors through GAC and those reactors without conductive materials (t-student analyses $p < 0.05$). As shown in Figure 1A, from the tenth day until the fifteenth day of experimentation, N0 and N10 consistently produced methane over 10 L CH$_4$ kg^{-1}VS day^{-1}, while N methane production was under 10 L CH$_4$ kg^{-1}VS day^{-1}. Moreover, on one occasion during the same period, N0 and N10 actually reached more than 22 L CH$_4$ kg^{-1}VS day^{-1}, while N did not produce values above 10 L CH$_4$ kg^{-1}VS day^{-1}.

As highlighted in Figure 1B, the cumulative methane production results draws focus to the significant differences between N, N0, and N10 (t-student analysis $p < 0.05$). The final methane generation for N, N0 and N10 reached values of 222 ± 23 L CH$_4$ kg^{-1}VS, 297 ± 10 L CH$_4$ kg^{-1}VS and 342 ± 29 L CH$_4$ kg^{-1}VS, respectively. N0 methane production was 34% higher than N, however the difference between N10 and N soared staggeringly to register an increase of 54%. Additionally, results show that the N10 methane volume generated was 15% higher than N0.

Methane production was predicted relative to biochemical oxygen demand (BOD) concentration. Nejayote BOD is around 5870 ± 1900 mg L^{-1} [3,5,53,54]. Theoretically, 1 kg of BOD produces 350 L methane [55]. Thus, the methane production from nejayote expected through anaerobic digestion process without DIET promotion was 73.96 mL CH$_4$ (C$_6$H$_{12}$O$_6$ (DBO) → 3 CO$_2$ + 3 CH$_4$) [55]. In this study, real methane production in N reactors was 69.12 ± 8.96 mL CH$_4$. Likewise, through the DIET process an increase between 20–33% in the production of methane has been reported due to a greater recovery of electrons in the reduction of CO$_2$ to CH$_4$ through VFA (acetate, propionate, and butyrate) and alcohol (ethanol) degradation [18,25,38,56–58]. According to these increments, the methane production promoting DIET expected was in the range of 88.75–98.36 mL CH$_4$. In this study, real methane production for N0 and N10 was 92.71 ± 5.38 and 106.69 ± 8.15 mL CH$_4$, respectively. These values are in the expected range.

The first difference in the biogas composition came after the sixth day, when N10 methane percentage was recorded as 5% higher than N and N0, as illustrated in Figure 1C. This difference remained constant between N10 and N0 throughout the BMP test.

Between days eleven and eighteen, N0 methane composition was 5% higher than N. Thus, N10 was 10% higher than N. The maximum methane concentration was reached on the fifteenth day with values of 64%, 71%, and 77% for N, N0, and N10, respectively. After the eighteenth day, there was no further biogas production in all of the reactors, and therefore the biogas composition plateaued to show similar levels of methane concentration in all reactors.

Once the BMP test was completed, the contents of the different reactors were analyzed to determine pH, redox potential, VFA, and COD, as shown in Table 1. There were no significant differences in pH values between the three assays. A huge reduction of redox potential was produced in reactors with GAC, with the measured value for N being recorded as −222 ± 7 mV, when compared with −466 ± 1 mV and −471 ± 2 mV for N0 and N10, respectively. VFA composition was formed by acetic acid 203 mg·L^{-1} and butyric acid 90 mg·L^{-1} in N reactors, while in N0 and N10 assays the VFA composition was made up solely of acetic acid because butyric acid was only present in N reactors. COD concentration was higher in N reactors than in N0 and N10, with values of 532 ± 10 mg·L^{-1}, 307 ± 21 mg·L^{-1}, and 218 ± 14 mg·L^{-1}, respectively.

Table 1. Reactors analysis after BMP test.

Parameter	N	N0	N10
pH	7.3 ± 0.11	7.2 ± 0.03	7.3 ± 0.02
Redox potential (mV)	−223 ± 5	−467 ± 1	−471 ± 2
Acetic acid (mg·L^{-1})	203 ± 12	139 ± 9	109 ± 8
Butyric acid (mg·L^{-1})	91 ± 5		
COD (mg·L^{-1})	532 ± 10	307 ± 21	218 ± 14

Control reactor (N), Reactor with granular activated carbon (N0), reactor with GAC added 10 days before undertaking biochemical methane potential test.

3.3. Microbial Community Analysis

As shown in Figure 2, the four main bacteria phylum in all of the samples were *Actinobacteria*, *Bacteroidetes*, *Firmicutes*, and *Proteobacteria*. *Firmicutes* was the dominant phylum in S, comprising almost 50% of the total bacteria. In sludges S0 and S10, there was a lower percentage of *Firmicutes*, forming 26% and 25% of the total bacteria, respectively. Furthermore, the percentage of *Firmicutes* was even lower in C0 and C10, measuring 15% and 16%, respectively. In contrast, *Bacteroidetes* follows the opposite trend to *Firmicutes*. *Bacteroidetes* had its highest percentage in C0 and C10, attached to the carbon, reaching 48% in C0 and 43% in C10, while in S it accounted for 18% of the microbial community. The *Proteobacteria* phylum presented a difference between the samples: it was detected at 17% in the C10 biofilm; whilst, for the other samples, it never reached 10%.

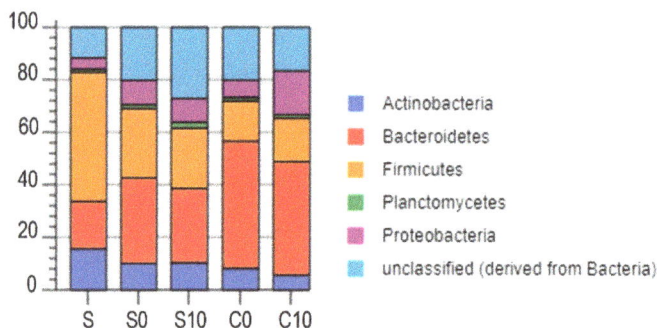

Figure 2. Bacteria phylum relative abundance. Phylum level with relative abundance lower than 1% were included in unclassified group. Sludge from control reactor (S), sludge from reactor with granular activated carbon (GAC) (S0), sludge from reactor with GAC biofilm developed before undertaking the BMP test (S10), GAC biofilm from reactor with GAC (C0), and GAC biofilm from reactor with GAC biofilm developed before undertaking the BMP test (C10).

Figure 3A illustrates the classification of bacteria by genus proportion. *Clostridium* anaerobic fermentative bacteria was highly developed in S, comprising 46% of the bacterial community, whereas it was less than 21% in S0 and S10, and interestingly, it accounted for 7% of the bacterial community attached to GAC in C0 and C10. *Parabacteroides* were developed in GAC biofilm and the sludge of GAC reactors with a relative abundance of 8%, 13%, 12%, and 14% for S0, S10, C0, and C10, respectively; however, there was no presence detected in S. *Parabacteroides* that is is able to produce VFA as an end product of fermentation. This, in turn, can be oxidized by exoelectrogenic bacteria. Additionally, an enrichment of this genus in microbial fuel cells [31] and microbial electrolysis cells [32] has been observed.

Geobacter, an exoelectrogenic bacteria widely accepted as an exoelectrogen DIET partner [7,8], was only detected in the C10 sample, where the biofilm was developed 10 days before undertaking the BMP test. The amount of *Geobacter* in C10, whilst not dominant, remains a substantive 14% of the overall bacteria community and 82% of *Proteobacteria* phylum. The most dominant genus on the surface of both GAC samples C0 and C10 was *Prolixibacter*, with 31% and 21%, respectively. *Prolixibacter* had a relative abundance of 16%, 18%, and 11% in S, S0, and S10, respectively. It must be mentioned that *Prolixibacter* has been reported as a possible exoelectrogenic bacteria in microbial electrolysis cells [33] and in marine-sediment fuel cells [34].

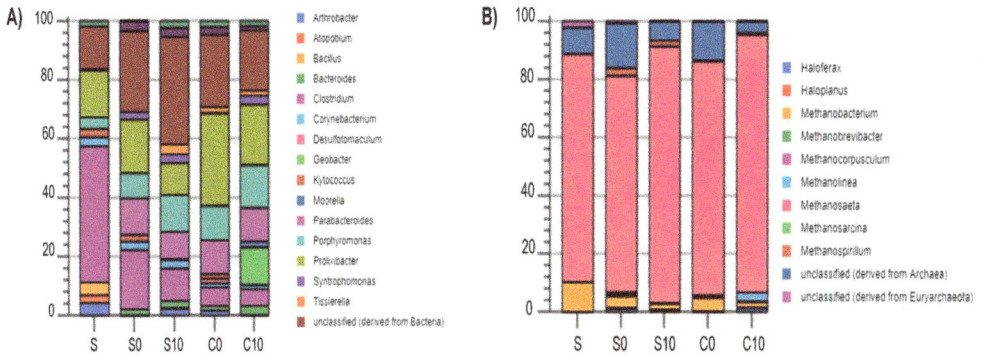

Figure 3. Bacteria (**A**) and Archaea community structure (**B**) at genus level. Genus level with relative abundance lower than 1% were included in unclassified groups. Sludge from control reactor (S), sludge from reactor with granular activated carbon (GAC) (S0), sludge from reactor with GAC biofilm developed before undertaking the BMP test (S10), GAC biofilm from reactor with GAC (C0) and GAC biofilm from reactor with GAC biofilm developed before undertaking the BMP test (C10).

A low enrichment for *Syntrophomonas* and *Bacteroides* was detected in the microbial community analysis. Both showed a relative abundance of 3% in C0 and C10, but contrastingly they were not detected in S. Both genera have only been reported as exoelectrogen DIET partners in studies with complex waste [19], as similar to nejayote.

In this study, the total amount of electroactive bacteria (*Geobacter, Proxilibacter, Bacteroides,* and *Syntrophomonas*) was 41%, 37%, 17%, 22%, and 16% for C10, C0, S10, S0, and S, respectively. This indicates an exoelectrogenic bacteria enrichment of 24% in C10 when compared with the control reactor, and an 18% higher electroactive bacteria community in C0 than S.

The archaeal community was dominated by *Methanosaeta* in all of the reactors, attached to the carbons and suspended in all sludges. It was most prevalent in C10, reaching 89% of the archaeal community, while in S, its percentage was 78%. *Methanosaeta* has been established as one of the main methanogens able to accept electrons in DIET [7]. *Methanolinea,* another methanogen associated with electrotrophic capacity in DIET [35], only appeared on the carbon surface of C10 with a relative abundance of less than 5% of the archaeal community.

The similiarities of the microbial community structures in the five samples were analyzed at the genus level by hierarchically clustered heat analysis (Figure 4). The heatmap includes 17 genera. C10 is the sample with the highest concentration of different genera. The communities attached to GAC, C10, and C0 clustered together finally. These samples share a high similitude when compared with sludge samples (S, S0, S10).

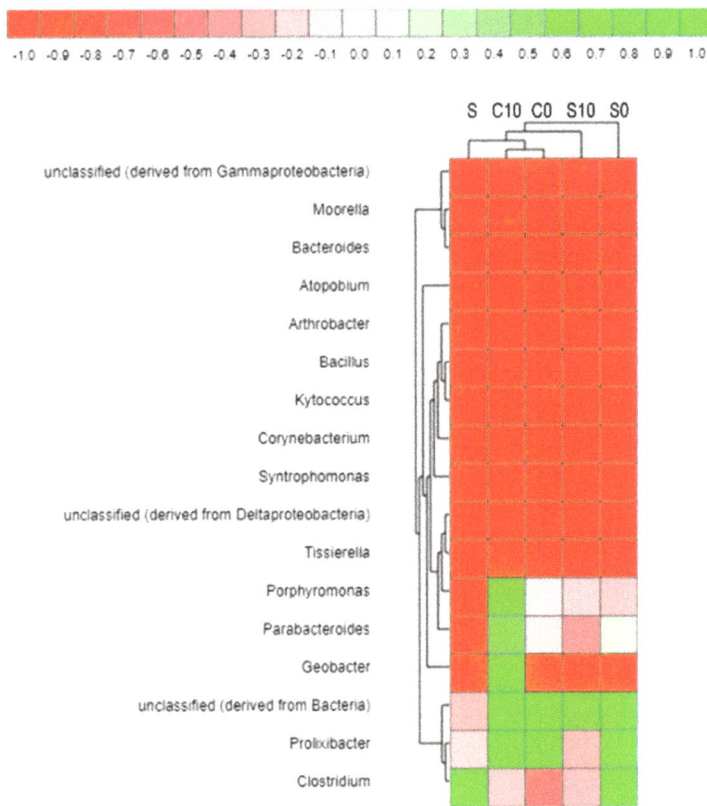

Figure 4. Heatmap of bacteria at genus level. Genus level with relative abundance lower than 1% were included in unclassified groups. Sludge from control reactor (S), sludge from reactor with granular activated carbon (GAC) (S0), sludge from reactor with GAC biofilm developed before undertaking the BMP test (S10), GAC biofilm from reactor with GAC (C0), and GAC biofilm from reactor with GAC biofilm developed before undertaking the BMP test (C10). Heatmap showing the 17 genera with significant difference of relative abundances among five samples. Heatmap is color-coded based scale (from red (−1) less abundance to green (+1) more abundance).

3.4. Scanning Electron Microscopy (SEM)

Through SEM, it was possible to examine the carbon surface and its porosity free of microbial communities before adding GAC to the reactors (Figure 5A). Additionally, microbial communities attached to GAC are clearly visible in Figure 5B. The exoelectrogen partner *Geobacter* was identified in the microbiological analysis attached to the GAC in sample C10 with a relative abundance of 14% in the bacterial community. Geobacter has been described as rod-shaped bacteria with a length between 2–4 μm [59]. In Figure 5C, bacteria can be visualized with the same morphological characteristics described above. This, coupled with the high percentage of *Geobacter* attached to GAC, makes it plausible to identify them in the Figure 5C. *Methanosaeta* is a rod-shaped archaea with average dimensions between 2 and 6 μm in length by single cell. Cells are enclosed inside an annular, striated sheathed structure, and they are separated by partitions forming a filamentous structure. *Methanosaeta* filaments can reach 100 μm of length [60]. In Figure 5D, it is possible to see *Methanosaeta* communities. Besides *Methanosaeta* was the dominant archaea in the microbial communities analysis, making its identification easy by SEM.

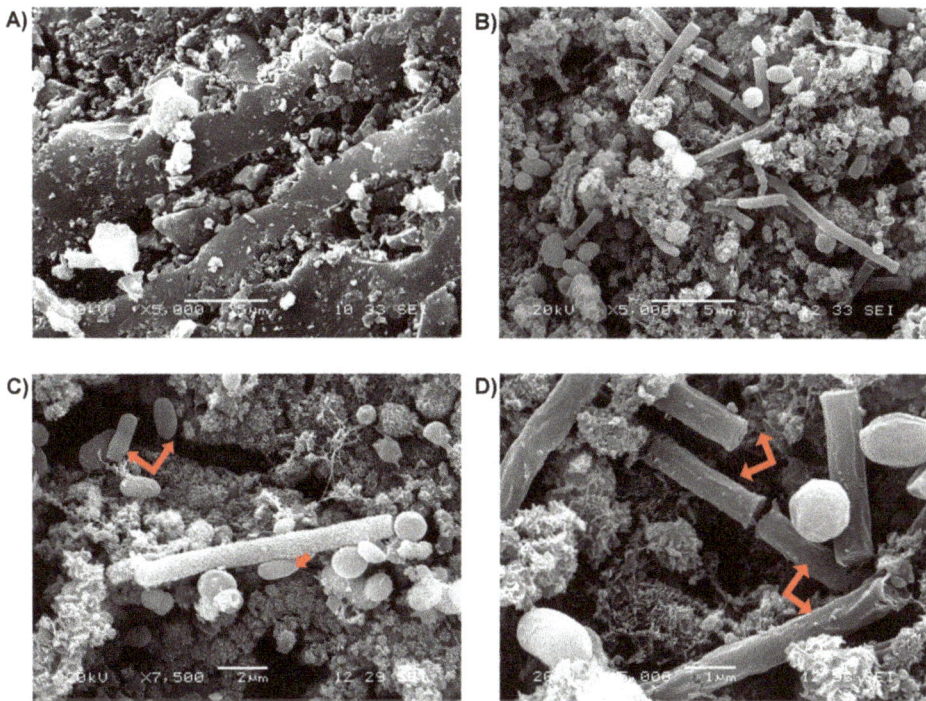

Figure 5. Scanning electron micrographs of the granular activated carbon surface (**A**), biofilm attached granular activated carbon (**B**), Geobacter communities attached to granular activated carbon (**C**), and Methanosaeta communities attached to granular activated carbon (**D**) after biochemical methane potential test (30 days).

4. Discussion

4.1. Nejayote and Inoculum

Table 2 shows the results of the physicochemical characterization of nejayote, as compared with the results that were obtained in other studies. As it is a wastewater that may come from an industrial process or from a more traditional process, which is not always done in the same way, the nejayote presents a high level of heterogeneity in the results of its composition, as seen in the wide range that is given for the majority of parameters. According to Ibarra-Mendívil et al. [61], different lime concentrations of 0%, 0.5%, and 1% were used for nixtamalization and pH values of 4.75, 7.72, and 11.01 were obtained, respectively. Small modifications in proportions of lime, cooking, and rest times, and different varieties of corn can give rise to notable variations in the composition of nejayote. The presence of lime in the process leads to the nejayote having a basic pH. This study is in the range expected with a value recorded of 10.2.

Table 2. Nejayote characterization compared with other studies.

Parameter	Results	Other Studies
pH	10.2	6.3–11.6 [1,3–5,62,63]
COD (mg·L^{-1})	15,433 ± 826	3430–40,058 [1,4,5,63]
NH$_3$-N (mg·L^{-1})	4.65 ± 0.05	2 [1]
TN (mg·L^{-1})	95.33 ± 3.7	209–428 [1,5]
SO$_4$$^{-2}$ (mg·L^{-1})	22.5 ± 2.5	13 [1]
PO$_4$$^{-3}$ (mg·L^{-1})	58.75 ± 1.2	7.6–1321 [1,5]
TA (mgCaCO$_3$·L^{-1})	1799 ± 116	5768 [4]
TSS (mg·L^{-1})	2676 ± 512	1810–8340 [3,4]
TS%	1.22 ± 0.01	0.34–2.5 [1,4,63]
VS%	0.84 ± 0.01	0.24–1.55 [1,63]
VS/TS	0.69 ± 0.01	0.70 [63]

The COD concentration of nejayote is within the range observed in the different studies (Table 2), with a value of 15,433 ± 826 mg·L^{-1}, coming mainly from the pericarp tissue of corn. The content in TN was 95.33 ± 3.7 mg·L^{-1}, lower than the composition analyzed in other studies, although this concentration is not sufficient to cause inhibition in anaerobic digestion. Concentrations between 1.7 and 14 g·L^{-1} cause a 50% reduction in methane production [64]. Both nitrogen and phosphorus are necessary for the good performance of an anaerobic reactor, since they are indispensable nutrients for bacteria. Concentrations lower than 0.3 mg·L^{-1} of phosphorus prevent the formation of the microbial community [65]. The nejayote that is characterized in this study has a concentration of 58.75 ± 1.25 mg·L^{-1}, which will aid a good performance in the BMP test. The concentration of SO$_4$ is 22.5 ± 2.5 mg·L^{-1}, which is a little higher than the previously reported 13 mg·L^{-1} (Table 2). The COD/SO$_4$ ratio is 686, so the presence of sulfur-reducing products and possible inhibition effects in the anaerobic digestion process due to the byproducts generated by them will be non-existent [66]. The percentages in both VS and TS are within the range referenced in Table 2. The values are 0.85 ± 0.01% VS and 1.22 ± 0.01% TS.

The inoculum has an alkalinity of 1652.5 ± 147 mg CaCO$_3$·L^{-1}. This contribution of alkalinity to the medium will absorb possible acidification processes that occur during anaerobic digestion due to the formation of VFAs [67]. The VS percentage of the inoculum is 2.67% ± 0.1. This percentage aids the good performance in the anaerobic digestion process [68].

4.2. BMP Test

The results of this study clearly demonstrate that generating a previous GAC biofilm before starting the wastewater anaerobic digestion enhances methane biogas production and results in an overall improved anaerobic digestion performance. The syntrophic relationship between exoelectrogenic bacteria and electrotrophic archaea is essential to enable DIET. Recently, it has been widely recognised that exoelectrogenic bacteria can only oxidize alcohols and VFA to send electrons to reduce carbon dioxide to methane within DIET [24,52]. The advantages of promoting DIET instead of IHT in methane production yield do not become obviously apparent until there are increased alcohols and VFA concentrations and they are available to be oxidized by exoelectrogenic bacteria. While examining the results of the study, it is important to state that the first difference between control reactors and GAC reactors in methane production did not appear until the seventh day in N0 and N10 (Figure 1A), due to the insufficient concentration of VFA and alcohols, which are available in the medium to be oxidized. In previous studies undertaken in batch assays with conductive materials, the difference in methane production emerges earlier due to the substrate employed. In these studies, the carbon sources used were synthetic wastewater, which has alcohols and VFA readily available from the beginning of experimentation [36,38]. Nejayote, on the other hand, is a complex waste that requires hydrolysis and acidogenesis steps before VFA and alcohols are available, which explains the delay in DIET.

After the BMP test was concluded, the COD concentration was higher in N reactors than N0 and N10 reactors. This fact explains part of the difference in accumulated methane production. Butyrate was detected in N medium, whilst there was no presence of butyrate in N0 and N10 reactors. The total VFA concentration of N was more than double that of N0 and N10, which can, in part, go some way towards explaining the difference between the methane potentials in the tests; however, it can not be held accountable for all of the differences. These differences in methane yield could also be due to some remnant organic matter that has not been yet converted in VFA in N reactors [49].

In Table 3, the results are compared with studies that were developed with conductive materials. It is noted in Table 3 that the percentage in the increase of methane production is within plausible values. An increase of 54% when the biofilm is previously generated in GAC and 34% without the previous formation of GAC biofilm are reasonable results.

Table 3. Methane production increase with conductive materials.

Reactor Volume (mL)	Substrate	Conductive Material	Methane Production Increase (%)	Time (Days)	Reference
120	Nejayote	GAC	54	30	This study
120	Nejayote	GAC	34	30	This study
250	Ethanol	Graphene	25	12	[38]
500	Synthetic wastewater	GAC	86	43	[17]
120	Synthetic wastewater	Magnetite	32	20	[56]
250	Glucose	Graphene	51	15	[36]
250	Sludge treatment plant	GAC	17	20	[33]

4.3. GAC Conductivity

DIET was stimulated by the addition of GAC in the reactors. GAC works as a support for microorganism communities and as electron conductor between exoelectrogenic bacteria and electrotrophic archaea. DIET using stainless steel as conductive support had a kinetic advantage that is 108 times greater than IHT to compete for the electron donor [69]. Stainless steel conductivity is 667 $\mu S \cdot cm^{-1}$, while the GAC conductivity employed in this study was 3600 $\mu S \cdot cm^{-1}$. Similar conductive materials employed before, such as magnetite [56] and biochar [30], possessed conductivities of 160 $\mu S \cdot cm^{-1}$ and 5 $\mu S \cdot cm^{-1}$, respectively. GAC has substantially higher values in terms of conductivity; making it an ideal material to use given its surface area, its high conductivity, and it is economically low cost [28,29].

4.4. Redox Potential

One of the most important effects, which GAC causes in the improvement of the methanogenic performance, is the change that it produces in redox potential. Redox potential must be under -200 mV for methanogenic activity to be possible. When redox potential is higher than -200 mV, this activity is virtually negligible [70]. All of the reactors in this study were under this barrier. As shown in the results, redox potential changed from -222 ± 7 mV to -471 ± 2 mV from control reactors N to N0 and N10 reactors. GAC had a direct effect on reducing redox potential by generating a reductive atmosphere, which is supported by evidence in other studies [71] where carbon nanotubes caused a reduction in redox potential in an abiotic environment. The potential energy difference ($\Delta E°$) between the electron donor and electron acceptor is directly proportional to the energy (ATP) that a microbe could employ from the metabolic process [72]. Consequently, a high reducing micro-environment, produced by the GAC presence in the anaerobic reactor, expedites electron donation to provide a thermodynamic driving force for accomplishing the electrophilic attack in terms of reduction process, increasing carbon dioxide reduction to methane, which could be one of the key points for improving the reactor performance [71].

4.5. Relationship Between Redox Potential, Methane Production and Archaea Communities

There was evidence that low redox potential stimulated the syntrophic relationship between exoelectrogenic bacteria and electrotrophic methanogens, resulting in elevated levels in methane concentration, which can be observed in Figure 1C. A previous study observed that a pure methanogenic archaeal culture with carbon materials had higher methane yield than a control reactor without carbon material, due to lower redox potential [71]. That means that, without an exoelectrogenic partner to send electrons through DIET to electrotrophic methanogens, methane yield is improved due to redox potential reduction. In a pure culture of hydrogenotrophic methanogens, the redox potential was under -450 mV, causing an increase in its population [70]. This change may be a result of the fact that it requires less thermodynamic energy to undertake an electrophilic attack by these methanogens to reduce carbon dioxide to methane. The low redox potential that was obtained in this study (-471 mV), by adding GAC, caused an increase of 10% in the acetoclastic methanogen (*Methanosaeta*) population.

4.6. Microbial Analysis

Only *Methanosaeta* and *Methanosarcina* are able to generate methane from acetate [73]. Additionally, *Methanosaeta* is recognised as an electron acceptor partner in DIET, reducing carbon dioxide to methane, hence in the control reactor, *Methanosaeta* played a vital role as acetate consumer to produce methane. However, in N0 and N10 reactors, due to its attachment to the GAC surface, *Methanosaeta* worked as an electron acceptor to reduce carbon dioxide to methane. *Methanobacterium* is a hydrogenotrophic archaea that tended to be in suspended solution instead of being attached to the carbon surface [74], although *Methanobacterium* has been detected in conductive support as an electron acceptor in DIET [44]. *Methanobacterium* had a relative abundance over 10% in the N reactor, whereas when detected in the sludges of N0 and N10, it was consistently under 5%. In this study, *Methanobacterium* was responsible for reducing hydrogen partial pressure through IHT in control reactor N. This is because *Methanobacterium* needs to be attached to conductive material to accept electrons. *Methanolinea*, which is a hidrogenotrophic archaea, was only detected in C10 samples with a proportion of 3%. *Methanolinea* and *Methanobacterium* can accept electrons when they are attached to a conductive support; therefore, they are able to work as electrotrophic archaea in DIET [17]. In this study, the two different hidrogenotrophic archaea played distinct roles: one (*Methanobacterium*) was suspended in the bulk and participated in IHT by controlling hydrogen partial pressure, and other (*Methanolinea*) was attached to the GAC surface and accepted electrons to reduce carbon dioxide to methane.

A variety of electroactive bacteria were enriched in the GAC surface; *Geobacter, Bacteroides,* and *Syntrophomonas*. All of these bacteria genera have been acknowledged as participants in DIET as electron donors [40,58]. *Syntrophomonas* and *Bacteroides* were detected with a relative abundance of 2% and 3% for C0 and C10, respectively. However, there was no presence of *Syntrophomonas* and *Bacteroides* in N reactors. Previous studies have shown that *Bacteroides* have the potential to grow significantly on the anode of a bioelectrochemical system and on the surface of carbon cloth in anaerobic reactors. This genus is able to extracellularly transfer electrons to ferric iron and donate electrons to electrotrophic methanogens via DIET [40,75]. There is no evidence that *Syntrophomonas* are able to transport electrons extracellularly, but notable growth has been reported when acting as a DIET partner in anaerobic systems [16].

Geobacter were only detected in C10 with a relative abundance of 13%. It must be concluded that this was because the biofilm was developed before introducing nejayote into the reactor and *Geobacter* could then grow under favorable conditions. The absence of *Geobacter* in C0 can be explained by the fact that *Geobacter* cannot efficiently degrade complex organic waste [76]. An additional supposition is that *Geobacter* is not viable in complex waste under hard conditions, as relative high salinity environment [27,40]. *Geobacter* enrichment was produced during the ten days before undertaking the BMP test in N10 reactors. This enrichment was not possible in N0 reactors due to the fact that GAC was added to the reactors at the same time as nejayote, and other exoelectrogenic bacteria coming from nejayote or inoculum, overtaking *Geobacter* and adapting better to the new environmental conditions.

Other enrichment produced in the BMP test involved *Prolixibacter*. This genus has not been reported before in any study that is related with DIET in anaerobic digestion as exoelectrogenic bacteria. The capacity of *Prolixibacter* to transport electrons extracellularly has not yet been concluded, but it has been reported as a fermentative bacterium in sediment fuel cells that could be able to transfer electrons extracellularly [77]. In addition, its enrichment in microbial electrolysis cells has been reported [78].

The GAC biofilm developed previously in N10 reactors had two positive effects in the exoelectrogenic bacteria community. C10 exoelectrogenic community was 4% higher than the C0 exoelectrogenic community and the *Geobacter* genus only appeared in C10 samples.

Microbial analysis has shown a relative abundance of 12% and 14% of *Parabacteroides* attached to N0 GAC and N10 GAC, respectively, while it was not detected in the N reactor. This genus is known for its capacity to degrade polysaccharides to acetate [79]. The main function of these fermentative bacteria was to convert organic matter to compounds that were available to be oxidized by exoelectrogenic bacteria. *Parabacteroides* played an important role in DIET, producing the essential VFA required by the exoelectrogenic bacteria to degrade and donate electrons to electrotrophic methanogens. *Parabacteroidetes* growth has been reported in microbial fuel cells [80], assuming the same role as in this study, however it has not been reported in studies that are related with the promotion of DIET in anaerobic reactors. This is a result of the fact that most of these studies employ synthetic wastewater that is easily converted to methane. One of the difficulties previously detected in DIET was the hydrolysis and acidogenesis limit of VFA and alcohol concentration that was required to take advantage of DIET instead of IHT [52]. *Parabacteroidetes* growth in the conductive material greatly assists in resolving this issue, by generating the crucial carbon source for exoelectrogenic bacteria to oxidize. The electroactive community in C10 was 8% higher than in C0 and 38% higher than S.

5. Conclusions

GAC addition had a positive impact on the BMP test. GAC had four main resulting effects on the anaerobic digestion process; increased methane production, redox potential reduction, DIET promotion, and electroactive biofilm development. A low redox potential made an electrotrophic attack to reduce carbon dioxide to methane easier. Additionally, DIET was promoted instead of IHT resulting in higher methane concentration and greater methane production. Furthermore, the electroactive community population intensified when the biofilm was developed 10 days before the assay, and the reactors with previous biofilm obtained a better yield than those without. This study was the first step to a better understanding of the relationship between DIET promotion and a complex wastewater, such as nejayote. Undertaking a study using bigger reactors is necessary to further support these promising results, in order to fully realise the potential of testing these conclusions on a large scale.

Author Contributions: D.V. design of experiments, performed physico-chemical parameter analysis and BMP tests, participated in the evaluation of microorganisms communities and drafted the manuscript. C.R. helped co-ordinate the project and revised the manuscript. B.C.-C. developed the DNA extraction protocol, participated in the evaluation of microorganisms communities and revised the manuscript. J.A.D.-M. performed VFA analysis and participated in physico-chemical parameter analysis and BMP test. R.T.-T. developed the DNA extraction protocol and performed MG RAST analysis. A.C.-V. participated in developing the DNA extraction protocol and performed DNA extraction. L.A.-G. designed and coordinated this work alongside helping to draft the manuscript. All authors read and approved the final manuscript.

Funding: Mexican Council for Science (CONACYT) for the financial support granted to carry out this study through grant 738499 awarded for doctoral studies.

Acknowledgments: The authors thank Q.I Tanit Toledano Thompson for technical assistance in SEM micrographs. Also we would like to thank M.S. Miguel Alonso Tzec Simá for technical assistance in DNA extraction.

Conflicts of Interest: The authors declare that they have no competing interests.

References

1. España-Gamboa, E.; Domínguez-Maldonado, J.A.; Tapia-Tussell, R.; Chale-Canul, J.S.; Alzate-Gaviria, L. Corn industrial wastewater (nejayote): A promising substrate in Mexico for methane production in a coupled system (APCR-UASB). *Environ. Sci. Pollut. Res.* **2017**, *25*, 712–722. [CrossRef] [PubMed]

2. Salmerón-Alcocer, A.; Rodríguez-Mendoza, N.; Pineda-Santiago, V.; Cristiani-Urbina, E.; Juárez-Ramírez, C.; Ruiz-Ordaz, N.; Galíndez-Mayer, J. Aerobic treatment of maize-processing wastewater (nejayote) in a single-stream multi-stage bioreactor. *J. Environ. Eng. Sci.* **2003**, *2*, 401–406. [CrossRef]

3. Rosentrater, K.A. A review of corn masa processing residues: Generation, properties, and potential utilization. *Waste Manag.* **2006**, *26*, 284–292. [CrossRef] [PubMed]

4. Valderrama-Bravo, C.; Gutiérrez-Cortez, E.; Contreras-Padilla, M.; Rojas-Molina, I.; Mosquera, J.C.; Rojas-Molina, A.; Beristain, F.; Rodríguez-García, M.E. Constant pressure filtration of lime water (nejayote) used to cook kernels in maize processing. *J. Food Eng.* **2012**, *110*, 478–486. [CrossRef]

5. Meraz, K.A.S.; Vargas, S.M.P.; Maldonado, J.T.L.; Bravo, J.M.C.; Guzman, M.T.O.; Maldonado, E.A.L. Eco-friendly innovation for nejayote coagulation–flocculation process using chitosan: Evaluation through zeta potential measurements. *Chem. Eng. J.* **2016**, *284*, 536–542. [CrossRef]

6. García-Zamora, J.L.; Sánchez-González, M.; Lozano, J.A.; Jáuregui, J.; Zayas, T.; Santacruz, V.; Hernández, F.; Torres, E. Enzymatic treatment of wastewater from the corn tortilla industry using chitosan as an adsorbent reduces the chemical oxygen demand and ferulic acid content. *Process Biochem.* **2015**, *50*, 125–133. [CrossRef]

7. Argun, M.S.; Argun, M.E. Treatment and alternative usage possibilities of a special wastewater: Nejayote. *J. Food Process Eng.* **2018**, *41*, e12609. [CrossRef]

8. Desloover, J.; De Clippeleir, H.; Boeckx, P.; Du Laing, G.; Colsen, J.; Verstraete, W.; Vlaeminck, S.E. Floc-based sequential partial nitritation and anammox at full scale with contrasting N_2O emissions. *Water Res.* **2011**, *45*, 2811–2821. [CrossRef] [PubMed]

9. Massara, T.M.; Komesli, O.T.; Sozudogru, O.; Komesli, S.; Katsou, E. A Mini Review of the Techno-environmental Sustainability of Biological Processes for the Treatment of High Organic Content Industrial Wastewater Streams. *Waste Biomass Valoriz.* **2017**, *8*, 1665–1678. [CrossRef]

10. Alexandropoulou, M.; Antonopoulou, G.; Lyberatos, G. Food industry waste's exploitation via anaerobic digestion and fermentative hydrogen production in an up-flow column reactor. *Waste Biomass Valoriz.* **2016**, *7*, 711–723. [CrossRef]

11. Veluchamy, C.; Kalamdhad, A.S. Biochemical methane potential test for pulp and paper mill sludge with different food/microorganisms ratios and its kinetics. *Int. Biodeterior. Biodegradation* **2017**, *117*, 197–204. [CrossRef]

12. Raposo, F.; Fernández-Cegrí, V.; De la Rubia, M.A.; Borja, R.; Béline, F.; Cavinato, C.; Demirer, G.; Fernández, B.; Fernández-Polanco, M.; Frigon, J.C. Biochemical methane potential (BMP) of solid organic substrates: Evaluation of anaerobic biodegradability using data from an international interlaboratory study. *J. Chem. Technol. Biotechnol.* **2011**, *86*, 1088–1098. [CrossRef]

13. Kafle, G.K.; Chen, L. Comparison on batch anaerobic digestion of five different livestock manures and prediction of biochemical methane potential (BMP) using different statistical models. *Waste Manag.* **2016**, *48*, 492–502. [CrossRef] [PubMed]

14. Hansen, T.L.; Schmidt, J.E.; Angelidaki, I.; Marca, E.; la Cour Jansen, J.; Mosbæk, H.; Christensen, T.H. Method for determination of methane potentials of solid organic waste. *Waste Manag.* **2004**, *24*, 393–400. [CrossRef] [PubMed]

15. Stams, A.J.M.; Plugge, C.M. Electron transfer in syntrophic communities of anaerobic bacteria and archaea. *Nat. Rev. Microbiol.* **2009**, *7*, 568–577. [CrossRef] [PubMed]

16. Zhao, Z.; Zhang, Y.; Quan, X.; Zhao, H. Evaluation on direct interspecies electron transfer in anaerobic sludge digestion of microbial electrolysis cell. *Bioresour. Technol.* **2016**, *200*, 235–244. [CrossRef] [PubMed]

17. Lee, J.-Y.; Lee, S.-H.; Park, H.-D. Enrichment of specific electro-active microorganisms and enhancement of methane production by adding granular activated carbon in anaerobic reactors. *Bioresour. Technol.* **2016**, *205*, 205–212. [CrossRef] [PubMed]

18. Rotaru, A.-E.; Shrestha, P.M.; Liu, F.; Shrestha, M.; Shrestha, D.; Embree, M.; Zengler, K.; Wardman, C.; Nevin, K.P.; Lovley, D.R. A new model for electron flow during anaerobic digestion: Direct interspecies electron transfer to Methanosaeta for the reduction of carbon dioxide to methane. *Energy Environ. Sci.* **2014**, *7*, 408–415. [CrossRef]

19. Baek, G.; Kim, J.; Kim, J.; Lee, C. Role and Potential of Direct Interspecies Electron Transfer in Anaerobic Digestion. *Energies* **2018**, *11*, 107. [CrossRef]

20. Dubé, C.-D.; Guiot, S.R. Direct Interspecies Electron Transfer in Anaerobic Digestion: A Review. In *Biogas Science and Technology*; Guebitz, M.G., Bauer, A., Bochmann, G., Gronauer, A., Weiss, S., Eds.; Springer International Publishing: Cham, Switzerland, 2015; pp. 101–115. ISBN 978-3-319-21993-6.

21. Liu, F.; Rotaru, A.-E.; Shrestha, P.M.; Malvankar, N.S.; Nevin, K.P.; Lovley, D.R. Promoting direct interspecies electron transfer with activated carbon. *Energy Environ. Sci.* **2012**, *5*, 8982–8989. [CrossRef]

22. Chen, S.; Rotaru, A.-E.; Shrestha, P.M.; Malvankar, N.S.; Liu, F.; Fan, W.; Nevin, K.P.; Lovley, D.R. Promoting Interspecies Electron Transfer with Biochar. *Sci. Rep.* **2014**, *4*, 5019. [CrossRef] [PubMed]

23. Zhao, Z.; Zhang, Y.; Holmes, D.E.; Dang, Y.; Woodard, T.L.; Nevin, K.P.; Lovley, D.R. Potential enhancement of direct interspecies electron transfer for syntrophic metabolism of propionate and butyrate with biochar in up-flow anaerobic sludge blanket reactors. *Bioresour. Technol.* **2016**, *209*, 148–156. [CrossRef] [PubMed]

24. Lovley, D.R. Happy together: Microbial communities that hook up to swap electrons. *ISME J.* **2016**, *11*, 327–336. [CrossRef] [PubMed]

25. Chen, S.; Rotaru, A.-E.; Liu, F.; Philips, J.; Woodard, T.L.; Nevin, K.P.; Lovley, D.R. Carbon cloth stimulates direct interspecies electron transfer in syntrophic co-cultures. *Bioresour. Technol.* **2014**, *173*, 82–86. [CrossRef] [PubMed]

26. Dang, Y.; Sun, D.; Woodard, T.L.; Wang, L.-Y.; Nevin, K.P.; Holmes, D.E. Stimulation of the anaerobic digestion of the dry organic fraction of municipal solid waste (OFMSW) with carbon-based conductive materials. *Bioresour. Technol.* **2017**, *238*, 30–38. [CrossRef] [PubMed]

27. Dang, Y.; Holmes, D.E.; Zhao, Z.; Woodard, T.L.; Zhang, Y.; Sun, D.; Wang, L.-Y.; Nevin, K.P.; Lovley, D.R. Enhancing anaerobic digestion of complex organic waste with carbon-based conductive materials. *Bioresour. Technol.* **2016**, *220*, 516–522. [CrossRef] [PubMed]

28. Hesas, R.H.; Arami-Niya, A.; Daud, W.M.A.W.; Sahu, J.N. Preparation and characterization of activated carbon from apple waste by microwave-assisted phosphoric acid activation: Application in methylene blue adsorption. *BioResources* **2013**, *8*, 2950–2966.

29. Arami Niya, A.; Daud, W.M.A.W.; Mjalli, F.S.; Abnisa, F.; Shafeeyan, M.S. Production of microporous palm shell based activated carbon for methane adsorption: Modeling and optimization using response surface methodology. *Chem. Eng. Res. Des.* **2012**, *90*, 776–784. [CrossRef]

30. Zhao, Z.; Zhang, Y.; Yu, Q.; Dang, Y.; Li, Y.; Quan, X. Communities stimulated with ethanol to perform direct interspecies electron transfer for syntrophic metabolism of propionate and butyrate. *Water Res.* **2016**, *102*, 475–484. [CrossRef] [PubMed]

31. Parker, R.; Tinsley, C.J. Electrical conduction in magnetite. *Phys. Status Solidi* **1976**, *33*, 189–194. [CrossRef]

32. Guo, K.; Soeriyadi, A.H.; Feng, H.; Prévoteau, A.; Patil, S.A.; Gooding, J.J.; Rabaey, K. Heat-treated stainless steel felt as scalable anode material for bioelectrochemical systems. *Bioresour. Technol.* **2015**, *195*, 46–50. [CrossRef] [PubMed]

33. Yang, Y.; Zhang, Y.; Li, Z.; Zhao, Z.; Quan, X.; Zhao, Z. Adding granular activated carbon into anaerobic sludge digestion to promote methane production and sludge decomposition. *J. Clean. Prod.* **2017**, *149*, 1101–1108. [CrossRef]

34. Mo, J.; Steen, S.M.; Zhang, F.-Y.; Toops, T.J.; Brady, M.P.; Green, J.B. Electrochemical investigation of stainless steel corrosion in a proton exchange membrane electrolyzer cell. *Int. J. Hydrogen Energy* **2015**, *40*, 12506–12511. [CrossRef]

35. Hu, Q.; Sun, D.; Ma, Y.; Qiu, B.; Guo, Z. Conductive polyaniline nanorods enhanced methane production from anaerobic wastewater treatment. *Polymer* **2017**, *120*, 236–243. [CrossRef]

36. Tian, T.; Qiao, S.; Li, X.; Zhang, M.; Zhou, J. Nano-graphene induced positive effects on methanogenesis in anaerobic digestion. *Bioresour. Technol.* **2017**, *224*, 41–47. [CrossRef] [PubMed]

37. Jing, Y.; Wan, J.; Angelidaki, I.; Zhang, S.; Luo, G. iTRAQ quantitative proteomic analysis reveals the pathways for methanation of propionate facilitated by magnetite. *Water Res.* **2017**, *108*, 212–221. [CrossRef] [PubMed]

38. Lin, R.; Cheng, J.; Zhang, J.; Zhou, J.; Cen, K.; Murphy, J.D. Boosting biomethane yield and production rate with graphene: The potential of direct interspecies electron transfer in anaerobic digestion. *Bioresour. Technol.* **2017**, *239*, 345–352. [CrossRef] [PubMed]

39. Xu, H.; Wang, C.; Yan, K.; Wu, J.; Zuo, J.; Wang, K. Anaerobic granule-based biofilms formation reduces propionate accumulation under high H2 partial pressure using conductive carbon felt particles. *Bioresour. Technol.* **2016**, *216*, 677–683. [CrossRef] [PubMed]

40. Lei, Y.; Sun, D.; Dang, Y.; Chen, H.; Zhao, Z.; Zhang, Y.; Holmes, D.E. Stimulation of methanogenesis in anaerobic digesters treating leachate from a municipal solid waste incineration plant with carbon cloth. *Bioresour. Technol.* **2016**, *222*, 270–276. [CrossRef] [PubMed]

41. Zhang, J.; Lu, Y. Conductive Fe_3O_4 nanoparticles accelerate syntrophic methane production from butyrate oxidation in two different lake sediments. *Front. Microbiol.* **2016**, *7*, 1316. [CrossRef] [PubMed]

42. Yan, W.; Shen, N.; Xiao, Y.; Chen, Y.; Sun, F.; Tyagi, V.K.; Zhou, Y. The role of conductive materials in the start-up period of thermophilic anaerobic system. *Bioresour. Technol.* **2017**, *239*, 336–344. [CrossRef] [PubMed]

43. Yamada, C.; Kato, S.; Ueno, Y.; Ishii, M.; Igarashi, Y. Conductive iron oxides accelerate thermophilic methanogenesis from acetate and propionate. *J. Biosci. Bioeng.* **2015**, *119*, 678–682. [CrossRef] [PubMed]

44. Zhuang, L.; Tang, J.; Wang, Y.; Hu, M.; Zhou, S. Conductive iron oxide minerals accelerate syntrophic cooperation in methanogenic benzoate degradation. *J. Hazard. Mater.* **2015**, *293*, 37–45. [CrossRef] [PubMed]

45. Barua, S.; Dhar, B.R. Advances towards understanding and engineering direct interspecies electron transfer in anaerobic digestion. *Bioresour. Technol.* **2017**, *244*, 698–707. [CrossRef] [PubMed]

46. American Public Health Association (APHA). *Standard Methods for the Examination of Water and Wastewater*; APHA: Washington, DC, USA, 2005.

47. Poggi-Varaldo, H.M.; Valdés, L.; Esparza-Garcia, F.; Fernández-Villagómez, G. Solid substrate anaerobic co-digestion of paper mill sludge, biosolids, and municipal solid waste. *Water Sci. Technol.* **1997**, *35*, 197–204. [CrossRef]

48. Wang, B.; Björn, A.; Strömberg, S.; Nges, I.A.; Nistor, M.; Liu, J. Evaluating the influences of mixing strategies on the Biochemical Methane Potential test. *J. Environ. Manag.* **2017**, *185*, 54–59. [CrossRef] [PubMed]

49. Valero, D.; Montes, J.A.; Rico, J.L.; Rico, C. Influence of headspace pressure on methane production in Biochemical Methane Potential (BMP) tests. *Waste Manag.* **2016**, *48*, 193–198. [CrossRef] [PubMed]

50. Bowler, N. Four-point potential drop measurements for materials characterization. *Meas. Sci. Technol.* **2011**, *22*, 12001. [CrossRef]

51. Wirth, R.; Kovács, E.; Maróti, G.; Bagi, Z.; Rákhely, G.; Kovács, K.L. Characterization of a biogas-producing microbial community by short-read next generation DNA sequencing. *Biotechnol. Biofuels* **2012**, *5*, 41. [CrossRef] [PubMed]

52. Zhao, Z.; Li, Y.; Quan, X.; Zhang, Y. Towards engineering application: Potential mechanism for enhancing anaerobic digestion of complex organic waste with different types of conductive materials. *Water Res.* **2017**, *115*, 266–277. [CrossRef] [PubMed]

53. Trejo-González, A.; Fería-Morales, A.; Wild-Altamirano, C. The role of lime in the alkaline treatment of corn for tortilla preparation. *Adv. Chem.* **1982**, *198*, 245–263.

54. González-Martinez, S. Biological treatability of the wastewaters from the alkaline cooking of maize (Indian corn). *Environ. Technol.* **1984**, *5*, 365–372. [CrossRef]

55. Eddy, M. *Wastewater Engineering: Treatment, Disposal and Reuse*; McGraw-Hill: New York, NY, USA, 1991.

56. Cruz Viggi, C.; Rossetti, S.; Fazi, S.; Paiano, P.; Majone, M.; Aulenta, F. Magnetite particles triggering a faster and more robust syntrophic pathway of methanogenic propionate degradation. *Environ. Sci. Technol.* **2014**, *48*, 7536–7543. [CrossRef] [PubMed]

57. Liu, Y.; Zhang, Y.; Zhao, Z.; Ngo, H.H.; Guo, W.; Zhou, J.; Peng, L.; Ni, B.-J. A modeling approach to direct interspecies electron transfer process in anaerobic transformation of ethanol to methane. *Environ. Sci. Pollut. Res.* **2016**, 1–9. [CrossRef] [PubMed]

58. Rotaru, A.-E.; Shrestha, P.M.; Liu, F.; Markovaite, B.; Chen, S.; Nevin, K.P.; Lovley, D.R. Direct interspecies electron transfer between Geobacter metallireducens and Methanosarcina barkeri. *Appl. Environ. Microbiol.* **2014**, *80*, 4599–4605. [CrossRef] [PubMed]

59. Lovley, D.R.; Phillips, E.J.P. Novel mode of microbial energy metabolism: Organic carbon oxidation coupled to dissimilatory reduction of iron or manganese. *Appl. Environ. Microbiol.* **1988**, *54*, 1472–1480. [PubMed]

60. Kamagata, Y.; Kawasaki, H.; Oyaizu, H.; Nakamura, K.; Mikami, E.; Endo, G.; Koga, Y.; Yamasato, K. Characterization of three thermophilic strains of Methanothrix ("Methanosaeta") thermophila sp. nov. and rejection of Methanothrix ("Methanosaeta") thermoacetophila. *Int. J. Syst. Evol. Microbiol.* **1992**, *42*, 463–468. [CrossRef] [PubMed]

61. Ibarra-Mendívil, M.H.; Gallardo-Navarro, Y.T.; Torres, P.I.; Ramírez Wong, B. Effect of processing conditions on instrumental evaluation of nixtamal hardness of corn. *J. Texture Stud.* **2008**, *39*, 252–266. [CrossRef]

62. Rosentrater, K.A.; Flores, R.A.; Richard, T.L.; Bern, C.J. Physical and nutritional properties of corn masa by-product streams. *Appl. Eng. Agric.* **1999**, *15*, 515–523. [CrossRef]

63. Krishnan, R.; Ríos, R.; Salinas, N.; Durán-de-Bazúa, C. Treatment of Maize Processing Industry Wastewater by Percolating Columns. *Environ. Technol.* **1998**, *19*, 417–424. [CrossRef]

64. Chen, Y.; Kurt, S.; Creamer, J.J.C. Inhibition of anaerobic digestion process: A review. *Bioresour. Technol.* **2007**, *99*, 4044–4064. [CrossRef] [PubMed]

65. Singh, R.P.; Kumar, S.; Ojha, C.S.P. Nutrient requirement for UASB process: A review. *Biochem. Eng. J.* **1999**, *3*, 35–54. [CrossRef]

66. Omil, F.; Lens, P.; Visser, A.; Hulshoff Pol, L.W.; Lettinga, G. Long-term competition between sulfate reducing and methanogenic bacteria in UASB reactors treating volatile fatty acids. *Biotechnol. Bioeng.* **1998**, *57*, 676–685. [CrossRef]

67. Gerardi, M.H. *The Microbiology of Anaerobic Digesters*; John Wiley & Sons: Hoboken, NJ, USA, 2003; ISBN 0471468959.

68. Lettinga, G.; Hobma, S.W.; Pol, L.W.H.; De Zeeuw, W.; De Jong, P.; Grin, P.; Roersma, R. Design operation and economy of anaerobic treatment. *Water Sci. Technol.* **1983**, *15*, 177–195. [CrossRef]

69. Li, Y.; Zhang, Y.; Yang, Y.; Quan, X.; Zhao, Z. Potentially direct interspecies electron transfer of methanogenesis for syntrophic metabolism under sulfate reducing conditions with stainless steel. *Bioresour. Technol.* **2017**, *234*, 303–309. [CrossRef] [PubMed]

70. Hirano, S.; Matsumoto, N.; Morita, M.; Sasaki, K.; Ohmura, N. Electrochemical control of redox potential affects methanogenesis of the hydrogenotrophic methanogen Methanothermobacter thermautotrophicus. *Lett. Appl. Microbiol.* **2013**, *56*, 315–321. [CrossRef] [PubMed]

71. Salvador, A.F.; Martins, G.; Melle-Franco, M.; Serpa, R.; Stams, A.J.M.; Cavaleiro, A.J.; Pereira, M.A.; Alves, M.M. Carbon nanotubes accelerate methane production in pure cultures of methanogens and in a syntrophic coculture. *Environ. Microbiol.* **2017**, *19*, 2727–2739. [CrossRef] [PubMed]

72. Thrash, J.C.; Coates, J.D. Direct and indirect electrical stimulation of microbial metabolism. *Environ. Sci. Technol.* **2008**, *42*, 3921–3931. [CrossRef] [PubMed]

73. Zinder, S.H. Physiological ecology of methanogens. In *Methanogenesis*; Springer: Boston, MA, USA 1993; pp. 128–206.

74. Xu, S.; He, C.; Luo, L.; Lü, F.; He, P.; Cui, L. Comparing activated carbon of different particle sizes on enhancing methane generation in upflow anaerobic digester. *Bioresour. Technol.* **2015**, *196*, 606–612. [CrossRef] [PubMed]

75. Wang, A.; Liu, L.; Sun, D.; Ren, N.; Lee, D.-J. Isolation of Fe (III)-reducing fermentative bacterium Bacteroides sp. W7 in the anode suspension of a microbial electrolysis cell (MEC). *Int. J. Hydrogen Energy* **2010**, *35*, 3178–3182. [CrossRef]

76. Wu, D.; Wang, T.; Huang, X.; Dolfing, J.; Xie, B. Perspective of harnessing energy from landfill leachate via microbial fuel cells: Novel biofuels and electrogenic physiologies. *Appl. Microbiol. Biotechnol.* **2015**, *99*, 7827–7836. [CrossRef] [PubMed]

77. Holmes, D.E.; Nevin, K.P.; Woodard, T.L.; Peacock, A.D.; Lovley, D.R. Prolixibacter bellariivorans gen. nov., sp. nov., a sugar-fermenting, psychrotolerant anaerobe of the phylum Bacteroidetes, isolated from a marine-sediment fuel cell. *Int. J. Syst. Evol. Microbiol.* **2007**, *57*, 701–707. [CrossRef] [PubMed]

78. Huang, L.; Jiang, L.; Wang, Q.; Quan, X.; Yang, J.; Chen, L. Cobalt recovery with simultaneous methane and acetate production in biocathode microbial electrolysis cells. *Chem. Eng. J.* **2014**, *253*, 281–290. [CrossRef]

79. Hodgson, D.M.; Smith, A.; Dahale, S.; Stratford, J.P.; Li, J.V.; Grüning, A.; Bushell, M.E.; Marchesi, J.R.; Avignone Rossa, C. Segregation of the anodic microbial communities in a microbial fuel cell cascade. *Front. Microbiol.* **2016**, *7*, 699. [CrossRef] [PubMed]

80. Toczyłowska-Mamińska, R.; Szymona, K.; Król, P.; Gliniewicz, K.; Pielech-Przybylska, K.; Kloch, M.; Logan, B.E. Evolving Microbial Communities in Cellulose-Fed Microbial Fuel Cell. *Energies* **2018**, *11*, 124. [CrossRef]

energies MDPI

Article

Nitric Acid Pretreatment of Jerusalem Artichoke Stalks for Enzymatic Saccharification and Bioethanol Production

Urszula Dziekońska-Kubczak *, Joanna Berłowska, Piotr Dziugan, Piotr Patelski, Katarzyna Pielech-Przybylska and Maria Balcerek

Institute of Fermentation Technology and Microbiology, Faculty of Biotechnology and Food Sciences, Lodz University of Technology, Wolczanska 171/173, 90-924 Lodz, Poland; joanna.berlowska@p.lodz.pl (J.B.); piotr.dziugan@p.lodz.pl (P.D.); piotr.patelski@p.lodz.pl (P.P.); katarzyna.pielech-przybylska@p.lodz.pl (K.P.-P.); maria.balcerek@p.lodz.pl (M.B.)
* Correspondence: urszula.dziekonska-kubczak@p.lodz.pl; Tel.: +48-042-631-3473

Received: 8 June 2018; Accepted: 13 August 2018; Published: 17 August 2018

Abstract: This paper evaluated the effectiveness of nitric acid pretreatment on the hydrolysis and subsequent fermentation of Jerusalem artichoke stalks (JAS). Jerusalem artichoke is considered a potential candidate for producing bioethanol due to its low soil and climate requirements, and high biomass yield. However, its stalks have a complexed lignocellulosic structure, so appropriate pretreatment is necessary prior to enzymatic hydrolysis, to enhance the amount of sugar that can be obtained. Nitric acid is a promising catalyst for the pretreatment of lignocellulosic biomass due to the high efficiency with which it removes hemicelluloses. Nitric acid was found to be the most effective catalyst of JAS biomass. A higher concentration of glucose and ethanol was achieved after hydrolysis and fermentation of 5% (w/v) HNO_3-pretreated JAS, leading to 38.5 g/L of glucose after saccharification, which corresponds to 89% of theoretical enzymatic hydrolysis yield, and 9.5 g/L of ethanol. However, after fermentation there was still a significant amount of glucose in the medium. In comparison to more commonly used acids (H_2SO_4 and HCl) and alkalis (NaOH and KOH), glucose yield (% of theoretical yield) was approximately 47–74% higher with HNO_3. The fermentation of 5% nitric-acid pretreated hydrolysates with the absence of solid residues, led to an increase in ethanol yield by almost 30%, reaching 77–82% of theoretical yield.

Keywords: Jerusalem artichoke; lignocellulose; acid pretreatment; nitric acid; alkali pretreatment; enzymatic hydrolysis; ethanol fermentation

---

1. Introduction

According to forecasts by the International Energy Agency (IEA), as the population increases (1.3-fold) between 2009 and 2050, energy consumption will grow even more quickly (2-fold), reaching 15–18 billion tons of oil equivalent (TOE) in 2035 [1,2]. Thus far, world energy demand has been met mainly by burning fossil fuels. However, due to the depletion of coal, oil, and natural gas reserves, as well as increasing public awareness of the environmental impact of emissions, more attention is being focused on developing renewable energy sources, such as biofuels [3]. First-generation biofuels are mostly produced from starch- and sugar-based biomass, derived from food crops grown on agricultural land using standard processes. This can affect food supply and prices. Interest has therefore been growing in second generation biofuels, produced using different feedstocks.

Second generation bioethanol production requires energy crops with high biomass yields per unit land area, resistance to changing climatic conditions, diseases and pests, and the ability to grow in poor soil conditions. Jerusalem artichoke (JA) corresponds to all these requirements [4,5]. An herbaceous perennial plant belonging to the *Asteraceae* (also called *Compositae*) family, JA consists mainly of small underground tubers with 1-3 m tall stalks. Originally used in food and feed, JA tubers contain 75–79% water, 15–16% carbohydrates (in the form of inulin), and 2–3% protein [6,7]. In the past few decades, JA has been cultivated extensively for fructose, oligofructose, and inulin extraction (as functional food ingredients, and more recently to produce bioethanol from the tuber inulin [8–11]. It is also possible to obtain other valuable biochemicals from lignocellulosic biomass, including biobutanol [12] and dimethylfuran [13]. Jerusalem artichoke tubers are considered an attractive feedstock, for producing these chemicals [14,15].

Much less attention has been given to the opportunities for producing bioethanol from JA stalks (JAS). Like all lignocellulosic material, JAS consist mainly of cellulose, hemicellulose, and lignin, bound together in a complex matrix which makes the raw material resistant to enzymatic hydrolysis [16,17]. Appropriate pretreatment must therefore be applied, to break down this recalcitrant structure and make the cellulose more accessible to hydrolyzing enzymes [16]. Many pretreatment methods exist, including physical, chemical, physicochemical, biological, and combined methods. However, the best method for pretreating JAS has not yet been found. Only a few papers describe bioethanol production from JAS. Kim et al. [18] and Kim and Kim [19] suggest using whole JA plants for bioethanol production. Using SSF and CBP processes, respectively, these authors treated the tubers as a source of nitrogen for *Kluyveromyces marxianus* yeasts. Khatun et al. [20] used inulinase producing *Saccharomyces cerevisiae* yeasts, for the fermentation of JAS pretreated with NaOH. Song et al. [21] investigated the effect of hydrogen peroxide-acetic acid (HPAC) pretreatment. All these studies have had a significant impact on research into bioethanol production from JAS. However, many issues regarding the pretreatment of JAS remain unresolved.

Dilute acid pretreatment is the most commonly used method for preparing lignocellulosic biomass and has great potential for industrial applications. This method is recommended by the National Renewable Energy Laboratory (NREL), as it enables the removal and recovery of most of the hemicelluloses, in the form of dissolved sugars, so higher yields of glucose can be achieved [22]. The removal of hemicelluloses facilitates access to the cellulose. However, this method does not result in significant delignification [23,24]. Dilute sulfuric acid, has been widely used to pretreat various types of lignocellulosic biomass [25–27]. Much less attention has been given to hydrochloric, and especially nitric acid. In comparison to sulfuric acid, nitric acid causes less equipment corrosion, and has been found to be much faster and effective for hydrolysis of lignocellulosic biomass [28–30]. Unlike acid pretreatment, alkaline pretreatment removes lignin, as well as acetyl and uronic acid groups present in hemicelluloses. It thus leads to the solubilization of lignocellulosic complexes. The main advantage of alkaline methods is that the reaction is performed under normal pressure and temperature, so less energy input is required [31]. Alkaline treatment has been applied with success to various lignocellulosic feedstocks, including switchgrass [32] and sweet sorghum bagasse [33], enabling a high rate of cellulose digestibility.

Given the advantages of JA as a feedstock for biofuel production, and the limited research regarding the use of JAS in second-generation processes, this paper assessed the impact of various pretreatments on this type of biomass. Each kind of lignocellulosic biomass requires the selection of the most appropriate pretreatment method. This study investigated the effect of different alkaline (sodium and potassium hydroxide) and acidic (sulfuric, hydrochloric and nitric acid) pretreatments, on the yields of subsequent enzymatic hydrolysis and fermentation. Based on our previous research [34], as well as on literature review [35,36], the effect of the acids and alkalis was evaluated under uniform conditions, i.e., 121 °C for 60 min; as these conditions were found to be sufficient to release the structure of the lignocellulosic complex, and thus enhance enzymatic hydrolysis. It was assumed that the use a of a less severe method of pretreatment would also result in the production of fewer inhibitors. To the best knowledge of the authors, this was the first study to evaluate the effect on JAS biomass, of pretreatment with hydrochloric and nitric acids or potassium hydroxide.

2. Results and Discussions

2.1. Chemical Composition of Raw and Pretreated Jerusalem Artichoke Stalks

The composition of whole plant JA biomass is highly variable and depends on climatic and cultivation conditions [18]. To evaluate the potential of JAS as a feedstock for second generation bioethanol production, the biomass was analyzed for total solids [37], cellulose [38], hemicellulose [39], and lignin content [40]. The raw JAS were composed of $89.05 \pm 0.55\%$ dry matter, $34.95 \pm 2.59\%$ d.m. cellulose, $12.69 \pm 3.23\%$ d.m. hemicellulose, and $20.24 \pm 2.13\%$ d.m. lignin. The content of cellulose, the most important component, was comparable to that of other feedstock commonly used for bioethanol production, e.g., 36.3% for corn stover [41], 36.9% for sweet sorghum bagasse [33], 35.8% for rice straw [42], and 38.7% for wheat straw [43].

To increase the susceptibility of lignocellulosic biomass to enzymatic hydrolysis, appropriate pretreatment methods must be applied. In our study, the effect of different concentrations of acids (2% and 5% w/v HCl, H_2SO_4, HNO_3) and alkalis (2% and 5% w/v NaOH, KOH) on JAS was evaluated by analyzing the solid and liquid fractions obtained after each pretreatment. The solid residues were analyzed after washing for cellulose, hemicellulose, and lignin content. The dry weight loss was also measured, to calculate the percentage of recovered solids. In the liquid fraction, the concentrations of inhibitory compounds (formic acid, acetic acid, 5-hydroxymethylfurfural—HMF, furfural, and total phenolics) and of sugars released during pretreatment (cellobiose, glucose, xylose, and arabinose) were measured using the HPLC method. The rate of hemicellulose and cellulose hydrolysis was calculated, based on the concentrations of xylose and glucose in the hydrolysates and the initial hemicellulose and cellulose contents in the raw JAS. The results are presented in Tables 1–3.

Table 1. Solids recovery and chemical composition of pretreated Jerusalem artichoke stalks (JAS).

Pretreatment Method	Solid Recovery (%)	Cellulose	Hemicellulose	Lignin
		(% d.m.)		
2% HCl	$74.75 \pm 2.13e$ [1]	$56.03 \pm 1.37bcd$	$5.41 \pm 0.41bcd$	$23.66 \pm 1.66bcd$
5% HCl	$59.51 \pm 2.01ab$	$54.53 \pm 1.99bc$	$2.31 \pm 0.22abc$	$25.74 \pm 1.20d$
2% H_2SO_4	$73.52 \pm 1.82d$	$53.09 \pm 2.46b$	$4.93 \pm 1.02bcd$	$24.98 \pm 2.35cd$
5% H_2SO_4	$60.27 \pm 1.98ab$	$42.16 \pm 1.80a$	$1.91 \pm 0.23ab$	$27.41 \pm 1.47d$
2% HNO_3	$56.62 \pm 1.18a$	$60.31 \pm 1.67cde$	$5.78 \pm 0.27cd$	$22.96 \pm 1.09bcd$
5% HNO_3	$66.27 \pm 1.42c$	$77.27 \pm 1.32f$	$1.31 \pm 0.16a$	$20.70 \pm 2.02bc$
2% NaOH	$74.83 \pm 2.65d$	$59.30 \pm 1.97cde$	$9.64 \pm 0.66e$	$6.47 \pm 1.22a$
5% NaOH	$66.29 \pm 2.03c$	$62.75 \pm 2.34e$	$8.17 \pm 0.70de$	$4.77 \pm 0.36a$
2% KOH	$72.48 \pm 1.66d$	$56.28 \pm 1.98bcd$	$10.43 \pm 1.50e$	$6.90 \pm 0.93a$
5% KOH	$64.86 \pm 2.02bc$	$59.98 \pm 2.35de$	$8.70 \pm 0.61de$	$5.23 \pm 0.64a$

[1] Different lower-case letters in the columns indicate a significant difference ($p < 0.05$), as analyzed by the Tukey's post-hoc test.

Table 2. Concentration of sugars in liquid fraction after acid or alkaline pretreatment of JAS, with hemicellulose and cellulose solubilization rate.

Pretreatment Method	Cellobiose	Glucose	Xylose	Arabinose	Hemicellulose Solubilization Rate (%)	Cellulose Solubilization Rate (%)
	(g/L)					
2% HCl	0.41 ± 0.04bcd [1]	2.20 ± 0.19a	9.38 ± 0.81b	0.53 ± 0.05ab	68.76 ± 5.92b	5.64 ± 0.49a
5% HCl	0.61 ± 0.12de	2.73 ± 0.53a	11.96 ± 1.32bc	0.84 ± 0.16c	87.70 ± 7.04bc	7.02 ± 1.36a
2% H$_2$SO$_4$	0.64 ± 0.09e	2.63 ± 0.39a	10.84 ± 1.21bc	0.47 ± 0.07ab	79.52 ± 6.72bc	6.76 ± 1.00a
5% H$_2$SO$_4$	0.54 ± 0.06cde	2.80 ± 0.30a	11.35 ± 0.90bc	0.68 ± 0.07bc	83.23 ± 5.81bc	7.19 ± 0.76a
2% HNO$_3$	0.09 ± 0.01a	2.45 ± 0.22a	9.41 ± 0.85b	0.55 ± 005ab	69.01 ± 6.26b	6.30 ± 0.57a
5% HNO$_3$	1.12 ± 0.10f	5.70 ± 0.52b	13.20 ± 1.19c	1.29 ± 0.12d	96.77 ± 8.75c	14.66 ± 1.33b
2% NaOH	0.42 ± 0.05bcd	2.95 ± 0.36a	n.d. [2]	0.62 ± 0.07abc	0.00 ± 0.00a	7.60 ± 0.92a
5% NaOH	0.23 ± 0.01ab	2.12 ± 0.07a	0.29 ± 0.01a	0.56 ± 0.02ab	2.15 ± 0.07a	5.45 ± 0.17a
2% KOH	0.37 ± 0.03bc	2.51 ± 0.18a	0.13 ± 0.01a	0.43 ± 0.03a	0.95 ± 0.07a	6.45 ± 0.47a
5% KOH	0.30 ± 0.02abc	2.99 ± 0.17a	0.31 ± 0.02a	0.52 ± 0.03abc	2.26 ± 0.13a	7.69 ± 0.44a

[1] Different lower-case letters in the columns indicate a significant difference ($p < 0.05$), as analyzed by the Tukey's post-hoc test. [2] n.d.—not detected.

Table 3. Concentration of inhibitors in liquid fraction after acid or alkaline pretreatment of JAS.

Pretreatment Method	Formic Acid	Acetic Acid	TPC [1]	HMF [2]	Furfural
	(g/L)			(mg/L)	
2% HCl	0.19 ± 0.02ab [3]	1.16 ± 0.10a	0.69 ± 0.06b	201.61 ± 17.34c	95.81 ± 8.24a
5% HCl	0.32 ± 0.06abc	2.92 ± 0.57b	0.55 ± 0.05a	136.62 ± 26.55bc	523.79 ± 81.79c
2% H$_2$SO$_4$	0.13 ± 0.02a	2.80 ± 0.41b	0.87 ± 0.12c	739.26 ± 108.93d	95.49 ± 14.07a
5% H$_2$SO$_4$	0.36 ± 0.04bc	3.25 ± 0.34bc	1.08 ± 0.08de	220.65 ± 23.36c	293.09 ± 31.03b
2% HNO$_3$	0.48 ± 0.04c	1.76 ± 0.16a	0.51 ± 0.06a	153.95 ± 13.97bc	8.33 ± 0.75a
5% HNO$_3$	1.18 ± 0.11d	3.19 ± 0.29bc	0.46 ± 0.04a	14.00 ± 1.26a	41.63 ± 3.76a
2% NaOH	1.16 ± 0.14d	3.42 ± 0.41bc	1.39 ± 0.11f	1.42 ± 0.17a	0.98 ± 0.12a
5% NaOH	2.86 ± 0.09f	5.02 ± 0.16de	1.12 ± 0.05e	2.46 ± 0.07a	16.43 ± 0.52a
2% KOH	0.97 ± 0.07d	4.07 ± 0.30cd	1.23 ± 0.10e	1.04 ± 0.07a	1.94 ± 0.08a
5% KOH	1.97 ± 0.11e	6.05 ± 0.34e	1.25 ± 0.11ef	2.92 ± 0.17a	10.17 ± 0.57a

[1] TPC—total phenolics concentration. [2] HMF—hydroxymethylfurfural. [3] Different lower-case letters in the columns indicate a significant difference ($p < 0.05$), as analyzed by the Tukey's post-hoc test.

Analysis of the solid fractions revealed that when acid pretreatment was applied, the hemicellulose content in the biomass decreased significantly, from 12.69% d.m. to 1.31% d.m. (nitric acid pretreatment), while the cellulose and lignin content increased. This could be explained by the fact that the main reaction which occurred during pretreatment in acidic environments was the decomposition of hemicelluloses, especially the xylan fraction. As a result, higher recovery of hemicellulosic sugars is possible, and the cellulose present in the solid fraction is more susceptible to enzymatic hydrolysis [44,45]. The highest cellulose content in the biomass (77.27% d.m.) was achieved when 5% nitric acid was applied (Table 1). Under the same conditions, 96% of the hemicelluloses were hydrolyzed (Table 2). However, biomass recovery was significantly lower ($p < 0.05$) (66% for 2% w/v HNO_3 and 57% for 5% w/v HNO_3) than when other pretreatment methods were used. Rodríguez-Chong et al. [28] reported that nitric acid exhibits much higher efficiency for hemicellulose removal, and the whole process is faster than when hydrochloric or sulfuric acids are used. Under optimal conditions (i.e., 6% HNO_3, 122 °C, 9.3 min), after pretreatment these authors obtained a liquor containing 18.6 g xylose/L, 2.87 g glucose/L, 2.04 g arabinose/L, 0.9 g acetic acid/L, and 1.32 g furfural/L. Zhang et al. [29] evaluated the effect of low concentrations of nitric acid on the hydrolysis of hemicelluloses. The optimum conditions were reported as 0.6% HNO_3, 150 °C for 1 min, releasing 22.01 g xylose/L, 1.91 g glucose/L, 2.90 g arabinose/L, 2.42 g acetic acid/L, and 0.21 g furfural/L. However, these authors did not perform enzymatic hydrolysis of the solids obtained, so the effects on cellulose saccharification are still unclear.

The recovery of solids after pretreatment with HCl or H_2SO_4, NaOH or KOH at 2% concentration, was approximately 72–74%. Increasing the acid or alkali concentration to 5% resulted in solids recovery in the range of 59–66%. In the liquid fraction obtained after acid pretreatment, glucose was present at a similar level (2.2–2.8 g/L), regardless of the type of acid used, except for 5% w/v nitric acid, with which 5.70 g/L of glucose was obtained (corresponding to 14.66% of cellulose hydrolysis) (Table 2). After acid pretreatment, large amounts of furan derivatives were found in the liquid fraction, reaching 0.74 g/L for HMF and 0.52 g/L for furfural (Table 3). Herrera et al. [46] obtained 2.5 g/L and 1.0 g/L of acetic acid and furfural, respectively, in hydrolysate from 2% w/v HCl-pretreated sorghum straw at 100 °C. When alkaline treatment is applied, the main reactions are saponification and solvation, causing lignin to be removed from the lignocellulosic matrix. Unlike acid treatment, the deployment of alkaline reagents leads to lower hemicellulose hydrolysis and lower production of inhibitory compounds [45]. In our study, the application of NaOH or KOH resulted in the removal of 66–76% of the lignin, whilst the cellulose and hemicellulose contents were approximately 56–63% d.m. and 8–10% d.m., respectively (Table 1).

The concentration of glucose in the liquid fraction was at a similar level to that with acid pretreatment ($p > 0.05$). However, the concentrations of xylose and furan derivatives were significantly lower, by up to 0.31 g/L and 0.98–16.43 mg/L, respectively (Tables 2 and 3). Nevertheless, the concentration of aliphatic carboxylic acids was higher than in the acid-treated samples ($p < 0.05$), due to the removal of acetic and other uronic acids, which were substituted for hemicelluloses because of alkaline treatment [47]. The concentration of phenolic compounds was also significantly higher ($p < 0.05$) in alkaline-treated samples than in acid-treated ones, and ranged from 1.12 to 1.39 g/L. The application of acidic pretreatment led to the release of 0.46 to 1.08 g/L of TPC. Moreover, there were significant differences ($p < 0.05$) in obtained level of TPC depending on acid used, with the lowest concentration of phenols obtained for nitric acid pretreatment, and the highest for sulfuric acid pretreatment. However, as reported by Kim et al [48], phenolic compounds are released from biomass during enzymatic hydrolysis, regardless of whether the pretreatment step was applied or not. Phenolic compounds are strong inhibitors, which are proven to inhibit enzymes used in the saccharification step, as well as being responsible for their deactivation. Ximenes et al. [49,50], have shown that phenolic compounds are important inhibitors and deactivators of cellulolytic enzymes, especially β-glucosidase, with the strongest negative effect exhibited by a polymeric compound—tannin acid. Authors imply that the removal of inhibitors, in particular, phenolic compounds, prior to the saccharification process, ought to be performed to ensure high enzyme activity and to achieve better hydrolysis yield.

Taking into consideration the results obtained in our study, the pretreatment using nitric acid, resulted in significantly lower ($p < 0.05$) concentrations of inhibitors in liquid fractions. Therefore, acid treatment using nitric acid is a promising pretreatment method for JAS biomass, enabling a large cellulose fraction to be obtained from the solid material.

2.2. Effect of Pretreatment on Enzymatic Hydrolysis

To evaluate the susceptibility of acid and alkali pretreated JAS to enzymatic hydrolysis, the biomass obtained after each pretreatment was subjected to saccharification at 50 °C for 72 h. During that time, samples were collected every 24 h and the concentrations of glucose and other saccharides (cellobiose, xylose, and arabinose) were assessed. The results are presented in Figure 1 and Table 4. The hydrolysis of untreated JAS resulted in the formation of a small amount of glucose (approximately 2.5 g/L). This concentration was achieved after 24 h of hydrolysis and remained at an almost constant level until the end of the process. The best results from among the acid pretreatment methods, were obtained when nitric acid was used. Not only did it provide the highest glucose yield after hydrolysis (89% of the theoretical yield) (Table 4), but more than 92% of this glucose was formed within the first 48 h of hydrolysis (Figure 1). A similar situation was observed in the case of 5% (w/v) HCl. However, the total amount of glucose and the glucose yield were significantly lower ($p < 0.05$) (17 g/L and 55.5%, respectively). The application of 2% (w/v) HCl, or H_2SO_4 (at both concentrations—2% and 5% (w/v)), required more than 48 h of hydrolysis (Figure 1). The application of sulfuric acid resulted in the formation of 12.5 and 13.5 g/L of glucose, respectively, when 2% and 5% (w/v) of reagent was used. Increasing the acid concentration did not lead to a significant increase in released glucose ($p > 0.05$). However, taking into consideration the cellulose content in pretreated biomass, the yield of cellulose hydrolysis was significantly ($p < 0.05$) higher when 5% (w/v) H_2SO_4 was applied (58% of the theoretical yield), as compared with 2% (w/v) H_2SO_4 (42% of the theoretical yield) (Table 4).

An inverse relation was observed with regard to xylose formation. When 2% (w/v) sulfuric acid was used, the concentration of released xylose was 1.90 g/L, in comparison to 1.28 g/L for 5% (w/v) acid, which was probably caused by greater hemicellulose decomposition during pretreatment with higher acid concentrations (Table 4). The application of hydrochloric acid showed a similar pattern to H_2SO_4 for glucose formation, i.e., increasing the concentration of HCl did not cause a significant increase. When 2% (w/v) nitric acid was used, the concentration of glucose was significantly higher than for other acidic reagents ($p < 0.05$) and reached 17.87 g/L (54% yield), after 72 h of enzymatic hydrolysis. However, the use of a higher concentration of HNO_3 (i.e., 5% w/v) resulted in a more than two-fold increase in the amount of released glucose, up to 38.47 g/L with 89% cellulose hydrolysis efficiency. Acid treatment converted most hemicelluloses into monomeric sugars, leading to a higher rate of cellulose bioconversion. Martin et al. [51] achieved almost 60% cellulose conversion after enzymatic hydrolysis of *Jatropha curcas* shells, pretreated with 1.5% sulfuric acid at 110 °C for 60 min.

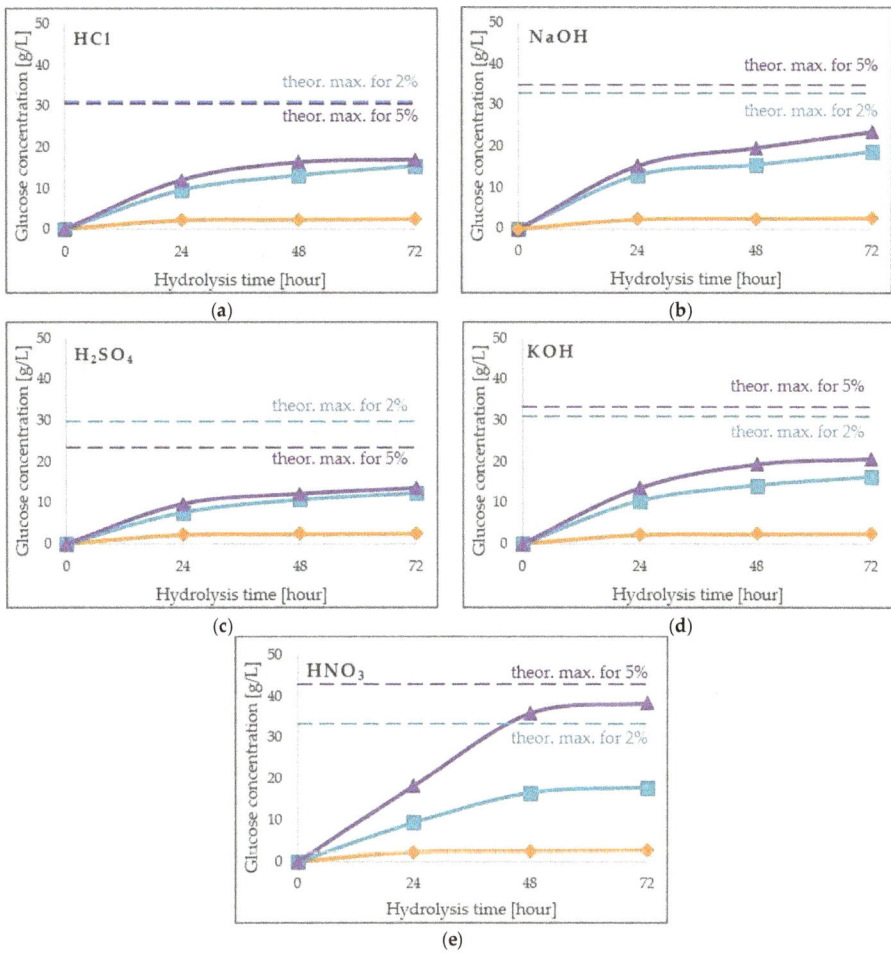

Figure 1. Formation of glucose during enzymatic hydrolysis of JAS pretreated with (**a**) hydrochloric acid, (**b**) sodium hydroxide, (**c**) sulfuric acid, (**d**) potassium hydroxide and (**e**) nitric acid at concentration of 2% (blue square) or 5% (purple triangle), in relation to untreated JAS (orange diamond). The dotted lines indicate the theoretical maximum glucose concentration that can be achieved for 2% (blue) and 5% (purple) acid/alkali concentrations.

Alkali treatment leads to the degradation of side ester and glycoside chains, as well as to cellulose swelling and decrystalization [52,53]. This enables better access by cellulolytic enzymes. In the present study, two alkaline reagents, sodium and potassium hydroxide, were used. The results showed that the application of NaOH resulted in higher glucose concentrations (18.85 g/L and 23.63 g/L, respectively, for 2% and 5% (*w/v*) NaOH), when compared with KOH (16.33 g/L and 20.67 g/L) (Table 4). However, after pretreatment using potassium hydroxide, the maximum level of glucose was obtained in a shorter time. The amount of cellulose converted to glucose was also significantly higher ($p < 0.05$) after NaOH treatment, than after pretreatment with KOH. Moreover, a significant amount of xylose, ranging from 4.99 g/L to 8.47 g/L, was released via enzymatic saccharification of JAS after alkali pretreatment. NaOH was found to be an effective method, for the pretreatment of Coastal Bermuda grass, at 121 °C with 0.75% NaOH for 15 min [54]. Under these conditions, conversion efficiencies for glucan and xylan, were 90.43% and 65.11%, respectively.

Table 4. Concentration of monomeric sugars and glucose yield after hydrolysis of acid- or alkali-pretreated JAS.

Pretreatment Method	Cellobiose	Glucose	Xylose	Arabinose	Glucose Yield (%)
	(g/L)				
Untreated	0.01 ± 0.001a [1]	2.62 ± 0.36a	1.42 ± 0.28ab	0.05 ± 0.01c	13.27 ± 1.21a
2% HCl	0.05 ± 0.01ab	15.70 ± 1.84b	1.53 ± 0.45ab	0.02 ± 0.00ab	50.81 ± 1.71c
5% HCl	0.05 ± 0.02ab	16.99 ± 2.50bc	1.54 ± 0.16ab	0.01 ± 0.01a	55.57 ± 1.81cde
2% H$_2$SO$_4$	0.02 ± 0.01a	12.47 ± 0.88b	1.90 ± 0.29ab	n.d. [2]	41.64 ± 1.48b
5% H$_2$SO$_4$	0.04 ± 0.01ab	13.55 ± 1.99b	1.28 ± 0.14a	0.01 ± 0.00a	57.64 ± 1.55de
2% HNO$_3$	0.06 ± 0.10ab	17.87 ± 1.09bc	2.56 ± 0.46bc	0.01 ± 0.00a	53.90 ± 1.33cd
5% HNO$_3$	0.21 ± 0.05b	38.47 ± 2.97d	2.68 ± 0.09c	n.d.	89.32 ± 2.24g
2% NaOH	0.10 ± 0.04ab	18.85 ± 2.88bc	6.83 ± 0.44e	0.04 ± 0.01c	57.53 ± 1.67de
5% NaOH	0.10 ± 0.03ab	23.63 ± 4.07c	4.99 ± 0.60d	n.d.	67.60 ± 1.51f
2% KOH	0.01 ± 0.00a	16.33 ± 1.64b	6.69 ± 0.68e	0.02 ± 0.00ab	52.14 ± 1.94c
5% KOH	0.01 ± 0.00ab	20.67 ± 2.52bc	8.47 ± 1.03e	0.02 ± 0.00ab	62.27 ± 1.68ef

[1] Different lower-case letters in the columns indicate a significant difference ($p < 0.05$), as analyzed by the Tukey's post-hoc test. [2] n.d.—not detected.

2.3. Concentration of Inhibitors and Fermentation of Hydrolysates

The next step in the process of obtaining bioethanol, after enzymatic hydrolysis of cellulose, is microbial fermentation of the released glucose to ethanol, conducted mainly by *Saccharomyces cerevisiae* (*S. cerevisiae*) yeasts. Unfortunately, these microbes are very sensitive to the presence in the hydrolysates, of products from the degradation of lignocellulose. These products include a wide range of substances formed either from the decomposition of sugars, i.e., furan derivatives and aliphatic carboxylic acids, or from lignin, i.e., phenolic compounds [45]. The furan derivatives (furfural and HMF), are formed by the dehydration of pentose and hexose sugars, respectively, and are responsible for reducing yeast growth, as well as ethanol yield and productivity. Acetic acid is formed because of the hydrolysis of the acetyl groups in hemicellulose, whilst formic acid is a product of the degradation of furan aldehydes [45,55]. A variety of phenolic compounds are formed from lignin, when pretreatment is applied. They can affect the growth of microorganisms, as well as fermentation efficiency, and their toxicity is most probably related with specific functional groups [55].

After pretreatment, the biomass was washed thoroughly with water to remove inhibiting compounds. However, there was a risk that small amounts of inhibiting compounds could remain bound to solid fractions and affect downstream processes. To determine the concentration of inhibitors in the hydrolysates, HPLC analysis was performed. The results are summarized in Table 5. In the raw JAS hydrolysates, only acetic acid was present, at a concentration of 1.11 mg/L. No formic acid, furan derivatives, or phenolic compounds were detected. The concentration of aliphatic carboxylic acids in hydrolysates obtained from pretreated biomass was in the range of 2.57–22.23 mg/L and 8.31–37.44 mg/L, respectively, for formic and acetic acids. The only exceptions were the hydrolysates obtained from HNO$_3$-pretreated biomass, in which the concentration of acetic acid was significantly higher ($p < 0.05$) than in hydrolysates of JAS treated with other chemicals, reaching 128.58 mg/L (2 mM). However, this concentration is still not considered to be toxic to *S. cerevisiae*. In fact, according to Larsson et al. [56], below 100 mM acetic acid can have a stimulatory effect on ethanol fermentation. Furan compounds, are the other large group of inhibitory compounds present in lignocellulosic hydrolysates. It has been reported that, as with carboxylic aliphatic acids, moderate amounts of furan aldehydes in fermented media can enhance the fermentation efficiency of recombinant xylose-utilizing *S. cerevisiae* [55]. In our study, the concentrations of HMF and furfural in hydrolysates, were in the ranges of 1.68–12.29 mg/L and 2.83–7.78 mg/L, respectively.

Table 5. Fermentation inhibitors in acid and alkali pretreated JAS hydrolysates.

Pretreatment Method	Formic Acid	Acetic Acid	HMF [1]	Furfural	TPC [2]
			(mg/L)		
Untreated	n.d. [3]	1.11 ± 0.09a	n.d.	n.d.	n.d.
2% HCl	6.41 ± 0.65bc [4]	29.99 ± 3.05f	2.87 ± 0.22ab	7.78 ± 0.48e	162.11 ± 19.15cde
5% HCl	8.26 ± 0.66cd	35.04 ± 2.25g	3.40 ± 0.38b	6.68 ± 0.44de	178.09 ± 13.34ef
2% H$_2$SO$_4$	12.29 ± 0.66e	12.36 ± 0.82bc	12.29 ± 1.41d	5.40 ± 0.42bcd	173.83 ± 12.56de
5% H$_2$SO$_4$	8.11 ± 0.55cd	37.44 ± 1.34g	5.26 ± 0.66c	6.18 ± 0.72cde	155.96 ± 15.42cd
2% HNO$_3$	14.60 ± 0.76f	128.58 ± 2.98i	2.54 ± 0.25ab	4.53 ± 0.12b	196.91 ± 8.27f
5% HNO$_3$	9.48 ± 1.04d	68.76 ± 2.97h	n.d.	4.93 ± 0.90bc	80.63 ± 8.11a
2% NaOH	19.62 ± 1.11g	14.09 ± 1.27cd	n.d.	6.39 ± 0.41cde	128.04 ± 18.20b
5% NaOH	22.23 ± 0.72h	8.31 ± 0.65b	n.d.	2.83 ± 0.31a	161.34 ± 14.13cde
2% KOH	2.57 ± 0.25a	18.56 ± 1.08de	2.75 ± 0.35ab	7.15 ± 0.73e	151.21 ± 19.11c
5% KOH	5.37 ± 0.99b	20.32 ± 1.68e	1.86 ± 0.24a	6.86 ± 0.75de	170.12 ± 17.15cde

[1] HMF—hydroxymethylfurfural. [2] TPC—total phenolics concentration. [3] n.d.—not detected. [4] Different lower-case letters in the columns indicate a significant difference ($p < 0.05$), as analyzed by the Tukey's post-hoc test.

The concentration of phenolic compounds ranged from 80.63 to 196.91 g/L, respectively, when 5% w/v and 2% w/v HNO$_3$ were applied. In hydrolysates obtained with the use of other reagents, the TPC value was between 128 and 178 mg/L. Despite the significant differences between the samples prepared using different pretreatment methods, no direct correlation was observed between the concentrations of furan or phenolic compounds, and the initial pretreatment methods used. This implied that washing the solids before enzymatic hydrolysis, had successfully removed all analyzed inhibitors which had formed during the pretreatment step.

To evaluate the susceptibility of the hydrolysates to be converted into bioethanol, fermentations were performed using commercial distiller's yeast at a dose of 0.5 g/L of hydrolysate, supplemented with ammonium phosphate (0.3 g/L) as a nutrient. The results are presented in Figures 2 and 3. The application of dilute sulfuric acid, resulted in the formation at most of 5 g/L of ethanol (Figure 2). It was also observed that when a higher concentration of acid was used, the yield of ethanol decreased, from 80% to 51% respectively, for 2% and 5% (w/v) H$_2$SO$_4$ (Figure 3). This was despite the fact that the initial level of sugars in the hydrolysates was at a similar level. Moreover, not all the available glucose was converted by the yeast, following pretreatment with 5% (w/v) H$_2$SO$_4$. Pretreatment with HCl or KOH (at concentrations of both 2% and 5%), as well as 2% NaOH, resulted in similar ethanol concentrations ($p > 0.05$), ranging from 6.2 g/L to 7.2 g/L, with corresponding ethanol yields in the rage of 68–78% of the theoretical yield. The utilization of glucose during fermentation in these samples was almost complete. Pretreatment with nitric acid had a significantly different impact on the fermentation process, compared with the other tested chemicals. The application of 5% (w/v) HNO$_3$, led to the formation of 38.5 g/L of glucose during saccharification. However, only 9.5 g/L of ethanol was obtained after fermentation, which was only 46% of the theoretical yield. Moreover, the amount of glucose that remained in the medium after fermentation was significant (11.5 g/L), which may imply that there were still some unidentified inhibitors present in hydrolysate.

As shown in Table 5, the concentrations of acetic and formic acids, as well as of furfural and HMF, were not at levels considered toxic to yeasts. The concentration of phenolic compounds was also relatively low. However, the low yield of ethanol may have been caused by the presence of inhibitory compounds, which were released from the pretreated biomass when the ethanol was produced, or due to synergic interactions between inhibitors. In addition, the exact composition of phenolic compounds in tested hydrolysates is unknown. Despite the low TPC level, the concentration of individual phenols can be close to inhibitory to the yeast fermentation process. Nitric acid is a strong oxidizing agent, and for that reason, some compounds that are products of the oxidative reactions of commonly found inhibitors may have been present in the medium [57]. Luo et al. [58] found 2-furancarboxylic acid, 2-furanacetic acid, and 5-hydroxymethylfurancarboxylic acid, of which furfural and HMF are precursors, in nitric-acid hydrolysate of aspen chips. Several phenolic compounds are also formed during the degradation of lignin, and some of these could have remained bound to the biomass. Phenolic compounds are much stronger inhibitors of *S. cerevisiae* than aliphatic acids or furans, so even at low concentrations, they can decrease the ethanol yield significantly. Modelska et al. [59] found the minimal inhibitory concentration (MIC) of vanillin to be 0.25% (2.5 g/L). Adeboye et al. [60] reported that the most toxic inhibitor of *S. cerevisiae* Ethanol Red, was 4-hydroxy-3-methoxycinnaldehydyde (coniferyl aldehyde), for which the toxicity limit was as low as 1.8 mM (0.32 g/L). The results presented in this paper, are in line with those of other authors working with dilute nitric acid pretreatment. Following the fermentation of wheat straw pretreated with 1% nitric acid at 130 °C, Tutt et al. [61] obtained an ethanol yield of 95 g/kg of biomass, which corresponded to 59.2% fermentation efficiency.

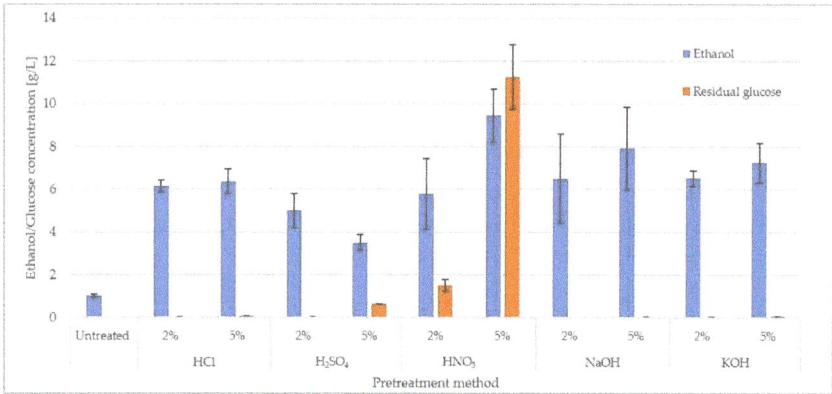

Figure 2. Concentration of ethanol in hydrolysates pretreated using different reagents after 72 h of fermentation and concentration of residual glucose after fermentation. Different lower-case letters, indicate significant differences ($p < 0.05$) between the mean values of the ethanol concentration (Tukey's post-hoc test). Different capital letters, indicate significant differences ($p < 0.05$) between the mean values of the residual glucose concentration (Tukey's post-hoc test).

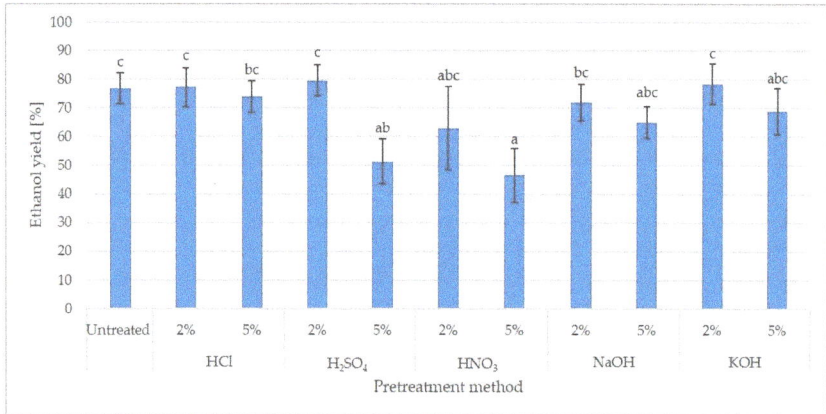

Figure 3. Yield of ethanol (% of theoretical) from hydrolysates pretreated using different reagents, after 72 h of fermentation. Different lower-case letters, indicate significant differences ($p < 0.05$) between the mean values of the ethanol yield (Tukey's post-hoc test).

Fermentation of Nitric Acid Pretreated Hydrolysates—The Effect of Yeast Strain, Inoculum Size and Presence of Solids Residues

To increase the ethanol yield in nitric acid pretreated JAS hydrolysates, additional fermentation trials were performed. The impact of a yeast strain (Thermosacc Dry, Ethanol Red), and applied yeast inoculum (0.5 and 1.5 g/L) were assessed. In addition, since all previous experiments were performed with the presence of solid residues during the fermentation process, the effect of separation of unhydrolyzed biomass before fermentation was also evaluated. The results are presented in Figure 4.

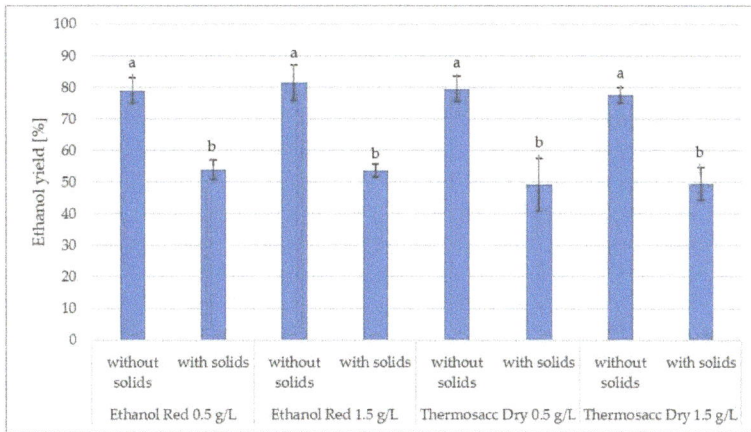

Figure 4. Yield of ethanol (% of theoretical) from nitric acid pretreated hydrolysates fermented with the use of different yeast strains and inoculum. Different lower-case letters, indicate significant differences ($p < 0.05$) between the mean values of the ethanol yield (Tukey's post-hoc test).

The obtained results showed that the ethanol yield was not affected by the yeast strain or the inoculum used. The fermentation trials of hydrolysates in the presence of solids, with use of Thermosacc Dry and Ethanol Red, both at a dose of 0.5 and 1.5 g/L, resulted in similar ($p > 0.05$) ethanol yield ranging from 49.22% to 53.85% of the theoretical yield. According to Azhar et al. [62], the concentration of inoculum has an impact on sugars consumption rate and ethanol productivity, but not on the final ethanol yield. The separation of unhydrolyzed solids before fermentation, resulted in a significant ($p < 0.05$) increase in fermentation efficiency, reaching 77.61–81.51% of the theoretical yield. However, the obtained yield was not influenced by yeast strain or inoculum size under applied conditions. Usually, the fermentation of lignocellulosic hydrolysates is carried out in the presence of solids, which contain mostly lignin [63]. Nonetheless, as reported by Monot et al. [64], it is a good practice to remove the solid parts before fermentation, unless hydrolysis and fermentation processes occur simultaneously. As shown in this paper, it is very important when fermentation of nitric acid pretreated biomass is carried out. The separation of solids before the fermentation allowed us to obtain almost 30% higher ethanol yield. This may have been the result of the presence of some toxic compounds, in the unhydrolyzed solid. These compounds may have been released from biomass when ethanol, even at low concentration, appeared in the medium.

3. Materials and Methods

3.1. Materials

3.1.1. Plant Biomass

Jerusalem artichoke (*Helianthus tuberosus* L.) stalks were harvested from organic 'Żeń-szeń' breeding plants (Kalinówka, Poland), towards the end of October 2017, and delivered fresh. The stalks were air dried and chopped into 0.2–0.5 cm lengths before chemical pretreatment.

3.1.2. Enzymes and Chemicals

Enzymatic hydrolysis was carried out with use of Cellic CTec2 (Novozymes A/S, Basgsværd, Denmark), a blend of cellulases, β-glucosidases, and hemicellulases with reported activities of 120 FPU/mL cellulase, 2371 U/g β-glucosidase, and 161 mg/g protein [65].

Hydrochloric acid (38% w/w), nitric acid (65% w/w), sulfuric acid (95% w/w), sodium hydroxide (pellets), and potassium hydroxide (pellets), all from Chempur (Piekary Slaskie, Poland) were used to prepare the pretreatment solutions. Glucose, xylose, arabinose, cellobiose, acetic acid, formic acid, HMF, and furfural from Sigma Aldrich (St. Louis, MO, USA) were used as standards for high-performance liquid chromatography (HPLC) analysis. All the reagents were at least analytical grade, and all the aqueous solutions were prepared using high-purity deionized water (18.2 MΩ·cm, Simplicity Millipore Waters, Milford, MA, USA). Penicillin and streptomycin (Sigma Aldrich, St. Louis, MO, USA) were used to prevent microbial growth during saccharification.

3.1.3. Yeasts

Fermentation was conducted using the commercial dry distiller's yeasts *Saccharomyces cerevisiae* strains: Thermosacc Dry (Lallemand Ethanol Technology, Montreal, QC, Canada) and Ethanol Red (Fermentis Ltd., Marcq-en-Baroeul, France). Both selected yeast strains, were declared by the producers to have the average number of live cells of 2×10^{10} per gram and approx. 95% of dry matter. Depending on experiments (see below), yeast dose was 0.5 or 1.5 g/L of hydrolysate. Before the inoculation of hydrolysates, dry yeast was hydrated in 10 times the volume of water, at approximate 35 °C for 15 min, to ensure higher activity and homogenization [66].

3.2. Pretreatment

The biomass pretreatments were carried out with 10% w/v dry solid loadings. For acid pretreatment, JAS was soaked in 150 mL of a 2% or 5% (w/v) solution of sulfuric acid, hydrochloric acid or nitric acid, whilst for alkali pretreatment, 150 mL of a 2% or 5% (w/v) solution of sodium hydroxide or potassium hydroxide was used. The samples were autoclaved at 121 °C for 60 min. After pretreatment, the samples were filtered, and the liquid fraction was analyzed after neutralization using the HPLC method to determine the individual sugars, as well as the products of lignocellulose degradation. The solid biomass was washed thoroughly with water until a neutral pH (7.0 ± 0.2) value was achieved. The content of cellulose, hemicellulose, and lignin was then measured, before the solid biomass was subjected to enzymatic hydrolysis.

3.3. Enzymatic Hydrolysis and Fermentation

The pretreated material was subjected to enzymatic hydrolysis with 5% w/v solid loading, based on dry matter, in 100 mL of 0.05 M citric buffer solution (pH 4.8). Cellic CTec2 preparation was used with a dose of 10 FPU/g cellulose (15.43 mg protein/g cellulose). A mixture of penicillin and streptomycin (50 U/mL and 50 µg/mL, respectively) was added. Hydrolysis experiments were performed at 50 °C for 72 h. Samples were taken every 24 h, and the supernatant was analyzed for monomeric sugar concentrations and lignocellulose degradation products using HPLC. After hydrolysis, the whole hydrolysate was used in the fermentation stage. The dry distiller's yeast

S. cerevisiae Thermosacc Dry was added to each sample at a dose of 0.5 g/L, along with ammonium phosphate as a source of nutrients (0.3 g/L). Fermentation trials were conducted for 72 h at 32 °C, in 250 mL flat-bottomed flasks equipped with fermentation tubes. At the same time, untreated JAS was enzymatically hydrolyzed and fermented as a reference.

The additional fermentation trials of nitric acid pretreated hydrolysates were conducted with the use of *S. cerevisiae* Thermosacc Dry and Ethanol Red strains, at a standard dose of 0.5 as well as 1.5 g/L. Fermentations were carried out directly in whole hydrolysate (including solids), or after separation of solid residues (fermentation of liquids only). The other parameters of the fermentation process (nitrogen nutrient, temperature, and time) remained unchanged.

3.4. Analysis and Calculations

The raw and pretreated JAS, were analyzed to determine their content of total solids according to National Renewable Energy Laboratory (NREL) protocol [37]; cellulose according to the Kurschner-Hoffer method [38]; hemicellulose according to the Ermakov method [39]; and lignin according to the NREL protocol [40]. The concentrations of sugars, organic acids, ethanol, and furan aldehydes (HMF, furfural) were analyzed using an HPLC system (Agilent 1260 Infinity, Agilent Technologies, Santa Clara, CA, USA). The compounds were separated using a Hi-Plex H column (7.7 × 300 mm, 8 µm) (Agilent Technologies, Santa Clara, CA, USA), equipped with a refractive index detector (RID). The column and detector temperatures were maintained at 60 °C and 55 °C, respectively. As a mobile phase, 0.05 M H_2SO_4 was used at a flow rate of 0.7 mL/min, with a sample volume of 20 µL. The samples were filtered through 0.22 µm syringe filters to remove any particulate matter. The total concentration of phenolic compounds was determined using a modified Folin-Ciocalteu method, as described by Antolak et al. [67]. Gallic acid (at concentrations of 0–250 mg/L) was used to prepare a standard curve. Samples of enzymatic hydrolysates were analyzed directly, whilst samples of a liquid fraction after pretreatment were diluted ten times before analysis.

The recovery of solid after pretreatment was calculated based on the following equation:

$$\text{Solid recovery} = \frac{\text{dry weight of biomass after pretreatment}}{\text{dry weight of untreated biomass}} \times 100\% \tag{1}$$

The hemicellulose solubilization rate was calculated based on the following equation:

$$\text{Hemicellulose solubilization} = \frac{\text{Xylose concentration in liquid fraction} \left[\frac{g}{L}\right] \times 0.88}{\text{Hemicellulose in untreated biomass} \left[\frac{g}{L}\right]} \times 100\% \tag{2}$$

The cellulose solubilization rate was calculated based on the following equation:

$$\text{Cellulose solubilization} = \frac{\text{Glucose concentration in liquid fraction} \left[\frac{g}{L}\right] \times 0.90}{\text{Cellulose in untreated biomass} \left[\frac{g}{L}\right]} \times 100\% \tag{3}$$

The glucose yield (cellulose hydrolysis efficiency) after enzymatic hydrolysis, was calculated based on the following equation:

$$\text{Glucose yield} = \frac{\text{Glucose concentration in hydrolysate} \left[\frac{g}{L}\right] \times 0.90}{\text{Cellulose in pretreated biomass} \left[\frac{g}{L}\right]} \times 100\% \tag{4}$$

The ethanol yield, after fermentation, was calculated based on the following equation:

$$\text{Ethanol yield} = \frac{\text{Ethanol concentration} \left[\frac{g}{L}\right]}{\text{Glucose in hydrolysate} \left[\frac{g}{L}\right] \times 0.511} \times 100\% \tag{5}$$

All tests were performed in triplicate. Statistical analysis was performed using Statistica 10 software (Tibco Software, Palo Alto, CA, USA), with the results expressed as a mean ± standard deviation. The data were analyzed using a one-way or two-way ANOVA, followed by the Tukey's post-hoc test, with a significance level of 0.05.

4. Conclusions

Pretreatment of JAS, using acid and alkaline reagents, successfully enhances the efficiency of enzymatic hydrolysis. The highest concentration of glucose (38.5 g/L, 89% of theoretical yield), after 72 h of enzymatic hydrolysis was obtained from material pretreated with 5% (w/v) HNO_3. However, due to the fact that pretreatments were performed under the same conditions (121 °C for 1 h), there is a possibility that other acid (besides HNO_3) and alkali reagents can reveal higher efficiency, when optimal conditions for them are used.

The highest ethanol concentration (9.5 g/L) was produced when 5% (w/v) HNO_3 was applied. However, almost 30% of the available glucose remained after fermentation, and the yield of ethanol was low (46% of the theoretical yield). Separation of unhydrolyzed solids before fermentation, allows obtaining of significantly higher ethanol yield (77–82% of the theoretical). Regardless of using for fermentation *S. cerevisiae* yeast strain (Ethanol-Red or Thermosacc Dry), an increase in inoculum size (from 0.5 to 1.5 g/L) did not influence the obtained ethanol yield.

Author Contributions: Conceptualization, U.D.-K., J.B. and P.D.; Formal analysis, U.D.-K. and K.P.-P.; Funding acquisition, J.B. and P.D.; Investigation, U.D.-K., P.P. and K.P.-P.; Methodology, U.D.-K. and P.P.; Project administration, J.B. and P.D.; Resources, J.B. and P.D.; Validation, U.D.-K. and M.B.; Writing—original draft, U.D.-K. and P.P.; Writing—review & editing, U.D.-K. and M.B.

Funding: This research was funded by the Polish National Center for Research and Development under grant number BIOSTRATEG2/296369/5/NCBR/2016.

Conflicts of Interest: The authors declare no conflict of interest.

References

1. Eyidogan, M.; Kilic, F.C.; Kaya, D.; Coban, V.; Cagman, S. Investigation of Organic Rankine Cycle (ORC) technologies in Turkey from the technical and economic point of view. *Renew. Sustain. Energy Rev.* **2016**, *58*, 885–895. [CrossRef]
2. Zecchina, A. Energy sources and carbon dioxide waste. *Rend. Fis. Acc. Lincei* **2014**, *25*, 113–117. [CrossRef]
3. Rastogi, M.; Shrivastava, S. Recent advances in second generation bioethanol production: An insight to pretreatment, saccharification and fermentation processes. *Renew. Sustain. Energ. Rev.* **2017**, *80*, 330–340. [CrossRef]
4. Wilson, P.; Glithero, N.J.; Ramsden, S.J. Prospects for dedicated energy crop production and attitudes towards agricultural straw use: The case of livestock farmers. *Energy Policy* **2014**, *74*, 101–110. [CrossRef] [PubMed]
5. Long, X.H.; Shao, H.B.; Liu, L.; Liu, L.P.; Liu, Z.P. Jerusalem artichoke: A sustainable biomass feedstock for biorefinery. *Renew. Sustain. Energ. Rev.* **2016**, *54*, 1382–1388. [CrossRef]
6. Yang, L.; He, Q.S.; Corscadden, K.; Udenigwe, C.C. The prospects of Jerusalem artichoke in functional food ingredients and bioenergy production. *Biotechnol. Rep.* **2015**, *5*, 77–88. [CrossRef] [PubMed]
7. Negro, M.J.; Ballesteros, I.; Manzanares, P.; Oliva, J.M.; Sáez, F.; Ballesteros, M. Inulin containing biomass for ethanol production: Carbohydrate extraction and ethanol fermentation. *Appl. Biochem. Biotechnol.* **2006**, *132*, 922–932. [CrossRef]
8. Panchev, I.; Delchev, N.; Kovacheva, D.; Slavov, A. Physicochemical characteristics of inulins obtained from Jerusalem artichoke (*Helianthus tuberosus* L.). *Eur. Food Res. Technol.* **2011**, *233*, 889–896. [CrossRef]
9. Roberfroid, M.B. Functional food: Concepts and application to inulin and oligofructose. *Br. J. Nutr.* **2002**, *87*, S139–S143. [CrossRef] [PubMed]
10. Wang, Y.Z.; Zou, S.M.; He, M.L.; Wang, C.H. Bioethanol production from the dry powder of Jerusalem artichoke tubers by recombinant *Saccharomyces cerevisiae* in simultaneous saccharification and fermentation. *J. Ind. Microbiol. Biotechnol.* **2015**, *42*, 543–551. [CrossRef] [PubMed]

11. Matías, J.; Encinar, J.M.; González, J.; González, J.F. Optimisation of ethanol fermentation of Jerusalem artichoke tuber juice using simple technology for a decentralised and sustainable ethanol production. *Energy Sustain. Dev.* **2015**, *25*, 34–39. [CrossRef]

12. Guan, W.; Xu, G.; Duan, J.; Shi, S. Acetone–Butanol–Ethanol Production from Fermentation of Hot-Water-Extracted Hemicellulose Hydrolysate of Pulping Woods. *Ind. Eng. Chem. Res.* **2018**, *57*, 775–783. [CrossRef]

13. Braun, M.; Antonietti, M. A continuous flow process for the production of 2,5-dimethylfuran from fructose using (non-noble metal based) heterogeneous catalysis. *Green Chem.* **2017**, *19*, 3813–3819. [CrossRef]

14. Hailong, C.; Qishun, L.; Shuguang, L.; Zhongbao, Z.; Yuguang, D. Helianthus tuberosus—A good kind of biomass source for dimethylfuran production. *J. Biotechnol.* **2008**, *136*, 271–272. [CrossRef]

15. Sarchami, T.; Rehmann, L. Optimizing Acid Hydrolysis of Jerusalem Artichoke-Derived Inulin for Fermentative Butanol Production. *Bioenergy Res.* **2015**, *8*, 1148. [CrossRef]

16. Kumar, P.; Barrett, D.M.; Delwiche, M.J.; Stroeve, P. Methods for pretreatment of lignocellulosic biomass for efficient hydrolysis and biofuel production. *Ind. Eng. Chem. Res.* **2009**, *48*, 3713–3729. [CrossRef]

17. Xiao, W.; Yin, W.; Xia, S.; Ma, P. The study of factors affecting the enzymatic hydrolysis of cellulose after ionic liquid pretreatment. *Carbohydr. Polym.* **2012**, *87*, 2019–2023. [CrossRef]

18. Kim, S.; Park, J.M.; Kim, C.H. Ethanol production using whole plant biomass of Jerusalem artichoke by *Kluyveromyces marxianus* CBS1555. *Appl. Biochem. Biotechnol.* **2013**, *169*, 1531–1545. [CrossRef] [PubMed]

19. Kim, S.; Kim, C.H. Evaluation of whole Jerusalem artichoke (*Helianthus tuberosus* L.) for consolidated bioprocessing ethanol production. *Renew. Energy* **2014**, *65*, 83–91. [CrossRef]

20. Khatun, M.M.; Li, Y.H.; Liu, C.G.; Zhao, X.Q.; Bai, F.W. Fed-batch saccharification and ethanol fermentation of Jerusalem artichoke stalks by an inulinase producing *Saccharomyces cerevisiae* MK01. *RSC Adv.* **2015**, *5*, 107112–107118. [CrossRef]

21. Song, Y.; Wi, S.G.; Kim, H.M.; Bae, H.-J. Cellulosic bioethanol production from Jerusalem artichoke (*Helianthus Tuberosus* L.) using hydrogen peroxide-acetic acid (HPAC) pretreatment. *Bioresour. Technol.* **2016**, *214*, 30–36. [CrossRef] [PubMed]

22. Yang, B.; Wyman, C. Pretreatment: The key to Unlocking Low-Cost Cellulosic Ethanol. *Biofuels Bioprod. Biorefin.* **2008**, *2*, 26–40. [CrossRef]

23. Mosier, N.; Wyman, C.; Dale, B.; Elander, R.; Lee, Y.Y.; Holtzapple, M.; Ladisch, M. Features of promising technologies for pretreatment of lignocellulosic biomass. *Bioresour. Technol.* **2005**, *96*, 673–686. [CrossRef]

24. Sannigrahi, P.; Kim, D.H.; Jung, S.; Ragauskas, A. Pseudo lignin and pretreatment chemistry. *Energy Environ. Sci.* **2011**, *4*, 1306–1310. [CrossRef]

25. Digman, M.F.; Shinners, K.J.; Casler, M.D. Optimizing on-farm pretreatment of perennial grasses for fuel ethanol production. *Bioresour. Technol.* **2010**, *101*, 5305–5314. [CrossRef] [PubMed]

26. Karapatsia, A.; Pappas, I.; Penloglou, G.; Kotrotsiou, O.; Kiparissides, C. Optimization of dilute acid pretreatment and enzymatic hydrolysis of Phalaris aquatica L. lignocellulosic biomass in batch and fed-batch processes. *Bioenergy Res.* **2017**, *10*, 225–236. [CrossRef]

27. Deshavath, N.N.; Dasu, V.V.; Goud, V.V.; Rao, P.S. Development of dilute sulfuric acid pretreatment method for the enhancement of xylose fermentability. *Biocatal. Agric. Biotechnol.* **2017**, *11*, 224–230. [CrossRef]

28. Rodíguez-Chong, A.; Ramírez, J.A.; Garrote, G.; Vázquez, M. Hydrolysis of sugar cane bagasse using nitric acid: A kinetic assessment. *J. Food Eng.* **2004**, *61*, 143–152. [CrossRef]

29. Zhang, R.; Lu, X.; Sun, Y.; Wang, X.; Zhang, S. Modeling and optimization of dilute nitric acid hydrolysis on corn stover. *J. Chem. Technol. Biotechnol.* **2011**, *86*, 306–314. [CrossRef]

30. Kim, I.; Lee, B.; Park, J.; Choi, S.; Han, J. Effect of nitric acid on pretreatment and fermentation for enhancing ethanol production of rice straw. *Carbohydr. Polym.* **2014**, *99*, 563–567. [CrossRef] [PubMed]

31. Maurya, D.P.; Singla, A.; Negi, S. An overview of key pretreatment processes for biological conversion of lignocellulosic biomass to bioethanol. *3 Biotech* **2015**, *5*, 597–609. [CrossRef] [PubMed]

32. Guan, W.; Shi, S.; Blersch, D. Effects of Tween 80 on fermentative butanol production from alkali-pretreated switchgrass. *Biochem. Eng. J.* **2018**, *135*, 61–70. [CrossRef]

33. Umagiliyage, A.L.; Choudhary, R.; Liang, Y.; Haddock, J.; Watson, D.G. Laboratory scale optimization of alkali pretreatment for improving enzymatic hydrolysis of sweet sorghum bagasse. *Ind. Crops Prod.* **2015**, *74*, 977–986. [CrossRef]

34. Berłowska, J.; Pielech-Przybylska, K.; Balcerek, M.; Dziekońska-Kubczak, U.; Patelski, P.; Dziugan, P.; Kręgiel, D. Simultaneous saccharification and fermentation of sugar beet pulp for efficient bioethanol production. *BioMed Res. Int.* **2016**, *2016*, 3154929. [CrossRef]

35. Gonzales, R.R.; Sivagurunathan, P.; Kim, S.H. Effect of severity on dilute acid pretreatment of lignocellulosic biomass and the following hydrogen fermentation. *Int. J. Hydrogen Energy* **2016**, *41*, 21678–21684. [CrossRef]

36. McIntosh, S.; Vancov, T. Enhanced enzyme saccharification of Sorghum bicolor straw using dilute alkali pretreatment. *Bioresour. Technol.* **2010**, *101*, 6718–6727. [CrossRef] [PubMed]

37. Sluiter, A.; Hames, B.; Hyman, D.; Payne, C.; Ruiz, R.; Scarlata, C.; Sluiter, J.; Templeton, D. *Determination of Total Solids in Biomass and Total Dissolved Solids in Liquid Process Samples*; Technical Report No. NREL/TP-510-42621; NREL: Golden, CO, USA, 2008.

38. Kurschner, K.; Hoffer, A. Cellulose and cellulose derivatives. *Fresenius J. Anal. Chem.* **1993**, *92*, 145–154. [CrossRef]

39. Arasimovich, V.V.; Ermakov, A.I. Measurement of the total content of hemielluloses. In *Methods for Biochemical Studies of Plants*; Ermakov, A.I., Ed.; Agropromizdat: Saint Petersburg, Russia, 1987; pp. 164–165.

40. Sluiter, A.; Hames, B.; Ruiz, R.; Scarlata, C.; Sluiter, J.; Templeton, D.; Crocker, D. *Determination of Structural Carbohydrates and Lignin in Biomass*; Technical Report No. NREL/TP-510-42618; NREL: Golden, CO, USA, 2012.

41. Ma, S.; Wang, H.; Wang, Y.; Bu, H.; Bai, J. Bio-hydrogen production from cornstalk wastes by orthogonal design method. *Renew. Energy* **2011**, *36*, 709–713. [CrossRef]

42. Imman, S.; Arnthong, J.; Burapatana, V.; Champreda, V.; Laosiripojana, N. Fractionation of rice straw by a single-step solvothermal process: Effects of solvents, acid promoters, and microwave treatment. *Renew. Energy* **2015**, *83*, 663–673. [CrossRef]

43. Valdez-Vazquez, I.; Pérez-Rangel, M.; Tapia, A.; Buitrón, G.; Molina, C.; Hernández, G.; Amaya-Delgado, L. Hydrogen and butanol production from native wheat straw by synthetic microbial consortia integrated by species of Enterococcus and Clostridium. *Fuel* **2015**, *159*, 214–222. [CrossRef]

44. Hendriks, A.T.W.M.; Zeeman, G. Pretreatments to enhance the digestibility of lignocellulosic biomass. *Bioresour. Technol.* **2009**, *100*, 10–18. [CrossRef] [PubMed]

45. Jönsson, L.J.; Martín, C. Pretreatment of lignocellulose: formation of inhibitory by-products and strategies for minimizing their effects. *Bioresour. Technol.* **2016**, *199*, 103–112. [CrossRef] [PubMed]

46. Herrera, A.; Tellez-Luis, S.J.; Gonzalez-Cabriales, J.J.; Ramírez, J.A.; Vazquez, M. Effect of the hydrochloric acid concentration on the hydrolysis of sorghum straw at atmospheric pressure. *J. Food Eng.* **2004**, *63*, 103–109. [CrossRef]

47. Chang, V.S.; Holtzapple, M.T. Fundamental factors affecting biomass enzymatic reactivity. *Appl. Biochem. Biotechnol.* **2000**, *84–86*, 5–37. [CrossRef]

48. Kim, D.; Orrego, D.; Ximenes, E.A.; Ladisch, M.R. Cellulose conversion of corn pericarp without pretreatment. *Bioresour. Technol.* **2017**, *245*, 511–517. [CrossRef] [PubMed]

49. Ximenes, E.; Kim, Y.; Mosier, N.; Dien, B.; Ladisch, M. Inhibition of cellulases by phenols. *Enzym. Microb. Technol.* **2010**, *46*, 170–176. [CrossRef]

50. Ximenes, E.; Kim, Y.; Mosier, N.; Dien, B.; Ladisch, M. Deactivation of cellulases by phenols. *Enzym. Microb. Technol.* **2011**, *48*, 54–60. [CrossRef] [PubMed]

51. Martín, C.; García, A.; Schreiber, A.; Puls, J.; Saake, B. Combination of water extraction with dilute-sulphuric acid pretreatment for enhancing the enzymatic hydrolysis of *Jatropha curcas* shells. *Ind. Crop. Prod.* **2015**, *64*, 233–241. [CrossRef]

52. Cheng, Y.S.; Zheng, Y.; Yu, C.W.; Dooley, T.M.; Jenkins, B.M.; Gheynst, J.S.V. Evaluation of high solids alkaline pretreatment of rice straw. *Appl. Biochem. Biotechnol.* **2010**, *162*, 1768–1784. [CrossRef] [PubMed]

53. Ibrahim, M.M.; El-Zawawy, W.K.; Abdel-Fattah, Y.R.; Soliman, N.A.; Agblevor, F.A. Comparison of alkaline pulping with steam explosion for glucose production from rice straw. *Carbohydr. Polym.* **2011**, *83*, 720–726. [CrossRef]

54. Wang, Z.; Keshwani, D.R.; Redding, A.P.; Cheng, J.J. Sodium hydroxide pretreatment and enzymatic hydrolysis of coastal Bermuda grass. *Bioresour. Technol.* **2010**, *101*, 3583–3585. [CrossRef] [PubMed]

55. Jönsson, L.J.; Björn, A.; Nilvebrant, N.O. Bioconversion of Lignocellulose: Inhibitors and Detoxification. *Biotechnol. Biofuels* **2013**, *6*, 16. [CrossRef] [PubMed]

56. Larsson, S.; Palmqvist, E.; Hahn-Hägerdal, B.; Tengborg, C.; Stenberg, K.; Zacchi, G.; Nilvebrant, N.O. The generation of fermentation inhibitors during dilute acid hydrolysis of softwood. *Enzym. Microb. Technol.* **1999**, *24*, 151–159. [CrossRef]

57. Farías-Sánchez, J.C.; Velázquez-Valadez, U.; Pineda-Pimentel, M.G.; López-Miranda, J.; Castro-Montoya, A.J.; Carrillo-Parra, A.; Vargas-Santillán, A.; Rutiaga-Quiñones, J.G. Simultaneous Saccharification and Fermentation of Pine Sawdust (*Pinus pseudostrobus* L.) Pretreated with Nitric Acid and Sodium Hydroxide for Bioethanol Production. *Bioresources* **2016**, *12*, 1052–1063. [CrossRef]

58. Luo, C.; Brink, D.L.; Blanch, H.W. Identification of potential fermentation inhibitors in conversion of hybrid poplar hydrolysate to ethanol. *Biomass Bioenergy* **2002**, *22*, 125–138. [CrossRef]

59. Modelska, M.; Berlowska, J.; Kregiel, D.; Cieciura, W.; Antolak, H.; Tomaszewska, J.; Binczarski, M.; Szubiakiewicz, E.; Witonska, I.A. Concept for Recycling Waste Biomass from the Sugar Industry for Chemical and Biotechnological Purposes. *Molecules* **2017**, *22*, 1544. [CrossRef] [PubMed]

60. Adeboye, P.T.; Bettiga, M.; Olsson, L. The chemical nature of phenolic compounds determines their toxicity and induces distinct physiological responses in *Saccharomyces cerevisiae* in lignocellulose hydrolysates. *AMB Express* **2014**, *4*, 46. [CrossRef] [PubMed]

61. Tutt, M.; Kikas, T.; Olt, J. Influence of different pretreatment methods on bioethanol production from wheat straw. *Agron. Res.* **2012**, *S1*, 269–276.

62. Azhar, S.H.M.; Abdulla, R.; Jambo, S.A.; Marbawi, H.; Gansau, J.A.; Faik, A.A.M.; Rodrigues, K.F. Yeasts in sustainable bioethanol production: A review. *Biochem. Biophys. Rep.* **2017**, *10*, 52–61. [CrossRef]

63. Häggström, C.; Rova, U.; Brandberg, T.; Hodge, D. Integration of Ethanol Fermentation with Second Generation Biofuels Technologies. In *Biorefineries: Integrated Biochemical Processes for Liquid Biofuels*; Qureshi, N., Hodge, D., Vertès, A., Eds.; Elsevier: Amsterdam, The Netherlands, 2014; Chapter 8; pp. 161–187. ISBN 978-0-444-59498-3.

64. Monot, F.; Duplan, J.L.; Alazard-Toux, N.; His, S. Biofuels. In *Renewable Energies*; Sabonnadière, J.C., Ed.; ISTE Ltd.: London, UK; John Wiley & Sons, Inc.: Hoboken, NY, USA, 2009; Chapter 10; pp. 329–395. ISBN 9781848211353.

65. Cannella, D.; Jørgensen, H. Do new cellulolytic enzyme preparations affect the industrial strategies for high solids lignocellulosic ethanol production? *Biotechnol. Bioeng.* **2014**, *111*, 59–68. [CrossRef] [PubMed]

66. Balcerek, M.; Pielech-Przybylska, K.; Dziekońska-Kubczak, U.; Patelski, P.; Strąk, E. Fermentation results and chemical composition of agricultural distillates obtained from rye and barley grains and the corresponding malts as a source of amylolytic enzymes and starch. *Molecules* **2016**, *21*, 1320. [CrossRef] [PubMed]

67. Antolak, H.; Czyzowska, A.; Sakač, M.; Mišan, A.; Đuragić, O.; Kregiel, D. Phenolic compounds contained in little-known wild fruits as antiadhesive agents against the beverage-spoiling bacteria *Asaia* spp. *Molecules* **2017**, *22*, 1256. [CrossRef] [PubMed]

energies

MDPI

Article

Bamboo Fiber and Sugarcane Skin as a Bio-Briquette Fuel

Anna Brunerová [1,*], **Hynek Roubík** [2] and **Milan Brožek** [1]

1   Faculty of Engineering, Department of Material Science and Manufacturing Technology, Czech University of
    Life Sciences Prague, Kamýcká 129, 165 00 Prague, Czech Republic; brozek@tf.czu.cz
2   Faculty of Tropical AgriSciences, Department of Sustainable Technologies, Czech University of Life Sciences
    Prague, Kamýcká 129, 165 00 Prague, Czech Republic; roubik@ftz.czu.cz
*   Correspondence: brunerova@tf.czu.cz; Tel.: +420-737-077-949

Received: 16 July 2018; Accepted: 16 August 2018; Published: 21 August 2018

Abstract: The present study deals with the issue of bio-briquette fuel produced from specific agriculture residues, namely bamboo fiber (BF) and sugarcane skin (SCS). Both materials originated from Thừa Thiên Huế province in central Vietnam and were subjected to analysis of their suitability for such a purpose. A densification process using a high-pressure briquetting press proved its practicability for producing bio-briquette fuel. Analysis of fuel parameters exhibited a satisfactory level of all measured quality indicators: ash content A_c (BF—1.16%, SCS—8.62%) and net calorific value NCV (BF—16.92 MJ·kg^{-1}, SCS—17.23 MJ·kg^{-1}). Equally, mechanical quality indicators also proved satisfactory; bio-briquette samples' mechanical durability DU occurred at an extremely high level (BF—97.80%, SCS—97.70%), as did their bulk density ρ (BF—986.37 kg·m^{-3}, SCS—1067.08 kg·m^{-3}). Overall evaluation of all observed results and factors influencing the investigated issue proved that both waste biomass materials, bamboo fiber and sugarcane skin, represent suitable feedstock materials for bio-briquette fuel production, and produced bio-briquette samples can be used as high-quality fuels.

Keywords: waste biomass; Vietnam; solid biofuel; calorific value; mechanical durability

1. Introduction

With the increasing prices and various environmental impacts created by the use of fossil fuels, the importance of biofuel production has subsequently increased. Such production has reached unprecedented volumes over the last 20 years [1]. In addition, the world's population is expected to continue growing, with the total population calculated to be almost ten billion by mid-2050 [2]. This population growth will lead to a deepening requirement for energy, and therefore, global energy demand will subsequently increase. Energy is considered a key source for the future and plays an important role in socioeconomic development because affordable energy is an essential ingredient for such development [3,4].

With deforestation comprising a major problem in many parts of the developing world, there is increasing demand for fuelwood for household cooking. This phenomenon particularly affects remote rural communities that have no access to fuels such as liquid petroleum gas (LPG) and that depend substantially on burning locally-collected biomass [5]. Such a demand for fuel could be covered by bio-briquettes, which may provide necessary energy from waste materials because biomass is globally recognized as a renewable and sustainable energy source [6]. In addition, agricultural residues have recently been posited as a major fuel source for many potential bio-energy projects in developing countries [1].

Biomass is now considered a primary global energy resource, providing 14% of the world's energy needs. It comprises at least one third of energy consumption in some developing countries. In addition, biomass combustion can be considered CO_2 neutral because during production, it removes CO_2 from the atmosphere by photosynthesis and is later released during combustion [7]; this aspect was however contradicted in Cherubini et al. [8], who analyzed CO_2 emissions from biomass combustion for bioenergy.

1.1. Bamboo and Sugarcane as a Source of Herbaceous Biomass

Bamboo (*Bambusoideae* spp.) and sugarcane plants (*Saccharum officinarum*) are members of the grass family *Poaceae* (also known as *Gramineae*). Both grasses are widely spread in tropical and subtropical climate regions. As a strongly growing species, their cultivation provides several advantages. Bamboo plants are the fastest growing plant species on Earth (approximately 91–122 cm per day); thus, significant potential for herbaceous biomass production is indicated. Moreover, the subfamily Bambusoideae contains 1100–1500 different bamboo species, with their classification and identification being complicated; approximately 69 naturally-occurring species were monitored in Vietnam alone [9,10].

Historically, bamboo is a traditional and widely-used plant in Asia; its utilization has lasted for centuries in different industries, for building various parts of houses, furniture production, as well as medicinal, food [10], clothing and home craft purposes, among others [11]. Concurrently, sugarcane plants cultivated for agricultural purposes provide 70% of sugar worldwide (having the most calories per unit area of cultivation of any plant), with a large amount of the herbaceous waste biomass being left behind after plant processing, for example sugarcane skin [12]. Such material is one of the by-products of the sugarcane processing industry [13]. Today, biological residues from sugarcane plants are a particular point of interest of different scientific fields [14,15]; thus, its potential seems significant. However, sugarcane skin reuse in Southeast Asia, namely in Vietnam, is not ensured; the open field burning or ejection into municipal waste is a common practice in rural areas when individuals or small processing plants are treating sugarcane crop. In an attempt to meet increasing energy demands and maintain an appropriate waste management strategy, waste biomass originated from processing of both mentioned grasses could be reused and utilized; for example, as a feedstock material for various biotechnological processes [13,16,17]. Bamboo plants commonly provide waste cuttings of bamboo logs or bamboo fiber (BF), while sugarcane processing results in the production of sugarcane bagasse. Utilization of such residues is not secured, and they are often left behind as useless waste or burned without a purpose. Occasionally, bamboo fiber is used as a fire starter and sugarcane skin for combustion purposes.

1.2. Biomass Availability in Vietnam

In Vietnam, biomass is considered an important source of energy, comprising approximately 90% of domestic energy consumption in rural areas [18], as well as being an important source of energy for small industries and farms, often located in rural areas [19]. Further, it has significant potential as a renewable energy source, even though the agricultural sector currently accounts only for approximately 20% of the country's GDP [19]. Agricultural residues, as described in Schirmer [19], and animal waste, as described in Roubík et al. [20], are offering significant potential for the generation of electricity, sources that have so far been insufficiently tapped. Currently, biomass is mainly used by households, and waste biomass is usually not used at all, or very ineffectively and with potential harmful consequences.

Agricultural residues (including bamboo fiber and sugarcane skin) are important sources of biomass in Vietnam and can be taken from residues left directly on fields or from the processing of agricultural products. The wider use of such residues may be relatively difficult because this agricultural waste usually occurs locally (for example, rice husk residues are produced in large quantities only in the local rice mills). As mentioned by Schirmer [19], there is a need for a reliable supply of biomass, its efficient collection, transportation and storage, meaning well-established supply chains, which currently do not exist.

Agricultural Waste Utilization through Briquetting

One technology for the utilization of agricultural wastes is biomass briquetting [21,22], which involves the densification of the biomass through the use of pressure [23,24]. The advantages of briquetting include the following: increased bulk density of the material, making it easier to transport and store, higher energy content per unit volume and the production of homogenous product fuel from various raw materials [22,25]. Therefore, as concluded by Chen et al. [23], densified solid biofuel (meaning bio-briquettes) could be an important route for efficient utilization of agro-residues.

2. Methodology

The present section details research activities performed for the collection of selected waste biomass (both materials represented by herbaceous biomass) in raw form and its initial processing in the target area of Huế city and surrounding villages in Thừa Thiên Huế province, central Vietnam. There is a detailed description of the investigated waste biomass laboratory preparation (drying, grinding) and laboratory testing of its fuel properties. Further, the utilization of selected waste biomass as a feedstock material for bio-briquette fuel production using a high-pressure briquetting press is described, together with the determination of their suitability for such a purpose. Finally, the complete evaluation of stated mechanical and chemical quality indicators of produced bio-briquette samples is performed.

In general, the production of bio-briquette fuel for commercial purposes is conducted to mandatory technical standards (stated by the country of bio-briquette fuel production) ensuring the quality and safety of such products. Thus, all procedures of experimental testing were performed in accordance to related European technical mandatory standards stated by the European Committee for Standardization, namely: EN 14918 (2010) [26], ISO 1928 (2010) [27], EN 15234-1 (2011) [28], EN ISO 16559 (2014) [29], EN ISO 17225-1 (2015) [30], EN ISO 17831-2 (2015) [31], EN ISO 18122 (2015) [32], EN ISO 18134-2 (2015) [33], EN ISO 16948 (2016) [34] and EN ISO 18123 (2016) [35].

2.1. Materials and Samples

Both investigated waste materials, bamboo fiber (*Bambusoideae* spp.) and sugarcane skin (*Saccharum officinarum*), were identified as herbaceous biomass due to their membership in the grass family *Poaceae*. Collection activities were performed at small processing plants in central Vietnam in June of 2017. Bamboo fiber samples originated from rural areas in Thừa Thiên Huế province; the collections were performed during several field trips; thus, samples originated from different villages, whereas sugarcane skin samples originated from one specific processing plant in Huế city. Nevertheless, unprocessed raw sugarcane stalks were harvested in surrounding rural areas of Huế city. The target area of the materials' collection is illustrated in Figure 1.

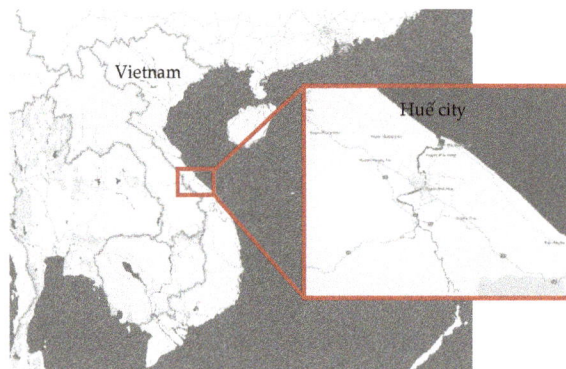

Figure 1. Area of sample collection: Huế district, Thừa Thiên Huế province.

Both investigated materials were chosen because they represented commodities frequently processed in large quantities in the target area, and their processing results in the production of a significant amount of waste biomass.

Bamboo stalks were already sundried prior to the fiber being manually separated in processing plants; thus, final waste biomass occurred in ideal conditions for combustion purposes (low level of moisture content); as shown in Figure 2.

(a)   (b)

Figure 2. Bamboo processing in Thừa Thiên Huế province, central Vietnam: (**a**) manual removing of fiber; (**b**) produced waste biomass: bamboo fibers.

In contrast, sugarcane stalks occurred in the raw state before their skin was manually removed in the processing plant; thus, produced waste biomass occurred in its initial high moisture content, as shown in Figure 3.

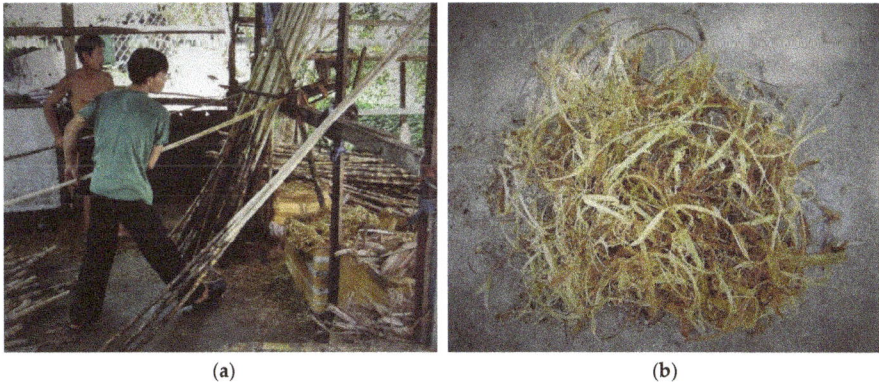

(a)   (b)

Figure 3. Sugarcane processing in Huế city, central Vietnam: (**a**) manual removing of skin; (**b**) produced waste biomass: sugarcane skin.

The drying process of sugarcane skin was thus necessary prior to its subsequent utilization, and therefore, the material was sundried in open fields in local conditions. Thereby, the drying process was ensured without the need for any other energy source investment. Moreover, local insects were not interested in such materials, which represents a significant advantage if considering that such materials contained residual sugars. Unfortunately, precise initial moisture content of both materials was not determined due to the fact that primary research activities were performed in conditions of rural areas of Vietnam without proper measuring equipment. Further, both materials were properly stabilized and stored in special hermetically-sealed bags for transportation.

After the materials' transportation to the laboratories at the target destination of Prague, Czech Republic, both materials were appropriately prepared for subsequent experimental measurements; their particle size and moisture content were modified to the required form. Primarily, both materials were dried in a laboratory dryer LAC, Type S100/03 (Rajhrad, Czech Republic), for 24 hours at a temperature of 105 °C until their moisture content was constant. Disintegration of feedstock materials was also performed using a grinding hammer mill Taurus, Type VM 7,5 (Chrudim, Czech Republic), with a vertical shaft with eight hammers as a working unit (see Figure 4).

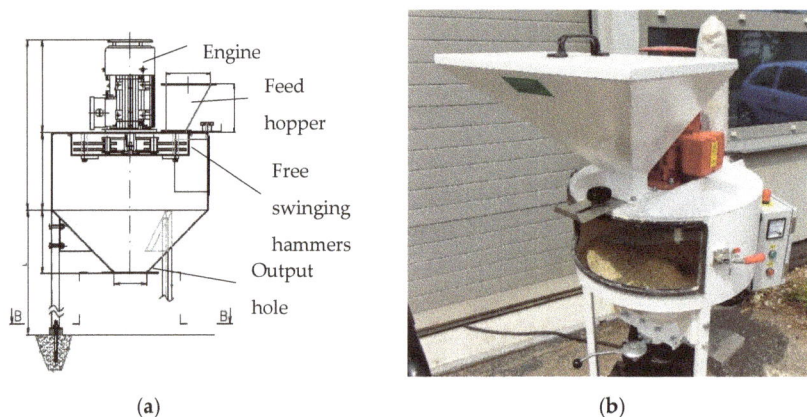

Figure 4. The disintegration device used, the hammer grinding mill: (a) scheme; (b) in practice.

A sieve with holes of 8 mm in diameter was used to unify the faction of both disintegrated feedstock materials. The final form of properly prepared samples is expressed in Figure 5.

Figure 5. The form of samples prepared for chemical and mechanical analysis: (a) bamboo fiber; (b) sugarcane skin.

Nevertheless, the bamboo material occurred in the fibrous form, and hence, the fibers that fell through sieve holes were prevalently slightly longer than 8 mm. In contrast, sugarcane skin (SCS) material also contained tiny particles in the form of dust; thus, the prevalent particle size was smaller than 8 mm. Due to such differences between the investigated materials, brief microscopic measurements were performed to describe the particle size and shapes of investigated materials (measurement scale 5 mm). A stereoscopic microscope Arsenal, Type 347 SZP 11-T Zoom (Prague, Czech Republic) was used for image analysis; the resulting images are shown in Figure 6.

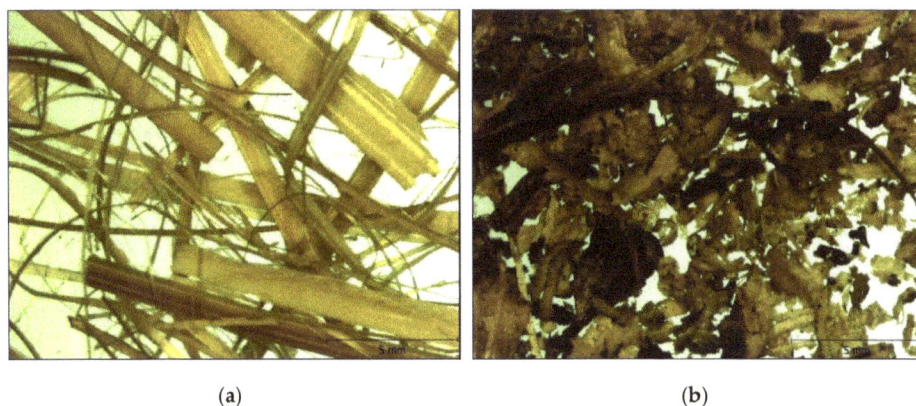

(a)   (b)

Figure 6. Microscopic analysis of investigated samples: (**a**) bamboo fiber; (**b**) sugarcane skin.

Despite the fact that both investigated materials were represented by the top layer of plant stem (the epidermis), the variability in materials' behavior during disintegration and particle size were detected during microscopic analyses (Figure 6); such variability was related to the taxonomy of processed plants.

2.2. Fuel Analysis of Feedstock Materials

Samples were subjected to analysis of their fuel properties, which defined their basic chemical parameters, energy potential and elementary composition. Such measurements were performed to determine the materials' suitability for the process of direct combustion. At least two or three repetitions of all of the following tests were performed for all investigated samples to ensure the correctness of measurement processes and observed data; data were recorded and evaluated using supporting software. Final data exhibited in the Results and Discussion Section are expressed as average values of all observed data with standard deviations.

2.2.1. Basic Parameters and Energy Potential (M_c, A_c, GCV, NCV)

Samples were initially milled into the required particle size (<0.1 mm) using a cutting mill. Moisture M_c (%) and ash content A_c (%) analysis was performed with a thermogravimetric analyzer LECO, Type TGA 701 (Saint Joseph, MO, USA). Experimental tests complied with related standards, namely EN 18134-2 (2015): Solid biofuels—Determination of moisture content—Oven dry method—Part 2: Total moisture-Simplified method and ISO 18122 (2015): Solid biofuels—Determination of ash content.

The principal parameters of fuel properties, gross calorific value (GCV) (MJ·kg^{-1}) and net calorific value NCV (MJ·kg^{-1}) express the energy potential of the investigated materials. Gross calorific values (GCV) (MJ·kg^{-1}) were primarily measured using an isoperibol calorimeter LECO, Type AC 600 (Saint Joseph, USA) and secondarily analyzed by related software. Net calorific values (NCV) (MJ·kg^{-1}) were calculated using relations between those two parameters. The complete process of analyses used standards EN 14918 (2010): Solid biofuels—Determination of calorific value and ISO 1928 (2010): Solid mineral fuels—Determination of gross calorific value by the bomb calorimetric method and calculation of NCV.

2.2.2. Elementary Composition (*C*, *H*, *N*, *S*, *O*)

Materials content of carbon *C* (%), hydrogen *H* (%), nitrogen *N* (%), sulfur *S* (%) and oxygen *O* (%) represented the last measured fuel parameters. Experimental testing was performed using laboratory equipment LECO, Type CHN628+S (Saint Joseph, MO, USA), while the testing process corresponded to requirements stated by related standard EN ISO 16948 (2016): Solid biofuels—Determination of total content of carbon, hydrogen and nitrogen.

2.3. Mechanical Analysis of Bio-Briquette Samples

When the suitability of investigated materials for the process of direct combustion was proven, they were utilized as a feedstock for bio-briquette fuel production. Subsequently, the produced bio-briquette samples were subjected to the determination of their mechanical quality represented by the following indicators: bulk density ρ, mechanical durability *DU* and rupture force *RF*. Specific methods of the mentioned experimental measurements are described in the following sections. The present section is divided into four sections: The first section is related to the production factors of the densification process and to the entire process of bio-briquette sample production. The remaining three sections describe bio-briquette sample quality testing and determination of specific mechanical quality indicator results.

2.3.1. Bio-Briquette Sample Production

As mentioned previously, high-pressure briquetting technology was used for densification of two different types of herbaceous waste biomass: bamboo fiber and sugarcane skin. Namely, a laboratory hydraulic piston high-pressure briquetting press Briklis, Type BrikStar 30-12 (Malšice, Czech Republic), was used as a pressing device (shown in Figure 7).

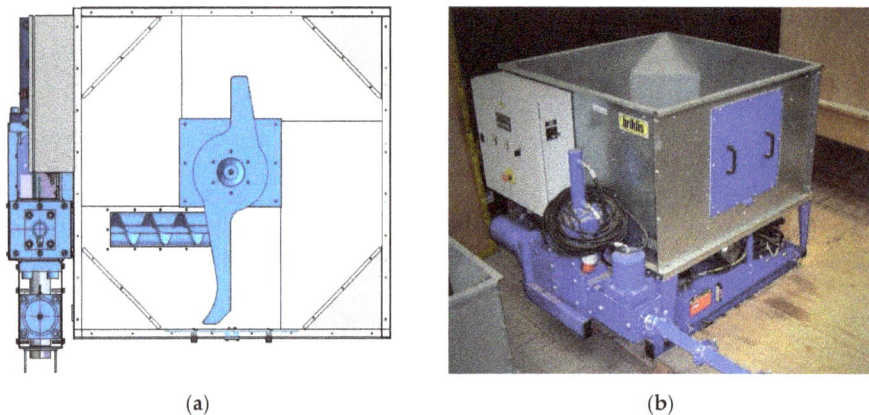

(a)   (b)

Figure 7. Briklis high-pressure briquetting press: (**a**) scheme; (**b**) in practice.

This press was operated with an automatic setting during feedstock compression with a specific focus on ensuring a similar level of bulk density ρ of all produced bio-briquette samples. However, such settings resulted in different lengths of final products. The device is equipped with a pressing chamber and matrix with a diameter of 50 mm, thus producing bio-briquette samples (expressed in Figure 8) of a diameter equal to approximately 50 mm; the dimensions of bio-briquette samples are provided in Table 1.

(a)   (b)

Figure 8. Produced bio-briquette fuel samples from: (**a**) bamboo fiber; (**b**) sugarcane skin.

Table 1. Basic mechanical properties of investigated briquette samples (±standard deviation).

Sample	Weight (g)	Height (mm)	Diameter (mm)
Bamboo fiber	79.68 ± 20.71	37.73 ± 10.61	51.55 ± 0.37
Sugarcane skin	120.61 ± 12.85	53.35 ± 5.55	51.96 ± 5.40

In general, the dimensions of bio-briquette fuel influence its final mechanical quality. Thus, its monitoring represents an important production parameter that needs to be considered.

2.3.2. Bulk Density ρ

The first investigated indicator, bulk density ρ, describes both the final mechanical quality of produced bio-briquette samples, as well as the ability of tested feedstock materials to be densified. Thus, it evaluates the efficiency of the densification process in the case of such specific feedstock materials. Calculation of bio-briquette sample bulk density ρ (kg·m^{-3}) used its dimensions, namely its volume V (m^3) and mass m (kg), and the following Equation (1).

$$\rho = \frac{m}{V}$$ (1)

where ρ is the bulk density (kg·m^{-3}), V the bio-briquette samples' volume (m^3) and m the bio-briquette sample mass (kg).

2.3.3. Mechanical Durability DU

The produced bio-briquette samples were subjected particularly to the determination of mechanical durability DU (%), which is the main indicator of bio-briquette fuel mechanical quality and which is defined by standard EN ISO 17831-2 (2015): Solid biofuels—Determination of mechanical durability of pellets and briquettes—Part 2: Briquettes. This standard states several quality levels of bio-briquette fuel and establishes the lowest acceptable level ($DU > 90\%$) for commercial bio-briquette fuel production. Experimental observation and measurements were performed repeatedly using an electrically-powered special dust–proof rotating drum (Figure 9) with a rectangular steel partition. The principle of such a test consists of projected impacts of tested bio-briquette samples in the rotating drum; thereby, the samples prove their durability.

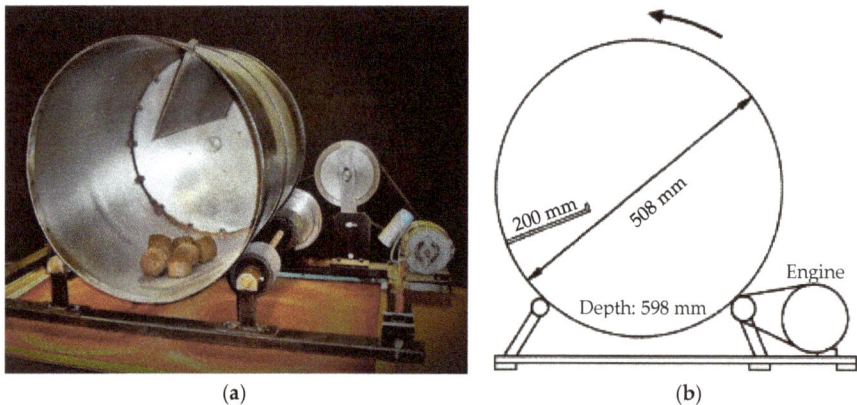

(a)   **(b)**

Figure 9. Equipment used for the determination of mechanical durability: (**a**) in practice; (**b**) scheme.

In practice, several groups of bio-briquette samples of a specific weight (2 ± 0.1 kg) were separately placed inside the drum and subjected to controlled impacts for a specific time (4 minutes 17 seconds), which was equal to the specific number of rotations (105 ± 5 rotations). Subsequently, bio-briquette samples were weighed prior to and after such testing, and their final resistance was determined by Equation (2).

$$DU = \frac{m_a}{m_e} \cdot 100 \tag{2}$$

where DU is mechanical durability (%), m_e the sample weight before testing (g) and m_a the sample weight after testing (g).

2.3.4. Rupture Force *RF*

The methodology of the last tested mechanical quality indicator, the rupture force *RF*, does not correspond to any mandatory technical standard. It was developed on the basis of previous scientific research focused on the pressing technology and measurement of the physical and mechanical properties of pressed items [36–38]. The stated methodology determined the hardness of tested bio-briquette samples, thus simulating their possible damage in practice.

The principle of measurement lies in the plate-loading test (Figure 10), which was performed using a universal hydraulic machine WPM, Type ZDM 5 (Leipzig, Germany) with a loading speed v equal to 20 mm·min^{-1}. The measurement ended when bio-briquette samples disintegrated due to the influence of force loading; thus, the maximal force loaded to the samples before their disintegration was noted. The number of tested bio-briquette samples was equal to 64 in the case of bamboo fiber and to 57 in the case of sugarcane skin.

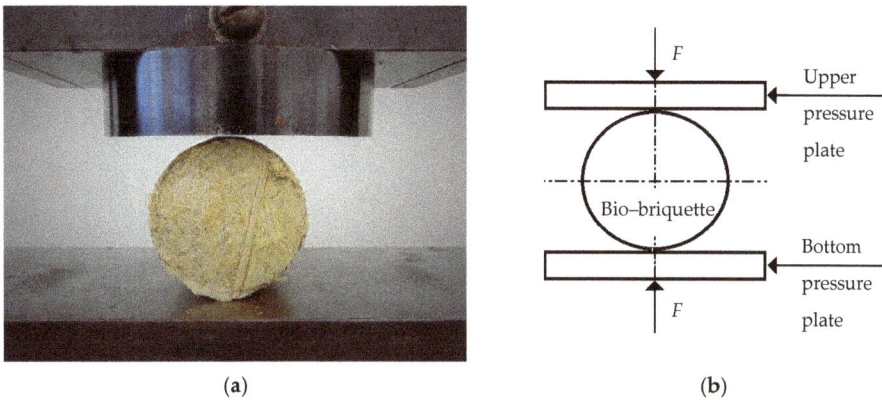

Figure 10. The equipment used for the determination of rupture force *RF*: (**a**) in practice; (**b**) plate, loading test principle (*F* compressive force (N)).

3. Results and Discussion

The complex evaluation of the investigated bio-briquette production efficiency was based on the set of quality indicators of both feedstock materials and produced bio-briquette samples. All chosen indicators described the suitability of the investigated bio-briquette fuel for commercial production. Therefore, required indicator levels occurred at a high level in an attempt to ensure the highest quality of produced bio-briquette fuel. Nevertheless, mandatory requirements for solid biofuel commercial production differ within each country; the obtained results were compared with the European standards.

The present research aimed to answer the question of whether the chosen waste materials were suitable for bio-briquette production. In general, it was a complex process containing several necessary steps and could be influenced by many factors. Thus, before answering the main question posed, the following specific issues must be answered: (a) Is there a sufficient amount of such waste materials? Is there any functional existing way of its subsequent utilization? (b) Are those materials suitable for direct combustion processes? Is their fuel analysis positive for such purposes? (c) Are the materials suitable for drying and disintegration processes or do such steps involve difficulties? Is it possible to produce bio-briquette fuel from those materials? What is the final mechanical quality of such biofuel?

Answers to all of these questions are noted in the following sections or were explained in the Methodology Section.

3.1. Feasibility and Practicability of Production

The conditions of investigated materials production and potential reuse were primarily monitored in the target area. Both materials originated from plants that abundantly grow and are processed in the target area, with the only difference being that sugarcane plants are purposely cultivated as an agriculture crop, while bamboo plants grow wildly throughout the target area. Specifically, processed bamboo culms originated from the forests in the surrounding areas.

Both plants were nevertheless processed in the target area at a large-scale. Sugarcane plants were used for juice production by individual sellers; plants stems were skinned before squeezing of juice, which resulted in the production of sugarcane skin. Dried bamboo stems were manually skinned by small storekeepers and subsequently used for the production of various commercial items. As observed, both produced waste materials, bamboo fiber and sugarcane skin, were not reused properly in the target area (Figure 11).

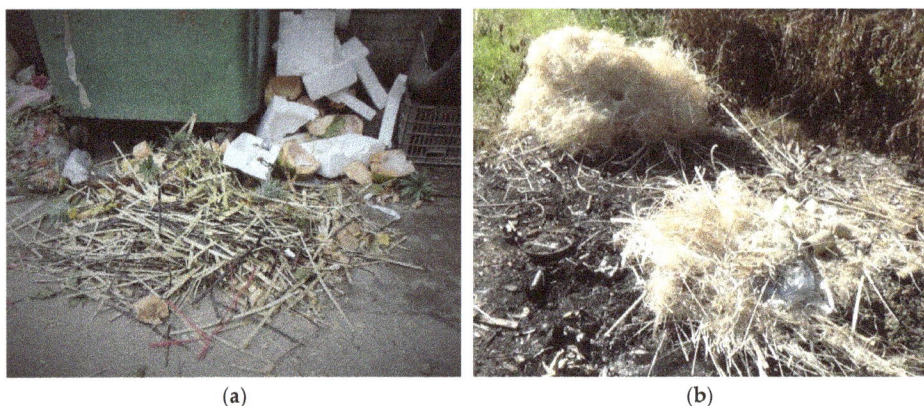

(a)   (b)

Figure 11. Untapped waste biomass potential in the target area: (**a**) sugarcane skin in municipal waste; (**b**) bamboo fiber in a fireplace.

Bamboo processing plants were located in rural areas. Thus, bamboo fiber was burned as waste without any purpose in the fields or individuals used it as a fire lighter at a small-scale. Sugarcane plants were processed directly in the streets of Huế city, thus leading to part of the produced sugarcane skin ending up in garbage bins for municipal waste (Figure 11). It is worth mentioning that the generation and amount of monitored waste materials were considerable and daily. In contrast, subsequent utilization of materials was rarely observed.

Bamboo plant populations or their cultivation were difficult to monitor because prevalent parts of the population are growing wildly. Focused on sugarcane plants, specific data of its production in Vietnam were available and are shown in Table 2. Meanwhile, according to a statistical database of the United Nations Food and Agriculture Organization (FAO), world sugarcane crop cultivation in 2016 provided the following data: harvested area 26,774,304 ha, production quantity 1,890,661,751 t and yield 706,148 hg·ha^{-1}.

Table 2. Sugarcane production in Vietnam in 2010–2014.

Year	Harvested Area (ha)	Production Quantity (t)	Yield (hg·ha^{-1})
2016	256,322	16,313,145	636,432
2015	284,262	18,337,227	645,083
2014	304,969	19,822,851	649,996
2013	310,264	20,131,088	648,757

t, tonne; ha, hectare; hg, hectogram. Source: FAO, 2018.

As observed in this table, sugarcane production in Vietnam has occurred at a high level in recent years. Thus, the production of sugarcane residues has also increased. Sugarcane plants are prevalently used for production of juice or sugar in Vietnam, and thus, sugarcane skin represents a considerable percentage ratio of produced waste biomass.

3.2. Fuel Analysis of Feedstock Materials

The present section describes the chemical parameters of the investigated materials, the required levels of which must be respected in practice, otherwise their burning (in the form of bio-briquette fuel) could cause environmental pollution. With respect to the data expressed in Table 3, it is clear that moisture content M_c (%) during experimental measurements occurred at a suitable level for bio-briquette production (i.e., $M_c < 15\%$) [30]. Mentioned values of moisture content M_c do not represent a material's moisture content in their initial form (at the moment of creation or collection) due to the necessary treatment within their transportation. However, as mentioned in the Methodology Section, bamboo fiber was removed from already sundried culms; thus, their (required) lower level of moisture content M_c was expected. Such a factor evaluated the amount of energy input positively within the entire process of bio-briquette production, which in contrast to others, also contains feedstock drying. Moreover, if moisture content M_c exceeds the suitable level, it complicates the densification process or makes it completely impossible to realize it.

Table 3. Basic chemical parameters of waste biomass samples (in w.b.) (± standard deviation).

Sample	M_c (%)	A_c (%)	GCV (MJ·kg^{-1})	NCV (MJ·kg^{-1})
Bamboo fiber	7.62 ± 0.47	1.16 ± 0.13	18.15 ± 0.03	16.92
Sugarcane skin	8.05 ± 0.52	8.62 ± 0.18	18.47 ± 0.50	17.23

M_c, moisture content; A_c, ash content; GCV, gross calorific value; NCV, net calorific value; w.b., wet basis.

When comparing the values of ash content A_c (%), it is clear that bamboo fiber exhibited a good result, while sugarcane skin exhibited a result that was at the margin of tolerance (i.e., A_c < 10%) [30]. Such a result represents a disadvantage of sugarcane material; however, it is still accepted by the mandatory requirements. Commonly, when a high level of ash content is detected, it can be caused by contamination of material due to external impurities (dust, soil). Nevertheless, sugarcane skin samples tested in the present research did not contain any impurities. Table 4 provides a comparison of fuel property data obtained within the present research with data from other studies.

Table 4. Fuel properties of bamboo and sugarcane residues.

	A_c (%)	GCV (MJ·kg^{-1})	Research
Bamboo culm	3.74	19.62	[39]
	3.70	18.32	[40]
Sugarcane bagasse	6.16	18.20	[41]
	11.27	17.33	[42]

A_c, ash content; GCV, gross calorific value.

A higher level of ash content A_c is commonly related to complications during biomass burning, which can result in low burning efficiency or damage to burning devices. In general, both materials represented herbaceous biomass, which prevalently exhibited a higher level of ash content A_c in comparison to wood biomass [43,44]. However, the results of bamboo fiber samples were very good, thus proving the significant potential of such material, which were qualitatively comparable with wood biomass [45,46].

Energy potential expressed by calorific value is the most important indicator of fuel chemical quality, which indicates the amount of energy released from fuel during burning [47]. The obtained *NVC* data for both tested materials exhibited a high level of such an indicator; thus, these results were satisfactory. For commercial sale, bio-briquette fuel must exhibit *NVCs* of at least 14.50 MJ·kg^{-1} for non-woody bio-briquette fuel and at least 15.00 MJ·kg^{-1} for woody bio-briquette fuel [30]. Every biomass kind exhibits different levels of calorific value, as well as the requirements on each specific biomass kind of bio-briquette fuel differing (Table 5).

Table 5. Comparison of various biomass kind energy potential.

Biomass Kind	Feedstock Material	NCV (MJ·kg^{-1})	Research
Woody	Vine pruning	19.19	[48]
	Pine sawdust	18.14	[46]
	Sycamore wood	15.62	[49,50]
	Wild cherry wood	15.55	
Herbaceous	Banana leaf	17.76	[41]
	Wheat straw	17.30	[45]
	Bamboo leaf	16.71	[51]
	Rice straw	15.07	[41]
Fruit	Grape seeds	20.39	[52]
	Oil palm empty fruit bunch	18.16	[53]
	Banana peel	18.89	[54]
	Durian peel	17.61	[55]
Aquatic	Water hyacinth	16.65	[41]
Mixed	Rice husk, oil palm sludge	21.68	[56]
	Tropical hardwood sawdust	18.94	[57]
	Rice straw, sugarcane leaves	17.83	

NCV, net calorific value.

The results of elementary composition (Table 6) primarily proved the higher level of oxygen content O (%) in the case of bamboo fiber, which is undesirable, while results for sugarcane skin were at a satisfactory level. In general, biomass elementary composition can influence final calorific value of produced biofuel, as well as influencing biofuel behavior during combustion. Higher levels of oxygen O (%) influence consumption of air during biofuel burning and production of flue gas [58].

Table 6. Elementary composition of waste biomass samples (in w.b.) (±standard deviation).

Sample	C (%)	H (%)	N (%)	S (%)	O (%)
Bamboo fiber	46.31 ± 0.14	6.19 ± 0.03	0.44 ± 0.02	0.04 ± 0.03	42.15
Sugarcane skin	45.85 ± 0.11	6.36 ± 0.02	0.96 ± 0.03	0.18 ± 0.01	34.02

C, carbon; H, hydrogen; N, nitrogen; S, sulfur; O, oxygen; w.b., wet basis.

Observed values were also converted into dry ash-free state to express results without the influence of ash presence (Table 7).

Table 7. Composition of samples in dry ash-free state (d.a.f.).

Sample	C (%)	H (%)	N (%)	S (%)	O (%)	GCV (MJ·kg^{-1})	NCV (MJ·kg^{-1})
Bamboo Fiber	51.01	5.82	0.49	0.04	42.65	20.00	18.73
Sugarcane Skin	54.95	6.46	1.15	0.21	37.23	22.14	20.73

C, carbon; H, hydrogen; N, nitrogen; S, sulfur; O, oxygen; GCV, gross calorific value; NCV, net calorific value; d.a.f., dry ash-free state.

3.3. Mechanical Analysis of Briquette Samples

Both investigated materials proved their suitability for bio-briquette production. For the densification process, it was possible to produce bio-briquette samples from them. Such a statement may not be a matter of course because the success of bio-briquette sample production is not guaranteed in advance.

Specific data evaluating the efficiency of the densification process were observed immediately following bio-briquette sample production. Thus, their bulk density ρ (kg·m^{-3}) was stated. The results of the first investigated mechanical quality indicator are noted in Table 8, together with the results of other investigated indicators.

Table 8. Mechanical quality indicators of the investigated bio-briquette samples (±standard deviation).

Sample	ρ (kg·m^{-3})	*DU* (%)	*RF* (N·mm^{-1})
Bamboo fiber	986.37 ± 181.18	97.80 ± 0.04	143.30 ± 1.70
Sugarcane skin	1067.08 ± 39.08	97.70 ± 0.08	46.50 ± 0.80

ρ, bulk density; *DU*, mechanical durability; *RF*, rupture force.

Observed bulk density ρ data proved satisfactory results in both cases, which indicates high quality bio-briquette fuel. As noted previously, bio-briquette fuel mechanical quality increases with increasing bulk density ρ [49,59,60] because it indicates bio-briquette fuels' longer burning time and larger amount of produced heat [61]. According to other published research, the level of bulk density ρ of high quality bio-briquette fuels should occur at approximately 1000 kg·m^{-3} [56,62,63].

The results of the next investigated indicator, bio-briquette sample *DU* (%), were obtained during experimental laboratory testing. A statement regarding bio-briquette fuel *DU* (%), also termed abrasion resistance, is necessary and required within commercial production and represents a major deciding factor [49,62,64]; thus, its performance was a pivotal point of mechanical analysis. Data noted in Table 8 expressed extremely good results; measured *DU* exceeded 97% for both materials, while the mandatory required level is a DU $\geq$ 90% [31]. Figure 12 illustrates the conditions of bio-briquette samples after abrasion testing.

(a)   (b)

Figure 12. Bio-briquette fuel samples after mechanical durability testing: (**a**) bamboo fiber; (**b**) sugarcane skin.

As observed in Figure 12, tested bio-briquette samples were in very good condition; only the edges of their bodies were abraded. Such positive results were caused by the structural characteristics of the materials, which were able to create strong bonds between particles.

Such excellent results correspond to the highest level of *DU* stated by related standards (i.e., DU $\geq$ 95%). Table 9 shows a comparison of the result values of other investigated bio-briquette fuels sorted according to specific mechanical durability levels.

Table 9. Mechanical durability *DU* of different bio-briquette fuel kinds.

DU (%)	Feedstock Material	Research
<90	Big bluestem sawdust	[59]
	Canola straw	
	Switchgrass	[65]
	Corn stover	
>90	Pine sawdust	[66]
	Canary grass	[67]
	Rice husk	[64]
	Corncob	
≥95	Eucalyptus sawdust, paper	[68]
	Soybean stalk	[69]
	Cotton stalk	[70]
	Digestate	[71]

DU, mechanical durability.

The level of *DU* is prevalently influenced by the specifications of pressed feedstock materials, but also, a considerable influence of the forming pressure used was proven. *DU* clearly increased with increasing forming pressure applied by pressing the briquetting press into the feedstock material [65]. Applying such a process in practice could improve the unsuitable lower levels of other specific bio-briquette's *DU*.

The last monitored indicator was rupture force *RF*, which is not defined by any mandatory standard. Thus, the evaluation of the observed results was performed only by comparison between the investigated materials. Data in Table 9 indicate the marked differences observed. BF bio-briquette samples had an *RF* = 143.30 N·mm^{-1}, while sugarcane skin bio-briquette samples exhibited an *RF* equal to 46.50 N·mm^{-1}. The conditions of bio-briquette samples after *RF* testing are visible in Figure 13.

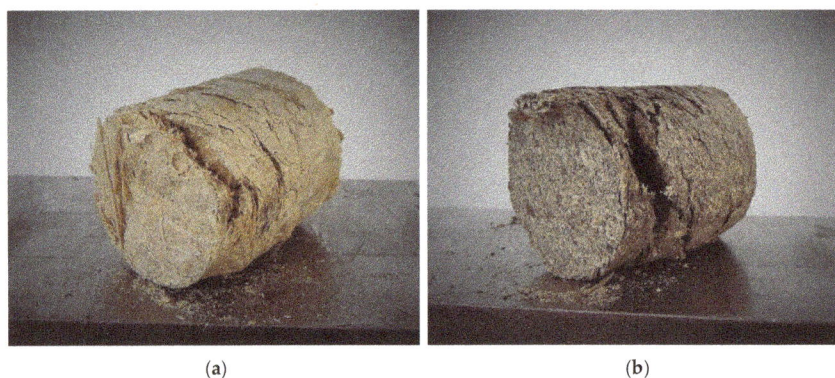

(a)   (b)

Figure 13. Bio-briquette fuel samples after rupture force testing: (a) bamboo fiber; (b) sugarcane skin.

The considerably better strength of bamboo fiber bio-briquette samples was caused by the positive behavior of fibers during pressing; above that, fibers made strong bonds already during feedstock preparation. Thus, it can be concluded that fibrous materials have significant potential for bio-briquette production. In addition, the overall evaluation of both tested materials proved satisfactory when compared with other published data [72–74].

4. Conclusions

The main question investigated in the present research was to determine whether specific agriculture waste residues, namely bamboo fibers and sugarcane skin, are suitable for bio-briquette fuel production. The answer to this question, based on all observed results, is 'Yes'. Both investigated materials proved suitable as complete feedstocks for bio-briquette fuel production. Monitoring of their production (quantity) and poor practice in terms of their reuse (in local conditions of the target area) indicated high potential of such waste biomass for any further meaningful utilization.

Fuel parameter analysis found satisfactory levels of all tested quality indicators; the only identified limitation was observed in the case of sugarcane skin ash content (A_c = 8.05%), but even that result is acceptable. Comparatively good results were obtained during determination of bio-briquette fuel mechanical quality indicators. Extremely high levels of mechanical durability (DU > 97%) strongly supported the statement that utilization of investigated materials for bio-briquette fuel production will result in high quality biofuel production.

In general, all measurements and tests performed within the present research provided satisfactory results, and hence, the utilization of bamboo fiber and sugarcane skin for bio-briquette fuel production can be highly recommended.

It is hoped that this research will also extend knowledge regarding appropriate waste management principles and reuse of all potential waste biomass for the production of environmentally-friendly solid biofuels.

Author Contributions: A.B. conceived of and designed the experiments, performed the sample collection and experiments, analyzed the results and wrote the majority of the paper. H.R. performed the sample collection and contributed to analyzing the results and to the writing of the paper. M.B. performed the experiments and provided comments on the paper. All authors have read and approved the final manuscript.

Acknowledgments: This research was supported by the Internal Grant Agency of the Czech University Life Sciences Prague 20173005 (31140/1313/3108) and by the Internal Grant Agency of the Faculty of Engineering, Czech University of Life Sciences Prague, Number 2018:31140/1312/3111.

Conflicts of Interest: The authors declare no conflict of interest.

Nomenclature

A_c	Ash content (%)
ρ	Bulk density of bio-briquette samples (kg·m^{-3})
C	Carbon (%)
DU	Mechanical durability (%)
F	Compressive force (N)
GCV	Gross calorific value (MJ·kg^{-1})
H	Hydrogen (%)
m	Mass of bio-briquette samples (kg)
m_a	Mass of bio-briquette samples after DU testing (g)
M_c	Moisture content (%)
m_e	Mass of bio-briquette samples before DU testing (g)
N	Nitrogen (%)
NCV	Net calorific value (MJ·kg^{-1})
O	Oxygen (%)
RF	Rupture force (N·mm^{-1})
S	Sulfur (%)
V	Volume of bio-briquette samples (m^3)
v	Loading speed (mm·min^{-1})

References

1. Popp, J.; Lakner, Z.; Harangi-Rákos, M.; Fári, M. The effect of bioenergy expansion: Food, energy, and environment. *Renew. Sustain. Energy Rev.* **2014**, *32*, 559–578. [CrossRef]

2. World Population Prospect. 2017. Available online: http://www.un.org/en/development/desa/population/publications/index.shtml (accessed on 28 May 2018).

3. Owusu, P.A.; Asumadu-Sarkodie, S. A review of renewable energy sources, sustainability issues and climate change mitigation. *Cogent Eng.* **2016**, *3*, 1167990. [CrossRef]

4. Mboumboue, E.; Njomo, D. Potential contribution of renewables to the improvement of living conditions of poor rural households in developing countries: Cameroon's case study. *Renew. Sustain. Energy Rev.* **2016**, *61*, 266–279. [CrossRef]

5. Liao, H.; Tang, X.; Wei, Y.M. Solid fuel use in rural China and its health effects. *Renew. Sustain. Energy Rev.* **2016**, *60*, 900–908. [CrossRef]

6. Liu, Z.; Mi, B.; Jiang, Z.; Fei, B.; Cai, Z.; Liu, X. Improved bulk density of bamboo pellets as biomass for energy production. *Renew. Energy* **2016**, *86*, 1–7. [CrossRef]

7. Werther, J.; Saenger, M.; Hartge, E.-U.; Ogada, T.; Siagi, Z. Combustion of agricultural residues. *Prog. Energy Combust. Sci.* **2000**, *26*, 1–27. [CrossRef]

8. Cherubini, F.; Peters, G.P.; Berntsen, T.; Stromman, A.H.; Hertwich, E. CO_2 emissions from biomass combustion for bioenergy: Atmospheric decay and contribution to global warming. *Glob. Chang. Biol. Bioenergy* **2011**, *3*, 413–426. [CrossRef]

9. Ohrnberger, D. *The Bamboos of the World: Annotated Nomenclature and Literature of the Species and the Higher and Lower Taxa*; Elsevier Press: New York, NY, USA, 1999.

10. Akinlabi, E.T.; Anane-Fenin, K.; Akwada, D.R. (Eds.) *Bamboo: The Multipurpose Plant*; Springer International Publishing: Cham, Switzerland, 2017; pp. 1–37. Available online: https://www.springer.com/la/book/9783319568072 (accessed on 6 July 2018).

11. Owen, A. Bamboo!! Improving island economy and resilience with Guam College students. *J. Mar. Isl. Cult.* **2015**, *4*, 65–75. [CrossRef]

12. Hofsetz, K.; Silva, M.A. Brazilian sugarcane bagasse: Energy and non-energy consumption. *Biomass Bioenergy* **2012**, *46*, 564–573. [CrossRef]

13. Pandey, A.; Soccol, C.R.; Nigam, P.; Soccol, V.T. Biotechnological potential of agro-industrial residues. I: Sugarcane bagasse. *Bioresour. Technol.* **2000**, *74*, 69–80. [CrossRef]

14. Ramirez, J.A.; Brown, R.; Rainey, T.J. Techno-economic analysis of the thermal liquefaction of sugarcane bagasse in ethanol to produce liquid fuels. *Appl. Energy* **2018**, *224*, 184–193. [CrossRef]

15. Xiong, W. Bagasse composites: A review of material preparation, attributes, and affecting factors. *J. Thermoplast. Compos. Mater.* **2018**, *31*, 1112–1146. [CrossRef]

16. Brienzo, M.; Siqueira, A.F.; Milagres, A.M.F. Search for optimum conditions of sugarcane bagasse hemicellulose extraction. *Biochem. Eng. J.* **2009**, *46*, 199–204. [CrossRef]

17. Jayapal, N.; Samanta, A.K.; Kolte, A.P.; Senani, S.; Sridhar, M.; Suresh, K.P.; Sampath, K.T. Value addition to sugarcane bagasse: Xylan extraction and its process optimization for xylooligosaccharides production. *Ind. Crops Prod.* **2013**, *42*, 14–24. [CrossRef]

18. Zwebe, D. Biomass Business Opportunities Viet Nam, SNV Netherlands Development Organization Vietnam. 2012. Available online: https://english.rvo.nl/sites/default/files/2013/12/Biomass_Opportunities_Viet_Nam.pdf (accessed on 10 July 2018).

19. Schirmer, M. Biomass and waste as a renewable and sustainable energy source in Vietnam. *J. Vietnam. Environ.* **2014**, *6*, 4–12. [CrossRef]

20. Roubík, H.; Mazancová, J.; Phung, L.D.; Dung, D.V. Quantification of biogas potential from livestock waste in vietnam. *Agron. Res.* **2017**. Available online: http://agronomy.emu.ee/wp-content/uploads/2017/05/Vol15nr2_Roubik.pdf (accessed on 29 May 2018).

21. Maninder; Kathuria, R.S.; Grover, S. Using agricultural residues as a biomass briquetting: An alternative source of energy. *IOSR J. Electr. Electron. Eng.* **2012**, *1*, 11–15. Available online: http://www.iosrjournals.org/iosr-jeee/Papers/vol1-issue5/C0151115.pdf (accessed on 29 May 2018).

22. Brand, M.A.; Jacinto, R.C.; Antunes, R.; da Cunha, A.B. Production of briquettes as a tool to optimize the use of waste from rice cultivation and industrial processing. *Renew. Energy* **2017**, *111*, 116–123. [CrossRef]

23. Chen, L.; Xing, L.; Han, L. Renewable energy from agro-residues in China: Solid biofuels and biomass briquetting technology. *Renew. Sustain. Energy Rev.* **2009**, *13*, 2689–2695. [CrossRef]

24. Zhang, G.; Sun, Y.; Xu, Y. Review of briquette binders and briquetting mechanism. *Renew. Sustain. Energy Rev.* **2018**, *82*, 477–487. [CrossRef]

25. Arelano, G.M.T.; Kato, Y.S.; Bacani, F.T. Evaluation of Fuel Properties of Charcoal Briquettes Derived from Combinations of Coconut Shell, Corn Cob and Sugarcane Bagasse. DLSU Research Congress, Manila, Philippines. 2015. Available online: http://www.dlsu.edu.ph/conferences/dlsu_research_congress/2015/proceedings/SEE/033-SEE_Bacani_FT.pdf (accessed on 12 July 2018).

26. EN 14918. *Solid Biofuels—Determination of Calorific Value*; ISO: Geneva, Switzerland, 2010.

27. ISO 1928. *Solid Mineral Fuels—Determination of Gross Calorific Value by the Bomb Calorimetric Method, and Calculation of Net Calorific Value*; ISO: Geneva, Switzerland, 2010.

28. EN 15234-1. *Solid Biofuels—Fuel Quality Assurance—Part 1: General Requirements*; ISO: Geneva, Switzerland, 2011.

29. EN ISO 16559. *Solid Biofuels—Terminology, Definitions and Descriptions*; ISO: Geneva, Switzerland, 2014.

30. EN ISO 17225-1. *Solid Biofuels—Fuel Specifications and Classes—Part 1: General Requirements*; ISO: Geneva, Switzerland, 2015.

31. EN ISO 17831-2. *Solid Biofuels—Determination of Mechanical Durability of Pellets and Briquettes—Part 2: Briquettes*; ISO: Geneva, Switzerland, 2015.

32. ISO 18122. *Solid Biofuels—Determination of Ash Content*; ISO: Geneva, Switzerland, 2015.

33. EN 18134-2. *Solid Biofuels—Determination of Moisture Content—Oven Dry Method—Part 2: Total Moisture—Simplified Method*; ISO: Geneva, Switzerland, 2015.

34. EN ISO 16948. *Solid Biofuels—Determination of Total Content of Carbon, Hydrogen and Nitrogen*; ISO: Geneva, Switzerland, 2016.

35. EN ISO 18123. *Solid Biofuels—Determination of the Content of Volatile Matter*; ISO: Geneva, Switzerland, 2016.

36. Seifi, M.R. The moisture content effect on some Physical and Mechanical Properties of Corn. *J. Agric. Sci.* **2010**, *2*, 125–134. [CrossRef]

37. Altuntas, E.; Yıldız, M. Effect of moisture content on some physical and mechanical properties of faba bean (*Vicia faba* L.) grains. *J. Food Eng.* **2007**, *78*, 174–183. [CrossRef]

38. Yahya, A.; Hamdan, K.; Ishola, T.A.; Suryanto, H. Physical and Mechanical Properties of *Jatropha curcas* L. Fruits from Different Planting Densities. *J. Appl. Sci.* **2013**, *13*, 1004–1012. [CrossRef]

39. Dietenberger, R. Eigenschaftsuntersuchungen von Bambus Zur Energetischen Verwertung [Analysis of Bamboo Characteristics for Energy Recovery]. Ph.D. Thesis, University of Freiburg, Freiburg, Germany, 2009.

40. Engler, B.; Schoenherr, S.; Zhong, Z.; Becker, G. Suitability of Bamboo as an Energy Resource: Analysis of Bamboo Combustion Values Dependent on the Culm's Age. *Int. J. For. Eng.* **2012**, *23*, 114–121. [CrossRef]

41. Phichai, K.; Pragrobpondee, P.; Khumpart, T.; Hirunpraditkoon, S. International Prediction Heating Values of Lignocellulosics from Biomass Characteristics. *J. Chem. Mol. Eng.* **2013**, *7*, 532–535. Available online: https://waset.org/Publications/prediction_heating_values_of_lignocellulosics_from_biomass_characteristics/16408 (accessed on 12 July 2018).

42. Guar, S.; Reed, T.B. *Thermal data for Natural and Synthetic Fuels*; Marcel Dekker, Inc.: New York, NY, USA, 1998.

43. Gürdil, G.A.K.; Selvi, K.C.; Malaťák, J.; Pinar, Y. Biomass Utilization for Thermal Energy. *Am. Agric. Mech. Asia Afr. Lat. Am.* **2009**, *40*, 80–85. Available online: https://www.researchgate.net/publication/290695799_Biomass_Utilization_for_Thermal_Energy (accessed on 10 May 2018).

44. Johansson, L.S.; Leckner, B.; Gustavsson, L.; Cooper, D.; Tullin, C.; Potter, A. Emission characteristics of modern and old-type residential boilers fired with wood logs and wood pellets. *Atmos. Environ.* **2004**, *38*, 4183–4195. [CrossRef]

45. McKendry, P. Energy production from biomass (part 1): Overview of biomass. *Bioresour. Technol.* **2002**, *83*, 37–46. [CrossRef]

46. Stolarski, M.J.; Szczukowski, S.; Tworkowski, J.; Krzyzaniak, M.; Gulczyński, P.; Mleczek, M. Comparison of Quality and Production Cost of Briquettes Made from Agricultural and Forest Origin Biomass. *Renew. Energy* **2013**, *57*, 20–26. [CrossRef]

47. Tumuluru, J.S.; Sokhansanj, S.; Wright, C.T.; Boardman, R.D.; Yancey, N. A Review on Biomass Classification and Composition, Co-Firing Issues and Pretreatment Methods. 2011. Available online: https://s3.amazonaws.com/academia.edu.documents/43368530/A_review_on_biomass_classification_and_c20160304-21801-1dvq0xm.pdf?AWSAccessKeyId=AKIAIWOWYYGZ2Y53UL3A&Expires=1531690192&Signature=qz4Y%2FadxiRqfCDsquKYyE6dKRDE%3D&response-content-disposition=inline%3B%20filename%3DA_review_on_biomass_classification_and_c.pdf (accessed on 15 May 2018).

48. Nasser, R.A.; Salem, M.Z.M.; Al-Mefarrej, H.A.; Abdel-Aal, M.A.; Soliman, S.S. Fuel Characteristics of Vine Prunings (*Vitis vinifera* L.) as a Potential Source for Energy Production. 2014. Available online: https://bioresources.cnr.ncsu.edu/wp-content/uploads/2016/06/BioRes_09_1_482_Nasser_SMAS_Fuel_Charac_Vine_Prunings_Alt_Wood_Source_4840.pdf (accessed on 11 July 2018).

49. Karunanithy, C.; Wang, Y.; Muthukumarappan, K.; Pugalendhi, S. Physiochemical characterization of briquettes made from different feedstocks. *Biotechnol. Res. Int.* **2012**, 165202. [CrossRef] [PubMed]

50. Telmo, C. and Lousada, J. The Explained Variation by Lignin and Extractive Contents on Higher Heating Value of Wood. *Biomass Bioenergy* **2011**, *35*, 1663–1667. [CrossRef]

51. Brunerová, A.; Roubík, H.; Brožek, M.; Velebil, J. Agricultural residues in Indonesia and Vietnam and their potential for direct combustion: With a focus on fruit processing and plantation crops. *Agron. Res.* **2018**, *16*, 656–668. [CrossRef]

52. Gravalos, I.; Xyradakis, P.; Kateris, D.; Gialamas, A.T. An Experimental Determination of Gross Calorific Value of Different Agroforestry Species and Bio-Based Industry Residues. *Nat. Resour.* **2016**, *7*, 57–68. [CrossRef]

53. Nasrin, A.B.; Choo, Y.M.; Lim, W.S.; Joseph, L.; Michael, S.; Rohaya, M.H.; Astimar, A.A.; Loh, S.K. Briquetting of Empty Fruit Bunch Fibre and Palm Shell as a Renewable Energy Fuel. *J. Eng. Appl. Sci.* **2011**, *6*, 446–451. [CrossRef]

54. Wilaipon, P.; Trirattansirichai, K.; Tangchaichit, K. Moderate Die–Pressure Banana–Peel Briquettes. *Int. J. Renew. Energy* **2007**, *2*, 53–58. Available online: http://www.sert.nu.ac.th/IIRE/V2N1(6).pdf (accessed on 10 July 2018).

55. Brunerová, A.; Roubík, H.; Brožek, M.; Herák, D.; Šleger, V.; Mazancová, J. Potential of Tropical Fruit Waste Biomass for Production of Bio-Briquette Fuel: Using Indonesia as an Example. *Energies* **2017**, *10*, 2119. [CrossRef]

56. Obi, O.F.; Okongwu, K.C.H. Characterization of fuel briquettes made from a blend of rice husk and palm oil mill sludge. Biomass Convers. *Biorefinery* **2016**, *6*, 449–456. [CrossRef]

57. Jittabut, P. Physical and Thermal Properties of Briquette Fuels from Rice Straw and Sugarcane Leaves by Mixing Molasses. *Energy Procedia* **2015**, *79*, 2–9. [CrossRef]

58. Demirbas, A. Combustion characteristics of different biomass fuels. *Prog. Energy Combust. Sci.* **2004**, *30*, 219–230. [CrossRef]

59. Tumuluru, J.S.; Tabil, L.G.; Song, Y.; Iroba, K.L.; Meda, V. Impact of process conditions on the density and durability of wheat, oat, canola, and barley straw briquettes. *Bioenergy Res.* **2015**, *8*, 388–401. [CrossRef]

60. Lindley, J.A.; Vossoughi, M. Physical properties of biomass briquettes. *Trans. ASABE* **1989**, *32*, 361–366. [CrossRef]

61. Obernberger, I.; Thek, G. Physical characterisation and chemical composition of densified biomass fuels with regard to their combustion behaviour. *Biomass Bioenergy* **2004**, *27*, 653–669. [CrossRef]

62. Kaliyan, N.; Morey, R.V. Factors affecting strength and durability of densified biomass products. *Biomass Bioenergy* **2009**, *33*, 337–359. [CrossRef]

63. Wakchaure, G.C.; Mani, I. Effect of binders and pressures on physical quality of some biomass briquettes. *J. Agric. Eng.* **2009**, *46*, 24–30. Available online: https://www.researchgate.net/profile/G_C_Wakchaure/publication/235944947_Effect_of_Binders_and_Pressures_on_Physical_Quality_of_Some_Biomass_Briquettes/links/02e7e51499eca04887000000.pdf (accessed on 10 July 2018).

64. Muazu, R.I.; Stegemann, J.A. Effects of operating variables on durability of fuel briquettes from rice husks and corn cobs. *Fuel Process. Technol.* **2015**, *133*, 137–145. [CrossRef]

65. Kaliyan, N.; Morey, R.V. Densification characteristics of corn stover and switchgrass. *Trans. ASABE* **2009**, *52*, 907–920. [CrossRef]

66. Li, Y.D.; Liu, H. High-pressure densification of wood residues to form an upgraded fuel. *Biomass Bioenergy* **2000**, *19*, 177–186. [CrossRef]

67. Repsa, E.; Kronbergs, E.; Pudans, E. Durability of compacted energy crop biomass. *Eng. Rural. Dev.* **2014**, *13*, 436–439. Available online: http://www.tf.llu.lv/conference/proceedings2014/Papers/74_Repsa_E.pdf (accessed on 10 July 2018).

68. Wachira, G.G.; Gitau, A.N.; Kimani, M.W.; Njoroge, B.N.K. Mechanical Properties of Saw Dust Briquettes of EucalyptusTree Species of Different Binders and Press Machines. *Int. J. Emerg. Technol. Adv. Eng.* **2015**, *5*, 532–538. Available online: https://pdfs.semanticscholar.org/100b/555644b59e9473181650d73509c067a7454a.pdf (accessed on 11 July 2018).

69. Rajkumar, D.; Venkatachalam, P. Physical properties of agro residual briquettes produced from Cotton, Soybean and Pigeon pea stalks. *Int. J. Power Eng. Energy* **2013**, *4*, 414–417. Available online: http://infomesr.org/attachments/JO-P-0039.pdf (accessed on 10 July 2018).

70. Eissa, A.H.A.; Alghannam, A.R.O. Quality Characteristics for Agriculture Residues to Produce Briquette. *World Acad. Sci. Eng. Technol.* **2013**, *78*, 166–171. Available online: https://www.researchgate.net/profile/Ayman_Amer_Eissa/publication/257357240_WASET_AyA2013v78-29/links/0046352500e5a91346000000.pdf (accessed on 10 July 2018).

71. Brunerová, A.; Pecen, J.; Brožek, M.; Ivanova, T. Mechanical durability of briquettes from digestate in different storage conditions. *Agron. Res.* **2016**, *14*, 327–336. Available online: http://agronomy.emu.ee/wp-content/uploads/2016/05/Vol14-_nr2_Brunerova.pdf#abstract-4144 (accessed on 12 July 2018).

72. Brunerová, A.; Brožek, M.; Müller, M. Utilization of waste biomass from post–harvest lines in the form of briquettes for energy production. *Agron. Res.* **2017**, *15*, 344–358. Available online: http://agronomy.emu.ee/wp-content/uploads/2017/05/Vol15nr2_Brunerova.pdf (accessed on 10 July 2018).

73. Brožek, M. Evaluation of selected properties of briquettes from recovered paper and board. *Res. Agric. Eng.* **2015**, *61*, 66–71. [CrossRef]

74. Brožek, M. The effect of moisture of the raw material on the properties briquettes for energy use. *Acta Univ. Agric. Silvic. Mendel. Brun.* **2016**, *64*, 1453–1458. [CrossRef]

![energies logo] *energies*

MDPI

Article

Biodiesel by Transesterification of Rapeseed Oil Using Ultrasound: A Kinetic Study of Base-Catalysed Reactions

José María Encinar [1,*], Ana Pardal [2], Nuria Sánchez [1] and Sergio Nogales [1]

[1]   Department of Chemical Engineering and Physical Chemistry, Extremadura University,
      Avenida de Elvas s/n., 06071 Badajoz, Spain; nuriass@unex.es (N.S.); senogalesd@unex.es (S.N.)
[2]   Department of Applied Sciences and Technology, IP Beja, Rua Pedro Soares s/n., 7800 Beja, Portugal;
      anap@ipbeja.pt
*   Correspondence: jencinar@unex.es; Tel.: +34-924-289-672

Received: 27 July 2018; Accepted: 23 August 2018; Published: 25 August 2018

Abstract: The objective of this work was to study the acceleration that ultrasound causes in the rate of biodiesel transesterification reactions. The effect of different operating variables, such as ultrasound power, catalyst (KOH) concentration and methanol:oil molar ratio, was studied. The evolution of the process was followed by gas chromatography, determining the concentration of methyl esters at different reaction times. The biodiesel was characterized by its density, viscosity, saponification and iodine values, acidity index, water content, flash and combustion points, cetane index and cold filter plugging point (CFPP), according to EN 14214 standard. High methyl ester yield and fast reaction rates were obtained in short reaction times. Ultrasound power and catalyst concentration had a positive effect on the yield and the reaction rate. The methanol:oil molar ratio also increased the yield of the reaction, but negatively influenced the process rate. The reaction followed a pseudo-first order kinetic model and the rate constants at several temperatures were determined. The activation energy was also determined using the Arrhenius equation. The main conclusion of this work is that the use of ultrasound irradiation did not require any additional heating, which could represent an energy savings for biodiesel manufacture.

Keywords: fatty acid methyl ester; catalyst; viscosity; iodine value; acidity index

---

1. Introduction

The future development of world economy makes finding renewable sources of energy that can replace fossil fuels necessary. For years, biodiesel has been a real alternative to fossil fuels used in internal combustion engines [1]. As it is known, biodiesel is a fuel consisting of monoalkyl esters of long-chain fatty acids (FAME) derived from renewable lipid feedstocks, and it is generally produced via transesterification [2,3]. A lot of different raw materials have been used to obtain biodiesel. Edible vegetable oils such as canola and soybean oil in the USA, palm oil in Malaysia or rapeseed oil in Europe have been used for biodiesel production and found to be good substitutes for diesel. Non-edible vegetable oils, such as *Pongamia pinnata* (karanja or honge), *Jatropha curcas* (jatropha or ratanjyote), *Citrus reticulata* (mandarin) and *Madhuca iondica* (mahua) have also been found to be suitable for biodiesel production [4–6]. Concerning the different types of vegetable oils and their composition, fatty acids with high unsaturation levels usually imply lower fluidity at low temperatures, leading to solidification. Thus, oils with a high ratio of monounsaturated fatty acid (and a low ratio of polyunsaturated fatty acids) usually show good performance at low temperatures. As rapeseed oil has such a composition, that is the reason why it is frequently used for biodiesel production in Europe. Rapeseed oil usually contains around 98% of triglycerides. The main components of this oil are oleic acid (monounsaturated), that is present in more than 60%, and linoleic acid (doubly unsaturated) that exceeds 20% [4]. For these reasons, rapeseed oil was chosen to carry out this work.

Although there are other possibilities, transesterification has been reported as the most common way to produce biodiesel from vegetable oil. Alcohols such as methanol or ethanol are the more frequently used alcohols. A catalyst is necessary to increase the reaction rate and the conversion yield. The catalysts used can be homogeneous and heterogeneous and they can be acid or basic in nature. Generally, basic homogenous catalysis is the preferred option to provide high reaction rates. Additionally, basic catalysts are readily available and very cheap. In the literature there is comprehensive information about the catalytic processes involved [3,7–9].

The mechanism of the transesterification of vegetable oils by means of basic catalysis is well known and published [10,11]. The mechanism comprises four stages. In the first step the base (catalyst) reacts with the alcohol, giving an alkoxide and the protonated catalyst. The second step consists in the nucleophilic attack of the alkoxide at the carbonyl group of the triglyceride, generating the alkyl ester and the corresponding anion of the diglyceride (third step). For the final step, this diglyceride deprotonates the catalyst, making it active and able to react with another alcohol, starting a new transtererification cycle. Diglycerides and monoglycerides are equally converted (to a mixture of alkyl esters and glycerol) by this mechanism.

The transesterification reaction is initially heterogeneous because methanol is only partially miscible with triglycerides at room temperature. For this reason, the reaction initially is slow and only takes place at the alcohol-oil interphase, and the process is dependent of mass transfer. When the stirring rate is high, emulsions are usually generated. These emulsions are caused, mainly, by the intermediate monoglycerides and diglycerides, which have both polar (hydroxyl groups) and non-polar (hydrocarbon chains) parts. Therefore, when a critical concentration of these intermediates is exceeded, emulsion takes place. Due to the low miscibility between methanol, vegetable oils and methyl esters, this emulsion is not stable and breaks. The use of ultrasonic irradiation could avoid this problem. Ultrasound produces special chemical and physical effects due to the collapse of cavitation bubbles. Low frequency sound waves can produce emulsions between immiscible liquids, being useful for the transesterification of triglycerides with alcohol [12–14]. Ultrasound has several effects on transesterification: acoustic streaming mixing or changes in sound pressure, leading to fast movement of fluids and cavitation bubbles that apply negative pressure gradients on liquids. The most important effect is the formation and collapse of cavitation bubbles, providing high temperature and pressure, with a significant influence on reaction rates, mass transfer and catalytic surface areas. As a consequence, the use of sonochemical reactors can favor the chemical reaction and propagation by way of enhanced mass transfer and interphase mixing between the phases and also can lower the requirements of the operating conditions (in terms of temperature and pressure) [15].

The main advantages of ultrasonic irradiation process are: shorter reaction times, lower alcohol/oil molar ratios, less energy consumption (50%), lower concentration of catalyst, higher reaction rate and conversion, improved yield, simpler separation and purification processes, and higher quality glycerol production [7,9,11,14].

In contrast, this process has some disadvantages such as: the reaction temperature was slightly higher for long reactions and the ultrasonic power must be under control due to the possibility of soap formation in fast reactions [16]. Nevertheless, as it is indicated, energetically the process is highly favorable, because of the formation of micro jets and neither localized temperature increases no agitation or heating are required to produce biodiesel with ultrasound application [17], therefore, according to the literature, the beneficial effect of ultrasound is due to the generation of a fine emulsion between methanol and fatty acids, increasing the surface area for the necessary chemical reactions. No adverse effects, such as the generation of free radicals and the subsequent chain reactions caused by them, were observed [18]. Thus, the use of ultrasound could imply an improvement in biodiesel production, increasing the yield of the product.

Therefore, it could be said that ultrasound radiation provides energy to form a mixture and overcome the energy activation barrier that is necessary for the process. On the other hand, high temperatures and pressures can be achieved locally, which could imply an intense mixture between the reagents. All these circumstances favor the reaction progress.

Considering the abovementioned facts, the objective of this work was to study the KOH-catalysed transesterification reaction of rapeseed oil using an ultrasonic mixing technique, determining the suitable reaction conditions to carry out the process in a short time and with low energy consumption. Also, a kinetic study of the reaction in the presence of ultrasonic radiation was carried out, determining the parameters necessary for the reactor design. In addition, the obtained results were compared to the previously obtained ones in the transesterification of rapeseed oil in the absence of radiation [19,20].

2. Materials and Methods

The raw material (rapeseed oil) was provided by the Research Center "La Orden-Valdesequera" (Badajoz, Spain), Section of Non-Edible Crops. The rapeseed oil was characterized by density, viscosity, water content, acid, iodine and saponification values and fatty acid profile (Table 1).

Table 1. Rapeseed oil fatty acid profile and properties.

Fatty Acid	Percentage (%)
C16:0 Palmitic	3.5
C18:0 Stearic	0.9
C18:1 Oleic	64.4
C18:2 Linoleic	22.3
C18:3 Linolenic	8.2
Other minority acids	0.7
Properties	
Density at 15 °C (kgm^{-3})	906.2
Viscosity at 40 °C (cSt)	36.3
Water content (wt, %)	0.8
Acidity index ($mg_{KOH} \cdot g_{oil}^{-1}$)	2.7
Iodine value ($g_{I2} \cdot 100 \ g^{-1}$)	113.5
Saponification value ($mg_{KOH} \cdot g_{oil}^{-1}$)	194.7

Potassium hydroxide (KOH, pellets GR for analysis), used as a catalyst, was supplied by Merck (Darmstadt, Germany) and methanol, 99%, was supplied by Panreac (Castellar del Vallés, Spain). The other chemicals were obtained commercially (Merck) and were of analytical grade. The experimental design is shown in Figure 1. The transesterification was carried out in a 500 mL spherical reactor, with a temperature sensor, sampling outlet and condensation systems, using a sonicator (Digital Sonifier, model 450, Branson, MO, USA), as shown in Figure 2. This sonicator has a fixed working frequency of 20 kHz and a power of 400 W, with adjustable levels for the latter between 0 and 100%.

Firstly, the reactor was charged with oil. Different amounts of catalyst were dissolved in different amounts of methanol and the resulting solution was added to the reactor. At this point, the sonicator was placed at the different power values and then the reaction started, taking place for 15–20 min. The different reaction conditions are specified in Table 2.

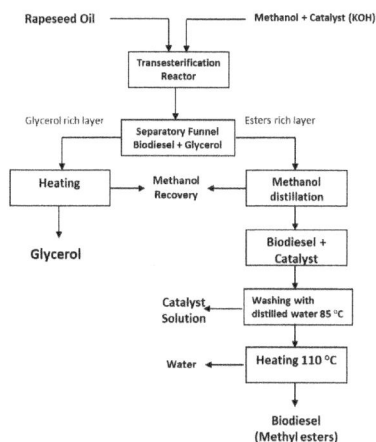

Figure 1. Schematic flow diagram of the biodiesel production from rapeseed oil.

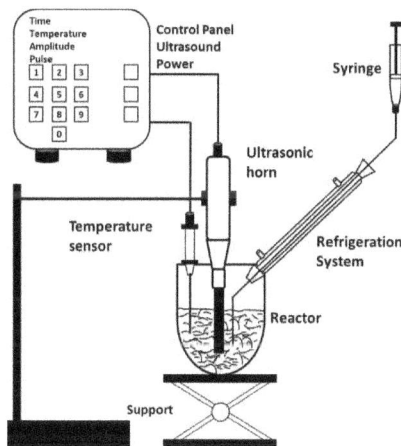

Figure 2. Schematic representation of the experimental setup.

Table 2. Experimental conditions. Properties of biodiesel produced from rapeseed oil and comparison with the EN 14214 standard.

Run	MeOH:Oil Molar Ratio	[a] KOH (%)	[b] Ultrasound Power, %, (W)	T_{max}, (°C)	[c] Yield, %	Density 15 °C, kg/m³	Viscosity 40 °C, cSt
1	9:1	0.7	20, (80 W)	49 (±2)	55.6 (±1.1)	879.4 (±2.1)	8.8 (±0.3)
2	9:1	0.7	40, (160 W)	60 (±2)	86.6 (±0.9)	872.0 (±2.0)	4.9 (±0.2)
3	9:1	0.7	60, (240 W)	64 (±1)	92.0 (±0.7)	867.9 (±1.7)	4.8 (±0.1)
4	9:1	0.7	80, (320 W)	68 (±1)	94.9 (±0.9)	866.3 (±1.2)	4.4 (±0.2)
5	9:1	0.7	100, (400 W)	70 (±2)	96.6 (±0.8)	861.9 (±1.5)	4.7 (±0.2)
6	9:1	0.3	100, (400 W)	69 (±1)	41.1 (±1.6)	888.5 (±2.1)	13.9 (±0.5)
7	9:1	0.5	100, (400 W)	70 (±2)	94.0 (±1.0)	866.2 (±1.3)	4.8 (±0.2)
8	9:1	1.0	100, (400 W)	70 (±2)	93.6 (±1.0)	866.5 (±0.9)	4.4 (±0.1)
9	3:1	0.7	100, (400 W)	80 (±1)	70.8 (±1.3)	875.5 (±1.4)	6.9 (±0.2)
10	6:1	0.7	100, (400 W)	75 (±1)	97.4 (±1.2)	870.9 (±0.8)	4.9 (±0.2)
11	12:1	0.7	100, (400 W)	70 (±1)	98.0 (±1.1)	866.5 (±1.3)	4.4 (±0.1)
12	9:1	0.7	0, (0 W)	55 (±2) (ISO)	61.2 (±1.4)	872.0 (±1.2)	7.6 (±0.2)
EN-14214	-	-	-	-	96.5	860–900	3.5–5.0

[a] The catalyst percentage is referred to the initial mass of oil; [b] The values in parentheses specify the applied power in watts; [c] The yield is referred to the mass of methyl esters in the total biodiesel mass.

After an appropriate reaction time, the mixture was placed in a separatory funnel for 24 h to ensure that the separation of biodiesel and glycerol phases was complete. The glycerol (bottom) phase was removed and left in a container. Methyl esters (biodiesel) were heated, at 85 °C to remove excess methanol. The remaining catalyst was extracted by successive rinses with distilled water. Finally, traces of water were eliminated by heating at 110 °C.

The evolution of the process (methyl ester content) was followed by gas chromatography on a VARIAN 3900 chromatograph (Varian, Palo Alto, California, USA), provided with an FID, employing a silica capillary column of 30 m length, 0.32 mm ID, and 0.25 mm film thickness. Heptane was used as a solvent, and the carrier gas was helium at a flow rate of 0.7 mL/min. The injector and the detector temperatures were kept at 270 °C and 300 °C, respectively. The temperature ramp started at 200 °C, and then increased 20 °C/min up to 220 °C. The calibration curve of the peak areas versus the quantity of biodiesel was linear. The samples were taken out from the reaction mixture, neutralized and heated to remove methanol, centrifuged for 5 min at 6000 rpm, and then analyzed by gas chromatography.

Thus, the yield was referred to the mass of FAME in the total biodiesel mass, as follows (Equation (1)):

$$Yield\ (\%) = \frac{\sum m_i}{m_T} * 100 \tag{1}$$

where m_i is the mass of each FAME, and m_T is the mass of biodiesel, both in mg. The analytical methods used to determine the characteristics of the biodiesel were basically those recommended by the European Organization for Normalization (CEN), for its use in motor vehicles. Also the recommended standards for the EN 14214 norm were employed. Details of the procedures used can be found in previous papers [10,19]. Most experiments were done in triplicate, and standard deviations are shown numerically in tables or as error bars in figures.

3. Results and Discussion

Before starting the study of the influence of variables, prior experiments in order to determine the optimal position of the probe in the reaction medium were carried out. In these experiments the methanol:oil molar ratio was 9:1, the concentration of catalyst (KOH) was 0.7% (w/w) and the power or amplitude of the probe was fixed at 40%. Three different positions were investigated: the probe in the oil phase, the probe in the methanol-oil interphase and the probe in the methanol phase. It was observed that the degree of conversion is small when the probe was in the oil phase. The conversion increased when the probe was placed in the methanol-oil interphase, reaching the top value when the probe is in the methanol phase. These results can be due to the extent cavitational intensity generated as a consequence of the presence of the probe in the oil phase, methanol phase or in the methanol-oil interphase. These circumstances affected the physicochemical properties, mainly viscosity, surface tension and density. As it is established in literature, methanol favors the generation of cavitation conditions and, as a consequence, the maximum conversion was obtained in this case [15]. In view of the above, the probe was placed in the methanol phase in all the subsequent experiments.

The generation of microturbulence for the cavitation bubbles located in the proximity of the interphase methanol-oil originated an emulsion of the two liquids. The dispersion of methanol in the oil depended on the intensity of the microturbulences that were generated by the cavitation bubbles in methanol and vice versa. The intensity of the microturbulences depended on the physical properties of the liquid medium such as density, viscosity and surface tension and also the amplitude of the acoustic waves driving the bubble motion. The methanol and oil phases have different physical properties and, therefore, the intensity of the microturbulences in the two phases was different. Consequently, the extension of the dispersion of methanol in oil cannot be the same as the dispersion of oil in methanol. The uniform dispersion of methanol in the oil phase originated the necessary interfacial area for the reaction to take place. This dispersion was produced, as it has been indicated, by the high level of microturbulences generated by the cavitation of the bubbles of methanol in the proximity with the interphase and, as it was manifested by Kalva [18], the dispersion increased with the methanol:oil molar ratio. As a consequence, the influence of this variable on the process should be notable.

Another aspect to consider is the reaction temperature. In the experiments, heating was not applied, but it was verified that the reaction temperature increased, due to irradiation, achieving maximum temperatures of up to 80 °C (run 9). Naturally, the experimental results showed that, for low temperatures, the extension of the conversion was also low and when temperature increased, the conversion also increased, showing, to a certain extent, a positive impact of temperature. But it is necessary to consider other factors. An increase in temperature implied an increase of the solubility of methanol in the oil phase and, also, an increase of the reaction rate. These two factors contributed to increase the conversion of the reaction. But at high temperatures, the extension of the cavitational effects was dampened and, on the other hand, methanol leaked of the reaction medium to surpass the boiling temperature. Hence, there was an optimal temperature that is dependent on all the previous aspects [15].

3.1. Effect of Ultrasound Power in Methyl Ester Conversion

In order to reduce the production costs of biodiesel it is necessary to optimize the amount of energy supplied to the reaction mixture. The objective was to obtain a maximum yield and a high formation rate with the lowest energy consumption. In this sense, it is known that when the intensity of the radiation (i.e., ultrasound power/irradiation area) increased, a more violent collapse of the bubbles of cavitation took place, producing a bigger mixing intensity in the methanol-oil interphase. These circumstances originated the formation of a very fine emulsion that favored the mass-transfer coefficient, causing a high formation of biodiesel [7].

Power had an influence on the size and number of bubbles, maximum live time, pressure that break these bubbles, elevation of the temperature inside the liquid, generation of cavitational effects and over the final collapse intensity. Also the intensity blending depended on the density of energy. Thus, generally, high levels of power dissipation promoted cavitational effects, but at times, these depended on the geometry of the reactor. For this reason, an optimal value of power dissipation is often observed due to phenomena of acoustic disengaging [21]. As a result, for high powers, a damping effect, which originates a decrease in the energy transfer and a low cavitational activity, was observed [15].

The results obtained in the study of the ultrasonic power are shown in Figure 3. The power was varied from 20 to 100% (experiments 1 to 5). Also, in order to draw a comparison, an experiment (run 12), carried out at 55 °C in isothermal conditions, with 0% of ultrasonic power, is enclosed. Also, in order to draw a comparison, an experiment (run 12), carried out at 55 °C in isothermal conditions, with 0% ultrasonic power, is enclosed. In this experiment, a heating power of 500 W was supplied in order to get the indicated conditions. The conditions of this experiment were chosen considering that are standard and common conditions of transesterification processes carried out by conventional heating methods. There was a positive effect of power, so that, for a given reaction time, the conversion was greater as the power was higher. The curves showed an induction period that diminished when power increased. Differences in conversion and induction times were smaller as the power level increased so, for powers above 60%, respective curves tended to an overlap situation. Thus, in the current situation, it took 15 min to achieve very high conversions (around 90%), for the experiments carried out with an ultrasound power between 160 and 400 W. An aspect that should be considered is the little variation of the maximum conversion with the applied power. In effect, runs 3, 4 and 5 lead to similar conversions even though the applied power differed considerably (60 to 100%). In this sense, the observations of Shing et al. [12] should be considered, showing that at higher amplitudes, ester yields were drastically reduced. This was attributable to cracking followed by FAME oxidation to aldehydes, ketones, and lower-chained organic fractions.

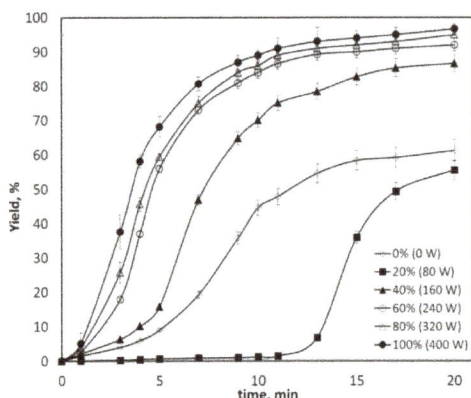

Figure 3. Effect of ultrasound power on the extent of conversion (MeOH:Oil 9:1, 0.7% KOH).

In addition, Figure 4 shows the evolution of temperature with reaction time for ultrasound powers between 40 and 100%. In all cases the same trend was observed: in the first minutes of reaction, the temperature increased very quickly, and later it stabilized at a constant value. The time required to achieve this constant value was between 3 and 4 min, and the maximum values of temperature were proportional to the applied power (60, 64, 68 and 70 °C for 40, 60, 80 and 100% powers respectively). Therefore, a balance between the heat generated as a consequence of the ultrasound radiation and the heat lost through the walls of the reactor took place. The result is a process that occurs near the isothermal regime. Another aspect to consider is the energy consumption. Ultrasonic powers of 160–240 W gave 90% yields. These results are in accordance with other research data, with similar power ranges and over 90% yields [22]. However, thermal powers of 500 W only produced a 55% yield. This implies a considerable energy savings.

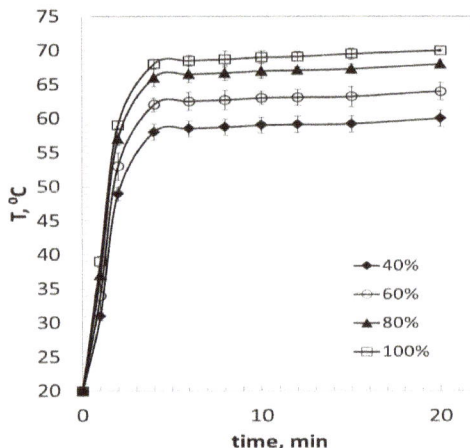

Figure 4. Temperature vs time. Effect of ultrasound power (methanol:oil 9:1, 0.7% KOH).

3.2. Effect of Catalyst Amount on Methyl Ester Conversion

Figure 5 shows the influence of the catalyst amount (run 5 to 8). The positive effect of this variable on the reaction rate of the process can be easily seen.

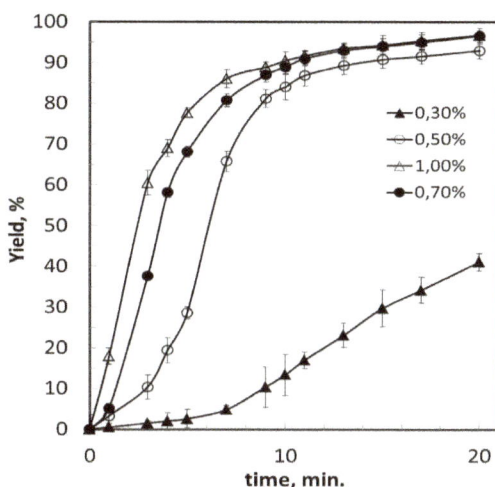

Figure 5. Effect of catalyst concentration on conversion (MeOH:Oil 9:1, 100% ultrasound power).

For very low concentrations, there was a long induction period. At high concentrations this induction period did not exist. For concentrations of catalyst under 0.7%, the conversion increased with concentration. Higher values of concentration lead to very similar conversions (run 5 and 8). This resulted in catalyst savings, which is positive from an economic point of view, and also facilitated the final product washing and/or the recovery of the catalyst, which is positive from a technical and environmental point of view.

Like in other works, it is possible to see two areas. In the first one, which can extend up to 7–12 min, a positive influence on the amount of catalyst on the rate of formation of methyl esters was observed. Indeed, as the amount of catalyst increased, it reached peak production sooner, that is, the asymptotic zone of the curve. The second area of the curve (the asymptote) was very similar in all cases (except for 0.3%).

Except for the experiment carried out with 0.3% catalyst (run 6) the short time necessary to achieve a maximum conversion was noticeable. In this sense, the ultrasound helps reduce the amount of catalyst needed due to the chemical activity due to cavitation. Thus, the necessary quantity of catalyst to achieve a given conversion was smaller when the reaction was carried out in the presence of ultrasound. A benefit was the resulting increase of the glycerol purity [23].

3.3. Effect of Methanol:Oil Ratio on Methyl Ester Conversion

One of the most important variables affecting the yield of esters is the alcohol:triglyceride molar ratio. As it is known, the stoichiometric ratio for transesterification requires three moles of alcohol and one mole of triglyceride to yield three moles of fatty acid alkyl esters and one mole of glycerol. However, transesterification is an equilibrium reaction in which a large excess of alcohol is required to drive the reaction to the right (products). However, the high molar ratio of alcohol to vegetable oil interferes with the separation of glycerin because there was an increase in solubility. When glycerin remained in solution, it helped to drive the equilibrium to the left (reactive), lowering the yield of esters [24]. Consequently, the methanol:oil molar ratio is a variable that should be optimized.

Figure 6 shows the results obtained (experiments 5, 9 to 11). Competing considerations are observed. On one hand, the higher initial reaction rate happened in the experiment with the lowest ratio of 3:1 and the lowest in the experiment with 12:1. On the other hand, the maximum conversion achieved was superior with increasing the methanol: oil molar ratio.

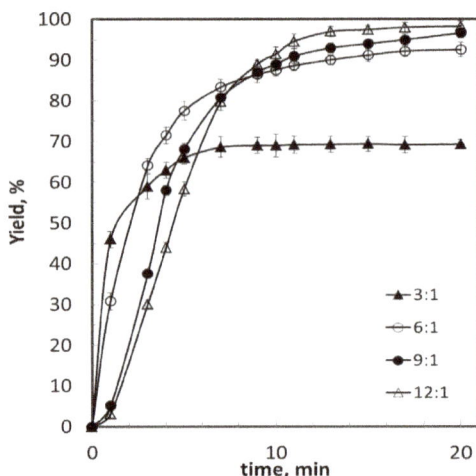

Figure 6. Effect of methanol:oil molar ratio on the extent of conversion (0.7% KOH, 100% ultrasound power).

This fact suggests that oil and methanol easily became an emulsion when there was a small amount of methanol, resulting in a higher rate of methanolysis. This implies that, under ultrasonic irradiation, small droplets of methanol were generated rapidly and easily achieved emulsion formation in the oil phase, leading to quick emulsion formation. In contrast, the droplet size of the emulsion was greater when there was a higher amount of methanol, and this resulted in a slower initial reaction rate. Furthermore, when there was a larger excess of methanol, the probability of the small size droplets of methanol encountering each other to agglomerate into larger droplets was larger than when there was a smaller excess of methanol. This trend towards large droplet formation may result in a slow transesterification reaction rate [25].

With an increase in the molar ratio, the quantity of methanol in the reaction mixture increased, which mainly affected the cavitation intensity. Excess of methanol provided additional cavitation events in the reactor, leading to the formation of enhanced emulsion quality (smaller drop sizes), providing additional area for the reaction and hence increased conversion [26].

Also, in connection with the previous figure, it can be noted that in the experiment with a molar ratio of 3:1, the reaction was stopped at 9 min due to the fast and excessive increase in temperature (80 °C at the end), which made the control of this variable difficult. In addition to the above, the low effect of temperature in ultrasonic system is related to the collapse of bubbles caused by cavitations that produced intense local heating and high pressures, with very short lifetimes, which have a much higher effect than elevating the temperature of the liquid media. Furthermore, temperature affected the vapor pressure, surface tension, and viscosity of the liquid medium. While higher temperature increased the number of bubble cavitations, the collapse of bubbles was cushioned by the higher vapor pressure, reducing the effect of ultrasound application in the reaction [27].

3.4. Biodiesel Properties

Table 2 shows the yield, density and viscosity obtained for all the experiments. Table 3 shows other parameters of the biodiesel corresponding with the two experiments with the highest yield in methyl esters. For comparison, in Tables 2 and 3, the values of the EN-14214 standard have been added. As it can be seen in Table 2, only in some experiments, it was possible to achieve the minimum conversion to methyl esters, required by the standard EN 14214. However, most of the experiments reached values close to the requirements. Only in experiments in which the applied power, or catalyst concentration, or the molar ratio of methanol: oil were minimal, was the conversion low.

Table 3. Properties of biodiesel produced from rapeseed oil in the best conditions of this study.

Parameter	Run 5	Run 10	EN-14214
FAME content, %	96.6 (±0.8)	97.4 (±1.2)	96.5
Water content, %	0.06 (±0.01)	0.06 (±0.01)	<0.05
Saponification value, $g_{KOH} \cdot g_{oil}^{-1}$	190.6 (±0.8)	188.9 (±0.6)	
Iodine value, wt %	106.9 (±0.3)	106.1 (±0.1)	≤120
Acidity Index, $mg_{KOH} \cdot g_{oil}^{-1}$	0.49 (±0.02)	0.48 (±0.01)	≤0.5
CFPP, °C	−7 (±1)	−7 (±1)	
Flash point, °C	178 (±1)	175 (±1)	≥120
Combustion point, °C	185 (±1)	189 (±1)	
Cetane index	44.1 (±0.2)	43.8 (±0.3)	
Density (kgm^{-3})	861.9 (±1.1)	870.9 (±0.8)	860–900
Viscosity at 40 °C (cSt)	4.7 (±0.2)	4.9 (±0.2)	3.5–5.0

In relation to density, the same ranges between 861.9 and 888.5 kg/m^3 were found, within the limits specified by the standard. Viscosity varied between 4.4 and 13.9 cSt. Except for the three experiments with low conversion, the presented viscosity values are in accordance with the standard. One overall effect was the relationship between conversion and viscosity: the more viscosity the less conversion. Consequently, most of the biodiesels obtained complied with the EN-14214 specifications, especially in those cases where the conversions obtained were high. For the other parameters (see Table 3), no significant differences were observed when high conversions were reached. Furthermore, it can be seen that, in general, the values of the standard were met.

3.5. Kinetic Study

As it has been indicated previously, the physical mechanism responsible for the beneficial action of ultrasound is the formation of a fine emulsion between oil and methanol that enhances the interface area for the reaction. No chemical effect of ultrasound, i.e., production of radical species and induction/acceleration of the reaction by these species, seemed to play any role [18]. In addition, the ultrasound radiation allowed one to get high temperature process in a short time. These temperatures (Figure 4) reached a maximum value (3 or 4 min) and at a later time they were constant. Therefore, the reaction, in the presence of ultrasound radiation, was similar to an isothermal regime process.

Besides, the transesterification reaction was complex, because secondary reactions of saponification or neutralization can happen. Additionally, the initial heterogeneous character of the reaction made the process difficult [28,29]. In our case, rapeseed oil was refined, and the free acidity was small, that is why saponification or neutralization reactions did not happen. The concentration of the catalyst did not change to get the asymptotic zone of the curves, and we assumed that the reaction-formations of monoglycerides, diglycerides and triglycerides were very fast and that there was no intermediate of reaction. That is, the transesterification reaction was considered as an only and global reaction, therefore, considering the above and in line with the literature [30], the reaction would be represented by Equation (2), where TG is the triglyceride, MeOH is methanol, ME is methyl-esters and G is glycerin.

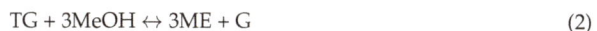

$$TG + 3MeOH \leftrightarrow 3ME + G \qquad (2)$$

The reaction (Equation (2)) is reversible, but in practice an excess of methanol was used and it can be considered irreversible. In addition, the kinetic study only was carried out until the curve showed the asymptotic trend. Under these conditions the inverse reaction lacks importance. The reaction rate is expressed by Equation (3), where α and β are the orders of reaction in relation to triglycerides and methanol respectively, k' is a constant that includes the effect of the catalyst, and t is the reaction time:

$$-d[TG]/dt = k'[TG]^{\alpha}[MeOH]^{\beta} \qquad (3)$$

Equation (3) can be rewritten like Equation (4), where k regroups k' and the concentration of methanol. This concentration can be considered constant because a very large concentration of methanol was used:

$$-d[TG]/dt = k[TG]^{\alpha} \tag{4}$$

Expressing the concentration of triglycerides in terms of conversion (X), Equation (4) takes the appearance of Equation (5), where TG_0 represents the initial concentration of triglycerides:

$$dX/dt = k[TG_0](1 - X)^{\alpha} \tag{5}$$

The integration of (5), with the most widely-used hypothesis in literature (pseudo first order kinetic model), that is, with $\alpha = 1$, leads to Equation (6):

$$\ln(1 - X) = -kt \tag{6}$$

Equation (1) has been applied to Experiments 2, 3, 4 and 5, in which the ultrasound power varied. In these experiments, the methanol:oil molar ratio was always 9:1 and the catalyst concentration 0.7%. In Table 4, the pseudo kinetic constants and R^2 coefficient, obtained by means of regression analysis of Equation (1) are shown.

Table 4. Kinetic analysis of the transesterification process.

Temperature, °C	Ultrasound Power (%)	Pseudo First-Order Kinetic, min^{-1}	R^2
60	40	0.1669	0.98
64	60	0.1872	0.98
68	80	0.2103	0.98
70	100	0.2170	0.99

The temperature shown for each experiment was the temperature that the system reached after 3 or 4 min (final temperature). The high values of R^2 confirmed that this system followed a pseudo-first kinetic model. The relationship among the specific reaction rate constant (k), absolute temperature (T) and activation energy (Ea) is given by Arrhenius Equation (7), where A is the frequency factor and R is the universal gas constant:

$$k = A\exp(-Ea/RT) \tag{7}$$

The linear regression of Equation (7) provides the value of the activation energy. The best fit is shown in Equation (8). These data lead to a value of Ea = 25.51 kJ/mol. This value was relatively small, which reveals a great catalytic activity of the basic catalysts (KOH):

$$\ln k = -3068.5/T + 7.428; R^2 = 0.993 \tag{8}$$

The activation energy was similar to that obtained in rapeseed oil transesterification in the presence of co-solvents (21.88 kJ/mol) [31] and slightly lower than in the case of *Jatropha curcas* transesterification by ultrasound (31.29 and 57.33 kJ/mol) [32]. On the other hand, the activation energy in other catalysis-base transesterification through thermal heat were superior. For instance, for palm oil transesterification an activation energy of 105 kJ/mol was found [33].

4. Conclusions

The suitability of the use of ultrasound for rapeseed oil transesterification was studied. The main findings of this research were:

Ultrasonic irradiation facilitated the rapeseed transesterification, since high biodiesel yields were achieved after short reaction times (20 min). It was observed, during experimentation, that the temperature of the reaction mixture increased, as a result of ultrasound radiation.

The use of ultrasound radiation did not need additional heating, what can suppose an energy savings. Nevertheless, it is necessary to consider the energy consumed in the generation of ultrasound.

The influence of the ultrasound power and the catalyst concentration was positive, so the yield of the process and the reaction rate increased as these variables increased. The methanol:oil molar ratio also lead to a bigger yield of the process, but the lower values of the methanol:oil molar ratio lead to the biggest reaction rate.

The obtained results showed that ultrasonic powers of 320 W, a catalyst concentration of 0.7% and a methanol:oil molar ratio of 9:1 are enough to achieve biodiesel yields of 95%. These conditions can be considered adequate to carry out the process. The characteristics of the biodiesel, determined by the EN 14214 standard, revealed that, in general, they met the established limits. The final product had similar properties to a diesel-oil.

The transesterification reactions followed a pseudo-first order kinetic model and the rate constants at several temperatures were determined. Also, the activation energy was determined by the Arrhenius equation.

Author Contributions: Conceptualization, J.M.E., A.P. and N.S.; Methodology, J.M.E., A.P., N.S. and S.N.; Validation, J.M.E., A.P., N.S. and S.N.; Formal Analysis, A.P. and N.S.; Investigation, A.P. and N.S.; Resources, J.M.E., A.P., N.S. and S.N.; Data Curation, J.M.E., A.P., N.S. and S.N.; Writing-Original Draft Preparation, J.M.E., A.P., N.S. and S.N.; Writing-Review & Editing, J.M.E. and S.N.; Supervision, J.M.E. and S.N.

Acknowledgments: The authors would like to thank the "Junta de Extremadura" ("Ayudas para la realización de actividades de investigación y desarrollo tecnológico, de divulgación y de transferencia de conocimiento por los Grupos de Investigación de Extremadura") and the FEDER "Fondos Europeos de Desarrollo Regional" for the financial support received.

Conflicts of Interest: The authors declare no conflict of interest.

References

1. Caldeira-Pires, A.; da Luz, S.M.; Palma-Rojas, S.; Oliveira Rodrigues, T.; Chaves Silverio, V.; Vilela, F.; Barbosa, P.C.; Alves, A.M. Sustainability of the biorefinery industry for fuel production. *Energies* **2013**, *6*, 329–350. [CrossRef]

2. Guo, M.; Song, M.; Buhain, J. Bioenergy and biofuels: History, status, and perspective. *Renew. Sustain. Energy Rev.* **2015**, *42*, 712–725. [CrossRef]

3. Ma, F.; Hanna, M.A. Biodiesel production: A review. *Bioresour. Technol.* **1999**, *70*, 1–15. [CrossRef]

4. Pinzi, S.; Garcia, I.L.; Lopez-Gimenez, F.J.; Luque de Castro, M.D.; Dorado, G.; Dorado, M.P. The Ideal Vegetable Oil-based Biodiesel Composition: A Review of Social, Economical and Technical Implications. *Energy Fuel* **2009**, *23*, 2325–2341. [CrossRef]

5. Jurac, Z.; Zlatar, V. Optimization of raw material mixtures in the production of biodiesel from vegetable and used frying oils regarding quality requirements in terms of cold flow properties. *Fuel Proc. Technol.* **2013**, *106*, 108–113. [CrossRef]

6. Azad, A.K. Biodiesel from Mandarin Seed Oil: A surprising source of alternative fuel. *Energies* **2017**, *10*, 1689. [CrossRef]

7. Mahamuni, N.N.; Adewuyi, Y.G. Optimization of the Synthesis of Biodiesel via Ultrasound-Enhanced Base-Catalyzed Transesterification of Soybean Oil Using a Multifrequency Ultrasonic Reactor. *Energy Fuels* **2009**, *23*, 2757–2766. [CrossRef]

8. Demirbas, A. Progress and recent trends in biodiesel fuels. *Energy Convers. Manag.* **2009**, *50*, 14–34. [CrossRef]

9. Ramachandran, K.; Suganya, T.; Nagendra Gandhi, N.; Renganathan, S. Recent developments for biodiesel production by ultrasonic assist transesterification using different heterogeneous catalyst: A review. *Renew. Sustain. Energy Rev.* **2013**, *22*, 410–418. [CrossRef]

10. Encinar, J.M.; González, J.F.; Pardal, A. Transesterification of castor oil under ultrasonic irradiation conditions. Preliminary results. *Fuel Proc. Technol.* **2012**, *103*, 9–15. [CrossRef]

11. Singh, A.K.; Fernando, S.D.; Hernandez, R. Base-Catalyzed Fast Transesterification of Soybean Oil Using Ultrasonication. *Energy Fuels* **2007**, *21*, 1161–1164. [CrossRef]

12. Hanh, H.D.; Dong, N.T.; Okitsu, K.; Nishimura, R.; Maeda, Y. Biodiesel production through transesterification of triolein with various alcohols in an ultrasonic field. *Renew. Energy* **2009**, *34*, 766–768. [CrossRef]

13. Ji, J.; Wang, J.; Li, Y.; Yu, Y.; Xu, Z. Preparation of biodiesel with the help of ultrasonic and hydrodynamic cavitation. *Ultrasonics* **2006**, *44*, e411–e414. [CrossRef] [PubMed]

14. Veljković, V.B.; Avramović, J.M.; Stamenković, O.S. Biodiesel production by ultrasound-assisted transesterification: State of the art and the perspectives. *Renew. Sustain. Energy Rev.* **2012**, *16*, 1193–1209. [CrossRef]

15. Hingu, S.M.; Gogate, P.R.; Rathod, V.K. Synthesis of biodiesel from waste cooking oil using sonochemical reactors. *Ultrason. Sonochem.* **2010**, *17*, 827–832. [CrossRef] [PubMed]

16. Talebian-Kiakalaieh, A.; Amin, N.A.S.; Mazaheri, H. A review on novel processes of biodiesel production from waste cooking oil. *Appl. Energy* **2013**, *104*, 683–710. [CrossRef]

17. Santos, F.F.P.; Rodrigues, S.; Fernandes, F.A.N. Optimization of the production of biodiesel from soybean oil by ultrasound assisted methanolysis. *Fuel Proc. Technol.* **2009**, *90*, 312–316. [CrossRef]

18. Kalva, A.; Sivasankar, T.; Moholkar, V.S. Physical mechanism of ultrasound-assisted synthesis of biodiesel. *Ind. Eng. Chem. Res.* **2009**, *48*, 534–544. [CrossRef]

19. Encinar, J.M.; Pardal, A.; Martínez, G. Transesterification of rapeseed oil in subcritical methanol conditions. *Fuel Process. Technol.* **2012**, *94*, 40–46. [CrossRef]

20. Martínez, G.; Sánchez, N.; Encinar, J.M.; González, J.F. Fuel properties of biodiesel from vegetable oils and mixtures. Influence of methyl esters distribution. *Biomass Bioenergy* **2014**, *63*, 22–32. [CrossRef]

21. Gole, V.L.; Gogate, P.R. A review on intensification of synthesis of biodiesel from sustainable feed stock using sonochemical reactors. *Chem. Eng. Process.* **2012**, *53*, 1–9. [CrossRef]

22. Sheng Ho, W.W.; Kiat Ng, H.; Gan, S. Advances in ultrasound-assisted transesterification for biodiesel production. *Appl. Therm. Eng.* **2016**, *100*, 553–563.

23. Kumar, D.; Kumar, G.; Poonam; Singh, C.P. Fast, easy ethanolysis of coconut oil for biodiesel production assisted by ultrasonication. *Ultrason. Sonochem.* **2010**, *17*, 555–559. [CrossRef] [PubMed]

24. Meher, L.C.; Vidya Sagar, D.; Naik, S.N. Technical aspects of biodiesel production by transesterification—A review. *Renew. Sustain. Energy Rev.* **2006**, *10*, 248–268. [CrossRef]

25. Thanh, L.T.; Okitsu, K.; Sadanaga, Y.; Takenaka, N.; Maeda, Y.; Bandow, H. Ultrasound-assisted production of biodiesel fuel from vegetable oils in a small scale circulation process. *Bioresour. Technol.* **2010**, *101*, 639–645. [CrossRef] [PubMed]

26. Deshmane, V.G.; Gogate, P.R.; Pandit, A.B. Ultrasound-Assisted Synthesis of Biodiesel from Palm Fatty Acid Distillate. *Ind. Eng. Chem. Res.* **2009**, *48*, 7923–7927. [CrossRef]

27. Santos, F.F.P.; Malveira, J.Q.; Cruz, M.G.A.; Fernandes, F.A.N. Production of biodiesel by ultrasound assisted esterification of Oreochromis niloticus oil. *Fuel* **2010**, *89*, 275–279. [CrossRef]

28. Veljkovic, V.B.; Stamenkovic, O.S.; Todorovic, Z.B.; Lazic, M.L.; Skala, D.U. Kinetics of sunflower oil methanolysis catalyzed by calcium oxide. *Fuel* **2009**, *88*, 1554–1562. [CrossRef]

29. Stamenkovic, O.S.; Todorovic, Z.B.; Lazic, M.L.; Veljkovic, V.B.; Skala, D.U. Kinetics of sunflower oil methanolysis at low temperatures. *Bioresour. Technol.* **2008**, *99*, 1131–1140. [CrossRef] [PubMed]

30. Ramezami, K.; Rowshanzamir, S.; Eikani, M.H. Castor oil transesterification reaction: A kinetic study and optimization of parameters. *Energy* **2010**, *35*, 4142–4148. [CrossRef]

31. Encinar, J.M.; Pardal, A.; Sánchez, N. An improvement to the transesterification process by the use of co-solvents toproduce biodiesel. *Fuel* **2016**, *166*, 51–58. [CrossRef]

32. Choudhury, H.A.; Srivastava, P.; Moholkar, V.S. Single-step ultrasonic synthesis of biodiesel from crude Jatropha curcas oil. *AIChE J.* **2014**, *60*, 1572–1581. [CrossRef]

33. Permsuwan, A.; Tippayawong, N.; Tanongkiat, T.; Thararux, C.; Wangkarn, S. Reaction Kinetics of Transesterification between Palm Oil and Methanol under Subcritical Conditions. *Energy Sci. Technol.* **2011**, *2*, 35–42.

energies

MDPI

Article

A Study of the Production and Combustion Characteristics of Pyrolytic Oil from Sewage Sludge Using the Taguchi Method

Guan-Bang Chen [1],*, Jia-Wen Li [2], Hsien-Tsung Lin [2], Fang-Hsien Wu [2] and Yei-Chin Chao [2],*

[1] Research Center for Energy Technology and Strategy, National Cheng Kung University, Tainan 701, Taiwan
[2] Department of Aeronautics and Astronautics, National Cheng Kung University, Tainan 701, Taiwan;
     z9809143@email.ncku.edu.tw (J.-W.L.); P48021027@mail.ncku.edu.tw (H.-T.L.);
     z10602031@email.ncku.edu.tw (F.-H.W.)
*   Correspondence: gbchen@mail.ncku.edu.twmailto (G.-B.C.); ycchao@mail.ncku.edu.tw (Y.-C.C.);
     Tel.: +886-6-2757575 (ext. 51030) (G.-B.C.); +886-6-2757575 (ext. 63690) (Y.-C.C.)

Received: 19 July 2018; Accepted: 24 August 2018; Published: 28 August 2018

Abstract: Sewage sludge is a common form of municipal solid waste, and can be utilized as a renewable energy source. This study examines the effects of different key operational parameters on sewage sludge pyrolysis process for pyrolytic oil production using the Taguchi method. The digested sewage sludge was provided by the urban wastewater treatment plant of Tainan, Taiwan. The experimental results indicate that the maximum pyrolytic oil yield, 10.19% (18.4% on dry ash free (daf) basis) by weight achieved, is obtained under the operation conditions of 450 °C pyrolytic temperature, residence time of 60 min, 10 °C/min heating rate, and 700 mL/min nitrogen flow rate. According to the experimental results, the order of sensitivity of the parameters that affect the yield of sludge pyrolytic oil is the nitrogen flow rate, pyrolytic temperature, heating rate and residence time. The pyrolysis and oxidation reactions of sludge pyrolytic oil are also investigated using thermogravimetric analysis. The combustion performance parameters, such as the ignition temperature, burnout temperature, flammability index and combustion characteristics index are calculated and compared with those of heavy fuel oil. For the blend of sludge pyrolytic oil with heavy fuel oil, a synergistic effect occurs and the results show that sludge pyrolytic oil significantly enhances the ignition and combustion of heavy fuel oil.

Keywords: sewage sludge; pyrolytic oil; Taguchi method; thermogravimetric analysis; synergistic effect

1. Introduction

Even now, a large part of the global energy supply still depends on fossil fuels, resulting in rapid depletion of these resources and enormous GHG (greenhouse gas) emissions. In order to mitigate the problems associated with this fossil fuel depletion and climate warming, it is necessary to either improve the efficiency of fossil fuel utilization [1,2], or partially replace the use of such fossil fuels with zero or neutral carbon footprint alternative energy supplies [3,4]. Among the alternative fuels that are now being considered, biomass is widely recognized as a promising, eco-friendly source of renewable energy, which has the advantage of being readily available around the world.

Biomass derives from botanical or biological sources, or from a combination of these. Ordinary sources of biomass include agricultural solid waste, forestry residues, municipal waste, energy crops, and biological waste. Sewage sludge, the major municipal waste from the wastewater treatment plants, is a complicated mixture of undigested organics, inorganic materials, and moisture. The traditional treatment of sewage sludge includes agricultural use, incineration, and landfill. However, these methods have several shortcomings. With regard to agricultural use, sewage sludge contains pathogenic bacteria, viruses, and parasitic helminths, which can harm the health of humans, animals, and plants. For incineration, the heating value of dewatered sewage sludge is low. For landfill, sludge contains a certain amount of organic substance, which generates a biogas rich in methane. Moreover, the cost of this approach is also increasing, because it is more and more difficult to find appropriate areas for landfill. In addition, with recent increases in the sewage treatment rate, the volume of sludge has grown rapidly, and thus alternative methods of sludge management are urgently needed. Thermochemical processes, such as pyrolysis or gasification, have been studied and recommended as suitable alternatives [5]. Thermochemical processes can produce energy from the sludge and have little influence on the environment.

For conversion into bio-fuel from biomass, biochemical conversion (BCC) and thermochemical conversion (TCC) are the two general broad approaches. BCC often uses fermentation and biological conversion processes. Direct combustion, gasification, pyrolysis, and liquefaction are generally used for TCC [6,7]. The resulting products can be used directly as fuels, or further processed for specialty chemicals [8,9].

Among the various thermochemical conversion processes, thermal pyrolysis has attracted more interest with regard to yielding liquid fuel products from various biomass species, such as woody biomass [10], bagasse [11], straws [12], miscanthus [13], castor meal [14], oil palm fiber [15], fruit waste [16], and municipal solid waste [6,17]. Even for waste oil, pyrolysis approaches offer some advantages over existing methods (transesterification, hydrotreating, gasification, solvent extraction, and membrane technology) in producing useful pyrolysis products for future reuse [18]. Products from pyrolysis processes depend strongly on the heating rate, heating temperature, gas residence time, particle size of biomass, and water content of the biomass. In fact, great work related to the conversion of biomass to liquid fuels was performed during the oil crisis of the 1970s [19], and most studies on sludge liquid fuel production are based on pyrolysis for lignocellulosic materials. In general, in order to increase the yield of bio-oil, the biomass in the fast pyrolysis process is rapidly heated to a high temperature (425–600 °C) and has a short gas residence time (<3 s) in the absence of oxygen [20].

The review work of sewage sludge pyrolysis by Fonts et al. [21] placed specific emphasis on liquid sewage sludge pyrolytic fuel production, and detailed comparisons of pyrolytic oils between sewage sludge and lignocellulose were also made. Several studies have performed sewage sludge pyrolysis to produce liquid fuel, which is composed of aqueous and organic phase compounds [22–27]. Some papers emphasize the influence of operational conditions and composition on the sludge pyrolysis product, especially the liquid yield. There are few studies related to the maximum yield of sludge pyrolytic oil and its combustion characteristics. Therefore, in this study we examined the influence of different key parameters on sewage sludge pyrolysis process for pyrolytic oil using the Taguchi method. As an example, the method was applied in determining the pyrolysis process parameters of the digested sewage sludge provided by the urban wastewater treatment plant of Tainan, Taiwan. The combustion characteristics of the resulting pyrolytic oil are also investigated by thermogravimetric analysis. The experimental results are also compared with those for heavy fuel oil, since the co-firing of sludge pyrolytic oil with heavy fuel oil is a promising potential alternative for practical applications.

2. Methodology

2.1. Experimental Setup

The feedstock used for the studies is typical digested sewage sludge from a wastewater treatment plant in Tainan, Taiwan (Figure 1). Even though the sludge is dewatered in the wastewater treatment plant, its water content is still higher than 30%. Therefore, before the pyrolysis experiment, the sewage sludge was preheated to 110 °C in an oven until the water content of the sewage sludge was lower than 10% (see Table 1). The drying step is important and can even affect the conversion of a waste into other products with high added value [28]. Figure 2a depicts the experimental thermal pyrolysis apparatus. In the experiment, the dried sewage sludge was mixed and ground to pass through a 100-mesh sieve, and 90 g of sewage sludge were packed in a cylindrical holder made of quartz and put into the furnace. The covers of the cylindrical holder were made of stainless steel with many holes. For pyrolysis, the tubular furnace was first vacuumed, and then flushed with N_2 to ambient pressure. The furnace was then heated to the preset temperature and maintained at the temperature for a designated residence time. The residence time here is different from the gas residence time, which represents the residence time for pyrolytic vapor inside the pyrolysis furnace. In addition, the gas residence time is inversely proportional to the nitrogen flow rate. The heated carrier gas can flow into the holder and interact with the sludge. The generated volatile gas was collected and delivered to a condensing system kept at 25 °C to form a liquid product. Figure 2b shows the thermal pyrolysis operating process in a fixed-bed tubular furnace and it must undergo two steps: being heated up from room temperature to the target temperature at a specific heating rate and holding at the targeted temperature for a designated residence time. The production of pyrolytic oil occurs in a certain conditions and is influenced by the combination of four operational parameters: heating rate, target pyrolytic temperature, residence time, and nitrogen flow rate. The Taguchi method will be used to analyze the influence of this parameter combination on the pyrolytic oil production from sewage sludge using a fixed-bed tubular furnace. After the experiment, the condensing bottle was held for 24 h, and the pyrolytic oil was separated from the aqueous matter using a separatory funnel. This pyrolytic oil has a high heating value, and was the target product in the study. As to the aqueous product, it could be used as a source for fertilizers or chemicals due to the nitrogen-containing compounds [29], or as a source of triacetonamine [30].

Figure 1. Photograph of sewage sludge, (a) sludge from the wastewater treatment plant and (b) sludge after preheating.

Table 1. Proximate analysis of sewage sludge.

Proximate Analysis	Sewage Sludge	Sewage Sludge [21]
Moisture (wt %)	6.94	1.5–7.1
Volatiles (wt %)	45.45	38.3–66.8
Ash (wt %)	37.67	22.6–52.0
Fixed carbon (wt %)	9.94	0.8–19.7

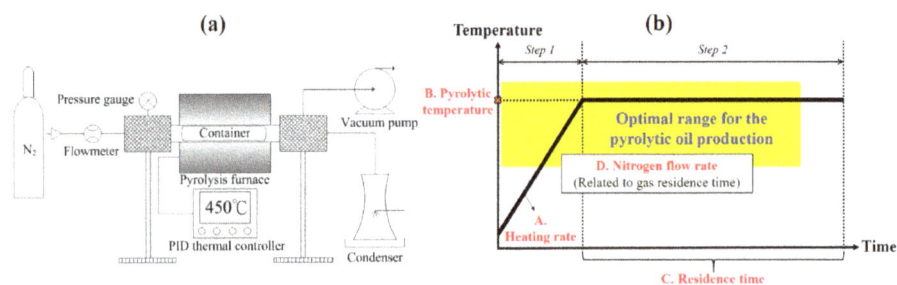

Figure 2. Schematic diagram of (**a**) experimental setup and (**b**) operating process for thermal pyrolysis.

Simultaneous thermogravimetric analysis (TGA) and differential scanning calorimetry (DSC) measurement were performed using the thermal analyzer (PerkinElmer, Waltham, MA, USA: STA 8000). The temperature range was 30–1200 °C, with a heating rate of 10 °C/min. A 145-μL alumina crucible was used to hold the testing sample, and the flow rate of air or nitrogen environmental gas was 50 mL/min.

A portable pH meter (ROCKER, New Taipei City, Taiwan: EC-210) was used to measure the pH value of sludge pyrolytic oil, with an accuracy of about 0.01 ± 1 digit. The pH value of sludge pyrolytic oil can be measured directly because of the low water content. In addition, a bomb calorimeter (Parr Instrument, Moline, IL, USA: Model 6200) was used to measure the heating values of different samples, while an elemental analyzer (Elementar, Langenselbold, Hesse, Germany: vario EL III) was used to analyze the C, H, O, N and S composition of the sewage sludge, pyrolytic sludge oil and heavy fuel oil. The chemical compounds contained in the sludge pyrolytic oil were analyzed by Agilent 7890A-GC and Agilent 5975C-MSD (Agilent Technologies, Santa Clara, CA, USA).

2.2. Taguchi Method

The Taguchi method was used to find the maximum yield of sludge pyrolytic oil from the pyrolysis experiments. The feature of the Taguchi method uses an orthogonal array experimental design with a simple analysis of variance. In Taguchi method, the variables are assigned to each column, called orthogonal arrays for the design of experiments [31], of a matrix related to experimental parameters and experiment levels. In general, the Taguchi method is not aimed at finding the optimal conditions as a full factorial method, but it can analyze the best trend with fewer experimental data, and thus is commonly applied to optimize industrial processes.

Taguchi method was previously applied to analyze caster meal pyrolysis by Chen et al. and this group [14]. The Taguchi method uses the S/N ratio, the signal to noise ratio , to measure the quality characteristics deviating from the desired value [32]. Different strategic categories of nominal—the best (NTB), smaller the better (STB), and larger the better (LB)—are used to describe the S/N ratio characteristics [33]. In this study, for example, to obtain the maximum yield of pyrolytic oil, the "larger the better" characteristic must be adopted. The expression of S/N ratio is shown as follows:

$$S/N_{LB} = -10\log\left(\frac{1}{N}\sum_{i=1}^{N}\frac{1}{y_i^2}\right) = -10\log\left[\frac{1}{\bar{y}^2}(1 + \frac{3s^2}{\bar{y}^2})\right], \tag{1}$$

where N is the test number, y_i is the value of the pyrolytic oil yield in the ith test, and $\bar{y}$ is the average yield of pyrolytic oil. The procedure for the experimental design with the Taguchi method can be found in previous work [14].

2.3. Characteristic Combustion Parameters

Several characteristic combustion parameters can be deduced from a TG-DTG curve to assess the combustion properties of fuels. This study will make use of the ignition temperature (T_i), burnout temperature (T_e), combustion characteristics index (S), and flammability index (C) to evaluate the combustion characteristics of sludge pyrolysis oil mixed with different percentages of heavy fuel oil. The weight loss curve is used to define the ignition temperature. There are many studies concerning the definition of the ignition temperature using TG-DTG curves [34–36], and the method proposed by Tognotti et al. [34] is widely used to predict the ignition temperature for different fuels. In this, the weight loss curve of fuel in the air atmosphere is overlapped with that in inert gas atmosphere (N_2), and the ignition temperature (T_i) corresponds to the first bifurcation point in the TG curves [37]. In addition, the burnout temperature (T_e) is defined as the temperature corresponding to 99% conversion of the fuel in the TG curve of air atmosphere.

Since the ignition and burnout temperatures show only a single combustion property of the fuel, an integrated index is required for the global combustion performance of fuels. Therefore, the combustion characteristic index (S) and flammability index (C) are also used for benchmarking the combustibility of fuels [38]. The combustion characteristic index (S) was first proposed by Cheng et al. [39]. The larger the combustion characteristics index, the better the combustion characteristics of the fuel. The combustion rate is expressed by the Arrhenius Law, as follows:

$$dW/d\tau = A\exp(-E/RT),\tag{2}$$

where $dW/d\tau$ is the combustion rate, A is a pre-exponential factor, E is the activation energy, and T is the absolute temperature.

Take the differential of Equation (2) with temperature and rearrange it as follows:

$$\frac{R}{E} \times \frac{d}{dT}\left(\frac{dW}{d\tau}\right) = \frac{dW}{d\tau} \times \frac{1}{T^2}.\tag{3}$$

At ignition temperature T_i, Equation (3) is multiplied by $\frac{(dW/d\tau)_{max}(dW/d\tau)_{mean}}{(dW/d\tau)_{T=T_i}T_e}$:

$$\frac{R}{E} \times \frac{d}{dT}\left(\frac{dW}{d\tau}\right)_{T=Ti} \frac{(dW/d\tau)_{max}}{(dW/d\tau)_{T=Ti}} \times \frac{(dW/d\tau)_{mean}}{T_e} = \frac{(dW/d\tau)_{max} \times (dW/d\tau)_{mean}}{T_i^2 \times T_e},\tag{4}$$

where $(dW/d\tau)_{max}$ is the maximum combustion rate, which can be obtained from the peak of DTG curve. $(dW/d\tau)_{mean}$ is the mean combustion rate, which can be obtained from the mean of DTG curve. The first term of Equation (4) on the left-hand side represents the reaction strength of fuel combustion, which is related to the activation energy. The second term stands for the change rate of fuel combustion rate at the ignition temperature and the third term refers to the ratio of maximum combustion rate to the combustion rate at the ignition temperature. These two terms are related to the ignition temperature. The last term is the ratio of mean combustion rate to the burnout temperature; the larger the value, the shorter the burnout time of the fuel.

The right-hand side of Equation (4) is the combustion characteristic index (S):

$$S = \frac{(dW/d\tau)_{max} \times (dW/d\tau)_{mean}}{T_i^2 \times T_e}.\tag{5}$$

The flammability index combines the influence of maximum combustion rate and the ignition temperature. It can reflect the difficulty of fuel combustion and the burning out speed. The fuel with a large flammability index will have better combustion stability and the flammability index (C) of the sample can be expressed as follows:

$$C = (dW/d\tau)_{\max}/T_i^2. \tag{6}$$

3. Results and Discussion

3.1. Thermogravimetric Analysis of Sewage Sludge

Table 1 shows the approximate analysis of the sewage sludge used in this study, and it is composed of 6.94 wt % moisture, 45.45 wt % volatiles, 9.94 wt % fixed carbon, and 37.67 wt % ash. This composition is well within the range of the different kinds of sewage sludge analyzed by Fonts et al. [21].

Figure 3 shows the TGA thermograph of sewage sludge. The heating rate is 10 °C/min and the nitrogen flow rate is 50 mL/min. The purple curve shows thermogravimetry (TG), the blue curve shows differential of thermogravimetry (DTG), the red curve shows differential scanning calorimetry (DSC), while the green curve represents TG in air atmosphere. The mass loss of sewage sludge increases along with the temperature, and the major reactions occur between 120 °C and 500 °C.

For the purple TG curve in Figure 3, there are three stages of weight loss. The first stage has a weight loss of 9.33% in the temperature range of 30–200 °C, due to the removal of the water and organics with low boiling points. In the second stage, there are two distinct peaks at the temperatures of 273 °C and 323 °C in the DTG curve. About 33.57% of the original sewage sludge is lost, and this is mainly from the decomposition of aliphatic compound protein and carbohydrates contained in the sewage sludge [40–43]. In the third stage, the weight loss tends to be mild when the temperature exceeds 500 °C, which might be from the decomposition of residual organic matter and the inorganic materials, such as calcium carbonate [44]. Finally, the residue is about 46.79% of the original weight of the sewage sludge when the temperature reaches 1000 °C.

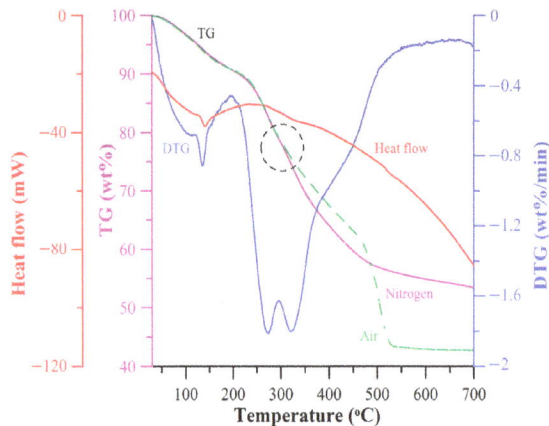

Figure 3. Thermogravimetric analysis results for sewage sludge at a heating rate of 10 °C/min and nitrogen flow rate of 50 mL/min.

3.2. Pyrolytic Parameters for Sewage Sludge

3.2.1. The Influence of Nitrogen Flow Rate on the Pyrolytic Oil Yield

Inert carrying gases such as nitrogen and carbon dioxide are often used in the pyrolysis process. Inert gas can keep the reactor in an oxygen-free state, and causes the volatiles to move quickly away from the system and thus avoid condensing again. Pütün et al. investigated the effects of the sweeping gas flow rate (N_2) on pyrolysis products, and found that increasing the sweeping gas will reduce the gas residence time and prevent the pyrolytic products from the secondary thermal decomposition, thus maximizing the liquid yield [45].

The influence of the N_2 flow rate on the sludge pyrolytic oil was also investigated in this study, as shown in Figure 4. The conditions were fixed as a pyrolytic temperature of 500 °C, heating rate of 20 °C/min and residence time of 60 min. The pyrolytic oil yield increases with the N_2 flow rate in the range of 300–700 mL/min. However, it then decreases again after the N_2 flow rate surpasses 700 mL/min. Higher liquid yield is usually obtained in a certain range of gas residence time. According to the volume of the quartz holder in this study, the filled volume of 90 g sludge and the nitrogen flow rate of 700 mL/min, the gas residence time is between 2.5 s and 3 s. It is under the range preferred for fast pyrolysis to obtain more oil product [20]. When the nitrogen flow rate is less than 700 mL/min, the gas residence time is increased and this gas residence time may cause condensable gases secondary decomposition, which is not conducive to oil product. Nevertheless, the variation in the nitrogen flow rate would also vary the degree of dilution of the pyrolysis vapors. When the dilution of the vapors increases, the vapor pressure decreases, making it more difficult to reach saturation pressure. This affects the condensation and thus the liquid yield. It is speculated this reason make sludge pyrolytic oil decrease after the nitrogen flow rate exceeds 700 mL/min. Therefore, the nitrogen flow rate was chosen as 300, 500, 700 and 900 mL/min in the following work.

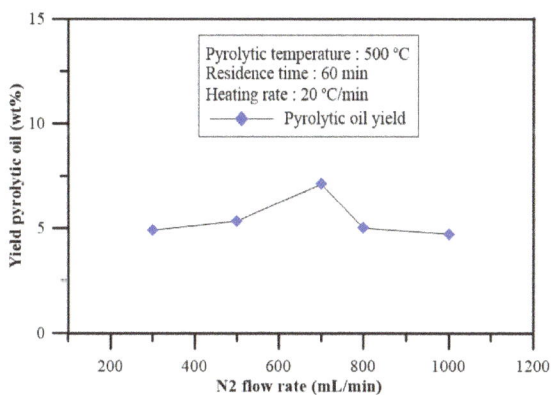

Figure 4. The effect of nitrogen flow rate on the yield of pyrolytic oil.

3.2.2. The Optimal Conditions for Sludge Pyrolysis

The optimal pyrolysis conditions for the maximum yield of sludge pyrolytic oil are investigated in the fixed-bed reactor using the Taguchi method. Table 2 gives the pyrolysis parameters and their levels. Four pyrolysis parameters are the heating rate, pyrolysis temperature, residence time, and nitrogen flow rate. These parameters are assumed to be independent and each parameter has four levels. According to the number of parameters and their levels, the $L_{16}(4^5)$ orthogonal array will be selected. Table 3 shows the experimental layout using an $L_{16}(4^5)$ orthogonal array and the original 256 experimental conditions for the full factorial method are thus diminished significantly to only 16. Table 4 reveals the mass balances of all 16 experimental conditions, including aqueous, oil, and solid yields. Gas yield was obtained by calculation since it was not collected. It shows that sewage sludge char accounts for 50–60% of total product, and the liquid yield is in the range of 18–27%. To assure the reliability of the results, each experiment was performed three times to take the average.

Table 2. Pyrolysis control factors and their levels.

	Level 1	Level 2	Level 3	Level 4
Heating rate (°C/min)	10	20	30	40
Temperature (°C)	450	500	550	600
Residence time (min)	30	60	90	120
N_2 flow rate (mL/min)	300	500	700	900

Table 3. Experimental layout using an $L_{16}(4^5)$ orthogonal array.

Experiment No.	Heating Rate (°C/min)	Temperature (°C)	Residence Time (min)	N_2 Flow Rate (mL/min)	Cooling Temperature (°C)
1	10	450	30	300	25
2	10	500	60	500	25
3	10	550	90	700	25
4	10	600	120	900	25
5	20	450	60	700	25
6	20	500	30	900	25
7	20	550	120	300	25
8	20	600	90	500	25
9	30	450	90	900	25
10	30	500	120	700	25
11	30	550	30	500	25
12	30	600	60	300	25
13	40	450	120	500	25
14	40	500	90	300	25
15	40	550	60	900	25
16	40	600	30	700	25

Table 4. The mass balances of all 16 experimental conditions.

Experiment No.	Aqueous (wt %)	Oil (wt %)	Solid (wt %)	Gas (by diff.) (wt %)
1	11.42	8.71	57.02	22.85
2	12.67	8.88	55.09	23.36
3	15.06	9.87	54.48	20.60
4	13.42	7.29	52.18	27.11
5	14.93	8.76	54.92	21.40
6	14.09	7.91	53.37	24.63
7	14.31	8.62	52.45	24.62
8	9.44	9.22	54.10	27.24
9	13.41	8.31	55.80	22.49
10	12.12	9.32	54.56	24.00
11	12.93	6.59	53.10	27.38
12	13.76	7.58	52.91	25.75
13	14.70	9.20	54.43	21.66
14	13.11	5.43	54.30	27.17
15	18.41	9.22	53.33	19.03
16	15.67	7.88	51.82	24.63

After obtaining the experimental yield of sludge pyrolytic oil, the corresponding S/N ratio can be calculated using Equation (1). Since the experimental design is orthogonal, the effect of each pyrolysis parameter at different levels can be separated out. The mean S/N ratio for the 16 experimental conditions is computed and the mean S/N ratio for each level of the other pyrolysis parameters can then be calculated. The S/N response for the mean S/N ratio of the pyrolysis parameters at each level are shown in Table 5 and it is used to determine the maximum yield of sludge pyrolytic oil.

Table 5. S/N response table for sludge pyrolytic oil.

	L1	L2	L3	L4	Max-Min
(A) Heating rate	−21.27	−21.29	−22.06	−22.20	0.929
(B) Temperature	−21.17	−22.25	−21.43	−21.98	1.074
(C) Residence time	−22.23	−21.32	−21.93	−21.34	0.910
(D) N_2 flow rate	−22.55	−21.52	−20.99	−21.77	1.564

Based on the experimental results and the calculated S/N ratio, the optimal operation conditions are determined as shown in Figure 5. In Figure 5, these four pyrolysis parameters are labelled as "A" for heating rate, "B" for pyrolytic temperature, "C" for residence time, and "D" for nitrogen flow rate. Although the results for pyrolytic temperature and residence time are zigzag, several studies concerning the Taguchi method also show the same phenomenon for S/N ratios [46–48]. It can be deduced from Figure 5 that the optimal pyrolysis conditions are A1, B1, C2 and D3, which represent a heating rate of 10 °C/min, pyrolytic temperature of 450 °C, residence time of 60 min, and nitrogen flow rate of 700 mL/min. It should be mentioned that fast pyrolysis, which represents a pyrolysis reaction carried out at high heating rates, moderate temperatures and short gas residence times, is usually used to obtain high yield of pyrolysis oil. There are three main kinds of technologies for fast pyrolysis: ablative pyrolysis, fluidized bed pyrolysis, and vacuum pyrolysis. However, in the study, a tubular furnace, a type of fixed-bed reactor was used. The results show that oil yield was affected by the combination of four operational parameters. The heating rate is not the dominant parameter due that it was restricted by the pyrolytic temperature. Among these parameters, the nitrogen flow rate has the most obvious effect, which is followed by the pyrolytic temperature and the residence time has the least effect. The nitrogen flow rate (inversely proportional to the gas residence time) has a more significant effect on the oil yield than heating rate.

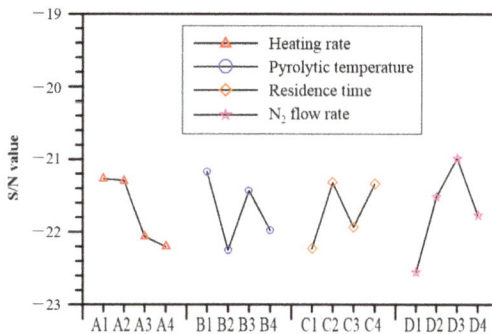

Figure 5. S/N response for sludge pyrolytic oil.

3.2.3. Confirmation Experiment

In the Taguchi method, the confirmation experiment is the last step and the aim of this is to validate the yield of sludge pyrolytic oil. After obtaining the optimum conditions and predicting the response under these conditions, a new experiment was performed with the optimum levels of these pyrolysis parameters. The theoretical optimal S/N ratio can be estimated as:

$$S/N_{opt} = S/N_{AV} + (A1 − S/N_{AV}) + (B1 − S/N_{AV}) + (C2 − S/N_{AV}) + (D3 − S/N_{AV}) = −19.63, \quad (7)$$

where S/N_{AV} is the total mean of the S/N ratio and S/N_{opt} is the S/N ratio at the optimal level.

From the experimental results of the standard $L_{16}(4^5)$ orthogonal array, the theoretical optimization S/N value is -19.63, and it can be deduced that the theoretical maximum pyrolytic oil yield is 0.1044 g/g sewage sludge (18.85% on dry ash free (daf) basis). The real yield of the confirmation experiment is 0.1019 g/g-sewage sludge (18.4% on dry ash free (daf) basis). The difference is only 0.0025 g/g sewage sludge, and thus real experimental result is very close to the theoretical value. In addition, under the optimal conditions, the solid product is about 54.83%, aqueous product is about 15% and by mass balance, the gas product is 19.98%. Although the solid product is not suitable to be used as fuel due to high content of inorganic matter (ash), it can be used to produce adsorbents for disposing of pollutants such as H_2S and NO_x [21]. The aqueous product could be used for the production of fertilizer since it has high content of nitrogen-containing compounds. As to the gas product, it could be recycled and used for the drying of feedstock.

3.3. Elemental Analysis and Heating Values of the Different Fuels

Table 6 shows the elemental analysis and heating value for the sewage sludge, sludge pyrolytic oil and heavy fuel oil on the wet basis. The sludge pyrolytic oil is obtained using the optimal conditions from the Taguchi method. The results show that the dried sewage sludge contained 1.28% sulfur and the sludge pyrolytic oil contained 1.25% sulfur. Compared to the sulfur content of heavy oil, it can be seen that when using sludge pyrolytic oil as fuel more care must be taken with regard to the SO_X emissions. However, since the combustion of pure sludge pyrolytic oil appears to be impractical because of the yield, it is more feasible to co-combust this and other fuels, such as heavy fuel oil. In this case, the sulfur content would drop significantly and thus decrease the sulfur oxide emissions. As to the high nitrogen content in the sludge pyrolytic oil, it is usually related to fuel NO_x during the combustion process. Therefore, the blending ratio of sludge pyrolytic oil and heavy fuel oil should be also well selected to have less SO_x and NO_x emissions. Sludge pyrolytic oil also has a higher oxygen content than heavy fuel oil. A high oxygen content directly affects the energy density, and the resulting heating value is usually lower than that of heavy fuel oil. Although the heating value of sludge pyrolytic oil is only about 2/3 that of heavy fuel oil (28.66 MJ/kg vs. 44.87 MJ/kg), it is still significantly higher than the common bio-oil from lignocellulose. In addition, it does not have the corrosive character that is often seen in other pyrolysis liquids obtained from lignocellulosic biomass, as the pH value of sludge pyrolytic oil is between 8.86 and 9.67 in this study. Accordingly, these advantageous properties make sludge pyrolytic oil favorable as fuel for the co-combustion with other fossil fuels, like heavy fuel oil. In addition, sludge pyrolytic oil contains a high nitrogen and sulfur content, primarily due to the decomposition of protein contained within the dried sewage sludge [21].

Table 6. Elemental analysis and heating values of different fuels.

Samples	N%	C%	H%	O%	S%	HHV(MJ/kg)
Sewage sludge	4.14	22.62	4.93	35.29	1.28	10.18
Sludge pyrolytic oil	5.92	58.44	9.08	18.98	1.25	28.66
Heavy fuel oil	0.46	86.7	11.81	0.26	0.28	44.87

Table 7 compares the properties of diesel, heavy fuel oil, and sludge pyrolytic oil. The optimal combination conditions obtained from the Taguchi method is used for the yield of sludge pyrolytic oil. The test standards used in this study for different properties are also shown in Table 7. The density of sludge pyrolytic oil is larger than that of diesel and heavy fuel oils and it is a little higher than the density of water. The viscosity of sludge pyrolytic oil is between that of diesel and heavy fuel oils. The flash point of sludge pyrolytic oil is 80 °C, which makes it safe to use as a fuel. The pour point of sludge pyrolytic oil is lower than that of heavy fuel oil and it still retains mobility at low temperatures.

Table 7. Properties of different kinds of liquid fuels.

Properties	Diesel	Heavy Fuel Oil	Sludge Oil (Pyrolysis)	Test Method
Density (g/cm^3, @15 °C)	0.8335	0.9533	1.04	ASTM D4052
Viscosity (mm^2/s @40 °C)	3.024	130.3	64.86	ASTM D445
Flash Point (°C)	52–80	110	80	ASTM D93
Pour Point (°C)	−9	12	6	ASTM D97
Heating Value (MJ/kg)	42.50	44.87	28.66	ASTM D240

The bio-oil characterization was carried out using GC/MS spectrometry at the optimal conditions for the maximum yield of the pyrolytic oil. Figure 6 shows the GC/MS mass spectra of sludge pyrolytic oil, and there are more than 100 peaks corresponding to different organic compounds detected. However, many of them cannot be identified. The major compounds are shown in Table 8 and the area percentage in the tables represents the area fraction under the peak for that identified component. The highest peak areas of the identified compounds were for *n*-hexadecanoic acid, cholest-4-ene, octadecanoic acid, hexadecanamide, oleic acid, 4-methyl phenol, toluene, indole, and so on. Oleic acid is a monounsaturated fatty acid, and can be found in a variety of animal and vegetable materials. The saturated form of this acid is stearic acid, and is used in Lorenzo's oil. Indole is generally used in the production of synthetic jasmine oil. Huang et al. also investigated bio-oil from pyrolysis of granular sewage sludge [49]. In their study, the GC-MS results of bio-oil also indicated *n*-hexadecanoic acid ($C_{16}H_{32}O_2$) is the major component at the pyrolysis temperature of 500 °C, which is close to our operation condition. The *n*-hexadecanoic acid accounts for a large proportion of the known ingredients in the sludge pyrolytic oil. The other identical components in both studies include octadecanoic acid ($C_{18}H_{36}O_2$), tetradecanoic acid ($C_{14}H_{28}O_2$) and hexadecanamide ($C_{16}H_{33}NO$). In general, carboxylic acid is the most compound type of the known ingredients.

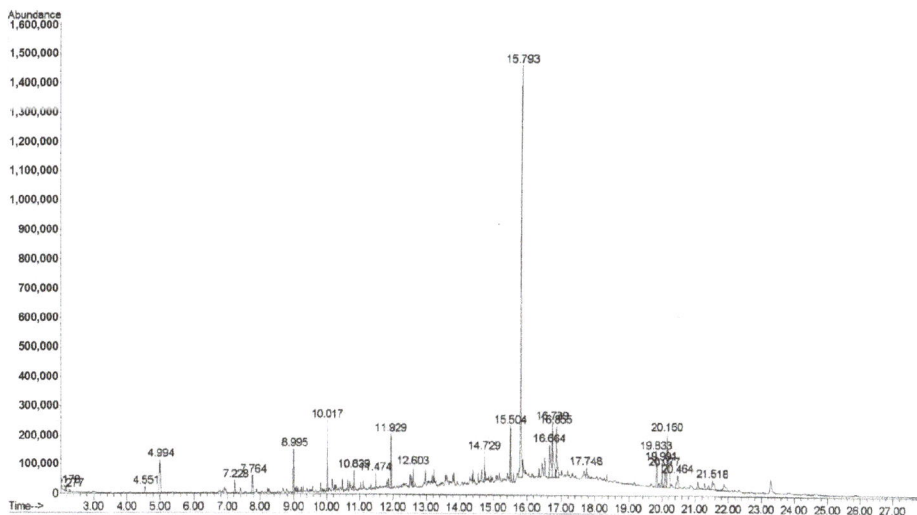

Figure 6. GC-MS mass spectra for sludge pyrolytic oil.

Table 8. GC-MS analysis of sludge pyrolytic oil.

Peak	Retention Time	Chemical Name	Mol Formula	Area%
1	2.178	acetic acid	$C_2H_4O_2$	0.40%
2	2.277	2-butanone	C_4H_8O	0.18%
3	4.551	Pyrrole	C_4H_5N	0.75%
4	4.994	Toluene	C_7H_8	3.63%
5	7.228	ethyl benzene	C_8H_{10}	0.93%
6	7.764	styrene	C_8H_8	1.47%
7	8.995	phenol	C_6H_6O	3.07%
8	10.017	4-methyl phenol	C_7H_8O	4.41%
9	10.839	4-ethyl phenol	$C_8H_{10}O$	1.23%
10	11.474	benzene propane nitrile	-	0.87%
11	11.929	indole	C_8H_7N	3.18%
12	12.603	3-methyl indole	C_9H_9N	1.31%
13	14.729	tetradecanoic acid	$C_{14}H_{28}O_2$	1.83%
14	15.504	pentadecanenitrile	$C_{15}H_{29}N$	3.17%
15	15.793	*n*-hexadecanoic acid	$C_{16}H_{32}O_2$	31.90%
16	16.664	oleic acid	$C_{18}H_{34}O_2$	4.86%
17	16.739	octadecanoic acid	$C_{18}H_{36}O_2$	5.89%
18	16.855	hexadecanamide	$C_{16}H_{33}NO$	4.97%
19	17.748	octadecanamide	$C_{18}H_{37}NO$	1.06%
20	19.833	Unknown	-	3.18%
21	19.991	cholest-2-ene	$C_{27}H_{46}$	2.08%
22	20.077	cholest-3-ene	$C_{27}H_{46}$	1.69%
23	20.16	cholest-4-ene	$C_{27}H_{46}$	9.42%
24	20.464	cholesta-3.5-dienl	-	2.10%
25	21.518	unknown	-	6.41%

3.4. Combustion Characteristic Parameters

Thermogravimetric analysis was performed for sludge pyrolytic oil (SPO), heavy fuel oil (HFO), 20% SPO/80% HFO, and 50% SPO/50% HFO, and the results were used to calculate the ignition temperature, burnout temperature, combustion characteristic index (S) and flammability index (C). In these experiments, the samples were inserted into an alumina crucible and heated up to 1200 °C at the heating rate of 10 °C/min. A nitrogen flow rate of 50 mL/min is used for the pyrolysis process and an air flow rate of 50 mL/min is used for the oxidation reaction.

Figure 7 shows the TGA burning profiles of different blending ratio for sludge pyrolytic oil/heavy fuel oil, and the TG pyrolysis curve for sludge pyrolytic oil with N_2 carrying gas is also shown to identify the ignition temperature. In Figure 7a, the main reaction for pure sludge pyrolytic oil occurs between 300 and 600 °C. The oxidation processes can be divided into three stages, as follows. (1) The stage of moisture release: the first peak of DTG curves occurs between 30 and 136 °C, corresponding to the endothermic reaction in the DSC curve. This mainly represents the heat adsorption by water and the organic matter with low boiling point. The weight loss in this stage is about 21.63 wt %. (2) The volatilization and oxidation stage of the light volatile substance, which occurs between 136 °C and 471 °C. The DSC curve indicates an exothermic reaction in this stage, and the weight loss is about 65.21 wt %. (3) The stage of nonvolatile combustion, which represents the combustion of heavy organic matter after 471 °C, and there is a peak in the DTG curve at the temperature of 565 °C, corresponding to a higher exothermic peak in the DSC curve. The weight loss in this stage is about 11.67 wt %. Finally, the remaining substance after oxidation is about 1.16 wt %.

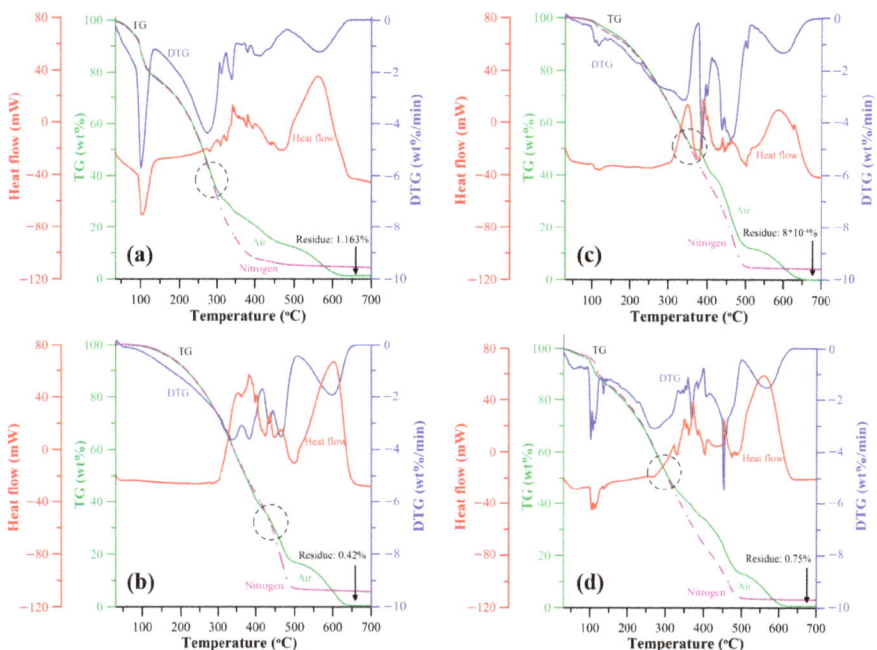

Figure 7. TGA burning profiles of sludge pyrolytic oil at a heating rate of 10 °C/min and air flow rate of 50 mL/min. (**a**) Pure sludge pyrolytic oil; (**b**) pure heavy fuel oil; (**c**) 20% sludge pyrolytic oil and 80% heavy fuel oil; (**d**) 50% sludge pyrolytic oil and 50% heavy fuel oil. (Green curve: TG; blue curve: DTG; red curve: DSC; purple curve: TG in the pyrolysis process).

In Figure 7b, the main reaction for pure heavy fuel oil occurs between 300 and 650 °C. The oxidation processes have two stages, as follows. (1) The volatilization and oxidation stage of light volatile substances and there are several peaks in the DTG curve between 300 and 510 °C. The DSC curve indicates an exothermic reaction in this stage, and the weight loss is about 83.41 wt %. (2) The stage of nonvolatile combustion represents the combustion of heavy organic matter after 510 °C, and there is a peak in the DTG curve at the temperature of about 600 °C, corresponding to a higher exothermic peak in the DSC curve. The weight loss in this stage is about 16.06 wt %. Finally, the remaining residue is about 0.42 wt %.

For the blends, Figure 7c shows the TGA burning profiles of 20% sludge pyrolytic oil/80% heavy fuel oil. The TG curve decreases faster than that of pure heavy oil due to the mixture of sludge pyrolytic oil, which has more volatile organic compounds. Similar to the case of pure sludge pyrolytic oil, the reaction processes can be divided into three stages, as follows. (1) The stage of moisture release, when the first peak of the DTG curves occurs between 30 and 150 °C and the weight loss in this stage is about 5.17 wt %. (2) The volatilization and oxidation stage of the light volatile substances, which occurs between 150 and 510 °C; the overall trend is the same as for the heavy oil. The DSC curve indicates an exothermic reaction in this stage, and the weight loss is about 82.97 wt %. (3) The stage of nonvolatile combustion, which represents the combustion of heavy organic matter above 510 °C. There is a peak in the DTG curve at the temperature of about 600 °C, corresponding to a higher exothermic peak in the DSC curve, and the weight loss in this stage is about 11.87 wt %. There is almost no residue left after oxidation.

When the blending ratio increases to 50% as shown in Figure 7d, The TG curve only decreases slower than that of pure sludge pyrolytic oil in these cases, and this is primarily due to the mixture of sludge pyrolytic oil, which has more volatile organic compounds. The reaction processes can also be divided into three stages, as follows. (1) The stage of moisture release, when the first peak of the DTG curves occurs between 30 and 145 °C, and the weight loss in this stage is about 14.23 wt %. (2) The volatilization and oxidation stage of the light volatile substances, which occurs between 145 and 500 °C and can be further divided into two parts. The first is close to the oxidation of sludge pyrolytic oil and is between 145 and 415 °C, while the second is close to the oxidation of heavy fuel oil and is between 415 and 500 °C. The DSC curve indicates an exothermic reaction in this stage, and the weight loss is about 72.58 wt %. (3) The stage of nonvolatile combustion, which represents the combustion of heavy organic matter above 500 °C. There is a peak in the DTG curve at the temperature of 566 °C, corresponding to a higher exothermic peak in the DSC curve, and the weight loss in this stage is about 12.3 wt %. After oxidation, the residue is about 0.75 wt %.

From Figure 7, these results show that for the blends the second DTG peak, which represents the combustion of volatile substances and determines the ignition temperature, shifts to a lower temperature when compared to the case of pure heavy fuel oil. As can be seen in the TGs and DTGs for the blends of sludge pyrolytic oil and heavy fuel oil, there is an obvious interaction between the blend components. Table 9 shows the combustion characteristic parameters of sludge pyrolytic oil, heavy fuel oil, and the blends. Sludge pyrolytic oil has a lower ignition temperature (274 °C) than heavy fuel oil (434 °C), since it has more volatile matter. In addition, a greater percentage of sludge pyrolytic oil in the blend will reduce the ignition temperature to a greater extent. We also found similar trends for combustion characteristics of a single droplet in the suspended droplet experimental system [50]. A higher blending ratio of sludge pyrolytic oil will make ignition easier. However, the burnout temperature is around 600 °C in all these cases because the combustion characteristics of heavy organic matter (the last DTG peak) are similar for these oils. The last DTG peak derived from the oxidation of heavy organic matter determines the burnout temperature. Table 9 also shows that the maximum combustion rate of sludge pyrolytic oil is higher than that of heavy fuel oil. An interesting finding is that the maximum combustion rate in the blend is even higher than for pure sludge pyrolytic oil. This indicates that there is an interaction between these blend components, and synergistic effects can be perceived. In addition, as the blending ratio of sludge pyrolytic oil increases, the maximum combustion rate also moves to a lower temperature and represents the domination shifts from heavy organic matter to light volatile substances. Finally, the flammability index and combustion characteristics index in the blend are higher than those of heavy fuel oil. Since the sludge pyrolytic oil has a higher volatile content than heavy fuel oil, adding a certain amount of sludge pyrolytic oil will promote the ignition performance of heavy fuel oil, which is conductive to the stability of heavy fuel oil combustion. The co-combustion of sludge pyrolytic oil and heavy fuel oil has better combustion characteristics than seen with pure heavy fuel oil. Moreover, it is even better than pure sludge pyrolytic oil, especially in the case of 50% sludge pyrolytic oil mixed with 50% heavy fuel oil.

Table 9. The combustion characteristic parameters of sludge pyrolytic oil, heavy fuel oil, and the blends.

Fuel	$(dW/d\tau)_{max}$	$(dW/d\tau)_{mean}$	T_i (°C)	T_e (°C)	$S \times 10^7$	$C \times 10^5$
Heavy fuel oil	3.517	1.544	434	612	0.470	1.863
20%SPO + 80%HFO	5.003	1.783	311	601	1.531	5.157
50%SPO + 50%HFO	5.464	1.562	295	600	1.631	6.257
Sludge pyrolytic oil	4.360	1.162	274	605	1.114	5.832

4. Conclusions

The Taguchi method was used in this study to optimize the pyrolysis of sewage sludge for obtaining the maximum pyrolytic oil yield. The combustion characteristic parameters are evaluated based on a thermogravimetric analysis to explore the combustion characteristics of sludge pyrolytic oil and heavy fuel oil. The findings of this study are summarized as follows.

(1) The effective sequence of pyrolytic parameters for the yield of sludge pyrolytic oil is nitrogen flow rate, pyrolytic temperature, heating rate, and residence time. The best operating conditions for sewage sludge pyrolysis are a heating rate of 10 °C/min, pyrolysis temperature of 450 °C, residence time of 60 min, and nitrogen flow rate of 700 mL/min. Under these conditions, the obtained pyrolytic oil yield is 10.19% (18.4% on dry ash free (daf) basis), which is very close to 10.44% (18.85% on dry ash free (daf) basis.), the ideal value from the Taguchi method.

(2) From the thermogravimetric analysis of sludge pyrolytic oil, the combustion performance parameters, such as the ignition temperature, burnout temperature, flammability index, and combustion characteristics index, are calculated and compared with those of heavy fuel oil. Since the sludge pyrolytic oil has volatile substances with lower boiling temperatures, it has a lower ignition temperature and better combustion characteristics than heavy fuel oil.

(3) For the blends of sludge pyrolytic oil and heavy fuel oil, the maximum combustion rate, the flammability index, and combustion characteristics index all increase markedly along with the amount of sludge pyrolytic oil. Sludge pyrolytic oil significantly enhances the combustion of heavy fuel oil, with the mix producing even better results than those seen with pure sludge pyrolytic oil (synergistic effects).

Author Contributions: G.-B.C. and Y.-C.C. contributed to the concept, results explanation and writing of the paper. J.-W.L. performed the experiments and analyzed the data. H.-T.L. and F.-H.W. contributed to experimental method and paper review.

Funding: This research was funded by the Ministry of Science and Technology of Republic of China grant number MOST 104-ET-E-006-001-ET.

Conflicts of Interest: The authors declare no conflict of interest.

References

1. Chen, G.B.; Li, Y.H.; Cheng, T.S.; Chao, Y.C. Chemical effect of hydrogen peroxide addition on characteristics of methane-air combustion. *Energy* **2013**, *55*, 564–570. [CrossRef]

2. Starikovskiy, A.; Aleksandrov, N. Plasma-assisted ignition and combustion. *Prog. Energy Combust. Sci.* **2013**, *39*, 61–110. [CrossRef]

3. Kaminski, W.; Marszalek, J.; Ciolkowska, A. Renewable energy source-dehydrated ethanol. *Chem. Eng. J.* **2008**, *13*, 95–102. [CrossRef]

4. Nigam, P.S.; Singh, A. Production of liquid biofuels from renewable resources. *Porg. Energy Combust. Sci.* **2011**, *37*, 52–68. [CrossRef]

5. Werther, J.; Ogada, T. Sewage sludge combustion. *Prog. Energy Combust. Sci.* **1999**, *25*, 55–116. [CrossRef]

6. Cantrell, K.B.; Ducey, T.; Ro, K.S.; Hunt, P.G. Livestock waste-to-bioenergy generation opportunities. *Bioresour. Technol.* **2008**, *99*, 7941–7953. [CrossRef] [PubMed]

7. Hossain, F.M.; Kosinkova, J.; Brown, R.J.; Ristovski, Z.; Hankamer, B.; Stephens, E.; Rainey, T.J. Experimental investigations of physical and chemical properties for microalgae HTL bio-crude using a large batch reactor. *Energies* **2017**, *10*, 467. [CrossRef]

8. Kim, K.H.; Lee, E.Y. Environmentally-Benign Dimethyl Carbonate-Mediated Production of Chemicals and Biofuels from Renewable Bio-Oil. *Energies* **2017**, *10*, 1790. [CrossRef]

9. Rowhani, A.; Rainey, T.J. Scrap tyre management pathways and their use as a fuel—A review. *Energies* **2016**, *9*, 888. [CrossRef]

10. Karagoz, S.; Bhaskar, T.; Muto, A.; Sakata, Y.; Oshiki, T.; Kishimoto, T. Low-temperature catalytic hydrothermal treatment of wood biomass: Analysis of liquid products. *Chem. Eng. J.* **2005**, *108*, 127–137. [CrossRef]

11. Asadullah, M.; Rahman, M.A.; Ali, M.M.; Motin, M.A.; Sultan, M.B.; Alam, M.R. Production of bio-oil from fixed bed pyrolysis of bagasse. *Fuel* **2007**, *86*, 2514–2520. [CrossRef]

12. Ferreira, L.C.; Nilsen, P.J.; Fdz-Polanco, F.; Perez-Elvira, S.I. Biomethane potential of wheat straw: Influence of particle size, water impregnation and thermal hydrolysis. *Chem. Eng. J.* **2014**, *242*, 254–259. [CrossRef]

13. Lewandowski, I.; Clifton-Brown, J.C.; Scurlock, J.M.O.; Huisman, W. Miscanthus: European experience with a novel energy crop. *Biomass Bioenergy* **2000**, *19*, 209–227. [CrossRef]

14. Chen, G.L.; Chen, G.B.; Li, Y.H.; Wu, W.T. A study of thermal pyrolysis for castor meal using the Taguchi method. *Energy* **2014**, *71*, 62–70. [CrossRef]

15. Chen, W.H.; Lin, B.J. Characteristics of products from the pyrolysis of oil palm fiber and its pellets in nitrogen and carbon dioxide atmospheres. *Energy* **2016**, *94*, 569–578. [CrossRef]

16. Lam, S.S.; Liew, R.K.; Lim, X.Y.; Ani, F.N.; Jusoh, A. Fruit waste as feedstock for recovery by pyrolysis technique. *Int. Biodeterior. Biodegrad.* **2016**, *113*, 325–333. [CrossRef]

17. Islam, M.N.; Beg, M.R.A.; Islam, M.R. Pyrolytic oil from fixed bed pyrolysis of municipal solid waste and its characterization. *Renew. Energy* **2005**, *30*, 413–420. [CrossRef]

18. Lam, S.S.; Liew, R.K.; Jusoh, A.; Chong, C.T.; Ani, F.N.; Chase, H.A. Progress in waste oil to sustainable energy, with emphasis on pyrolysis techniques. *Renew. Sustain. Energy Rev.* **2016**, *53*, 741–753. [CrossRef]

19. Mohan, D.; Pittman, C.U.; Steele, P.H. Pyrolysis of wood/biomass for bio-oil: A critical review. *Energy Fuels* **2006**, *208*, 48–89. [CrossRef]

20. Basu, P. *Biomass Gasification, Pyrolysis and Torrefaction Practial Design and Theory*, 2nd ed.; Academic Press: Cambridge, MA, USA, 2013.

21. Fonts, I.; Gea, G.; Azuar, M.; Ábrego, J.; Arauzo, J. Sewage sludge pyrolysis for liquid production: A review. *Renew. Sustain. Energy Rev.* **2012**, *16*, 2781–2805. [CrossRef]

22. Inguanzo, M.; Domínguez, A.; Menéndez, J.A.; Blanco, C.G.; Pis, J.J. On the pyrolysis of sewage sludge: The influence of pyrolysis conditions on solid, liquid and gas fractions. *J. Anal. Appl. Pyrolysis* **2002**, *63*, 209–222. [CrossRef]

23. Kim, Y.; Parker, W. A technical and economic evaluation of the pyrolysis of sewage sludge for the production of bio-oil. *Bioresour. Technol.* **2008**, *99*, 1409–1416. [CrossRef] [PubMed]

24. Domínguez, A.; Menéndez, J.A.; Inguanzo, M.; Bernad, P.L.; Pis, J.J. Gas chromatographic-mass spectrometric study of the oil fractions produced by microwave-assisted pyrolysis of different sewage sludges. *J. Chromatogr. A* **2003**, *1012*, 193–206. [CrossRef]

25. Domínguez, A.; Menéndez, J.A.; Inguanzo, M.; Pis, J.J. Investigations into the characteristics of oils produced from microwave pyrolysis of sewage sludge. *Fuel Process Technol.* **2005**, *86*, 1007–1020. [CrossRef]

26. Domínguez, A.; Menéndez, J.A.; Inguanzo, M.; Pis, J.J. Production of bio-fuels by high temperature pyrolysis of sewage sludge using conventional and microwave heating. *Bioresour. Technol.* **2006**, *97*, 1185–1193. [CrossRef] [PubMed]

27. Sánchez, M.E.; Menéndez, J.A.; Domínguez, A.; Pis, J.J.; Martínez, O.; Calvo, L.F.; Bernad, P.L. Effect of pyrolysis temperature on the composition of the oils obtained from sewage sludge. *Biomass Bioenergy* **2009**, *33*, 933–940. [CrossRef]

28. Jeguirim, M.; Dutourné, P.; Zorpas, A.A.; Limousy, L. Olive mill wastewater: From a pollutant to green fuels, agricultural water source and bio-fertilizer—Part 1. The drying kinetics. *Energies* **2017**, *10*, 1423. [CrossRef]

29. Azuara, M.; Ábrego, J.; Fonts, I.; Gea, G.; Murillo, M.B. Ammonia Content of Bottom Phase Liquid from Pyrolyisis of Sewage Sludge in a Bubbling Fluidized Bed. In Proceedings of the 18th European Biomass Conference and Exhibition, Lyon, France, 3–7 May 2010.

30. Cao, J.P.; Zhao, X.Y.; Morishita, K.; Li, L.Y.; Xiao, X.B.; Obara, R.; Wei, X.Y.; Takarada, T. Triacetonamine formation in a bio-oil from fast pyrolysis of sewage sludge using acetone as the absorption solvent. *Bioresour. Technol.* **2010**, *101*, 4242–4245. [CrossRef] [PubMed]

31. Tang, B. Orthogonal array-based hypercube. *J. Am. Statist. Assoc.* **1993**, *88*, 1392–1397. [CrossRef]

32. Yang, W.H.; Tarng, Y.S. Design optimization of cutting parameters for turning operations based on the Taguchi method. *J. Mater. Process. Technol.* **1998**, *84*, 122–129. [CrossRef]

33. Ghani, J.A.; Choudhury, I.A.; Hassan, H.H. Application of Taguchi method in the optimization of end milling parameters. *J. Mater. Process. Technol.* **2004**, *145*, 84–92. [CrossRef]

34. Tognotti, L.; Malotti, A.; Petarca, L.; Zanelli, S. Measurement of ignition temperature of coal particles using a thermogravimetric technique. *Combust. Sci. Technol.* **1985**, *15*, 15–28. [CrossRef]

35. Crelling, J.C.; Hippo, E.J.; Woerner, B.A.; West, D.P., Jr. Combustion characteristics of selected whole coals and macerals. *Fuel* **1992**, *71*, 151–158. [CrossRef]

36. Huang, X.; Jiang, X.; Han, X.; Wang, H. Combustion characteristics of fine- and micro-pulverizes coal in the misture of O_2/CO_2. *Energy Fuels* **2008**, *22*, 3756–3762. [CrossRef]

37. Wall, T.F.; Gupta, R.P.; Gururajan, V.S.; Zhang, D. The ignition of coal particles. *Fuel* **1991**, *70*, 1011–1016. [CrossRef]

38. Jin, Y.; Li, Y.; Liu, F. Combustion effects and emission characteristics of SO_2, CO, NO_x and heavy metals during co-combustion of coal and dewatered sludge. *Front. Environ. Sci. Eng.* **2015**, *10*, 201–210. [CrossRef]

39. Cheng, J.Y.; Sun, X.X. Determination of the devolatilization index and combustion characteristic index of pulverized coals. *Power Eng.* **1987**, *7*, 13–18.

40. Wang, Y.; Chen, G.; Li, Y.; Yan, B.; Pan, D. Experimental study of the bio-oil production from sewage sludge by supercritical conversion process. *Waste Manag.* **2013**, *33*, 2408–2415. [CrossRef] [PubMed]

41. Thipkhunthod, P.; Meeyoo, V.; Rangsunvigit, P.; Kitiyanan, B.; Siemanond, K.; Rirksomboon, T. Pyrolytic characteristics of sewage sludge. *Chemosphere* **2006**, *64*, 955–962. [CrossRef] [PubMed]

42. Gómez-Rico, M.F.; Font, R.; Fullana, A.; Martín-Gullón, I. Thermogravimetric study of different sewage sludges and their relationship with the nitrogen content. *J. Anal. Appl. Pyrolysis* **2005**, *74*, 421–428. [CrossRef]

43. Kristensen, E. Characterization of biogenic organic natter by stepwise thermogravimetry (STG). *Biogeochemistry* **1990**, *9*, 135–139. [CrossRef]

44. Li, J.J.; Qi, B.Y.; Li, A.M.; Duan, Y.; Wang, Z. Thermal analysis and products distribution of dried sewage sludge pyrolysis. *J. Anal. Appl. Pyrolysis* **2014**, *105*, 43–48.

45. Pütün, E. Catalytic pyrolysis of biomass: Effects of pyrolysis temperature, sweeping gas flow rate and MgO catalyst. *Energy* **2010**, *35*, 2761–2766.

46. Chen, W.H.; Chen, C.J.; Hung, C.I. Taguchi approach for co-gasification optimization of torrefied biomass and coal. *Bioresour. Technol.* **2013**, *144*, 615–622. [CrossRef] [PubMed]

47. Das, S.P.; Gupta, A.; Das, D.; Goyal, A. Enhanced bioethanol production from water hyacinth (*Eichhornia crassipes*) by statistical optimization of fermentation process parameters using Taguchi orthogonal array design. *Int. Biodeterior. Biodegrad.* **2016**, *109*, 174–184. [CrossRef]

48. Pattanaik, A.; Satpathy, M.P.; Mishra, S.C. Dry sliding wear behavior of epoxy fly ash composite with Taguchi optimization. *Eng. Sci. Technol. Int. J.* **2016**, *19*, 710–716. [CrossRef]

49. Huang, F.; Yu, Y.; Huang, H. Temperature influence and distribution of bio-oil from pyrolysis of granular sewage sludge. *J. Anal. Appl. Pyrolysis* **2018**, *130*, 36–42. [CrossRef]

50. Li, J.W. A Study of Thermal Process for Sewage Sludge Pyrolytic Oil and its Combustion Characteristics. Master's Thesis, National Cheng Kung University, Tainan, Taiwan, 2016.

energies

MDPI

Article

Combined Ball Milling and Ethanol Organosolv Pretreatment to Improve the Enzymatic Digestibility of Three Types of Herbaceous Biomass

Seong Ju Kim [1], Byung Hwan Um [2], Dong Joong Im [3], Jin Hyung Lee [4] and Kyeong Keun Oh [3,5,*]

[1] Department of Biomolecular and Chemical Engineering, Hankyong National University, Anseong, Gyeonggi 17579, Korea; ksj1756@hknu.ac.kr
[2] Department of Chemical Engineering and Research Center of Chemical Technology, Hankyong National University, Anseong, Gyeonggi 17579, Korea; bhum11@hknu.ac.kr
[3] R&D Center, SugarEn Co., Ltd., Yongin, Gyeonggi 16890, Korea; Biol@daum.net
[4] Korea Institute of Ceramic Engineering and Technology, Jinju, Gyeongnam 52851, Korea; leejinh1@kicet.re.kr
[5] Department of Chemical Engineering, Dankook University, Yongin, Gyeonggi 16890, Korea
* Correspondence: kkoh@dankook.ac.kr; Tel.: +82-31-8005-3548

Received: 31 August 2018; Accepted: 14 September 2018; Published: 16 September 2018

Abstract: A combined ball milling and ethanol organosolv process is proposed for the pretreatment of three types of herbaceous biomass, giant miscanthus, corn stover, and wheat straw. The combined pretreatment was effective at both removing lignin and increasing the glucan content. After 120 min pretreatment, the glucan content increased to 63.09%, and 55.89% of the acid-insoluble lignin was removed from the giant miscanthus sample. The removal of cellulose, hemicellulose, and acetyl groups were correlated with the removal of lignin. The pretreatment of corn stover showed the highest removal of cellulose, but this was dependent on the removal of acid-insoluble lignin. The slope of the regression lines, which shows the correlation between the removal of lignin and cellulose, was lower than other correlations. The changes in biomass size were analyzed using size distribution graphs. With increasing pretreatment time, the particle size reduction improved in the three types of herbaceous biomass. Because of the combined physicochemical pretreatment, the enzymatic digestibility improved, and a maximum of 91% glucan digestibility was obtained from the pretreated corn stover when 30 FPU/g-glucan enzyme was added. Finally, compositional analysis of the recovered lignin from the remaining black liquor was investigated.

Keywords: combined pretreatment; ball mill; ethanol organosolv; herbaceous biomass; lignin recovery

---

1. Introduction

Sugars derived from herbaceous biomass are important platform chemicals and can be further converted to high-value-added chemicals and biofuels via thermochemical or biological routes [1,2]. Pretreatment is an essential step in producing sugars from herbaceous biomass. Pretreatment reduces the amount of the structural polymer lignin, as well as the crystalline cellulose structure, thus increasing the accessibility of enzymes to the holocellulose [3,4]. Recently, lignin has emerged as a renewable source of aromatics for the chemical industry [5]. Therefore, both carbohydrate and lignin fractionation should be considered when pretreatment steps are applied to herbaceous biomass.

Pretreatment consists of biological, chemical, or physical processes [6,7]. Recently, combined physical and chemical pretreatment processes have been proposed to enhance enzymatic hydrolysis [8–12]. These combined methods increase enzymatic accessibility and reduce energy consumption [13].

Ball milling is one physical pretreatment method. Previous studies have reported that ball milling increases the accessible surface area and pore size and reduces the crystallinity and the polymerization degree of cellulose [14–17]. Because chemical catalysts are not used, there is no significant change in the chemical composition [18]. As a chemical pretreatment approach, organosolv is a promising method for the alcoholysis of lignin from herbaceous biomass [19]. This pretreatment employs organic solvents, mainly alcohols, or their aqueous solutions, which can be later reused and recovered [20,21]. The advantage of using organosolv is that highly pure lignin is obtained as a solid from the liquid phase; this can be used for high-value products such as adhesives, fibers, and biodegradable polymers [20,22] or thermal fuel [23]. During the organosolv process, cellulose is recovered as a solid phase, whereas most of the lignin and partial hemicellulose is dissolved in the organic solvent as a liquid phase. The recovered solid, rich in cellulose, is readily hydrolyzed by enzymatic digestion because the lignin or hemicellulose has been removed, and this material can be used for further processing [24–27]. Finally, the fractionation of lignin and partial hemicellulose reduces the recalcitrance of the material to digestion and increases the accessibility of the material to enzymes, thereby enhancing enzymatic digestibility [28]. However, the two-stage pretreatment (physical and chemical) requires considerable energy, as well as additional reactors. Previous studies have reported that pretreatment combined with mechanical size reduction is effective in reducing the energy consumption in biorefinery plants [14,29–33].

In this study, we propose a novel pretreatment process where ball mill grinding and the ethanol-based organosolv process are conducted simultaneously. In this process, ball milling is used as a physical treatment to reduce the biomass size and improve enzymatic accessibility. Ethanol organosolv was used as a chemical treatment to fractionate the lignin, the major obstacle to cellulose digestion. In this study, the effects of the combined physicochemical pretreatment on fermentable sugar production were investigated using three types of herbaceous biomass.

2. Materials and Methods

2.1. Raw Materials

Three herbaceous biomasses namely, giant miscanthus (GM), corn stover (CS), and wheat straw (WS) were supplied by the Rural Development Administration (Wanju, Republic of Korea). The samples were air dried at room temperature and used directly in the pretreatment processes. The moisture contents of GM, CS, and WS were 2.92%, 3.08%, and 3.20%, based on the total wet biomass weight, respectively. The chemical compositions of the raw materials are shown in Table 1.

Table 1. Compositions of the three types of herbaceous biomass used in this study. GM: Giant Miscanthus; CS: corn stover; WS: wheat straw.

Compositions (%)		GM	CS	WS
Carbohydrates	Glucan	43.77 ± 0.07	30.30 ± 0.26	31.42 ± 0.44
	Xylan	21.22 ± 0.08	16.58 ± 0.19	21.38 ± 0.09
	Mannan	-	0.60 ± 0.21	-
	Galactan	-	1.53 ± 0.47	1.56 ± 0.14
	Arabinan	0.50 ± 0.01	0.82 ± 0.02	1.00 ± 0.02
	Sub Total	65.49	49.83	55.36
Lignin	Acid soluble	0.96 ± 0.01	1.47 ± 0.03	1.11 ± 0.02
	Acid insoluble	19.55 ± 0.01	12.38 ± 0.09	13.47 ± 0.24
	Sub Total	20.51	13.85	14.58
Extractive	Water	4.96 ± 0.71	21.14 ± 0.12	17.11 ± 0.33
	Ethanol	1.47 ± 0.11	3.71 ± 0.22	2.77 ± 0.27
	Sub Total	6.43	24.85	19.88
Acetyl group		3.88 ± 0.02	2.76 ± 0.06	3.02 ± 0.06
Ash		2.18 ± 0.03	7.92 ± 0.05	7.36 ± 0.10
Total		98.49	99.21	100.2

2.2. Combined Physicochemical Pretreatment

The combined ball milling and ethanol organosolv pretreatment was performed in a rotary-pressured type reactor (Hanul Engineering Co., Gunpo, Korea). For the pretreatment, 370 g of dried herbaceous biomass and 11 kg of alumina balls (Ø = 10 mm) were added to the reactor with 3.7 L of 60% (*v*/*v*) ethanol solution (1:10:30 (*w*/*v*/*w*) for biomass, solvent, and ball, respectively). The rotary-pressure-type reactor was kept at 170 °C for 30–120 min and rotated at 50 rpm. After reaction, it was cooled to 80 °C, and the mixture was separated into solid biomass, lignin-containing liquid, and alumina balls. The treated solid biomass was washed with distilled water and stored until use. After pretreatment, the particle size distribution of the pretreated biomass was determined by the shaking sieve method after drying at 45 °C overnight. The sieves used were US sieve numbers 6, 14, 20, 40, 80, 170, and 270 (sieve opening sizes: 3.35, 1.4, 0.85, 0.425, 0.180, 0.09, and 0.053 mm, respectively). Lignin was recovered from the remaining black liquor by 3-fold dilution with distilled water. After 24 h, the precipitated lignin was collected by filtration.

2.3. Enzymatic Digestibility

Enzymatic digestibility was analyzed according to the National Renewable Energy Laboratory (NREL) standard procedures [34]. A commercial cellulase (Cellic CTec 2, Novozymes Korea Ltd., Seoul, Korea) was used. The enzyme hydrolysis was performed at 50 °C and pH 4.8 (50 mM sodium citrate buffer) while the solution was shaken at 150 rpm. The quantity of cellulase loaded was 15 or 30 FPU (filter paper unit)/g-glucan. Experiments were conducted in 250-mL Erlenmeyer flasks with a total working volume of 100 mL, and the glucan concentration was maintained at 1.0% (*w*/*v*). Samples were taken periodically (0, 3, 6, 9, 12, 24, 48, 72, and 96 h) and boiled for 5 min to deactivate the enzymes, followed by filtration through a 0.45 μm nylon membrane filter for glucose content analysis using a high-performance liquid chromatography (HPLC) column (Bio-Rad Laboratories, Hercules, CA, USA). All experiments were performed in triplicate for error analysis. Finally, the glucan digestibility was calculated using Equation (1).

$$\text{Lignin removal percentage (\%)} = 100 - m_{\text{Re}}/m_{\text{L}} \times 100 \tag{1}$$

where m_{Re} refers to the dry weight mass of remaining lignin after composition analysis of solid residues and m_{L} represents the mass weight of lignin in the original loaded biomass (m_{L} = the loaded biomass mass weight × the lignin content) [35].

2.4. Compositional Analysis of the Solid and Liquid Phases

The compositions of the solid and liquid samples were determined according to analytical procedures of the NREL (NREL/TP-510-42623 for structural carbohydrates and lignins; NREL/TP-510-42618 for sugars in the liquids or in the hydrolyzates) [36,37]. The sugars were determined using a HPLC (Agilent 1260 Infinity, Agilent Technologies, Santa Clara, CA, USA) equipped with a refractive index detector (Agilent 1260 Infinity, Agilent Technologies, Santa Clara, CA, USA). A Bio-Rad Aminex HPX-87H column (300 mm length × 7.8 mm internal diameter) and a Cation H micro-guard cartridge (30 mm length × 4.6 mm internal diameter, Bio-Rad Laboratories Inc., Hercules, CA, USA) were used for sugar, organic acid, and decomposition product analysis. The samples were filtered using syringe filters having a 0.45-μm pore size before analysis, and 5 mM sulfuric acid was used as the mobile phase. The temperature of the column was set at 65 °C and the flow rate of the mobile phase was 0.5 mL/min. The remaining solid was filtered, dried, and burned in a muffle furnace at 575 °C for lignin analysis. The ash content of the samples was determined by complete combustion in a muffle furnace (DMF12, Romax, Co., Seoul, Korea) equipped with a temperature controller and by running a temperature ramp program according to the NREL/TP-510-42622 method. The remaining residue in the crucible was taken as the ash content [38]. The compositional analysis was performed in triplicate.

3. Results

3.1. Compositions of Herbaceous Biomasses Used in This Study

The compositions of the herbaceous biomass samples used in this study are presented in Table 1. GM had the highest carbohydrate and lignin content among the three types of herbaceous biomass: 43.8% glucan, 21.22% xylan, mannan, and galactan (XMG), and 20.51% lignin (acid soluble + acid insoluble). In contrast, CS and Wheat straw (WS) contained approximately 50% carbohydrates and 14% lignin. However, the extractive content of GM was much lower than those of CS and WS. The differences in biomass composition could be due to the characteristics of each herbaceous biomass source. These differences in composition result in unique physical/chemical properties. Therefore, it is important to find the optimum pretreatment conditions that are tailored to the properties of each biomass type.

3.2. Composition Changes after Combined Physicochemical Pretreatment

The goals of pretreatment are to reduce the lignin content to improve the accessibility of the enzyme and to increase the glucan and XMG content. Wanting to optimize the combined physicochemical pretreatment conditions, we investigated the conditions where lignin removal was maximal while the cellulose content was maintained.

Table 2 presents the compositional changes with increasing reaction time of the GM sample after pretreatment. The combined pretreatment by ball mill and ethanol organosolv was effective in both removing acid-insoluble lignin (AIL) and increasing glucan content in GM (Table 2). With increasing reaction time, the glucan content increased to 63.09% after 120 min reaction. Removal of acid-insoluble lignin increased with reaction time. The XMG contents decreased when the reaction time increased from 90 to 120 min (Table 2), and the acetyl group contents also decreased. However, the AIL content did not decrease but the glucan content increased by about 6% when the remaining solid is considered.

Table 2. Changes in composition of giant miscanthus after the combined ball mill and organosolv pretreatment performed at 170 °C at 50 rpm rotation. XMG: xylan, mannan, and galactan; AIL: acid-insoluble lignin.

Time (min)	Solid Remaining (%)	Glucan (%)	XMG [1] (%)	Acetyl Group (%)	AIL (%)	AIL Removal Percentage (%)
Initial	100	43.77	21.22	3.88	19.53	
30	79.25	50.01	23.03	3.73	16.28	33.95
60	72.29	52.22	21.79	3.36	15.89	41.18
90	65.41	56.90	20.72	3.04	14.30	52.11
120	57.44	63.09	15.14	2.10	15.00	55.89

Table 3 presents the changes in the composition of corn stover before and after the combined physicochemical pretreatment with increasing reaction time. The AIL removal yield of corn stover linearly increased to 52.04% during 90 min reaction. The acetyl group content continuously decreased from 2.76% to 1.65% and glucan content steadily increased over 120 min reaction. However, the XMG content changed from 23.89% to 22.08% from 30 to 90 min of reaction time, respectively. The reason for this change could be the removal of extractives by ethanol. As shown for giant miscanthus, the AIL removal slightly increased to 53.91% after 120 min reaction when compared to that after 90 min reaction.

Table 3. Change in composition of corn stover after the combined ball mill and organosolv pretreatment performed at 170 °C with 50 rpm rotation.

Time (min)	Solid Remaining (%)	Glucan (%)	XMG (%)	Acetyl Group (%)	AIL (%)	AIL Removal Percentage (%)
Initial	100	30.30	18.71	2.76	12.23	
30	56.98	46.51	23.89	2.62	14.61	31.92
60	48.32	48.58	22.91	2.18	14.97	40.84
90	42.24	51.73	22.08	1.93	13.89	52.04
120	39.28	53.16	20.27	1.65	14.35	53.91

Table 4 shows the compositional changes of wheat straw. The removal percentage of AIL reached 48.39% after 120 min, which is less than those of GM and CS. Like the other biomass types, the XMG and acetyl group contents steadily decreased but the glucan content increased with reaction time, up to 120 min. The AIL content did not seem to change significantly. However, considering the relative content in the remaining solid, the AIL content was greatly decreased.

Table 4. Change in composition of wheat straw after the combined ball mill and organosolv pretreatment performed at 170 °C with 50 rpm rotation.

Time (min)	Solid Remaining (%)	Glucan (%)	XMG (%)	Acetyl Group (%)	AIL (%)	AIL Removal Percentage (%)
Initial	100	31.42	22.94	3.02	14.61	
30	64.47	41.36	27.83	2.61	14.53	35.90
60	58.07	44.93	26.76	2.45	15.19	39.62
90	52.06	48.78	24.13	2.29	15.55	44.58
120	50.70	50.15	22.71	1.97	14.87	48.39

In previous studies, the pretreatment of lignocellulosic biomasses was carried out using acidic catalysts. El Hage et al. obtained 62% AIL removal of *Miscanthus* samples when using 44% EtOH with 0.5% H_2SO_4 catalyst [39]. Huijgen et al. removed lignin from wheat straw using an acid catalyst and obtained a 56% removal yield [40]. However, using an acid catalyst caused the dissolution of glucan into the liquid, and the acidic solution corroded the pretreatment equipment. In contrast, the combined physicochemical pretreatment used in this study did not cause a loss of glucan and showed good lignin removal yields.

After 30 min reaction time, of the three herbaceous biomass types, the highest percentage of remaining solid was found for GM. The percentages of remaining solid were 79.25%, 56.98%, and 64.47% for GM., CS, and WS, respectively. The remaining solid could be related to the extractive content. The extractive contents (water + ethanol) were 6.43%, 24.85%, and 19.88% for GM, CS, and WS, respectively, after 30 min pretreatment. The extractives were drawn into the liquid by ethanol during the combined physicochemical pretreatment. The remaining solid further decreased with reaction time, and the lowest value was 39.28% in the case of CS because it had the highest extractive content and, thus, underwent the highest removal of the main components.

3.3. Correlation of Lignin Removal with the Removal of Other Compounds

Using the combined physicochemical pretreatment, lignin was mostly removed, and cellulose/hemicellulose were partially removed. The robustness (or recalcitrance) of the lignocellulosic biomass is attributable to the crosslinking between the polysaccharides (cellulose and hemicellulose) and lignin via ester and ether linkages [41]. During the combined physicochemical pretreatment, the polysaccharides were first disconnected from lignin. The newly freed celluloses became dissolved in the ethanol and were broken down into monosaccharides. Simultaneously, lignin and hemicellulose were concurrently fractionated. The correlation between lignin removal with the removal of the various components is presented in Figure 1. It was found that the removal of cellulose, hemicellulose, and acetyl groups is correlated with lignin removal. As shown by the regression lines in Figure 1a, the dissolution of cellulose into the liquor occurred simultaneously with the removal of AIL. With increasing AIL removal, the cellulose dissolution also increased. The regression line for CS has a different slope compared to those of the other herbaceous biomass samples (Figure 1). CS showed higher removal of cellulose depending on the removal of AIL. For example, the percentages of removed cellulose were 66%, 35%, and 29% for CS, WS, and GM, respectively, at the 100% AIL removal point on the regression lines (Figure 1a). The correlation between lignin removal and hemicellulose removal is different from the correlation between the removal of lignin and cellulose. The slopes of the regression lines, which show the correlation between the removal of lignin and hemicellulose, is higher than that of lignin and cellulose because the hemicellulose is more easily removed because of the weak and amorphous structure. From our results, we determined that hemicellulose was rapidly removed when the AIL was dissolved during the combined ball milling and organosolv pretreatment. The correlation between the removal of lignin and acetyl groups also showed a similar pattern to that of hemicellulose. The cleavage of acetyl groups occurred rapidly, and the acetyl groups were also removed when AIL was removed (Figure 1c). Using the combined physicochemical pretreatment, the removal of crosslinking between lignin and the other compounds (cellulose, hemicellulose, and acetyl group) occurred initially. Subsequently, the components were dissolved and removed. However, complete AIL removal was difficult because of the strong crosslinks between the polysaccharides and lignin.

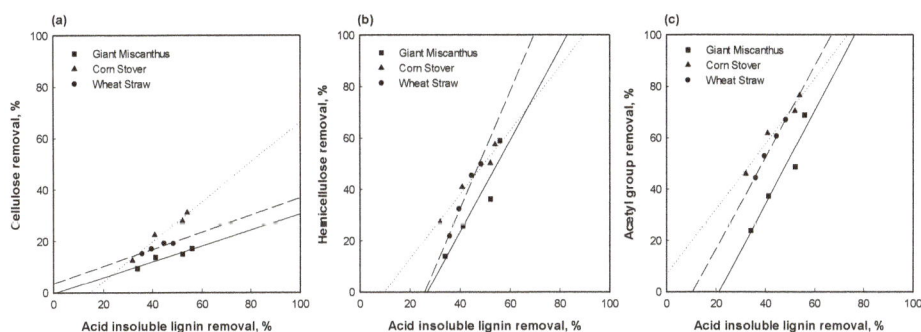

Figure 1. Correlation between the removal of lignin and other compounds: (**a**) cellulose; (**b**) hemicellulose; and (**c**) acetyl group by the combined physicochemical pretreatment.

3.4. Changes in Biomass Particle Size

The change in the size of the biomass particles with pretreatment time was investigated and expressed by calculating the number of particles that passed through sieves of different mesh sizes. All untreated herbaceous biomass samples showed positive skewness (Figure 2). With increasing pretreatment time, the size distribution curves became negatively skewed. The particle sizes after 30 and 60 min pretreatment did not significantly change when compared to those of the untreated samples (data not shown). However, significant changes to the size distribution curves were found after 90 min pretreatment time. Many untreated GM particles were collected after sieving through a 20-mesh sieve. With increasing pretreatment time, the relative amount of GM increased between the 40- and 170-mesh sieves. After 120 min pretreatment, a significant increase was found for the 40-mesh and 170-mesh sieves, indicating that the GM particles had become finer. Figure 2b shows that the relative amount of CS particles passing through the 80-mesh sieve drastically increased from 0.02% to 0.4% after 120 min pretreatment. Meanwhile, the relative number of CS particles larger than 20 mesh was reduced by the pretreatment. Similar results were also found for WS (Figure 2c). In this study, the ball milling process ground the biomass because of the friction between the ball and the biomass. Based on our results, ball milling was effective in pretreating the three types of herbaceous biomass used in this study.

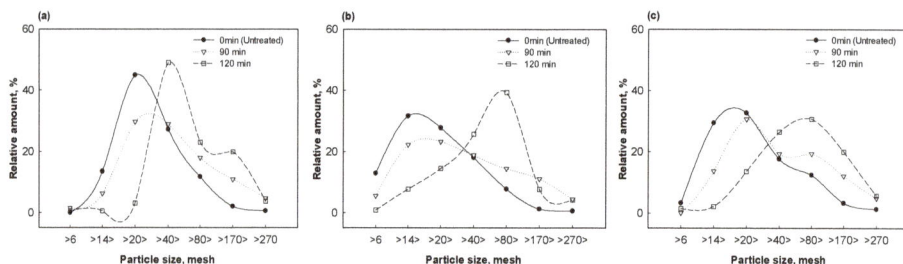

Figure 2. Biomass particle size distributions before and after the combined physicochemical pretreatment at different reaction times: (**a**) GM; (**b**) CS; and (**c**) WS.

3.5. Enzymatic Digestibility of Pretreated Herbaceous Biomasses

To evaluate the effect of pretreatment on cellulose accessibility, the enzyme digestibility was compared before and after the combined physicochemical pretreatment. Untreated GM showed very low glucan digestibility, less than 10%. Even when the enzyme dosage was doubled, the glucose production did not increase significantly. However, the combined physicochemical pretreatment significantly improved the glucan digestibility. For the pretreated samples, a five-fold increase in glucan digestibility was obtained compared to that of the untreated sample (Figure 3a). In addition, the glucan digestibility increased with increasing enzyme dosage. Untreated CS showed relatively higher glucan digestibility than the other two types of herbaceous biomass. However, the glucan digestibility drastically increased after the combined physicochemical pretreatment. Specifically, treated CS showed 91% glucan digestibility when the enzyme dose was 30 FPU/g-glucan (Figure 3b). Untreated WS showed less than 20% glucan digestibility, even after the addition of 30 FPU/g-glucan (Figure 3c). The treated WS showed 79% and 88% glucan digestibility for enzyme doses of 15 and 30 FPU/g-glucan, respectively, as shown in Figure 3c. These results indicate that the combined physicochemical pretreatment used in this study increased the cellulose surface area available for reaction with cellulase, improving enzyme accessibility.

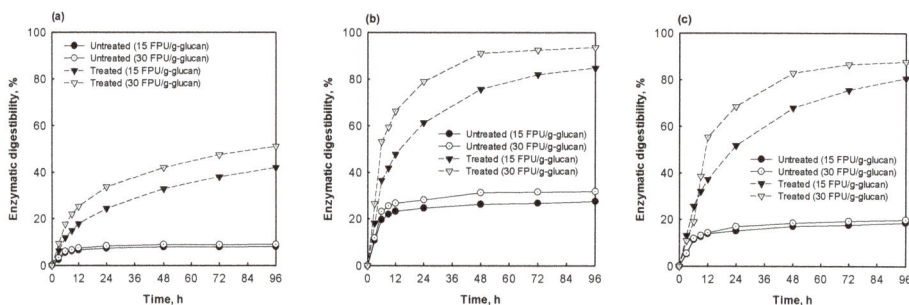

Figure 3. Glucan digestibility before and after pretreatment depending on enzyme dosage: (**a**) GM; (**b**) CS; and (**c**) WS.

3.6. Composition of Recovered Lignin from the Black Liquor

Lignin is byproduct in this process, but it is a very useful compound having biopolymer applications [42]. With increasing pretreatment time from 30 to 120 min in GM, the recovered lignin (AIL and acid-soluble lignin (ASL)) increased from 57.61% to 80.26% (Table 5). In contrast, the contents of glucan and XMG decreased. The contents of recovered lignin were 71.29% and 64.86% for CS and WS, respectively, after 120 min pretreatment. With increasing pretreatment time, the quantity of recovered lignin increased significantly for GM but increased only slightly for CS and WS.

Table 5. Composition of lignin obtained from the black liquor depending on pretreatment time. ASL: acid-soluble lignin; AIA: acid-insoluble ash.

Biomass	Time (min)	AIL (%)	ASL (%)	AIA (%)	Glucan (%)	XMG (%)	Acetyl Group (%)
GM	30	55.62	1.99	5.39	14.57	12.60	2.59
	60	59.02	1.85	5.41	10.74	13.00	2.65
	90	62.27	2.06	6.57	11.34	13.96	2.71
	120	77.94	2.32	1.58	2.66	11.87	2.73
CS	30	59.37	3.07	2.90	5.83	5.56	1.14
	60	64.28	3.29	2.34	5.76	6.23	1.37
	90	56.34	3.78	2.32	6.14	7.28	1.42
	120	67.81	3.48	2.39	4.68	6.28	1.24
WS	30	50.91	3.65	2.97	10.47	12.24	1.93
	60	54.42	2.78	4.30	10.10	11.44	1.69
	90	55.38	3.03	3.22	8.16	12.97	1.97
	120	61.66	3.20	2.32	5.94	12.24	1.90

4. Conclusions

A combined ball milling and ethanol organosolv pretreatment was effective in both increasing the glucan content and removing the acid-insoluble lignin. With increasing pretreatment time, the glucan content and lignin recovery were improved in all three types of herbaceous biomass. Lignin removal was correlated with the removal of cellulose, hemicellulose, and acetyl groups but the slope of the plot of cellulose removal to lignin removal was lower than those of the others. In summary, the combined pretreatment significantly improved the enzyme digestibility of all herbaceous biomass samples tested in this study.

Author Contributions: S.J.K., B.H.U., D.J.I., J.H.L., and K.K.O. contributed equally to this work. S.J.K. and D.J.I. contributed to the experimental process for combined fractionation process. B.H.U. and K.K.O. contributed to the project administration and experimental design. In addition, J.H.L. contributed to providing methodology and data validation. All the authors contributed to the writing and review of this document.

Funding: This work was funded by a grant the Korea Institute of Energy Technology Evaluation and Planning (KETEP) (Grant No. 2015010091990).

Conflicts of Interest: The authors declare no conflict of interest.

References

1. Wettstein, S.G.; Alonso, D.M.; Gürbüz, E.I.; Dumesic, J.A. A roadmap for conversion of lignocellulosic biomass to chemicals and fuels. *Curr. Opin. Chem. Eng.* **2012**, *1*, 218–224. [CrossRef]
2. Sheldon, R.A. Green and sustainable manufacture of chemicals from biomass: State of the art. *Green Chem.* **2014**, *16*, 950–963. [CrossRef]
3. Sun, Y.; Cheng, J. Hydrolysis of lignocellulosic materials for ethanol production: A review. *Bioresour. Technol.* **2002**, *83*, 1–11. [CrossRef]
4. Kumar, P.; Barrett, D.M.; Delwiche, M.J.; Stroeve, P. Methods for pretreatment of lignocellulosic biomass for efficient hydrolysis and biofuel production. *Ind. Eng. Chem. Res.* **2009**, *48*, 3713–3729. [CrossRef]
5. Cazacu, G.; Capraru, M.; Popa, V.I. Advances concerning lignin utilization in new materials. In *Advances in Natural Polymers*; Springer: Berlin, Germany, 2013; pp. 255–312.

6. Menon, V.; Rao, M. Trends in bioconversion of lignocellulose: Biofuels, platform chemicals & biorefinery concept. *Prog. Energy Combust. Sci.* **2012**, *38*, 522–550. [CrossRef]

7. Taherzadeh, M.J.; Karimi, K. Pretreatment of lignocellulosic wastes to improve ethanol and biogas production: A review. *Int. J. Mol. Sci.* **2008**, *9*, 1621–1651. [CrossRef] [PubMed]

8. Zakaria, M.R.; Hirata, S.; Hassan, M.A. Combined pretreatment using alkaline hydrothermal and ball milling to enhance enzymatic hydrolysis of oil palm mesocarp fiber. *Bioresour. Technol.* **2014**, *169*, 236–243. [CrossRef] [PubMed]

9. Kim, S.M.; Dien, B.S.; Tumbleson, M.E.; Rausch, K.D.; Singh, V. Improvement of sugar yields from corn stover using sequential hot water pretreatment and disk milling. *Bioresour. Technol.* **2016**, *216*, 706–713. [CrossRef] [PubMed]

10. Barakat, A.; Chuetor, S.; Monlau, F.; Solhy, A.; Rouau, X. Eco-friendly dry chemo-mechanical pretreatments of lignocellulosic biomass: Impact on energy and yield of the enzymatic hydrolysis. *Appl. Energy* **2014**, *113*, 97–105. [CrossRef]

11. Yuan, Z.; Long, J.; Wang, T.; Shu, R.; Zhang, Q.; Ma, L. Process intensification effect of ball milling on the hydrothermal pretreatment for corn straw enzymolysis. *Energy Convers. Manag.* **2015**, *101*, 481–488. [CrossRef]

12. Deng, A.; Ren, J.; Wang, W.; Li, H.; Lin, Q.; Yan, Y.; Sun, R.; Liu, G. Production of xylo-sugars from corncob by oxalic acid-assisted ball milling and microwave-induced hydrothermal treatments. *Ind. Crops Prod.* **2016**, *79*, 137–145. [CrossRef]

13. Huo, D.; Fang, G.; Yang, Q.; Han, S.; Deng, Y.; Shen, K.; Lin, Y. Enhancement of eucalypt chips' enzymolysis efficiency by a combination method of alkali impregnation and refining pretreatment. *Bioresour. Technol.* **2013**, *150*, 73–78. [CrossRef] [PubMed]

14. Da Silva, A.S.; Inoue, H.; Endo, T.; Yano, S.; Bon, E.P.S. Milling pretreatment of sugarcane bagasse and straw for enzymatic hydrolysis and ethanol fermentation. *Bioresour. Technol.* **2010**, *101*, 7402–7409. [CrossRef] [PubMed]

15. Inoue, H.; Yano, S.; Endo, T.; Sakaki, T.; Sawayama, S. Combining hot-compressed water and ball milling pretreatments to improve the efficiency of the enzymatic hydrolysis of eucalyptus. *Biotechnol. Biofuels* **2008**, *1*, 2. [CrossRef] [PubMed]

16. Lin, Z.; Huang, H.; Zhang, H.; Zhang, L.; Yan, L.; Chen, J. Ball milling pretreatment of corn stover for enhancing the efficiency of enzymatic hydrolysis. *Appl. Biochem. Biotechnol.* **2010**, *162*, 1872–1880. [CrossRef] [PubMed]

17. Silva, G.G.D.; Couturier, M.; Berrin, J.-G.; Buléon, A.; Rouau, X. Effects of grinding processes on enzymatic degradation of wheat straw. *Bioresour. Technol.* **2012**, *103*, 192–200. [CrossRef] [PubMed]

18. Lee, J.H.; Kwon, J.H.; Kim, T.H.; Choi, W. Impact of planetary ball mills on corn stover characteristics and enzymatic digestibility depending on grinding ball properties. *Bioresour. Technol.* **2017**, *241*, 1094–1100. [CrossRef] [PubMed]

19. Zhang, K.; Pei, Z.; Wang, D. Organic solvent pretreatment of lignocellulosic biomass for biofuels and biochemicals: A review. *Bioresour. Technol.* **2016**, *199*, 21–33. [CrossRef] [PubMed]

20. Pan, X.; Arato, C.; Gilkes, N.; Gregg, D.; Mabee, W.; Pye, K.; Xiao, Z.; Zhang, X.; Saddler, J. Biorefining of softwoods using ethanol organosolv pulping: Preliminary evaluation of process streams for manufacture of fuel-grade ethanol and co-products. *Biotechnol. Bioeng.* **2005**, *90*, 473–481. [CrossRef] [PubMed]

21. Zhao, X.; Cheng, K.; Liu, D. Organosolv pretreatment of lignocellulosic biomass for enzymatic hydrolysis. *Appl. Microbiol. Biotechnol.* **2009**, *82*, 815. [CrossRef] [PubMed]

22. Zakzeski, J.; Bruijnincx, P.C.A.; Jongerius, A.L.; Weckhuysen, B.M. The catalytic valorization of lignin for the production of renewable chemicals. *Chem. Rev.* **2010**, *110*, 3552–3599. [CrossRef] [PubMed]

23. De la Torre, M.; Moral, A.; Hernández, M.; Cabeza, E.; Tijero, A. Organosolv lignin for biofuel. *Ind. Crops Prod.* **2013**, *45*, 58–63. [CrossRef]

24. Chang, V.S.; Holtzapple, M.T. Fundamental factors affecting biomass enzymatic reactivity. In *Twenty-first Symposium on Biotechnology for Fuels and Chemicals*; Humana Press: Totowa, NJ, USA, 2000; pp. 5–37.

25. Lee, D.; Yu, A.H.C.; Saddler, J.N. Evaluation of cellulase recycling strategies for the hydrolysis of lignocellulosic substrates. *Biotechnol. Bioeng.* **1995**, *45*, 328–336. [CrossRef] [PubMed]

26. Öhgren, K.; Bura, R.; Saddler, J.; Zacchi, G. Effect of hemicellulose and lignin removal on enzymatic hydrolysis of steam pretreated corn stover. *Bioresour. Technol.* **2007**, *98*, 2503–2510. [CrossRef] [PubMed]

27. Zheng, Y.; Zhang, S.; Miao, S.; Su, Z.; Wang, P. Temperature sensitivity of cellulase adsorption on lignin and its impact on enzymatic hydrolysis of lignocellulosic biomass. *J. Biotechnol.* **2013**, *166*, 135–143. [CrossRef] [PubMed]

28. Koo, B.-W.; Kim, H.-Y.; Park, N.; Lee, S.-M.; Yeo, H.; Choi, I.-G. Organosolv pretreatment of Liriodendron tulipifera and simultaneous saccharification and fermentation for bioethanol production. *Biomass Bioenergy* **2011**, *35*, 1833–1840. [CrossRef]

29. Barakat, A.; De Vries, H.; Rouau, X. Dry fractionation process as an important step in current and future lignocellulose biorefineries: A review. *Bioresour. Technol.* **2013**, *134*, 362–373. [CrossRef] [PubMed]

30. Mathew, A.K.; Chaney, K.; Crook, M.; Humphries, A.C. Alkaline pre-treatment of oilseed rape straw for bioethanol production: Evaluation of glucose yield and pre-treatment energy consumption. *Bioresour. Technol.* **2011**, *102*, 6547–6553. [CrossRef] [PubMed]

31. Lee, S.-H.; Teramoto, Y.; Endo, T. Enzymatic saccharification of woody biomass micro/nanofibrillated by continuous extrusion process I—Effect of additives with cellulose affinity. *Bioresour. Technol.* **2009**, *100*, 275–279. [CrossRef] [PubMed]

32. Hideno, A.; Inoue, H.; Tsukahara, K.; Fujimoto, S.; Minowa, T.; Inoue, S.; Endo, T.; Sawayama, S. Wet disk milling pretreatment without sulfuric acid for enzymatic hydrolysis of rice straw. *Bioresour. Technol.* **2009**, *100*, 2706–2711. [CrossRef] [PubMed]

33. Zhu, J.Y.; Pan, X.J. Woody biomass pretreatment for cellulosic ethanol production: Technology and energy consumption evaluation. *Bioresour. Technol.* **2010**, *101*, 4992–5002. [CrossRef] [PubMed]

34. Selig, M.; Weiss, N.; Ji, Y. *Enzymatic Saccharification of Lignocellulosic Biomass: Laboratory Analytical Procedure*; National Renewable Energy Laboratory: Golden, CO, USA, 2008.

35. Zhang, L.; Yan, L.; Wang, Z.; Laskar, D.D.; Swita, M.S.; Cort, J.R.; Yang, B. Characterization of lignin derived from water only and dilute acid flowthrough pretreatment of poplar wood at elevated tempeatures. *Biotechnol. Biofuels* **2015**, *8*, 203–217. [CrossRef] [PubMed]

36. Sluiter, A.; Hames, B.; Ruiz, R.; Scarlata, C.; Sluiter, J.; Templeton, D.; Crocker, D. *Determination of Structural Carbohydrates and Lignin in Biomass: Laboratory Analytical Procedure*; National Renewable Energy Laboratory: Golden, CO, USA, 2008.

37. Sluiter, A.; Hames, B.; Ruiz, R.; Scarlata, C.; Sluiter, J.; Templeton, D. *Determination of Sugars, Byproducts, and Degradation Products in Liquid Fraction Process Samples: Laboratory Analytical Procedure*; National Renewable Energy Laboratory: Golden, CO, USA, 2012.

38. Sluiter, A.; Hames, B.; Ruiz, R.; Scarlata, C.; Sluiter, J.; Templeton, D. *Determination of Ash in Biomass: Laboratory Analytical Procedure*; National Renewable Energy Laboratory: Golden, CO, USA, 2008.

39. El Hage, R.; Chrusciel, L.; Desharnais, L.; Brosse, N. Effect of autohydrolysis of Miscanthus x giganteus on lignin structure and organosolv delignification. *Bioresour. Technol.* **2010**, *101*, 9321–9329. [CrossRef] [PubMed]

40. Huijgen, W.J.J.; Smit, A.T.; De Wild, P.J.; Den Uil, H. Fractionation of wheat straw by prehydrolysis, organosolv delignification and enzymatic hydrolysis for production of sugars and lignin. *Bioresour. Technol.* **2012**, *114*, 389–398. [CrossRef] [PubMed]

41. Jeffries, T.W. Biodegradation of lignin-carbohydrate complexes. In *Physiology of Biodegradative Microorganisms*; Springer: Dordrecht, The Netherlands, 1991; pp. 163–176.

42. Zadeh, E.M.; O'Keefe, S.F.; Kim, Y.-T. Utilization of lignin in biopolymeric packaging films. *ACS Omega* **2018**, *3*, 7388–7398. [CrossRef]

energies

MDPI

Article

Biodiesel Production from a Novel Nonedible Feedstock, Soursop (*Annona muricata* L.) Seed Oil

Chia-Hung Su *, Hoang Chinh Nguyen, Uyen Khanh Pham, My Linh Nguyen and Horng-Yi Juan

Graduate School of Biochemical Engineering, Ming Chi University of Technology, New Taipei City 24301, Taiwan; d10522811@mail.ntust.edu.tw (H.C.N.); d913612@alumni.nthu.edu.tw (U.K.P.); m07138213@mail2.mcut.edu.tw (M.L.N.), hyjuan@mail.mcut.edu.tw (H.-Y.J.)
* Correspondence: chsu@mail.mcut.edu.tw; Tel.: +886-22-908-9899 (ext. 4665)

Received: 11 August 2018; Accepted: 24 September 2018; Published: 26 September 2018

Abstract: This study investigated the optimal reaction conditions for biodiesel production from soursop (*Annona muricata*) seeds. A high oil yield of 29.6% (w/w) could be obtained from soursop seeds. Oil extracted from soursop seeds was then converted into biodiesel through two-step transesterification process. A highest biodiesel yield of 97.02% was achieved under optimal acid-catalyzed esterification conditions (temperature: 65 °C, 1% H_2SO_4, reaction time: 90 min, and a methanol:oil molar ratio: 10:1) and optimal alkali-catalyzed transesterification conditions (temperature: 65 °C, reaction time: 30 min, 0.6% NaOH, and a methanol:oil molar ratio: 8:1). The properties of soursop biodiesel were determined and most were found to meet the European standard EN 14214 and American Society for Testing and Materials standard D6751. This study suggests that soursop seed oil is a promising biodiesel feedstock and that soursop biodiesel is a viable alternative to petrodiesel.

Keywords: *Annona muricata*; biodiesel production; seed oil; soursop; two-step process

1. Introduction

Fossil fuel depletion and environmental concerns have stimulated the search for alternative fuels from renewable sources. Biodiesel, a biomass-derived fuel, is renewable, exhibits superior combustion properties, and is completely suitable for diesel engines [1,2]. Furthermore, the use of biodiesel results in relatively low environmental pollution because biodiesel is sulfur free and emits minimal carbon monoxide and hydrocarbons [3–5]. Because of these merits, biodiesel has been developed worldwide to replace petrodiesel.

Biodiesel has been mainly produced from edible oil using an acid, alkali, or enzyme catalyst [6,7]. Nevertheless, the use of edible oil as a feedstock increases the production cost of biodiesel [8], thus limiting the commercialization of biodiesel. Furthermore, the use of edible feedstock for fuel purpose may cause adverse effects on food supply [9–11]; therefore, alternative feedstocks must be identified for biodiesel synthesis. Numerous cheap and nonedible feedstocks, including microalgae oil [12–14], Jatropha oil [15,16], waste cooking oil [17–19], insect fat [20–22], Chinese tallow tree seed oil [23], tobacco seed oil [24], sweet basil seed oil [25], *Brucea javanica* seed oil [26], spent coffee grounds [27], and food waste [28] have been investigated as vital feedstocks for biodiesel synthesis. Two-step transesterification (acid-catalyzed esterification followed by alkali-catalyzed transesterification) is a promising method to produce biodiesel from high free fatty acid oils [29,30]. The acid oils (fatty acid content >1%, w/w) should be esterified using an acid catalyst to lower the oil acidity before applying an alkali catalyst to transesterify the oil into biodiesel [31–33]. This two-step process not only minimizes soap formation but also enhances biodiesel yield [31,34,35].

Soursop (*Annona muricata* L.), which belongs to the Annonaceae family, is an economically critical crop worldwide [36,37]. *A. muricata* is native to North and South America and is popularly distributed in the tropical and subtropical areas of Western Africa, Central America, the Caribbean, and the Asian continent [37–39]. The *A. muricata* tree is approximately 5–8-m tall with low branches [39], and the trees yield up to 10 tons of fruit per hectare [38]. The oval or heart-shaped fruit is 15–30-cm in length, 10–20-cm in width, and, on average, weighs up to 4.0 kg [36,39]. The edible white pulp of soursop fruit comprises about 80% water, 18% carbohydrate, 1% protein, and 1% fiber content and contains beneficial vitamins [40]. The mesocarp contains numerous black seeds, which are each approximately 2 cm long and 1 cm wide [36]. The soursop is mainly cultivated for its fruit, which is used in fresh and processed forms in the production of juice, ice cream, sherbet, beverages, and candy [37,41,42]. The use of soursop fruit in food production results in various waste materials, including seeds which account for 5–8.5% of the fruit [37,38,43]. The seeds are usually discarded and cannot be used as animal feed because they contain toxic substances such as annonacin and acetogenins [37,43,44]. Studies have shown that oil comprises up to 40% of the soursop seed [37]. The seed oil mainly comprises palmitic acid, oleic acid, and linoleic acid [37,45]. This composition is similar to that of other biodiesel feedstocks [24,29,31]. Therefore, the nonedible soursop seed oil is a promising and cheap biodiesel feedstock. In addition, the use of this seed for biodiesel synthesis can resolve the problematic surplus of the seed in the food industry. Only few studies have reported the potential use of this seed oil for biodiesel production [46,47]. However, no optimization study on reaction conditions has investigated for biodiesel production from *A. muricata* seed oil.

This study optimized the reaction factors for producing biodiesel from *A. muricata* seed oil. Because of the presence of a high level of free fatty acid (FFA) in the seed oil, a two-step process was used to convert FFAs and triglycerides into biodiesel. The effects of reaction factors (molar ratio of methanol to oil, temperature, catalyst amount, and reaction time) on esterification and transesterification were investigated to optimize the reaction conditions. The biodiesel's properties were finally characterized according to the American Society for Testing and Materials (ASTM) methods.

2. Materials and Methods

2.1. Materials

Ripe soursop fruits were purchased from Thu Duc Agromarket (Ho Chi Minh City, Vietnam). The seeds were removed from the fruit and air-dried at room temperature for 3 days. The soursop seed kernels were then separated from the hulls, ground with a blender (EUPA TSK-935BAP, Tsann Kuen Enterprise Co., Ltd., Taipei, Taiwan), and stored at room temperature. Methanol, sulfuric acid, sodium hydroxide, n-hexane, and other reagents used in this study were of analytical grade (≥99.0% purity) and were obtained from Tedia Company, Inc. (Fairfield, CT, USA).

2.2. Extraction of Crude Oil

Soursop seed powder was immersed in n-hexane (1:4, w/v) at room temperature for 2 days, and stirred to extract the oil from soursop seed. After extraction, the hexane layer was separated from solid residue by filtration (Advantec No. 5C filter paper). The n-hexane was then removed using a R300 Buchi Rotary Evaporator (Büchi Labortechnik, Flawil, Switzerland), and the soursop seed crude oil was obtained. The crude oil's properties, such as acidity, saponification, and iodine values, were measured using the standard method [48,49].

2.3. Production of Biodiesel through Two-Step Process

2.3.1. Esterification Step

An H_2SO_4-catalyzed pretreatment was employed to reduce the oil's acidity and convert its FFAs into biodiesel. To study the influence of reaction factors on esterification, several experimental trials were conducted in a sealed reactor with stirring under different conditions: molar ratios of methanol to oil (4:1–12:1), temperatures (45–85 °C), catalyst amounts (0.25–2.0%), and reaction times (30–150 min). After each reaction, the samples were withdrawn to evaluate the FFA conversion.

2.3.2. Transesterification Step

The oil pretreated through H_2SO_4-catalyzed esterification was used for the transesterification step. The esterified reaction mixture was kept in a funnel for phase separation. After two phases were completely separated, the crude oil and biodiesel (upper layer) was poured into a sealed reactor and subsequently transesterified into biodiesel using NaOH as catalyst. A set of experiments with various methanol to oil molar ratios (4:1–12:1), temperatures (45–85 °C), catalyst amounts (0.4–1.2%), and reaction times (15–75 min) were studied for their effects on the conversion yield. After each reaction, the reactor was placed at room temperature for phase separation. The mixture's upper layer containing biodiesel was collected to determine the biodiesel yield.

2.4. Analysis

The acid value (AV) of the oil sample was measured using a titration method reported previously [50,51]. The FFA conversion was then calculated as follows:

$$\text{FFA conversion (\%)} = \frac{AV_1 - AV_2}{AV_1} \times 100 \tag{1}$$

where AV_1 is the initial acid value, and AV_2 is the acid value after the esterification reaction.

The composition of the biodiesel was quantified using a Shimadzu GC-2014 gas chromatograph system (Shimadzu Corp., Kyoto, Japan) equipped with a Stabilwax capillary column (Restek Corp., Bellefonte, PA, USA) and a flame ionization detector (Shimadzu Corp., Kyoto, Japan) according to the procedure reported in our previous study [10]. The fatty acid profiles of the soursop biodiesel were characterized based on Supelco 37 Component FAME Mix reference standards (Sigma-Aldrich Corp., St. Louis, MO, USA). The biodiesel content was quantified by comparing the peak areas of fatty acid methyl esters with those of the internal standard, methyl pentadecanoate. The soursop biodiesel yield was then calculated as follows [10]:

$$\text{Biodiesel yield (\%)} = \frac{A_{\text{sample}}}{A_{\text{standard}}} \times \frac{W_{\text{standard}}}{W_{\text{sample}}} \times \frac{W_{\text{biodiesel}}}{W_{\text{oil}}} \times 100 \tag{2}$$

where A_{sample} is peak area of biodiesel sample, A_{standard} is peak area of international standard, W_{sample} is weigh of biodiesel sample, W_{standard} is weight of internal standard, $W_{\text{biodiesel}}$ is weight of total biodiesel product, and W_{oil} is weight of oil used.

The acid value, viscosity, sulfur content, water content, ester content, cetane index, density, and flash point of the produced biodiesel were determined using the ASTM D664, D445, D5453, D95, D7371, D613, D1480, and D93 methods, respectively [52].

3. Results and Discussion

3.1. Properties of Soursop Seed Oil

Table 1 shows the characteristics of soursop seed oil. The oil extracted from soursop seeds reached the yield of 29.6%, demonstrating the soursop seed's high oil content and subsequent potential as an oil source. The saponification value of soursop seed oil was 244.7 mg KOH/g, indicating that the average molecular weight of soursop seed oil was 884.4 g/mol. The acid value of the soursop seed oil was 54.4 mg KOH/g, indicating a high FFA content. Biodiesel production processes must be refined to maximize the value of materials and minimize costs [25,53,54]. To maximize the biodiesel yield from oils with high FFA levels, esterification must be performed to reduce the level of FFAs prior to transesterification [31,34,35]. Therefore, the two-step process of acid-catalyzed esterification followed by alkali-transesterification was selected for biodiesel synthesis from soursop seed oil in this study.

Table 1. Properties of crude soursop seed oil.

Fat Yield (%)	Acid Value (mg KOH/g)	Saponification Value (mg KOH/g)
29.6 ± 0.2	54.4 ± 0.4	244.7 ± 1.6

3.2. Conversion of FFAs into Biodiesel through Acid-Catalyzed Esterification

3.2.1. Effect of Methanol to Oil Molar Ratio

Esterification pretreatment enhances the biodiesel yield by reducing the oil's acidity and converting its FFAs into biodiesel [29,51]. To optimize reaction conditions, this study investigates the influences of methanol to oil molar ratio, temperature, H_2SO_4 amount, and reaction time on the FFAs conversion. First, esterification was performed at 75 °C with 1% H_2SO_4 (w/w) and various methanol to oil molar ratios (4:1–12:1) for 60 min. As shown in Figure 1a, FFA conversion was greater at higher methanol to oil molar ratios. This result corresponds with those of other studies [35,55]. The molar ratio of methanol to oil is a critical factor affecting the efficiency of esterification reactions. A high methanol to oil molar ratio is required to drive esterification reactions toward completion [55]. In this work, the highest FFA conversion occurred at the methanol:oil molar ratio of 12:1. However, the FFA conversion had insignificant differences between the methanol:oil molar ratios of 12:1 and 10:1. Therefore, the methanol:oil molar ratio of 10:1 was chosen for the next experiments.

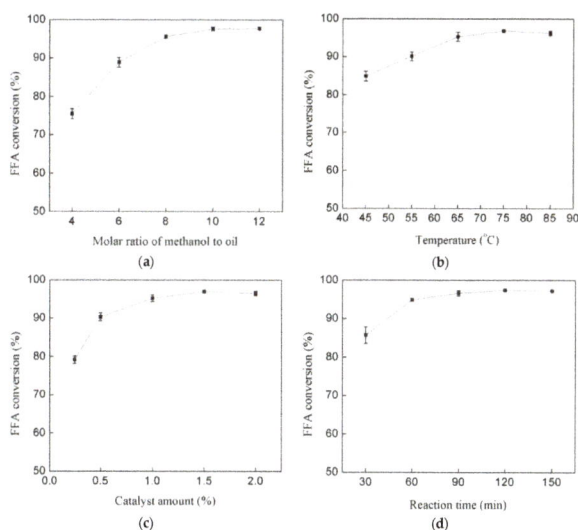

Figure 1. Effects of (**a**) molar ratio of methanol to oil (with a fixed temperature of 75 °C, 1% H_2SO_4, and a reaction time of 60 min); (**b**) temperature (with a fixed methanol:oil molar ratio of 10:1, 1% H_2SO_4, and a reaction time of 60 min); (**c**) catalyst amount (with a fixed methanol:oil molar ratio of 10:1, a temperature of 65 °C, and a reaction time of 60 min); and (**d**) reaction time (with a fixed methanol:oil molar ratio of 10:1, a temperature of 65 °C, and 1% H_2SO_4) on FFA conversion in soursop seed oil.

3.2.2. Effect of Temperature

To investigate the impact of temperature on the efficiency of esterification, the reaction was performed at various temperatures (45–85 °C), whilst keeping the other factors constant. As shown in

Figure 1b, FFA conversion was enhanced from 84.86 to 96.16% when the temperature was increased from 45 to 85 °C. This result is reasonable because a high temperature enhances the reaction rate [51,56]. However, no significant differences were found in the proportion of FFA conversion at temperatures of 65, 75, and 85 °C. Therefore, to reduce energy consumption, 65 °C was selected as the optimal temperature for the esterification reaction.

3.2.3. Effect of Catalyst Amount

The results in Figure 1c reflect the effect of sulfuric acid amount on FFA conversion. FFA conversion significantly improved when the catalyst levels increased from 0.25 to 1.0%. Nevertheless, increasing the catalyst load to 1.5% resulted in only a slight increase in FFA conversion, and a 2% catalyst load caused a slight decrease in conversion efficiency. This slight decrease at a 2% catalyst load is similar to of the results of other studies [35,55]. Excess H_2SO_4 catalyst can activate the polymerization of unsaturated FFA, causing the product's darkened color due to the oxidation and decarboxylation of FFA [55]. Therefore, 1.0% H_2SO_4 was chosen as the optimal catalyst amount for further experiments.

3.2.4. Effect of Reaction Time

Various reaction times (30–150 min) were tested for esterification performed at 65 °C with a methanol:oil molar ratio of 10:1 and 1% H_2SO_4 (w/w). As shown in Figure 1d, FFA conversion increased from 85.71 to 96.61% when increasing reaction time from 30 to 90 min. Increases in reaction times beyond 90 min resulted in insignificant increases in FFA conversion, indicating that reactions reached equilibrium at 90 min. In conclusion, the optimal conditions for the H_2SO_4-catalyzed esterification were determined to be a methanol:oil molar ratio: 10:1, a temperature: 65 °C, 1% H_2SO_4 (w/w), and 90 min. These conditions were thus used in this study for the esterification step in biodiesel production.

3.3. Conversion of Triglyceride into Biodiesel through Alkali-Catalyzed Transesterification

3.3.1. Effect of Methanol to Oil Molar Ratio

The esterified oil was used as the material for producing biodiesel through alkali-catalyzed transesterification. To optimize transesterification conditions, the influences of methanol to oil molar ratio, temperature, NaOH amount, and reaction time on the biodiesel yield were examined. First, transesterification was performed at 65 °C with 0.8% NaOH (w/w) and various methanol to oil molar ratios (4:1–12:1) for 30 min. As can be seen from Figure 2a, the biodiesel yield increased from 81% to 96.37% when the molar ratio of methanol to oil was increased from 4:1 to 8:1. Nevertheless, a higher methanol to oil molar ratio caused a decrease in the biodiesel yield. This result is consistent with that reported in the study of biodiesel synthesis from *Jatropha curcus* seed oil [31] and *Croton megalocarpus* oil [57]. A high level of methanol may have increased the glycerol solubility in the solution, driving the equilibrium to a reverse reaction and thus lowering the biodiesel yield [57]. Therefore, this study selected the methanol:oil molar ratio of 8:1 as the optimal reactant ratio for transesterification.

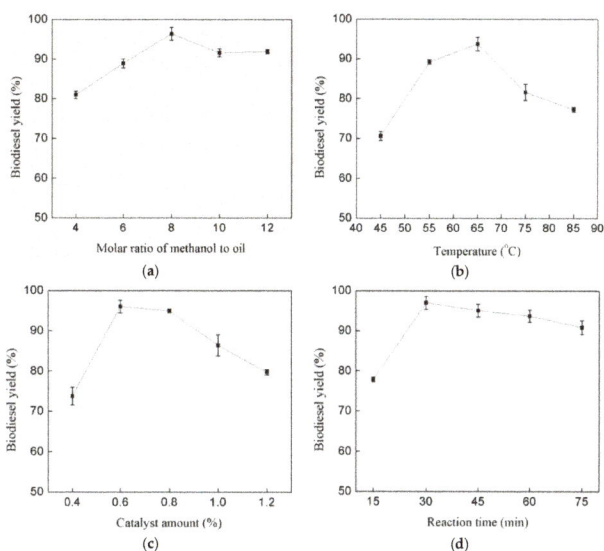

Figure 2. Effects of (**a**) molar ratio of methanol to oil (with a fixed temperature of 65 °C, 0.8% NaOH, and a reaction time of 30 min); (**b**) temperature (with a fixed methanol:oil molar ratio of 8:1, 0.8% NaOH, and a reaction time of 30 min); (**c**) catalyst amount (with a fixed methanol:oil molar ratio of 8:1, a temperature of 65 °C, and a reaction time of 30 min); and (**d**) reaction time (with a fixed methanol:oil molar ratio of 8:1, a temperature of 65 °C, and 0.6% NaOH) on transesterification of soursop seed oil.

3.3.2. Effect of Temperature

To investigate the influence of temperature on the biodiesel yield, transesterification was carried out at various temperatures (45–85 °C) with a methanol to oil molar ratio, NaOH amount, and reaction time maintained at 8:1, 0.8%, and 30 min, respectively. As can be seen from Figure 2b, the biodiesel yield increased when temperature was increased from 45 to 65 °C. Nevertheless, the biodiesel yield reduced at temperatures greater than 65 °C. This result is in agreement with those of other studies [57,58]. A high temperature may have enhanced side reactions, including saponification, thus resulting in a lower biodiesel yield [57]. Based on this result, 65 °C was chosen as the optimal temperature for the transesterification.

3.3.3. Effect of Catalyst Amount

Catalyst amount is a critical factor affecting the efficiency of transesterification. In this study, various NaOH amounts were tested with other factors maintained as constant to evaluate the influence of NaOH amount on the biodiesel yield. Results revealed that the biodiesel yield increased when the amount of catalyst increased from 0.4 to 0.6% (Figure 2c). However, when the catalyst amount is higher than 0.6%, the biodiesel yield reduced. Excess catalyst favored the saponification reaction, enhancing the formation of an emulsion and gel, thus lowering the biodiesel yield [31,55,57]. Therefore, 0.6% NaOH was selected for use in further experiments.

3.3.4. Effect of Reaction Time

Finally, the influence of reaction time on the biodiesel yield was examined using the optimal reaction conditions obtained in previous experiments. Transesterification was performed at 65 °C with a methanol:oil molar ratio of 8:1, 0.6% NaOH (w/w), and various reaction times (15–75 min). As indicated in Figure 2d, the biodiesel yield increased from 77.85 to 97.02% when the reaction time increased from 15 to 30 min. A reaction time longer than 30 min caused a slight decrease in the biodiesel yield. Therefore, a reaction time of 30 min was determined as sufficient for the transesterification reaction. In conclusion, the highest biodiesel yield of 97.02% was achieved under the following optimal transesterification conditions: 65°C, 0.6% NaOH (w/w), 30 min, and methanol:oil molar ratio of 8:1.

3.4. Fatty Acid Profiles of Soursop Biodiesel

Table 2 illustrates the fatty acid profiles of soursop biodiesel in comparison with rapeseed biodiesel. Nine fatty acid methyl esters were identified in the soursop biodiesel, among which oleic acid methyl ester (43.68%), linoleic acid methyl ester (32.45%), and palmitic acid methyl ester (18.14%) were present in the highest amounts. The synthesized biodiesel was found to comprise 78.07% unsaturated fatty acids and 21.93% saturated ones. The saturated fatty acid level in soursop biodiesel was higher than that of rapeseed biodiesel (4.3%) [51], indicating a higher cetane index for the soursop biodiesel. This is because the high level of saturated fatty acid increases the cetane index of a fuel [59]. In addition, because saturated fatty acid methyl esters exhibit higher oxidative stability than unsaturated ones [4,60], the soursop biodiesel can have more oxidative stability than rapeseed biodiesel. These results suggest that soursop seed oil is a suitable feedstock for biodiesel synthesis.

Table 2. Fatty acid methyl ester compositions of soursop biodiesel compared with rapeseed biodiesel.

Composition	Rapeseed Biodiesel [a] (%)	Soursop Biodiesel [b] (%)
Palmitic acid methyl ester (C16:0)	3.5	18.14
Palmitoleic acid methyl ester (C16:1)	na [c]	0.81
Stearic acid methyl ester (18:0)	0.8	3.79
Oleic acid methyl ester (C18:1)	64.4	43.68
Linoleic acid methyl ester (C18:2)	22.3	32.45
Linolenic acid methyl ester (C18:3)	8.2	1.13

[a] Data obtained from Reference [51]; [b] This study; [c] na = none reported.

3.5. Properties of Soursop Biodiesel

The soursop biodiesel's properties were characterized using ASTM standard methods and were compared with the corresponding properties of rapeseed biodiesel [51]. As indicated in Table 3, most soursop biodiesel's properties were similar to those of rapeseed biodiesel. Remarkably, most properties of synthesized biodiesel, sulfur content (0.04%), ester content (98.6%), viscosity (5.5 mm^2/s), water content (300 mg/kg), cetane number (53), density (868 kg/m^3), and flash point (123°C), met the standards ASTM D6751 [52] and EN 14,214 [61]. These results indicate that the synthesized biodiesel may serve as an alternative to petrodiesel. Moreover, the high ester content in the biodiesel indicates that the conditions identified in this study are optimal for the esterification and transesterification reactions. However, the acid value of synthesized biodiesel was 0.8, which was higher than the standards EN 14,214 and ASTM D6751: this could be due to the presence of free fatty acid in the biodiesel product. A further purification step is therefore required in order to reduce the acid value [62].

Table 3. Soursop biodiesel's properties compared with those of rapeseed biodiesel, the standards ASTM D6751, and EN 14214.

Properties	ASTM Method	ASTM D6751 [a]	EN 14214	Rapeseed Biodiesel [b]	This Study
Acid value (mg KOH/g)	D664	<0.5	<0.5	0.31	<0.8
Sulfur content (wt. %)	D5453	<0.05	<0.05	<0.01	0.04
Ester content (%)	D7371	na [c]	>96.5	na [c]	98.6
Viscosity at 40 °C (mm^2/s)	D445	1.9–6.0	3.5–5.0	6.35	5.5
Water content (mg/kg)	D95	na [c]	<500	300	300
Cetane number	D613	>47	>51	45	53
Density (kg/m^3)	D1480	na [c]	860–900	880	868
Flash point (closed cup) (°C)	D93	100–170	>120	na [c]	123

[a] Data obtained from Reference [52]; [b] Data were obtained from Reference [51]; [c] na = none reported.

3.6. The Feasibility of Soursop Seed Oil as Biodiesel Feedstock

With an increasing demand for renewable energy, biodiesel has been widely produced to replace petrodiesel. To reduce the production cost, various non-edible feedstocks including microbial oil [12–14], waste cooking oil [17–19], insect fat [20–22], and plant seed oil [23–25], have been studied for biodiesel production because of their low-price. However, the availability of those oils is still a major concern for large-scale production [10]. Therefore, efforts have been made to search for new low-cost biodiesel feedstocks. In recent years, the use of plant seed obtained from fruit production industry for biodiesel production has attracted much attention due to its low cost and availability [25,26]. In the state of Bahia (Brazil), approximately 20 thousand tons of soursop fruit are produced annually, thus producing about 1.7 thousand tons of soursop seed each year [37]. Since soursop is an economically important crop worldwide [37,41], the global food production from soursop fruit can cause a problematic surplus of this seed. To add value to the soursop seeds, the effective method of recycling these seeds is being investigated. In the present study, soursop seeds were used as a non-edible feedstock for biodiesel production—a solution to the soursop seed disposal problem. The results suggested that soursop seed can be a potential biodiesel feedstock along with other non-edible plant seed oils.

4. Conclusions

This paper investigates the use of soursop seed oil for biodiesel production. In the current study, the reaction conditions of acid-catalyzed esterification and alkali-catalyzed transesterification were optimized to maximize biodiesel yield. Under optimized conditions, 97.02% biodiesel yield was obtained. The properties of the soursop biodiesel were determined and were found to meet the ASTM D6751 and EN 14214. The results of this study suggest that soursop seed oil is a potential biodiesel feedstock and the soursop biodiesel can serve as an alternative for petrodiesel.

Author Contributions: Conceptualization, C.-H.S. and H.C.N.; methodology, C.-H.S. and H.C.N.; validation, H.-J.Y. and U.K.P.; formal analysis, U.K.P and M.L.N.; investigation, U.K.P. and M.L.N; resources, C.-H.S.; writing—original draft preparation, C.-H.S. and H.C.N.; writing—review and editing, C.-H.S.; supervision, C.-H.S.; project administration, C.-H.S.; funding acquisition, C.-H.S.

Funding: This research was funded by Ministry of Science and Technology (MOST) of Taiwan, grant number 106-2221-E-131-028.

Conflicts of Interest: The authors declare no conflicts of interest.

References

1. Mosarof, M.; Kalam, M.; Masjuki, H.; Alabdulkarem, A.; Ashraful, A.; Arslan, A.; Rashedul, H.; Monirul, I. Optimization of performance, emission, friction and wear characteristics of palm and *Calophyllum inophyllum* biodiesel blends. *Energy Convers. Manag.* **2016**, *118*, 119–134. [CrossRef]

2. Damanik, N.; Ong, H.C.; Tong, C.W.; Mahlia, T.M.I.; Silitonga, A.S. A review on the engine performance and exhaust emission characteristics of diesel engines fueled with biodiesel blends. *Environ. Sci. Pollut. Res.* **2018**, *25*, 15307–15325. [CrossRef] [PubMed]

3. Anwar, M.; Rasul, M.G.; Ashwath, N. Production optimization and quality assessment of papaya (*Carica papaya*) biodiesel with response surface methodology. *Energy Convers. Manag.* **2018**, *156*, 103–112. [CrossRef]

4. Knothe, G. "Designer" biodiesel: Optimizing fatty ester composition to improve fuel properties. *Energy Fuels* **2008**, *22*, 1358–1364. [CrossRef]

5. Abdul Malik, M.S.; Shaiful, A.I.M.; Mohd Ismail, M.S.; Mohd Jaafar, M.N.; Mohamad Sahar, A. Combustion and emission characteristics of coconut-based biodiesel in a liquid fuel burner. *Energies* **2017**, *10*, 458. [CrossRef]

6. Leung, D.Y.; Wu, X.; Leung, M. A review on biodiesel production using catalyzed transesterification. *Appl. Energy* **2010**, *87*, 1083–1095. [CrossRef]

7. Bhuyan, M.S.U.S.; Alam, A.H.M.A.; Chu, Y.; Seo, Y.C. Biodiesel production potential from littered edible oil fraction using directly synthesized S-TiO$_2$/MCM-41 catalyst in esterification process via non-catalytic subcritical hydrolysis. *Energies* **2017**, *10*, 1290. [CrossRef]

8. Mardhiah, H.H.; Ong, H.C.; Masjuki, H.; Lim, S.; Lee, H. A review on latest developments and future prospects of heterogeneous catalyst in biodiesel production from non-edible oils. *Renew. Sustain. Energy Rev.* **2017**, *67*, 1225–1236. [CrossRef]

9. Martindale, W.; Trewavas, A. Fuelling the 9 billion. *Nat. Biotechnol.* **2008**, *26*, 1068–1070. [CrossRef] [PubMed]

10. Nguyen, H.C.; Liang, S.H.; Doan, T.T.; Su, C.H.; Yang, P.C. Lipase-catalyzed synthesis of biodiesel from black soldier fly (*Hermetica illucens*): Optimization by using response surface methodology. *Energy Convers. Manag.* **2017**, *145*, 335–342. [CrossRef]

11. Tuntiwiwattanapun, N.; Monono, E.; Wiesenborn, D.; Tongcumpou, C. In-situ transesterification process for biodiesel production using spent coffee grounds from the instant coffee industry. *Ind. Crops Prod.* **2017**, *102*, 23–31. [CrossRef]

12. El Shimi, H.I.; Moustafa, S.S. Biodiesel production from microalgae grown on domestic wastewater: Feasibility and Egyptian case study. *Renew. Sustain. Energy Rev.* **2018**, *82*, 4238–4244. [CrossRef]

13. Chia, S.R.; Ong, H.C.; Chew, K.W.; Show, P.L.; Phang, S.M.; Ling, T.C.; Nagarajan, D.; Lee, D.J.; Chang, J.S. Sustainable approaches for algae utilisation in bioenergy production. *Renew. Energy* **2018**, *129*, 838–852. [CrossRef]

14. Ghorbani, A.; Rahimpour, M.R.; Ghasemi, Y.; Raeissi, S. The biodiesel of microalgae as a solution for diesel demand in Iran. *Energies* **2018**, *11*, 950. [CrossRef]

15. Kamel, D.A.; Farag, H.A.; Amin, N.K.; Zatout, A.A.; Ali, R.M. Smart utilization of jatropha (*Jatropha curcas* Linnaeus) seeds for biodiesel production: Optimization and mechanism. *Ind. Crops Prod.* **2018**, *111*, 407–413. [CrossRef]

16. Lin, J.J.; Chen, Y.W. Production of biodiesel by transesterification of Jatropha oil with microwave heating. *J. Taiwan Inst. Chem. Eng.* **2017**, *75*, 43–50. [CrossRef]

17. Milano, J.; Ong, H.C.; Masjuki, H.H.; Silitonga, A.S.; Chen, W.H.; Kusumo, F.; Dharma, S.; Sebayang, A.H. Optimization of biodiesel production by microwave irradiation-assisted transesterification for waste cooking oil-*Calophyllum inophyllum* oil via response surface methodology. *Energy Convers. Manag.* **2018**, *158*, 400–415. [CrossRef]

18. Hossain, M.N.; Siddik Bhuyan, M.S.U.; Alam, A.H.M.A.; Seo, Y.C. Biodiesel from hydrolyzed waste cooking oil using a S-ZrO$_2$/SBA-15 super acid catalyst under sub-critical conditions. *Energies* **2018**, *11*, 299. [CrossRef]

19. Poudel, J.; Karki, S.; Sanjel, N.; Shah, M.; Oh, S.C. Comparison of biodiesel obtained from virgin cooking oil and waste cooking oil using supercritical and catalytic transesterification. *Energies* **2017**, *10*, 546. [CrossRef]

20. Nguyen, H.C.; Liang, S.H.; Chen, S.S.; Su, C.H.; Lin, J.H.; Chien, C.C. Enzymatic production of biodiesel from insect fat using methyl acetate as an acyl acceptor: Optimization by using response surface methodology. *Energy Convers. Manag.* **2018**, *158*, 168–175. [CrossRef]

21. Zheng, L.; Hou, Y.; Li, W.; Yang, S.; Li, Q.; Yu, Z. Exploring the potential of grease from yellow mealworm beetle (*Tenebrio molitor*) as a novel biodiesel feedstock. *Appl. Energy* **2013**, *101*, 618–621. [CrossRef]

22. Surendra, K.; Olivier, R.; Tomberlin, J.K.; Jha, R.; Khanal, S.K. Bioconversion of organic wastes into biodiesel and animal feed via insect farming. *Renew. Energy* **2016**, *98*, 197–202. [CrossRef]

23. Barekati-Goudarzi, M.; Muley, P.D.; Clarens, A.; Nde, D.B.; Boldor, D. Continuous microwave-assisted in-situ transesterification of lipids in seeds of invasive Chinese tallow trees (*Triadica sebifera* L.): Kinetic and thermodynamic studies. *Biomass Bioenergy* **2017**, *107*, 353–360. [CrossRef]

24. García-Martínez, N.; Andreo-Martínez, P.; Quesada-Medina, J.; de los Ríos, A.P.; Chica, A.; Beneito-Ruiz, R.; Carratalá-Abril, J. Optimization of non-catalytic transesterification of tobacco (*Nicotiana tabacum*) seed oil using supercritical methanol to biodiesel production. *Energy Convers. Manag.* **2017**, *131*, 99–108. [CrossRef]

25. Amini, Z.; Ong, H.C.; Harrison, M.D.; Kusumo, F.; Mazaheri, H.; Ilham, Z. Biodiesel production by lipase-catalyzed transesterification of *Ocimum basilicum* L. (sweet basil) seed oil. *Energy Convers. Manag.* **2017**, *132*, 82–90. [CrossRef]

26. Hasni, K.; Ilham, Z.; Dharma, S.; Varman, M. Optimization of biodiesel production from *Brucea javanica* seeds oil as novel non-edible feedstock using response surface methodology. *Energy Convers. Manag.* **2017**, *149*, 392–400. [CrossRef]

27. Tuntiwiwattanapun, N.; Tongcumpou, C. Sequential extraction and reactive extraction processing of spent coffee grounds: An alternative approach for pretreatment of biodiesel feedstocks and biodiesel production. *Ind. Crops Prod.* **2018**, *117*, 359–365. [CrossRef]

28. Sakuragi, K.; Li, P.; Otaka, M.; Makino, H. Recovery of bio-oil from industrial food waste by liquefied dimethyl ether for biodiesel production. *Energies* **2016**, *9*, 106. [CrossRef]

29. Chen, L.; Liu, T.; Zhang, W.; Chen, X.; Wang, J. Biodiesel production from algae oil high in free fatty acids by two-step catalytic conversion. *Bioresour. Technol.* **2012**, *111*, 208–214. [CrossRef] [PubMed]

30. Silva, L.N.; Cardoso, C.C.; Pasa, V.M. Production of cold-flow quality biodiesel from high-acidity on-edible oils—Esterification and transesterification of Macauba (*Acrocomia aculeata*) oil using various alcohols. *BioEnergy Res.* **2016**, *9*, 864–873. [CrossRef]

31. Berchmans, H.J.; Hirata, S. Biodiesel production from crude *Jatropha curcas* L. seed oil with a high content of free fatty acids. *Bioresour. Technol.* **2008**, *99*, 1716–1721. [CrossRef] [PubMed]

32. Suresh, R.; Antony, J.V.; Vengalil, R.; Kochimoolayil, G.E.; Joseph, R. Esterification of free fatty acids in non-edible oils using partially sulfonated polystyrene for biodiesel feedstock. *Ind. Crops Prod.* **2017**, *95*, 66–74. [CrossRef]

33. Nguyen, H.C.; Huong, D.T.M.; Juan, H.Y.; Su, C.H.; Chien, C.C. Liquid lipase-catalyzed esterification of oleic acid with methanol for biodiesel production in the presence of superabsorbent polymer: Optimization by using response surface methodology. *Energies* **2018**, *11*, 1085. [CrossRef]

34. Çaylı, G.; Küsefoğlu, S. Increased yields in biodiesel production from used cooking oils by a two step process: Comparison with one step process by using TGA. *Fuel Process. Technol.* **2008**, *89*, 118–122. [CrossRef]

35. Hayyan, A.; Alam, M.Z.; Mirghani, M.E.; Kabbashi, N.A.; Hakimi, N.I.N.M.; Siran, Y.M.; Tahiruddin, S. Reduction of high content of free fatty acid in sludge palm oil via acid catalyst for biodiesel production. *Fuel Process. Technol.* **2011**, *92*, 920–924. [CrossRef]

36. Coria-Téllez, A.V.; Montalvo-Gónzalez, E.; Yahia, E.M.; Obledo-Vázquez, E.N. *Annona muricata*: A comprehensive review on its traditional medicinal uses, phytochemicals, pharmacological activities, mechanisms of action and toxicity. *Arab. J. Chem.* **2016**, *11*, 662–691. [CrossRef]

37. Schroeder, P.; do Nascimento, B.P.; Romeiro, G.A.; Figueiredo, M.K.K.; da Cunha Veloso, M.C. Chemical and physical analysis of the liquid fractions from soursop seed cake obtained using slow pyrolysis conditions. *J. Anal. Appl. Pyrolysis* **2017**, *124*, 161–174. [CrossRef]

38. Fasakin, A.; Fehintola, E.; Obijole, O.; Oseni, O. Compositional analyses of the seed of sour sop, *Annona muricata* L., as a potential animal feed supplement. *Sci. Res. Essays* **2008**, *3*, 521–523.

39. Moghadamtousi, S.Z.; Fadaeinasab, M.; Nikzad, S.; Mohan, G.; Ali, H.M.; Kadir, H.A. *Annona muricata* (Annonaceae): A review of its traditional uses, isolated acetogenins and biological activities. *Int. J. Mol. Sci.* **2015**, *16*, 15625–15658. [CrossRef] [PubMed]

40. de Lima, M.C.; Alves, R. Soursop (*Annona muricata* L.). In *Postharvest Biology and Technology of Tropical and Subtropical Fruits: Mangosteen to White Sapote*; Yahia, E., Ed.; Woodhead Publishing: Cambridge, UK, 2011; pp. 363–392.

41. Gajalakshmi, S.; Vijayalakshmi, S.; Devi Rajeswari, V. Phytochemical and pharmacological properties of *Annona muricata*: A review. *Int. J. Pharm. Pharm. Sci.* **2012**, *4*, 3–6.

42. Awan, J.; Kar, A.; Udoudoh, P. Preliminary studies on the seeds of *Annona muricata* Linn. *Plant Foods Hum. Nutr.* **1980**, *30*, 163–168. [CrossRef]

43. Badrie, N.; Schauss, A.G. Soursop (*Annona muricata* L.): Composition, nutritional value, medicinal uses, and toxicology. In *Bioactive Foods in Promoting Health*; Watson, R.R., Preedy, V.R., Eds.; Academic Press: Oxford, UK, 2010; pp. 621–643.

44. Zafra-Polo, M.C.; González, M.C.; Estornell, E.; Sahpaz, S.; Cortes, D. Acetogenins from Annonaceae, inhibitors of mitochondrial complex I. *Phytochemistry* **1996**, *42*, 253–271. [CrossRef]

45. Elagbar, Z.A.; Naik, R.R.; Shakya, A.K.; Bardaweel, S.K. Fatty acids analysis, antioxidant and biological activity of fixed oil of *Annona muricata* L. seeds. *J. Chem.* **2016**, *6*. [CrossRef]

46. Phoo, Z.W.M.M.; Ilham, Z.; Goembira, F.; Razon, L.; Saka, S. Physico-chemical properties of biodiesel from various feedstocks. In *Zero-Carbon Energy Kyoto 2012*; Yao, T., Ed.; Springer: Berlin, Germany, 2013; pp. 113–121.

47. Schroeder, P.; dos Santos Barreto, M.; Romeiro, G.A.; Figueiredo, M.K.K. Development of energetic alternatives to use of waste of *Annona muricata* L. *Waste Biomass Valorization* **2018**, *9*, 1459–1467. [CrossRef]

48. Su, C.H.; Fu, C.C.; Gomes, J.; Chu, I.; Wu, W.T. A heterogeneous acid-catalyzed process for biodiesel production from enzyme hydrolyzed fatty acids. *AIChE J.* **2008**, *54*, 327–336. [CrossRef]

49. Vicente, G.; Martınez, M.; Aracil, J. Integrated biodiesel production: A comparison of different homogeneous catalysts systems. *Bioresour. Technol.* **2004**, *92*, 297–305. [CrossRef] [PubMed]

50. Su, C.H. Recoverable and reusable hydrochloric acid used as a homogeneous catalyst for biodiesel production. *Appl. Energy* **2013**, *104*, 503–509. [CrossRef]

51. Li, Q.; Zheng, L.; Cai, H.; Garza, E.; Yu, Z.; Zhou, S. From organic waste to biodiesel: Black soldier fly, *Hermetia illucens*, makes it feasible. *Fuel* **2011**, *90*, 1545–1548. [CrossRef]

52. Annual Book of ASTM Standards. Available online: https://www.astm.org/BOOKSTORE/BOS/index.html (accessed on 26 September 2018).

53. Ramadhas, A.; Jayaraj, S.; Muraleedharan, C. Characterization and effect of using rubber seed oil as fuel in the compression ignition engines. *Renew. Energy* **2005**, *30*, 795–803. [CrossRef]

54. Emil, A.; Yaakob, Z.; Kumar, M.S.; Jahim, J.M.; Salimon, J. Comparative evaluation of physicochemical properties of Jatropha seed oil from Malaysia, Indonesia and Thailand. *J. Am. Oil Chem. Soc.* **2010**, *87*, 689–695. [CrossRef]

55. Ramadhas, A.S.; Jayaraj, S.; Muraleedharan, C. Biodiesel production from high FFA rubber seed oil. *Fuel* **2005**, *84*, 335–340. [CrossRef]

56. Nguyen, H.C.; Liang, S.H.; Li, S.Y.; Su, C.H.; Chien, C.C.; Chen, Y.J.; Huong, D.T.M. Direct transesterification of black soldier fly larvae (*Hermetia illucens*) for biodiesel production. *J. Taiwan Inst. Chem. Eng.* **2018**, *85*, 165–169. [CrossRef]

57. Kafuku, G.; Mbarawa, M. Biodiesel production from *Croton megalocarpus* oil and its process optimization. *Fuel* **2010**, *89*, 2556–2560. [CrossRef]

58. Meng, X.; Chen, G.; Wang, Y. Biodiesel production from waste cooking oil via alkali catalyst and its engine test. *Fuel Process. Technol.* **2008**, *89*, 851–857. [CrossRef]

59. Yang, S.; Li, Q.; Gao, Y.; Zheng, L.; Liu, Z. Biodiesel production from swine manure via housefly larvae (*Musca domestica* L.). *Renew. Energy* **2014**, *66*, 222–227. [CrossRef]

60. Ramos, M.J.; Fernández, C.M.; Casas, A.; Rodríguez, L.; Pérez, Á. Influence of fatty acid composition of raw materials on biodiesel properties. *Bioresour. Technol.* **2009**, *100*, 261–268. [CrossRef] [PubMed]

61. Automotive Fuels-Fatty Acid Methyl Esters (FAME) for Diesel Engines-Requirements and Test Methods. Available online: http://agrifuelsqcs-i.com/attachments/1598/en14214.pdf (accessed on 25 September 2018).

62. Bala, V.S.S.; Thiruvengadaravi, K.V.; Kumar, P.S.; Premkumar, M.P.; Kumar, M.H.; Sivanesan, S. Removal of free fatty acids in *Pongamia pinnata* (Karanja) oil using divinylbenzene-styrene copolymer resins for biodiesel production. *Biomass Bioenergy* **2012**, *37*, 335–341. [CrossRef]

energies

MDPI

Article

Optimisation of Second-Generation Biodiesel Production from Australian Native Stone Fruit Oil Using Response Surface Method

Mohammad Anwar [1], Mohammad G. Rasul [1], Nanjappa Ashwath [2] and Md Mofijur Rahman [1,*]

[1] School of Engineering and Technology, Central Queensland University, North Rockhampton, Queensland 4702, Australia; m.anwar@cqu.edu.au (M.A.); m.rasul@cqu.edu.au (M.G.R.)

[2] School of Health, Medical and Applied Sciences, Central Queensland University, North Rockhampton, Queensland 4702, Australia; n.ashwath@cqu.edu.au

* Correspondence: m.rahman@cqu.edu.au or m.anwar@cqu.edu.au; Tel.: +617-4930-6371

Received: 13 August 2018; Accepted: 25 September 2018; Published: 26 September 2018

Abstract: In this study, the production process of second-generation biodiesel from Australian native stone fruit have been optimised using response surface methodology via an alkali catalysed transesterification process. This process optimisation was performed varying three factors, each at three different levels. Methanol: oil molar ratio, catalyst concentration (wt %) and reaction temperature were the input factors in the optimisation process, while biodiesel yield was the key model output. Both 3D surface plots and 2D contour plots were developed using MINITAB 18 to predict optimum biodiesel yield. Gas chromatography (GC) and Fourier transform infrared (FTIR) analysis of the resulting biodiesel was also done for biodiesel characterisation. To predict biodiesel yield a quadratic model was created and it showed an R^2 of 0.98 indicating the satisfactory performance of the model. Maximum biodiesel yield of 95.8% was obtained at a methanol: oil molar ratio of 6:1, KOH catalyst concentration of 0.5 wt % and a reaction temperature of 55 °C. At these reaction conditions, the predicted biodiesel yield was 95.9%. These results demonstrate reliable prediction of the transesterification process by Response surface methodology (RSM). The results also show that the properties of the synthesised Australian native stone fruit biodiesel satisfactorily meet the ASTM D6751 and EN14214 standards. In addition, the fuel properties of Australian native stone fruit biodiesel were found to be similar to those of conventional diesel fuel. Thus, it can be said that Australian native stone fruit seed oil could be used as a potential second-generation biodiesel source as well as an alternative fuel in diesel engines.

Keywords: response surface methodology; RSM; second-generation biodiesel; stone fruit; optimisation; biodiesel testing; transesterification

1. Introduction

Global climate change and the resulting desire for renewable energy sources has generated the interests for using biofuel in the transport sector [1]. Due to the higher production of biofuel in recent years, it currently contributes 1.5% global transportation fuel. It has been reported that nearly 40% of the total worldwide biofuel supply comes from emerging and developing countries. However, the expansion of biofuel production around the world has raised major concerns, for example the existence of several first-generation biofuels. Biofuels that are produced from edible sources are termed first-generation biofuels [2], and these have been increasingly questioned over some concerns such as food-fuel controversy, environmental pollution, and climate change. The increasing concern regarding the sustainability of several first-generation biofuels has led to investigations into the potential of producing biodiesel from non-food crops which are termed as second-generation biodiesel.

The potential benefits offered by the second-generation biodiesels are that they consume waste oils, make use of abandoned land and do not compete with food crops [3].

In addition, second-generation biofuel from locally available sources can play a great role in economic development of rural and emerging region of a country [4]. Despite significant socio-economic advantages and continuous support from government and non-government organisations, the market for biofuel production around the world has not expanded very much over the last few years. Many countries have announced second-generation biofuel support policies, e.g., the United States has adopted the policies to produce 60 billion litres by 2022 and the European Union set their target to use 20% renewable energy in the transport sector by 2020 [5]. Both the US and EU policies could play an important role for the worldwide biofuel development because of their market size and considerable amount of biofuel imports. In addition, the Australian Federal Government and its State Governments have developed relevant policies to promote a sustainable biofuel industry to ensure Australian's long-term energy security. Leading oil companies such as Caltex, Shell, BP, and Exxon Mobile are also coming forward in second-generation biofuel research with more investment. A few plants with research activities are going to be established soon in emerging countries.

Biodiesel is one of the biofuels and has proved its potential as an alternative fuel worldwide. Biodiesel is biodegradable, renewable and environmentally friendly [6,7]. The feedstock selection of biodiesel is very important as 75% of the total cost of biodiesel production is associated with obtaining feedstocks alone. A high oil yield of any feedstock ensures a commercial scale biodiesel production at reasonable prices [8]. Feedstock security of supply, feedstock cost of supply and feedstock storage are the important factors to consider when choosing the biodiesel feedstock [9]. In addition, biodiesel should be produced from the feedstock that is consistently available, economically viable and locally available. Currently, the major feedstocks for biodiesel production in Australia are waste cooking oil, animal tallow, macadamia, beauty leaf, canola and mustard oils [10]. However, stone fruit such as *Prunus armeniaca* L. is widely cultivated in Australia, and it yields 22–38% of kernels which contain up to 54.2% oil. Australia produces about 100,000 tons of summer stone-fruit from October to April each year and, in 2008, about 16,917 tons of *Prunus armeniaca* L. fruit were produced from all six mainland states in Australia. This could therefore be a potential second-generation biodiesel feedstock in Australia. The main aim of this research was to investigate and optimise the production process of second-generation biofuel from this Australian native feedstock as the research on it is still far behind that into other feedstocks. This biodiesel could overcome the limitations associated with first-generation biodiesels and be used as an alternative to conventional fossil fuels.

2. Literature Review

Stone fruit is similar to a small peach, generally 1.5–2.5 cm in diameter, with its colour varying from yellow to orange or red. Its single seed is enclosed in a hard stony shell. During fruit processing, the seeds are discarded due to the presence of hydro-cyanic acid [11]. To utilise this waste product, it is important to optimise the procedures involved in oil extraction and its conversion into biodiesel.

Many researchers [12–17] have optimised the production of biodiesel from different first- and second-generation feedstocks using various methods. For example, Saydu et al. [13] optimised the process of biodiesel production from hazelnut and sunflower oil using single step transesterification with methanol, and employing potassium hydroxide as a catalyst. Razack and Duraiarasan [12] optimised the waste cooking oil biodiesel production process using response surface methodology using encapsulated mixed enzyme as a catalyst. Dharma et al. [17] optimised the biodiesel production process of *Jatropha curcas* and *Ceiba pentandra* oil using response surface methodology as also did Ong et al. [16] for the *Calophyllum inophyllum* biodiesel production process.

A few studies have been done on the optimisation of the stone fruit oil (SFO) biodiesel production process but none of them used any statistical modelling. For instance, Gumus et al. [18] used alkali transesterification with methanol and potassium hydroxide catalyst for producing SFO methyl ester. Abdelrahman [19] produced SFO biodiesel via alkali transesterification with 0.75% potassium hydroxide catalyst and at a methanol: oil molar ratio of 6:1. Faizan et al. [20] showed that the wild *Prunus Armeniaca* L. oil can be transesterified by a single step process via the use of sodium hydroxide catalyst at a methanol: oil molar ratio of 6:1, and reported a biodiesel yield of 93%. Ashok et al. [21] performed single step alkali transesterification using 1% potassium hydroxide as a catalyst at 55 °C and 60 min reaction time with a constant stirring at 400 rpm and obtained a biodiesel yield of 96.5%. Thus, many process parameters, including reaction temperature, catalyst type and catalyst concentration, type of alcohol used, the oil to methanol molar ratio, reaction time and agitation speed have been found to influence the optimum transesterification process [22–26]. From the above literature, it is obvious that no/limited investigation has been done on the optimisation of second-generation biodiesel production process from *Prunus Armeniaca* L. oil using any statistical modelling. Thus, this study has explored optimisation of the biodiesel production processes from Australian native stone fruit oil using response surface methodology.

3. Materials and Methods

3.1. Materials

Stone fruit (*Prunus armeniaca* L. species) seed oil was purchased from a local producer named Chromium Group Pty Ltd. of Eumundi, Queensland, Australia. The chemicals used in this study were methanol (99.9% purity), potassium hydroxide (KOH pellets, 99% purity) and sodium hydroxide (NaOH pellets, 99% purity). All were of analytical reagent grade (AR) and were procured from the School of Engineering and Technology, Central Queensland University, Rockhampton, Australia. A three-neck laboratory reactor (1 L) along with a reflux condenser and a thermocouple placed on a magnetic heater/stirrer were used in the SFO biodiesel conversion experiments. In this experiment, methanol, KOH, NaOH and Whatman 541 grade filter paper (pore size 22 μm) were used.

3.2. Oil Extraction

The stone fruits were collected and the fleshy parts were separated manually for drying purposes. The seeds were separated for kernel collection and oil extraction. For easy breaking of the hard shell, seeds were softened by immersing in water for 10–20 min. The broken shell can be used as fertilizer or firewood after the oil extraction process [27]. Kernels were separated from the broken shells and were crushed using a pestle and mortar and sieved through a 40 mesh or 0.8 mm sieve [19,28]. The ground kernel was placed in a Soxhlet apparatus and the oil was extracted using petroleum ether (40–60 °C) over 6–8 h until the extraction was completed [19]. After oil extraction, the petroleum ether was evaporated using a rotary evaporator at 25 °C [29]. The oil was placed in an oven at 60 °C for 60 min to remove the remaining solvent. The oil was filtered using Whatman 541 filter paper. After filtration, the SFO was kept in a sealed container for characterisation. The oil yield was calculated using Equation (1).

$$\text{Oil Yield (\%wt/wt dry kernels)} = \frac{\text{weight of oil extracted (g)}}{\text{weight of dry kernels used (g)}} \times 100 \tag{1}$$

3.3. Biodiesel Production

The acid value of raw SFO was determined as 1.65 mg KOH/g. After transesterification using KOH catalyst transesterification, the acid value of SFO biodiesel was found to be 0.25 mg KOH/g. This trial experiment suggested that only the single stage alkyl catalyst transesterification process was satisfactory for SFO biodiesel production. Thus, in each experiment, the experiment was performed by reacting a known quantity of SFO with methanol and the catalyst.

Initially, the SFO was poured into a three-neck laboratory reactor and heated to the desired temperature. The measured quantities (molar basis) of methanol and catalyst (KOH) were poured into a separate beaker and stirred vigorously using a magnetic stirrer at 50 °C at 600 rpm for 10 min to produce methoxide. This solution was slowly poured into the three-neck reactor containing SFO. The blend was agitated continuously at 600 rpm and the temperature and reaction time were varied as per the experimental design. At the end of the transesterification reaction, the blend was transferred into a separating funnel. Although the separation of glycerol and biodiesel occurred instantaneously, the funnel was left undisturbed for 24 h. Two separate liquid phases were formed, with the top layer being methyl ester (biodiesel) and the bottom layer of red viscous glycerol and impurities. The bottom layer was drained off while the top layer was collected and washed with warm (50 °C) distilled water. This moist biodiesel was then heated to 110 °C for 15 min to remove residual water that would have been retained by the biodiesel during the washing process. The Whatman® qualitative Grade 1 filter paper was used to filter the biodiesel and finally stored in an airtight container at room temperature until its characterisation. Biodiesel yield was calculated using Equation (2) and its composition was determined using a Gas Chromatogram [22]. The graphical illustration of the SFO production process is shown in Figure 1.

$$\text{SFO Biodiesel Yield} = \text{FAME percent from GC analysis} \times \frac{\text{weight of SFO biodiesel}}{\text{weight of Stone fruit oil}} \qquad (2)$$

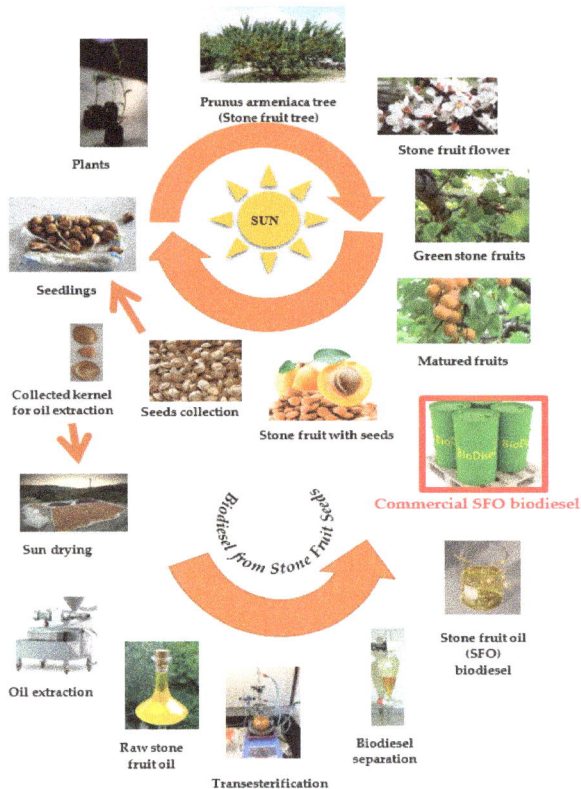

Figure 1. Graphical representation of producing biodiesel from SFO biodiesel.

3.4. Physicochemical Characterisation of SFO

This section discusses the international standards used in characterising the SFO biodiesel and its composition via Gas Chromatography (GC) and Fourier Transform Infrared (FTIR) spectroscopy.

3.4.1. Fuel Properties

The physicochemical properties and fatty acid compositions of crude SFO and SFO biodiesel were tested according to ASTM and EN standards. The properties studied were density at 15 °C (ASTM D1298), kinematic viscosity at 40 °C (ASTM D445), acid value (ASTM D664), calorific value (ASTM D240), flash point (ASTM D93) and oxidation stability (ASTM D2274). The fatty acid compositions were determined using a gas chromatograph according to EN 14103.

Fuel properties calculated based on the fatty acid composition of the SFO biodiesel were cetane number (CN), saponification value (SV), iodine value (IV), long-chain saturated factor (LCSF) and degree of unsaturation (DU). The numerical calculations were determined using the following equations [30]:

$$CN = 46.3 + \left(\frac{5458}{SV}\right) - (0.225 \times IV) \tag{3}$$

$$SV = \sum \frac{(560 \times A_i)}{MW_i} \tag{4}$$

$$IV = \sum \frac{(254 \times D \times A_i)}{MW_i} \tag{5}$$

$$LCSF = 0.1 \times (C16:0 \text{ wt } \%) + 0.5 \times (C18:0 \text{ wt } \%) + 1 \times (C20:0 \text{ wt } \%)$$
$$+ 1.5 \times (C22:0 \text{ wt } \%) + 2.0 \times (C24:0 \text{ wt } \%) \tag{6}$$

$$DU = \sum MUFA + (2 \times PUFA) \tag{7}$$

where D indicates the number of double bonds, A_i is the percentage of each fatty acid in the FAME, and MW_i is the molecular mass of each component. MUFA denotes monounsaturated fatty acid and PUFA refers to polyunsaturated fatty acid. The degree of unsaturation was calculated using both MUFA and PUFA concentrations. Fatty acids of C16.0, C18.0, C20.0, C22.0 and C24.0 stand for palmitic acid, stearic acid, arachidic acid, behenic acid and lignoceric acid, respectively, and were used for measuring the long chain saturated factor.

3.4.2. Gas Chromatography

According to EN14103, a gas chromatograph (GC) (Thermo Scientific Trace 1310 GC) was used to determine the fatty acid composition of the SFO. 25 mg of the SFO biodiesel was dissolved in high purity hexane (10 mL). Then, this solution was poured to 2 mL auto-sampler vials. The equipment for the GC test included Thermo Scientific Trace 1310 GC with a split/split less (SSL) injector, flame ionisation detector, and TriPlus auto-sampler. At 240 °C, a 1 µL sample was injected in split mode (40:1) by maintaining a constant helium flow of 1.2 mL/min. The conditions for separating FAMEs were: using a BPX-70 column (60 m × 250 µm × 0.25 µm film) with a temperature program: 110 °C (4 min); 10 °C/min, 150 °C; 3.9 °C/min, 230 °C (5 min). In this study, SFO biodiesel individual components were identified by retention time compared to a standard FAME mixture that had certified concentrations, namely Supelco CRM18920 (FAME C8-C22). Chromeleon 7.2 software was used for data acquisition and processing.

3.4.3. Fourier Transform Infrared (FTIR) Spectroscopy

The various functional groups present in the crude oil and the biodiesel sample were determined with Fourier transform infrared (FTIR) spectroscopy. The Spectrum 100 FTIR spectrometer with a universal Attenuated total reflectance (ATR) sampling accessory (Perkin Elmer, Melbourne, Australia) was used to record ATR infrared spectra. SFO biodiesel samples were placed directly on the ATR window at approximately 40% transmission to record the spectra with four scans, 4000–650 cm^{-1}. After ATR correction, Spectrum 6.2.0 software was used to acquire data and processing.

3.5. Design of Experiments

Box-Behnken is one of the most commonly used responses surface methodology designs. This design was used for designing and statistical analysis of this experiment. The Box-Behnken design matrix was utilised to find the optimum conditions for maximum biodiesel yield production. The experimental optimisation was achieved via analysis of variance (ANOVA) using Minitab 18 software. The effects of process factors such as methanol: oil molar ratio, KOH catalyst concentration, and reaction temperature were tested. Using these three factors at three levels required a total of 15 runs for identifying the optimum conditions for transesterification. The coded symbols, ranges, and levels of the investigated factors are listed in Table 1. The design matrix for the three factors was varied at three levels, namely −1, 0 and +1. The range levels of the factors investigated were chosen by considering the initial tests carried out on the effect of individual factors on biodiesel yield as well as the operating limits of the biodiesel production process conditions as evidenced from the literature.

Table 1. Experimental range and levels coded for independent factors.

Factors/Variables	Unit	Symbol Coded	Range and Levels		
			−1	0	+1
Methanol: Oil ratio	mol/mol	M	4:1	5:1	6:1
KOH catalyst concentration	wt %	C	0.5	1.00	1.5
Temperature	°C	T	45	55	65

Methanol: oil molar ratio ranged from 4:1 to 6:1, catalyst concentrations were 0.5–1.5% by weight of oil and the reaction temperature was varied from 45 °C to 65 °C (boiling point of methanol). Once the experiments were completed, the response factor (biodiesel yield) was applied in a full quadratic model to correlate the response factor to the independent factors. The general form of the full quadratic model is shown in Equation (8).

$$Y = P_0 + P_1Q_1 + P_2Q_2 + P_3Q_3 + P_{1,2}Q_1Q_2 + P_{1,3}Q_1Q_3 + P_{2,3}Q_2Q_3 + P_{1,1}Q_1^2 + P_{2,2}Q_2^2 + P_{3,3}Q_3^2 \quad (8)$$

where Y is the response factor (biodiesel yield, %); P_0 is a constant; P_1, P_2, and P_3 are regression coefficients; $P_{1,1}$, $P_{1,2}$, $P_{1,3}$, $P_{2,2}$, $P_{2,3}$, and $P_{3,3}$ are quadratic coefficient; and Q_1, Q_2, and Q_3 are independent variables.

4. Results and Discussion

This section includes the results of the characterisation of both crude SFO and SFO biodiesel, fatty acid compositions of SFO biodiesel, optimisation of reaction conditions by response surface methodology and response surface plots for SFO biodiesel production.

4.1. Characterisation of Crude SFO

The properties of crude stone fruit seed oil used in this study were evaluated prior to the optimisation process. Physicochemical properties are the most important features to check the quality of any crude oil. The SFO was characterised by viscosity, density, specific gravity, acid value, calorific value, saponification number and iodine value. The properties of SFO from this study along with those from other studies and those of petro diesel were compared and are presented in Table 2. The density of the oil was found to be 910 kg/m^3 which matches with that reported in the literature. Again, the acid value of SFO was determined to be 1.65 mg KOH/g, indicating the presence of low levels of free fatty acids in the oil. The kinematic viscosity of the oil was found to be 34.54 m^2/s and the calorific value was 38.45 MJ/kg, which is within the values found in the literature [18]. Based on above results, it is clear that Australian native SFO oil have similar fuel properties including fatty acid, calorific value and viscosity with the data of other researchers, thus it is expected that Australian native SFO may serve as a good feedstock for biodiesel production.

Table 2. Physical and chemical properties of SFO.

Properties	Units	SFO This Study	SFO [18]	SFO [21]	SFO [11]	Petro Diesel
Kinematic Viscosity @ 40 °C	m^2/s	34.54	34.82	20.53	26.22	3.23
Density	kg/m^3	910	920	913	916.6	827.2
Specific Gravity @ 15 °C	g/cm^3	0.91	0.91	0.91	0.91	0.83
Acid value	mg KOH/g	1.65		2.60	0.68	0.05
Calorific value	MJ/kg	38.4	39.6	31.5		45.3
Saponification number	mg KOH/g	173		188	187	
Iodine value	mgI$_2$/100 g	103		90	101	

4.2. Properties and Qualities of SFO Biodiesel

The physical and chemical properties of SFO biodiesel from this study, along with results from other researchers' work on SFO biodiesel, are compared with other non-edible biodiesels and petrodiesel in Table 3. It was found that all the properties and qualities of the SFO biodiesel fulfilled the international standards (USA ASTM D6751 and European Union EN14214). Many researchers [31–34] showed that densities of biodiesels do not vary considerably, as the density of methanol and oil are close to the density of produced biodiesel, which usually varies between 850 and 900 kg/m^3. The density of SFO and papaya seed oil (PSO) was found to be 855 and 840 kg/m^3 respectively, whereas that of petrodiesel was found to be 827.2 kg/m^3. Densities of other SFO biodiesels also matched with the international standards. Karanja biodiesel has a density of 931 kg/m^3, which is outside the ASTM and EN standards specification, thus limiting its efficiency of fuel atomisation in airless combustion systems [35]. However, other biodiesels (Table 3) have slightly higher densities than petrodiesel fuel, but they are within the range of the international standards. The viscosity of SFO biodiesel was determined to be 4.26 mm^2/s and other biodiesels (except Karanja) ranged from 1.9–6.0 mm^2/s and also fulfil the requirements of the standards. The viscosity of SFO biodiesels of other studies was found to be within the range as well. The acid values of biodiesels (except Neem) were also in line with requirements of ASTM and EN biodiesel standards which are less than 0.5 mg KOH/g. Higher acid values can cause corrosion of IC engines and internal metal parts. Cetane number is an important fuel property for diesel engines. A higher speed diesel engine works more efficiently with a fuel with a higher cetane number. A lower cetane number fuel has longer ignition delays providing more time to complete the combustion process. The cetane number of biodiesel increases with an increase in fatty acid proportion. Longer fatty acid chains and higher saturated fatty acid content will lead to a higher cetane number [36,37]. Moringa biodiesel has the highest cetane number of 67.1 compared with all other biodiesels in Table 3 and is within the international standard limit. All calorific values are lower than those of petrodiesel fuel (45.3 MJ/kg). The SFO biodiesel calorific value was found to be 39.64 MJ/kg in this study, thus meeting the minimum EN standard requirement of 35 MJ/kg. All other SFO biodiesels have similar calorific values as well. The flashpoint of this SFO biodiesel was found to be 105 °C, whereas the ASTM standard specifies 100–170 °C and petrodiesel fuel is 68.5 °C. This suggests that SFO biodiesel fuel is safer to handle and store than petroleum diesel. The iodine value for this SFO was recorded as 104.7 mgI$_2$ which met the range of the EN standard. The higher the iodine value, the more unsaturated double bonds are present in the methyl ester, leading to better biodiesel fuel quality. Biodiesel with higher oxidation stability is preferable as low oxidation stability can affect the quality of biodiesel [38]. This SFO biodiesel has an oxidation stability of 7.15 h, which falls above the minimum values of both the ASTM (minimum 3 h) and EN standards (minimum 6 h).

Some biodiesels with poor oxidation stability such as Tobacco, Cottonseed, Jatropha and Moringa biodiesel can be easily remedied by adding antioxidants.

Table 3. Comparison of SFO biodiesel with other non-edible biodiesels.

Non-Edible Biodiesels	Density (kg/m³)	Viscosity at 40 °C, mm²/s	Acid Value, mg KOH/g	Cetane Number (CN)	Calorific Value, MJ/Kg	Flash Point, °C	Iodine Value (IV) mgI₂/100 g	Oxidation Stability (OS), h
SFO this study	855.0	4.26	0.25	50.45	39.64	105	104.70	7.15
SFO [18]	884.3	4.92			39.95	111		
SFO [21]	857.0	5.20	0.32	58.70	38.93	180	100.70	6.30
SFO [11]	879.4	4.21	0.08		39.12	170	100.66	
Petro diesel [39]	827.2	3.23	0.05	48.00	45.30	68.5	38.3	39.0
Tobacco [23,38]	888.5	4.23		51.60		165.4	136	0.80
PSO [39]	840.0	3.53	0.42	48.29	38.49	112	115.89	5.61
Jatropha [38]	879.5	4.80	0.40	51.60	39.23	135	104	2.30
Rapeseed [23]	882.0	4.43		54.40	37.00	170		7.60
Cottonseed [23]	875.0	4.07	0.16	54.13	40.43	150		1.83
Neem [23,38]	868.0	5.21	0.65		39.81	76		7.10
Karanja [38]	931.0	6.13	0.42	55.00	43.42	95		
Moringa [23]	883.0	5.00	0.18	67.1		160	74	2.3
ASTM D6751	880.0	1.9~6.0	maximum 0.5	minimum 47		93~170		minimum 3
EN14214	860~900	3.5~5.0	maximum 0.5	minimum 51	35	>120	maximum 120	minimum 6

4.3. The Fatty Acid Composition of SFO Biodiesel

The fatty acid composition of any biodiesel feedstock is an important fuel property. The fatty acid composition is highly dependent on the quality of the feedstock, its growth condition and the geographic location in which the plant has grown. The chromatogram of the SFO biodiesel produced in this study shows the existence of derivatives of C16:0 (palmitic acid), C18:0 (Stearic acid), C18:1 (oleic acid), C18:2 (linoleic acid), C18:3 (linolenic acid), and C22:1 (behenic acid) in Figure 2. GC chromatogram of SFO biodiesel ensured the formation of methyl ester.

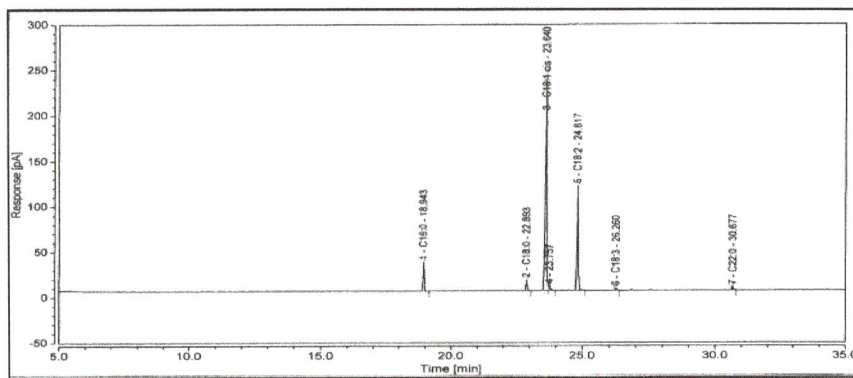

Figure 2. GC Chromatogram of SFO biodiesel.

The fatty acid compositional analysis of SFO biodiesel produced in this study is shown in Table 4, which indicates a high level (89.7%) of unsaturated fatty acids made up of both polyunsaturated and monounsaturated fatty acids. Saturated fatty acids such as palmitic acid, stearic acid, and behenic acid were found to be present at 5.85%, 2.51%, and 0.66%, respectively. Monounsaturated fatty acids (MUFA) such as oleic acid were found to be the dominant fatty acids (63.84%), whereas polyunsaturated fatty acids (PUFA) such as linoleic acid and linolenic acid were 25.34% and 0.51%, respectively. Pedro et al. [40] indicated that the degree of unsaturation of biodiesel did not significantly affect the engine performance and the start of injection, but it had a significant influence on combustion characteristics and emissions. The degree of unsaturation was recorded as 115.54% in this study.

Altun [41] indicated that degree of unsaturation and cetane number of biodiesel could highly influence NO_x formation in biodiesel-fuelled diesel engines. Generally, higher degrees of saturation relate to higher cetane numbers of biodiesel. Unsaturated biodiesel produces higher NO_x and lowers HC emissions than saturated biodiesels [40]. Furthermore, higher degrees of unsaturation in crude oil results in the production of less viscous biodiesel.

Table 4. The fatty acid composition of SFO.

Fatty Acids.	Formula	Molecular Weight	Structure	wt %
Palmitic	$C_{16}H_{32}O_2$	256	16:0	5.85
Stearic	$C_{18}H_{36}O_2$	284	18:0	2.51
Oleic	$C_{18}H_{34}O_2$	282	18:1	63.8
Linoleic	$C_{18}H_{32}O_2$	280	18:2	25.3
Linolenic	$C_{18}H_{30}O_2$	278	18:3	0.51
Behenic	$C_{22}H_{44}O_2$	340	22:0	0.66
Others				1.29
Total Saturated Fatty Acids (SFA)				9.02
Total Monounsaturated Fatty Acids (MUFA)				63.84
Total Polyunsaturated Fatty Acids (PUFA)				25.85
The degree of Unsaturation (DU)				115.5
Long Chain Saturated Factor (LCSF)				2.83

The distribution of the main fatty acids of some non-edible biodiesel feedstocks (including the SFO in this study and those from other research works on SFO) are shown in Table 5. Gas chromatography (GC) analysis of those biodiesels shows that the most abundant fatty acids are oleic acid, linoleic acid, and palmitic acid. The SFO biodiesel produced in this study shows similar results compared with those of previous research [18,21,28,42,43] on SFO biodiesel fuel. The presence of higher MUFA and PUFA can contribute to lower oxidative stability. As mentioned earlier, low oxidative stability could affect the quality of produced biodiesel. This low oxidation has a negative impact on both kinematic viscosity and acid value [44]. However, the presence of high contents of MUFA and PUFA ensures biodiesel with good fuel flow properties (especially in cold prone countries) compared with saturated fatty acids (SFA).

Table 5. Variations in the main fatty acid compositions of selected biodiesel feedstocks.

Non-Edible Biodiesels	Fatty Acids (% *w/w*)						Ref.
	C16:0	C18:0	C18:1	C18:2	C18:3	Others	
SFO this study	5.85	2.51	63.84	25.34	0.51	1.95	
SFO	5.62	1.27	67.31	24.68	0.08	1.04	[18]
SFO	4.20	2.32	71.00	20.15	1.20	1.13	[21]
SFO	3.87	0.92	67.21	27.12	0.11	0.77	[28]
SFO	3.79	1.01	65.23	28.92	0.14	0.91	[42]
SFO			66.20	28.20		5.60	[43]
PSO	6.07	3.13	47.73	37.25	1.78	4.04	[39]
Tobacco	8.90	3.50	14.10	70.10	1.00	2.40	[45]
Jatropha	16.20	8.20	38.40	36.80	0.40	0	[46]
Rapeseed	3.49	0.85	64.40	22.30	8.23	0.73	[47]
Cottonseed	28.70	0.90	13.00	57.40	0	0	[48]
Neem	12.00	10.00	61.00	16.00	0	1.00	[49]
Karanja	9.80	6.20	72.20	11.80	0	0	[50]
Moringa	6.50	6.00	72.20	1.0	0.65	13.65	[51]

The comparison of different non-edible biodiesels and SFO biodiesel fatty acid compositions are shown in Figure 3. It can be seen that MUFA-oleic acid (C18:1) dominates most of the feedstocks (except tobacco and cottonseed), followed by PUFA-linoleic acid (C18:2) and SFA-palmitic acid (C16:0). Both tobacco and cottonseed had high proportions of PUFA and it can be seen in Table 3 that their oxidation stabilities are very poor (0.8 h and 1.83 h, respectively). Oxidation occurs due to the presence of high proportions of unsaturated fatty acid chains and double bonds, i.e., PUFA in the parent molecule reacts with oxygen as soon as it is exposed to air [23,52]. Therefore, biodiesels with high linoleic acid (C18:2) and linolenic acids (C18:3) such as tobacco and cottonseed tend to have lower oxidation stabilities.

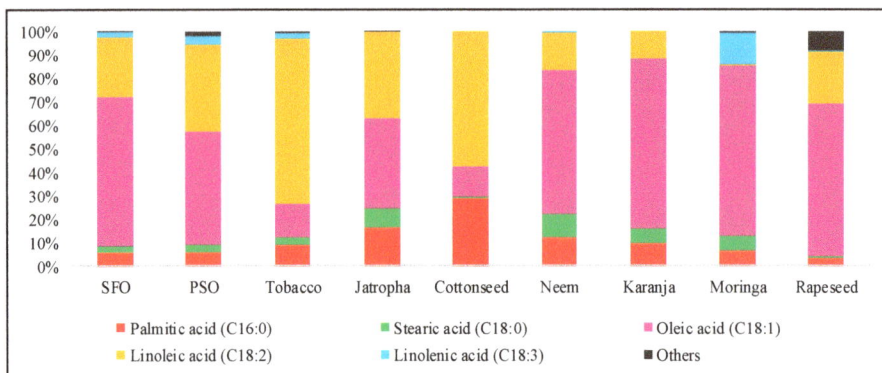

Figure 3. Comparison of the fatty acid composition of SFO and other non-edible vegetable oils.

Figure 4 shows the FTIR spectrum of the SFO methyl ester produced in this study. The advantages of using the FTIR method compared with GC is the ability to analyse whole samples, including precipitated fractions, without any further preparation [53].

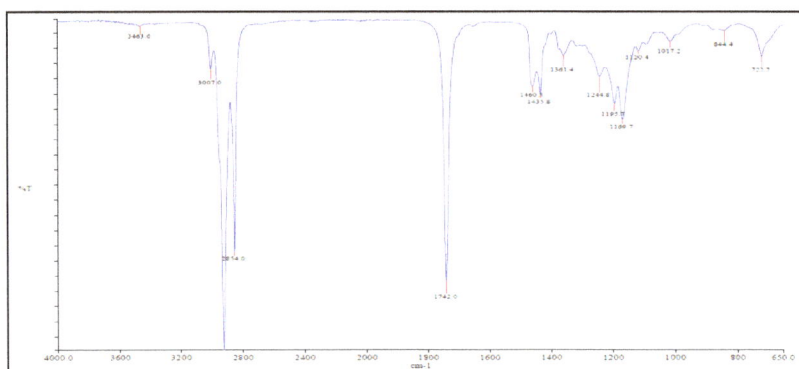

Figure 4. Fourier transforms infrared (FTIR) spectrum of SFO biodiesel.

The most important functional groups, wave number, band assignment and absorption intensity of absorption peaks detected in the FTIR spectrum of the SFO methyl ester are presented in Table 6. The peak at 1435.8 cm^{-1} corresponds to the asymmetric stretching of –CH$_3$ present in the SFO biodiesel sample, which is shown in Figure 4. The peak in the region of 2800–3000 cm^{-1} represents the CH$_3$ asymmetric stretching vibration. The peak of stretching of the carbonyl group (–C=O) is 1742 cm^{-1} located in the region of 1800–1700 cm^{-1} which is common for esters. The fingerprint region of 1500–900 cm^{-1} is the major spectrum from the SFO methyl ester which has a peak at 1244.8 cm^{-1}, corresponding to the bending vibration of –CH$_3$ [5]. These results reflect the conversion of triglycerides to methyl ester.

Table 6. Functional groups of SFO biodiesel detected in the FTIR spectrum.

Wavenumber (cm^{-1})	Group Attribution	Vibration Type	Functional Groups	Absorption Intensity
2924	=C–H	Asymmetric stretching vibration	Alkyl	Strong
2854	–CH$_2$	Symmetric stretching vibration	Aromatic	Strong
1742	–C=O	Stretching	Carbonyl	Strong
1460.5	–CH$_2$	Shear-type vibration	Alkanes	Weak
1244.8	–CH$_3$	Bending vibration	Alkanes	Weak
1195.7	C–O–C	Anti-symmetric stretching vibration	Ethers	Middling
1169.7	C–O–C	Anti-symmetric stretching vibration	Ethers	Middling
1120.4	C–O–C	Anti-symmetric stretching vibration	Ethers	Weak
1017.2	C–O–C	Anti-symmetric stretching vibration	Ethers	Weak
722.7	–CH$_2$	Plane rocking vibration	Aromatic	Weak

4.4. Optimisation of Reaction Conditions by RSM

The results of the Box–Behnken design model to optimise biodiesel production process parameters are shown in Table 7. In the transesterification experiments, the SFO biodiesel yield ranged from 75.2% to 95.8%. This design matrix also shows the experimental run order, experimental yields and predicted yields. These results show that the biodiesel yield varies with the production process. To avoid systematic errors, all run orders were randomised.

Table 7. Experimental matrix and results for Box-Behnken design model. The combination in italics shows the best combination for SFO biodiesel production.

Exp. Number	Run Order	M	C	T	Methanol: Oil (Molar Ratio)	KOH (wt %)	Temp (°C)	SFO Biodiesel Yield (%) Experimental	SFO Biodiesel Yield (%) Predicted
1	7	0	−1	−1	5	0.5	45	86.65	85.27
2	13	*1*	*−1*	*0*	*6*	*0.5*	*55*	*95.75*	*95.89*
3	14	0	0	0	5	1	55	91.65	92.98
4	12	−1	0	−1	4	1	45	80.34	80.85
5	15	1	0	−1	6	1	45	87.71	88.95
6	6	−1	−1	0	4	0.5	55	75.24	76.12
7	9	1	1	0	4	1.5	55	82.27	82.13
8	11	0	0	0	5	1	55	93.65	92.98
9	5	0	0	0	5	1	55	93.65	92.98
10	8	0	1	−1	5	1.5	45	84.58	84.21
11	4	1	1	0	6	1.5	55	82.11	81.23
12	3	1	0	1	6	1	65	89.33	88.82
13	2	−1	1	1	4	1	65	79.30	78.06
14	1	0	−1	1	5	0.5	65	86.71	87.08
15	10	0	1	1	5	1.5	65	78.10	79.48

The predicted biodiesel yield values were generated from a quadratic regression model as obtained from Minitab software version 18.0 through response surface methodology (RSM) statistical analysis of the experimental data. The Minitab 18 program was used to calculate the effects of each parameter and its interactions with other parameters. The response parameter (biodiesel yield %) was correlated with other parameters using a full quadratic regression model shown in Equation (9). The model represents SFO biodiesel predicted yield (Y) as a function of methanol: oil molar ratio (M), catalyst concentration (wt %) (C) and reaction temperature (T).

$$Y = 92.983 + 4.719M - 2.161C - 0.730T - 4.490M^2 - 4.650C^2 - 4.323T^2 - 5.168MC \\ + 0.665MT - 1.635CT \tag{9}$$

Figure 5a shows the line of a perfect fit with points corresponding to zero error as indicated by the comparison of experimental and predicted biodiesel yields. Points closer to the straight line indicate a good agreement between the experimental and predicted values. Figure 5b shows that there is an adequate correlation between RSM predicted values and experimental values, which verifies the acceptability of the model. The model represents a relatively good description of the experimental data regarding the SFO yield.

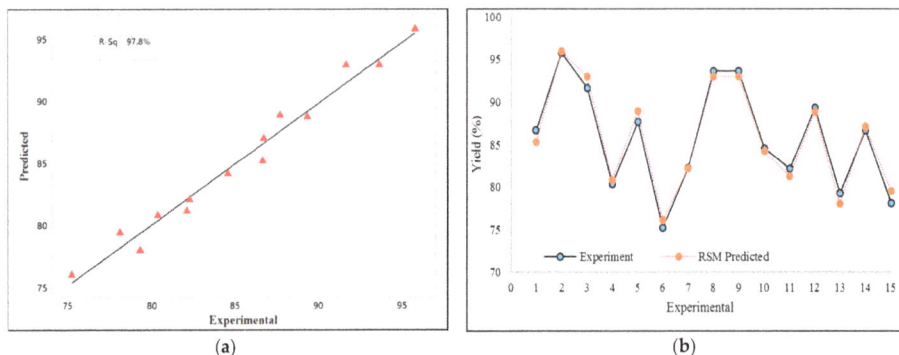

(a)   (b)

Figure 5. Biodiesel yield values (%): (**a**) Experimental versus (**b**) RSM predicted.

The linear, quadratic and interaction effects of the parameters were considered to investigate the impacts on the biodiesel yield. Table 8 displays the significance of those parameters in terms of the probability value (*p* value). It also summarises the resulting regression coefficients and computed T-values. In the model, positive coefficients M and MT showed a linear effect on biodiesel yield, whereas quadratic terms of M^2, C^2, T^2, MC, and CT had adverse effects that reduce the biodiesel yield. At the 95% confidence level, the *p* values were less than 0.05, indicating significant effects of those parameters. The analysis of variance (ANOVA) was used to determine the significance and fitness of the quadratic model.

Table 8. Regression coefficient of the predicted quadratic polynomial model.

Term.	Coefficients	Standard Errors	Computed T-Value	*p* Value
Constant	92.983	0.893	104.17	0.000
M	4.719	0.547	8.63	0.000
C	−2.161	0.547	−3.95	0.011
T	−0.730	0.547	−1.34	0.239
M × M	−4.490	0.805	−5.58	0.003
C × C	−4.650	0.805	−5.78	0.002
T × T	−4.323	0.805	−5.37	0.003
M × C	−5.168	0.773	−6.69	0.001
M × T	0.665	0.773	0.86	0.429
C × T	−1.635	0.773	−2.12	0.088

Table 9 shows the level of significance of individual terms and their interactions on the selected response. The quadratic regression model has higher F value (24.76) and lower *p* value (0.001) than significance level (*p* < 0.05), which indicates that the model is significant at the 99.9% confidence level. The *p* value represents the probability of error and is used to check the significance of each regression coefficient. The interaction effect of each cross product can be revealed through the *p* value. It is found, M (methanol: oil molar ratio), C (catalyst concentration), M^2 (quadratic effect of methanol amount), C^2 (quadratic effect of catalyst concentration), T^2 (quadratic effect of reaction temperature) and MC (methanol amount with catalyst) have significant effects on SFO biodiesel production. Among all other parameters, methanol: oil molar ratio (M) has the lowest p value (0.000) and highest F value (74.53). These results show that M is the most important parameter in SFO biodiesel production. According to the regression model in Equation (9), M has a positive effect and both C and reaction temperature (T) have negative effects on SFO biodiesel yield. This implies that increasing M will increase the speed of the transesterification process. However, increase in C and T will slow the speed of the transesterification reaction. The square term of T^2 was also significant although it has a smaller F value compared to its corresponding linear term which indicated its weaker influence in the model. Again, the ANOVA results showed that the linear term of T with p value was not significant (more than 0.05) and its quadratic term T^2 with p value was significant (less than 0.05). ANOVA also showed that both C and C^2 terms were significant with their F values (15.63 and 27) which indicated their medium effect in the model. The Lack of Fit was also determined for this regression model. F value and p value of Lack of Fit parameters were found to be 2.32 and 0.315 respectively. The p value (0.315) of the Lack of Fit parameter is greater than 0.050, which indicated the quadratic model has an insignificant Lack of Fit, i.e., the model sufficiently described a relationship between independent parameters such as M, C and T, with the dependent parameter (SFO biodiesel yield). The coefficient of determination (R^2) was employed to identify the quality of the model fitness. R^2 also indicates the good correlation between the independent parameters. In this study, R^2 was found to be 97.8% and the adjusted coefficient of determination (Adj. R^2) was 93.9%. This means that the model explains 97.8% of the variation in the experimental data. In conclusion, the regression model developed for SFO biodiesel yield was valid and showed a satisfactory experimental relationship between the response and parameters.

Table 9. ANOVA results for SFO biodiesel.

Source	Sum of Squares	Degree of Freedom	Mean Square	F-Value	p-Value	Remarks
Model	532.603	9	59.178	24.76	0.001	Significant
M-Methanol	178.123	1	178.123	74.53	0.000	Highly significant
C-Catalyst	37.364	1	37.364	15.63	0.011	Significant
T-Temperature	4.263	1	4.263	1.78	0.239	Not significant
M^2	74.447	1	74.447	31.15	0.003	Significant
C^2	79.847	1	79.847	27.00	0.002	Significant
T^2	69.004	1	69.004	28.87	0.003	Significant
MC	106.823	1	106.823	44.69	0.001	Significant
MT	1.769	1	1.769	0.74	0.429	Not significant
CT	10.693	1	10.693	4.47	0.088	Not significant
Lack of Fit	9.284	3	3.095	2.32	0.315	Not Significant
Pure Error	2.667	2	1.333			
Total	544.553	14				
$R^2 = 0.9781$			Adj $R^2 = 0.9386$			

4.5. Response Surface Plots for SFO Biodiesel Production

The interactive effect of the two factors on the transesterification process for biodiesel production is necessary for interpreting the impact of independent variables used in the optimisation process. Figures 6–8 show both surface plots and contour plots of SFO biodiesel yield obtained by the regression model in Equation (9). Surface plots were produced by plotting three-dimensional (3D) surface curves against any two independent variables while keeping the other variables fixed at their medium values. The interaction effect of the two parameters plotted while the third parameter was fixed at a medium value in the contour plot. Contour plots can identify the variation in biodiesel yield with any change in experimental conditions.

4.5.1. Interaction Effect of Methanol: Oil Molar Ratio and Catalyst Concentration

Figure 6a shows the 3D response surface for SFO biodiesel yield production as a function of methanol: oil molar ratio, and KOH catalyst concentration under the current conditions of Box-Behnken design matrix. With an increase of methanol: oil molar ratio up to 6:1 (highest) and 0.5 wt % of catalyst concentration (lowest), biodiesel yield percentage increases. The maximum SFO biodiesel yield of 95.8% was found for KOH 0.5 wt % (Run 13). Table 7 design matrix indicated that highest KOH concentration at 1.5 wt % and unchanged methanol: oil ratio at 6:1 resulted in lower SFO biodiesel yield to 82.1% (Run 4). When the methanol:oil molar ratio remains unchanged at 6:1, and the catalyst concentration is at the highest value of 1.5 wt %, the SFO biodiesel decreases to 82.1% (Run 4). When the methanol: oil molar ratio was reduced to 4:1 (lowest level), and with the highest value of catalyst concentration of 1.5 wt %, the biodiesel yield was found to be 82.3% (Run 9). Again, at methanol: oil molar ratio of 6:1, and with the mid-value of catalyst concentration of 1wt %, the yield was found to be 89.3% (Run 3). On the other hand, when the methanol: oil molar ratio was reduced to 4:1, and with the mid-value of catalyst concentration of 1 wt %, the yield dropped to 80.3% (Run 12). Methanol: oil molar ratio affected total biodiesel yield production. ANOVA from Table 9 confirmed that both M, C and MC interaction were significant. The 2D contour plot with MC interaction along with biodiesel yield is shown in Figure 6b. It is easy to identify the optimum operating conditions and the related response values (yield) through the 2D contour plot. Therefore, both M and C are significant factors for higher biodiesel yield. Although literature showed that higher amount of C could result in less biodiesel yield production, and eventually produce emulsion and phase separation [54].

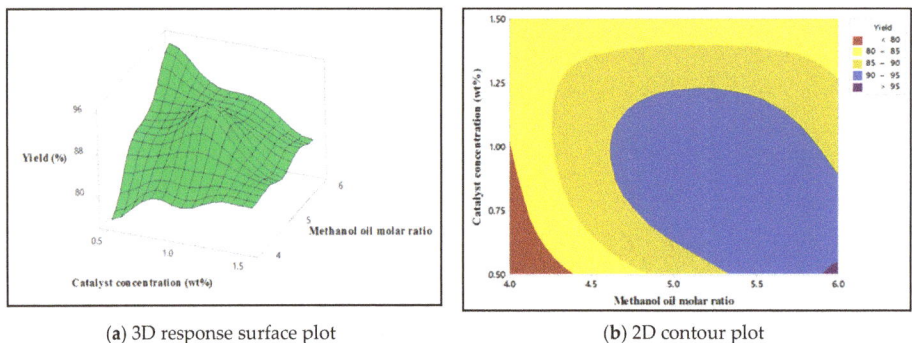

(a) 3D response surface plot

(b) 2D contour plot

Figure 6. Interaction effect of methanol: oil molar ratio (M) and catalyst concentration (C) on the SFO biodiesel yield.

4.5.2. Interaction Effect of Catalyst Concentration and the Reaction Temperature

The interaction effect of catalyst concentration, C and reaction temperature, T on the SFO biodiesel yield in 3D surface plots is shown in Figure 7a. An increase in T to mid-level (55 °C), and the low-level of C (0.5 wt %) can enhance the SFO biodiesel yield up to 95.8% (Run 13), considering M of 6:1 and reaction time of 60 min constant, which is presented in Table 7 of the Box–Behnken design matrix. It is found that increasing the T to 65 °C and C to 1 wt % resulted in the decline in biodiesel yield to 89.3% (Run 3). Similarly, increasing C to the highest level of 1.5 wt % and increasing T to the mid-level of 55 °C resulted in a stepwise decline in biodiesel yield to 82.1% (Run 11). Figure 7a displays the interaction between C and T on biodiesel yield production up to 92%, keeping the experimental conditions of T at 55 °C and C at 1 wt %. ANOVA results in Table 9 confirms that the interaction between C and T is not significant. Moreover, the higher amount of C and higher T might induce saponification of triglycerides as well as form soap at the end [55]. Figure 7b shows the 2D contour plots of CT interaction, which is not significant for SFO biodiesel yield production.

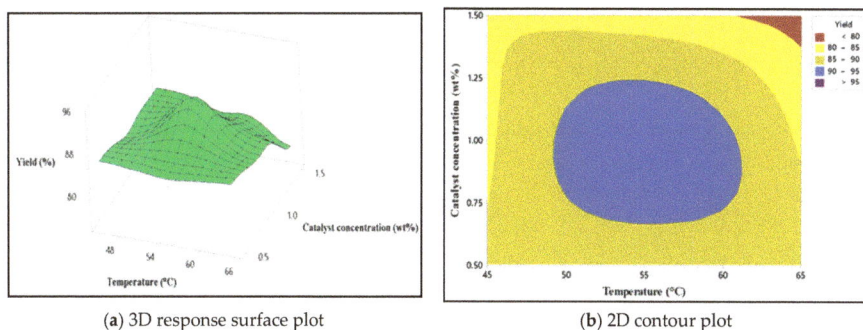

(**a**) 3D response surface plot   (**b**) 2D contour plot

Figure 7. Interaction effect of reaction temperature (T) and catalyst concentration (C) on SFO biodiesel yield.

4.5.3. Interaction Effect of Reaction Temperature and Methanol: Oil Molar Ratio

Figure 8a shows the 3D surface plot of reaction temperature, T and methanol: oil molar ratio, M with SFO biodiesel yield. The T at mid-level (55 °C) and mid-level M of 5:1 shows the maximum yield. Table 7 Box–Behnken design matrix shows that, at the mid-level T and mid-level M, a biodiesel yield of 93% can be achieved. However, at mid-level T and the highest level M, biodiesel yield was optimised and found to be 95.8%. The overall biodiesel yield decreased significantly to 89.3% when T reached 65 °C (Run 3). Any change in T either by an increase or decrease from its mid-level (55 °C) resulted in reduced biodiesel yield. The optimum M was found to be 6:1 and any decreasing the molar ratio (<6) lowered the biodiesel yield. Figure 8b shows the contour plot of interaction between M and T. It shows that, at the mid-level of both T and M, biodiesel yield is maximum. Therefore, ANOVA results in Table 9 confirms that both T and TM are not significant in SFO biodiesel production.

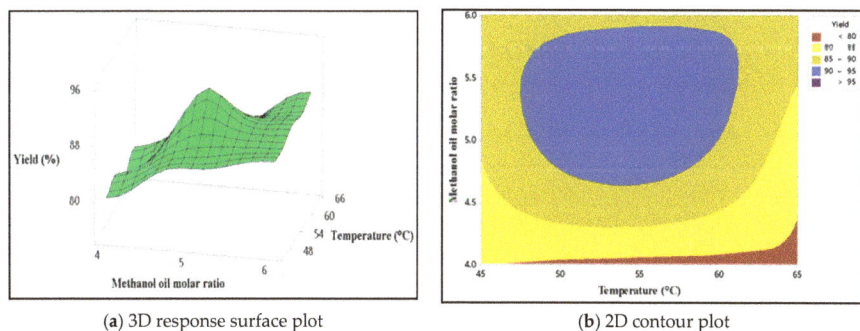

(**a**) 3D response surface plot   (**b**) 2D contour plot

Figure 8. Interaction effect of methanol: oil molar ratio (M) and reaction temperature (T) on SFO biodiesel yield.

From ANOVA results in Table 9, it is found that M is the only highly significant process factor that affects the production of SFO biodiesel. Interaction of MC is also a significant process factor for biodiesel production. Both the M and MC have higher F values of 74.5 and 44.7, respectively. Therefore, the optimum reaction conditions are M of 6:1, KOH C of 0.5 wt % and a T of 55 °C and the optimum SFO biodiesel yield is predicted to be 95.9%. To check the validity of the regression model (Equation (9)), experiments were carried out under predicted optimum conditions. The results of the experimental values under the optimum conditions indicated the highest (95.8%) conversion of SFO to

SFO biodiesel. This matches very closely with the predicted value (95.9%). Finally, this small degree of error (<0.5%) indicates the high accuracy of the model.

5. Conclusions

A response surface methodology-based Box–Behnken design matrix was employed to achieve the optimum operating parameters for second-generation biodiesel production from SFO. Three major parameters were varied individually within different ranges to anticipate biodiesel yield in that matrix. Based on the results, optimum operating parameters for transesterification of stone fruit seed oil were found to be methanol: oil molar ratio of 6:1, catalyst concentration 0.5 wt %, and a reaction temperature of 55 °C, considering both reaction time and reaction agitation speed were fixed at 60 min and 600 rpm. The maximum biodiesel yield under such conditions was 95.8%, which also confirmed the RSM model prediction of 95.9%. ANOVA statistics of this study confirmed that methanol: oil molar ratio has the most significant effect on the stone fruit biodiesel yield, whereas catalyst concentration and reaction temperature does not seem to have any significant impact. The results show a significant improvement in fuel properties of stone fruit biodiesel with kinetic viscosity 4.26 (mm^2/s), density 0.855 (kg/m^3), acid value 0.25 (mg/KOH/g), flash point 105 (°C), cloud point −4 (°C), pour point −8 (°C), higher heating value 39.04 (MJ/kg), cetane number 50.45 and oxidation stability 7.15 (h), all of which meet both the ASTM D6751 and EN14214 standards. In conclusion, stone fruit oil is a potential for biodiesel production, and this environment-friendly biodiesel can be used as an alternative to diesel fuel.

Author Contributions: M.A. collected oil, produced and characterised biodiesel, optimised the biodiesel, and drafted the manuscript; M.G.R. contributed to the experimental design and thoroughly revised the paper; N.A. helped revise and improve the paper; and M.M.R. checked and revised the manuscript.

Funding: This research received no external funding.

Acknowledgments: The authors would like to acknowledge Tim Mcsweeney, Adjunct Research Fellow, Tertiary Education Division at Central Queensland University, Australia for his contribution in proofreading of this article.

Conflicts of Interest: The authors declare no conflict of interest.

References

1. Rahman, M.M.; Rasul, M.G.; Hassan, N.M.S.; Azad, A.K.; Uddin, M.N. Effect of small proportion of butanol additive on the performance, emission, and combustion of Australian native first-and second-generation biodiesel in a diesel engine. *Environ. Sci. Pollut. Res.* **2017**, *24*, 22402–22413. [CrossRef] [PubMed]

2. Rahman, M.; Rasul, M.; Hassan, N. Study on the tribological characteristics of Australian native first generation and second generation biodiesel fuel. *Energies* **2017**, *10*, 55. [CrossRef]

3. Bhuiya, M.M.K.; Rasul, M.G.; Khan, M.M.K.; Ashwath, N.; Azad, A.K.; Hazrat, M.A. Second generation biodiesel: Potential alternative to-edible oil-derived biodiesel. *Energy Procedia* **2014**, *61*, 1969–1972. [CrossRef]

4. Bhuiya, M.M.K.; Rasul, M.G.; Khan, M.M.K.; Ashwath, N.; Azad, A.K.; Hazrat, M.A. Prospects of 2nd generation biodiesel as a sustainable fuel—Part 2: Properties, performance and emission characteristics. *Renew. Sustain. Energy Rev.* **2016**, *55*, 1129–1146.

5. Popp, J.; Lakner, Z.; Harangi-Rákos, M.; Fári, M. The effect of bioenergy expansion: Food, energy, and environment. *Renew. Sustain. Energy Rev.* **2014**, *32*, 559–578. [CrossRef]

6. NguyenThi, T.; Bazile, J.-P.; Bessières, D. Density measurements of waste cooking oil biodiesel and diesel blends over extended pressure and temperature ranges. *Energies* **2018**, *11*, 1212. [CrossRef]

7. Yusop, A.; Mamat, R.; Yusaf, T.; Najafi, G.; Yasin, M.; Khathri, A. Analysis of particulate matter (pm) emissions in diesel engines using palm oil biodiesel blended with diesel fuel. *Energies* **2018**, *11*, 1039. [CrossRef]

8. Ali, O.; Mamat, R.; Najafi, G.; Yusaf, T.; Safieddin Ardebili, S. Optimization of biodiesel-diesel blended fuel properties and engine performance with ether additive using statistical analysis and response surface methods. *Energies* **2015**, *8*, 14136–14150. [CrossRef]

9. Silitonga, A.S.; Atabani, A.E.; Mahlia, T.M.I.; Masjuki, H.H.; Badruddin, I.A.; Mekhilef, S. A review on prospect of jatropha curcas for biodiesel in Indonesia. *Renew. Sustain. Energy Rev.* **2011**, *15*, 3733–3756. [CrossRef]

10. Rahman, M.; Rasul, M.; Hassan, N.; Hyde, J. Prospects of biodiesel production from macadamia oil as an alternative fuel for diesel engines. *Energies* **2016**, *9*, 403.

11. Sharma, R.; Gupta, A.; Abrol, G.S.; Joshi, V.K. Value addition of wild apricot fruits grown in north-west himalayan regions: A review. *J. Food Sci. Technol.* **2012**, *51*, 2917–2924. [CrossRef] [PubMed]

12. Razack, S.A.; Duraiarasan, S. Response surface methodology assisted biodiesel production from waste cooking oil using encapsulated mixed enzyme. *Waste Manag.* **2016**, *47*, 98–104. [CrossRef] [PubMed]

13. Saydut, A.; Erdogan, S.; Kafadar, A.B.; Kaya, C.; Aydin, F.; Hamamci, C. Process optimization for production of biodiesel from hazelnut oil, sunflower oil and their hybrid feedstock. *Fuel* **2016**, *183*, 512–517. [CrossRef]

14. Martínez, S.L.; Romero, R.; Natividad, R.; González, J. Optimization of biodiesel production from sunflower oil by transesterification using Na$_2$O/NaX and methanol. *Catal. Today* **2014**, *220*, 12–20. [CrossRef]

15. Kostić, M.D.; Djalović, I.G.; Stamenković, O.S.; Mitrović, P.M.; Adamović, D.S.; Kulina, M.K.; Veljković, V.B. Kinetic modeling and optimization of biodiesel production from white mustard (*Sinapis alba* L.) seed oil by quicklime-catalyzed transesterification. *Fuel* **2018**, *223*, 125–139. [CrossRef]

16. Ong, H.C.; Masjuki, H.H.; Mahlia, T.M.I.; Silitonga, A.S.; Chong, W.T.; Leong, K.Y. Optimization of biodiesel production and engine performance from high free fatty acid *Calophyllum inophyllum* oil in Ci diesel engine. *Energy Convers. Manag.* **2014**, *81*, 30–40. [CrossRef]

17. Dharma, S.; Masjuki, H.H.; Ong, H.C.; Sebayang, A.H.; Silitonga, A.S.; Kusumo, F.; Mahlia, T.M.I. Optimization of biodiesel production process for mixed *Jatropha curcas–Ceiba pentandra* biodiesel using response surface methodology. *Energy Convers. Manag.* **2016**, *115*, 178–190. [CrossRef]

18. Gumus, M.; Kasifoglu, S. Performance and emission evaluation of a compression ignition engine using a biodiesel (apricot seed kernel oil methyl ester) and its blends with diesel fuel. *Biomass Bioenergy* **2010**, *34*, 134–139. [CrossRef]

19. Fadhil, A.B. Evaluation of apricot (*Prunus armeniaca* L.) seed kernel as a potential feedstock for the production of liquid bio-fuels and activated carbons. *Energy Convers. Manag.* **2017**, *133*, 307–317. [CrossRef]

20. Ullah, F.; Nosheen, A.; Hussain, I.; Bano, A. Base catalyzed transesterification of wild apricot kernel oil for biodiesel production. *Afr. J. Biotechnol.* **2009**, *8*, 3309–3313.

21. Yadav, A.K.; Pal, A.; Dubey, A.M. Experimental studies on utilization of *Prunus armeniaca* L. (wild apricot) biodiesel as an alternative fuel for CI engine. *Waste Biomass Valoriz.* **2017**, 1–9. [CrossRef]

22. Hamze, H.; Akia, M.; Yazdani, F. Optimization of biodiesel production from the waste cooking oil using response surface methodology. *Process Saf. Environ. Prot.* **2015**, *94*, 1–10. [CrossRef]

23. Atabani, A.E.; Silitonga, A.S.; Badruddin, I.A.; Mahlia, T.M.I.; Masjuki, H.H.; Mekhilef, S. A comprehensive review on biodiesel as an alternative energy resource and its characteristics. *Renew. Sustain. Energy Rev.* **2012**, *16*, 2070–2093. [CrossRef]

24. Atadashi, I.M.; Aroua, M.K.; Abdul Aziz, A.R.; Sulaiman, N.M.N. The effects of water on biodiesel production and refining technologies: A review. *Renew. Sustain. Energy Rev.* **2012**, *16*, 3456–3470. [CrossRef]

25. Banerjee, A.; Chakraborty, R. Parametric sensitivity in transesterification of waste cooking oil for biodiesel production—A review. *Resour. Conserv. Recycl.* **2009**, *53*, 490–497. [CrossRef]

26. Yaakob, Z.; Mohammad, M.; Alherbawi, M.; Alam, Z.; Sopian, K. Overview of the production of biodiesel from waste cooking oil. *Renew. Sustain. Energy Rev.* **2013**, *18*, 184–193. [CrossRef]

27. Targais, K.; Stobdan, T.; Yadav, A.; Singh, S.B. Extraction of apricot kernel oil in cold desert Ladakh, India. *Indian J. Tradit. Knowl.* **2011**, *10*, 304–306.

28. Wang, L. Properties of manchurian apricot (*Prunus mandshurica* Skv.) and siberian apricot (*Prunus sibirica* L.) seed kernel oils and evaluation as biodiesel feedstocks. *Ind. Crops Prod.* **2013**, *50*, 838–843. [CrossRef]

29. Fan, S.; Liang, T.; Yu, H.; Bi, Q.; Li, G.; Wang, L. Kernel characteristics, oil contents, fatty acid compositions and biodiesel properties in developing siberian apricot (*Prunus sibirica* L.) seeds. *Ind. Crops Prod.* **2016**, *89*, 195–199. [CrossRef]

30. Rahman, M.M.; Hassan, M.H.; Kalam, M.A.; Atabani, A.E.; Memon, L.A.; Rahman, S.M.A. Performance and emission analysis of *Jatropha curcas* and *Moringa oleifera* methyl ester fuel blends in a multi-cylinder diesel engine. *J. Clean. Prod.* **2014**, *65*, 304–310. [CrossRef]

31. Kafuku, G.; Mbarawa, M. Alkaline catalyzed biodiesel production from moringa oleifera oil with optimized production parameters. *Appl. Energy* **2010**, *87*, 2561–2565. [CrossRef]

32. Alptekin, E.; Canakci, M. Determination of the density and the viscosities of biodiesel–diesel fuel blends. *Renew. Energy* **2008**, *33*, 2623–2630. [CrossRef]

33. Encinar, J.M.; González, J.F.; Rodríguez-Reinares, A. Biodiesel from used frying oil. Variables affecting the yields and characteristics of the biodiesel. *Ind. Eng. Chem. Res.* **2005**, *44*, 5491–5499. [CrossRef]

34. Graboski, M.S.; McCormick, R.L. Combustion of fat and vegetable oil derived fuels in diesel engines. *Prog. Energy Combust. Sci.* **1998**, *24*, 125–164. [CrossRef]

35. Silitonga, A.S.; Ong, H.C.; Mahlia, T.M.I.; Masjuki, H.H.; Chong, W.T. Biodiesel conversion from high FFA crude *Jatropha curcas*, *Calophyllum inophyllum* and *Ceiba pentandra* oil. *Energy Procedia* **2014**, *61*, 480–483. [CrossRef]

36. Silitonga, A.S.; Ong, H.C.; Masjuki, H.H.; Mahlia, T.M.I.; Chong, W.T.; Yusaf, T.F. Production of biodiesel from *Sterculia foetida* and its process optimization. *Fuel* **2013**, *111*, 478–484. [CrossRef]

37. Keera, S.T.; El Sabagh, S.M.; Taman, A.R. Transesterification of vegetable oil to biodiesel fuel using alkaline catalyst. *Fuel* **2011**, *90*, 42–47. [CrossRef]

38. Hasni, K.; Ilham, Z.; Dharma, S.; Varman, M. Optimization of biodiesel production from *Brucea javanica* seeds oil as novel non-edible feedstock using response surface methodology. *Energy Convers. Manag.* **2017**, *149*, 392–400. [CrossRef]

39. Anwar, M.; Rasul, M.G.; Ashwath, N. Production optimization and quality assessment of papaya (*Carica papaya*) biodiesel with response surface methodology. *Energy Convers. Manag.* **2018**, *156*, 103–112. [CrossRef]

40. Benjumea, P.; Agudelo, J.R.; Agudelo, A.F. Effect of the degree of unsaturation of biodiesel fuels on engine performance, combustion characteristics, and emissions. *Energy Fuels* **2011**, *25*, 77–85. [CrossRef]

41. Altun, Ş. Effect of the degree of unsaturation of biodiesel fuels on the exhaust emissions of a diesel power generator. *Fuel* **2014**, *117*, 450–457. [CrossRef]

42. Wang, L.; Yu, H. Biodiesel from siberian apricot (*Prunus sibirica* L.) seed kernel oil. *Bioresour. Technol.* **2012**, *112*, 355–358. [CrossRef] [PubMed]

43. Kate, A.E.; Lohani, U.C.; Pandey, J.P.; Shahi, N.C.; Sarkar, A. Traditional and mechanical method of the oil extraction from wild apricot kernel: A comparative study. *Res. J. Chem. Environ. Sci.* **2014**, *2*, 54–60.

44. Atabani, A.E.; César, A.D.S. *Calophyllum inophyllum* L.—A prospective non-edible biodiesel feedstock. Study of biodiesel production, properties, fatty acid composition, blending and engine performance. *Renew. Sustain. Energy Rev.* **2014**, *37*, 644–655. [CrossRef]

45. García-Martínez, N.; Andreo-Martínez, P.; Quesada-Medina, J.; de los Ríos, A.P.; Chica, A.; Beneito-Ruiz, R.; Carratalá-Abril, J. Optimization of non-catalytic transesterification of tobacco (*Nicotiana tabacum*) seed oil using supercritical methanol to biodiesel production. *Energy Convers. Manag.* **2017**, *131*, 99–108. [CrossRef]

46. Goyal, P.; Sharma, M.P.; Jain, S. Optimization of conversion of high free fatty acid *Jatropha curcas* oil to biodiesel using response surface methodology. *ISRN Chem. Eng.* **2012**, *2012*, 8. [CrossRef]

47. Lin, L.; Zhou, C.; Vittayapadung, S.; Shen, X.; Dong, M. Opportunities and challenges for biodiesel fuel. *Appl. Energy* **2011**, *88*, 1020–1031. [CrossRef]

48. Demirbaş, A. Biodiesel fuels from vegetable oils via catalytic and non-catalytic supercritical alcohol transesterifications and other methods: A survey. *Energy Convers. Manag.* **2003**, *44*, 2093–2109. [CrossRef]

49. Shankar, V.; Jambulingam, R. Waste crab shell derived CaO impregnated Na-ZSM-5 as a solid base catalyst for the transesterification of neem oil into biodiesel. *Sustain. Environ. Res.* **2017**, *27*, 273–278. [CrossRef]

50. Patel, R.L.; Sankhavara, C.D. Biodiesel production from karanja oil and its use in diesel engine: A review. *Renew. Sustain. Energy Rev.* **2017**, *71*, 464–474. [CrossRef]

51. Rashid, U.; Anwar, F.; Moser, B.R.; Knothe, G. Moringa oleifera oil: A possible source of biodiesel. *Bioresour. Technol.* **2008**, *99*, 8175–8179. [CrossRef] [PubMed]

52. Silitonga, A.S.; Masjuki, H.H.; Mahlia, T.M.I.; Ong, H.C.; Chong, W.T.; Boosroh, M.H. Overview properties of biodiesel diesel blends from edible and non-edible feedstock. *Renew. Sustain. Energy Rev.* **2013**, *22*, 346–360. [CrossRef]

53. Gupta, J.; Agarwal, M.; Dalai, A.K. Optimization of biodiesel production from mixture of edible and nonedible vegetable oils. *Biocatal. Agric. Biotechnol.* **2016**, *8*, 112–120. [CrossRef]

Energies **2018**, *11*, 2566

54. Silva, G.F.; Camargo, F.L.; Ferreira, A.L.O. Application of response surface methodology for optimization of biodiesel production by transesterification of soybean oil with ethanol. *Fuel Process. Technol.* **2011**, *92*, 407–413. [CrossRef]
55. Vicente, G.; Martínez, M.; Aracil, J. Optimisation of integrated biodiesel production. Part I. A study of the biodiesel purity and yield. *Bioresour. Technol.* **2007**, *98*, 1724–1733. [CrossRef] [PubMed]

energies

MDPI

Article

Effects of Organosolv Pretreatment Using Temperature-Controlled Bench-Scale Ball Milling on Enzymatic Saccharification of *Miscanthus × giganteus*

Tae Hoon Kim [1,2], Dongjoong Im [2], Kyeong Keun Oh [2,3,*] and Tae Hyun Kim [1,*]

[1] Department of Materials Science and Chemical Engineering, Hanyang University, Ansan, Gyeonggi-do 15588, Korea; thkim@sugaren.co.kr
[2] R&D Center, SugarEn Co., Ltd., Yongin, Gyeonggi-do 16890, Korea; Myred3@daum.net
[3] Department of Chemical Engineering, Dankook University, Yongin, Gyeonggi-do 16890, Korea
* Correspondence: hitaehyun@hanyang.ac.kr (T.H.K.); kkoh@dankook.ac.kr (K.K.O);
  Tel.: +82-31-400-5222 (T.H.K.); +82-31-330-5222 (K.K.O)

Received: 16 August 2018; Accepted: 1 October 2018; Published: 5 October 2018

Abstract: The effect of organosolv pretreatment was investigated using a 30 L bench-scale ball mill reactor that was capable of simultaneously performing physical and chemical pretreatment. Various reaction conditions were tried in order to discover the optimal conditions for the minimal cellulose loss and enhanced enzymatic digestibility of *Miscanthus × giganteus* (MG), with conditions varying from room temperature to 170 °C for reaction temperature, from 30 to 120 min of reaction time, from 30% to 60% ethanol concentration, and a liquid/solid ratio (L/S) of 10–20 under non-catalyst conditions. The pretreatment effects were evaluated by chemical compositional analysis, enzymatic digestibility test and X-ray diffraction of the treated samples. The pretreatment conditions for the highest glucan digestibility yield were determined as 170 °C, reaction time of 90 min, ethanol concentration of 40% and L/S = 10. With these pretreatment conditions, the XMG (xylan + mannan + galactan) fractionation yield and delignification were 84.4% and 53.2%, respectively. The glucan digestibility of treated MG after the aforementioned pretreatment conditions was 86.0% with 15 filter paper units (FPU) of cellulase (Cellic® CTec2) per g-glucan enzyme loading.

Keywords: lignocellulosic biomass; *Miscanthus*; mechanical pretreatment; organosolv pretreatment

1. Introduction

Lignocellulosic biomass is considered to be an abundant and attractive resource in the biorefinery system. However, lignocellulosic biomass is a complex organic polymer, which is composed mainly of carbohydrate polymers (cellulose and hemicellulose) and lignin. In addition, chemical and physical barriers such as lignin and the crystalline structure of cellulose inhibit the hydrolytic reaction of enzymes on the cellulose substrate. Thus, pretreatment is an indispensable process to enhance the enzymatic saccharification of cellulose, and various pretreatment methods have been attempted extensively. Pretreatment methods can be divided into two different categories: physical and chemical pretreatment methods and biological pretreatment methods. The purpose of pretreatment is to break down the structure of the lignocellulosic biomass to improve enzymatic digestibility and thus to enable the efficient bioconversion of monomeric sugars to biochemicals [1–3].

Chemical pretreatment generally uses various catalysts such as acids, alkalis, ionic liquids, oxidizing agents, or organic solvents. Among them, acid and alkali catalysts effectively affect the biomass by either reducing the degree of polymerization or removing lignin [3,4]. Despite its many advantages, chemical pretreatment has encountered major problems such as equipment corrosion, long reaction times, high input of solvents, and sugar degradation due to the high reactivity and sever reaction conditions [5,6]. In addition, the neutralization and washing of the treated solid is essential to remove the residual chemicals; therefore, a large amount of water is required. As one type of chemical pretreatment, the organosolv pretreatment method uses a variety of organic or aqueous solvents under catalytic or non-catalytic conditions. The solvents being used in the organosolv pretreatment include ethanol, methanol, acetone, formic acid, acetic acid, and glycerol [7,8]. In addition, acid catalysts such as sulfuric acid and hydrochloric acid are generally applied to lower the pH, because the carbohydrate–lignin complex efficiently cleaved at low pH. The organosolv pretreatment can effectively cleave the lignin-to-lignin and lignin-to-hemicellulose bonds, which can recover high purity lignin [8,9]. As a result, the organosolv-pretreated biomass with low hemicellulose and lignin contents can be expected to have increased digestibility. However, organosolv pretreatment usually requires a high reaction temperature range (170–210 °C), long reaction time (30–120 min), and high chemical loading (50–75% in the case of ethanol and 0.5–2.0 wt.% of acid catalyst (such as H_2SO_4 or HCl)) [10–13]. However, in case of non-acid catalytic conditions (organosolv-only without additional acid), it typically requires more severe reaction condition. For example, a study of J Wildschut [14] (using a wheat straw) showed very low enzymatic digestibility of 31.7% at 170 °C, 60 min, and 60% ethanol concentration. When the reaction temperature was increased by 30 °C (200 °C, 60 min, 60% ethanol concentration), the digestibility was only 44.4%. On the other hand, when the most severe reaction conditions (210 °C, 90 min, 50% ethanol concentration) were applied, the enzymatic digestibility increased to 85.9%.

Meanwhile, mechanical pretreatment methods commonly include grinding, milling, and comminution. They usually reduce the particle size and increase the surface area, which improves the contact surface of the biomass exposed to the enzyme [14]. Among the mechanical pretreatment methods, ball milling pretreatment severely reduces the crystallinity of cellulose by breaking hydrogen bonds in the substrate [15]. It increases the amorphous portion of cellulose, which improves the accessibility of the substrate to the enzyme [16]. In some studies on the tendency of biomass size reduction, the negative effect of milling on enzymatic digestibility was reported that the grinded MG (using hammer mill to 80 µm) resulted in only 22.5% digestibility [17]. In addition, ball mill pretreatment has previously received less attention because of its high energy consumption and long reaction time. However, integrating the mechanical and chemical pretreatment has the potential to overcome the shortcomings of chemical pretreatment method [18–20].

The main purpose of this study was to improve the pretreatment effects for *Miscanthus* × *giganteus* (MG) by combining mechanical and chemical pretreatment methods. The various reaction conditions (reaction temperature, time, and ethanol concentration) were evaluated to reduce the reaction time and decrease the reaction temperature of the organosolv pretreatment. Ethanol was used because ethanol is not only cheaper than other organic solvents, but it also has a low boiling point, which makes it easy to recover [21,22]. In general, organosolv pretreatment is supplemented with organic or inorganic acid to control the pH level for improved hydrolytic reaction, but this study was carried out under non-catalytic conditions to avoid the problems with an acid catalyst, such as equipment corrosion and the formation of toxic substances such as levulinic acid, 5-(hydroxymethyl)furfural (5-HMF) and furfural [1]. To the best of our knowledge, there have been few studies on the combined pretreatment with organosolv and ball milling in a large-scale reactor system. According to a study by Y. Teramoto [23] (ball milling and organosolv treatment were performed separately), enzymatic digestibility was improved to 67.4% under the pretreatment conditions of 200 °C, 60 min, and 75% ethanol concentration.

In this study, one of the representative mechanical pretreatments, the ball mill pretreatment with alumina balls, was applied. A 30-L reactor was custom designed to demonstrate the combined effects of chemical and physical pretreatment. In addition, this large-scale ball mill can be operated at a maximum temperature of 200 °C and pressure of 20 kgf/cm^2. Prior to testing the 30-L ball mill, small lab scale tests were performed.

2. Results and Discussion

2.1. Effects of Ethanol Concentration on Compositions of Pretreated Solids and Liquids Using Small Scale Batch Reactor

In a previous study of organosolv pretreatment, high ethanol concentration and loading were applied to achieve effective pretreatment effects; therefore, this method was expected to increase the overall processing cost. In general, the use of a lower ethanol concentration to reduce the processing cost of the organosolv pretreatment process. A preliminary experiment was conducted to determine the adequate ethanol concentration, and the chemical compositions of the organosolv-treated MG are shown in Figure 1. The reaction temperature and reaction time were fixed at 170 °C for 60 min, which were pre-selected based on the previous studies [10–13].

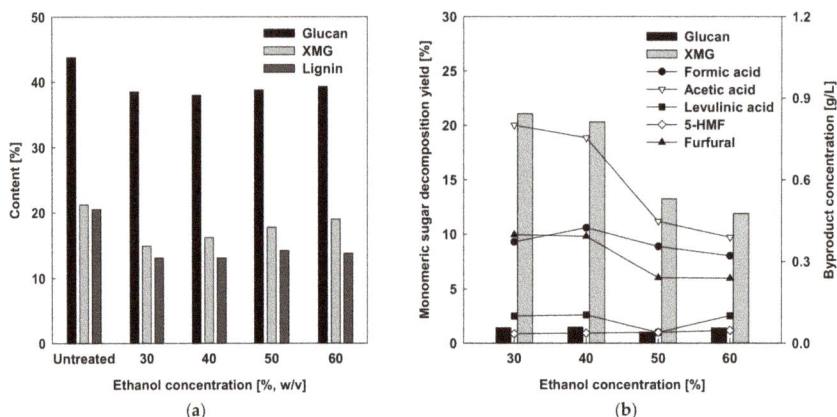

Figure 1. Effect of ethanol concentration on (**a**) solid content of untreated and pretreated *Miscanthus × giganteus* (MG); (**b**) monomeric sugar and byproduct concentrations of liquid hydrolysate. Reaction conditions: 170 °C, 60 min, liquid/solid (L/S) ratio = 10. Data in the figure are based on untreated biomass (wt.%).

As shown in Figure 1a, solubilizations of glucan (10.2–13.2%) were generally lowers than those (10.5–29.8%) of XMG (sum of xylan, mannan, and galactan). In addition, the glucan and XMG contents of the treated solids were more preserved as the ethanol concentration increased from 30% to 60%. The reason for this phenomenon was because of the different hydrogen ion concentrations in water produced at high temperatures.

A low ethanol concentration can produce more hydronium ion (H_3O^+) than a high ethanol concentration does due to the greater water content [24]. Therefore, the pH value of the low ethanol concentration becomes lower than that of the high ethanol concentration, and this promotes the depolymerization of hemicellulose [25]. In contrast, lignin content showed no important change as ethanol concentrations rose from 30% to 60%. The hemicellulose is bound with lignin in the cell wall of the biomass, which is more susceptible to hydrolysis under severe conditions [26,27]. Concentrated ethanol is known to be effective for lignin solubilization. However, reaction under harsh conditions is generally necessary for the effective cleavage of the lignin–hemicellulose complex, which was assumed to be related to pH. The extent of delignification with increased ethanol concentration was shown to be consistent at the tested ethanol concentration (Figure 1a). The tendency of sugars in the residual solid upon various ethanol concentrations was similar to that of the solubilized sugar in liquid hydrolyzate shown in Figure 1b. As shown in Figure 1b, the formations of XMG and byproducts (formic acid, acetic acid, levulinic acid, 5-HMF, and furfural) in liquid hydrolyzate were decreased with increased ethanol concentration. However, the amount of glucan decomposition was not affected by ethanol concentration. At 30% ethanol concentration, 21.1% of XMG was solubilized. Under the same conditions, byproduct was formed including 0.4 g/L of formic acid, 0.8 g/L of acetic acid, and 0.4 g/L of furfural (levulinic acid and 5-HMF were < 0.1 g/L). At 60% ethanol concentration, 11.9% of XMG was solubilized. Under these conditions, byproduct was formed and contained 0.3 g/L of formic acid, 0.4 g/L of acetic acid and 0.2 g/L of furfural (*c.f.* levulinic acid and 5-HMF were < 0.1 g/L). In general, organosolv pretreatment has more advantage for delignification with high ethanol concentration. Based on these results in Figure 1b, the ethanol concentration for maximum XMG solubilization was determined to be 40%, which can achieve maximum XMG solubilization and possibly more delignification.

2.2. Organosolv Pretreatment Using Bench-Scale Ball-Milling Reactor

2.2.1. Effects of Reaction Temperature and Time

The various effects of organosolv pretreatment were evaluated using a bench scale ball-milling reactor. Table 1 presents the effects of organosolv pretreatment on the chemical compositions of the treated solids and liquid hydrolyzate. As the reaction temperature increased from room temperature (RT) to 170 °C, the XMG removal and lignin removal (delignification) were highly increased for all pretreatment times. However, the changes in XMG solubilization from the treated solids at RT and 130 °C were inconsiderable at all reaction times.

Table 1. Effects of reaction temperature and time on chemical compositions of treated solids and liquid hydrolyzate.

Conditions		Treated Solids			Liquid Hydrolyzate		Total Cons. [1]	
Temp. [°C]	Time [min]	Glucan [wt.%]	XMG [2] [wt.%]	Lignin [3] [wt.%]	Glucan [wt.%]	XMG [wt.%]	Glucan [wt.%]	XMG [wt.%]
Untreated		43.8	21.2	20.5	-	-	-	-
RT [4]	30	43.5	20.9	20.2	0.4	0.5	99.7	99.1
	60	43.3	21.0	20.0	0.4	0.7	99.5	99.8
	120	43.0	20.7	20.3	0.6	0.8	98.9	98.5
130	30	41.7	20.1	18.3	0.7	2.1	96.0	97.0
	60	42.3	20.2	18.5	0.9	2.6	97.5	97.7
	120	42.7	20.0	16.7	0.8	3.4	98.5	97.7
150	30	42.1	19.0	15.4	1.1	8.4	97.3	98.0
	60	41.5	17.6	14.1	1.4	15.4	96.3	98.4
	120	40.7	14.4	12.3	2.0	30.0	95.1	97.8
170	30	41.2	10.1	10.6	1.9	40.9	96.0	88.3
	60	40.8	4.2	10.0	1.9	65.3	95.0	85.3
	120	35.4	2.0	7.8	4.1	74.1	85.0	83.6

[1] Total conservation: [(remaining (wt.%) in treated solid + fractionated (wt.%) in liquid)/untreated (wt.%) in MG] × 100; [2] XMG: xylan + mannan + galactan; [3] Lignin: acid-insoluble lignin; [4] room temperature (~25 °C). Reaction conditions: 40% ethanol concentration and L/S ratio = 10.

At 150 °C for 120 min, XMG removal was 32.1%, whereas the delignification was 40%. On the other hand, with the 170 °C treatment, the XMG removal and delignification differed considerably between the 30 min and 120 min treatments. At 170 °C for 120 min, the XMG removal and delignification yields were 90.6% and 62.0%, respectively, which were considerably increased from the 52.4% XMG removal and 48.3% delignification of the 30 min treatment. The glucan contents were preserved well at all treatment conditions (>80.8%).

The composition of liquid hydrolyzates had similar trends. As the reaction conditions became more severe, the XMG solubilization yields in the liquid hydrolyzate were proportionally increased. The solubilization yield of XMG was less than 30.0% at RT, 130 °C, and 150 °C, while substantial changes were observed at 170 °C treatment. The maximum recovery yield of XMG was 74.1% at 170 °C for 120 min.

The glucan and XMG contents were decreased as the reaction temperature and time became more severe. Under the most severe conditions of 170 °C and 120 min, total conservation of glucan and XMG were 85.0% and 83.6%, respectively. This suggested that appropriate reaction conditions must be selected to accomplish the effective fractionation of sugars and lignin.

Figure 2 summarizes the byproduct formations (mainly formic acid, acetic acid, and furfural) at two different reaction temperatures (150 °C and 170 °C). The byproduct formations (formic acid, acetic acid, levulinic acid, 5-HMF, and furfural) at RT and 130 °C were not substantial, which were less than 0.3 g/L (data not shown). Byproducts were produced by the decomposition of cellulose and hemicellulose, which typically occurred at the more severe reaction conditions [28]. Therefore, this is a good indicator to determine the severity of the reaction. At the most severe conditions (i.e., 170 °C for 120 min), the productions of formic acid, acetic acid, and furfural were 1.2 g/L, 3.5 g/L and 2.0 g/L, respectively.

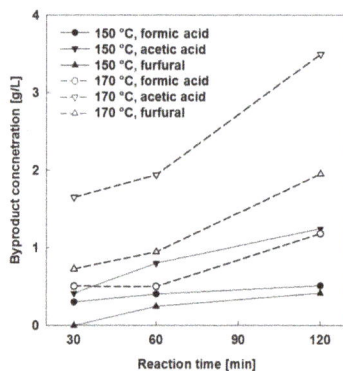

Figure 2. Effects of reaction temperature and time on concentrations of byproducts in liquid hydrolyzate. Reaction conditions: 40% ethanol concentration and 10 L/S.

2.2.2. Effects of Liquid/Solid (L/S) Ratio

The effects of ethanol concentrations (40% and 60%), L/S ratio (10 and 20), and reaction times (from 30 to 120 min) on the chemical compositions of treated solids at 170 °C are summarized in Table 2. The results indicate that the run No. 1 (ethanol concentration of 40% and L/S ratio of 10) conditions seem to be more severe than those of run No. 2 (ethanol concentration of 40% and L/S ratio of 20) and run No. 3 (ethanol concentration of 60% and L/S ratio of 10), because the retention of XMG in treated solid in run No. 1 was lower than those of the other runs. For the decomposition of sugar, in particular, XMG was depending on the pretreatment reaction severity, and the observed results in Table 2 were assumed to be related to the generation of acid products such as formic acid and acetic acid. Because the acetyl group of hemicellulose is cleaved as the reaction progresses, acetic acid is generated, which would drop the pH of the solution during the reaction and consequently increase the reaction severity, therefore promoting the hydrolysis of hemicellulose. Regardless of the amount generated, the acid concentration at an L/S of 10 is expected to be higher than that of 20. After treatment, the pH values of the liquid hydrolyzate of runs No. 1 and 2 at a reaction time of 120 min were 3.32 and 4.17, respectively. This means that L/S = 10 is a more severe condition than L/S = 20, and this can further promote the hydrolysis of MG during the reaction. On the other hand, the high ethanol concentration (run No. 3) resulted in the formation of a low concentration of hydronium ion; therefore, the hydrolysis effect was not improved. In addition, the delignifications of runs No. 1, 2, and 3 were 58.0%, 61.0%, and 62.0%, respectively. This indicated that a 40% ethanol concentration could effectively remove the lignin (~60%).

Table 2. Chemical compositions of solids and liquid hydrolyzates and total conservation yield of ball mill organosolv pretreatment under various reaction conditions.

No.	Conditions			Treated Solid			Liquid Hydrolyzate		Total Conservation	
	EtOH [%]	L/S [-]	Time [min]	Glucan [%]	XMG [%]	Lignin [%]	Glucan [%]	XMG [%]	Glucan [%]	XMG [%]
Untreated				43.8	21.2	20.5	-	-	-	-
1	40	10	30	41.2	10.1	10.6	1.9	40.9	96	88.3
			60	40.8	4.2	10	1.9	65.3	95	85.3
			90	37.5	3.3	9.6	1.8	68.6	87.6	84
			120	35.4	2	7.8	4.1	74.1	85	83.6
2	40	20	30	39.6	12.8	11.8	1.7	35.5	92.3	95.8
			60	38.8	8.8	9.1	2.3	56.5	91	98
			90	37.8	6.3	8.5	2.2	57.4	88.5	87.1
			120	36.6	5.3	8	2.4	59.2	86.1	84.1
3	60	10	30	42.3	18.3	12.9	1.8	9.8	98.5	95.8
			60	39.5	15.7	11.5	1.9	16.8	92.2	91
			90	38.8	13.6	9.4	2	24.5	90.6	88.4
			120	37.5	8.7	8.6	2.1	45.3	87.8	86.3

Total conservation: [(remaining (wt.%) in treated solid + fractionated (wt.%) in liquid)/untreated (wt.%) in MG] × 100; EtOH, ethanol; XMG: xylan + mannan + galactan; Lignin: acid-insoluble lignin. Reaction temperature: 170 °C.

2.2.3. Enzymatic Digestibility Testing of Pretreated Solids

The enzymatic digestibility tests were performed on untreated and treated MG, and the glucan digestibilities are summarized in Figure 3. Figure 3a compares the digestibility of untreated MG and treated MG under various reaction conditions with the fixed L/S ratio of 10, because L/S = 10 required less solvent than L/S = 20 and effective hydrolysis could be expected. As shown in Figure 3a, the digestibility at three treatment temperatures, RT, 130 °C and 150 °C, did not exceed 40% for 72 h of hydrolysis. The digestibility of the solids treated at 150 °C for 120 min was only 39.4% at 72 h of hydrolysis. However, the enzyme digestibility of the sample treated at 170 °C was increased to the extent to which it was similar to that using α-cellulose. The 72-h digestibility values of the samples treated for 60 and 120 min were 75.7% and 96.9%, respectively. In particular, the 72-h digestibility of the sample treated at 170 °C for 120 min was 8.9% higher than that 88.0% of α-cellulose. The 72-h digestibility values of untreated MG and α-cellulose as the reference and control were 6.5% and 88.0%, respectively.

The effects of ethanol concentration and L/S ratio on digestibility at the 90 min reaction time are presented in Figure 3b. The 72-h digestibility of treated MG with 60% ethanol was considerably lower than that of the treated solids with 40% ethanol. Although the digestibility of the two treatment conditions using 40% ethanol were close to 80% in 24 h, the hydrolysis rate was faster than that of α-cellulose (70.2%) with 15 filter paper units (FPU)/g-glucan enzyme loading. Considering the aforementioned results (Table 2 and Figure 3), the following conditions were selected for the effective pretreatment and fractionation of MG: 170 °C reaction temperature, 90 min reaction time, 40% ethanol concentration and 10 L/S ratio. The 72-h digestibility of the organosolv-treated sample (L/S of 10 with 40% ethanol concentration) was 86.0%. In addition, during pretreatment reaction at 170 °C (Table 2 and Figure 3b), it was observed that lignin removals were not substantial (<58.5% of delignification) at all tested conditions. It indicated that lignin could melt or dissolve and re-precipitate back on the solids, which was to suggest that removal of lignin to the liquid phase was only a partial measure of what is happening with the lignin fraction.

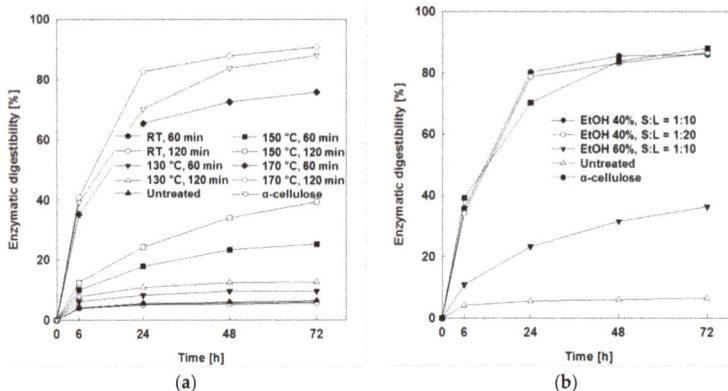

(a)   (b)

Figure 3. Enzymatic digestibility profiles of treated and untreated *Miscanthus* × *giganteus* at: (**a**) 40% ethanol concentration and L/S = 10 and (**b**) 170 °C for 90 min. RT (room temperature); EtOH (ethanol); L/S (liquid/solid); enzymatic hydrolysis conditions: 15 FPU of Cellic® CTec2/g-glucan, pH 4.8, 50 °C, and 150 rpm.

2.3. Crystallinity Index (CrI) of Untreated and Treated MG

The X-ray diffraction (XRD) patterns and *CrI* of untreated and treated MG at various reaction temperatures (RT, 130 °C, 150 °C, and 170 °C) are presented in Figure 4. Several studies have shown that as the reaction severity increases, the *CrI* of treated solid increases due to the removal of the amorphous parts (xylan and lignin) [29]. However, as shown in Figure 4, the crystallinities of the samples treated at RT (59.1) and 130 °C (62.9) were lower than that of the untreated sample (64.1). We speculated that the decrease in *CrI* for the samples treated at RT and 130 °C was because of the severe size reduction effect due to the ball milling pretreatment. In contrast, for the samples treated at 150 and 170 °C, the *CrI*s were 67.3 and 67.0, respectively. As the reaction temperature increased to >150 °C, XMG and lignin were removed and the *CrI* increased to more than that of the untreated sample, which may be due to the more removal of the amorphous portions.

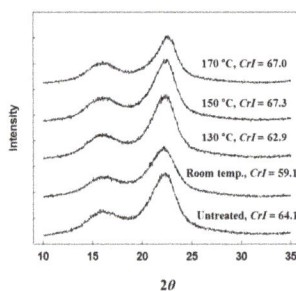

Figure 4. The diffraction patterns and *CrI*s (crystallinity indices) of untreated and treated *Miscanthus* × *giganteus* at different reaction temperatures. Reaction conditions: 60 min reaction time, 40% ethanol concentration and L/S = 10.

3. Materials and Methods

3.1. Materials

The air-dried MG, harvested in Wanju, Jeollabuk-do, Korea in 2016, was supplied by the National Institute of Crop Science (Wanju, Jeollabuk-do, Korea). The MG was ground (Blender 7012s, Warning Commercial Inc., Stamford, CT, USA) and screened to a nominal size of 14–45 mesh (from 0.36 to 1.4 mm). The average moisture content of the MG was determined to be 5.6% (by oven-drying at 105 °C for 24 h). The chemical composition of the untreated MG was determined using the laboratory analytical procedure (LAP) developed by the National Renewable Energy Laboratory (NREL, 2008).

The initial composition of the MG was: 43.8 wt.% glucan, 21.2 wt.% XMG, 0.5 wt.% arabinan, 3.9 wt.% acetyl group, 1.0 wt.% acid-soluble lignin, 19.6 wt.% acid-insoluble lignin, 1.5 wt.% organosolv extractives, 5.0 wt.% water-soluble extractives, and 2.2 wt.% ash ($n = 3$, standard deviations < 0.2). The mass closure of the untreated MG reached 98.5 wt.% (oven-dried biomass weight). The α-cellulose (catalog number C8002) and ethanol (95.0%, denatured, catalog number E0219) were purchased from Sigma-Aldrich Korea (Yongin, Gyeonggi-do, Korea) and Samchun Pure Chemical Co., Ltd. (Pyeongtaek, Gyeonggi-do, Korea), respectively.

3.2. Experimental Setup and Operation

3.2.1. Small Batch Organosolv Pretreatment Reactor

The small batch reactor (Figures 5 and A1) consisted of a control box, a reaction bath (molten salt bath and silicon oil bath), and a cooling bath (water bath). The bomb tubular reactor was made of SS-316L (internal diameter of 1.0 cm, length of 20.0 cm, and internal volume of 15.7 cm³). The temperature of the reaction bath was measured using a high temperature junction-type thermocouple (catalog number HY-72D, HANYOUNG NUX, Inchoen, Gyeonggi-do, Korea). The control box had a timer, a movement controller, and a digital temperature controller.

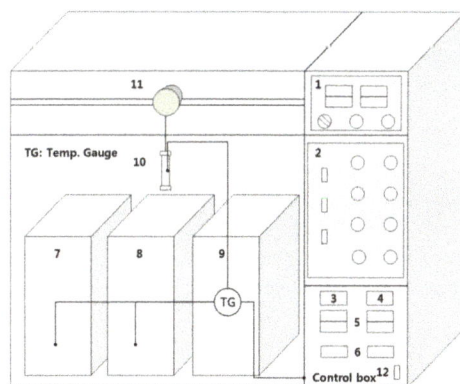

Figure 5. Schematic diagram of the small batch reactor setup: (1) timer and counter; (2) control box; (3) voltmeter; (4) ammeter; (5) thermo-controller; (6) temperature indicator; (7) salt bath; (8) silicon oil bath; (9) cooling bath; (10) batch reactor; (11) electric motor; (12) main switch.

First, 0.6 g (oven dry weight) MG and 6.0 mL ethanol (30%, 40%, 50% or 60%, *w/v*) were packed into the tubular-shaped batch reactor. The reactor was placed in a molten salt bath to preheat it to the target reaction temperature (170 °C). To shorten the preheating time, the molten salt bath was set to 50 °C higher than the target temperature. The preheating time was less than 1 min. When the target temperature was reached, the reactor was transferred to the silicon oil bath set at the target temperature (170 °C) targeted time (60 min). At the end of the reaction time, it was transferred to a water bath and rapidly cooled. After cooling, liquid samples were removed from the reactor then then diluted 3-fold with DI (deionized) water—it facilitated the lignin precipitation. This liquid sample was subjected to the evaporation (at 55 °C for 4 h), then acid hydrolysis was carried out to analyze the liquid sample by adding 72% sulfuric acid to a final sulfuric acid concentration of 4.0%. The solids removed from the reactor were washed with ethanol (30%, 40%, 50% or 60%, w/v) and water, and were then dried in a 45 °C convection oven. The dried solids were measured for weight loss and subjected to composition analysis. Each experiment was performed in triplicate.

3.2.2. Bench Scale Ball-Milling Pretreatment Reactor

A schematic diagram of the bench scale ball-milling reactor is shown in Figures 6 and A2. The main reactor was constructed using SS-316L (internal diameter of 30.0 cm, depth of 42.0 cm and internal volume of 30 L) with a wall thickness of 10.0 mm. The pressure, temperature, and rpm could be driven up to 20 kgf/cm^2, 200 °C, and 60 rpm, respectively. A metal spiral gasket was installed between the head and reactor at each of the two sides (front head and back head) to prevent pressure leakage. The front head was equipped with a hand hole 7.5 cm in diameter to feed in the reactant (biomass balls and ethanol) and to discharge the product. The reactor heating system was controlled of a mantle type electrical heater and capacity of heater was 10 kW. The motor (380 V, 3 Phase and 0.5 Hp) was installed to rotating for reactor and slip ring was installed with a rotating connector that could supply power or signals to the rotating equipment without twisting the line. The reactor was mounted on a frame with a turning roller to assist rotation and was equipped with a tilting system. The reactor was based on ball milling, but it was also designed to enable its operation under high temperature and high pressure. The alumina ball (HD, sphere type, 10 mm diameter and 3.6 g/cm^3 density) was purchased from NIKKATO Co., (Osaka, Japan). The alumina balls, MG, and ethanol (40% or 60%, w/v) were placed in the bench scale ball milling pretreatment reactor. The ratio of ball/MG/solvent was 30:1:10 ($w/w/v$). The reactor was preheated for 1 h until it reached the target temperature (130 °C, 150 °C or 170 °C), and the reactor was rotated during preheating. The pretreatment at RT was not preheated. When the target temperature was reached, the reaction was continued for a designated time (30, 60, 90 or 120 min). At the completion of the reaction, liquid samples were removed from the reactor then subjected to the evaporation (at 55 °C for 3 h), which was then diluted 3-fold with DI-water, which facilitated the lignin precipitation. This liquid sample was subjected to the evaporation (at 55 °C for 4 h), then acid hydrolysis was carried out to analyze the sugars and lignin in the liquid sample by adding 72% sulfuric acid to a final sulfuric acid concentration of 4.0%. The solids removed from the reactor were washed with ethanol (40% or 60%, w/v) and water and separated into two portions. One portion was dried in a 45 °C convection oven to determine weight loss (remaining solids) and subjected to composition analysis. The other portion was washed with DI-water and stored in a tightly sealed container until it underwent an enzymatic digestibility test. Each experiment was performed in duplicate.

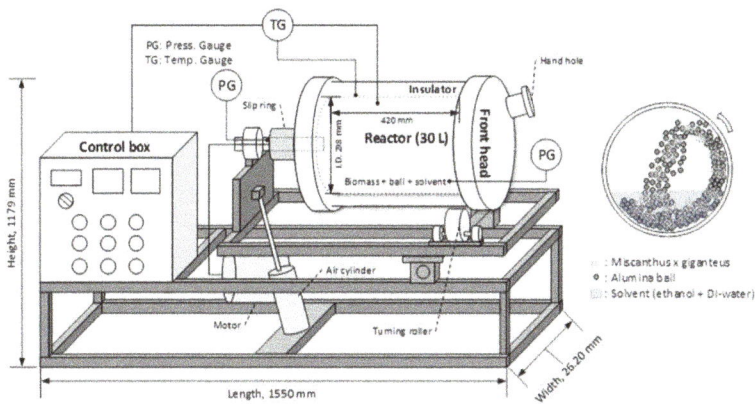

Figure 6. Schematic diagram of bench scale ball milling pretreatment reactor setup. PG (pressure gauge); TG (temperature gauge); DI (deionized) water.

3.3. Enzymatic Digestibility Test

The enzymatic digestibility of untreated and treated MG was determined according to the NREL Technical Report 510-42629 [30]. The tests were performed under the following conditions: 50 °C, 150 rpm, and pH 4.8 in a shaking incubator (BioFree, model BF-175SI, Co., Ltd., Dongdaemun-gu, Seoul, Korea). Enzyme loadings were 15 FPU/g-glucan of commercial cellulase enzyme Cellic CTec2 (Novozymes, A/S Bagsvaerd, Denmark). The initial glucan concentration was 1.0% (*w/v*) based on 100 mL (50 mM sodium citrate buffer) of total liquid in 250-mL Erlenmeyer flasks. To prevent microbial contamination, 1.0 mL of 20 mg/mL sodium azide was added. Samples were taken at the appropriate sampling times (6, 12, 24, 48 and 72 h) and analyzed for glucose content using high performance liquid chromatography (HPLC) with an HPX-87H column (cat. No. 125-0098, Bio-Rad Laboratories Inc., Hercules, CA, USA). The total glucose released after 72 h of hydrolysis was used to calculate the enzymatic digestibility for glucan as follows:

$$\text{Glucan digestibility (\%)} = \left[\frac{\text{Total released glucose (g)} \times 0.9}{\text{Initial glucan loading (g)}} \right] \times 100 \qquad (1)$$

where 0.9 is the conversion factor of glucose to the equivalent in glucan. Untreated MG and α-cellulose were put through the same procedure as the reference and the control, respectively.

3.4. Analytical Methods

The extractives, chemical compositions, and liquid fractions of the solid and liquid samples were determined by following the procedures of NREL-LAP [31]. Glucose, XMG and byproducts (formic acid, acetic acid, levulinic acid, 5-HMF, and furfural) were analyzed by HPLC. The HPLC system (1260 Infinity, Agilent Technologies Inc., Santa Clara, CA, USA) was used for the measurement of monomeric sugars and byproducts. The analytical column and detector used were an Aminex HPX-87H organic acid column (Bio-Rad) and a refractive index detector (1260 RID, Agilent Technologies Inc.), respectively. The mobile phase was 5 mM sulfuric acid in deionized water filtered through 0.2-μm filters and degassed by sonication. The operating conditions for the HPLC column were 65 °C with a mobile phase with a volumetric flow rate of 0.5 mL/min. External standards consisting of glucose, XMG, arabinose, formic acid, acetic acid, levulinic acid, 5-HMF, and furfural were used for quantification.

3.5. Crystallinity Measurement

Crystallinities of the untreated and treated MG were determined by XRD (Rigaku Co., Tokyo, Japan) operated at 40 kV and 40 mA. The samples were scanned at 4°/min (2θ = 10–35°, 0.02 increment). The crystallinity indices of the samples were calculated using the following equation [32].

$$CrI = \left[\frac{I_{002} - I_{18}}{I_{002}} \right] \times 100 \tag{2}$$

where I_{002} is the peak intensity corresponding to the 002 lattice plane of the cellulose molecule observed at 2θ equal to 22.5°, and I_{18} (at 2θ = 18°) is the peak intensity corresponding to amorphous cellulose.

4. Conclusions

In this study, we demonstrated that the combined mechanical and chemical pretreatment effectively reduced the recalcitrance of the MG. We also confirmed that the integration of mechanical and chemical pretreatment can be implemented in a large-scale reactor system. Reduced XMG, lignin, and cellulose crystallinity with minimal sugar loss could improve the enzymatic digestibility of the treated solids. For the improved enzymatic digestibility of glucan using commercial cellulase, pretreatment conditions were determined (170 °C reaction temperature, 90 min reaction time, 40% ethanol concentration, and 10 L/S ratio).

Author Contributions: T.H.K. as the first author conducted all experiments, summarized the data, and drafted the manuscript. D.J.I. as the co-author conducted a part of experiments and analyzed the data. T.H.K. and K.K.O. designed the reactor system as well as the overall study and experiments, interpreted the results, wrote the manuscript, and finalized the manuscript. T.H.K. and K.K.O., as the co-corresponding authors, contributed equally. All authors read and approved the final manuscript.

Funding: This work was supported by the New & Renewable Energy Core Technology Program of the Korea Institute of Energy Technology Evaluation and Planning, and was granted financial resources from the Ministry of Trade, Industry & Energy of the Republic of Korea (No. 20153030091050).

Conflicts of Interest: The authors declare no conflict of interest.

Appendix

Figure A1. Small batch reactor setup for organosolv only pretreatment.

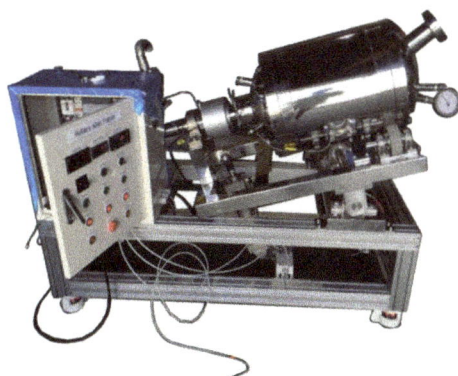

Figure A2. Bench scale ball milling pretreatment reactor setup for ball-milling organosolv pretreatment.

References

1. Chen, H.; Liu, J.; Chang, X.; Chen, D.; Xue, Y.; Liu, P.; Lin, H.; Han, S. A review on the pretreatment of lignocellulose for high-value chemicals. *Fuel Process. Technol.* **2017**, *160*, 196–206. [CrossRef]
2. Kim, T.H.; Kim, T.H. Overview of technical barriers and implementation of cellulosic ethanol in the US. *Energy* **2014**, *66*, 13–19. [CrossRef]
3. Zabed, H.; Sahu, J.N.; Boyce, A.N.; Faruq, G. Fuel ethanol production from lignocellulosic biomass: An overview on feedstocks and technological approaches. *Renew. Sustain. Energy Rev.* **2016**, *66*, 751–774. [CrossRef]
4. Szczodrak, J.; Fiedurek, J. Technology for conversion of lignocellulosic biomass to ethanol. *Biomass Bioenergy* **1996**, *10*, 367–375. [CrossRef]
5. Rao, L.V.; Goli, J.K.; Gentela, J.; Koti, S. Bioconversion of lignocellulosic biomass to xylitol: An overview. *Bioresour. Technol.* **2016**, *213*, 299–310.
6. Hassan, S.S.; Williams, G.A.; Jaiswal, A.K. Emerging Technologies for the Pretreatment of Lignocellulosic Biomass. *Bioresour. Technol.* **2018**, *262*, 310–318. [CrossRef] [PubMed]
7. Salapa, I.; Katsimpouras, C.; Topakas, E.; Sidiras, D. Organosolv pretreatment of wheat straw for efficient ethanol production using various solvents. *Biomass Bioenergy* **2017**, *100*, 10–16. [CrossRef]
8. Zhang, K.; Pei, Z.; Wang, D. Organic solvent pretreatment of lignocellulosic biomass for biofuels and biochemicals: A review. *Bioresour. Technol.* **2016**, *199*, 21–33. [CrossRef] [PubMed]
9. Pan, X.; Kadla, J.F.; Ehara, K.; Gilkes, N.; Saddler, J.N. Organosolv ethanol lignin from hybrid poplar as a radical scavenger: Relationship between lignin structure, extraction conditions, and antioxidant activity. *J. Agric. Food Chem.* **2006**, *54*, 5806–5813. [CrossRef] [PubMed]
10. Wildschut, J.; Smit, A.T.; Reith, J.H.; Huijgen, W.J. Ethanol-based organosolv fractionation of wheat straw for the production of lignin and enzymatically digestible cellulose. *Bioresour. Technol.* **2013**, *135*, 58–66. [CrossRef] [PubMed]
11. Pan, X.; Gilkes, N.; Kadla, J.; Pye, K.; Saka, S.; Gregg, D.; Ehara, K.; Xie, D.; Lam, D.; Saddler, J. Bioconversion of hybrid poplar to ethanol and co-products using an organosolv fractionation process: Optimization of process yields. *Biotechnol. Bioeng.* **2006**, *94*, 851–861. [CrossRef] [PubMed]
12. Geng, A.; Xin, F.; Ip, J.Y. Ethanol production from horticultural waste treated by a modified organosolv method. *Bioresour. Technol.* **2012**, *104*, 715–721. [CrossRef] [PubMed]
13. Sun, Y.C.; Wen, J.L.; Xu, F.; Sun, R.C. Structural and thermal characterization of hemicelluloses isolated by organic solvents and alkaline solutions from *Tamarix austromongolica*. *Bioresour. Technol.* **2011**, *102*, 5947–5951. [CrossRef] [PubMed]
14. Lee, J.H.; Kwon, J.H.; Kim, T.H.; Choi, W.I. Impact of planetary ball mills on corn stover characteristics and enzymatic digestibility depending on grinding ball properties. *Bioresour. Technol.* **2017**, *241*, 1094–1100. [CrossRef] [PubMed]

15. Avolio, R.; Bonadies, I.; Capitani, D.; Errico, M.E.; Gentile, G.; Avella, M. A multitechnique approach to assess the effect of ball milling on cellulose. *Carbohydr. Polym.* **2012**, *87*, 265–273. [CrossRef]

16. Zhao, H.; Kwak, J.H.; Wang, Y.; Franz, J.A.; White, J.M.; Holladay, J.E. Effects of crystallinity on dilute acid hydrolysis of cellulose by cellulose ball-milling study. *Energy Fuels* **2006**, *20*, 807–811. [CrossRef]

17. Khullar, E.; Dien, B.S.; Rausch, K.D.; Tumbleson, M.E.; Singh, V. Effect of particle size on enzymatic hydrolysis of pretreated Miscanthus. *Ind. Crops Prod.* **2013**, *135*, 11–17. [CrossRef]

18. Liu, H.; Pang, B.; Zhao, Y.; Lu, J.; Han, Y.; Wang, H. Comparative study of two different alkali-mechanical pretreatments of corn stover for bioethanol production. *Fuel* **2018**, *221*, 21–27. [CrossRef]

19. Shi, Z.; Liu, Y.; Xu, H.; Yang, Q.; Xiong, C.; Kuga, S.; Matsumoto, Y. Facile dissolution of wood pulp in aqueous NaOH/urea solution by ball milling pretreatment. *Ind. Crops Prod.* **2018**, *118*, 48–52. [CrossRef]

20. Dai, L.; Li, C.; Zhang, J.; Cheng, F. Preparation and characterization of starch nanocrystals combining ball milling with acid hydrolysis. *Carbohydr. Polym.* **2018**, *180*, 122–127. [CrossRef] [PubMed]

21. Zhao, X.; Cheng, K.; Liu, D. Organosolv pretreatment of lignocellulosic biomass for enzymatic hydrolysis. *Appl. Microbiol. Biotechnol.* **2009**, *82*, 815. [CrossRef] [PubMed]

22. Nitsos, C.; Rova, U.; Christakopoulos, P. Organosolv fractionation of softwood biomass for biofuel and biorefinery applications. *Energies* **2017**, *11*, 50. [CrossRef]

23. Teramoto, Y.; Tanaka, N.; Lee, S.H.; Endo, T. Pretreatment of eucalyptus wood chips for enzymatic saccharification using combined sulfuric acid-free ethanol cooking and ball milling. *Biotechnol. Bioeng.* **2008**, *99*, 75–85. [CrossRef] [PubMed]

24. Yan, J.; Joshee, N.; Liu, S. Utilization of hardwood in biorefinery: A kinetic interpretation of pilot-scale hot-water pretreatment of Paulownia elongata woodchips. *J. Biobased Mater. Bioenergy* **2016**, *10*, 339–348. [CrossRef]

25. Liu, S. Woody biomass: Niche position as a source of sustainable renewable chemicals and energy and kinetics of hot-water extraction/hydrolysis. *Biotechnol. Adv.* **2010**, *28*, 563–582. [CrossRef] [PubMed]

26. Yuan, T.Q.; You, T.T.; Wang, W.; Xu, F.; Sun, R.C. Synergistic benefits of ionic liquid and alkaline pretreatments of poplar wood. Part 2: Characterization of lignin and hemicelluloses. *Bioresour. Technol.* **2013**, *136*, 345–350. [CrossRef] [PubMed]

27. Weinwurm, F.; Turk, T.; Denner, J.; Whitmore, K.; Friedl, A. Combined liquid hot water and ethanol organosolv treatment of wheat straw for extraction and reaction modeling. *J. Clean. Prod.* **2017**, *165*, 1473–1484. [CrossRef]

28. Kim, T.H.; Jeon, Y.J.; Oh, K.K.; Kim, T.H. Production of furfural and cellulose from barley straw using acidified zinc chloride. *Korean J. Chem. Eng.* **2013**, *30*, 1339–1346. [CrossRef]

29. Kim, J.S.; Lee, Y.Y.; Kim, T.H. A review on alkaline pretreatment technology for bioconversion of lignocellulosic biomass. *Bioresour. Technol.* **2016**, *199*, 42–48. [CrossRef] [PubMed]

30. Selig, M.; Weiss, N.; Ji, Y. *Enzymatic Saccharification of Lignocellulosic Biomass*; NREL/TP-510–42629; National Renewable Energy Laboratory: Golden, CO, USA, 2018.

31. Sluiter, A.; Hames, B.; Ruiz, R.; Scarlata, C.; Sluiter, J.; Tmpleton, D.; Crocker, D. *Determination of Structural Carbohydrates and Lignin in Biomass*; NREL/TP-510–42618; National Renewable Energy Laboratory: Golden, CO, USA, 2012.

32. Cao, Y.; Tan, H. Study on crystal structures of enzyme-hydrolyzed cellulosic materials by X-ray diffraction. *Enzyme Microb. Technol.* **2005**, *36*, 314–317. [CrossRef]

Review

A Critical Analysis of Bio-Hydrocarbon Production in Bacteria: Current Challenges and Future Directions

Ziaur Rahman [1,*], **Javed Nawab** [2], **Bong Hyun Sung** [3] **and Sun Chang Kim** [4,*]

[1] Department of Microbiology, Abdul Wali Khan University Mardan, Mardan 23200, Pakistan
[2] Department of Environmental Sciences, Abdul Wali Khan University Mardan, Mardan 23200, Pakistan; javednawab11@yahoo.com
[3] Cell Factory Research Center, Korea Research Institute of Bioscience and Biotechnology, Daejeon 34141, Korea; bhsung@kribb.re.kr
[4] Department of Biological Sciences, Korea Advanced Institute of Science and Technology, Daejeon 34141, Korea
[*] Correspondence: ziamicrobiologist@gmail.com (Z.R.); sunkim@kaist.ac.kr (S.C.K.);
Tel.: +92-340-901-2316 (Z.R.); +82-42-350-2619 (S.C.K.)

Received: 10 September 2018; Accepted: 3 October 2018; Published: 6 October 2018

Abstract: As global fossil reserves are abruptly diminishing, there is a great need for bioenergy. Renewable and sustainable bioenergy products such as biofuels could fulfill the global energy demand, while minimizing global warming. Next-generation biofuels produced by engineered microorganisms are economical and do not rely on edible resources. The ideal biofuels are alcohols and n-alkanes, as they mimic the molecules in fossil fuels and possess high energy densities. Alcohols and n-alkane hydrocarbons (C_2 to C_{18}) have been produced using engineered microorganisms. However, it is difficult to optimize the complex metabolic networks in engineered microorganisms to obtain these valuable bio-hydrocarbons in high yields. Metabolic engineering results in drastic and adverse cellular changes that minimize production yield in microbes. Here, we provide an overview of the progress in next-generation biofuel (alcohols and n-alkanes) production in various engineered microorganisms and discuss the latest tools for strain development that improve biofuel production.

Keywords: microbial biofuel; metabolic engineering; alkanes; alcohols

---

1. Introduction

The global scarcity and predicted depletion of fossil fuel reserves is a threat to future energy demand [1–3]. Excess consumption of transportation fuel leads to ever-increasing global warming [1,4]. There is an urgent need to replace the current fossil fuel technology with renewable, sustainable, and environmentally friendly alternatives [5,6]. Biofuels are considered sustainable if they are produced from biomass through biological or non-biological methods. The ideal biofuels are bio-hydrocarbons and these have been suggested as replacements for fossil fuels. Bio-hydrocarbons produced in bacteria, such as alcohols (C_2–C_{18}) and n-alkanes (C_{10}–C_{20}), are of great importance [7–9]. These bio-hydrocarbons have advantages over fossil fuels, as they possess higher energy densities and emit fewer toxic chemicals into the environment [10,11]. These ready-to-use bio-hydrocarbons do not require the replacement of existing engines and are compatible with current technologies [12]. The current methods used for production of these biofuels are non-economical and are based on the conversion of plant and vegetable oil using the Fischer-Tropsch method [13,14]. The genetically engineered microorganisms could utilize cheaper raw materials such as cellulosic sugars and carbon dioxide to produce bio-hydrocarbons more economically.

Bio-hydrocarbons (alcohols and n-alkanes) are produced in engineered bacteria by modifying their central metabolism, including glycolytic, tricarboxylic, and fatty acid pathways (Figure 1). The current production yield of bio-hydrocarbons from modified bacteria is far below commercial scale due to the challenges of metabolic engineering. Current challenges include cellular toxicity, growth retardation, cofactor exhaustion and metabolic pathway imbalance. In this paper, the metabolic engineering of bacteria for bio-hydrocarbon production is reviewed. In addition, we discuss current problems and challenges in strain optimization for bio-hydrocarbon production and provide future directions for this technology.

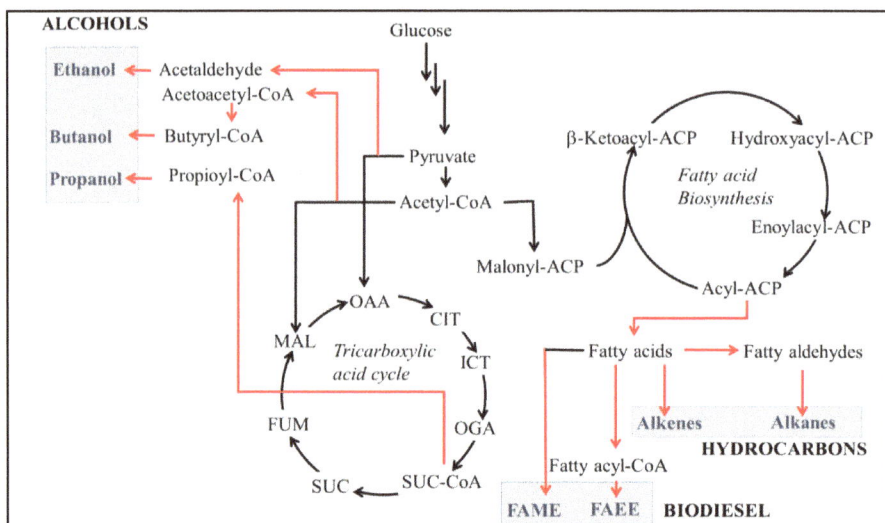

Figure 1. Metabolic pathways of short- and medium-chain alcohols and hydrocarbon production in an engineered bacterial cell. Glucose is metabolized to acetyl-CoA, a precursor of the tricarboxylic acid (TCA) cycle and fatty acid biosynthesis pathways. Short-chain alcohols (ethanol, propanol, butanol) are produced from acetyl-CoA and TCA. Fatty alcohols, fatty acid ethyl esters and *n*-alkanes are produced by utilizing fatty acyl-ACPs of the fatty acid machinery. The native pathways are shown in black, while heterologous pathways are shown in red. CoA: Coenzyme A; ACP: Acyl Carrier Protein; OAA: Oxalo Acetic Acid; CIT: Citrate; ICT: Iso Citrate; OGA: Oxo Glutaric Acid; SUC: Succinic Acid; FUM: Fumaric Acid; MAL: Malic Acid; FAME: Fatty Acid Methyl Ester; FAEE: Fatty Acid Ethyl Ester.

2. Engineering Metabolic Pathways for Hydrocarbon Production in Bacteria

Initial work on engineering microbes for biofuel production involved introducing the ethanol-producing pathway from *Zymomonas mobilis* into *Escherichia coli* [15]. The progress of most biofuels produced this way is discussed in detail in previous reviews [16–18]. In this section, we present an overview of alcohol and *n*-alkane biofuels produced by engineered microorganisms.

2.1. Alcohols Production in Bacteria

Next-generation alcohols are a group of compounds used in chemical industries and as transportation biofuels [19,20]. They can be grouped into short, medium, and long-chain alcohols, ranging from C_2 to C_6, C_6 to C_{12}, and C_{12} to C_{18} chain lengths, respectively [21–23]. They can be used as 100% solutions or blended with petroleum. Custom alcohols are currently produced by distillation during petroleum processing or chemically from fatty acids, which makes their production unviable in the long run. The sustainable approach to their production is to use fermenters, utilizing microorganisms that can easily convert cellulosic material into ready-to-use biofuels. Ethanol is a renewable liquid fuel and its annual production is greater than 100 billion liters, with an increase of 10% per year. More than 60% of ethanol is produced with fermentation technology using microorganisms [24].

From an engineering perspective, microbial strain selection is an important consideration. The best strain is one that suits and resists the characteristics of the fermenter environment, such as pH, temperature, acidity, agitation, and biofuel toxicity.

Wild-type *Escherichia coli* produces ethanol under fermentative conditions by converting pyruvate to acetyl-CoA and formate through the enzyme pyruvate formate lyase, encoded by *pfl* gene [15]. This process involves a two-step reaction: acetyl-CoA is converted to acetaldehyde, followed by ethanol synthesis. The overall reaction is catalyzed by a bifunctional aldehyde dehydrogenase (ADH), encoded by the *adhE* gene (Figure 1). Similarly, butanol is produced by the fermentative anaerobe, *Clostridium acetobutylicums*, through the acetone-butanol-ethanol (ABE) pathway [19]. The problem with wild-type strains is the low production titer. This is the main focus of most research into enhancing yields using metabolic engineering tools.

Pichia pastoris used for industrial production of chemicals was also engineered for butanol production [25]. The amino acid synthesis pathways of *Pichia pastoris*, specifically the endogenous L-valine biosynthesis pathway, was overexpressed to produce 2.22 g/L of butanol from glucose as a carbon source [25]. Furthermore, using cheaper raw materials for the production of renewable biofuel could minimize the final cost. Glycerol is one of the cost effective byproducts of the biodiesel production industry. Engineered microbes could utilize glycerol as a carbon source. To obtain the phenotype, Saini et al. (2017) rewired the central metabolism of *Escherichia coli* to utilize glycerol as a carbon source, introduced a butanol synthesis pathway, and obtained a titer of 6.9 g/L butanol from 20 g/L glycerol (Table 1) [26]. This highest titer was obtained because of forcing glycolytic flux, enabling gluconeogenic pathway, improving the breakdown of glycerol and moderately suppressing tricarboxylic acid pathway [26]. To enhance production yield, Liew et al. (2017) introduced a novel pathway enzyme, aldehyde/alcohol ferredoxin reductase (AOR), to replace the native bifunctional aldehyde dehydrogenase in *Clostridium autoethanogenum* [9]. The strain was engineered using the allelic mutagenesis approach and as a result, the final strain produced ethanol at a titer of up to 180% (0.28 g/L) from CO_2 as a carbon source in batch fermentation experiments [9].

Furthermore, the yeast strains were found to be promising for butanol production. *Pichia pastoris* used industrial chemical production and has also been engineered for butanol production [25]. The endogenous L-valine biosynthesis pathway of *P. pastoris* was overexpressed to produce 2.22 g/L of butanol from glucose as a carbon source [25]. Furthermore, using cheaper raw materials for renewable biofuel production could minimize the final cost. Glycerol, a byproduct of biodiesel production and other industrial processes, is a cost-effective raw material and engineered microbes could utilize it as a carbon source. Saini et al. (2017) rewired the central metabolism of *Escherichia coli* to utilize glycerol as the sole carbon source and introduced a butanol synthesis pathway that gives a production titer of 6.9 g/L of butanol from 20 g/L of glycerol (Table 1) [26]. This high titer was obtained by forcing glycolytic flux, enabling the gluconeogenic pathway, improving breakdown of glycerol, and moderately suppressing the tricarboxylic acid pathway [26]. The breakthrough report for butanol production was obtained using a *Clostridium tyrobutyricum* strain [27]. The complex regulatory pathway of butyrate production was manipulated by using the CRISPR-Cas system to delete *cat1* (regulatory component) and to integrate *adhE1* (alcohol producing) genes. The resultant strain achieved a record butanol titer of 26.2 g/L in batch fermentation using CO_2 as the carbon source (Table 1) [27].

2.2. Alkanes Production in Bacteria

High-energy renewable-fuel-like *n*-alkanes are direct replacements for petroleum fuel. These molecules range in chain length from C_8 to C_{18}. Chemically, *n*-alkanes are produced by the Fischer–Tropsch process, making them uneconomical. Biologically, *n*-alkanes are produced in *Escherichia coli* by modifying the lipid biosynthesis pathway and introducing aldehyde reductase (AAR) and aldehyde-deformylating oxygenase (ADO) from cyanobacteria [28]. Fatty acid synthesis plays a vital role in the production of *n*-alkanes. The prokaryotic fatty acid pathway is composed of individual components/enzymes. Acetyl-CoA enters the fatty acid pathway and a subsequent condensation reaction leads to the elongation of fatty acids. The alkane's production is directly related to fatty acid biosynthesis in bacteria. The fatty acid synthesis machinery of *E. coli* produces fatty acids from malonyl-CoA condensation. The first step in the fatty acid biosynthesis is carried out by acetyl-CoA carboxylase (ACC), a tetramer enzyme which produces malonyl-CoA from acetyl-CoA. The malonyl-CoA then transfers to acyl carrier protein (ACP) an elongation unit to get malonyl-ACP with the help of FabD, malonyl-CoA: ACP transcylase enzyme. The condensation cycle composes of four reactions catalyzed by FabH (3-oxooacyl-ACP synthase III), FabG (3-oxooacyl-ACP reductase), FabA/FabZ (enoyl-ACP synthase), and FabI (enoyl-ACP reductase), responsible for the production of 3-ketoacyl-ACP, 3-hydroxyacyl-ACP, enoyl-ACP, and acyl-ACP, respectively [16]. The fatty acids are then released from acyl carrier proteins with the help of thioesterases. Free fatty acids are then converted into fatty aldehydes and *n*-alkanes by AAR and ADO, respectively. The details of microbial biofuel production are shown in Table 1 and Figure 2.

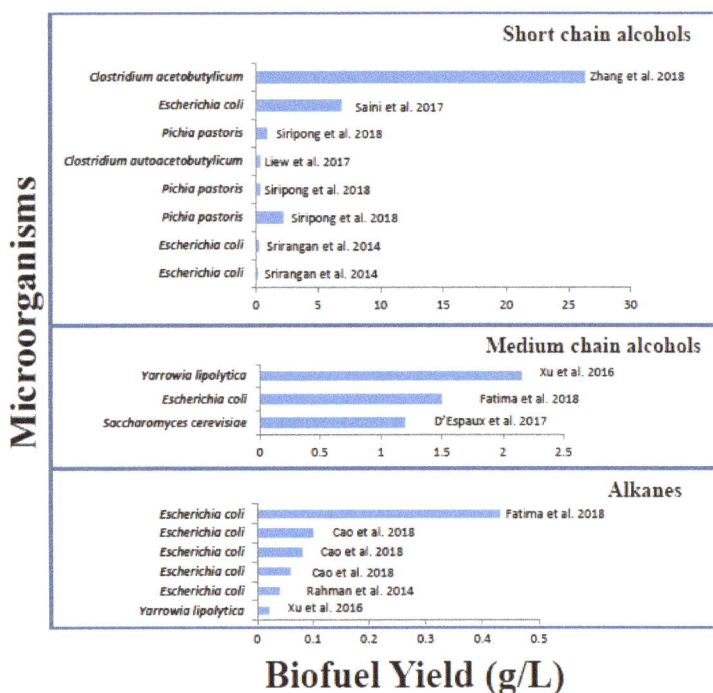

Figure 2. Recent studies on the production of short-chain alcohols, medium-chain alcohols and *n*-alkanes in microorganisms. The data represent batch fermentation experiments. For more detail, see Table 1.

3. Comparative Analysis of Bio-Hydrocarbons Production in Bacteria

Metabolic engineering plays a vital role in enhancing the production yield of bio-hydrocarbons. The ethanol produced in wild type *E. coli* with aldehyde dehydrogenase enzyme is far below the commercial threshold. To improve production yield, the genes of sleepy beauty mutase operon containing methyl malonyl CoA mutase, methyl malonyl CoA decarboxylase propinoyl CoA: Succinate transferase enzymes were expressed in *E. coli* that successfully produced 31 and 7 g/L ethanol and propanol, respectively in a fed-batch fermentation (Tables 1 and 2). The pathway for the production of propanol using sleeping beauty mutase (Sbm) was different when compared to propanol production using the keto acid pathway [21]. The highest production of propanol by Sbm pathway may be due to anaerobic conditions as compared to the aerobic keto acid pathway [21,29].

The C_4 alcohol (butanol) is considered better than ethanol in terms of hygroscopic properties and high energy densities [21]. The *Clostridium acetobutylicum* is a natural producer of butanol. The main problem in the wild type strain is butyrate production that lowers butanol synthesis. Zhang et al. (2018) speculated that by interrupting the butyrate synthesis genes, the butanol production may be enhanced. The *spoA* and *cat1* genes of butyrate synthesis were selectively deleted by using the CRISPR-Cas technology and as a result the highest titer of 26.2 g/L butanol in batch fermentation was achieved (Tables 1 and 2). The breakthrough report for butanol production was obtained in *Clostridium tyrobutyricum* strain [27]. The complex regulatory pathway of butyrate production was manipulated with the CRISPR-Cas system by deletion of *cat1* (regulatory component) and integration of *adhE1* (alcohol producing) genes [27]. The resultant strain achieved record production of butanol up to 26.2 g/L in a batch fermentation experiment using CO_2 as a carbon source (Figure 2) [27].

The fatty alcohol's production is linked with fatty acid synthesis pathways [16]. The fermentative products such as lactate, acetate, pyruvate, and formate are also synthesized in fatty alcohols producing strains that lower the final production titer [8]. To overcome these problems, Fatima et al. (2018) developed a constraint-based modeling approach and identified the competing pathways, such as formate, lactate, and acetate synthesis. By blocking these pathways, 1.5 and 12.5 g/L fatty alcohols were obtained in batch and fed-batch fermentation, respectively (Tables 1 and 2).

The alkane hydrocarbons are considered as a direct replacement for fossil petroleum [20]. The bio production of alkanes remains a major challenge because of slow acting aldehyde deformylating oxygenase, an alkane-producing enzymes (Figure 1) [28]. Currently, a number of efforts have been made to boost the alkane production but it is still far below the commercial threshold. The second problem is the toxicity of intermediate fatty aldehyde to the host cells. The toxicity of fatty aldehydes was overcome by applying DNA scaffolds to enzymes AAR and ADO and enhanced 8-fold alkanes production 0.04 g/L [28]. Cao et al. (2016) blocked the competing pathways such as endogenous reductases, and engineered the electron transfer system towards alkane-producing enzymes. The resultant strains produced 0.1 and 1.3 g/L alkanes in a batch and fed-batch fermentation, respectively [30]. The details about current production yields and experimental conditions can be found in Tables 1 and 2.

Table 1. Biofuel production in engineered bacteria.

Biofuel	Substrate	Microorganism	Metabolic Engineering	Yield (g/L)	References
Ethanol	Glycerol	*Escherichia coli*	Heterologous expression of sleeping beauty mutase operon	0.20 [a], 31 [b]	[29]
	CO_2	*Clostridium autoacetobutylicum*	Allelic mutagenesis of aldehyde dehydrogenase (*adhE*) and expression of aldehyde ferredoxin reductase (AOR)	0.28 [a]	[9]
Propanol	Glycerol	*Escherichia coli*	Heterologous expression of sleeping beauty mutase operon	0.15 [a], 7 [b]	[29]
Butanol	CO_2	*Clostridium acetobutylicum*	Targeted deletion of *sopA* and *cat1* (butyrate production) and integration of *adhE* gene by CRISPR-Cas9	26.2 [a]	[27]
	Glycerol	*Escherichia coli*	Enhanced glycolytic flux, prevention of glycerol breakdown	6.9 [a]	[26]
	2-keto isovalerate	*Pichia pastoris*	Expression of keto acid degradation pathway	0.28 [a]	[25]
	Glucose	*Pichia pastoris*	Overexpression of endogenous L-Valine biosynthesis pathway	0.89 [a]	[25]
		Pichia pastoris	Fine tuning of expression levels using episomal plasmids	2.22 [a]	[25]

Table 1. *Cont.*

Biofuel	Substrate	Microorganism	Metabolic Engineering	Yield (g/L)	References
Fatty alcohols	Glucose	*Escherichia coli*	In *silico* modelling of metabolic pathways for alkane production to remove bottlenecks identified in pentose phosphate pathway	1.5 [a], 12.5 [b]	[8]
	Glucose	*Yarrowia lipolytica*	Expression of alkane-producing pathway, consisting of aldehyde reductase (AAR) and aldehyde deformylating oxygenase (FAD)	2.15 [a]	[20]
	Lignocellulose	*Saccharomyces cerevisiae*	Expression of alkane-producing pathway Bottlenecks were identified by flux balance analysis and bacterial fatty aldehyde reductases (FARs) were replaced with mouse FARs	1.2 [a], 6 [b]	[7]
Alkanes	Glucose	*Escherichia coli*	Heterologous expression of alkane-producing pathway (AAR & aldehyde-deformylating oxygenase (ADO)) from cyanobacteria, with no further modification	0.06 [a]	[30]
		Escherichia coli	Expression of alkane-producing pathway Deletion of *adhE* (endogenous reductase)	0.08 [a]	[30]
		Escherichia coli	Expression of alkane-producing pathway and modification of electron transfer system	0.1 [a], 1.3 [b]	[30]
		Escherichia coli	Expression of alkane-producing pathway and DNA scaffolding of cyanobacterial AAR & ADO	0.04 [a]	[28]
		Yarrowia lipolytica	Expression of alkane-producing pathway (FAR & FAD), with no further modification	0.02 [a]	[20]
		Escherichia coli	In *silico* modelling of metabolic pathways for alkane production to help remove bottlenecks in pentose phosphate pathway	0.43 [a], 2.5 [b]	[8]

Notes: [a]: batch fermentation; [b]: fed-batch fermentation.

Table 2. Comparison of experimental parameters of biofuel production in bacteria.

Biofuel	Microorganism	Experimental Conditions						Yield (g/L)	References
		Substrate	Substrate Concentration	pH	Temperature	Cell Density Initial-Final	Reaction Time		[29]
Ethanol	*Escherichia coli*	Glycerol	30 g/L	7.0	30 °C	0.4–1.8 g/DCW/L	28 h	0.20 [a], 31 [b]	[9]
	Clostridium autoacetobutylicum	CO_2	50 kPa	5.8	37 °C	0.2–1.8 $O.D_{600}$	30 h	0.28 [a]	[29]
Propanol	*Escherichia coli*	Glycerol	30 g/L	7.0	30 °C	1.8 g/DCW/L	28 h	0.15 [a], 7 [b]	[27]
Butanol	*Clostridium acetobutylicum*	CO_2	-	6.0	37 °C	-	30 h	26.2 [a]	[26]
	Escherichia coli	Glycerol	30 g/L	7.0	37 °C	0.2–7.0 $O.D_{550}$	40 h	6.9 [a]	[25]
	Pichia pastoris	2-keto isovalerate	4 g/L	-	30 °C	0.05–12 $O.D_{550}$	72 h	0.28 [a]	[25]
	Pichia pastoris	Glucose	20 g/L	-	30 °C	0.05–1.2 $O.D_{550}$	72 h	0.89 [a]	[25]
	Pichia pastoris	Glucose	20 g/L	-	30 °C	0.05–2.0 $O.D_{550}$	72 h	2.22 [a]	[8]
Fatty alcohols	*Escherichia coli*	Glucose	20 g/L	7.0	30 °C	0.02-$O.D_{550}$	48 h	1.5 [a], 12.5 [b]	[20]
	Yarrowia lipolytica	Glucose	100 g/L	5.5	28 °C	1–80 $O.D_{600}$	120 h	2.15 [a]	[7]
	Saccharomyces cerevisiae	Lignocellulose	20 g biomass	5.0	30 °C	1–25 $O.D_{600}$	240 h	1.2 [a], 6 [b]	[30]
Alkanes	*Escherichia coli*	Glucose	20 g/L	7.0	30 °C	0.5–4.0 $O.D_{600}$	50 h	0.06 [a]	[30]
	Escherichia coli	Glucose	20 g/L	7.0	30 °C	0.5–4.0 $O.D_{600}$	50 h	0.08 [a]	[30]
	Escherichia coli	Glycerol	140 g/L	7.0	30 °C	0.5–140 $O.D_{600}$	50 h	0.1 [a], 1.3 [b]	[28]
	Escherichia coli	Glucose	20 g/L	7.0	30 °C	0.4–3.0 $O.D_{600}$	48 h	0.04 [a]	[20]
	Yarrowia lipolytica	Glucose	100 g/L	5.5	28 °C	1–80 $O.D_{600}$	120 h	0.02 [a]	[8]
	Escherichia coli	Glucose	20 g/L	7.0	37 °C	0.02–102 $O.D_{550}$	72 h	0.43 [a], 2.5 [b]	[29]

Notes: [a]: batch fermentation; [b]: fed-batch fermentation.

4. Current Challenges in Improving Hydrocarbon Production

4.1. Promoters

Promoters are important genetic elements responsible for the controlled expression of genes on a chromosome or plasmid [31,32]. In wild-type cells, promoters are well organized and controlled by repressors and inducible factors. From a bioengineering perspective, promoters are critical for balancing the metabolic flux of heterologous and/or engineered pathways by controlling the expression levels of individual genes. Therefore, choosing the correct promoter is an important step in engineering prokaryotic and eukaryotic cells. Uncontrolled gene expression in a heterologous host exerts a metabolic burden and results in the exhaustion of cellular biomolecules that are essential for cellular growth, leading to undesired physiological changes (Figure 3) [33]. Therefore, the promoter needs to be properly employed for smooth growth and productivity of the desired products. For example, when a target protein is toxic to the cell, a low-strength promoter is needed. If the expression of multiple genes is not optimized or if the catalytic activity of an enzyme is greater than the corresponding enzymes of the same metabolic pathway, it is recommended to balance the pathway with native/synthetic promoters to optimize flux towards the product. A wide range of promoters have been characterized and are available for research and academic purposes [34]. Historically, natural constitutive and inducible promoters have been used in this research. However, more recently, novel artificial promoters have been designed by modifying the −35 (TTGACA) and −10 (TATA) sequences. The expression patterns of these novel promoters are diverse [34]. Furthermore, two modes of transcription initiation have been reported. In the positive control mode, the interaction between regulatory proteins and their corresponding recognition sequences switches on transcription. In contrast, the negative control mode inhibits transcription initiation by the interaction of regulatory proteins with their regulatory regions. These synthetic promoters have been studied with the help of reporter genes coding for fluorescent proteins (GFP, RFP), luciferase, and galactosidase [35]. The promoters can be switched on/off using both chemical and physical stimuli. For instance, the Lac promoter is controlled by the LacI repressor and the IPTG inducer. The Tet promoter is controlled by the TetR repressor and the Tc inducer [36]. Therefore, selecting the appropriate promoter is critical for controlling gene expression levels and optimizing pathway flux. This can be achieved using either single-promoter-single-gene or single-promoter-multiple-gene expression systems. In addition, certain enzymes are required at different times of fermentation. Heat-inducible phage promoters and constitutive expression systems of Bacillus subtilis have proven promising for industrial and batch fermentation scales [37]. Currently, synthetic or artificial promoters are used in bacterial, yeast and mammalian cells. Basic questions about balancing pathway flux for biofuel production have been answered using synthetic promoters [38–40]. Another bottleneck in strain development for biofuel production is the combinatorial assembly of building blocks (promoters, genes, and terminators). These issues have mostly been solved using computational approaches, such as the DNA fragment compiler software, BioPartsBuilder [41].

4.2. Gene Copy Number

Gene expression level can also be regulated by controlling gene copy number (GCN). Changes in microbial GCN have been successful at balancing the flow of metabolites towards biofuels in engineered pathways. Enzymes in metabolic pathways have different catalytic turnover rates. Therefore, if their expression is not normalized according to their catalytic efficiencies, physiological disturbances such as toxicity and cell death can occur [33]. The strategies used to control the desired gene expression levels through GCN by selecting low-, medium-, and high-copy-number plasmids are discussed in the literature [42]. Gene expression levels can also be controlled by knocking down genes or integrating them into a chromosome, leading to a decrease or increase in GCN, respectively [43]. Furthermore, the strategy to engineer high GCN plasmids has been successful in the fission yeast, *Saccharomyces cerevisiae*. In *S. cerevisiae*, CEN/ARS a centromere based plasmid has been engineered to obtain plasmid diversity with low, medium and high copy number versions [44]. For the development of Escherichia coli strains, plasmids with low, medium and high copy number per cell have been developed. Based on the ColE1 replicon, 15 to 20 plasmid copies per cell can be maintained. To obtain higher gene expression levels, 200+ copies of the pMB1 replicon in the pUC plasmid have been used [45]. More recently, traditional gene cloning approaches are being replaced with novel self-cloning techniques, based on transcription activator-like effector nuclease (TALEN) in *Pseudochoricystis ellipsoidea* [46].

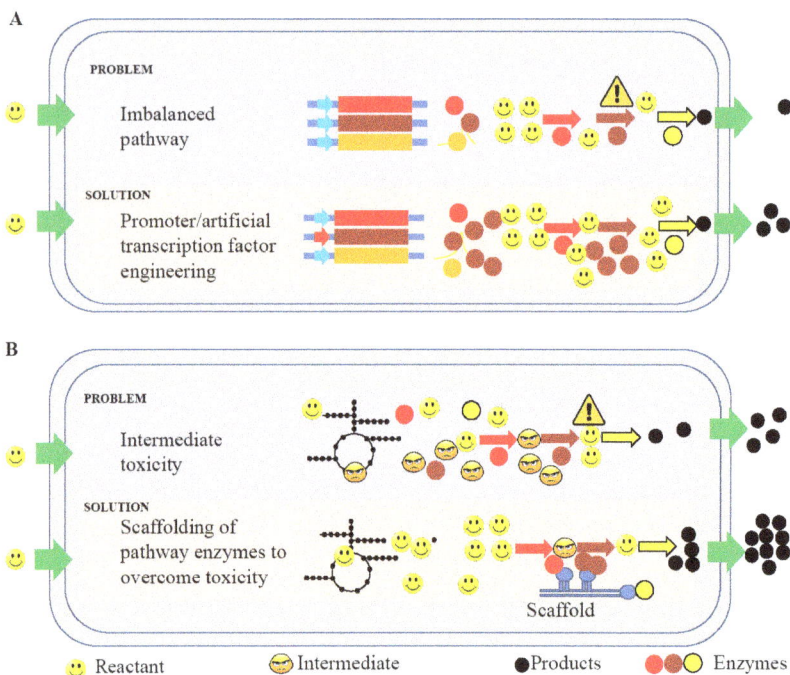

Figure 3. Graphical representation of improving bio-hydrocarbon production in bacteria. The figure shows challenges and possible solutions in transcription (**A**) and metabolite toxicity (**B**) in newly engineered strains.

4.3. Artificial Transcription Factor (ATF) Engineering

Artificial transcription factors (ATF) control genome-wide gene expression levels in engineered host cells to balance metabolic pathways (Figure 3). Transcription factors are composed of a DNA binding domain and an effector domain, both of which are required for regulating the expression of downstream genes. Specifically, the DNA binding domain is responsible for binding to the target regulatory site of a gene and the effector domain facilitates the initiation of downstream gene expression by RNA polymerase. Most ATFs are composed of zinc fingers (ZF) and sigma factors. ZF proteins bind specifically to DNA and function as regulatory factors. ATFs recognize specific DNA sequences, typically near the transcription start site of the target gene. Numerous methods have been developed for the construction of ATFs [47]. At present, ATFs are mainly used in industrial microorganisms to improve their tolerance to environmental stresses and to enhance the yield of chemicals and fuels. The ATF-based strain engineering approach has proven successful in improving tolerance to various environmental and physical stresses [48]. The T2 ATF has been shown to generate thermo tolerance and butanol tolerance phenotypes [48]. Furthermore, DNA microarray analysis of a T2 ATF strain showed up-regulation of the outer membrane protein gene, *ompW* and down-regulation of the *marRAB* operon. The exact function of OmpW is not known, but a previous publication has suggested that it has an important role during environmental stress and under physically varied conditions in industrial bioreactors [48]. Lee et al. (2011) engineered an ATF library by fusing ZFs with a cyclic adenosine mono phosphate receptor protein (CRP) and screened for butanol tolerance in *E. coli* [49]. Wild-type *E. coli* could not tolerate 1% (v/v) butanol, but a butanol-tolerant strain with the ATF was able to tolerate 1.5% (v/v) butanol (Lee et al. 2011). Similarly, lignocellulosic properties of fungal species were studied for efficient degradation of plant cell walls. Ling et al. (2011) engineered a transcription factor from *Saccharomyces cerevisiae* to generate an *n*-alkanes-resistant phenotype by site-directed mutagenesis of the drug resistant transcription factors (PDR), PDR1P and PDR3P. Point mutations were generated (PDR1, F815S and PDR3, Y276H), which resulted in high tolerance to C_{10}-alkanes (Ling et al. 2015). More recently, an acetic-acid-resistant phenotype was also generated using an ATF that successfully recognized novel genes responsible for the tolerance phenotype [50].

4.4. Ribosome Engineering

Ribosome engineering is also considered a useful tool for modulating gene expression levels. Ochi (2007) showed that the gene for ribosomal proteins, *rpsL*, could be engineered to dramatically turn on previously switched off genes [51]. Ribosomes, therefore, may be a useful tool for the production of microbes for biofuel production. The principle of ribosomal gene modification is to mimic the natural antibiotic resistance system developed by bacteria [52].

Ribosome engineering has proven to be a successful approach for the production of primary and secondary metabolites. Tanaka modified the *rpsL* gene in *Propionibacterium shermanii*, resulting in more than fivefold enhanced production of vitamin B12 [53]. The challenge with ribosome engineering is growth retardation of engineered strains, which may be due to the unavailability of free ribosomes for cellular growth. In addition, ribosome modification in *Bacillus subtilis* resulted in up to a 30% increase in amylase and protease production in the stationary phase [54]. Besides ribosome modification, the rapid availability of ribosomes for desired metabolic pathways has also proven to be beneficial for the production of antibiotics. This approach, which involves the introduction of a ribosome release factor, has been applied for improved antibiotic production in *Streptomyces* [55,56].

4.5. Transfer RNA Engineering

Translation is the final step in the gene expression process. Translation requires energy and involves several steps: initiation, elongation and termination. The process is tightly regulated by a number of factors, such as the synthesis and degradation of the messenger RNA (mRNA), activation of transfer RNA (tRNA), and the coupling of tRNA with ribosomes. To obtain strains with high biofuel production, individual components involved in the translation process such as mRNA, tRNA, ribosomes, and various proteins required for translation can be modulated.

Biofuel-producing phenotypes have problems with fuel toxicity, as most proteins are denatured by the hydrophobic biofuel. This type of issue may be solved using tRNA engineering approaches. Recently, modified tRNAs were designed with different codon specificities to incorporate unnatural amino acids into proteins, resulting in the formation of novel proteins with altered or enhanced properties, such as resistance to toxic substances or enhanced catalytic efficiency [57].

Protein synthesis is based on the universal genetic code recognized by specific tRNAs. The term "tRNA engineering" refers to the chemical modification of tRNAs with unnatural amino acids that can then be incorporated into proteins [58]. Recently, in vivo modification of tRNA was demonstrated using a special aminoacyl tRNA synthetase to esterify unnatural amino acids to an engineered tRNA, i.e., tRNApyl [59,60]. More than 30 novel amino acids have been introduced into proteins using this technology [60]. Similarly, pyrrolysine, an unnatural amino acid analogue of lysine with a methyl-pyrroline moiety, was first identified in the active site of the enzyme monomethylamine methyltransferase from *Methanosarcina barkeri* [61]. Furthermore, it was found that pyrrolysine incorporation was due to tRNApyl containing the CUA anticodon, complementary to UAG bases. The uncharged tRNApyl is activated by pyrrolysyl-tRNA synthetase (PylRS).

This technique can also be used to fine-tune certain enzymes to improve their function or pharmacological properties, e.g., in vivo incorporation of p-azido-L-phenylalanine for Phe170 or Phe281 in *E. coli* urate-oxidase [57]. Engineering proteins for functional metabolic pathways using this strategy may improve the productivity of biofuel in microbes. The described technologies also provide a path forward for biotech industries to modify their targets for improved activity [57].

Finally, advances in tRNA engineering broaden the field of phenotype engineering by allowing the incorporation of unnatural amino acids into structural and functional proteins of the cell. However, engineering aminoacyl tRNA synthetases and screening natural systems still present challenges for tRNA engineering.

4.6. Cofactor Regeneration

Cofactors such as NADH, NADPH, GTP, ATP, FAD, and FADH are important partners of catalytic reactions and enzymatic pathways. Therefore, the availability of these molecules is vital for a chemical reaction to proceed and improving their availability can enhance the efficiency of the pathway. The continuous supply of ATP in an engineered microbial host can boost the desired products [62]. The glutathione metabolic pathway requires ATP and blocking the ATP-consuming *nlpD* gene improves glutathione production [62]. Wang et al. (2013) showed up to a 90% production yield of 2,3-butanediol by improving NADH availability through overexpression of NADH-generating glucose and formate dehydrogenase enzymes [63]. Similar approaches have been applied in the yeast *Pichia pastoris* where blocking NADH-consuming pathways led to a boost in butanediol production [64].

4.7. Scaffold System

Nature has provided a sophisticated mechanism for enhancing pathway catalysis by using scaffolding enzymes. Many organisms have evolved multifunctional enzyme complexes such as fatty acid, polyketide, and alkaloid synthases for the biosynthesis of fatty acids polyketides, and alkaloids, respectively [65–67]. Applying this approach has led to the efficient production of artemisinin, an anti-malarial drug; *n*-alkanes; and biodiesel [16,28,68–70]. In our previous study, we demonstrated the potential of a DNA scaffold system for *n*-alkane production in *E. coli*. The critical enzymes of the *n*-alkane pathway, acyl-ACP reductase (AAR) and ADO, were linked to four zinc finger DNA-binding domains, capable of binding to a 12 bp sequence on the DNA scaffold. This resulted in an eightfold increase in *n*-alkane production [16]. Design in synthetic biology is usually inspired by the spatial organization of enzymes observed in nature. This close approximation of enzymes to their desired distances and angles is made possible using modular scaffolds. The concept of scaffolding involves the use of small building blocks to allow the channeling of intermediates during a biochemical reaction (Figure 3). The advantage of such a system is that it prevents the diffusion of intermediates of sequential enzymatic reactions to nearby competing pathways, minimizing transit time and toxicity of certain intermediates [16,28,70,71]. Types of scaffolding include the immobilization, co-localization, compartmentalization, and spatial organization of enzymes on nucleic acids (DNA/RNA) and protein-protein scaffolding. During immobilization and co-localization, the enzymes of interest are targeted to the membrane via signal peptides as leading sequences. Enzyme scaffolds on nucleic acids are constructed by tagging nucleic acid binding domains onto the desired enzymes so that they land on the sequence of interest. Protein scaffolds can be constructed either as a fusion protein of two or more enzymes to form a large single multi-enzyme protein or by fusing an individual enzyme with protein binding domains that eventually lead to a complex of proteins or scaffolds. Coiled-coil proteins have been used to create this type of protein scaffold [72].

Energies **2018**, *11*, 2663

5. Conclusions and Future Directions

The global energy crisis and ever-increasing global warming may be addressed by decreasing reliance on fossil fuels. Alcohols and *n*-alkanes are currently the best replacements for petroleum fuel. The production yields of these biofuels are far below commercial level due to technical problems in strain optimization. In this review, gene expression optimization strategies such as the use of promoters, artificial transcription factors and gene copy number alterations were discussed. In addition, strategies for cofactor regeneration and scaffolding of pathway enzymes for metabolic pathway optimization were also discussed. Future tools must focus on rapid screening technologies for the selection of single microbial cells in real time to boost biofuel production.

Author Contributions: Z.R. conceived of this work. J.N. contributed to the manuscript text and figures. B.H.S. and S.C.K. supervised this work.

Funding: This work was supported by grants from the Higher Education Commission of Pakistan (Grant No. 5197/KPK/NRPU/R&D/HEC/2016), the Intelligent Synthetic Biology Center of Global Frontier Project through the National Research Foundation funded by the Ministry of Science & ICT of Korea, and the Research Initiative Program of KRIBB.

Acknowledgments: We are very thankful to Raham Sher, Department of Biotechnology, AWKUM for providing insights about global warming.

Conflicts of Interest: The authors have no conflicts of interest to declare.

References

1. Goldthau, A. The G20 must govern the shift to low-carbon energy. *Nature* **2017**, *546*, 203–205. [CrossRef] [PubMed]
2. Balch, J.K.; Nagy, R.C.; Archibald, S.; Bowman, D.M.; Moritz, M.A.; Roos, C.I.; Scott, A.C.; Williamson, G.J. Global combustion: The connection between fossil fuel and biomass burning emissions (1997–2010). *Philos. Trans. R. Soc. Lond. B Biol. Sci.* **2016**, *371*. [CrossRef] [PubMed]
3. Den, W.; Sharma, V.K.; Lee, M.; Nadadur, G.; Varma, R.S. Lignocellulosic biomass transformations via greener oxidative pretreatment processes: Access to energy and value-added chemicals. *Front. Chem.* **2018**, *6*, 141. [CrossRef] [PubMed]
4. Nejat, P.; Jomehzadeh, F.; Taheri, M.M.; Gohari, M.; Majid, M.Z.A. A global review of energy consumption, CO_2 emissions and policy in the residential sector (with an overview of the top ten CO_2 emitting countries). *Renew. Sustain. Energy Rev.* **2015**, *43*, 843–862. [CrossRef]
5. Shields-Menard, S.A.; Amirsadeghi, M.; French, W.T.; Boopathy, R. A review on microbial lipids as a potential biofuel. *Bioresour. Technol.* **2018**, *259*, 451–460. [CrossRef] [PubMed]
6. Dresselhaus, M.; Thomas, I. Alternative energy technologies. *Nature* **2001**, *414*, 332. [CrossRef] [PubMed]
7. d'Espaux, L.; Ghosh, A.; Runguphan, W.; Wehrs, M.; Xu, F.; Konzock, O.; Dev, I.; Nhan, M.; Gin, J.; Reider Apel, A.; et al. Engineering high-level production of fatty alcohols by *Saccharomyces cerevisiae* from lignocellulosic feedstocks. *Metab. Eng.* **2017**, *42*, 115–125. [CrossRef] [PubMed]
8. Fatma, Z.; Hartman, H.; Poolman, M.G.; Fell, D.A.; Srivastava, S.; Shakeel, T.; Yazdani, S.S. Model-assisted metabolic engineering of *Escherichia coli* for long chain alkane and alcohol production. *Metab. Eng.* **2018**, *46*, 1–12. [CrossRef] [PubMed]
9. Liew, F.; Henstra, A.M.; Kpke, M.; Winzer, K.; Simpson, S.D.; Minton, N.P. Metabolic engineering of *Clostridium autoethanogenum* for selective alcohol production. *Metab. Eng.* **2017**, *40*, 104–114. [CrossRef] [PubMed]
10. Demirbas, A. Biofuels sources, biofuel policy, biofuel economy and global biofuel projections. *Energy Convers. Manag.* **2008**, *49*, 2106–2116. [CrossRef]
11. Pandey, R.K.; Tewari, L. Mycotechnology for lignocellulosic bioethanol production: An emerging approach to sustainable environment. In *Microbial Biotechnology in Environmental Monitoring and Cleanup*; IGI Glob: Hershey, PA, USA, 2018; pp. 28–43. [CrossRef]
12. Babu, V.; Murthy, M. Butanol and pentanol: The promising biofuels for CI engines–A review. *Renew. Sustain. Energy Rev.* **2017**, *78*, 1068–1088.

13. No, S.-Y. Application of hydrotreated vegetable oil from triglyceride based biomass to CI engines—A review. *Fuel* **2014**, *115*, 88–96. [CrossRef]

14. Van Vliet, O.P.; Faaij, A.P.; Turkenburg, W.C. Fischer—Tropsch diesel production in a well-to-wheel perspective: A carbon, energy flow and cost analysis. *Energy Convers. Manag.* **2009**, *50*, 855–876. [CrossRef]

15. Ohta, K.; Beall, D.S.; Mejia, J.P.; Shanmugam, K.T.; Ingram, L.O. Genetic improvement of *Escherichia coli* for ethanol production: Chromosomal integration of *Zymomonas mobilis* genes encoding pyruvate decarboxylase and alcohol dehydrogenase ii. *Appl. Environ. Microbiol.* **1991**, *57*, 893–900. [PubMed]

16. Rahman, Z.; Rashid, N.; Nawab, J.; Ilyas, M.; Sung, B.H.; Kim, S.C. *Escherichia coli* as a fatty acid and biodiesel factory: Current challenges and future directions. *Environ. Sci. Pollut. Res. Int.* **2016**, *23*, 12007–12018. [CrossRef] [PubMed]

17. Lu, X. A perspective: Photosynthetic production of fatty acid-based biofuels in genetically engineered *Cyanobacteria*. *Biotechnol. Adv.* **2010**, *28*, 742–746. [CrossRef] [PubMed]

18. Zhou, Y.J.; Buijs, N.A.; Siewers, V.; Nielsen, J. Fatty acid-derived biofuels and chemicals production in *Saccharomyces cerevisiae*. *Front Bioeng. Biotechnol.* **2014**, *2*, 32. [CrossRef] [PubMed]

19. Lin, Y.L.; Blaschek, H.P. Butanol production by a butanol-tolerant strain of *Clostridium acetobutylicum* in extruded corn broth. *Appl. Environ. Microbiol.* **1983**, *45*, 966–973. [PubMed]

20. Xu, P.; Qiao, K.; Ahn, W.S.; Stephanopoulos, G. Engineering *Yarrowia lipolytica* as a platform for synthesis of drop-in transportation fuels and oleochemicals. *Proc. Natl. Acad. Sci. USA* **2016**, *113*, 10848–10853. [CrossRef] [PubMed]

21. Atsumi, S.; Cann, A.F.; Connor, M.R.; Shen, C.R.; Smith, K.M.; Brynildsen, M.P.; Chou, K.J.; Hanai, T.; Liao, J.C. Metabolic engineering of *Escherichia coli* for 1-butanol production. *Metab. Eng.* **2008**, *10*, 305–311. [CrossRef] [PubMed]

22. Rutter, C.D.; Rao, C.V. Production of 1-decanol by metabolically engineered *Yarrowia lipolytica*. *Metab. Eng.* **2016**, *38*, 139–147. [CrossRef] [PubMed]

23. Elgaali, H.; Hamilton-Kemp, T.; Newman, M.; Collins, R.; Yu, K.; Archbold, D. Comparison of long-chain alcohols and other volatile compounds emitted from food-borne and related gram positive and gram negative bacteria. *J. Basic Microbiol.* **2002**, *42*, 373–380. [CrossRef]

24. Kohse-Hoinghaus, K.; Osswald, P.; Cool, T.A.; Kasper, T.; Hansen, N.; Qi, F.; Westbrook, C.K.; Westmoreland, P.R. Biofuel combustion chemistry: From ethanol to biodiesel. *Angew. Chem. Int. Ed. Engl.* **2010**, *49*, 3572–3597. [CrossRef] [PubMed]

25. Siripong, W.; Wolf, P.; Kusumoputri, T.P.; Downes, J.J.; Kocharin, K.; Tanapongpipat, S.; Runguphan, W. Metabolic engineering of *Pichia pastoris* for production of isobutanol and isobutyl acetate. *Biotechnol. Biofuels.* **2018**, *11*, 1. [CrossRef] [PubMed]

26. Saini, M.; Wang, Z.W.; Chiang, C.J.; Chao, Y.P. Metabolic engineering of *Escherichia coli* for production of n-butanol from crude glycerol. *Biotechnol. Biofuels.* **2017**, *10*, 173. [CrossRef] [PubMed]

27. Zhang, J.; Zong, W.; Hong, W.; Zhang, Z.T.; Wang, Y. Exploiting endogenous crispr-cas system for multiplex genome editing in *Clostridium tyrobutyricum* and engineer the strain for high-level butanol production. *Metab. Eng.* **2018**, *47*, 49–59. [CrossRef] [PubMed]

28. Rahman, Z.; Sung, B.H.; Yi, J.Y.; Bui le, M.; Lee, J.H.; Kim, S.C. Enhanced production of *n*-alkanes in *Escherichia coli* by spatial organization of biosynthetic pathway enzymes. *J. Biotechnol.* **2014**, *192 Pt A*, 187–191. [CrossRef]

29. Srirangan, K.; Liu, X.; Westbrook, A.; Akawi, L.; Pyne, M.E.; Moo-Young, M.; Chou, C.P. Biochemical, genetic, and metabolic engineering strategies to enhance coproduction of 1-propanol and ethanol in engineered *Escherichia coli*. *Appl. Microbiol. Biotechnol.* **2014**, *98*, 9499–9515. [CrossRef] [PubMed]

30. Cao, Y.X.; Xiao, W.H.; Zhang, J.L.; Xie, Z.X.; Ding, M.Z.; Yuan, Y.J. Heterologous biosynthesis and manipulation of alkanes in *Escherichia coli*. *Metab. Eng.* **2016**, *38*, 19–28. [CrossRef] [PubMed]

31. Garcia, H.G.; Brewster, R.C.; Phillips, R. Using synthetic biology to make cells tomorrow's test tubes. *Integr. Biol.* **2016**, *8*, 431–450. [CrossRef] [PubMed]

32. Smanski, M.J.; Zhou, H.; Claesen, J.; Shen, B.; Fischbach, M.A.; Voigt, C.A. Synthetic biology to access and expand nature's chemical diversity. *Nat. Rev. Microbiol.* **2016**, *14*, 135–149. [CrossRef] [PubMed]

33. Wu, G.; Yan, Q.; Jones, J.A.; Tang, Y.J.; Fong, S.S.; Koffas, M.A.G. Metabolic burden: Cornerstones in synthetic biology and metabolic engineering applications. *Trends Biotechnol.* **2016**, *34*, 652–664. [CrossRef] [PubMed]

34. Jensen, P.R.; Hammer, K. Artificial promoters for metabolic optimization. *Biotechnol. Bioeng.* **1998**, *58*, 191–195. [CrossRef]

35. Polli, F.; Meijrink, B.; Bovenberg, R.A.L.; Driessen, A.J.M. New promoters for strain engineering of *Penicillium chrysogenum*. *Fungal. Genet. Biol.* **2016**, *89*, 62–71. [CrossRef] [PubMed]

36. Lutz, R.; Bujard, H. Independent and tight regulation of transcriptional units in *Escherichia coli* via the lacr/o, the tetr/o and arac/i1-i2 regulatory elements. *Nucleic. Acids. Res.* **1997**, *25*, 1203–1210. [CrossRef] [PubMed]

37. Ho, K.M.; Lim, B.L. Co-expression of a prophage system and a plasmid system in *Bacillus subtilis*. *Protein Expr. Purif.* **2003**, *32*, 293–301. [CrossRef] [PubMed]

38. Dehli, T.; Solem, C.; Jensen, P.R. Tunable promoters in synthetic and systems biology. *Subcell. Biochem.* **2012**, *64*, 181–201. [PubMed]

39. Song, Y.; Nikoloff, J.M.; Fu, G.; Chen, J.; Li, Q.; Xie, N.; Zheng, P.; Sun, J.; Zhang, D. Promoter screening from *Bacillus subtilis* in various conditions hunting for synthetic biology and industrial applications. *PLoS ONE* **2016**, *11*. [CrossRef] [PubMed]

40. Rudge, T.J.; Brown, J.R.; Federici, F.; Dalchau, N.; Phillips, A.; Ajioka, J.W.; Haseloff, J. Characterization of intrinsic properties of promoters. *ACS Synth. Biol.* **2016**, *5*, 89–98. [CrossRef] [PubMed]

41. Yang, K.; Stracquadanio, G.; Luo, J.; Boeke, J.D.; Bader, J.S. Biopartsbuilder: A synthetic biology tool for combinatorial assembly of biological parts. *Bioinformatics* **2016**, *32*, 937–939. [CrossRef] [PubMed]

42. Stewart, G.S.; Lubinsky-Mink, S.; Jackson, C.G.; Cassel, A.; Kuhn, J. Phg165: A PBR322 copy number derivative of puc8 for cloning and expression. *Plasmid* **1986**, *15*, 172–181. [CrossRef]

43. Atkinson, M.R.; Savageau, M.A.; Myers, J.T.; Ninfa, A.J. Development of genetic circuitry exhibiting toggle switch or oscillatory behavior in *Escherichia coli*. *Cell* **2003**, *113*, 597–607. [CrossRef]

44. Lian, J.; Jin, R.; Zhao, H. Construction of plasmids with tunable copy numbers in *Saccharomyces cerevisiae* and their applications in pathway optimization and multiplex genome integration. *Biotechnol. Bioeng.* **2016**, *113*, 2462–2473. [CrossRef] [PubMed]

45. Shizuya, H.; Birren, B.; Kim, U.J.; Mancino, V.; Slepak, T.; Tachiiri, Y.; Simon, M. Cloning and stable maintenance of 300-kilobase-pair fragments of human DNA in *Escherichia coli* using an f-factor-based vector. *Proc. Natl. Acad. Sci. USA* **1992**, *89*, 8794–8797. [CrossRef] [PubMed]

46. Kasai, Y.; Oshima, K.; Ikeda, F.; Abe, J.; Yoshimitsu, Y.; Harayama, S. Construction of a self-cloning system in the unicellular green alga *Pseudochoricystis ellipsoidea*. *Biotechnol. Biofuels*. **2015**, *8*, 94. [CrossRef] [PubMed]

47. Sera, T. Zinc-finger-based artificial transcription factors and their applications. *Adv. Drug. Deliv. Rev.* **2009**, *61*, 513–526. [CrossRef] [PubMed]

48. Lee, J.Y.; Sung, B.H.; Yu, B.J.; Lee, J.H.; Lee, S.H.; Kim, M.S.; Koob, M.D.; Kim, S.C. Phenotypic engineering by reprogramming gene transcription using novel artificial transcription factors in *Escherichia coli*. *Nucleic. Acids. Res.* **2008**, *36*. [CrossRef] [PubMed]

49. Lee, J.Y.; Yang, K.S.; Jang, S.A.; Sung, B.H.; Kim, S.C. Engineering butanol-tolerance in *Escherichia coli* with artificial transcription factor libraries. *Biotechnol. Bioeng.* **2011**, *108*, 742–749. [CrossRef] [PubMed]

50. Ma, C.; Wei, X.; Sun, C.; Zhang, F.; Xu, J.; Zhao, X.; Bai, F. Improvement of acetic acid tolerance of *Saccharomyces cerevisiae* using a zinc-finger-based artificial transcription factor and identification of novel genes involved in acetic acid tolerance. *Appl. Microbiol. Biotechnol.* **2015**, *99*, 2441–2449. [CrossRef] [PubMed]

51. Ochi, K. From microbial differentiation to ribosome engineering. *Biosci. Biotechnol. Biochem.* **2007**, *71*, 1373–1386. [CrossRef] [PubMed]

52. Santos, C.N.; Stephanopoulos, G. Combinatorial engineering of microbes for optimizing cellular phenotype. *Curr. Opin. Chem. Biol.* **2008**, *12*, 168–176. [CrossRef] [PubMed]

53. Tanaka, Y.; Kasahara, K.; Izawa, M.; Ochi, K. Applicability of ribosome engineering to vitamin B12 production by *Propionibacterium shermanii*. *Biosci. Biotechnol. Biochem.* **2017**, *81*, 1636–1641. [CrossRef] [PubMed]

54. Kurosawa, K.; Hosaka, T.; Tamehiro, N.; Inaoka, T.; Ochi, K. Improvement of alpha-amylase production by modulation of ribosomal component protein S12 in *Bacillus subtilis* 168. *Appl. Environ. Microbiol.* **2006**, *72*, 71–77. [CrossRef] [PubMed]

55. Hosaka, T.; Xu, J.; Ochi, K. Increased expression of ribosome recycling factor is responsible for the enhanced protein synthesis during the late growth phase in an antibiotic-overproducing *Streptomyces coelicolor* ribosomal rpsl mutant. *Mol. Microbiol.* **2006**, *61*, 883–897. [CrossRef] [PubMed]

56. Janosi, L.; Shimizu, I.; Kaji, A. Ribosome recycling factor (ribosome releasing factor) is essential for bacterial growth. *Proc. Natl. Acad. Sci. USA* **1994**, *91*, 4249–4253. [CrossRef] [PubMed]

57. Chen, M.; Cai, L.; Fang, Z.; Tian, H.; Gao, X.; Yao, W. Site-specific incorporation of unnatural amino acids into urate oxidase in *Escherichia coli*. *Protein Sci.* **2008**, *17*, 1827–1833. [CrossRef] [PubMed]

58. Rothschild, K.J.; Gite, S. Trna-mediated protein engineering. *Curr. Opin. Biotechnol.* **1999**, *10*, 64–70. [CrossRef]

59. Gladyshev, V.N.; Jeang, K.T.; Stadtman, T.C. Selenocysteine, identified as the penultimate c-terminal residue in human t-cell thioredoxin reductase, corresponds to tga in the human placental gene. *Proc. Natl. Acad. Sci. USA* **1996**, *93*, 6146–6151. [CrossRef] [PubMed]

60. Yanagisawa, T.; Ishii, R.; Fukunaga, R.; Kobayashi, T.; Sakamoto, K.; Yokoyama, S. Multistep engineering of pyrrolysyl-trna synthetase to genetically encode n(epsilon)-(o-azidobenzyloxycarbonyl) lysine for site-specific protein modification. *Chem. Biol.* **2008**, *15*, 1187–1197. [CrossRef] [PubMed]

61. Burke, S.A.; Lo, S.L.; Krzycki, J.A. Clustered genes encoding the methyltransferases of methanogenesis from monomethylamine. *J. Bacteriol.* **1998**, *180*, 3432–3440. [PubMed]

62. Hara, K.Y.; Shimodate, N.; Hirokawa, Y.; Ito, M.; Baba, T.; Mori, H.; Mori, H. Glutathione production by efficient atp-regenerating *Escherichia coli* mutants. *FEMS Microbiol. Lett.* **2009**, *297*, 217–224. [CrossRef] [PubMed]

63. Wang, Y.; Li, L.; Ma, C.; Gao, C.; Tao, F.; Xu, P. Engineering of cofactor regeneration enhances (2S,3S)-2,3-butanediol production from diacetyl. *Sci. Rep.* **2013**, *3*, 2643. [CrossRef] [PubMed]

64. Geier, M.; Brandner, C.; Strohmeier, G.A.; Hall, M.; Hartner, F.S.; Glieder, A. Engineering *Pichia pastoris* for improved NADH regeneration: A novel chassis strain for whole-cell catalysis. *Beilstein. J. Org. Chem.* **2015**, *11*, 1741–1748. [CrossRef] [PubMed]

65. Ishikawa, M.; Tsuchiya, D.; Oyama, T.; Tsunaka, Y.; Morikawa, K. Structural basis for channelling mechanism of a fatty acid beta-oxidation multienzyme complex. *EMBO J.* **2004**, *23*, 2745–2754. [CrossRef] [PubMed]

66. Jorgensen, K.; Rasmussen, A.V.; Morant, M.; Nielsen, A.H.; Bjarnholt, N.; Zagrobelny, M.; Bak, S.; Moller, B.L. Metabolon formation and metabolic channeling in the biosynthesis of plant natural products. *Curr. Opin. Plant. Biol.* **2005**, *8*, 280–291. [CrossRef] [PubMed]

67. Pfeifer, B.A.; Khosla, C. Biosynthesis of polyketides in heterologous hosts. *Microbiol. Mol. Biol. Rev.* **2001**, *65*, 106–118. [CrossRef] [PubMed]

68. Akhtar, M.K.; Dandapani, H.; Thiel, K.; Jones, P.R. Microbial production of 1-octanol: A naturally excreted biofuel with diesel-like properties. *Metab. Eng. Commun.* **2015**, *2*, 1–5. [CrossRef] [PubMed]

69. Martin, V.J.; Pitera, D.J.; Withers, S.T.; Newman, J.D.; Keasling, J.D. Engineering a mevalonate pathway in *Escherichia coli* for production of terpenoids. *Nat. Biotechnol.* **2003**, *21*, 796–802. [CrossRef] [PubMed]

70. Yu, P.; Chen, X.; Li, P. Enhancing microbial production of biofuels by expanding microbial metabolic pathways. *Biotechnol. Appl. Biochem.* **2017**, *64*, 606–619. [CrossRef] [PubMed]

71. Jones, J.A.; Toparlak, O.D.; Koffas, M.A. Metabolic pathway balancing and its role in the production of biofuels and chemicals. *Curr. Opin. Biotechnol.* **2015**, *33*, 52–59. [CrossRef] [PubMed]

72. Peacock, A.F. Recent advances in designed coiled coils and helical bundles with inorganic prosthetic groups-from structural to functional applications. *Curr. Opin. Chem. Biol.* **2016**, *31*, 160–165. [CrossRef] [PubMed]

![energies logo] *energies*

MDPI

Article

Electrochemical Hydrogenation of Acetone to Produce Isopropanol Using a Polymer Electrolyte Membrane Reactor

Chen Li [1], Ashanti M. Sallee [2], Xiaoyu Zhang [3,* and Sandeep Kumar [1,*

[1]   Department of Civil & Environmental Engineering, Old Dominion University, Norfolk, VA 23529, USA; cxxli001@odu.edu
[2]   Department of Chemistry, Hampton University, Hampton, VA 23668, USA; amsallee12@gmail.com
[3]   Department of Mechanical & Aerospace Engineering, Old Dominion University, Norfolk, VA 23529, USA
*   Correspondence: x1zhang@odu.edu (X.Z.); skumar@odu.edu (S.K.); Tel.: +1-757-683-4913 (X.Z.); +1-757-683-3898 (S.K.)

Received: 17 August 2018; Accepted: 2 October 2018; Published: 10 October 2018

Abstract: Electrochemical hydrogenation (ECH) of acetone is a relatively new method to produce isopropanol. It provides an alternative way of upgrading bio-fuels with less energy consumption and chemical waste as compared to conventional methods. In this paper, Polymer Electrolyte Membrane Fuel Cell (PEMFC) hardware was used as an electrochemical reactor to hydrogenate acetone to produce isopropanol and diisopropyl ether as a byproduct. High current efficiency (59.7%) and selectivity (>90%) were achieved, while ECH was carried out in mild conditions (65 °C and atmospheric pressure). Various operating parameters were evaluated to determine their effects on the yield of acetone and the overall efficiency of ECH. The results show that an increase in humidity increased the yield of propanol and the efficiency of ECH. The operating temperature and power supply, however, have less effect. The degradation of membranes due to contamination of PEMFC and the mitigation methods were also investigated.

Keywords: acetone; electrochemical hydrogenation; isopropanol; membrane contamination; polymer electrolyte membrane; relative humidity

---

1. Introduction

Propanol is an important organic raw material in chemical production, two isomers 1-propanol and isopropanol are widely used in the paint, medicine and pesticide industries [1]. Compared to 1-propanol, isopropanol has more extensive and important applications. Along with ethanol, n-butanol, and methanol, isopropanol belongs to the group of alcohol solvents, about 6.4 million tons of which were utilized worldwide in 2011 [2]. Isopropanol is primarily produced by combining water and propene in a hydration reaction, through either an indirect or direct process. In an indirect process, propene reacts with sulfuric acid and forms a mixture of sulfate esters. Subsequent hydrolysis of those esters by steam produces isopropanol. In a direct hydration process, propene reacts with water or steam at high pressure (200–300 atm) and high temperatures (230–270 °C), in the presence of solid or supported acidic catalysts [3,4]. Isopropanol is produced by a direct combination of propene and water. Both processes require intensive energy input and use of corrosive chemicals.

Thermal hydrogenation of acetone is a relatively new and more advanced method to produce isopropanol, where acetone is hydrogenated either in the liquid or gas phase over a Raney nickel or copper and chromium oxide mixture [5]. Compared to the aforementioned conventional methods, thermal hydrogenation can be carried out at a lower temperature (75 °C) with up to 35% yield rate. However, an elevated temperature (350–400 °C) is still required to enable fully activated catalysts. In addition, handling corrosive chemicals remains a problem [6].

Electrochemical hydrogenation (ECH) provides a more energy efficient and environment-friendly method of upgrading organics, by integrating both the electrochemical and catalytic methods [7]. The overall reaction mechanism of ECH of an unsaturated organic molecule is suggested as the following Equations (1)–(5) [8–13].

Anode side: $H_2 \rightarrow 2(H^+) + 2e^-$; $H_2O \rightarrow 1/2O_2 + 2(H^+) + 2e^-$

Cathode side:

(1) Protons react with electrons and generate $M(H)_{ads}$ (M is an adsorption site):

$$(H^+) + e^- + M \rightarrow M(H)_{ads} + H_2O \tag{1}$$

(2) An organic molecule Y = Z is adsorbed by an adsorption site M:

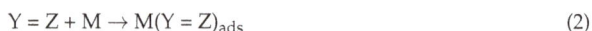

$$Y = Z + M \rightarrow M(Y = Z)_{ads} \tag{2}$$

(3) $M(H)_{ads}$ reacts with the adsorbed organic molecule:

$$M(Y = Z)_{ads} + 2M(H)_{abs} \rightarrow M(YH\text{-}ZH)_{abs} + 2M \tag{3}$$

(4) A hydrogenated product is generated:

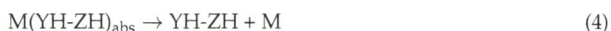

$$M(YH\text{-}ZH)_{abs} \rightarrow YH\text{-}ZH + M \tag{4}$$

(5) H_2 gas is also produced:

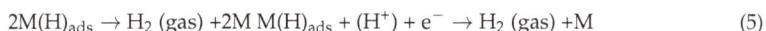

$$2M(H)_{ads} \rightarrow H_2 \text{ (gas)} + 2M \quad M(H)_{ads} + (H^+) + e^- \rightarrow H_2 \text{ (gas)} + M \tag{5}$$

In the process, chemisorbed hydrogen $M(H)_{ads}$ is generated in situ on the electrocatalyst surface through either hydrogen pumping or water electrolysis and reacts with the adsorbed organics (Equation (1)). Note that hydrogenation also competes with hydrogen gas evolution (Equation (5)), which results in a decrease of the current efficiency. ECH allows the reactions (Equations (2)–(4)) to happen at lower temperatures and ambient pressure. Compared to conventional hydrogenation methods, ECH mainly uses electrical energy and all the reactions take place in a mild operating condition (i.e., low temperature and atmospheric pressure). Therefore, intense thermal energy input is not required in ECH. ECH also uses either water or H_2 to supply protons, which eliminate the need for any reducing agent. ECH can be conducted onsite using fuel cell stacks and renewable power sources to produce hydrogen enriched compounds. This will avoid or minimize the storage and transportation of corrosive and hazardous chemicals.

ECH has been widely used to upgrade unsaturated compounds to corresponding saturated chemicals, such as furfural [14–20], aromatic compounds [7,13,21–29], soybean oil [30], edible oil [31,32], levulinic acid [33–35], lactic acid [36], acetaldehyde [37], ethanol [37], acetylene [38], bio-oil [39,40], cyclohexane [9], glucose [41], and lignin [42]. In all those cases, reactions take place under mild conditions with temperatures below 100 °C and atmospheric pressure. A maximum current efficiency up to 45%, which is defined as the efficiency of electrogenerated H_2 addition to unsaturated bonds, was reported [30–32,38]. The yield of the electrocatalytic hydrogenation of organic molecules is directly related to the processes described in Equations (1)–(5), which are determined by the capabilities of catalysts. Compared to nickel (Ni), copper (Cu) and lead (Pb), activated carbon fibers supported platinum (Pt) demonstrated the best catalytic activities for upgrading various organics, such as furfural [20] and acetaldehyde [37]. The electrochemical conversion rate is also affected by the nature of electrodes [43], current density [44], temperature [26,38,40,45], solvent compositions [9,12,45], solution pH [9,14,21,40,46], and chemical potential [30,39,40]. The most commonly used ECH reactor is H-type cell [7,10,14,15,19,20,22,23,27,39,42,44,46]. A typical H-type cell consists of two electrode chambers, between which a cation exchange membrane is sandwiched. Compared to the traditional homogenous electrolyte method of the single chamber cell [14,24,33,34,36,37], the basic anolyte and the acidic/neutral catholyte method were applied in many H-type cells [7,14,40] to promote the proton transmission efficiency.

Hydrogenation of unsaturated compounds using polymer electrolyte membrane fuel cell (PEMFC) reactors was also reported [18,29–32]. PEMFC is a type of low-temperature fuel cell that takes its name from ion conductive polymer membrane used as the electrolyte [47]. A typical PEMFC assembly consists of an ion exchange membrane, two electrodes made of carbon layer loaded with Pt, and two gas diffusion media. The reactor provides gas distribution, current collection, temperature control, and mechanical support of the PEMFC assembly [47]. Alfonso et al. electrocatalytically hydrogenated acetophenone by H_2 using a PEMFC reactor. They reported the selectivity of produced 1-phenylethanol around 90% with only methylbenzene and hydrogen as by-products [29]. Green et al. reported that the main products from ECH of furfural were furfural alcohol (54–100% selectivity) and tetrahydrofurfuryl alcohol (0–26% selectivity). A higher production rate was achieved by feeding pure hydrogen gas than that from electrolysis of water [18]. Pintauro et al. and An et al. both studied ECH of soybean oil in a PEMFC reactor at 60–90 °C and atmospheric pressure [30–32]. Pintauro et al. proved that a bimetallic cathode (Pd/Co or Pd/Fe) could increase the yield rate of the ECH process [30]. An et al. proved Pd-black cathode worked significantly better than Pt and the best current efficiency could reach 41% [31,32]. There are many advantages of using a PEMFC reactor for ECH [48]: First, compared to H-Cell and one chamber cell reactor, the PEMFC reactor has a smaller internal resistance, due to its highly conductive and thin membrane electrolyte assembly (MEA), resulting in significantly less electric energy loss. In addition, the energy consumption can be further reduced if protons are supplied from hydrogen oxidation, rather than electrolysis. Second, since PEMFCs can be easily scaled up by simply stacking them, the space-time yield of ECH using PEMFC reactors is superior to the other methods. PEMFC can be easily applied in space limited area such as transportation, stationary, and portable/micro power generation sectors [49].

In this paper, ECH of acetone to produce isopropanol was demonstrated using a PEMFC reactor at ambient pressure. Various factors that impact the yield of propanol were investigated, including current density, temperature, relative humidity (RH), and membrane degradation. The main objective of this work was to evaluate the appropriate pathways of ECH of bio-oil components using a PEMFC reactor.

2. Materials and Method

2.1. Material and Experimental Setup

The experiments were performed using a standard PEMFC hardware (Scribner Associates Inc., USA) with an active area of 25 cm^2. Such standard PEMFC hardware has been widely used for PEMFC evaluation tests [29,50]. Commercially available MEAs were purchased from Ion Power Inc., New Castle, DE, USA. The MEAs consist of Nafion® 117 membranes sandwiched with porous carbon-based electrodes, each of which has a Pt loading of 0.3 mg/cm^2. Micro-porous carbon papers (SIGRACET® 10BC) were trimmed and used as gas diffusion layers (GDLs) for both electrodes. Teflon gaskets were used to seal around the assembly. A pair of graphite bipolar plates with flow patterns were used to distribute flows and enclose the assembly. All temperatures were acquired via K-type thermocouples (OMEGA, USA). A fuel cell test station (850e, Scribner Associates Inc., Southern Pines, NC, USA) was used to control temperatures, flow rates, and humidity. It was also used for data acquisition.

Ultra-high purity (99.999%) N_2 (Airgas, Radnor, PA, USA), and filtered shop air were connected to the fuel cell test station and supplied to the PEMFC through the purge line. They were used for purging system and making current and voltage curve, respectively, and were cut off while ECH experiment was running. Ultra-high purity (99.999%) H_2 (Airgas, USA) and Deionized (DI) water tank were purged through the anode side of the reactor, using the fuel cell test station. H_2 was the electrons donor and deionized (DI) water tank was used for adding humidity to H_2. Acetone (Fisher

Scientific, Hampton, NH, USA, Certified ACS Reagent Grade) was injected into the cathode side of the reactor by a syringe pump (MTI Corporation, Richmond, CA, USA, EQ-300sp-LD). A direct current (DC) power supply (Tektronix, Beaverton, OR, USA, PWS 4205) was used to supply DC power for ECH. The positive probe was connected to the anode, while the negative probe was connected to the cathode. While ECH was running, electrons were deprived of H_2 and transferred to the cathode trough the DC power supply. Acetone was the electron acceptor and reacted with the produced protons. The products from the cathode outlet were collected by airbags (Tedlar®), which were held by water bath. Room temperature water bath was used for condensing the unreacted acetone and products. The cell temperature was controlled by the fuel cell test station. The exhaust gases and byproducts from the anode were condensed in a knock-out bottle prior to venting out. The schematic diagram of the whole experimental setup is shown in Figure 1.

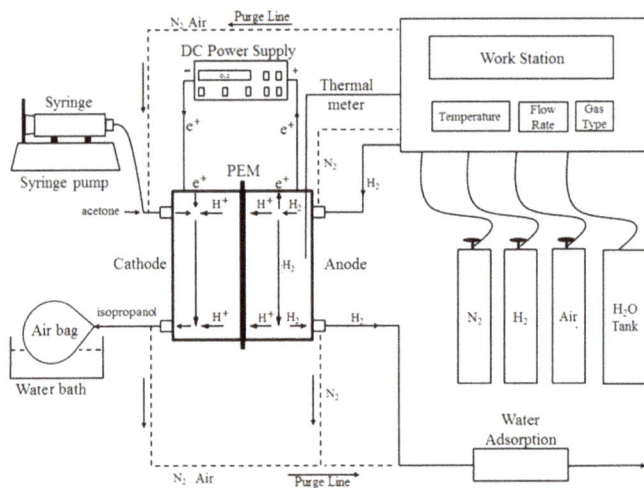

Figure 1. Schematic diagram of the experimental setup.

2.2. Electrochemical Hydrogenation and Characterization

All the PEMFCs were assembled in accordance with the standard assembly procedure of fuel cell hardware [51]. Prior to each ECH experiment, a new membrane was conditioned based on the standard protocol [51] for at least 24 h, until it reached a fully functional state. Current–voltage (I-V) sweeps were performed to establish the baseline data for the following ECH experiments.

In each ECH experiment, the acetone and H_2 flow rates were controlled at 6 mL/h and 0.25 slpm, respectively. The voltage was consistent for each reaction; the current was recorded every 5 min. Four different factors, namely cell temperature, RH, supplied voltage, and membrane degradation, were investigated to identify the optimized operating conditions. The operating conditions were in the range of 55–80 °C, 35–90% and 0.01–0.02 V, respectively. I-V scans were conducted before and after ECH to characterize the membrane degradation.

The flow chart of the ECH experimental operation is shown in Figure 2. Initially, three collection methods, including the dead end, partially confined and an open end were evaluated to identify the best means for accurate collection of products. In the partially confined method, products were collected by airbags with pressure relief valves, which prevents pressure buildup while trapping the products. In the other two collection methods, although the products were condensed by room temperature water bath, dead end still resulted in too much back pressure accumulation, whereas open end failed to collect enough products for analysis. Therefore, partially confined airbags were used for all the experiments. The pressure differential of two electrodes can be controlled by airbags confined extent. Green et al. proved that a suitable pressure differential between the anode and cathode could decrease cross-over and increase conversion [52]. Dadda et al. believed that water transport in the membrane of a PEMFC is influenced by a convective force, resulting from a pressure gradient [53]. Many researchers also point out the importance of the flow pressure [54].

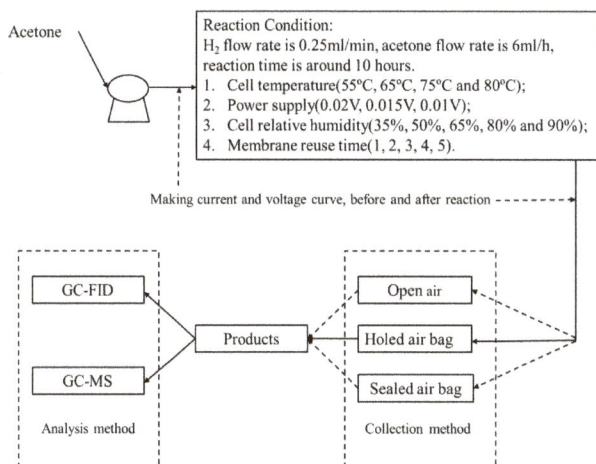

Figure 2. Flowchart of the experiment.

The collected samples were analyzed by gas chromatography–mass spectroscopy (Shimadzu, GC-MS-QP2010 SE), using a Shimadzu (SH-Rxi-5Sii) MS column (length: 30 m, inner diameter: 0.25 mm). Volume quantitative analysis was conducted by another gas chromatograph (SRI Instrument, 8610C), using a Restek (MXT-WAX) column (length: 30 m, inner diameter: 0.53 mm). Helium was used as a carrier gas for both gas chromatographs.

3. Results and Discussion

3.1. Product Characterization

Both the liquid and gaseous products were analyzed by GC-MS. Unreacted acetone (C_3H_6O), isopropanol (C_3H_8O), and diisopropyl ether ($C_6H_{14}O$) were detected in the liquid products. Unreacted acetone, diisopropyl ether, and isopropanol were found in the gaseous and liquid products.

The reactions on both electrodes are catalyzed by Pt. While applying a DC voltage, protons formed at the anode are electrochemically pumped to the cathode. The protons then react with acetone to produce isopropanol and diisopropyl ether. The most feasible reaction pathways at the cathode are shown below:

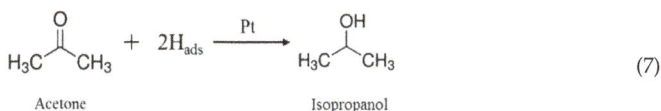

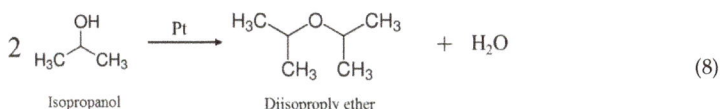

$$2H^+ + 2e^- \xrightarrow{Pt} 2H_{ads} \tag{6}$$

$$\tag{7}$$

$$\tag{8}$$

Electrosorbed hydrogen is formed on Pt surface by reduction of H^+ (Equation (6), where H_{ads} is the electrosorbed hydrogen). Hydrogenation of the C=O bond then proceeds as in catalytic hydrogenation through the reaction of the acetone with the electrosorbed hydrogen (Equation (7)). As a result, isopropanol, which is the main product, is generated. Two isopropanol molecules may also combine and free one water, generating diisoproply ether as a byproduct is formed (Equation (8)). Note that hydrogen gas can also be regenerated, which is an unfavorable electrochemical reaction during ECH. Hydrogen regeneration reduces the efficiency by electrochemically pumping useless hydrogen through the MEA, resulting in a reduction of the yield rate of products.

Each MEA had undergone at least three ECH experiments before replacement. To minimize the impacts of MEA degradation on ECH experiments, the results of the I-V scans, which were carried out prior to the experiments, were compared to the baseline performance of each MEA. If the I-V curve demonstrated an obvious deflection from the baseline performance, then the MEA needed to be replaced. The components detected in the products were isopropanol, diisopropyl ether, acetone, and water, with their volumetric percentages ranging 12–16%, 1–2%, 69–75%, and 11–14%, respectively. Note that abundant acetone was supplied to the cathode to prevent fuel starvation. As a result, the maximum conversion rate of acetone to isopropanol was 23%. The selectivity of isopropanol was calculated more than 90%. Acetone was also detected in the anode due to crossover, which is discussed below. The produced isopropanol can be easily separated from the mixture using extraction and distillation, which are two widely adopted methods in the industry [55] and therefore not discussed in detail here.

Three different control parameters, including RH, operating temperature, and input voltage, were assessed to identify the optimized operating conditions for ECH of acetone. All operating parameters used in the experiments are shown in the Table S1 in the Supplementary Materials.

RH is a very important parameter that affects the performance of PEMFCs [54,56–60]. Figure 3a shows a typical impact of RH on the product yield during ECH of acetone. It was obviously evidenced that higher humidity promoted higher yield of isopropanol. The composition of isopropanol in the products increased from 4.9% to 16.1%, while RH climbed from 35% to 90%. The reason humidity had such a significant impact is that the MEA usually uses a perfluorosulfonic acid membrane (e.g., Nafion®) as the electrolyte. A high or nearly saturated humidity (RH > 80%) is usually required to obtain practical performance because the conductivity of perfluorosulfonic acid membranes depends on the water content. Higher humidity means higher conductivity, consequently resulting in better performance [61–64]. Water management is critical for PEMFC operation. Sufficient water must be absorbed into the membrane to ionize the acid groups, whereas excess water can cause flooding issues and thus diminishing the performance [52]. The inlet RH of the electrodes must be controlled to prevent both membranes from drying out and electrode flooding. Although better performance is usually obtained by increasing RH, excess moisture may result in water flooding that hinders gas transport [65]. In the present experiments, RH was maintained between 35% and 90% to prevent either water starvation or over saturation. In fact, many researchers have investigated the mechanism of humidity influence. It is generally believed that RH can impact electro-osmotic drag, water diffusion, membrane ionic conductivity, and water back diffusion flux in the MEAs, which consequently influence the performance [66,67]. Elevated RH can greatly improve the PEMFC performance, through increasing the membrane conductivity [68,69], the catalyst activities [68,70], the electrode kinetics [71,72], and the mass transfer rates [73,74].

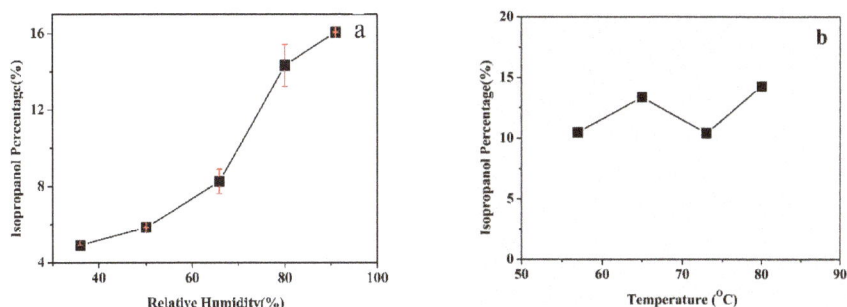

Figure 3. Isopropanol yield (percentage) change as a result of: (**a**) RH; and (**b**) temperature. H_2 and acetone flow rates were 0.25 L/min and 6 mL/h, respectively. (**a**) The ECH experiments were conducted five times at 65 °C. Each ECH experiment was repeated twice. (**b**) The ECH experiments were conducted four times at 80% RH.

Operating temperature is the second factor that was assessed in this study. Generally, the temperature was found to have a slightly positive impact on the product yield. As shown in Figure 3b, the isopropanol yield percentage of products varied between 9.3% and 14.3%, when the temperature increased from 55 °C to 80 °C. However, the temperature seemed to have minimal effect on the total efficiency. The results agreed with the findings from the literature. Singh et al. [25] investigated ECH of phenol by Pt accordance with increasing temperature. They believed that dehydrogenated phenol adsorbents easily block the active sites of Pt at higher temperatures. The ECH efficiency was claimed to be directly correlated with the adsorption properties of acetone, hydrogen, and propanol onto the Pt/C catalyst. Murillo and Chen [75] used temperature programmed desorption (TPD) to monitor the desorption and decomposition property of propanol in a wide temperature range on the Pt surface. According to their research, propanol decomposition peaked at 65 °C and 117 °C. In the present research, the operating temperature ranged from 55 °C to 80 °C, between which propanol decomposition could happen at a higher temperature (>80 °C). Decomposition of propanol was believed to cause the decrease of its yield. Therefore, increasing the operating temperature in the range does not necessarily result in an increase in the product yield.

Finally, the influence of applied voltage on the product yield was also investigated. Generally, the input voltage has no obvious impact on the product yield. In the experiments, the voltage ranged from 10 mV to 20 mV, with 5 mV increments. At 10 mV and 15 mV, the yield of isopropanol was 15.9% and 17.0%, respectively. Diisopropyl ether was not detected in either case. However, when the input voltage was increased to 0.02 V, the volumetric percentage of isopropanol produced was up to 16%, and about 1% diisopropyl ether was detected.

3.2. System Analysis

Selectivity, H_2 utilization, and current efficiency were selected to evaluate the hydrogenation efficiency. Selectivity represents the yield of desirable products. As the major product, higher isopropanol selectivity was pursued. The selectivity is calculated based on the following equation [14,20], where acetone unreacted is excluded:

$$Selectivity = \frac{Moles\ of\ Desired\ Product}{Total\ Moles\ of\ Products} \times 100\% \tag{9}$$

In the present research, H_2 was supplied to the anode to produce protons for ECH reactions on the cathode. Due to gas diffusion resistance, gas crossover, and hydrogen regeneration on the cathode, some H_2 was wasted. The H_2 utilization is directly related to the overall ECH efficiency. Higher H_2 utilization percentage is desired since more hydrogen will be involved in the ECH process.

The actual amount of H_2 used to produce isopropanol can be derived from the amount of product. The H_2 utilization is calculated by the following equation:

$$H_2\ Utilization = \frac{Atomic\ Hydrogen\ Used\ for\ Faraday\ Current}{Total\ Atomic\ Hydrogen\ Supplied} \times 100\% \qquad (10)$$

During the ECH process, acetone reacts with $M(H)_{ads}$ to produce isopropanol and byproducts on the cathode. Concurrently, H_2 regeneration happens and is an unfavorable process simply because it wastes energy. The H_2 regeneration reaction is affected by supplied voltage, temperature, humidity, and catalyst. Hereby, current efficiency (shown below) is used as an important parameter to determine how efficient H_2 is used for the ECH process [14,20].

$$Current\ Efficiency = \frac{Current\ used\ for\ the\ ECH\ process}{Total\ Faraday\ current} \times 100\% \qquad (11)$$

The total efficiency is defined by the H_2 utilization multiplying the current efficiency, as shown in the following equation:

$$Total\ Efficiency = H_2\ Utilization \times Current\ Efficiency \qquad (12)$$

Figure 4a shows the impact of temperature on H^+ utilization, current efficiency, and total efficiency in a typical set of experiments. As the operating temperature increased from 50 °C to 80 °C, H^+ utilization increased from 1.5% to 6.0%, whereas current efficiency decreased from 28% to 18.5%. Temperature affects H^+ utilization and current efficiency differently. Higher ionic mobility and catalytic activities are achieved with higher operating temperatures, resulting in higher H_2 utilization. Consequently, the electrochemical conversion and reaction rates increase with elevated temperatures [52]. However, Figure 4a indicates that, although elevated operating temperature enabled more hydrogen being involved in reactions, the yield of products did not increase or even decreased. That resulted in a loss of current efficiency, which means most extra protons produced were somehow wasted. The conclusion can also be evidenced by the curve of the total efficiency, which remained almost flat. Note that the total efficiency was low because abundant H_2 was supplied to the system to minimize the impact of fuel starvation and gas diffusion resistance. Practically, stoichiometric flow can be fed to the system based upon the actual current. In their experiments of ECH of acetone, Sara et al. observed that the current efficiency increased while the cell temperature increased from 25 °C to 50 °C, which seems to contradict our results [52]. However, in the present research, PEMFCs were operated in a recommended range between 50 °C and 80 °C to achieve the best performance. The reduction of current efficiency is believed mainly due to propanol decomposition, as mentioned in Section 3.1. Another minor reason was acetone vaporization, since the boiling point of acetone is 56 °C. Acetone gasification might have negative impact on the hydrogenation reactions on the cathode, due to increased pressure and thus higher diffusion resistance.

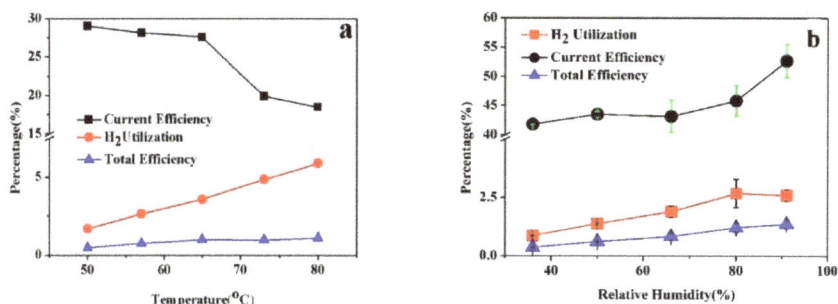

Figure 4. H_2 Utilization, Current Efficiency, and Total efficiency calculated at different: (**a**) temperatures; and (**b**) RH. H_2 and acetone flow rates were 0.25 L/min and 6 mL/h, respectively. (**a**) The ECH experiments were conducted five times at 80% RH; (**b**) The ECH experiments were conducted five times at 65 °C. Each ECH experiment was repeated twice.

Figure 4b shows the impact of RH on H_2 utilization, current efficiency, and total efficiency. The operating temperature was set at 65 °C, and the RH was controlled by setting the humidifier's temperature. It is seen that higher RH resulted in better efficiencies. As the RH ranged from 35% to 90%, the H_2 utilization increased from 0.9% to 2.8%, and the total efficiency increased from 0.4% to 1.5%. A sudden spike of the current efficiency was observed when ramping the RH from 80% to 91%. It can be concluded that higher RH is favorable for ECH of acetone and will result in a higher yield of products. It has been proved that higher ionic conductivity can be achieved when MEAs become more hydrated [68–70,76–79]. Practically, high RH is required to maintain the best fuel cell performance. The higher water content in the Nafion membrane will ease proton transport, i.e., reduce ionic resistance. As a result, more protons can be created and transported to the cathode for ECH. Typically, >80% RH is recommended [61–64], which explains why a spike in the current efficiency was observed when the RH surpassed 80%.

The present research shows that the performance of ECH of acetone is correlated with RH, input voltage, and temperature, in which RH has the most obvious effect. It is suggested that the optimized operating conditions are RH of 80% or more, the input voltage of 0.02 V or less, and temperature of between 50 °C and 55 °C. The obtained maximum H_2 utilization and maximum current efficiency achieved in the present experiments were 5.9% and 59.7%, respectively. To further increase those efficiencies, stoichiometric flow control is strongly recommended.

3.3. MEA Degradation

Long-term durability is one important factor that affects the practical applications of ECH using PEMFC reactors. Nowadays, commercial MEAs are fairly durable for their common roles as the power sources. The ECH process, however, involves organics that may contaminate MEAs and thus shorten their lifetime. To our best knowledge, very limited research has been conducted to evaluate the impacts of contaminants on the durability of ECH. In the past decade, extensive research has been carried out on mitigating contamination of PEMFCs from impurities, including CO, CO_2, H_2S, NO_x, SO_x, and hydrocarbons [80,81]. Impurities may contaminate one or more components of the MEA, resulting in performance degradation. Three major contamination effects were identified as the poisoning of the electrode catalysts, a decrease of the ionic conductivity, and an increase of the mass transfer resistance.

Additionally, the crossover is another factor that negatively impacts the PEMFC performance. Crossover of organic compounds during hydrogenation using PEMFCs has been reported [82,83]. One immediate drawback is the loss of fuel and/or products, which decreases the efficiency.

Furthermore, contaminants not only poison just one electrode but also may crossover and further poison the catalyst on the other electrode [12].

To investigate the impacts of MEA degradation on the present hydrogenation tests, polarization scans (V-I sweeps) were performed after each test [51]. The black curves in Figure 5 are the baseline data recorded for the fresh MEAs prior to ECH tests. After each test, pure N_2 was purged for at least 10 h to remove all the temporary contaminants. The effects of RH and temperature on the MEA degradation were also evaluated.

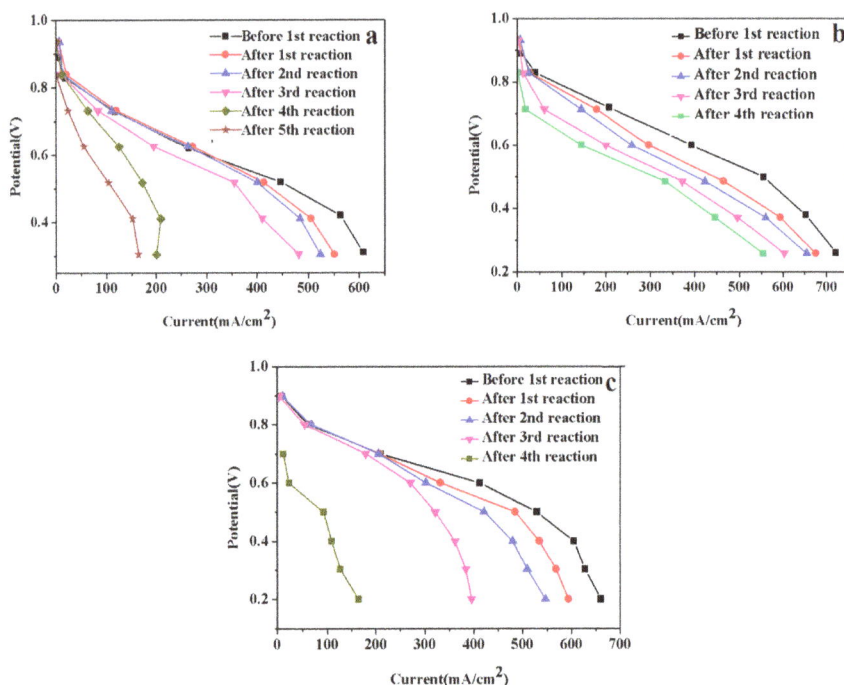

Figure 5. V-I scans performed during three sets of ECH experiments: (**a**) The ECH experiments were conducted five times at 65 °C and 80% RH; (**b**) the ECH experiments were conducted four times at 65 °C but different RH (65%, 65%, 50% and 50%, respectively); (**c**) the ECH experiments were conducted five times at 80% RH but different temperatures (80 °C, 73 °C, 57 °C, and 50 °C, respectively). The black curves are the baseline data recorded for the fresh MEAs prior to ECH tests. After each test, pure N_2 was purged for at least 10 h to remove all the temporary contaminants.

Figure 5a shows the results of five sets of V-I measurements performed on a PEMFC, which underwent five 10-h ECH experiments. Both the ECH experiments and V-I measurements shown in Figure 5a were conducted at 65 °C with 80% RH. It clearly shows that the MEA performance degraded as more ECH tests were conducted, especially after the third ECH experiment. The open circuit voltage (OCV) dropped a lot starting from the fourth V-I scan, which indicates that crossover became significant. It implied that pinholes might form due to degradation.

Figure 5b shows the effect of RH on the MEA degradation during ECH. For the ECH experiments conducted in Figure 5b, RH was reduced to 65% for the first and second tests and was further reduced to 50% for the third and fourth tests. To compare with the same baseline data (the black curve in Figure 5b), all V-I scans were performed using the same operating conditions that have been used for the baseline scan. The results illustrate that reducing RH was able to mitigate the degradation to some extent. It is believed that less RH resulted in less mass transport via the MEA, which eventually extended the lifetime of the catalysts. However, permanent damage to the MEA still existed, as seen from the general trend of V-I scans. Similar to the observations in Figure 5a, purging with pure N_2 could not remove permanent contaminants.

Finally, the effect of operating temperature on the MEA degradation was investigated, as shown in Figure 5c. For the ECH experiments conducted in Figure 5c, the operating temperatures were 80 °C, 73 °C, 57 °C, and 50 °C, while maintaining the same RH. Again, all V-I scans were performed using the same operating conditions as those used for the baseline scan (black curve in Figure 5c). The results show that temperature variation has no observable impact on MEA degradation. In other words, changing the operating temperature did not mitigate degradation.

Figure 6 shows that the trends of current efficiency, H_2 utilization, and total efficiency using the same MEA for several ECH experiments. Figure 6 shows that, as the MEA degraded, current efficiency, H_2 utilization, and total efficiency all decreased. Until a method of contamination mitigation is found, it is acceptable to use one MEA three times.

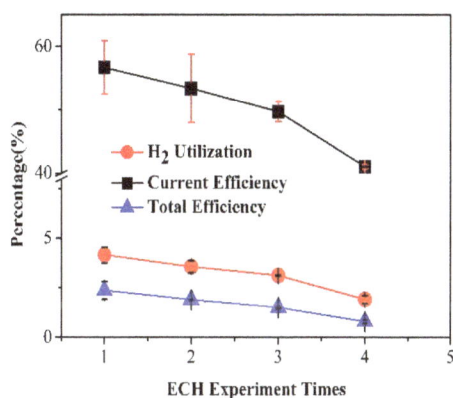

Figure 6. H_2 Utilization, Current Efficiency and Total efficiency as a function of ECH experiment times. The ECH experiments were conducted four times at 65 °C and 80% RH. H_2 and acetone flow rates were 0.25 L/min and 6 mL/h, respectively. After each test, pure N_2 was purged for at least 10 h to remove all the temporary contaminants. Each ECH experiment was repeated twice.

Usually, the lifespan of a PEMFC under steady-state operation can be very long, up to thousands of hours [84–89]. However, catalyst contamination is the major factor that diminishes the PEMFC performance and very likely results in significant degradation [90–92]. In fact, many organic compounds can contaminate the MEA. Those compounds include acetaldehyde, toluene, propane, vinyl acetate, methyl methacrylate, acetonitrile, dichloromethane, acetylene, chlorobenzene, formic acid, methanol, ethanol, phenol, butane, acetone, and naphthalene [93–95], and the list is expanding. The main reason that so many contaminants were found is that the catalysts used in common MEAs are Pt-based. Pt is a premium catalyst, but also sensitive to so many contaminants. The MEAs used in the present research contain pure Pt as the catalyst. Although developing non-Pt catalyst is beyond the scope of the present research, to further conduct durable ECH experiments, MEAs with contamination tolerant catalysts need to be used.

Reactant and product crossing over is another possible reason that caused the MEA degradation. Liquids that contained mainly acetone were detected at the anode side during the ECH experiments. Those liquids not only decrease the fuel utilization but also further contaminate the anode catalyst. Feasible solutions to this issue include adopting thicker MEAs, feeding gaseous feedstock instead of liquid, and using non-Pt catalyst [96].

In summary, to minimize the MEA contamination using the current setup, keeping low RH is suggested. To solve the problem essentially, novel non-Pt catalysts need to be developed, such as Pd- and Ni-based catalysts [97,98]. Even though a wide range of metals can be used as electrocatalysts at the cathode, those with the strong hydrogen absorption capability are desired.

4. Conclusions

Electrochemical hydrogenation of acetone using a PEMFC reactor was successfully demonstrated in the present research. The results proved that ECH can be a feasible way of hydrogenating acetone to produce isopropanol in mild conditions. In the experiments, the main product obtained was isopropanol with a selectivity of approximately 90%. A small amount (about 1%) diisopropyl ether was also obtained as a byproduct. The mild operation conditions, including low temperature and ambient pressure, are the greatest advantages of the proposed ECH method. The present research suggests that the optimized conditions for ECH of acetone using a PEMFC reactor include an operating temperature around 65 °C and relatively high RH.

Contamination impact using the PEMFC reactor during ECH was also investigated. It was concluded that organic compounds can contaminate the MEAs, resulting in serious degradation. However, methods to mitigate contamination are limited. The present research only demonstrated that lower RH could help reduce contamination. Eventually, novel non-Pt catalysts need to be developed for durable ECH process.

Supplementary Materials: The following are available online at http://www.mdpi.com/1996-1073/11/10/2691/s1, Table S1: Operating Parameters of the Experiments.

Author Contributions: Conceptualization, C.L. and S.K.; Methodology, X.Z.; Software, C.L. and A.M.S.; Validation, C.L., X.Z. and S.K.; Formal Analysis, C.L.; Investigation, C.L. and A.M.S.; Resources, X.Z. and S.K.; Data Curation, C.L.; Writing—Original Draft Preparation, C.L.; Writing—Review and Editing, X.Z. and S.K.; Visualization, C.L.; Supervision, S.K.; Project Administration, X.Z.; and Funding Acquisition, X.Z.

Funding: This research was supported by the United States National Science Foundation (#1560194) and the Old Dominion University.

Acknowledgments: The authors would like to acknowledge the technical assistance from Can Zhou, Qiang Tang, and Yazdanshenas Elias who helped with the experiment set up. The authors are also grateful to Ali Teymouri, Caleb Talbot, Adams Kameron, Alexander Asiedu, and Samaratunga Ashani for technical discussion and sample analysis.

Conflicts of Interest: The authors declare no conflict of interest.

References

1. World Health Organization. Guide to Local Production: WHO-Recommended Handrub Formulations. 2011, pp. 1123–1156. Available online: http://www.who.int/gpsc/1125may/Guide_to_Local_Production.pdf (accessed on 2 October 2018).

2. Ceresana Inc. Market Study Solvents, 4th ed. 2012, p. 10. Available online: https://www.ceresana.com/en/market-studies/chemicals/solvents/ (accessed on 2 October 2018).

3. Alt, C. *Ullmann's Encyclopedia of Industrial Chemistry*, 1th ed.; Wiley-VCH: New York, NY, USA, 2006; pp. 2345–2351.

4. Onoue, Y.; Mizutani, Y.; Akiyama, S.; Izumi, Y.; Ihara, H. Isopropyl alcohol by direct hydration of propylene. *Bull. Jpn. Pet. Inst.* **1973**, *15*, 50–55. [CrossRef]

5. John, E.L.; Loke, R.A. Isopropyl Alcohol. In *Kirk-Othmer Encyclopedia of Chemical Technology*, 5th ed.; John, E., Logsdon, L.R.A., Eds.; Wiley-VCH: New York, NY, USA, 2000; Volume 6, pp. 98–100. ISBN 9780471238966.

6. Yurieva, T.M.; Plyasova, L.M.; Makarova, O.V.; Krieger, T.A. Mechanisms for hydrogenation of acetone to isopropanol and of carbon oxides to methanol over copper-containing oxide catalysts. *J. Mol. Catal. A Chem.* **1996**, *113*, 455–468. [CrossRef]

7. Cyr, A.; Chiltz, F.; Jeanson, P.; Martel, A.; Brossard, L.; Lessard, J.; Ménard, H. Electrocatalytic hydrogenation of lignin models at Raney nickel and palladium-based electrodes. *Can. J. Chem.* **2000**, *78*, 307–315. [CrossRef]

8. Bockris, J. Electrochemical Processing. In *Comprehensive Treatise of Electrochemistry*, 1th ed.; Bockris, J., Brian, E.C., Yeager, E., White, R.E., Eds.; Springer Science & Business Media: New York, NY, USA, 2013; Volume 2, pp. 537–616. ISBN 1468437852.

9. Cirtiu, C.M.; Brisach-Wittmeyer, A.; Ménard, H. Electrocatalysis over Pd catalysts: A very efficient alternative to catalytic hydrogenation of cyclohexanone. *J. Catal.* **2007**, *245*, 191–197. [CrossRef]

10. Amouzegar, K.; Savadogo, O. Electrocatalytic hydrogenation of phenol on dispersed Pt: Reaction mechanism and support effect. *Electrochim. Acta* **1998**, *43*, 503–508. [CrossRef]

11. St-Pierre, G.; Chagnes, A.; Bouchard, N.-A.; Harvey, P.D.; Brossard, L.; Ménard, H. Rational design of original materials for the electrocatalytic hydrogenation reactions: Concept, preparation, characterization, and theoretical analysis. *Langmuir* **2004**, *20*, 6365–6373. [CrossRef] [PubMed]

12. Cirtiu, C.M.; Hassani, H.O.; Bouchard, N.-A.; Rowntree, P.A.; Ménard, H. Modification of the surface adsorption properties of alumina-supported Pd catalysts for the electrocatalytic hydrogenation of phenol. *Langmuir* **2006**, *22*, 6414–6421. [CrossRef] [PubMed]

13. Song, Y.; Chia, S.H.; Sanyal, U.; Gutiérrez, O.Y.; Lercher, J.A. Integrated catalytic and electrocatalytic conversion of substituted phenols and diaryl ethers. *J. Catal.* **2016**, *344*, 263–272. [CrossRef]

14. Li, Z.; Kelkar, S.; Lam, C.H.; Luczek, K.; Jackson, J.E.; Miller, D.J.; Saffron, C.M. Aqueous electrocatalytic hydrogenation of furfural using a sacrificial anode. *Electrochim. Acta* **2012**, *64*, 87–93. [CrossRef]

15. Chu, D.; Hou, Y.; He, J.; Xu, M.; Wang, Y.; Wang, S.; Wang, J.; Zha, L. Nano TiO_2 film electrode for electrocatalytic reduction of furfural in ionic liquids. *J. Nanopart. Res.* **2009**, *11*, 1805–1809. [CrossRef]

16. Parpot, P.; Bettencourt, A.; Chamoulaud, G.; Kokoh, K.; Belgsir, E. Electrochemical investigations of the oxidation–reduction of furfural in aqueous medium: Application to electrosynthesis. *Electrochim. Acta* **2004**, *49*, 397–403. [CrossRef]

17. Chamoulaud, G.; Floner, D.; Moinet, C.; Lamy, C.; Belgsir, E. Biomass conversion II: Simultaneous electrosyntheses of furoic acid and furfuryl alcohol on modified graphite felt electrodes. *Electrochim. Acta* **2001**, *46*, 2757–2760. [CrossRef]

18. Green, S.K.; Lee, J.; Kim, H.J.; Tompsett, G.A.; Kim, W.B.; Huber, G.W. The electrocatalytic hydrogenation of furanic compounds in a continuous electrocatalytic membrane reactor. *Green Chem.* **2013**, *15*, 1869–1879. [CrossRef]

19. Nilges, P.; Schroder, U. Electrochemistry for biofuel generation: Production of furans by electrocatalytic hydrogenation of furfurals. *Energy Environ. Sci.* **2013**, *6*, 2925–2931. [CrossRef]

20. Zhao, B.; Chen, M.; Guo, Q.; Fu, Y. Electrocatalytic hydrogenation of furfural to furfuryl alcohol using platinum supported on activated carbon fibers. *Electrochim. Acta* **2014**, *135*, 139–146. [CrossRef]

21. Ilikti, H.; Rekik, N.; Thomalla, M. Electrocatalytic hydrogenation of alkyl-substituted phenols in aqueous solutions at a Raney nickel electrode in the presence of a non-micelle-forming cationic surfactant. *J. Appl. Electrochem.* **2004**, *34*, 127–136. [CrossRef]

22. Laplante, F.; Brossard, L.; Ménard, H. Considerations about phenol electrohydrogenation on electrodes made with reticulated vitreous carbon cathode. *Can. J. Chem.* **2003**, *81*, 258–264. [CrossRef]

23. Song, Y.; Gutiérrez, O.Y.; Herranz, J.; Lercher, J.A. Aqueous phase electrocatalysis and thermal catalysis for the hydrogenation of phenol at mild conditions. *Appl. Catal. B Environ.* **2016**, *182*, 236–246. [CrossRef]

24. Santana, D.S.; Lima, M.V.F.; Daniel, J.R.R.; Navarro, M. Electrocatalytic hydrogenation of organic compounds using current density gradient and sacrificial anode of nickel. *Tetrahedron Lett.* **2003**, *44*, 4725–4727. [CrossRef]

25. Singh, N.; Song, Y.; Gutiérrez, O.Y.; Camaioni, D.M.; Campbell, C.T.; Lercher, J.A. Electrocatalytic Hydrogenation of Phenol over Platinum and Rhodium: Unexpected Temperature Effects Resolved. *ACS Catal.* **2016**, *6*, 7466–7470. [CrossRef]

26. Tountian, D.; Brisach-Wittmeyer, A.; Nkeng, P.; Poillerat, G.; Ménard, H. Effect of support conductivity of catalytic powder on electrocatalytic hydrogenation of phenol. *J. Appl. Electrochem.* **2009**, *39*, 411–419. [CrossRef]

27. Ilikti, H.; Rekik, N.; Thomalla, M. Electrocatalytic hydrogenation of phenol in aqueous solutions at a Raney nickel electrode in the presence of cationic surfactants. *J. Appl. Electrochem.* **2002**, *32*, 603–609. [CrossRef]

28. Tsyganok, A.; Holt, C.M.; Murphy, S.; Mitlin, D.; Gray, M.R. Electrocatalytic hydrogenation of aromatic compounds in ionic liquid solutions over WS2-on-glassy carbon and Raney nickel cathodes. *Fuel* **2012**, *93*, 415–422. [CrossRef]

29. Sáez, A.; García-García, V.; Solla-Gullón, J.; Aldaz, A.; Montiel, V. Electrocatalytic hydrogenation of acetophenone using a Polymer Electrolyte Membrane Electrochemical Reactor. *Electrochim. Acta* **2013**, *91*, 69–74. [CrossRef]

30. Pintauro, P.N.; Gil, M.P.; Warner, K.; List, G.; Neff, W. Electrochemical Hydrogenation of Soybean Oil with Hydrogen Gas. *Ind. Eng. Chem. Res.* **2005**, *44*, 6188–6195. [CrossRef]

31. An, W.; Hong, J.K.; Pintauro, P.N.; Warner, K.; Neff, W. The electrochemical hydrogenation of edible oils in a solid polymer electrolyte reactor. II. Hydrogenation selectivity studies. *J. Am. Oil Chem. Soc.* **1999**, *76*, 215–222. [CrossRef]

32. An, W.; Hong, J.K.; Pintauro, P.N.; Warner, K.; Neff, W. The electrochemical hydrogenation of edible oils in a solid polymer electrolyte reactor. I. Reactor design and operation. *J. Am. Oil Chem. Soc.* **1998**, *75*, 917–925. [CrossRef]

33. Xin, L.; Zhang, Z.; Qi, J.; Chadderdon, D.J.; Qiu, Y.; Warsko, K.M.; Li, W. Electricity Storage in Biofuels: Selective Electrocatalytic Reduction of Levulinic Acid to Valeric Acid or γ-Valerolactone. *ChemSusChem* **2013**, *6*, 674–686. [CrossRef] [PubMed]

34. Qiu, Y.; Xin, L.; Chadderdon, D.J.; Qi, J.; Liang, C.; Li, W. Integrated electrocatalytic processing of levulinic acid and formic acid to produce biofuel intermediate valeric acid. *Green Chem.* **2014**, *16*, 1305–1315. [CrossRef]

35. Nilges, P.; dos Santos, T.R.; Harnisch, F.; Schroder, U. Electrochemistry for biofuel generation: Electrochemical conversion of levulinic acid to octane. *Energy Environ. Sci.* **2012**, *5*, 5231–5235. [CrossRef]

36. Dalavoy, T.S.; Jackson, J.E.; Swain, G.M.; Miller, D.J.; Li, J.; Lipkowski, J. Mild electrocatalytic hydrogenation of lactic acid to lactaldehyde and propylene glycol. *J. Catal.* **2007**, *246*, 15–28. [CrossRef]

37. Lai, S.C.; Koper, M.T. Electro-oxidation of ethanol and acetaldehyde on platinum single-crystal electrodes. *Faraday Discuss.* **2009**, *140*, 399–416. [CrossRef]

38. Huang, B.; Durante, C.; Isse, A.A.; Gennaro, A. Highly selective electrochemical hydrogenation of acetylene to ethylene at Ag and Cu cathodes. *Electrochem. Commun.* **2013**, *34*, 90–93. [CrossRef]

39. Li, Z.; Kelkar, S.; Raycraft, L.; Garedew, M.; Jackson, J.E.; Miller, D.J.; Saffron, C.M. A mild approach for bio-oil stabilization and upgrading: Electrocatalytic hydrogenation using ruthenium supported on activated carbon cloth. *Green Chem.* **2014**, *16*, 844–852. [CrossRef]

40. Li, Z.; Garedew, M.; Lam, C.H.; Jackson, J.E.; Miller, D.J.; Saffron, C.M. Mild electrocatalytic hydrogenation and hydrodeoxygenation of bio-oil derived phenolic compounds using ruthenium supported on activated carbon cloth. *Green Chem.* **2012**, *14*, 2540–2549. [CrossRef]

41. Kwon, Y.; Koper, M. Electrocatalytic hydrogenation and deoxygenation of glucose on solid metal electrodes. *ChemSusChem* **2013**, *6*, 455–462. [CrossRef] [PubMed]

42. Lam, C.H.; Lowe, C.B.; Li, Z.; Longe, K.N.; Rayburn, J.T.; Caldwell, M.A.; Houdek, C.E.; Maguire, J.B.; Saffron, C.M.; Miller, D.J. Electrocatalytic upgrading of model lignin monomers with earth abundant metal electrodes. *Green Chem.* **2015**, *17*, 601–609. [CrossRef]

43. Moutet, J.-C. Electrocatalytic hydrogenation on hydrogen-active electrodes. A review. *Org. Prep. Proced. Int.* **1992**, *24*, 309–325. [CrossRef]

44. Mahdavi, B.; Chapuzet, J.M.; Lessard, J. The electrocatalytic hydrogenation of phenanthrene at Raney nickel electrodes: The effect of periodic current control. *Electrochim. Acta* **1993**, *38*, 1377–1380. [CrossRef]

45. Lipkowski, J.; Ross, P.N. *Electrocatalysis*, 1th ed.; John Wiley&Sons: New York, NY, USA, 1998; pp. 75–290. ISBN 0471246735.

46. Robin, D.; Comtois, M.; Martel, A.; Lemieux, R.; Cheong, A.K.; Belot, G.; Lessard, J. The electrocatalytic hydrogenation of fused poly cyclic aromatic compounds at Raney nickel electrodes: The influence of catalyst activation and electrolysis conditions. *Can. J. Chem.* **1990**, *68*, 1218–1227. [CrossRef]

47. Garraín, D.; Lechón, Y.; de la Rúa, C. Polymer electrolyte membrane fuel cells (PEMFC) in automotive applications: Environmental relevance of the manufacturing stage. *Smart Grid Renew. Energy* **2011**, *2*, 68. [CrossRef]

48. Montiel, V.; Sáez, A.; Expósito, E.; García-García, V.; Aldaz, A. Use of MEA technology in the synthesis of pharmaceutical compounds: The electrosynthesis of N-acetyl-l-cysteine. *Electrochem. Commun.* **2010**, *12*, 118–121. [CrossRef]

49. Wang, Y.; Chen, K.S.; Mishler, J.; Cho, S.C.; Adroher, X.C. A review of polymer electrolyte membrane fuel cells: Technology, applications, and needs on fundamental research. *Appl. Energy* **2011**, *88*, 981–1007. [CrossRef]

50. Zhang, X.; Pasaogullari, U.; Molter, T. Influence of ammonia on membrane-electrode assemblies in polymer electrolyte fuel cells. *Int. J. Hydrogen Energy* **2009**, *34*, 9188–9194. [CrossRef]

51. Florida Solar Energy Center, Test Protocol for Cell Performance Tests Performed Under DOE. 2009; pp. 19–20. Available online: https://www.energy.gov/sites/prod/files/2014/03/f10/htmwg_may09_pem_single_cell_testing.pdf (accessed on 8 October 2018).

52. Green, S.K.; Tompsett, G.A.; Kim, H.J.; Kim, W.B.; Huber, G.W. Electrocatalytic Reduction of Acetone in a Proton-Exchange-Membrane Reactor: A Model Reaction for the Electrocatalytic Reduction of Biomass. *ChemSusChem* **2012**, *5*, 2410–2420. [CrossRef] [PubMed]

53. Dadda, B.; Abboudi, S.; Zarrit, R.; Ghezal, A. Heat and mass transfer influence on potential variation in a PEMFC membrane. *Int. J. Hydrogen Energy* **2014**, *39*, 15238–15245. [CrossRef]

54. Zong, Y.; Zhou, B.; Sobiesiak, A. Water and thermal management in a single PEM fuel cell with non-uniform stack temperature. *J. Power Sources* **2006**, *161*, 143–159. [CrossRef]

55. Berg, L. Separation of Acetone from Isopropanol-Water Mixtures by Extractive Distillation. US 5897750, 1999. Available online: https://patents.justia.com/patent/5897750 (accessed on 8 October 2018).

56. Jian, Q.; Ma, G.; Qiu, X. Influences of gas relative humidity on the temperature of membrane in PEMFC with interdigitated flow field. *Renew. Energy* **2014**, *62*, 129–136. [CrossRef]

57. Anantaraman, A.; Gardner, C. Studies on ion-exchange membranes. Part 1. Effect of humidity on the conductivity of Nafion®. *J. Electroanal. Chem.* **1996**, *414*, 115–120. [CrossRef]

58. Sumner, J.; Creager, S.; Ma, J.; DesMarteau, D. Proton conductivity in Nafion® 117 and in a novel bis [(perfluoroalkyl) sulfonyl] imide ionomer membrane. *J. Electrochem. Soc.* **1998**, *145*, 107–110. [CrossRef]

59. Sone, Y.; Ekdunge, P.; Simonsson, D. Proton conductivity of Nafion 117 as measured by a four-electrode AC impedance method. *J. Electrochem. Soc.* **1996**, *143*, 1254–1259. [CrossRef]

60. Yuan, W.; Tang, Y.; Pan, M.; Li, Z.; Tang, B. Model prediction of effects of operating parameters on proton exchange membrane fuel cell performance. *Renew. Energy* **2010**, *35*, 656–666. [CrossRef]

61. Wong, K.; Loo, K.; Lai, Y.; Tan, S.-C.; Chi, K.T. A theoretical study of inlet relative humidity control in PEM fuel cell. *Int. J. Hydrogen Energy* **2011**, *36*, 11871–11885. [CrossRef]

62. Lee, C.-I.; Chu, H.-S. Effects of cathode humidification on the gas–liquid interface location in a PEM fuel cell. *J. Power Sources* **2006**, *161*, 949–956. [CrossRef]

63. Guvelioglu, G.H.; Stenger, H.G. Flow rate and humidification effects on a PEM fuel cell performance and operation. *J. Power Sources* **2007**, *163*, 882–891. [CrossRef]

64. Zhang, J.; Tang, Y.; Song, C.; Xia, Z.; Li, H.; Wang, H.; Zhang, J. PEM fuel cell relative humidity (RH) and its effect on performance at high temperatures. *Electrochim. Acta* **2008**, *53*, 5315–5321. [CrossRef]

65. Jeon, D.H.; Kim, K.N.; Baek, S.M.; Nam, J.H. The effect of relative humidity of the cathode on the performance and the uniformity of PEM fuel cells. *Int. J. Hydrogen Energy* **2011**, *36*, 12499–12511. [CrossRef]

66. Ikeda, T.; Koido, T.; Tsushima, S.; Hirai, S. MRI investigation of water transport mechanism in a membrane under elevated temperature condition with relative humidity and current density variation. *ECS Trans.* **2008**, *16*, 1035–1040.

67. Misran, E.; Hassan, N.S.M.; Daud, W.R.W.; Majlan, E.H.; Rosli, M.I. Water transport characteristics of a PEM fuel cell at various operating pressures and temperatures. *Int. J. Hydrogen Energy* **2013**, *38*, 9401–9408. [CrossRef]

68. Knights, S.D.; Colbow, K.M.; St-Pierre, J.; Wilkinson, D.P. Aging mechanisms and lifetime of PEFC and DMFC. *J. Power Sources* **2004**, *127*, 127–134. [CrossRef]

69. Abe, T.; Shima, H.; Watanabe, K.; Ito, Y. Study of PEFCs by AC Impedance, Current Interrupt, and Dew Point Measurements I. Effect of Humidity in Oxygen Gas. *J. Electrochem. Soc.* **2004**, *151*, A101–A105. [CrossRef]

70. Saleh, M.M.; Okajima, T.; Hayase, M.; Kitamura, F.; Ohsaka, T. Exploring the effects of symmetrical and asymmetrical relative humidity on the performance of H_2/air PEM fuel cell at different temperatures. *J. Power Sources* **2007**, *164*, 503–509. [CrossRef]

71. Xu, H.; Song, Y.; Kunz, H.R.; Fenton, J.M. Effect of elevated temperature and reduced relative humidity on ORR kinetics for PEM fuel cells. *J. Electrochem. Soc.* **2005**, *152*, A1828–A1836. [CrossRef]

72. Uribe, F.A.; Springer, T.E.; Gottesfeld, S. A microelectrode study of oxygen reduction at the platinum/recast—Nafion film interface. *J. Electrochem. Soc.* **1992**, *139*, 765–773. [CrossRef]

73. Jang, J.-H.; Yan, W.-M.; Li, H.-Y.; Chou, Y.-C. Humidity of reactant fuel on the cell performance of PEM fuel cell with baffle-blocked flow field designs. *J. Power Sources* **2006**, *159*, 468–477. [CrossRef]

74. Broka, K.; Ekdunge, P. Oxygen and hydrogen permeation properties and water uptake of Nafion® 117 membrane and recast film for PEM fuel cell. *J. Appl. Electrochem.* **1997**, *27*, 117–123. [CrossRef]

75. Murillo, L.E.; Chen, J.G. Adsorption and reaction of propanal, 2-propenol and 1-propanol on Ni/Pt (111) bimetallic surfaces. *Surf. Sci.* **2008**, *602*, 2412–2420. [CrossRef]

76. Takalloo, P.K.; Nia, E.S.; Ghazikhani, M. Numerical and experimental investigation on effects of inlet humidity and fuel flow rate and oxidant on the performance on polymer fuel cell. *Energy Convers. Manag.* **2016**, *114*, 290–302. [CrossRef]

77. Jiang, R.; Kunz, H.R.; Fenton, J.M. Investigation of membrane property and fuel cell behavior with sulfonated poly(ether ether ketone) electrolyte: Temperature and relative humidity effects. *J. Power Sources* **2005**, *150*, 120–128. [CrossRef]

78. Xu, H.; Kunz, H.R.; Fenton, J.M. Analysis of proton exchange membrane fuel cell polarization losses at elevated temperature 120 °C and reduced relative humidity. *Electrochim. Acta* **2007**, *52*, 3525–3533. [CrossRef]

79. Jiang, R.; Kunz, H.R.; Fenton, J.M. Influence of temperature and relative humidity on performance and CO tolerance of PEM fuel cells with Nafion®–Teflon®–Zr (HPO₄) 2 higher temperature composite membranes. *Electrochim. Acta* **2006**, *51*, 5596–5605. [CrossRef]

80. Wang, X.; Baker, P.; Zhang, X.; Garces, H.F.; Bonville, L.J.; Pasaogullari, U.; Molter, T.M. An experimental overview of the effects of hydrogen impurities on polymer electrolyte membrane fuel cell performance. *Int. J. Hydrogen Energy* **2014**, *39*, 19701–19713. [CrossRef]

81. Cheng, X.; Shi, Z.; Glass, N.; Zhang, L.; Zhang, J.; Song, D.; Liu, Z.-S.; Wang, H.; Shen, J. A review of PEM hydrogen fuel cell contamination: Impacts, mechanisms, and mitigation. *J. Power Sources* **2007**, *165*, 739–756. [CrossRef]

82. Benziger, J.; Nehlsen, J. A polymer electrolyte hydrogen pump hydrogenation reactor. *Ind. Eng. Chem. Res.* **2010**, *49*, 11052–11060. [CrossRef]

83. Lundin, M.D.; McCready, M.J. High pressure anode operation of direct methanol fuel cells for carbon dioxide management. *J. Power Sources* **2011**, *196*, 5583–5590. [CrossRef]

84. Scholta, J.; Berg, N.; Wilde, P.; Jörissen, L.; Garche, J. Development and performance of a 10 kW PEMFC stack. *J. Power Sources* **2004**, *127*, 206–212. [CrossRef]

85. Cleghorn, S.; Mayfield, D.; Moore, D.; Moore, J.; Rusch, G.; Sherman, T.; Sisofo, N.; Beuscher, U. A polymer electrolyte fuel cell life test: 3 years of continuous operation. *J. Power Sources* **2006**, *158*, 446–454. [CrossRef]

86. St-Pierre, J.; Wilkinsor, D.; Knights, S.; Bos, M. Relationships between water management, contamination and lifetime degradation in PEFC. *J. New Mater. Electrochem. Syst.* **2000**, *3*, 99–106.

87. Satyapal, S. Hydrogen & Fuel Cells Program Overview. In Proceedings of the Annual Merit Review and Peer Evaluation Meeting, Arlington, VA, USA, 7–11 June 2011.

88. Wipke, K.; Sprik, S.; Kurtz, J.; Ramsden, T.; Ainscough, C.; Saur, G. VII. 1 Controlled Hydrogen Fleet and Infrastructure Analysis. In Proceedings of the DOE Annual Merit Review and Peer Evaluation Meeting, Washington, DC, USA, 13–17 May 2012.

89. Wu, J.; Yuan, X.Z.; Martin, J.J.; Wang, H.; Zhang, J.; Shen, J.; Wu, S.; Merida, W. A review of PEM fuel cell durability: Degradation mechanisms and mitigation strategies. *J. Power Sources* **2008**, *184*, 104–119. [CrossRef]

90. Ferreira, P.; Shao-Horn, Y.; Morgan, D.; Makharia, R.; Kocha, S.; Gasteiger, H. Instability of Pt/C electrocatalysts in proton exchange membrane fuel cells a mechanistic investigation. *J. Electrochem. Soc.* **2005**, *152*, A2256–A2271. [CrossRef]

91. Xie, J.; Wood, D.L.; More, K.L.; Atanassov, P.; Borup, R.L. Microstructural changes of membrane electrode assemblies during PEFC durability testing at high humidity conditions. *J. Electrochem. Soc.* **2005**, *152*, A1011–A1020. [CrossRef]

92. Antolini, E. Formation, microstructural characteristics and stability of carbon supported platinum catalysts for low temperature fuel cells. *J. Mater. Sci.* **2003**, *38*, 2995–3005. [CrossRef]

93. Murugan, A.; Brown, A.S. Review of purity analysis methods for performing quality assurance of fuel cell hydrogen. *Int. J. Hydrogen Energy* **2015**, *40*, 4219–4233. [CrossRef]

94. St-Pierre, J.; Angelo, M.; Zhai, Y. Focusing research by developing performance related selection criteria for PEMFC contaminants. *ECS Trans.* **2011**, *41*, 279–286.

95. St-Pierre, J.; Zhai, Y.; Angelo, M.S. Effect of selected airborne contaminants on PEMFC performance. *J. Electrochem. Soc.* **2014**, *161*, F280–F290. [CrossRef]

Energies **2018**, *11*, 2691

96. Fukuzumi, S.; Lee, Y.M.; Nam, W. Mechanisms of two-electron versus four-electron reduction of dioxygen catalyzed by earth—Abundant metal complexes. *ChemCatChem* **2018**, *10*, 9–28. [CrossRef]

97. Chen, Y.; Hwang, C.; Liaw, C. One-step synthesis of methyl isobutyl ketone from acetone with calcined Mg/Al hydrotalcite-supported palladium or nickel catalysts. *Appl. Catal. A Gen.* **1998**, *169*, 207–214. [CrossRef]

98. Nikolopoulos, A.; Jang, B.-L.; Spivey, J. Acetone condensation and selective hydrogenation to MIBK on Pd and Pt hydrotalcite-derived MgAl mixed oxide catalysts. *Appl. Catal. A Gen.* **2005**, *296*, 128–136. [CrossRef]

energies

MDPI

Article

A Systematic Multivariate Analysis of *Carica papaya* Biodiesel Blends and Their Interactive Effect on Performance

Mohammad Anwar [1,*], **Mohammad G. Rasul** [1] **and Nanjappa Ashwath** [2]

[1] School of Engineering and Technology, Central Queensland University, North Rockhampton, QLD 4702, Australia; m.rasul@cqu.edu.au
[2] School of Health, Medical and Applied Sciences, Central Queensland University, North Rockhampton, QLD 4702, Australia; n.ashwath@cqu.edu.au
*   Correspondence: m.anwar@cqu.edu.au; Tel.: +61-7-4930-6371

Received: 11 October 2018; Accepted: 25 October 2018; Published: 26 October 2018

Abstract: This paper investigates the interactive relationship between three operating parameters (papaya seed oil (PSO) biodiesel blends, engine load, and engine speed) and four responses (brake power, BP; torque; brake specific fuel consumption, BSFC; and, brake thermal efficiency, BTE) for engine testing. A fully instrumented four cylinder four-stroke, naturally aspirated agricultural diesel engine was used for all experiments. Three different blends: B5 (5% PSO biodiesel + 95% diesel), B10 (10% PSO biodiesel + 90% diesel), and B20 (20% PSO biodiesel + 80% diesel) were tested. Physicochemical properties of these blends and pure PSO biodiesel were characterised, and the engine's performance characteristics were analysed. The results of the engine performance experiments showed that, in comparison with diesel, the three PSO biodiesel blends caused a slight reduction in BP, torque, and BTE, and an increase in BSFC. The analysis of variance and quadratic regression modelling showed that both load and speed were the most important parameters that affect engine performance, while PSO biodiesel blends had a significant effect on BSFC.

Keywords: diesel; *Carica papaya*; engine performance; biodiesel; characterisation

1. Introduction

Diesel fuel has successfully contributed to all sectors of human life, especially to transportation, industry, and agricultural sectors due to its availability, reliability, adaptability, higher combustion efficiency, and excellent handling/storage properties. However, fossil reserves (oil, gas, and coal) are limited, therefore, researchers have been exploring an alternative source for diesel for many years, and, over the last decade, biofuel (i.e., mainly biodiesel) has drawn massive attention for its excellent environmental and sustainability attributes [1,2]. Biodiesel is biodegradable, non-toxic, non-explosive, non-flammable, renewable, and an environmentally friendly (produces fewer emissions) fuel. Although the biodiesel energy content is about 10–12% less than diesel, it can be mixed with diesel at specific proportions to make the blended fuel properties close to diesel [3,4]. Furthermore, researchers are working hard to find suitable sustainable biodiesels and their blends, which can be used as fuel in unmodified diesel engines.

Numerous researchers have explored the production of biodiesel from different feedstocks focusing on non-edible oil, but very few investigations [5–10] have been performed on papaya seed oil (PSO) biodiesel. Among these, most of the articles deal with only the biodiesel production process when using different catalysts and methanol: oil molar ratios. Asokan et al. [2] examined engine performance in a single cylinder diesel engine fuelled with a mixture of papaya seed oil and watermelon seed oil and found that B20 performed close to diesel. Prabhakaran et al. [11] analysed engine performance of a single cylinder Kirloskar-TV1 diesel engine with PSO biodiesel blends and found B25 blend has lower BSFC than all other blends. Sundar Raj and Karthikayan [12] did diesel engine (single cylinder) performance analysis with PSO-diesel blends with/without additives and found PSO-diesel blends with additives have better combustion and emission characteristics. However, no other literature has been found on engine performance analysis of a fully instrumented four cylinder four-stroke diesel engine fuelled with PSO biodiesel.

PSO is non-edible and converting the waste product (seeds) of the fruit into biodiesel is a sensible option. *Carica papaya* was originally native to the tropics of the Americas and it is now primarily grown in tropical-subtropical climates of Asia (in particular, India with 42% of world production), South America (mainly Brazil), Africa (Nigeria and Congo), and Polynesia [13,14]. Therefore, the evaluation of papaya seed oil as biodiesel feedstock will contribute to the development of regional communities and their overall economy. In this study, papaya seed oil was used as feedstock to produce biodiesel and investigate its suitability as an alternative fuel source. Determining the fuel properties of PSO and its various blends with diesel were also undertaken. Finally, interactive relationships between several operating parameters (PSO biodiesel blends, load, and speed) and engine performance are evaluated and discussed.

2. Materials and Method

2.1. Materials

The raw PSO was obtained from a supplier in Eumundi, Queensland, Australia. The chemicals that were utilised for this study were methanol (99.9% purity), potassium hydroxide (KOH pellets, 99% purity), and sodium hydroxide (NaOH pellets, 99% purity). All chemicals were of analytical reagent grade (AR) and they were procured from the School of Engineering and Technology, Central Queensland University. A reflux condenser and a thermocouple fitted on 0.5 L and 1 L three-neck laboratory reactors were used for the PSO biodiesel conversion experiments.

2.2. Equipment List

The PSO properties were characterised for density, viscosity, refractive index, angular rotation, acid value, oxidation stability, and iodine value. These properties were measured before the biodiesel conversion and optimisation process. Table 1 summarises the equipment used in this study to measure the properties of PSO, PSO biodiesel, and the biodiesel-diesel blends, and the relevant standards applied. The properties studied were density at 15 °C, kinematic viscosity at 40 °C, acid value, calorific value, flash point, and oxidation stability. A gas chromatograph was used to determine the fatty acid compositions by EN 14103.

Table 1. Equipment used for measuring properties of papaya seed oil (PSO) and related products in this study.

Property	Equipment	Standard Applied	Accuracy
Kinematic viscosity	NVB classic (Normalab, France)	ASTM D445	$\pm 0.01 \text{ mm}^2/\text{s}$
Density	DM40 LiquiPhysics^{TM} density meter (Mettler Toledo, Switzerland)	ASTM D1298	$\pm 0.1 \text{ kg/m}^3$
Flash point	NPM 440 Pensky-Martens flash point tester (Normalab, France)	ASTM D93	± 0.1 °C
Acid value	Automation titration Rondo 20 (Mettler Toledo, Switzerland)	ASTM D664	± 0.001 mg KOH/g
Calorific value	6100EF semi-auto bomb calorimeter (Perr, USA)	ASTM D240	± 0.001 MJ/kg
Oxidation stability at 110 °C	873 Rancimat (Metrohm, Switzerland)	ASTM D2274	± 0.01 h
Refractive index	RM 40 Refractometer	-	± 0.0001

2.3. Biodiesel Production

The acid value of raw PSO was determined to be 0.98 mg KOH/g. After transesterification using the KOH catalyst, the acid value of PSO biodiesel was found to be 0.42 mg KOH/g. This trial experiment suggested that only the single-stage alkyl catalyst transesterification process was necessary for satisfactory PSO biodiesel production. Further, only a few researchers [6–10] have found that the single-stage transesterification process was sufficient for PSO biodiesel conversion. Based on these researches and the trial experiment, it was decided to use single-stage alkaline transesterification for biodiesel production. Thus, in each experiment, the conversion was performed by reacting a known quantity of PSO with methanol and the catalyst.

In each of these experiments, PSO was poured into a three-neck laboratory reactor and was heated to the preferred temperature. The catalyst (KOH) was dissolved in methanol in a separate beaker, and a magnetic stirrer was used to provide vigorous agitation at 600 rpm for 10 min at 50 °C to produce methoxide. This solution was then poured into the reactor that contained heated PSO to allow the transesterification reaction at a constant speed of agitation of 600 rpm. Other reaction conditions, such as temperature and time, were adjusted as per the individual experimental designs determined from the optimisation process of PSO biodiesel [14]. At the end of the transesterification reaction, the blend was poured into a separating funnel and left to settle 24 h for layer separation. Under gravity, two distinct liquid phases were formed; the upper layer was methyl ester (biodiesel), and the bottom dark brown layer consisted of glycerol and impurities. The glycerol and impurities were drawn off, whence the methyl ester layer was collected and washed with warm (50 °C) distilled water. The methyl ester was then heated to 110 °C for 15 min to remove any water that would have been retained from the washing process. A Whatman® qualitative Grade 1 filter paper was used to filter the methyl ester (i.e., biodiesel) and was stored at laboratory temperature for characterisation. Figure 1 shows the PSO production process.

Figure 1. PSO biodiesel production.

PSO biodiesel yield was calculated using Equation (1).

$$\text{PSO Biodiesel Yield} = \text{FAME percent(from GC analysis)} \times \frac{\text{weight of PSO biodiesel}}{\text{weight of PSO}} \quad (1)$$

2.4. Fatty Acid Composition

Fatty acids with a double bond are known as unsaturated fatty acids, while those without a double bond are called saturated fatty acids. High unsaturated fatty acid levels make biodiesel prone to autoxidation [14]. However, high unsaturated fatty acid levels do ensure good flow properties as compared with saturated fatty acids. The major drawback of high saturated fatty acid levels in biodiesel is its poor fuel filterability, particularly in cold weather conditions. The PSO biodiesel fatty acid composition was determined using a gas chromatograph (GC), (Thermo Trace 1310 GC) according to EN14103. Table 2 shows the operating condition details of the GC.

Table 2. Gas chromatograph (GC) operating conditions.

Property	Details
Brand, model	Thermo Scientific Trace 1310 GC
Carrier gas	Helium
Flow rates	Air: 350 mL/min, H_2: 35 mL/min, N_2: 30 mL/min
Detector temperature	240 °C
Column dimensions	60 m × 250 µm × 0.25 µm
Column head pressure	23.8 psi
Injector	Split injector, 40:1 ratio, split flow 48 mL/min, constant flow 1.2 mL/min, 1 µL injection volume
Temperature ramp 1	110 °C hold for 4 min
Temperature ramp 2	10 °C/min to 230 °C, hold for 3 min

The compositional analysis that is detailed in Table 3 shows that PSO biodiesel contains a high level (87.5%) of unsaturated fatty acid methyl esters (FAME) made up of polyunsaturated fatty acid (PUFA) at 39.03% and monounsaturated fatty acid (MUFA) at 48.49%. Amongst these, the dominant fatty acids were found to be oleic acid (C18:1) at 47.7% and linoleic acid (C18:2) at 37.3%. The saturated fatty acids included palmitic acid (C16:0) at 6%.

Table 3. PSO biodiesel fatty acid composition.

Fatty Acids	Formula	Molecular Weight	Structure	wt%
Palmitic	$C_{16}H_{32}O_2$	256	16:0	6.07
Stearic	$C_{18}H_{36}O_2$	284	18:0	3.13
Oleic	$C_{18}H_{34}O_2$	282	18:1	47.73
Linoleic	$C_{18}H_{32}O_2$	280	18:2	37.25
Linolenic	$C_{18}H_{30}O_2$	278	18:3	1.78
Eicosenoic	$C_{20}H_{38}O_2$	310	20:1	0.76
Behenic	$C_{22}H_{44}O_2$	340	22:0	0.68
Erucic	$C_{22}H_{42}O_2$	338	22:1	1.51
Others				1.09
Total Saturated Fatty Acids (SFA)				9.88
Total Monounsaturated Fatty Acids (MUFA)				48.49
Total Polyunsaturated Fatty Acids (PUFA)				39.03
Degree of Unsaturation (DU)				126.55
Long Chain Saturated Factor (LCSF)				3.19

2.5. Properties Analysis

The properties of the produced PSO methyl ester (i.e., pure biodiesel or B100) are compared with those of diesel (B0) and PSO biodiesel-diesel blends (B5, B10, and B20) in Table 4. The properties and qualities of PSO biodiesel and the diesel blends comply with the requirements of international standards ASTM D6751 and EN14214. The USA's ASTM D6751 ensures that the parameters of any pure biodiesel (B100) satisfy the standard before being used as a blend with diesel or pure fuel, whereas the European Union's EN14214 defines the minimum standards for FAME [14,15]. The PSO biodiesel was found to comply with both standards.

Table 4. Comparison of PSO biodiesel and blends with diesel.

Properties	Units	PSO	B100	B20	B10	B5	B0	ASTM D6751-2	EN14214-03
Density	Kg/m^3	885	840	829.76	828.48	827.84	827.2	870–890	860–900
Viscosity at 40 °C	mm^2/s	27.3	3.53	3.29	3.26	3.25	3.23	1.9–6.0	3.5–5.0
Acid value	mg KOH/g	0.98	0.42	0.12	0.09	0.07	0.05	max. [1] 0.5	max. 0.5
Cetane number (CN)	-	-	48.29	48.06	48.03	48.01	48.00	min. [2] 47	min. 51
Calorific value	MJ/kg	-	38.49	43.94	44.62	44.96	45.30	-	35.00
Flash point	(°C)	-	112	77.20	72.85	70.68	68.50	min. 93	>120
Iodine value (IV)	-	79.95	115.89	53.82	46.06	42.18	38.30	-	max. 120
Oxidation stability (OS)	Hour	77.97	5.61	32.32	35.66	37.33	39	min. 3	min. 6

[1] max. = maximum; [2] min. = minimum.

2.6. Fourier Transform Infrared (FTIR) Analysis

The various functional groups that are present in the pure PSO biodiesel sample were determined with FTIR spectroscopy. In this study, a Spectrum 100 series FTIR spectrometer with a universal Attenuated total reflectance (ATR) sampling accessory was used. Samples of the PSO biodiesel were inserted directly on the ATR window to record the spectra over the frequency range of 4000–650 cm^{-1} with four scans at approximately 40% transmission. Spectrum analysis program (Spectrum version 6.2.0, Perkin-Elmer Life and Analytical Sciences, Bridgeport, CT, USA) was used to acquire data and for processing. Figure 2 shows the FTIR spectrum of the pure PSO biodiesel (B100).

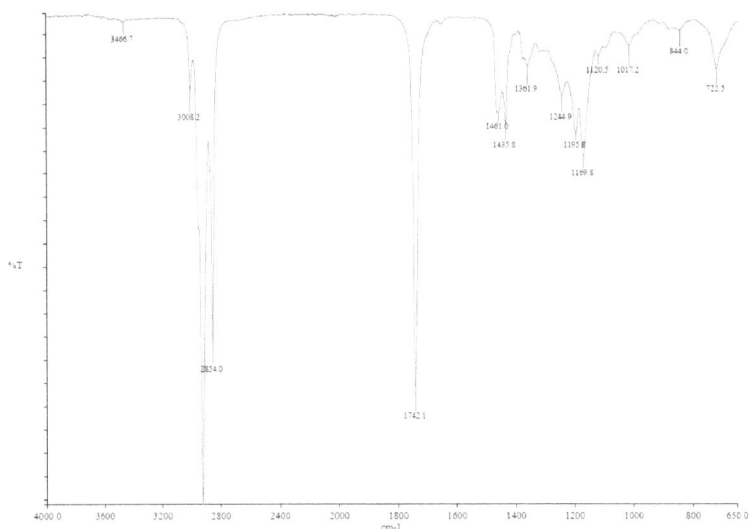

Figure 2. PSO methyl ester investigation Fourier Transform Infrared (FTIR) spectrum.

2.7. Blending of Biodiesel

The pure PSO biodiesel (B100) was mixed with pure diesel (B0) to prepare various blends. Those blends comprised homogeneous mixtures of 5% vol. of biodiesel with 95% vol. of diesel denoted by B5, 10% vol. of biodiesel with 90% vol. of diesel indicated by B10 and 20% vol. of biodiesel with 80% vol. of diesel meant by B20. The blends were prepared in a 2 L flask and were agitated at 2000 rpm with a magnetic stirrer for 60 min.

2.8. Experimental Setup for Engine Testing

This experimental investigation was carried out using pure diesel (B0) and PSO biodiesel-diesel blends (B5, B10, and B20) in an agricultural tractor four-stroke diesel engine with four cylinders (model: Kubota V3300) coupled with an eddy current dynamometer. Table 5 details the engine specification. This tractor engine was used for testing PSO biodiesel-diesel blends with regard to engine performance and characteristics of exhaust emissions. The engine performance and emissions data were collected under full load (100%) condition, keeping the throttle 100% wide open and varying speeds in the range from 1200 to 2400 rpm with an increment of 200 rpm. A photograph of the engine and test bed and a schematic of the setup of the engine along with the data acquisition system are shown in Figure 3. The engine is connected with the test bed along with instrumentation consoles that measure engine speed, torque, air, and fuel consumption and temperature. A Dynolog data acquisition system was used to convert the console-measured data to a display in the computer monitor. The exhaust gas emissions of NO_x and HC in ppm, and CO and CO_2 in vol. %, were measured with a CODA 5 gas analyser (CODA Products Pty Ltd., Hamilton NSW 2303, Australia). For measuring Particulate Matter (PM) emissions, a PM meter (MPM-4M) (MAHA Maschinenbau Haldenwang GmbH & Co. KG, Haldenwang, Germany) was used. Table 6 shows the specification of the gas analyser and PM meter. Before taking any data from the diesel engine, it was run with pure diesel (B0) for 20 min at full load to ensure that the engine was warmed up. The biodiesel-diesel blend (B5, B10, and B20) was then fuelled into the engine for analysis and data acquisition. At the end of any test or experiment with blended fuel, the engine was again flushed out with pure diesel to clear the fuel line of blended fuel and the injection system. All of the tests were repeated three times to minimise any possible error.

There are a large number of studies available in literatures on different biodiesels, and their engine performance and emission characteristics, but a very few researchers have analysed the engine performance or emission characteristics in terms of multivariate analysis. This study focuses on the multivariate analysis of PSO biodiesel blends and their interactive effect on engine performance.

Table 5. Details of the test engine.

Property	Apparatus, Model
Engine Model	Kubota V3300 Indirect injection
Type	Vertical, 4 cycle liquid cooled diesel
Number of cylinders	4
Total displacement (L)	3.318
Bore × Stroke (mm)	98 × 110
Combustion system	Spherical type (E-TVCS) (three vortex combustion system)
Intake system	Natural aspired
Rated power output (KW/rpm)	53.9/2600
Rated torque (Nm/rpm)	230/1400
Compression ratio	22.6
Emissions certification	Tier 2

(**a**) Photograph of test bed and diesel engine

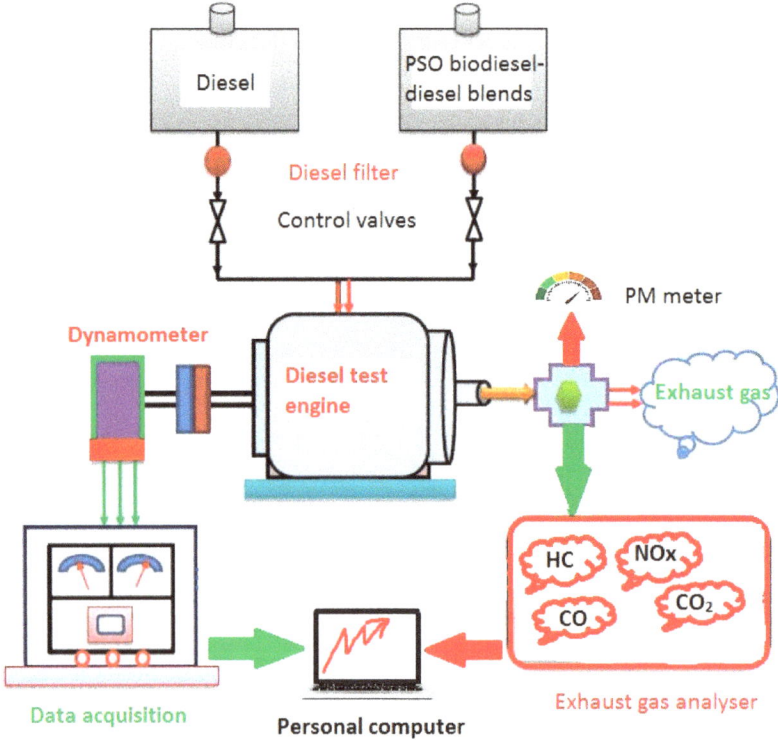

(**b**) Schematic of test procedure setup

Figure 3. Test engine set up.

Table 6. Exhaust gas analyser, PM meter specification, and error analysis.

Measured Gas	Measurement		
	Range	Resolution	Accuracy
HC	0–30,000 ppm (n-hexane)	1 ppm	±1 ppm
CO	0–15%	0.001%	±0.02%
CO_2	0–20%	0.001%	±0.3%
NO_x	0–5000 ppm	1 ppm	±1 ppm
Meter	Particle Size	Particle Concentration Range	Resolution
Particulate Matter	<100 nm to >10 μm	0.1 to >700 mg/m^3	±0.1 mg/m^3
Measurements	Accuracy	Relative Uncertainty (%)	Average Reading for B0
BP	±0.41 kW	0.0105	39.20
BSFC	±5 g/kWh	0.0195	256

3. Results and discussion

3.1. Characterisation of PSO Biodiesel-Diesel Blends

While the engine performance experiments in this study investigated B5, B10, and B20 biodiesel blends, the characteristics of a much broader series of PSO biodiesel-diesel blends has been investigated. A total of 10 PSO biodiesel-diesel blends (from B5-B90) were prepared, and their individual fuel properties of density, viscosity, flash point, calorific value, and oxidation stability have been analysed and are presented in Figure 4a–e.

Density is one of the vital properties of biodiesel that affects fuel atomisation efficiency in an airless combustion system [16]. According to relevant ASTM and EN standards, the density of biodiesel at 15 °C should be in the range of 860–900 kg/m^3. The densities of B5 (827.84 kg/m^3), B10 (828.48 kg/m^3), and B20 (829.76 kg/m^3) were very close to that of B0 diesel (827.20 kg/m^3). Generally, biodiesel has a slightly higher density than diesel. In addition, density increased as the percentage of PSO biodiesel in the blends increased (Figure 4a).

Some vegetable oils have a higher viscosity, which causes poor fuel flow. Raw PSO has a viscosity of 27.3 mm^2/s, which is 8–9 times higher than diesel (B0). However, after the transesterification process, the viscosity of PSO drops to an acceptable limit. Generally, the viscosity of biodiesel ranges from 1.9–6.0 mm^2/s, and all PSO biodiesel-diesel blends in Figure 4b fulfill this requirement. However, the viscosity of B5 (3.25 mm^2/s), B10 (3.26 mm^2/s), B20 (3.29 mm^2/s), and B30 (3.32 mm^2/s) were very close to B0 (3.23 mm^2/s). Therefore, those blends (B10–B30) can be used as a diesel engine fuel without any modifications to the engine.

According to ASTM D6751, the flashpoint should be 100–170 °C, and all of the biodiesel-diesel blends met that requirement as well. The flash point of B0 was recorded as 68.50 °C, whereas B5, B10, and B20 were found to be 70.68 °C, 72.85 °C, and 77.20 °C, respectively. Biodiesel tends to a higher flash point than diesel, and Figure 4c shows the pattern of the increased flash point with increased biodiesel blends. The more the flashpoint is above 66 °C is indicative of a safer fuel with better storage ability [1,17]. From that perspective, all PSO biodiesel-diesel blends can be stored safely and they can be used to fuel a diesel engine without any modifications.

Calorific value is another important property in the selection of any fuel. Figure 4d shows that pure PSO biodiesel and all of its blends have slightly lower calorific values than diesel although they are within the requirements of international standards. Biodiesel has nearly 10% more oxygen content than diesel, and less hydrogen-carbon content hence will produce less thermal energy [17]. Calorific values of B5 (44.96 MJ/kg), B10 (44.62 MJ/kg), and B20 (43.94 MJ/kg) blends are very close to B0 (45.30 MJ/kg).

Another vital fuel property is oxidation stability (OS), as it indicates the degree of oxidation that occurs during prolonged storage. Hasni et al. [18] mentioned that lower oxidation stability could adversely affect fuel quality. The higher the oxidation stability, the better the fuel quality. As the PSO biodiesel (unsaturated fatty acid) percentage increases, the OS decreases, which is shown in Figure 4e. As per European standard EN 590, the minimum OS of any fuel should be 20 h (indicated by the red line on the Figure 4e). PSO biodiesel-diesel blends B5 to B50 meet that standard. B5 has an OS value of 37.33 h and B50 is 22.31 h, while B0 has an OS value of 39 h. According to ASTM D6751 and EN14214, the minimum OS values are 3 and 6 h respectively. This study found the 100% PSO biodiesel has an OS value of 5.61 h, which falls within the ASTM and close to EN standards.

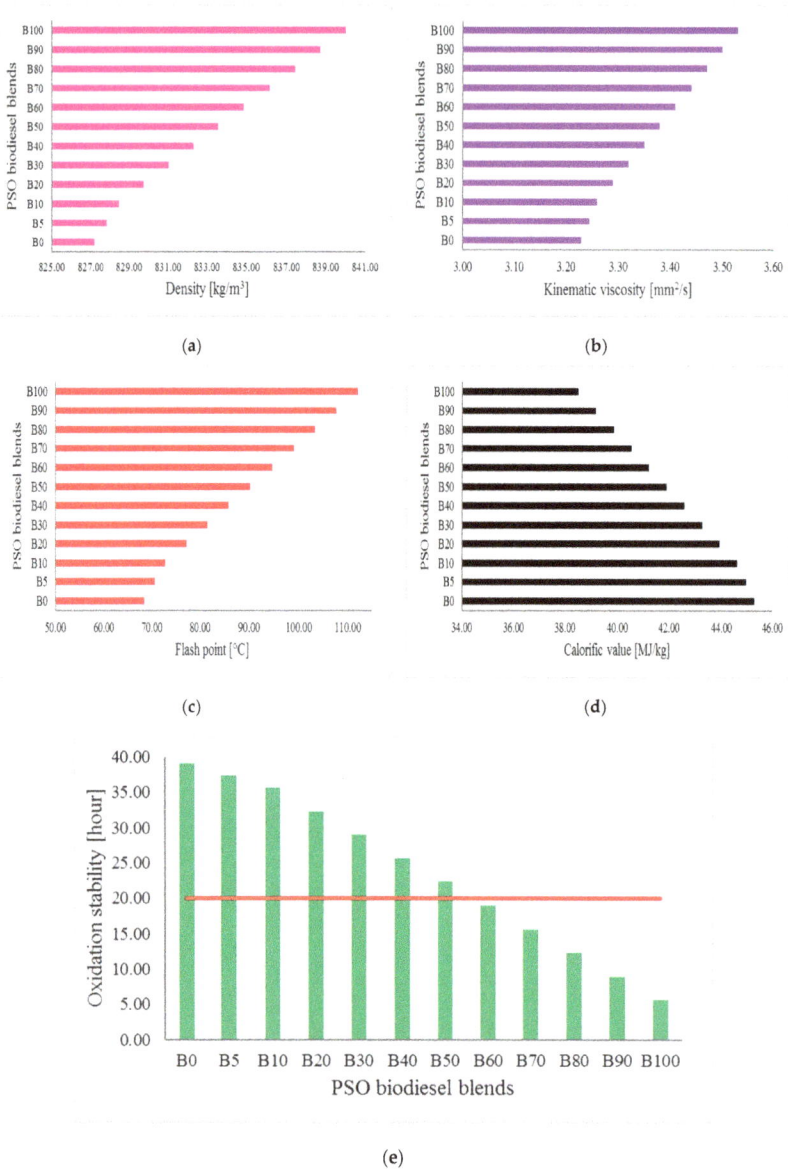

Figure 4. PSO biodiesel-diesel blending effects on: (**a**) density, (**b**) kinematic viscosity, (**c**) flash point, (**d**) calorific value, and (**e**) oxidation stability.

3.2. Analysis of Engine Performance

3.2.1. Brake Power (BP)

The brake power (BP) outputs from the diesel engine fuelled with diesel (B0) and PSO B5, B10, and B20 blends are shown in Figure 5a. It can be seen that the BP gradually increases with the increase of engine speed. Maximum BP was observed at 2400 rpm and was 45.9 kW, 45.4 kW, 45.08 kW, and 48.34 kW for B5, B10, B20, and B0, respectively. Several factors, such as calorific value and viscosity, can have the effect of uneven combustion that results in lowering the BP value. In this study, the viscosity of diesel (B0) and PSO biodiesel-diesel blends were very close. Only the difference in calorific values of blends and diesel caused the variations in BP values. Average BP value reductions for B5, B10, and B20 in comparison with B0 were 2.88%, 3.87%, and 5.13%, respectively.

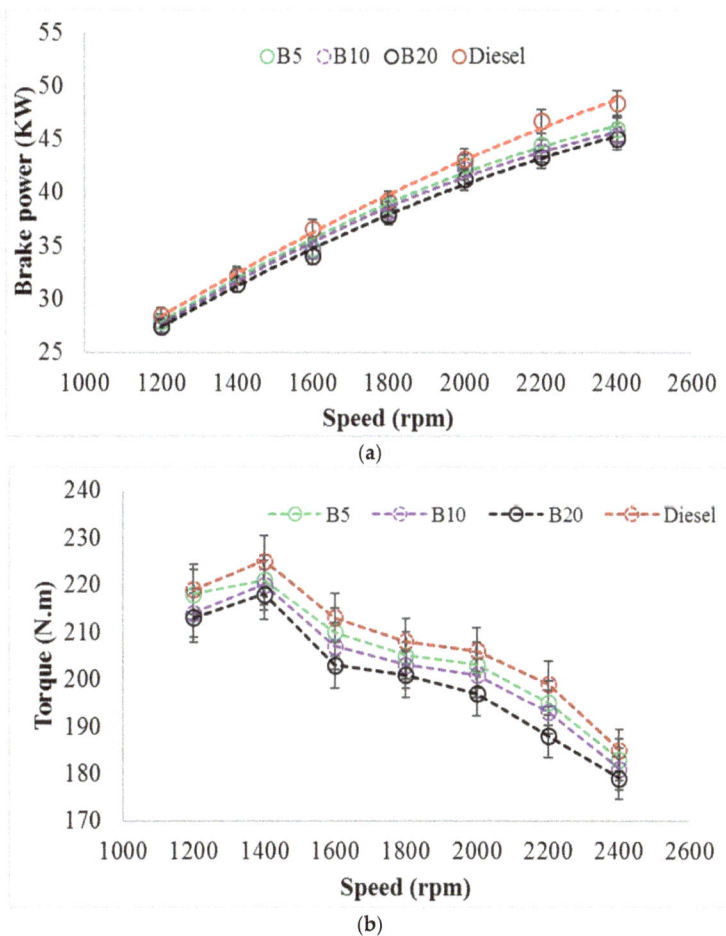

(a)

(b)

Figure 5. *Cont.*

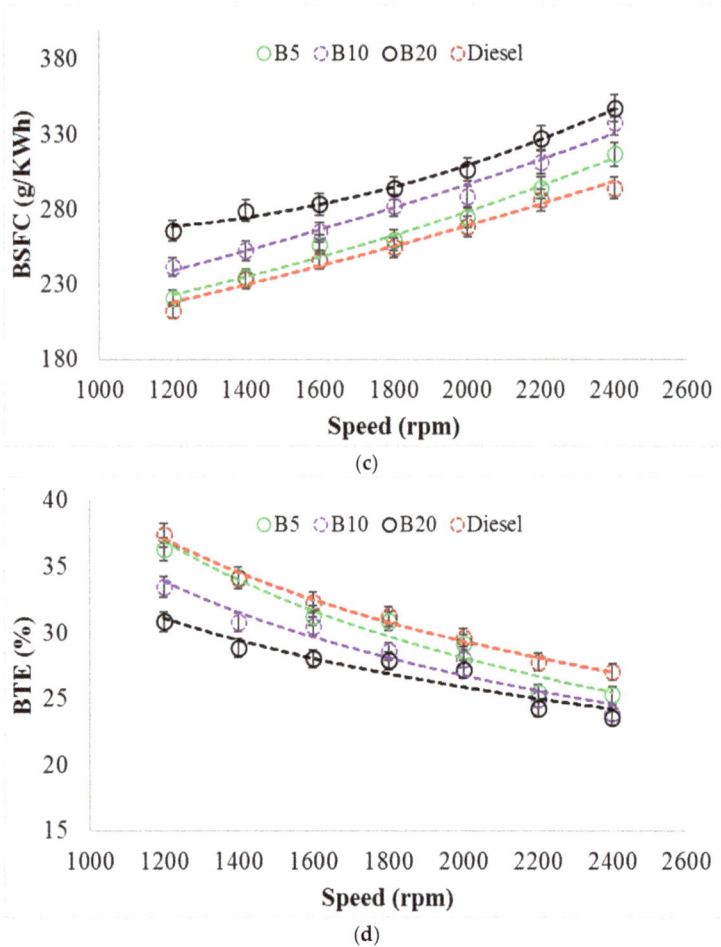

Figure 5. Variation of: (**a**) brake power, (**b**) torque, (**c**) brake specific fuel consumption (BSFC), and (**d**) brake thermal efficiency (BTE) for all PSO biodiesel-diesel blends and diesel with respect to engine speed at full load condition.

3.2.2. Torque

Figure 5b shows that torque increases initially with speed increase up to 1400 rpm when the maximum level is achieved and then decreases continuously until the maximum speed of 2400 rpm for all biodiesel blends and B0. B5 had the highest torque among the biodiesel blends due to its higher calorific value and lower density and viscosity. As expected, B0 (diesel) had the highest torque followed by B5, B10, and B20. The maximum torques were recorded at 1400 rpm, and these were 225 Nm, 221 Nm, 220 Nm, and 218 Nm for B0, B5, B10, and B20, respectively. The average torque reduction for B5, B10, and B20 as compared with B0 was 1.37%, 2.47%, and 3.85%, respectively.

3.2.3. Brake Specific Fuel Consumption (BSFC)

Brake specific fuel consumption (BSFC) is the ratio of fuel flow rate and brake power. According to Mofijur et al. [17], BSFC values depend on the relationship between the fuel injection system, fuel density, calorific value, and viscosity. Figure 5c shows the variation of BSFC for all PSO biodiesel-diesel blends and diesel with respect to engine speed. It is found that diesel has the lowest BSFC value in comparison with PSO biodiesel-diesel blends (B5, B10, and B20). The average BSFC value of B0 was measured as 256.5 g/kWh, whereas those for blends of B5, B10, and B20 were 265 g/kWh, 282.6 g/kWh, and 300.4 g/kWh, respectively. As mentioned earlier, the combined effects of fuel properties, such as density, viscosity, and calorific values of biodiesel may result in higher BSFC. Besides, biodiesel needs more fuel for producing the same power due to its lower calorific value in comparison with diesel. The average increase in BSFC values for B5, B10, and B20 as compared with B0 (diesel) were 3.35%, 10.16%, and 17.13%, respectively.

3.2.4. Brake Thermal Efficiency (BTE)

Brake thermal efficiency (BTE) is the ratio of the brake power and heat energy that is produced by fuel. Figure 5d shows that BTE decreases with the increase of biodiesel in the blends from B5 to B20. A higher BTE value (%) depends on some specific fuel properties such as higher calorific value, and lower density and viscosity. When compared with other blends, the properties of B5 were matched closely with diesel. The average BTE value of B5 was measured as 30.35%, whereas diesel was recorded as 31.33%. The lower calorific values and higher fuel consumptions of both B10 and B20 resulted in BTE values of 28.59% and 27.23%, respectively. The average reduction in BTE values for B5, B10, and B20 as compared with B0 (diesel) were 3.1%, 8.76%, and 13.1%, respectively.

3.3. Interaction Effects of Operating Parameters on PSO Engine Performance

The complex interaction effects of operating parameters, such as biodiesel blends, load, and speed, on each engine output response (BP, torque, BSFC, and BTE) could not be analysed independently. The significance of each of the various parameters in the model was obtained via analysis of variance (ANOVA). The experiments were carried out by use of the Box-Behnken response surface design. Minitab 18 was used to carry out the statistical analysis. Table 7 shows the factors and the range and levels of the investigated variables.

Table 7. Experimental range and levels coded for analysis of variance (ANOVA).

Factors	Unit	Symbol Coded	Range and Levels		
			−1	0	1
Biodiesel blends	%	BL	0	10	20
Load	%	LD	0	50	100
Speed	rpm	SP	1200	1800	2400

Once the experiments were completed, a full quadratic model was applied for the correlation of the response variable to the independent variables. The form of the full quadratic model is shown in Equation (2).

$$R = P_0 + P_1Q_1 + P_2Q_2 + P_3Q_3 + P_{1,2}Q_1Q_2 + P_{1,3}Q_1Q_3 + P_{2,3}Q_2Q_3 + P_{1,1}Q_1{}^2 + P_{2,2}Q_2{}^2 + P_{3,3}Q_3{}^2 \quad (2)$$

where R is the response factor; P_0 is a constant; P_1, P_2, P_3 are regression coefficients, $P_{1,1}$, $P_{1,2}$, $P_{1,3}$, $P_{2,2}$, $P_{2,3}$, and $P_{3,3}$ are quadratic coefficient; and, Q_1, Q_2, and Q_3 are independent variables.

Consideration was given to the linear, quadratic and combined effects of operating parameters to identify their impacts on the response. Each parameter's the significance was evaluated by the probability value (*p*-value) from ANOVA. At the 95% confidence level, the *p*-values less than 0.05 indicate a 'significant' effect of those parameters on the response. In other words, *p*-values more than 5% or 0.05 indicate 'not-significant' effects of those parameters on the response.

3.3.1. Effects of Biodiesel Blends, Load and Speed on Brake Power (BP)

The relationships between brake power and three operating parameters of biodiesel blends, load, and speed were analysed. A quadratic regression model based on the coded parameters with determined coefficients for statistical prediction as defined by Equation (3) was developed using Minitab 18 to predict BP (kW) as a function of biodiesel blends (BL), load (LD), and speed (SP).

$$BP = 20 - 0.018 \text{ BL} + 17.945 \text{ LD} + 4.41 \text{ SP} + 0.501 \text{ BL}^2 - 0.324 \text{ LD}^2 - 0.799 \text{ SP}^2 - $$
$$0.267 \text{ BL} \times \text{LD} - 2.177 \text{ BL} \times \text{SP} + 4.093 \text{ LD} \times \text{SP} \tag{3}$$

p-values from Table 8 show that the model is highly significant with an insignificant lack of fit. When considering the linear, quadratic, and combined effects, BL is not significant as a linear parameter; also, none of the quadratic terms is significant. Only the combined effects of LD and SP are highly significant, whereas BL and SP have a significant effect on BP. The ANOVA results in Table 8 also show that both LD and SP have the lowest *p*-values (<0.0001 each) and highest F-values (3307.29 and 199.74). According to the quadratic Equation (3), both LD and SP have positive effects on BP. This means that increasing the LD and SP will increase BP as well. The coefficient of determination (R^2) and the adjusted coefficient of determination (adj. R^2) were 99.86% and 99.61%, respectively, which indicates a high accuracy for the model.

Table 8. ANOVA results of the interactive effect of biodiesel blends, load, and speed on brake power.

Source	DF	Adj. SS	Adj. MS	F-Value	*p*-Value	Significant
Model	9	2821.89	313.54	402.53	<0.0001	Highly
Biodiesel blends (BL)	1	0.00	0.00	0.00	0.957	No
Load (LD)	1	2576.18	2576.18	3307.29	<0.0001	Highly
Speed (SP)	1	155.58	155.58	199.74	<0.0001	Highly
BL × BL	1	0.93	0.93	1.19	0.325	No
LD × LD	1	0.39	0.39	0.50	0.512	No
SP × SP	1	2.36	2.36	3.02	0.143	No
BL × LD	1	0.29	0.29	0.37	0.571	No
BL × SP	1	18.97	18.97	24.35	0.004	Yes
LD × SP	1	66.99	66.99	86.01	<0.0001	Highly
Lack-of-Fit	3	3.58	1.19	7.71	0.117	No
Pure Error	2	0.31	0.15	-	-	-
Total	14	2825.79	-	-	-	-
R^2 = 0.9986			Adj. R^2 = 0.9961			-

The ANOVA results in Table 8 for both LD × SP, and BL × SP interaction effects on BP are shown graphically in Figure 6. The three-dimensional (3D) surface plot and two-dimensional (2D) contour plot of LD and SP effects on BP are presented in Figure 6a,b respectively. BP (kW) increases with the increase of LD up to 100% and exceeds 40 kW with SP of 2000 rpm onwards. The maximum BP was found to be 45 kW at 2400 rpm. At 50% LD, the average BP value was recorded at about 20 kW with minimum effects from SP. However, the combined effects of LD and SP on BP are more significant with an increase of LD above 50%. Figure 6c,d present the 3D surface plot and 2D contour plot, respectively, which show only minor influences of changes of BL and SP on BP. It is therefore concluded that BL has a slight effect on BP, irrespective of SP changes.

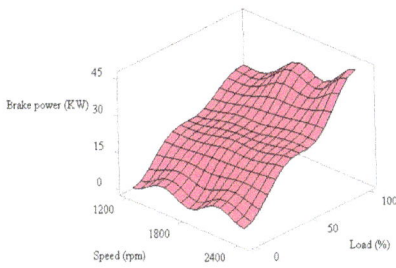

(**a**) Three-dimensional (3D) surface plot of load and speed on brake power (BP)

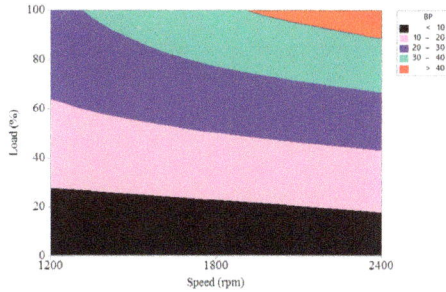

(**b**) Two-dimensional (2D) contour plot of load and speed on BP

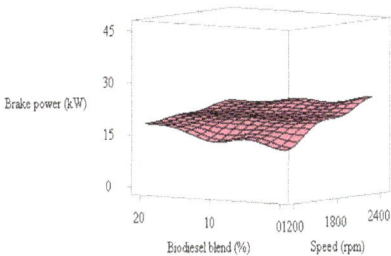

(**c**) 3D surface plot of biodiesel blends and speed on BP

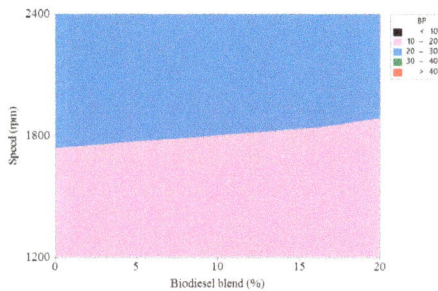

(**d**) 2D contour plot of biodiesel blends and speed on BP

Figure 6. Combined effects of biodiesel blends, load, and speed on BP (kW).

3.3.2. Effects of Biodiesel Blends, Load, and Speed on Torque

The relationships between torque and three operating parameters of biodiesel blend, load, and speed were analysed. Based on the coded parameters, a quadratic regression model with determined coefficients for statistical prediction, as defined by Equation (4), was developed using Minitab 18 to predict torque (N.m) as a function of biodiesel blend (BL), load (LD), and speed (SP).

$$T = 137.47 + 4.596\,BL + 95.884\,LD - 8.105\,SP - 16.6\,BL^2 - 15.06\,LD^2 - 18.13\,SP^2 -$$
$$0.53\,BL \times LD - 11.42\,BL \times SP - 12\,LD \times SP \tag{4}$$

p-values from Table 9 show that the model is highly significant with an insignificant lack of fit. When considering the linear, quadratic, and combined effects, only the combined effects of BL and LD are not significant. The ANOVA results in Table 9 show that LD is not significant, as it has the lowest *p*-value and highest F-value, while BL is found to be significant and all other parameters are highly significant. According to the quadratic Equation (4), both LD and BL have positive effects on torque. This means that increasing the LD and BL will increase torque as well. The coefficient of determination (R^2) and the adjusted coefficient of determination (adj. R^2) were 99.96% and 99.89%, respectively, which indicates the high accuracy of the model.

Table 9. ANOVA results of the interactive effect of biodiesel blends, load, and speed on torque.

Source	DF	Adj. SS	Adj. MS	F-Value	*p*-Value	Significant
Model	9	78,005.7	8667.3	1352.54	<0.0001	Highly
BL	1	169.0	169.0	26.37	0.004	Yes
LD	1	73,549.5	73,549.5	11,477.45	<0.0001	Highly
SP	1	525.5	525.5	82.01	<0.0001	Highly
BL × BL	1	1017.2	1017.2	158.73	<0.0001	Highly
LD × LD	1	837.8	837.8	130.73	<0.0001	Highly
SP × SP	1	1213.7	1213.7	189.40	<0.0001	Highly
BL × LD	1	1.1	1.1	0.18	0.691	No
BL × SP	1	521.7	521.7	81.41	<0.0001	Highly
LD × SP	1	575.5	575.5	89.81	<0.0001	Highly
Lack-of-Fit	3	30.4	10.1	12.13	0.077	No
Pure Error	2	1.7	0.8	-	-	-
Total	14	78,037.7	-	-	-	-
R^2 = 0.9996				Adj. R^2 = 0.9989		

The ANOVA results in Table 9 for LD × SP, and BL × SP interaction effects on torque are shown graphically in Figure 7. The 3D surface plot and 2D contour plot of LD and SP effects on torque are presented in Figure 7a,b, respectively. Torque (Nm) decreases with the increase of SP at 100% LD. The maximum torque was found to be 220 Nm at 1400 rpm. At 50% LD, the average torque was recorded as about 120 Nm with minimum effects from SP. However, the combined effects of LD and SP on torque are more significant with an increase of LD above 50%. Figure 7c,d present the 3D surface plot and 2D contour plot, respectively, which show only minor influences of changes of BL and SP on torque. It is therefore concluded that BL has a slight effect on torque, irrespective of SP changes.

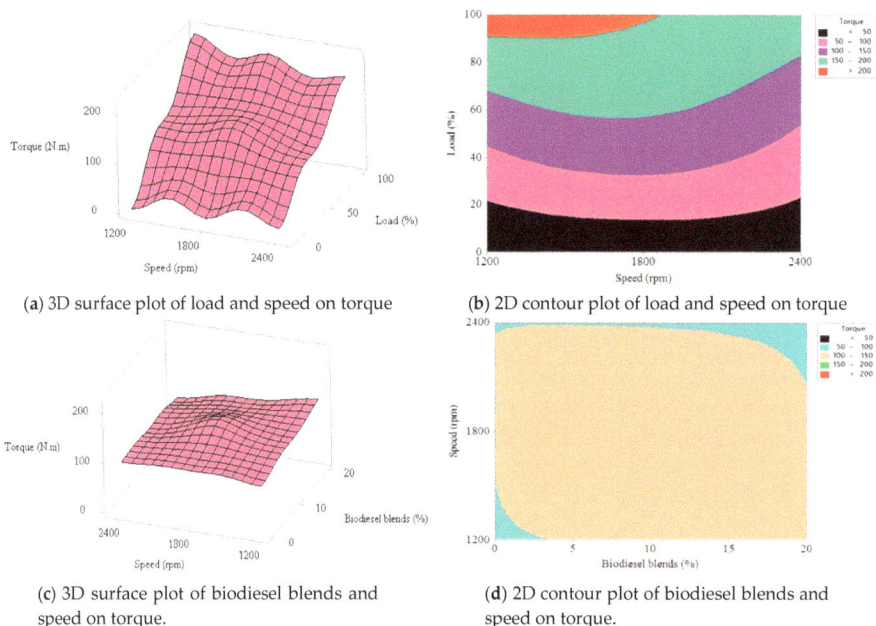

(**a**) 3D surface plot of load and speed on torque

(**b**) 2D contour plot of load and speed on torque

(**c**) 3D surface plot of biodiesel blends and speed on torque.

(**d**) 2D contour plot of biodiesel blends and speed on torque.

Figure 7. Combined effects of biodiesel blends, load, and speed on torque (Nm).

3.3.3. Effects of Biodiesel Blends, Load, and Speed on Brake Specific Fuel Consumption (BSFC)

The relationships between three operating parameters of biodiesel blend, load, and speed with brake specific fuel consumption (BSFC) were analysed. Based on the coded parameters, a quadratic regression model with determined coefficients for statistical prediction, as defined by Equation (5), was developed using Minitab 18 to predict BSFC (gm/kWh) as a function of biodiesel blends (BL), load (LD), and speed (SP).

$$BSFC = 252.17 + 138.13\ BL - 526.25\ LD - 30.43\ SP + 62.5\ BL^2 + 488.88\ LD^2 + 71.66\ SP^2 - 125.71\ BL \times LD - 38.55\ BL \times SP + 78.67\ LD \times SP \tag{5}$$

p-values in Table 10 show that the model is highly significant with an insignificant lack of fit. When considering the linear, quadratic, and combined effects, all of the parameters are highly significant. Among all parameters, LD is the parameter with the lowest *p*-value and highest F-value (20,429.23). According to the quadratic Equation (5), both LD and SP have a negative effect on BSFC. This means that increasing the LD will decrease BSFC, whereas decreasing the LD will increase BSFC. The SD parameter has less influence in changing BSFC. The coefficient of determination (R^2) and the adjusted coefficient of determination (adj. R^2) were 99.98% and 99.95%, respectively, which indicates the high accuracy of the model.

Table 10. ANOVA results of the interactive effect of biodiesel blends, load and speed on brake specific fuel consumption (BSFC).

Source	DF	Adj. SS	Adj. MS	F-Value	*p*-Value	Significant
Model	9	3,358,105	373,123	3440.61	<0.0001	Highly
BL	1	152,631	152,631	1407.43	<0.0001	Highly
LD	1	2,215,481	2,215,481	20,429.23	<0.0001	Highly
SP	1	7408	7408	68.31	<0.0001	Highly
BL × BL	1	14,424	14,424	133.00	<0.0001	Highly
LD × LD	1	882,480	882,480	8137.45	<0.0001	Highly
SP × SP	1	18,960	18,960	174.83	<0.0001	Highly
BL × LD	1	63,215	63,215	582.91	<0.0001	Highly
BL × SP	1	5944	5944	54.81	0.001	Highly
LD × SP	1	24,756	24,756	228.28	<0.0001	Highly
Lack-of-Fit	3	521	174	16.38	0.058	No
Pure Error	2	21	11	-	-	-
Total	14	3,358,647	-	-	-	-
$R^2 = 0.9998$				Adj. $R^2 = 0.9995$		

The ANOVA results in Table 10 for LD × SP, BL × LD, and BL × SP interaction effects on BSFC are shown graphically in Figure 8. The 3D surface plot and 2D contour plot of LD and SP effects on BSFC are presented in Figure 8a,b respectively. BSFC decreases with the increase of LD, irrespective of SP. The maximum BSFC was found to be 1450 gm/kWh at 0% LD and 1200 rpm SP. At 50% LD, the average BSFC was recorded at about 400 gm/kWh with almost no effect from changes in SP (rpm). Figure 8c,d present the 3D surface plot and 2D contour plot, respectively, of BL and LD effects on BSFC. At 0% LD, BL has a significant effect on BSFC. The 20% biodiesel blend (BL20) shows the maximum BSFC of 1450 gm/kWh at 0% LD. BSFC values decrease with the increase of LD, irrespective of BL. From Figure 8d, it can be seen that, at 36% LD, the 0–10% biodiesel blends (BL0, BL5, and BL10) have low BSFC values whereas, at 60% LD, BL20 also has achieved a lower BSFC. It is therefore concluded that BL has a minor effect on BSFC, irrespective of SP changes. Figure 8e,f show the 3D surface plot and 2D contour plot respectively of BL and SP effects on BSFC. BSFC values for BL0 have recorded for SP values in the range of 1200 to 2400 rpm and found to be less than 400 gm/kWh. Changes in both BL and SP influence BSFC values. From Figure 8f, it can be seen that both BL at 5% and BL at 10%

have low BSFC for SP values in the range of 1600 to 2000 rpm. Higher BL percentages with higher SP values (up to 2000 rpm) result in higher values of BSFC.

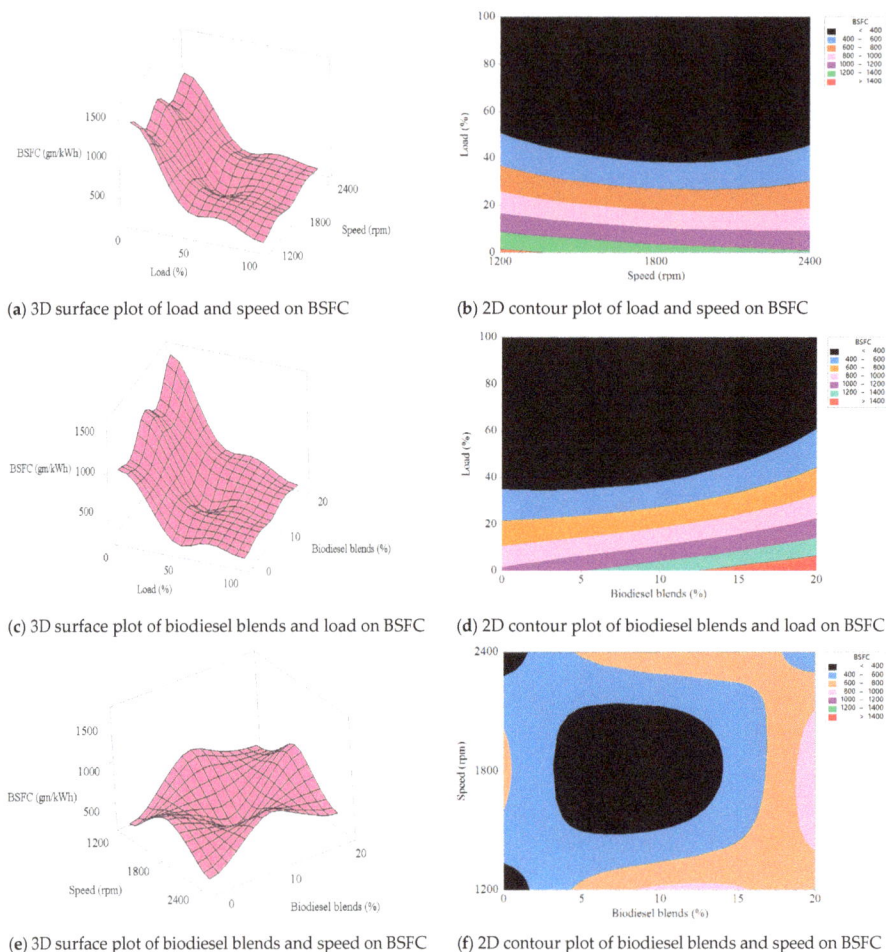

(**a**) 3D surface plot of load and speed on BSFC

(**b**) 2D contour plot of load and speed on BSFC

(**c**) 3D surface plot of biodiesel blends and load on BSFC

(**d**) 2D contour plot of biodiesel blends and load on BSFC

(**e**) 3D surface plot of biodiesel blends and speed on BSFC

(**f**) 2D contour plot of biodiesel blends and speed on BSFC

Figure 8. Combined effects of biodiesel blends, load, and speed on BSFC (gm/kWh).

3.3.4. Effects of Biodiesel Blends, Load, and Speed on Brake Thermal Efficiency (BTE)

The relationships between brake thermal efficiency (BTE) and three operating parameters of biodiesel blends, load, and speed were analysed. Based on the coded parameters, a quadratic regression model with determined coefficients for statistical prediction, as defined by Equation (6), was developed using Minitab 18 to predict BTE (%) as a function of biodiesel blends (BL), load (LD), and speed (SP).

$$BTE = 32.283 - 1.36\,BL + 11.752\,LD - 1.71\,SP - 2.795\,BL^2 - 12.05\,LD^2 - 1.9\,SP^2 + \\ 1.038\,BL \times LD + 0.267\,BL \times SP - 2.097\,LD \times SP \tag{6}$$

p-Values in Table 11 show that the model is highly significant with an insignificant lack of fit. When considering the linear, quadratic, and combined effects, only the BL × LD and BL × SP combined parameters are not significant; all other parameters are significant. Only the combined effects of LD × SP are significant, whereas the quadratic terms LD × LD, BL × BL, and SP × SP were all found to have significant effects on BTE. The ANOVA results in Table 11 also show that LD has both the lowest *p*-values (<0.0001) and highest F-value (940.76). According to the quadratic Equation (6), only LD has a positive effect on BTE. This means that increasing the LD will increase BTE as well. The coefficient of determination (R^2) and the adjusted coefficient of determination (adj. R^2) were 99.66% and 99.05%, respectively, which indicates the high accuracy of the model.

Table 11. ANOVA results of the interactive effect of biodiesel blends, load, and speed on brake thermal efficiency (BTE).

Source	DF	Adj. SS	Adj. MS	F-Value	*p*-Value	Significant
Model	9	1717.12	190.79	162.44	<0.0001	Highly
BL	1	14.80	14.80	12.60	0.016	Yes
LD	1	1104.97	1104.97	940.76	<0.0001	Highly
SP	1	23.39	23.39	19.92	0.007	Yes
BL × BL	1	28.85	28.85	24.57	0.004	Yes
LD × LD	1	536.17	536.17	456.49	<0.0001	Highly
SP × SP	1	13.34	13.34	11.35	0.020	Yes
BL × LD	1	4.31	4.31	3.67	0.114	Not
BL × SP	1	0.29	0.29	0.24	0.642	No
LD × SP	1	17.60	17.60	14.98	0.012	Yes
Lack-of-Fit	3	5.48	1.83	9.26	0.099	No
Pure Error	2	0.39	0.20	-	-	-
Total	14	1722.99	-	-	-	-
R^2 = 0.9966				Adj. R^2 = 0.9905		

The ANOVA results in Table 11 for LD × SP interaction effects on BTE are shown graphically in Figure 9. The 3D surface plot and 2D contour plot of LD and SP effects on BTE are presented in Figure 9a,b, respectively. BTE increases with the increase of LD with minimum impacts from SP. The maximum BTE was found to be 33.43% at 1200 rpm. However, the combined effects of LD and SP on BTE are affected by an increase of LD above 50%. It is therefore concluded that SP has a minor impact on BTE, irrespective of LD changes.

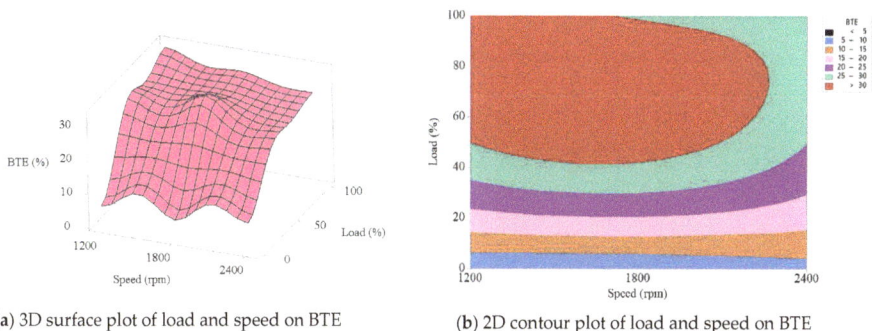

(a) 3D surface plot of load and speed on BTE

(b) 2D contour plot of load and speed on BTE

Figure 9. Combined effects of biodiesel blends, load, and speed on BTE (%).

4. Conclusions

This study investigated the effect of papaya seed oil (PSO) biodiesel in an agricultural diesel engine with various biodiesel-diesel blends and the resulting engine performance outcomes as compared with those from the use of a reference diesel. A response surface methodology was introduced to analyse and describe the performance of this engine. The results of this investigation can be summarised, as follows:

- PSO biodiesel-diesel blends (B5–B50) meet the European standard EN 590 for having minimum oxidation stability of 20 h.
- Over the entire speed range of 1200 to 2400 rpm, the average BP value reductions for B5, B10, and B20 in comparison with B0 were 2.88%, 3.87%, and 5.13%, respectively.
- The average torque reductions for B5, B10, and B20 in comparison with B0 were 1.37%, 2.47%, and 3.85%, respectively.
- The average increase in BSFC values for B5, B10, and B20 in comparison with B0 were 3.35%, 10.16%, and 17.13%, respectively.
- The average BTE value of B5 was measured as 30.35%, whereas B0 (diesel) was recorded as 31.33%.
- The interactive relationships between three operating parameters (biodiesel blends, load, and speed) and four responses (BP, torque, BSFC, and BTE) were analysed. ANOVA and a statistical regression model show that load and speed were the two most important parameters that affect all four responses. The biodiesel blends parameter had a significant effect on BSFC.

These results show that B5 and B10 PSO biodiesel-diesel blends can be used to fuel diesel engines without further engine modification. Therefore, it can be concluded that papaya seed oil can be considered as a promising source of biodiesel production. However, before recommending as a future alternative energy source in commercial scale, further research needs to be conducted in terms of engine emissions, in-cylinder pressure, burn rate data, combustion analysis, and tribological performance analysis.

Author Contributions: The contributions of each author are as follows: M.A. produced and characterised biodiesel-diesel blends, conducted and analysed engine performance, and drafted the manuscript; M.G.R. contributed to the experimental design and thoroughly revised the paper, and N.A. helped to revised and improved the paper.

Funding: This research received no external funding.

Acknowledgments: The authors would like to acknowledge Tim McSweeney, Adjunct Research Fellow, Tertiary Education Division at Central Queensland University, Australia for his contribution in proof reading of this article.

Conflicts of Interest: The authors declare no conflict of interest.

References

1. Bhuiya, M.M.K.; Rasul, M.G.; Khan, M.M.K.; Ashwath, N.; Azad, A.K.; Hazrat, M.A. Second Generation Biodiesel: Potential Alternative to-edible Oil-derived Biodiesel. *Energy Procedia* **2014**, *61*, 1969–1972. [CrossRef]
2. Asokan, M.A.; Prabu, S.S.; Kamesh, S.; Khan, W. Performance, combustion and emission characteristics of diesel engine fuelled with papaya and watermelon seed oil bio-diesel/diesel blends. *Energy* **2018**, *145*, 238–245. [CrossRef]
3. Tüccar, G.; Tosun, E.; Özgür, T.; Aydın, K. Diesel engine emissions and performance from blends of citrus sinensis biodiesel and diesel fuel. *Fuel* **2014**, *132*, 7–11. [CrossRef]
4. Ozsezen, A.N.; Canakci, M.; Turkcan, A.; Sayin, C. Performance and combustion characteristics of a DI diesel engine fueled with waste palm oil and canola oil methyl esters. *Fuel* **2009**, *88*, 629–636. [CrossRef]
5. Agunbiade, F.O.; Adewole, T.A. Methanolysis of Carica papaya Seed Oil for Production of Biodiesel. *J. Fuels* **2014**, *2014*, 904076. [CrossRef]
6. Wong, C.S.; Othman, R. Biodiesel Production by Enzymatic Transesterification of Papaya Seed Oil and Rambutan Seed Oil. *Int. J. Eng. Technol.* **2014**, *6*, 2773–2777.
7. Mohan, A.; Sen, S.S.S. Emission Analysis Of The Biodiesel from Papaya and Chicken Blends. *Int. J. Inform. Futurist. Res.* **2015**, *2*, 2258–2264.
8. De Melo, M.L.S.; Santos, N.A.; Rosenhaim, R.; Souza, A.G.; Filho, P.F.A. Use of thermal analysis techniques for evaluation of the stability and chemical properties of papaya biodiesel (*Carica papaya* L.) at low temperatures. *J. Therm. Anal. Calorim.* **2011**, *106*, 831–836. [CrossRef]
9. Daryono, E.D.; Sinaga, E.J. Rapid In Situ Transesterification of Papaya Seeds to Biodiesel with the Aid of Co-solvent. *Int. J. Renew. Energy Res.* **2017**, *7*, 379–385.

10. Anwar, M.; Rasul, M.G.; Ashwath, N. Optimization of biodiesel production process from papaya (*Carica papaya*) seed oil. In Proceedings of the 2017 IEEE 7th International Conference on Power and Energy Systems (ICPES), Toronto, ON, Canada, 1–3 November 2017.

11. Prabhakaran, P.; Saravanan, C.G.; Aalam, C.S. Effects of Papaya Methyl Ester on DI Diesel Engine Combustion, Emission and Performance Characteristics. *Int. Res. J. Eng. Technol.* **2016**, *3*, 319–325.

12. Sundar Raj, C.; Karthikayan, M. Effect of additive on the performance, emission and combustion characteristics of a diesel engine run by diesel-papaya methyl ester blends. *Int. J. Chem. Sci.* **2016**, *14*, 2823–2834.

13. Sharma, S.K.; Mitra, S.K.; Saran, S. *Papaya Production in India—History, Present Status and Future Prospects*; International Society for Horticultural Science (ISHS): Leuven, Belgium, 2016; Volume 1111, pp. 87–94.

14. Anwar, M.; Rasul, M.G.; Ashwath, N. Production optimization and quality assessment of papaya (*Carica papaya*) biodiesel with response surface methodology. *Energy Convers. Manag.* **2018**, *156*, 103–112. [CrossRef]

15. Silitonga, A.S.; Masjuki, H.H.; Mahlia, T.M.I.; Ong, H.C.; Chong, W.T.; Boosroh, M.H. Overview properties of biodiesel diesel blends from edible and non-edible feedstock. *Renew. Sustain. Energy Rev.* **2013**, *22* (Suppl. C), 346–360. [CrossRef]

16. Silitonga, A.S.; Ong, H.C.; Mahlia, T.M.I.; Masjuki, H.H.; Chong, W.T. Biodiesel Conversion from High FFA Crude Jatropha Curcas, Calophyllum Inophyllum and Ceiba Pentandra Oil. *Energy Procedia* **2014**, *61* (Suppl. C), 480–483. [CrossRef]

17. Mofijur, M.; Masjuki, H.H.; Kalam, M.A.; Atabani, A.E. Evaluation of biodiesel blending, engine performance and emissions characteristics of Jatropha curcas methyl ester: Malaysian perspective. *Energy* **2013**, *55*, 879–887. [CrossRef]

18. Hasni, K.; Ilham, Z.; Dharma, S.; Varman, M. Optimization of biodiesel production from Brucea javanica seeds oil as novel non-edible feedstock using response surface methodology. *Energy Convers. Manag.* **2017**, *149* (Suppl. C), 392–400. [CrossRef]

![energies logo] *energies*

MDPI

Article

Mathematical Modeling of Non-Premixed Laminar Flow Flames Fed with Biofuel in Counter-Flow Arrangement Considering Porosity and Thermophoresis Effects: An Asymptotic Approach

Mehdi Bidabadi [1], Peyman Ghashghaei Nejad [1], Hamed Rasam [1], Sadegh Sadeghi [1] and Bahman Shabani [2,*]

[1] School of Engineering, Iran University of Science and Technology, Narmak, Tehran 16846-13114, Iran; bidabadi@iust.ac.ir (M.B.); peymanghashghaie@yahoo.com (P.G.N.); h_rasam@mecheng.iust.ac.ir (H.R.); sadeghsadeghi@mecheng.iust.ac.ir (S.S.)

[2] School of Engineering, RMIT University, Melbourne, VIC 3083, Australia

* Correspondence: bahman.shabani@rmit.edu.au; Tel.: +61-(0)-3-9925-4353

Received: 2 October 2018; Accepted: 24 October 2018; Published: 29 October 2018

Abstract: Due to the safe operation and stability of non-premixed combustion, it can widely be utilized in different engineering power and medical systems. The current paper suggests a mathematical asymptotic technique to describe non-premixed laminar flow flames formed in organic particles in a counter-flow configuration. In this investigation, fuel and oxidizer enter the combustor from opposite sides separately and multiple zones including preheating, vaporization, flame and post-flame zones were considered. Micro-sized lycopodium particles and air were respectively applied as a biofuel and an oxidizer. Dimensionalized and non-dimensionalized mass and energy conservation equations were determined for the zones and solved by Mathematica and Matlab software by applying proper boundary and jump conditions. Since lycopodium particles have numerous spores, the porosity of the particles was involved in the equations. Further, significant parameters such as lycopodium vaporization rate and thermophoretic force corresponding to the lycopodium particles in the solid phase were examined. The temperature distribution, flame sheet position, fuel and oxidizer mass fractions, equivalence ratio and flow strain rate were evaluated for the counter-flow non-premixed flames. Ultimately, the thermophoretic force caused by the temperature gradient at different positions was computed for several values of porosity, fuel and oxidizer Lewis numbers.

Keywords: porosity; thermophoretic force; biomass fuel; non-premixed combustion; counter-flow structure; mathematical modeling

---

1. Introduction

Considering the progressive depletion of fossil fuels, the soaring costs of these fuels and the greenhouse gas (GHG) emissions associated with them, significant attention has been turned toward using alternatives to these fuels [1,2]. In recent years, bioenergy has been introduced as a reasonable form of renewable energy with a high degree of sustainability for use in various industrial and medical applications [3]. Bioenergy can readily be achieved through the combustion of biomass fuels [4]. On the basis of structure, combustion is categorized into premixed, partially premixed and non-premixed flames [5,6], among which non-premixed combustion is widely utilized in engineering combustion power systems due to its safe operation and stability [7]. So far, different experimental and analytical investigations have been performed by various researchers to specify the propagation of non-premixed flames through biofuels.

Joshi and Berlad [8] experimentally studied the behavior of premixed flames through a lycopodium-air mixture and reported the flame temperature distribution in the vertical position. In another investigation, Berlad and Joshi [9] used both analytical and experimental approaches to model the extinction and temperature distribution of premixed flames burning lycopodium dust clouds. Berlad and Tangirala [10] studied the effect of thermal radiation on the behavior and structure of lycopodium-fueled flames considering a small quantity of gravity force. The behavior of a triple-zone premixed flame containing uniformly-distributed lycopodium particles was developed by Seshadri, applying an asymptotic method [11]. Han et al. [12] experimentally perused flame propagation through a lycopodium–air mixture within a vertical duct and evaluated the weight loss of the lycopodium particles and flame velocity. In further experimental work, Han et al. [13] investigated the propagation of lycopodium dust particles in a premixed combustible system using particle image velocimetry (PIV). Xi et al. [14] used an experimental method to study the pulsation characteristics of a swirl non-premixed flame. Proust [15] experimentally investigated flame propagation through lycopodium-air and sulphur flour-air mixtures. Skjold et al. [16] examined turbulent flame propagation through lycopodium dust clouds at a constant pressure and made a comparison between the structures of the produced premixed and non-premixed flames. Shamooni et al. [17] employed a finite-rate scale similarity method to predict the combustion and heat release rates in non-premixed jet flames. A two-dimensional analytical model was presented by Rahbari et al. [18] for premixed flames propagating through a lycopodium-air mixture taking into account the particle size and equivalence ratio parameters. Bidabadi and Esmaeilnejad et al. [19] suggested an analytical triple-zone model for counter-flow premixed flames propagating through a lycopodium-air mixture. Bidabadi et al. [20] proposed a non-premixed triple-zone combustion model for lycopodium dust flames in a counter flow configuration. Xi et al. [21] experimentally and numerically simulated the flame shape and size for a high-pressure turbulent non-premixed swirl combustion. The combustion behavior of non-premixed micro-jet flames through a methane-air mixture in a co-flow configuration was reported by Li et al. [22]. Bidabadi et al. [23] theoretically investigated the effect of thermal resistance on the structure of the premixed combustion of micro-organic dust particles with air. Spijker and Raupenstrauch [24] performed a numerical investigation to reveal the effect of the inner structure of lycopodium particles on the propagation of premixed flames. Chen et al. [25] experimentally described the temperature distribution of the propagation of premixed flames through a methane-air mixture considering the porosity parameter. Bidabadi et al. [26] developed a method for the theoretical investigation of multi-zone counter-flow premixed flames burning lycopodium dust clouds considering the vaporization rate. Ji et al. [27] experimentally computed the maximum pressure and the maximum rate of pressure rise of several hybrid mixtures taking into account different venting diameters and static activation pressures. Di Benedetto et al. [28] employed a theoretical/numerical method to determine the minimum ignition temperature of polyethylene dust at several dust concentrations in air. Sanchirico et al. [29] evaluated the flammability and combustion behavior of several complex hybrid mixtures, e.g., mixtures of nicotinic acid, lycopodium and methane, at different dust concentrations. In another study, Sanchirico et al. [30] experimentally demonstrated the combustion and flammability behavior of combustible dust mixtures including lycopodium, Nicotinic acid and Ascorbic acid.

As reviewed of the literature, numerous theoretical and experimental examinations have been undertaken on the behavior of premixed flames propagating through organic fuels. Until now, no comprehensive mathematical models have been available in the literature on the behavior of multi-zone counter-flow non-premixed flames. In most of the previous studies, in which limited models were introduced, the shape of lycopodium particles was considered to be spherical while these particles consist of numerous minute spores that can affect the structure of premixed and non-premixed flames [12]. In addition, the thermophoretic force, caused by a temperature gradient for transferring micro-sized fuel particles, has not yet been investigated mathematically for non-premixed flames.

The current paper aimed to theoretically describe the structure of counter-flow non-premixed flames through organic particles taking into account the effects of fuel particle porosity and thermophoresis. The biofuel particles and oxidizer were separately injected into a combustor from opposite sides. Multiple zones composed of preheating, vaporization, flame and post-flame zones, are presented. Lycopodium particles and air were injected into the system as biomass and oxidizer, respectively. The governing equations, including mass and energy conservation equations, were derived for the above-mentioned zones. The impacts of fuel particle porosity, vaporization and thermophoresis on the structure of the flames are discussed. Changes in the flame temperature and the location of the flame sheet with fuel and oxidizer Lewis numbers were examined for several porosity factors. Additionally, the influence of the equivalence ratio on the temperature distribution was evaluated. Afterwards, the mass fractions and temperature distributions of the biofuel and oxidizer at different positions were obtained. Then, variations of the flow critical strain rate with oxidizer and fuel Lewis numbers are described. Finally, the thermophoretic force caused the temperature gradient of the solid particles was computed for several different porosity factors, fuel and oxidizer Lewis numbers.

2. Theoretical Modeling

2.1. Lycopodium Characteristics

In the last couple of decades, the accidental ignition of flammable organic and inorganic dusts, as well as knowledge of combustion and flammability, has gained much attention [31]. Lycopodium particles have been introduced as a reasonable reference for volatile biomass fuels due to its excellent dispersibility and flammability [12]. Lycopodium can efficiently be used as a reference fuel for testing different biomass-fueled systems and recognizing the combustion processes in the systems prior to industrial scale-up [12,29,30]. A lycopodium particle typically contains almost 50% fat oil, 2% sucrose and 24% sporopollenin [12]. One of the main advantages of this biofuel is that almost no solid residue is produced during the combustion process [12]. In this study, as can be found in the literature, it was assumed that a gaseous fuel with a certain chemical composition, gaseous methane, is produced through the vaporization of these volatile particles [11]. In this regard, the pyrolysis process was disregarded in the current study. It should be noted that lycopodium particles are mono-disperse in size [14] and contain numerous minute spores [12], as illustrated in Figure 1. Moreover, when the lycopodium particles were injected into a combustible system via a nozzle, they did not undergo breakage due to their sponge-like behavior (i.e., the numerous minute spores are "elastic") [31].

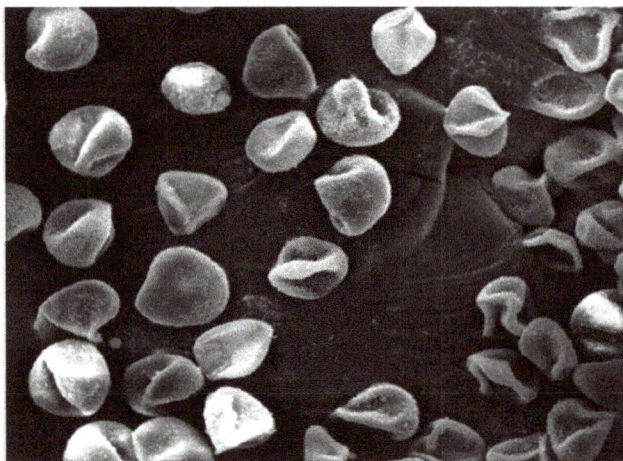

Figure 1. Scanning electron microscopy of lycopodium biofuel particles [22].

2.2. Porosity of Lycopodium Particles

As described in Figure 1, lycopodium particles are not completely spherical-shaped particles. Therefore, the porosity effect should be involved in the modeling of the flame structures fueled by these particles. Figure 2 shows the porosity of the lycopodium particles considered in this study.

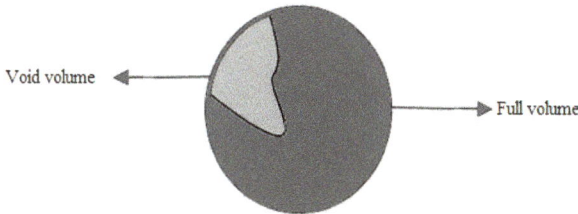

Figure 2. Schematic representation of the porous biofuel particle considered in the current analysis.

The volume porosity and porosity factor [32] are defined by the following equations, respectively:

$$\varepsilon = \frac{V_{void}}{V_{total}} \tag{1}$$

$$f_e = 1 - \varepsilon \tag{2}$$

where ε, V_{void} and V_{total} are the fraction of the empty volume, void volume and total volume, respectively.

2.3. Thermophoretic Force

The thermophoresis phenomenon, first introduced by Tyndall, is due to the temperature gradient that transfers micro-sized particles from a warmer zone to a colder zone [33–35]. This phenomenon is notably influenced by the particle size and mean free path of the molecules [36–38]. It should be pointed out that the ratio of the molecules mean free path to diameter is represented by the Knudsen number that can be obtained by the following equation [36]:

$$Kn = \frac{L}{r_p} \tag{3}$$

where r_p is the radius of the particles, and L is the mean free path of the molecules that can be defined as follows [36,38]:

$$L = \frac{\mu}{\varnothing \rho_g \overline{C}} \tag{4}$$

where $\varnothing$ is equal to 0.941 [38], and μ and ρ_g are the dynamic viscosity of the fuel particles and the density of the gas, respectively. $\overline{c}$ is the mean thermal velocity of the molecules that can be expressed as follows [36]:

$$\overline{c} = \sqrt{\frac{8RT}{\pi M}} \tag{5}$$

where T and M are the temperature and molecular mass, respectively. In addition, R is the universal gas constant, which is equal to 8.314 $\frac{J}{mol \cdot K}$.

To assess the thermophoretic force, the following correlation can be applied [36]:

$$F_{Th} = -3\pi\mu^2 r_p k_T \frac{\nabla T}{\rho_g T_\infty} \tag{6}$$

where k_T is calculated by the following correlation [36]:

$$k_T = \frac{2C_k\left(\frac{k_g}{k_p} + C_t Kn\right)}{(1 + 3C_m Kn)\left(1 + 2\frac{k_g}{k_p} + 2C_t Kn\right)} \tag{7}$$

where μ, r_p, k_g, k_p, ∇T and T_∞ are the dynamic viscosity of the fuel particles, the particle radius, the gas thermal conductivity, the solid particle thermal conductivity, the temperature gradient and the ambient temperature, respectively. C_k, C_t and C_m are the temperature jump, the temperature creep and the velocity jump coefficients [37]. Table 1 lists the constant parameters used in Equation (7).

Table 1. Constant parameters used in Equation (7) [36,37].

Parameter	Value
k_p	$1.446538 \times 10^{-4} \, \frac{kj}{m \cdot s \cdot K}$
k_g	$0.3468 \times 10^{-4} \, \frac{kj}{m \cdot s \cdot K}$
C_k	1.147
C_t	2.2
C_m	1.146

The Knudsen number can be obtained by the following correlation [36]:

Assuming small Knudsen numbers, the Talbot correlation (Equation (6)) can be rewritten as follows [36,38,39]:

$$F_{Th,continuum} = -6\pi\mu^2 r_p C_p \frac{k_g}{k_p + 2k_g} \frac{\nabla T}{\rho_g T_\infty} \tag{8}$$

where C_p is the specific heat of the lycopodium particles.

2.4. The Flame Structure

Figure 3 represents the structure of a non-premixed flame, in which the biofuel and oxidizer were entrained in a counter-flow configuration. Regarding this figure, the porous biofuel particles and oxidizer separately enter the combustor from $-\infty$ and $+\infty$ (opposed sides), respectively. In this analysis, a gaseous fuel with a specific chemical composition, gaseous methane, was produced from solid biofuel particles once they asymptotically crossed the vaporization front [11]. Subsequently, the gaseous fuel blended with the oxidizer within the reaction zone and a huge quantity of thermal bioenergy was released. In this investigation, an asymptotic approach was used to model the vaporization and the reaction processes. Besides, the momentum of the fuel and oxidizer streams was considered to be equal so a stagnation plane was placed in the middle of the fuel and oxidizer nozzles. The location of the stagnation plane was employed as the reference coordinate for measuring the location of the flame sheet and vaporization front, as shown in Figure 3. The mass particle concentration value at $-\infty$ was assumed to be less than $100 \, gr/cm^3$. Therefore, the initial position of the flame sheet was presumed to be on the left-hand side of the stagnation plane. It should be noted that the flame sheet position was the axial distance from the stagnation plane at which the flame front was detected.

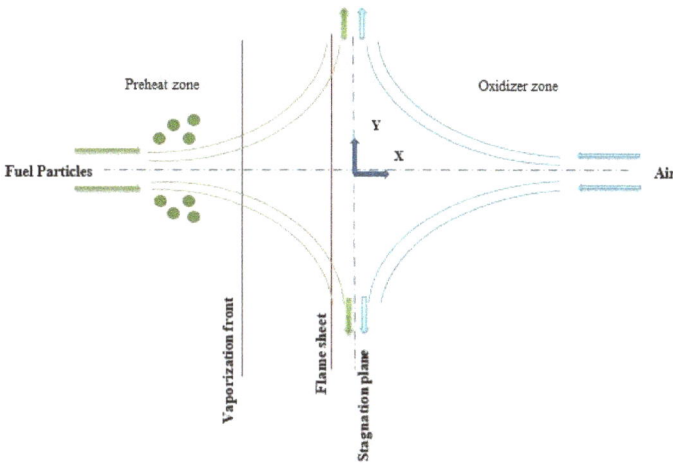

Figure 3. Schematic of the non-premixed flame in a counter-flow arrangement.

2.5. Mathematical Modeling of the Flame

In this paper, the vaporization rate was modeled as the amount of gaseous fuel mass per unit volume per unit time that can be described as follows [40]:

$$\omega_v = \frac{Y_s}{\tau_{vap}} H(T - T_v) \tag{9}$$

where Y_s, τ_{vap}, H, T and T_v are the mass fraction of solid particles, the constant vaporization characteristic time, the Heaviside function, the fuel temperature and the vaporization temperature of the particles. The Heaviside function is defined as follows:

$$H(T - T_v) = \begin{cases} 0 & T < T_v \\ 1 & T \geq T_v \end{cases} . \tag{10}$$

The thermal diffusivity (α) to mass diffusivity (D) ratio (i.e., Lewis number) is defined as follows [41]:

$$Le = \frac{\alpha}{D}. \tag{11}$$

Chemical kinetics were assumed to be a one-stage reaction and expressed as follows [42]:

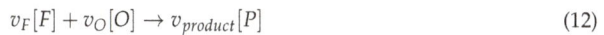

$$v_F[F] + v_O[O] \rightarrow v_{product}[P] \tag{12}$$

where $[F]$, $[O]$ and $[P]$ denote the fuel, oxidizer and products, respectively. v_F, v_O and v_P are the stoichiometric coefficients of the fuel, oxidizer and products, respectively. The one-dimensional velocity field of the particles was considered as follows:

$$u = -aX \tag{13}$$

where u and a represent the velocity in the X direction and the flow strain rate, respectively.

The following assumptions were considered in this analysis:

- For simplicity, it was assumed that values of density and specific heat were constant and the momentum of the fuel and oxidizer streams were the same.
- The vaporization process occurred in a very thin zone (asymptotic limit).

- Thermal radiation and heat losses were disregarded.
- Lycopodium particles were uniformly dispersed. In other words, the size and shape of the particles were assumed to be the same.
- A large Zeldowich number was presumed. Thus, the thickness of the flame zone would be too small.
- The ambient temperature $(T_{\pm\infty})$ was assumed to be 300 K.
- In order to analytically solve the coupled complex conservation equations of mass and energy in the considered zones, it was assumed that a gaseous fuel with a certain chemical composition evolved from the asymptotic vaporization of the lycopodium particles. Therefore, pyrolysis was disregarded as clearly considered in References [11,20,43].
- No chemical interaction occurred between the solid particles before the vaporization front.

2.5.1. Dimensionalized Governing Equations

The conservation of mass equation for a fuel in the gaseous phase was derived, considering the effect of porosity:

$$- aX\frac{dY_F}{dX} = D_F\frac{d^2Y_F}{dX^2} - f_e\frac{\omega_F}{\rho} + f_e\frac{\omega_v}{\rho} \tag{14}$$

where D_F, Y_F and ω_F are the fuel diffusivity, the gaseous fuel mass fraction and the chemical reaction rate, respectively. The chemical reaction rate was obtained using a one-step Arrhenius correlation [44,45]:

$$\omega_F = \beta\rho^2 v_F v_O \overline{Y_F Y_O} exp(-\frac{E}{RT}) \tag{15}$$

where β and E are the frequency constant and overall activation energy, respectively. $\overline{Y_F}$ and $\overline{Y_O}$ are expressed as follows [41]:

$$\overline{Y_F} = Y_F\frac{m}{v_F m_F} \tag{16}$$

$$\overline{Y_O} = Y_O\frac{m}{v_O m_O} \tag{17}$$

where Y_O and Y_F represent the oxidizer and fuel volume fractions. m_F and m_O are the fuel and oxidizer molecular weights, and m denotes the molecular weight of the mixture.

By neglecting the diffusion of the solid particles (no interaction exists between the solid particles), the mass conservation equation for the solid fuel particles can be presented as follows:

$$- aX\frac{dY_s}{dX} = -\omega_v \tag{18}$$

where Y_s and ω_v are the mass fraction and vaporization rate of the solid fuel particles, respectively.

The conservation equation of energy was obtained as follows:

$$- aX\frac{dT}{dX} = D_T\frac{d^2T}{dX^2} + f_e\omega_F\frac{Q}{\rho C} - f_e\omega_v\frac{Q_v}{C} \tag{19}$$

where Q is the released heat per unit mass of consumed fuel, Q_v is the latent heat of vaporization and $D_T = \frac{\lambda}{\rho C}$ is the thermal diffusivity coefficient. In Equation (19), C represents the specific heat capacity of the mixture that can be calculated using the following correlation [40,46,47]:

$$C = C_a + f_e\frac{4\rho_p}{3\rho}\pi r^3 C_p n_p \tag{20}$$

where ρ_p is the density of the solid particles and n_p is the particle number density per unit volume. The density of the mixture is stated as follows, taking into account the porosity of the particles [40]:

$$\rho = \rho_a + f_e \frac{4}{3} \pi r_p^3 n_p \rho_p. \tag{21}$$

The conservation of mass equation for the oxidizer is as follows:

$$-aX\frac{dY_O}{dX} = D_O \frac{d^2 Y_O}{dX^2} - f_e \vartheta \frac{w_F}{\rho} \tag{22}$$

where D_O and Y_O are the oxidizer mass diffusion and mass fraction, respectively.

2.5.2. Normalization of the Governing Equations

For normalizing the governing equations, the following parameters were employed [41]:

$$\theta = \frac{C(T - T_\infty)}{QY_{F-\infty}}, \quad y_F = \frac{Y_F}{Y_{F-\infty}}, \quad y_O = \frac{Y_O}{\vartheta Y_{F-\infty}}, \quad y_s = \frac{Y_s}{Y_{F-\infty}}, \quad x = \frac{X}{\sqrt{\frac{\lambda}{\rho C a}}} \tag{23}$$

where θ, y_F, y_O, y_s and x are the normalized forms of the temperature, the fuel mass fraction, the oxidizer mass fraction, the solid fuel mass fraction and position, respectively. $Y_{F-\infty}$ is the initial mass fraction of the fuel in the gaseous phase at $x \to -\infty$ (where fuel particles exit in the fuel nozzle). By applying the non-dimensionalized parameters (Equation (23)) into Equations (14), (18), (19) and (22) and considering the definition of a chemical reaction, the vaporization rate and Lewis number, the following normalized equations were obtained for the gaseous fuel mass conservation (Equation (24)), the solid fuel mass conservation (Equation (25)), the oxidizer mass conservation equation (Equation (26)) and the energy conservation equation (Equation (27)), respectively:

$$\frac{1}{Le_F}\frac{d^2 y_F}{dx^2} + x\frac{dy_F}{dx} + f_e \frac{y_s}{a\tau_{vap}}H(T - T_v) = f_e D_C y_F y_O \exp\left(-\frac{T_a}{T}\right) \tag{24}$$

$$x\frac{dy_s}{dx} = \frac{y_s}{a\tau_{vap}}H(T - T_v) \tag{25}$$

$$x\frac{dy_O}{dx} + \frac{1}{Le_O}\frac{d^2 y_O}{dx^2} = f_e D_C y_F y_O \exp\left(-\frac{T_a}{T}\right) \tag{26}$$

$$\frac{d^2\theta}{dx^2} + x\frac{d\theta}{dx} - f_e \frac{q}{a\tau_{vap}}y_s H(T - T_v) = -f_e D_C y_F y_O \exp\left(-\frac{T_a}{T}\right) \tag{27}$$

where T_a is the activation temperature, which is defined as $T_a = \frac{E}{R}$ (E is the overall activation energy), q is defined as $q = \frac{Q_v}{Q}$ and D_C is the Damkohler number, which was determined using the following equation [41]:

$$D_C = \rho B \vartheta_O Y_{F-\infty} / W_F a \tag{28}$$

where ϑ_O is the number of moles of oxygen participating in the reaction with one mole of fuel and W_F is the molecular weight of the fuel.

2.5.3. Boundary and Jump Conditions

In order to solve the non-dimensionalized governing equations in each of the zones, proper boundary and jump conditions were used. In this study, the following division was considered:

Preheat zone:	$R_1 : \{x \mid -\infty < x \le x_v\}$
Post vaporization zone:	$R_2 : \{x \mid x_v \le x \le x_f\}$
Oxidizer zone:	$R_3 : \{x \mid x_f \le x < \infty\}$

where x_v and x_f are locations of the vaporization front and flame sheet, respectively. As previously mentioned, solid particles turned into the gaseous phase asymptotically once they passed the vaporization front. Therefore, the following conditions were considered:

$$-\infty < x \le x_v: \quad y_s = 1$$
$$x_v \le x < \infty: \quad y_s = 0 \tag{29}$$

The following boundary conditions were applied in the analysis of the counter-flow non-premixed flame:

$$y_F = 0, \ y_O = 0, \ y_s = 1, \ \theta = 0 \quad @ \ x \to -\infty \tag{30}$$

$$y_F = 0, \ y_O = \alpha, \ y_s = 0, \ \theta = 0 \quad @ \ x \to +\infty \tag{31}$$

where α was obtained using the following correlation [38]:

$$\alpha = \frac{Y_{O,\infty}}{\vartheta Y_{F-\infty}} \tag{32}$$

The above boundary conditions were defined for the temperature and mass fraction of the fuel and oxidizer at $x = -\infty$ and $x = +\infty$. Nevertheless, to detect the positions of the vaporization front (x_v) and flame sheet (x_f), and the corresponding temperature and mass fractions, and enforce the continuity throughout the system, additional conditions were required at $x = x_v$ and $x = x_f$. It should be pointed out that at the flame sheet, $\theta = \theta_f$ and $y = y_F$, and at the vaporization front, $\theta = \theta_v$ and $y = y_{Fv}$, which must be solved. Since the current analysis was conducted based on an asymptotic concept, the vaporization and reaction processes occurred within a very thin zone. At the vaporization front, the convection and reaction terms were neglected compared to the diffusion and vaporization terms. Moreover, at the flame sheet position, the convection and vaporization terms can be disregarded against the reaction and diffusion terms. By using the aforementioned explanations and integrating Equations (24), (26) and (27) from x_v^- to x_v^+ and x_f^- to x_f^+, the following jump conditions were obtained at $x = x_v$ and $x = x_f$, respectively, as below:

$$-\frac{1}{Le_F}\left[\frac{dy_F}{dx}\right] = \left[\frac{d\theta}{dx}\right] = qx_v, \ \left[\frac{dy_O}{dx}\right] = 0, [y_F] = [y_O] = [\theta] = 0 \quad @ \ x = x_v \tag{33}$$

$$\frac{1}{Le_O}\left[\frac{dy_O}{dx}\right] = \frac{1}{Le_F}\left[\frac{dy_F}{dx}\right] = -\left[\frac{d\theta}{dx}\right], \ [y_F] = [y_O] = [\theta] = 0 \quad @ \ x = x_f \tag{34}$$

where [] (square brackets) is defined as $[] = ( )_+ - ( )_-$. No reaction occurred at the vaporization front. Therefore, the following governing equations were used for the modeling of the vaporization process:

$$\frac{d^2\theta}{dx^2} + x\frac{d\theta}{dx} - f_e\frac{q}{a\tau_{vap}}y_s H(T - T_v) = 0 \tag{35}$$

$$\frac{1}{Le_F}\frac{d^2y_F}{dx^2} + x\frac{dy_F}{dx} + f_e\frac{y_s}{a\tau_{vap}}H(T - T_v) = 0 \tag{36}$$

$$\frac{1}{Le_O}\frac{d^2y_O}{dx^2} + x\frac{dy_O}{dx} = 0. \tag{37}$$

By integrating the above equations from x_v^- to x_v^+, the following equations were achieved:

$$-\frac{1}{Le_F}\left[\frac{dy_F}{dx}\right] = \left[\frac{d\theta}{dx}\right] = qx_v \tag{38}$$

$$\left[\frac{dy_O}{dx}\right] = 0 \tag{39}$$

Equations (38) and (39) describe the jump conditions at the vaporization front (x_v). In order to obtain the jump condition at the flame sheet (x_f), the vaporization terms were neglected relative to the diffusion and reaction terms. Further, since the value of the Zeldovich number was presumed to be very large, the thickness of the reaction zone would be very small and the convection term can be neglected. Hence, the following equations were used to model the reaction process:

$$\frac{d^2\theta}{dx^2} = -f_e D_C y_F y_O \exp\left(-\frac{T_a}{T}\right) \tag{40}$$

$$\frac{1}{Le_F}\frac{d^2 y_F}{dx^2} = f_e D_C y_F y_O \exp\left(-\frac{T_a}{T}\right). \tag{41}$$

By aggregating Equations (40) and (41), the following equation was achieved:

$$\frac{d^2\theta}{dx^2} + \frac{1}{Le_F}\frac{d^2 y_F}{dx^2} = 0. \tag{42}$$

By performing a similar procedure for the oxidizer, the following equation was obtained:

$$\frac{1}{Le_O}\frac{d^2 y_O}{dx^2} = f_e D_C y_F y_O \exp\left(-\frac{T_a}{T}\right). \tag{43}$$

The following equation was determined by aggregating Equations (40) and (41):

$$\frac{d^2\theta}{dx^2} + \frac{1}{Le_O}\frac{d^2 y_O}{dx^2} = 0. \tag{44}$$

Finally, by integrating Equations (42) and (44) from x_f^- to x_f^+, the jump condition at the flame sheet was achieved as follows:

$$\frac{1}{Le_F}\left[\frac{dy_F}{dx}\right] = \frac{1}{Le_O}\left[\frac{dy_O}{dx}\right] = -\left[\frac{d\theta}{dx}\right]. \tag{45}$$

2.5.4. Solution of the Governing Equations

In this subsection, the solution of the non-dimensionalized governing equations is presented considering the boundary and jump conditions. It must be noted that for solving the complex equations, Mathematica and Matlab software were employed.

- Zone R_1: $-\infty \leq x \leq x_v$

  - Temperature distribution

In this zone, the temperature distribution was obtained by solving Equation (35):

$$\theta = C_1\sqrt{\frac{\pi}{2}}\, erf\left(\frac{x}{\sqrt{2}}\right) + C_2 \tag{46}$$

Considering the presented boundary conditions, C_1 and C_2 were calculated and the following function was achieved for the temperature distribution:

$$\theta = \theta_v \frac{erf\left(\frac{x}{\sqrt{2}}\right) + 1}{erf\left(\frac{x_v}{\sqrt{2}}\right) + 1}. \tag{47}$$

  - Oxidizer mass fraction

By solving Equation (37), the following equation was obtained for the oxidizer mass fraction:

$$y_O = C_1 \sqrt{\frac{1}{Le_O}} \sqrt{\frac{\pi}{2}} erf\left(\frac{x}{\sqrt{2}\sqrt{\frac{1}{Le_O}}}\right) + C_2. \tag{48}$$

By applying the boundary conditions, C_1 and C_2 were found to be zero and therefore the mass fraction of the oxidizer will be zero:

$$y_O = 0. \tag{49}$$

- Fuel mass fraction

By solving Equation (36), the following equation was obtained for the fuel mass fraction in this zone:

$$y_f = C_1 \sqrt{\frac{1}{Le_f}} \sqrt{\frac{\pi}{2}} erf\left(\frac{x}{\sqrt{2}\sqrt{\frac{1}{Le_f}}}\right) + C_2. \tag{50}$$

By using the boundary conditions presented in Equation (51), the fuel mass fraction function was calculated as presented in Equation (52).

$$x = -\infty \rightarrow y_F = 0, \ x = x_v \rightarrow y_F = y_{Fv} \tag{51}$$

$$y_f = \frac{y_{Fv}\left[erf\left(\frac{x}{\sqrt{2}\sqrt{\frac{1}{Le_f}}}\right) + 1\right]}{erf\left(\frac{x_v}{\sqrt{2}\sqrt{\frac{1}{Le_f}}}\right) + 1}. \tag{52}$$

- Zone R_2: $x_v \leq x \leq x_f$

  - Temperature distribution

As previously mentioned, the solid fuel particles were immediately turned into a gaseous fuel once they crossed the vaporization front. In this zone, the following boundary conditions were applied:

$$x = x_v \rightarrow \theta = \theta_v, \ x = x_f \rightarrow \theta = \theta_f. \tag{53}$$

By solving Equation (35) and using these boundary conditions, the temperature distribution of a gaseous fuel was obtained as follows:

$$\theta = \theta_v + \left(erf\left(\frac{x}{\sqrt{2}}\right) - erf\left(\frac{x_v}{\sqrt{2}}\right)\right) \frac{\theta_f - \theta_v}{erf\left(\frac{x_f}{\sqrt{2}}\right) - erf\left(\frac{x_v}{\sqrt{2}}\right)} \tag{54}$$

- Oxidizer mass fraction

In this zone, there was no oxidizer; thus, the mass fraction of the oxidizer was found to be zero.

$$y_O = 0. \tag{55}$$

- Fuel mass fraction

The fuel mass fraction was obtained by solving Equation (36) and applying the following boundary conditions:

$$x = x_v \rightarrow y_F = y_{Fv}, \ x = x_f \rightarrow y_F = 0 \tag{56}$$

$$y_F = y_{Fv} \frac{\left[erf\left(\frac{x}{\sqrt{2} \sqrt{\frac{1}{Le_f}}} \right) - erf\left(\frac{x_f}{\sqrt{2} \sqrt{\frac{1}{Le_f}}} \right) \right]}{\left[erf\left(\frac{x_v}{\sqrt{2} \sqrt{\frac{1}{Le_f}}} \right) - erf\left(\frac{x_f}{\sqrt{2} \sqrt{\frac{1}{Le_f}}} \right) \right]}. \tag{57}$$

- ■ Zone R_3: $x_f \leq x \leq +\infty$

  - ● Temperature distribution

In this zone, no gaseous fuel existed. The temperature distribution can be achieved by calculating Equation (35) and using the following boundary conditions:

$$x = x_f \rightarrow \theta = \theta_f, \ x = +\infty \rightarrow \theta = 0 \tag{58}$$

$$\theta = \theta_f \frac{erf\left(\frac{x}{\sqrt{2}} \right) - 1}{erf\left(\frac{x_f}{\sqrt{2}} \right) - 1}. \tag{59}$$

  - ● Oxidizer mass fraction

The oxidizer mass fraction was obtained by solving Equation (37), considering the below boundary conditions:

$$x = x_f \rightarrow y_O = 0, \ x = +\infty \rightarrow y_\alpha = 0 \tag{60}$$

$$y_O = \alpha \frac{\left[erf\left(\frac{x}{\sqrt{2} \sqrt{\frac{1}{Le_O}}} \right) - erf\left(\frac{x_f}{\sqrt{2} \sqrt{\frac{1}{Le_O}}} \right) \right]}{1 - erf\left(\frac{x_f}{\sqrt{2} \sqrt{\frac{1}{Le_O}}} \right)}. \tag{61}$$

  - ● Fuel mass fraction

In this zone, there was no fuel. Hence, the fuel mass faction was found to be zero:

$$y_F = 0. \tag{62}$$

It is notable that the values of the five parameters including the flame temperature, the fuel mass fraction, the oxidizer mass fraction and the location of the vaporization front and flame sheet were unknown.

2.6. Flame Zone Analysis

In order to study the flame zone, as previously expressed, it was assumed that the quantity of the Zeldovich number was very large so the thickness of the flame zone will be very small. In this regard, employing an asymptotic approach could be a promising technique for analyzing this zone. The Zeldovich number was calculated using the following correlation:

$$Ze = \frac{EQY_{F-\infty}}{RCT_f^2} \tag{63}$$

where E, R and T are the reaction activation energy, the universal gas constant and the flame temperature, respectively. To evaluate the critical extinction of the flame, a reduced Damkohler number was applied. Below is the critical extinction of the Damkohler number [41]:

$$\delta_{0E} \approx 2e(Z_f - 2Z_f^2 + 1.04Z_f^3 + 0.44Z_f^4) \tag{64}$$

where $Z_f = \frac{1}{2}erfc(\frac{x_f}{\sqrt{2}})$ and δ_0 are an order of unity that can be approximated using the following equation:

$$\delta = \delta_0 + O(\frac{1}{Ze}) \tag{65}$$

where δ is the reduced Damkohler number that can be calculated using the following equation:

$$\delta = \frac{8\pi \exp(x_f^2)DLe_OLe_fZ_f^2}{\alpha^2 F_{Of}^2 Z_e^3} \exp\left(-\frac{T_a}{T_f}\right) \tag{66}$$

where F_{Of} is defined as follows [38]:

$$F_{Of} \equiv F_O(x_f, Le_O) = \frac{t_{Of}}{t\sqrt{Le_{Of}}} \left(\frac{1 - \frac{0.276}{t_f} + \frac{2.15}{t_f^2}}{1 - \frac{0.276}{t_{Of}} + \frac{2.15}{t_{Of}^2}}\right) \tag{67}$$

$$t_{Of} = 1 + 0.33333 x_f \sqrt{Le_O} \tag{68}$$

$$t_f = 1 + 0.33333 x_f. \tag{69}$$

By substituting Equation (64) into Equation (66), the following correlation was obtained:

$$D_{OE} \approx \frac{\alpha^2 F_{Of}^2 Z_e^3 \exp\left(\frac{T_a}{T_f} - x_f^2 + 1\right)}{4\pi Le_O Le_f Z_f}(1 - 2Z_f + 1.04Z_f^2 + 0.44Z_f^3). \tag{70}$$

According to Equation (70), it can be readily implied that D_{OE} mainly depends on the flame sheet position (x_f) and the flame temperature (T_f). The ratio of a to a_0 (for which $Le_O = Le_F = 1$) is presented in the following:

$$\left(\frac{a}{a_0}\right)_{crit} = \frac{Le_O Le_f}{F_{Of}^2} \left(\frac{T_f}{T_f^0}\right)^6 \frac{\eta^2}{d_E} \exp\left[\frac{T_a}{T_f^0}\left(1 - \frac{T_f^0}{T_f}\right) + (x_f^2 - (x_f^0)^2)\right]. \tag{71}$$

Equation (71) describes the critical strain rate as a function of the fuel and oxidizer Lewis numbers. In this equation, η and d_E are expressed as follows [20]:

$$\eta = \frac{Z_f}{Z_f^0} \tag{72}$$

$$d_E = \eta \frac{1 - 2Z_f + 1.04Z_f^2 + 0.44Z_f^3}{1 - 2Z_{f^0} + 1.04(Z_{f^0})^2 + 0.44(Z_{f^0})^3}. \tag{73}$$

2.7. Calculation of the Thermophoretic Force

- Zone R_1: $-\infty \le x \le x_v$

By applying $\theta = \theta_v \frac{erf\left(\frac{x}{\sqrt{2}}\right)+1}{erf\left(\frac{x_v}{\sqrt{2}}\right)+1}$ into Equation (47), the thermophoretic forcewas obtained as follows:

$$T = \left(\left(erf\left(\frac{x}{\sqrt{2}}\right)+1\right) \times 1107.65\right) + 300 \tag{74}$$

$$\nabla T(x) = \frac{dT}{dx} = 1107.65\frac{\sqrt{2}\exp\left(-\frac{x^2}{2}\right)}{\pi^{\frac{1}{2}}} \tag{75}$$

$$F_{Th,continuum} = -3.3132 \times 10^{-7} \times \exp\left(-\frac{x^2}{2}\right) \tag{76}$$

According to the definition of the thermophoretic force, this force mainly depends on the temperature gradient and the radius of the fuel particles. As previously mentioned, after the vaporization front, the solid fuel particles were completely transformed into a gaseous fuel. In this regard, the radius of the particles and subsequently the value of the thermophoretic force will be almost zero.

3. Results and Discussion

The properties of the lycopodium biofuel used in the present study are listed in Table 2.

Table 2. Properties of the biofuel particles and oxidizer [15,41,48].

Property	Value	Property	Value
ρ_p	$1000 \frac{kg}{m^3}$	C_p	$5.67 \frac{kJ}{kg\cdot K}$
ρ_a	$1.2 \frac{kg}{m^3}$	C_a	$1.001 \frac{kJ}{kg\cdot K}$
Q	$64,895.4 \frac{kJ}{kg}$	q	0.4
r	$12 \mu m$	n	12 Giga
T_{in}	300 K	T_{vap}	650 K
v	2	$y_{O,+\infty}$	0.13

In this section, changes in flame temperature, the flame sheet position, the mass fractions of the biofuel and oxidizer, the flow strain rate and the thermophoretic force with Lewis number, the position and the equivalence ratio were studied. In order to follow the temperature and mass fraction distributions of the fuel and oxidizer, solutions of the mass and energy conservation equations were used considering the afore-presented boundary and jump conditions. As mentioned earlier, the locations of the flame sheet and vaporization front were measured relative to the location of the stagnation plane whose position was employed as the reference coordinate. As assumed in References [11,48], the fuel supplied by the vaporization of the lycopodium particles was gaseous methane. The complete reaction of gaseous methane with air is considered as follows:

$$CH_4 + 2(O_2 + 3.76N_2) \rightarrow CO_2 + 2H_2O + (7.52)N_2. \tag{77}$$

By taking into account the volatility of the biofuel, the effective gas phase equivalence ratio is defined as follows [11,48]:

$$\phi_u = \frac{17.18Y_{F-\infty}}{1 - Y_{F-\infty}} \tag{78}$$

where $Y_{F-\infty}$ was calculated using the following equation [11,48]:

$$Y_{F-\infty} = \frac{\frac{4}{3}\pi r_p{}^3 n_p \rho_p}{\rho}. \tag{79}$$

In order to include the effect of particle porosity, the following correlation was applied:

$$Y_{F-\infty}{}^* = f_e \frac{\frac{4}{3}\pi r_p{}^3 n_p \rho_p}{\rho} \tag{80}$$

Figure 4 shows the variations of flame temperature (K) with the fuel Lewis number for several porosity factors when the oxidizer Lewis number was unified. With regard to the definition of the Lewis number in Equation (11) (the ratio of thermal diffusivity to mass diffusivity), an increase of the Lewis number leads to a reduction in the fuel mass fraction. Therefore, an accessible fuel reaching the reaction zone declined, which leads to a decrease in flame temperature. As can be observed, an increment in the fuel Lewis number from 0.2 to 1.4 resulted in a gradual increase of the flame temperature between ~1355 and 1615 K for a lycopodium mass concentration of 1000 $\frac{kg}{m^3}$ and a unity value of the porosity factor. It is worth mentioning that decreasing the porosity factor caused a decline in the flame temperature.

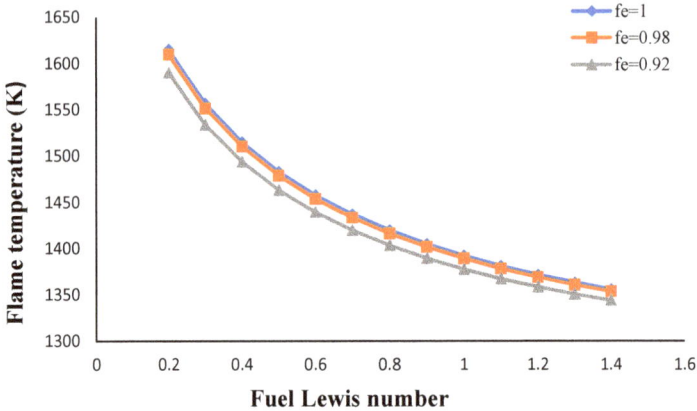

Figure 4. Flame temperature against the fuel Lewis number for several porosity factors.

Figure 5 shows the change in the flame sheet position with the fuel Lewis number for several porosity factors assuming that a unity value of the oxidizer Lewis number. Again, considering the definition of the Lewis number, increasing the Lewis number leads to a reduction in the fuel mass fraction and flame temperature. Thus, the amount of available fuel for the combustion process decreased leading to the movement of the flame sheet toward the fuel nozzle (left-hand side of the stagnation plane). According to Figure 5, increasing the fuel Lewis number from 0.2 to 1.4 changed the flame sheet position from 0.0967 to −0.1545 when the value of the porosity factor was unity.

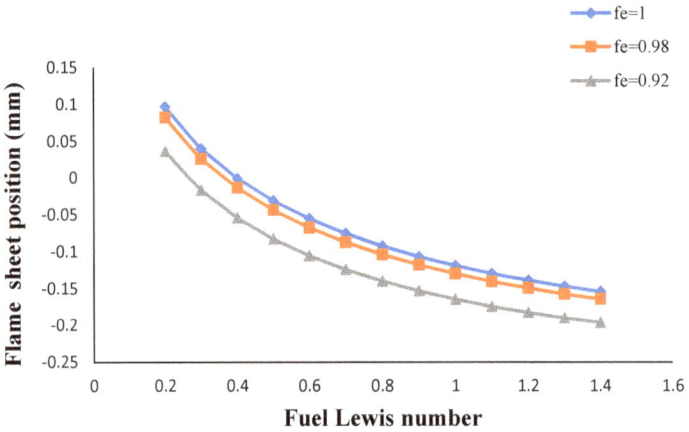

Figure 5. Flame sheet position against the fuel Lewis number for different porosity factors.

Figure 6 indicates a change in flame temperature with the oxidizer Lewis number for different porosity factors when the fuel Lewis number was unity. As it is seen in Figure 6, an increase of the oxidizer Lewis number caused a decline in the mass diffusivity of the oxidizer (according to the

Lewis number formula) leading to a decline in the flame temperature. For the considered range of the oxidizer Lewis number from 0.3 to 1, the flame temperature varied from 1955 K to 1391 K.

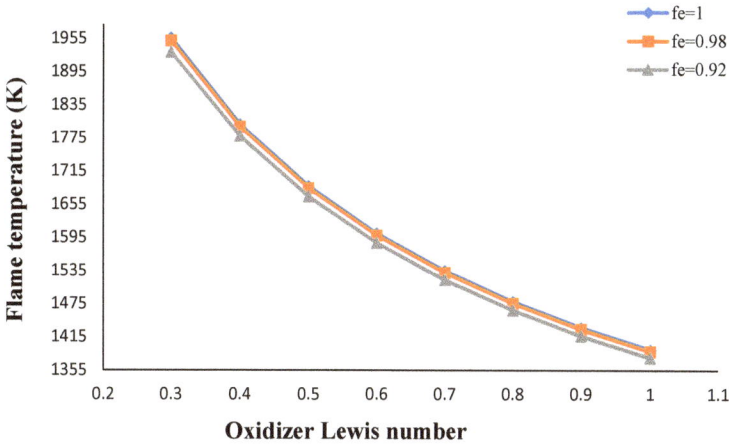

Figure 6. Flame temperature against the oxidizer Lewis number for several porosity factors.

Figure 7 represents the change in the flame sheet position with the oxidizer Lewis number for different porosity factors and a unity magnitude of the fuel Lewis number. According to the figure, an increment in the oxidizer Lewis number caused a rise in the value of the flame sheet position. In other words, the flame sheet shifted toward the oxidizer nozzle. Based on the Lewis number formula, increasing the oxidizer Lewis number reduces the mass diffusivity of oxidizer. Accordingly, the flame sheet moved toward the oxidizer zone. Regarding Figure 7, for the considered values of the oxidizer Lewis number from 0.3 to 1, the flame sheet position changed gradually from −0.533 to −0.119. Moreover, increasing the particle porosity factor increased the flame temperature and shifted the flame sheet toward the oxidizer zone. With an increasing porosity factor, the available fuel mass for the reaction decreased so the flame sheet would be closer to the fuel nozzle. For validation purposes, the current results were compared to the numerical results reported by Wang [42] for a methane-air mixture under the same conditions (the unity values of the porosity factor and the fuel Lewis number). As represented in Figure 7, there was consistency between the present analytical results and the data provided by Reference [42] for the case in which the porosity factor was unity.

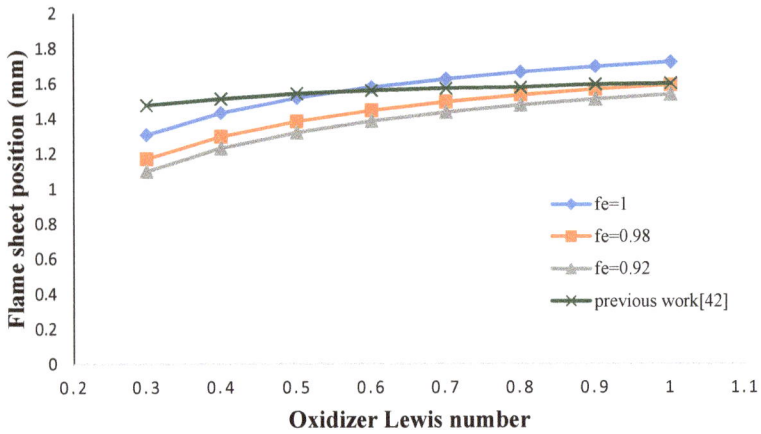

Figure 7. Flame sheet position against the oxidizer Lewis number for different porosity factors.

In Figure 8, the effect of the fuel Lewis number on the temperature distribution of the flame is shown for different mass particle concentrations. With regard to Figure 8, an increase of the fuel Lewis number resulted in a decrease of the flame temperature, which can readily be explained by the definition of the Lewis number. According to this figure, by increasing the fuel Lewis number from 0.5 to 1, the flame temperature varied from about 1380 to 1485 K. Furthermore, by increasing the value of the mass particle concentration, the amount of available fuel for the formation of the flame increased, resulting in an increase of the flame temperature.

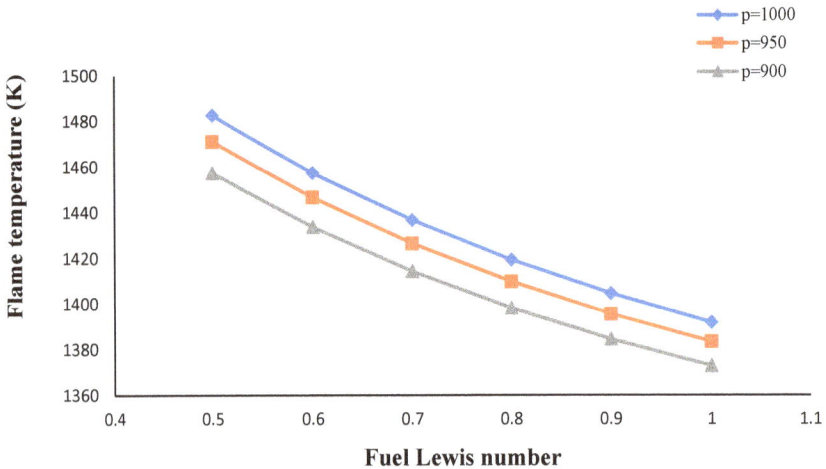

Figure 8. Flame temperature against the fuel Lewis number for several values of mass particle concentration.

Figure 9 depicts the effect of the fuel Lewis number on the location of the flame sheet considering several different mass particle concentrations. By increasing the fuel Lewis number from 0.5 to 1, the flame sheet position ranged from −0.096 to −0.01765 when the mass particle concentration was 900 kg/m^3. Increasing the Lewis number of the fuel decreased the amount of fuel mass approaching the reaction zone, which shifted the flame sheet toward the fuel nozzle. It is notable that the mass fraction of the fuel for the reaction process increased with an increase in the mass particle concentration. Therefore, further fuel mass diffusivity pushed the flame sheet toward the oxidizer zone.

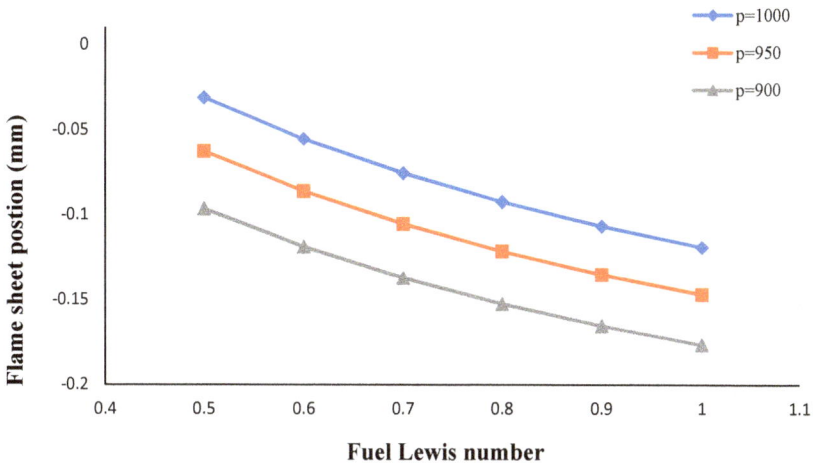

Figure 9. Flame sheet position against the fuel Lewis number for different mass particle concentrations.

The variation of the flame temperature with an effective equivalence ratio (ϕ_u) is drawn in Figure 10 for different lycopodium radii. As it is observed in Figure 10, an increase of the equivalence ratio resulted in an increment in the flame temperature. In addition, increasing the size of the fuel particles (from 8 to 18 μm) caused a reduction in the flame temperature. An enhancement of the particle size decreased the ratio of the particle surface to the particle volume. On this basis, a portion of released heat during the reaction was consumed for the preheating and vaporization of the lycopodium particles with larger diameters that led to a decrease in the flame temperature.

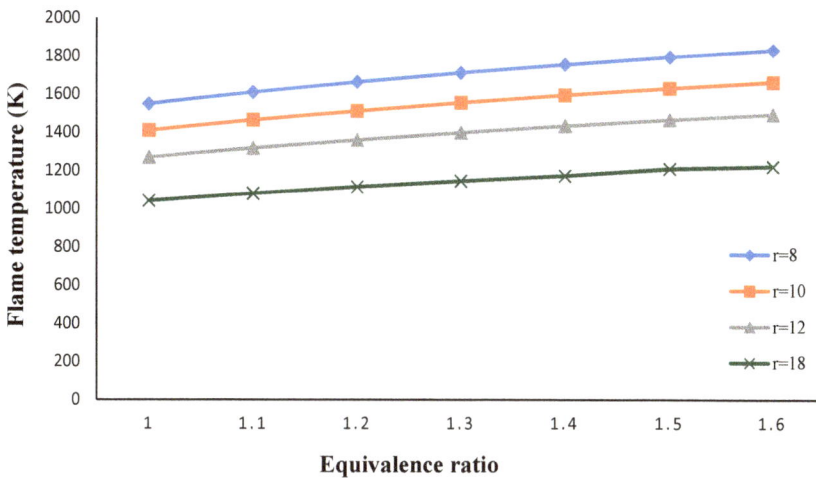

Figure 10. Flame temperature against the effective equivalence ratio (ϕ_u) considering several lycopodium radii.

Figure 11 represents the change in the temperature distribution with the flame position for various oxidizer Lewis numbers when the fuel Lewis number was unity. According to Figure 11, the temperature of the fuel particles grew until it approached the maximum temperature corresponding to the flame sheet position. The temperature of the oxidizer also grew until it reached the flame temperature (right-hand side of the stagnation plane). Reducing the oxidizer Lewis number from 1 to 0.4 caused a decrease in the flame temperature and a shift of the flame sheet toward the oxidizer nozzle.

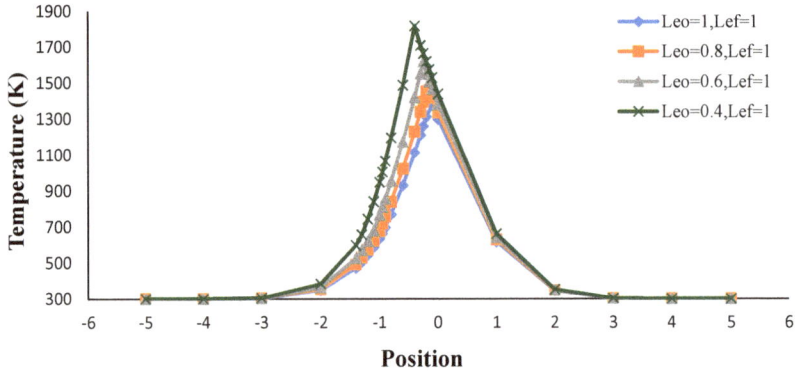

Figure 11. Temperature distribution against position for different oxidizer Lewis numbers.

In Figure 12, changes in the gaseous fuel and oxidizer mass fractions with the flame position are described for several different mass particle concentrations. Regarding Figure 12, the gaseous fuel mass fraction increased until it reached the vaporization front at which the mass fraction of gaseous fuel was the lowest. Afterwards, the gaseous fuel was oxidized and consumed in the reaction zone so its mass fraction reduced to zero. From the opposite side, the oxidizer flowed toward the reaction zone and was consumed for the generation of the flame. Therefore, its mass fraction decreased as it moved closer to the flame front. Similarly to the fuel mass fraction, the oxidizer was entirely consumed at the flame front position and its mass fraction became zero. It can also be implied that increasing the value of the porosity factor increased the gaseous fuel mass fraction. Therefore, the vaporization front moved toward the oxidizer nozzle.

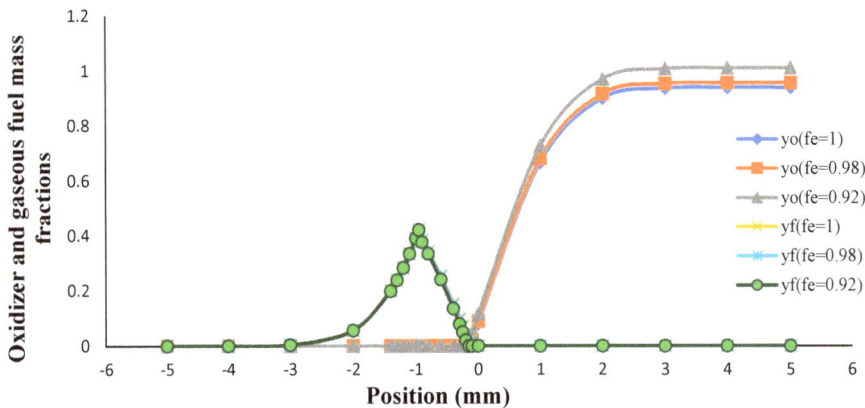

Figure 12. Oxidizer and gaseous fuel mass fractions against the position for different mass particle concentrations.

Figures 13 and 14 illustrate the effects of the fuel and oxidizer Lewis numbers on the critical strain rate ($\frac{a}{a_0}$) for different porosity factors. According to Figure 13, for unity values of the fuel Lewis number and the porosity factor, increasing the oxidizer Lewis number led to a decline in the ratio of $\frac{a}{a_0}$. Moreover, decreasing the value of the porosity factor decreased the value of x_f leading to a decrease of $\frac{a}{a_0}$. It is notable that by increasing the value of $\frac{a}{a_0}$, the flame sheet shifted toward the oxidizer zone. Figure 14 demonstrates the variation of $\frac{a}{a_0}$ with the fuel Lewis number considering different porosity factors when the oxidizer Lewis number was unity. Regarding this figure, the value of $\frac{a}{a_0}$ decreased as the fuel Lewis number increased. Decreasing the fuel Lewis number incremented the value of the flame sheet position (x_f) and so a remarkable enhancement in the value of the exponential term in Equation (71) that resulted in an increase in the value of $\frac{a}{a_0}$. In order to validate the results, a comparison was made between the results obtained for $\frac{a}{a_0}$ in this investigation and the results provided by Seshadri and Trevino [41] for the critical strain rate under the same conditions (T_a = 30,000 K and the oxidizer Lewis number was unity). With regard to Figure 14, a reasonable agreement exists between the compared results under the considered conditions.

Figures 15 and 16 delineate the impact of the oxidizer and fuel Lewis numbers on the thermophoretic force for the unity value of the porosity factor, respectively. According to these figures, until reaching the vaporization front, the thermophoretic force continuously reduced with the flame position. As the oxidizer and fuel Lewis numbers decreased, the maximum temperature occurring at the reaction zone increased. In this regard, a temperature gradient and subsequently the value of the thermophoretic force increased. Further, it can be seen that by decreasing the oxidizer and fuel Lewis numbers, the location of the minimum thermophoretic force moved toward the fuel nozzle. As there was no solid fuel after the vaporization front, the value of r_p and the thermophoretic force will be zero.

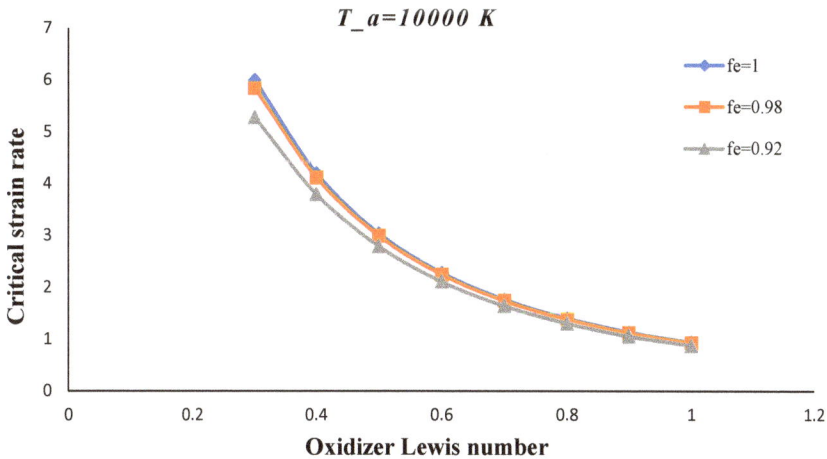

Figure 13. Critical flow strain rate against the oxidizer Lewis number for several porosity factors.

$$T\_a = 30000 \ K$$

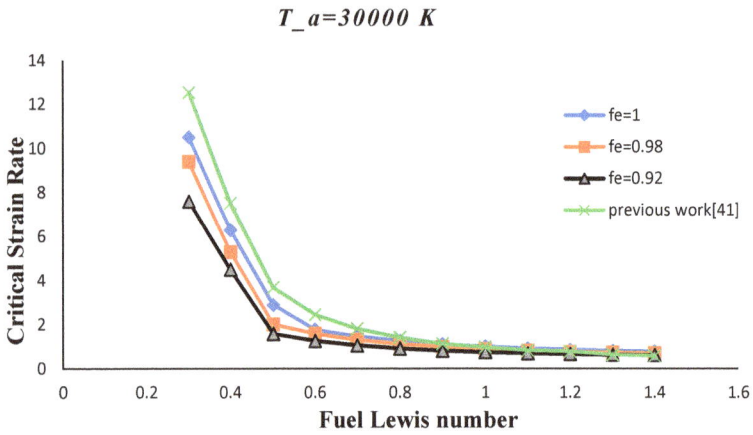

Figure 14. Critical flow strain rate against the fuel Lewis number for different porosity factors.

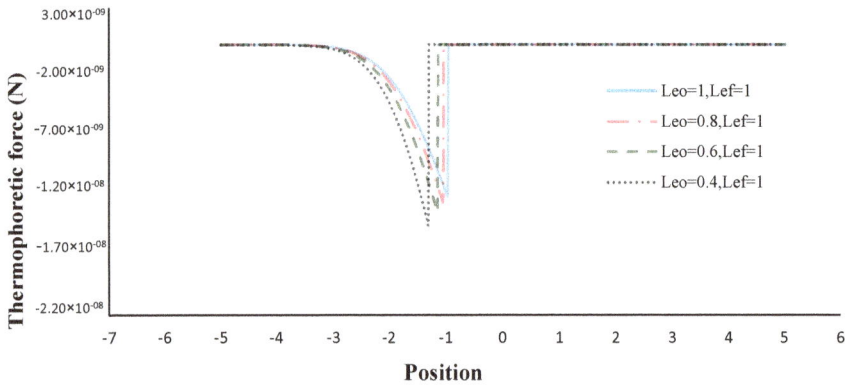

Figure 15. Thermophoretic force against the position for different oxidizer Lewis numbers.

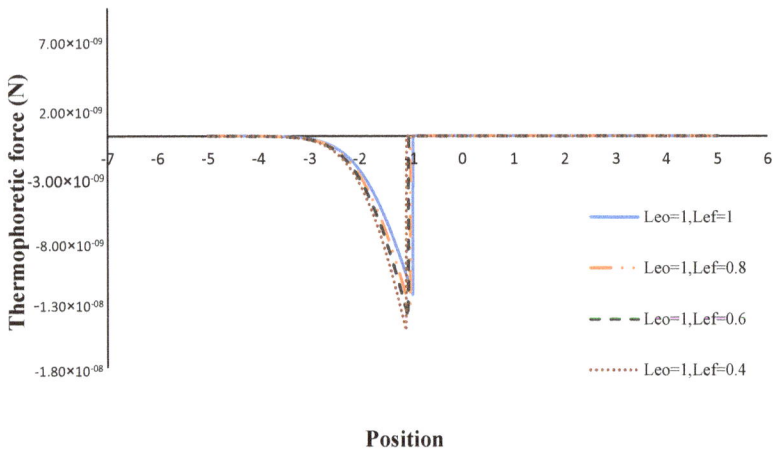

Figure 16. Thermophoretic force against position for different fuel Lewis numbers.

The impact of porosity on the thermophoretic force is illustrated in Figure 17 taking into account the unity values of the oxidizer and fuel Lewis numbers. According to this figure, an increase of the porosity factor enhanced the thermophoretic force. By increasing the porosity factor, the accessible fuel that approached the flame zone decreased leading to a decline in the value maximum flame temperature. In this regard, the temperature gradient term in Equation (8) decreased causing a decline in the value of the thermophoretic force. Similarly to Figures 15 and 16, since no solid fuel existed after the vaporization front, the value of r_p and then the thermophoretic force will be zero.

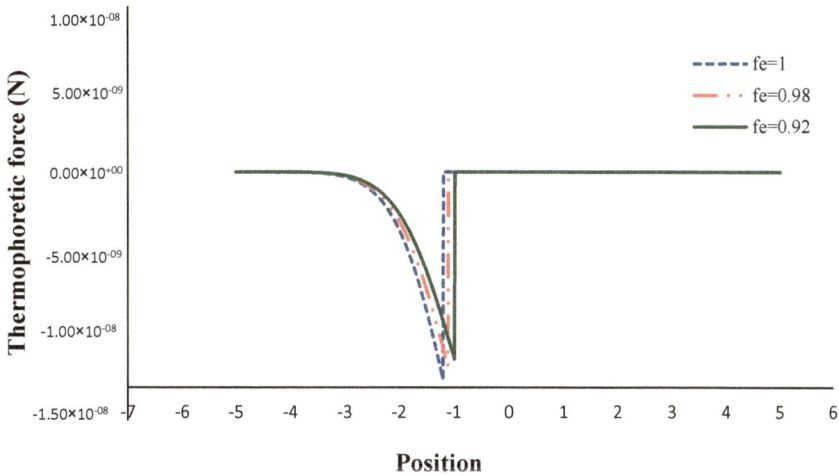

Figure 17. Thermophoretic force against position for different porosity factors.

4. Conclusions

Due to its appropriate dispersibility and flammability, lycopodium has been used as a reasonable reference organic fuel for testing different combustion systems and recognizing the processes occurring in the system prior to industrial scale-up. To provide helpful theoretical results about the effective parameters (i.e., the porosity and thermophoretic force), this paper used a mathematical approach to study the structure of multi-zone counter non-premixed flames fueled by lycopodium particles. Preheat, vaporization, flame and oxidizer zones were considered for the flame. Mass and energy equations were obtained for the zones. In this analysis, the porosity of the biofuel particles was also taken into account. Further, the vaporization rate and thermophoresis effects were modelled. Changes in the flame temperature and flame sheet position with fuel and oxidizer Lewis numbers were presented by considering different porosity factors. The impacts of the effective equivalence ratio, the flow critical strain rate, and the gaseous fuel and oxidizer mass fractions were evaluated. Finally, the thermophoretic force caused by the temperature gradient was obtained for different porosity factors, fuel and oxidizer Lewis numbers. For the case in which the oxidizer and fuel Lewis numbers were equal to 0.4 and 1, a maximum temperature of the flame was found to be ~1860 K. The maximum mass fraction of gaseous fuel that occurred at the vaporization front was found to be 0.44 when the value of the porosity factor was unity. In this analysis, for the unity value of the fuel Lewis number, the minimum value of the thermophoretic force was found to be -1.48×10^{-8} N, when the oxidizer Lewis number was 0.4. On the other hand, for the unity value of oxidizer Lewis number, the minimum value of the thermophoretic force was equal to -1.53×10^{-8}N at a fuel Lewis number of 0.4. By considering the porosity factor, the minimum value of the thermophoretic force occurred at a unity value of the porosity factor and its value was found to be -1.28×10^{-8} N. Studying the effects of oscillation and instability and calculating the quenching distance for counter-flow non-premixed combustion can be considered in future works. The main conclusions of the present study are summarized as follows:

- The flame temperature increased by decreasing the volume porosity, fuel and oxidizer Lewis numbers.
- The flame sheet position moved toward the fuel nozzle with an increasing volume porosity and fuel Lewis number.
- The flame sheet position shifted toward the fuel nozzle by decreasing the oxidizer Lewis number.
- The thermophoretic force increased by decreasing the volume porosity, fuel and oxidizer Lewis numbers.
- The critical strain rate increased by decreasing the volume porosity, fuel and oxidizer Lewis numbers.

Author Contributions: Project Administration, M.B.; Methodology and Analysis, P.G.N.; Methodology and Validation, H.R. and S.S.; Writing-Original, S.S.; Writing-Review & Editing, B.S.; Analysis-Review M.B. & B.S.

Funding: This research received no external funding.

Conflicts of Interest: The authors declare no conflicts of interest.

Nomenclature

a	Strain rate
$\overline{C}$	Mean thermal velocity of gas molecular
C_a	gaseous phase specific heat $\left(\frac{kJ}{kg\cdot K}\right)$
C_k	temperature jump coefficient
C_p	Solid particle specific heat $\left(\frac{kJ}{kg\cdot K}\right)$
C_t	Temperature creep coefficient
C_m	Velocity jump coefficient
C_s	Gas velocity discontinuities coefficient
D_C	Damkohler number
D_F	Mass diffusivity coefficient of gaseous fuel (m^2/s)
D_O	Mass diffusivity coefficient of oxidizer (m^2/s)
D_{0E}	Critical Damkohler number
D_T	Thermal diffusivity coefficient (m^2/s)
E	activation energy (kj)
$erfi(x)$	Error function
f_e	Porosity factor
H	Heavi side function
k_g	Gas thermal conductivity $\left(\frac{kJ}{m\cdot s\cdot K}\right)$
k_p	Lycopodium thermal conductivity $\left(\frac{kJ}{m\cdot s\cdot K}\right)$
k_T	Constant Defined in Equation (4)
L	Mean free path
Le	Lewis number
m	Mixture molecular mass ($\frac{kg}{mol}$)
m_f	Fuel molecular mass ($\frac{kg}{mol}$)
m_O	Oxygen molecular mass ($\frac{kg}{mol}$)
n_p	Number of particle per volume unit

Q	Reaction heat per unit of fuel mass $\left(\frac{kJ}{kg}\right)$
q	Dimensionless heat
R	Universal constant of gases $\left(\frac{m^3 Pa}{mol \cdot K}\right)$
r_p	Particle radius
T	Fuel temperature (K)
T_a	activation temperature (K)
T_f	Flame temperature (K)
T_v	Particle Start temperature of vaporization (K)
W_F	molecular weight of fuel
x	Dimension length
x_f	Flame position
x_v	Vaporization front position
Y_F	Gaseous fuel mass fraction
Y_O	Oxidizer mass fraction
Y_s	Particle mass fraction
y_F	Dimensionless fuel mass fraction
y_O	Dimensionless oxidizer mass fraction
y_s	Dimensionless mass fraction of solid particles
Z	Secondary coordinate axis
Ze	Zeldovich number

Greek symbols

α	Initial mass fraction of oxidizer
ε	Volume porosity
Θ	Dimensionless Temperature
λ	Thermal conductivity of fuel or oxidizer $\left(\frac{kJ}{m \cdot s \cdot K}\right)$
μ	Dynamic viscosity
v_F	Fuel stoichiometric coefficient
v_O	oxidizer stoichiometric coefficient
$v_{product}$	Product stoichiometric coefficient
ρ	Density $\left(\frac{kg}{m^3}\right)$
ρ_a	Gaseous phase density $\left(\frac{kg}{m^3}\right)$
ρ_p	Density of Solid particle $\left(\frac{kg}{m^3}\right)$
τ_{vap}	constant time characteristic of vaporization
ω_v	Particle vaporization rate $\left(\frac{kg}{m \cdot s^2}\right)$
ω_F	Rate of chemical reaction $\left(\frac{kg}{m \cdot s^2}\right)$
$\varnothing$	Constant and equal to 0.941

References

1. Caputo, A.C.; Palumbo, M.; Pelagagge, P.M.; Scacchia, F. Economics of biomass energy utilization in combustion and gasification plants: Effects of logistic variables. *Biomass Bioenergy* **2005**, *28*, 35–51. [CrossRef]
2. Wahlund, B.; Yan, J.; Westermark, M. Increasing biomass utilisation in energy systems: A comparative study of CO_2 reduction and cost for different bioenergy processing options. *Biomass Bioenergy* **2004**, *26*, 531–544. [CrossRef]
3. Demirbas, A.H.; Demirbas, I. Importance of rural bioenergy for developing countries. *Energy Convers. Manag.* **2007**, *48*, 2386–2398. [CrossRef]
4. Hall, D.O.; Scrase, J.I. Will biomass be the environmentally friendly fuel of the future? *Biomass Bioenergy* **1998**, *15*, 357–367. [CrossRef]
5. Hu, Y.; Kurose, R. Nonpremixed and premixed flamelets LES of partially premixed spray flames using a two-phase transport equation of progress variable. *Combust. Flame* **2018**, *188*, 227–242. [CrossRef]
6. Bhoi, P.R.; Channiwala, S.A. Emission characteristics and axial flame temperature distribution of producer gas fired premixed burner. *Biomass Bioenergy* **2009**, *33*, 469–477. [CrossRef]

7.	Akbarzadeh, M.; Birouk, M. Effect of Fuel Nozzle Geometry on the Stability of Non-Premixed Turbulent Methane Flame. In Proceedings of the 23rd ICDERS, Irvine, CA, USA, 24–29 July 2011.

8.	Josh, N.D.; Berlad, A.L. Gravitational effects on stabilized, premixed, lycopodium-air flames. *Combust. Sci. Technol.* **1986**, *47*, 55–68. [CrossRef]

9.	Berlad, A.L.; Joshi, N.D. Gravitational effects on the extinction conditions for premixed flames. *Acta Astronaut.* **1985**, *12*, 539–545. [CrossRef]

10.	Berlad, A.L.; Tangirala, V.; Facca, L.R. Radiative structures of lycopodium-air flames in low gravity. *J. Propuls. Power* **1991**, *7*, 5–8. [CrossRef]

11.	Seshadri, K.; Berlad, A.L.; Tangirala, V. The structure of premixed particle-cloud flames. *Combust. Flame* **1992**, *89*, 333–342. [CrossRef]

12.	Han, O.S.; Yashima, M.; Matsuda, T.; Matsui, H.; Miyake, A.; Ogawa, T. Behavior of flames propagating through lycopodium dust clouds in a vertical duct. *J. Loss Prev. Process. Ind.* **2000**, *13*, 449–457. [CrossRef]

13.	Han, O.S.; Yashima, M.; Matsuda, T.; Matsui, H.; Miyake, A.; Ogawa, T. A study of flame propagation mechanisms in lycopodium dust clouds based on dust particles' behavior. *J. Loss Prev. Process. Ind.* **2001**, *14*, 153–160. [CrossRef]

14.	Xi, Z.; Fu, Z.; Hu, X.; Sabir, S.; Jiang, Y. An experimental investigation on flame pulsation for a swirl non-premixed combustion. *Energies* **2018**, *11*, 1757. [CrossRef]

15.	Proust, C. Flame propagation and combustion in some dust-air mixtures. *J. Loss Prev. Process. Ind.* **2006**, *19*, 89–100. [CrossRef]

16.	Skjold, T.; Olsen, K.L.; Castellanos, D. A constant pressure dust explosion experiment. *J. Loss Prev. Process. Ind.* **2013**, *26*, 562–570. [CrossRef]

17.	Shamooni, A.; Cuoci, A.; Faravelli, T.; Sadiki, A. Prediction of Combustion and Heat Release Rates in Non-Premixed Syngas Jet Flames Using Finite-Rate Scale Similarity Based Combustion Models. *Energies* **2018**, *11*, 2464. [CrossRef]

18.	Rahbari, A.; Shakibi, A.; Bidabadi, M. A two-dimensional analytical model of laminar flame in lycopodium dust particles. *Korean J. Chem. Eng.* **2015**, *32*, 1798–1803. [CrossRef]

19.	Bidabadi, M.; Esmaeilnejad, A. An analytical model for predicting counterflow flame propagation through premixed dust micro particles with radiative heat loss. *J. Loss Prev. Process. Ind.* **2015**, *35*, 182–199. [CrossRef]

20.	Bidabadi, M.; Ramezanpour, M.; Poorfar, A.K.; Monteiro, E.; Rouboa, A. Mathematical Modeling of a Non-premixed Organic Dust Flame in a Counterflow Configuration. *Energy Fuels* **2016**, *30*, 9772–9782. [CrossRef]

21.	Xi, Z.; Fu, Z.; Hu, X.; Sabir, S.W.; Jiang, Y. An Investigation on Flame Shape and Size for a High-Pressure Turbulent Non-Premixed Swirl Combustion. *Energies* **2018**, *11*, 930. [CrossRef]

22.	Li, X.; Zhang, J.; Yang, H.; Jiang, L.; Wang, X.; Zhao, D. Combustion characteristics of non-premixed methane micro-jet flame in coflow air and thermal interaction between flame and micro tube. *Appl. Therm. Eng.* **2017**, *112*, 296–303. [CrossRef]

23.	Bidabadi, M.; Harati, M.; Afzalabadi, A.; Rahbari, A. Effect of Thermal Resistance on the Random Combustion of Micro-Organic Dust Particles. *J. Energy Eng.* **2017**, *144*, 04017073. [CrossRef]

24.	Spijker, C.; Raupenstrauch, H. Numerical investigation on inner particle effects in Lycopodium/Air dust deflagrations. *J. Loss Prev. Process. Ind.* **2017**, *49*, 870–879. [CrossRef]

25.	Chen, L.; Xia, Y.F.; Li, B.W.; Shi, J.R. Flame front inclination instability in the porous media combustion with inhomogeneous preheating temperature distribution. *Appl. Therm. Eng.* **2018**, *128*, 1520–1530. [CrossRef]

26.	Bidabadi, M.; Ebrahimi, F.; Bordbar, V. Modeling multi regional counter flow combustion of lycopodium dust cloud with considering radiative heat loss. *J. Cent. South Univ.* **2017**, *24*, 2638–2648. [CrossRef]

27.	Ji, W.; Yu, J.; Yu, X.; Yan, X. Experimental investigation into the vented hybrid mixture explosions of lycopodium dust and methane. *J. Loss Prev. Process. Ind.* **2018**, *51*, 102–111. [CrossRef]

28.	Di Benedetto, A.; Di Sarli, V.; Russo, P. On the determination of the minimum ignition temperature for dust/air mixtures. *Chem. Eng. Trans.* **2010**, *19*, 189–194.

29.	Sanchirico, R.; Russo, P.; Saliva, A.; Doussot, A.; Di Sarli, V.; Di Benedetto, A. Explosion of lycopodium-nicotinic acid–methane complex hybrid mixtures. *J. Loss Prev. Process. Ind.* **2015**, *36*, 505–508. [CrossRef]

30.	Sanchirico, R.; Russo, P.; Di Sarli, V.; Di Benedetto, A. On the explosion and flammability behavior of mixtures of combustible dusts. *Process. Saf. Environ. Prot.* **2015**, *94*, 410–419. [CrossRef]

31. Sanchirico, R.; Di Sarli, V.; Russo, P.; Di Benedetto, A. Effect of the nozzle type on the integrity of dust particles in standard explosion tests. *Powder Technol.* **2015**, *279*, 203–208. [CrossRef]

32. Kassebaum, J.L.; Chelliah, H.K. Oxidation of isolated porous carbon particles: Comprehensive numerical model. *Combust. Theory Model.* **2009**, *13*, 143–166. [CrossRef]

33. Comtois, P. John Tyndall and the floating matter of the air. *Aerobiologia* **2001**, *17*, 193–202. [CrossRef]

34. Davies, C.N. *Aerosol Science*; London Academic Press: London, UK, 1966.

35. Braun, D.; Libchaber, A. Trapping of DNA by thermophoretic depletion and convection. *Phys. Rev. Lett.* **2002**, *89*, 188103. [CrossRef] [PubMed]

36. Talbot, L.R.; Cheng, R.K.; Schefer, R.W.; Willis, D.R. Thermophoresis of particles in a heated boundary layer. *J. Fluid Mech.* **1980**, *101*, 737–758. [CrossRef]

37. Bakanov, S.P. The thermophoresis of solids in gases. *J. Appl. Math. Mech.* **2005**, *5*, 767–772. [CrossRef]

38. Rahbari, A.; Wong, K.F.; Vakilabadi, M.A.; Poorfar, A.K.; Afzalabadi, A. Theoretical investigation of particle behavior on flame propagation in lycopodium dust cloud. *J. Energy Resour. Technol.* **2017**, *139*, 012202. [CrossRef]

39. Bidabadi, M.; Natanzi, A.H.; Mostafavi, S.A. Thermophoresis effect on volatile particle concentration in micro-organic dust flame. *Powder Technol.* **2012**, *217*, 69–76. [CrossRef]

40. Haghiri, A.; Bidabadi, M. Modeling of laminar flame propagation through organic dust cloud with thermal radiation effect. *Int. J. Therm. Sci.* **2010**, *49*, 1446–1456. [CrossRef]

41. Seshadri, K.; Trevino, C. The influence of the Lewis numbers of the reactants on the asymptotic structure of counterflow and stagnant diffusion flames. *Combust. Sci. Technol.* **1989**, *64*, 243–261. [CrossRef]

42. Wang, H.Y.; Chen, W.H.; Law, C.K. Extinction of counterflow diffusion flames with radiative heat loss and nonunity Lewis numbers. *Combust. Flame* **2007**, *148*, 100–116. [CrossRef]

43. Bidabadi, M.; Panahifar, P.; Sadeghi, S. Analytical development of a model for counter-flow non-premixed flames with volatile biofuel particles considering drying and vaporization zones with finite thicknesses. *Fuel* **2018**, *231*, 172–186. [CrossRef]

44. Linan, A. The asymptotic structure of counterflow diffusion flames for large activation energies. *Acta Astronaut.* **1974**, *1*, 1007–1039. [CrossRef]

45. Fendell, F.E. Ignition and extinction in combustion of initially unmixed reactants. *J. Fluid Mech.* **1965**, *21*, 281–303. [CrossRef]

46. Huang, Y.; Risha, G.A.; Yang, V.; Yetter, R.A. Effect of particle size on combustion of aluminum particle dust in air. *Combust. Flame* **2009**, *156*, 5–13. [CrossRef]

47. Wichman, I.S.; Yang, M. Double-spray counterflow diffusion flame model. *Strain* **1998**, *2*, 373–398. [CrossRef]

48. Rockwell, S.R.; Rangwala, A.S. Modeling of dust air flames. *Fire Saf. J.* **2013**, *59*, 22–29. [CrossRef]

energies

MDPI

Article

Simultaneous Extraction and Emulsification of Food Waste Liquefaction Bio-Oil

David Längauer [1,2], Yu-Ying Lin [2], Wei-Hsin Chen [2,3,*], Chao-Wen Wang [2], Michal Šafář [1,2] and Vladimír Čablík [1]

[1] Institute of Environmental Engineering, Faculty of Mining and Geology, VŠB—Technical University of Ostrava, 17. listopadu 15, 708 33 Ostrava-Poruba, Czech Republic; david.langauer@vsb.cz (D.L.); michal.safar@vsb.cz (M.Š.); vladimir.cablik@vsb.cz (V.Č.)

[2] Department of Aeronautics and Astronautics, National Cheng Kung University, Tainan 701, Taiwan; z935309@gmail.com (Y.-Y.L.); kevin19940706@gmail.com (C.-W.W.)

[3] Research Center for Energy Technology and Strategy, National Cheng Kung University, Tainan 701, Taiwan

* Correspondence: weihsinchen@gmail.com or chenwh@mail.ncku.edu.tw; Tel.: +886-6-2004456

Received: 6 October 2018; Accepted: 31 October 2018; Published: 5 November 2018

Abstract: Biomass-derived bio-oil is a sustainable and renewable energy resource, and liquefaction is a potential conversion way to produce bio-oil. Emulsification is a physical upgrading technology, which blends immiscible liquids into a homogeneous emulsion through the addition of an emulsifier. Liquefaction bio-oil from food waste is characterized by its high pour point when compared to diesel fuel. In order to partially replace diesel fuel by liquefaction bio-oil, this study aimed to develop a method to simultaneously extract and emulsify the bio-oil using a commercial surfactant (Atlox 4914, CRODA, Snaith, UK). The solubility and stability of the emulsions at various operating conditions such as the bio-oil-to-emulsifier ratio (B/E ratio), storage temperature and duration, and co-surfactant (methanol) addition were analyzed. The results demonstrate that higher amounts of bio-oil (7 g) and emulsifier (7 g) at a B/E ratio = 1 in an emulsion have a higher solubility (66.48 wt %). When the B/E ratio was decreased from 1 to 0.556, the bio-oil solubility was enhanced by 45.79%, even though the storage duration was up to 7 days. Compared to the emulsion stored at room temperature (25 °C), its storage at 100 °C presented a higher solubility, especially at higher B/E ratios. Moreover, when methanol was added as a co-surfactant during emulsification at higher B/E ratios (0.714 to 1), it rendered better solubility (58.83–70.96 wt %). Overall, the emulsified oil showed greater stability after the extraction-emulsification process.

Keywords: emulsification; liquefaction; bio-oils; co-surfactant; surfactant; diesel

1. Introduction

In the 21st century, food wastes in excess of one billion metric tons are generated worldwide every year from various sources and such great amounts lead to a severe environmental problem. Failure to recycle or treat food wastes properly will lead to the emission of methane (CH_4) and carbon dioxide (CO_2) during bio-degradation. These greenhouse gases may cause human health and environmental problems [1].

In recent years, there has been an improvement in thermochemical conversion equipment, which can transform wastes into biofuels or bioenergy products [2,3]. Food waste is considered a valuable source of energy due to its high biodegradability and high organic matter content [4,5]. Nevertheless, there is a great challenge in the conversion of food wastes into biofuels by traditional thermochemical conversions such as pyrolysis and gasification. The high moisture content of food waste requires more time and energy for drying before the conversion proceeds.

Energies **2018**, *11*, 3031; doi:10.3390/en11113031   398   www.mdpi.com/journal/energies

Hydrothermal technology (HTL) provides opportunities for more energy efficient valorization of wet biomass in that water is used directly as a reaction medium, thus, bypassing the drying process [6,7]. After HTL is performed, a separation procedure is needed to obtain the oil phase, also known as the liquefaction bio-oil. The separation method includes using a gravimetric separation funnel or extraction with an organic solvent such as tetrahydrofuran (THF), ethyl acetate, and dichloromethane (DCM) [8,9]. The liquefaction bio-oil is a mixture of several hundred organic compounds such as acids, alcohols, aldehydes, amides, esters, ketones, phenols, and lignin oligomers [10,11]. The liquefaction bio-oil has undesirable features for industrial applications such as a high viscosity and high ash content. This implies that the bio-oil composition is complex, making it relatively difficult to control the performance during the combustion process.

Emulsification is an effective way for upgrading bio-oil and blending it with other fuels [12,13]. Emulsification technology is popular for upgrading pyrolysis oil. Since diesel and bio-oil have polar molecules, the emulsifier or surfactants can disperse them both to form a stable and homogeneous emulsion [12]. The emulsification of bio-oils or water in diesel using various emulsifiers is summarized in Table 1. Michio [14] studied the emulsion of diesel oil and pyrolysis oil from hardwood and investigated the effects of five operation parameters (bio-oil concentration, surfactant concentration, residence time, motor speed and emulsification temperature). The results were optimized by a statistical model and the dominant variables were identified for minimal stratification formation. These dominant parameters were bio-oil concentration, and surfactant and power input. Chiaramonti [15] studied diesel/bio-oil emulsions with four kinds of pyrolysis oils from Canadian oak, beech wood, California pine, and pine wood. The results suggested that it is necessary to use fresh bio-oil and micronize these products to obtain a stable emulsion. Guo [16] used a KDL-5 molecular distillation apparatus to separate the middle fraction and the heavy fraction of the crude bio-oil. The blending of the diesel and bio-oil was conducted by ultrasonic and ultrasonic-mechanical emulsification. The droplet size distribution was measured using a Malvern nanometer particle size analyzer for the three emulsions (heavy fraction with 40–300 nm, middle fraction with 15–60 nm and crude bio-oil with 8–25 nm) and had the same tendency with the stability time 14 min, 216 min and 31 days, respectively. As for the emulsion on the liquefaction bio-oil, Lijian. [17] employed micro-emulsion technologies to dissolve three kinds of sewage sludge liquefaction bio-oil extracted by different solvents into diesel. The experiments showed that the bio-oils obtained at a higher liquefaction temperature (360 °C) benefited the micro-emulsion.

Table 1. Literature of bio-oil and liquefaction oil in diesel emulsions.

Pyrolysis Bio-Oil Content	Mixing Method	Emulsifier and Co-Surfactant	Stability	Ref.
10–30 wt %	Micro-emulsifier (800–1750 rpm and 5–20 min)	Hypermer B246SF/2234 (Croda International)	3 to 42 days	[14]
25–75 wt %	Variable speed electrical motor (1100 W)	Polymeric surfactants/short chain additives (n-octanol)	3 days	[15]
10 wt %	Ultrasonic (20 kHz; 2 min) and Ultrasonic-mechanical (20 kHz; 2 min; and 5000 rpm; 5 min)	Span80 (sorbitan monooleate)/Tween 80 (Polyoxyethylenesorbitan monolaurate)/Span 85 (sorbitan trioleate)	27 to 31 days	[16]
2–40 wt %	Homogenizer (8800–16,700 rpm)	Span 80 (sorbitan monooleate)/Tween 80 (Polyoxyethylenesorbitan monolaurate)	5 days	[18]
Liquefaction Bio-Oil Content	**Mixing Method**	**Emulsifier or Co-Surfactant**	**Stability**	**Ref.**
5 wt %	Manual shaking (1 min) and centrifugation (6000 rpm; 20 min)	Span 80 (sorbitan monooleate)	-	[17]
5 to 30 wt %	Vortex mixer (3000 rpm; 2 min)	Atlox4914 (Croda International)/Methanol	7 days	This study

Despite some useful information provided from these studies, it should be underlined that there is still a lack of emulsification information for the liquefaction of bio-oil from food waste. Moreover, the mechanisms of extraction and emulsification are related to the surface tension, viscosity, and polarity between the molecules [15–17]. Thus, this study aimed to investigate the feasibility of the simultaneous emulsification-extraction of the food waste liquefaction bio-oil and diesel with the addition of an emulsifier. This process can combine liquefaction bio-oil extraction and emulsification in one step, which can reduce the experiment cost and time. The details of the proportion of the liquefaction bio-oil in the emulsion after mixing with diesel are also addressed.

2. Materials and Methods

2.1. Materials

The raw liquefaction bio-oil was produced from the hydrothermal liquefaction process of food waste, which was collected from the local food waste recycling field. It included rhizomes and peels of several kinds of vegetables and fruits, which were obtained from livelihood waste. Before the start of the liquefaction process, the size of the food waste (800 g) was reduced using a blender to intensify the reaction surface area. The food waste had a high moisture content (74%) for the liquefaction reaction, thus, dismissing additional water. Then, the food waste slurry was pretreated with a 5 wt % potassium carbonate (K_2CO_3, alkali catalyst) by heating the mixture to 100 °C for 1 h. Subsequently, the temperature was raised to 320 °C and held for 30 min as the conditions of liquefaction. The process yielded products at three different phases: gas, liquid, and solid. The water vapor in the reactor was removed by cooling to 105 °C under atmospheric pressure. The leftover liquid/solid mixture was heated to 100 °C for 4 h in the oven to remove the excess water content in the mixture. Finally, the remaining residue was considered as the raw liquefaction bio-oil having an average conversion rate of 40 wt %.

The diesel fuel used for experiments was commercially available and sourced from petrol station of China Petroleum Corporation (CPC). In the emulsification-extraction process, the liquefaction oil was dispersed into the diesel, leading to the bulk of the emulsion volume. The emulsifier used in the experiments was Atlox 4914 obtained from Croda International Plc. Methanol (>99 vol %) was used as the co-surfactant. The properties of the bio-oil, diesel, Atlox 4914, and co-surfactant such as density, viscosity, and higher heating value (HHV) are tabulated in Table 2.

Table 2. Properties of the bio-oil, diesel, surfactant, and co-surfactant.

Compound	Density (kg m^{-3})	Viscosity (mPa s)	HHV (MJ kg^{-1})
Bio-oil	1100.0	4916	31.41
Diesel	832.0	1.6–5.8	44.80
Atlox 4914	970.0	-	37.64
Methanol	791.0	0.533	20.40

2.2. The Procedure for Preparing the Sample

To determine the minimum volume required for the emulsifier, a fixed quantity of 20 g of diesel was mixed with various amounts of the emulsifier (based on B/E ratio) in a 50 mL centrifuge tube. A vortex mixer (Model: VM-2000, Digisystem Laboratory Instruments Inc., Taipei, Taiwan) was used to mix the diesel and emulsifier for 1 min at a speed of 3000 rpm. The liquefaction bio-oil was in the form of the solid phase at room temperature, so it was preheated to 100 °C in the oven to transform it into a liquid before the emulsification-extraction process. Thereafter, the liquid raw liquefaction bio-oil was added into the diesel-emulsifier mixture and mixed by the vortex mixer for 2 min at a speed of 3000 rpm. Then, the emulsion samples were stored at room temperature (25 °C). Samples, which needed higher-temperature condition, were stored in an oven at 100 °C. The co-surfactant could reduce the surface tension of bio-oil and enhance the performance of the emulsions [13]. Therefore, further emulsion experiments examined the effect of the adjunction of the co-surfactant by adding one more step in the process of sample preparation. After the emulsions of diesel, emulsifier, and bio-oil were prepared, methanol was added and mixed for 1 min at a speed of 3000 rpm. The detailed process is shown in Figure 1.

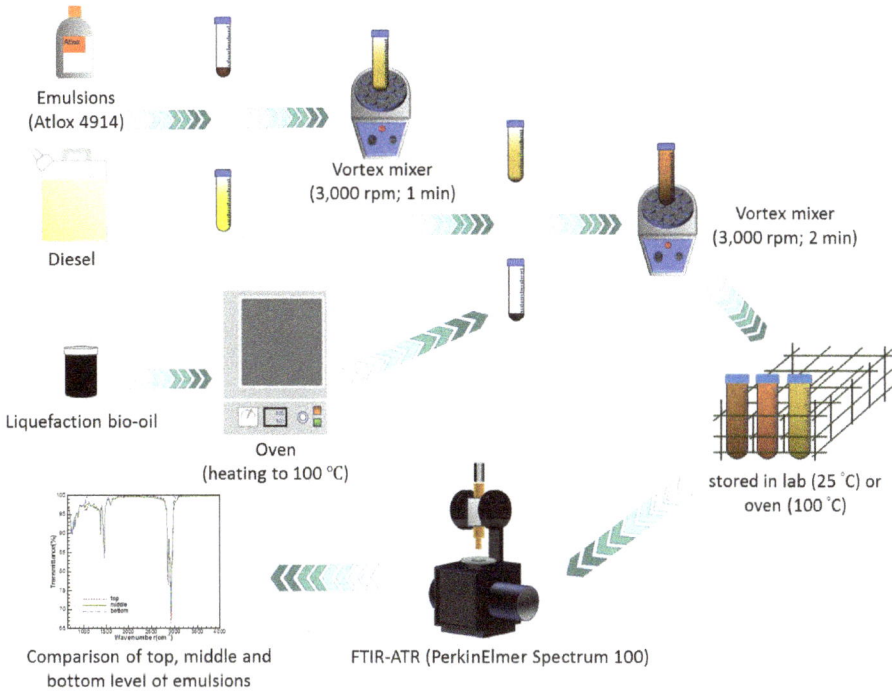

Figure 1. Scheme of emulsion preparation and stability analysis.

2.3. Analytical Methods

There are two directions to investigate the performance of the extraction-emulsification experiment: quantitative and qualitative methods. The quantitative research is to identify the amount of bio-oil dissolved in diesel. The deposited bio-oil is in the solid phase, so it is easy to separate it from the liquid emulsion and weigh the amount. In order to measure the extent of diesel-solvable bio-oil, a parameter called solubility was defined as

$$\text{Solubility } (\%) = \frac{m_{added\ bio-oil} - m_{insolvable\ bio-oil}}{m_{added\ bio-oil}} \times 100\% \tag{1}$$

where $m_{added\ bio-oil}$ stands for the added mass of bio-oil and $m_{insolvable\ bio-oil}$ denotes the mass of deposited bio-oil in the bottom of the tube.

Qualitative research is to evaluate the emulsion stability. Several methods can be used such as measuring the pH value, moisture content, the viscosity of the emulsion, or droplet size distribution in the emulsion [14,15]. However, there is no standard procedure for examining the long-term stability [15]. In this study, Fourier transform infrared spectroscopy (FTIR) coupled with attenuated total reflection (ATR) was used to examine the stability and homogeneity of the emulsions through the identification of the functional groups present in the emulsions [12,13,19]. In some cases, even though the emulsion seemed to be uniform by visual inspection, the FTIR spectra were different, revealing the phase separation in the emulsion. For a homogeneous emulsion, the FTIR spectra at different places would overlap. Therefore, the emulsion was divided into three regions, namely, the top, middle, and bottom liquid levels in the sample tube and the three regions were analyzed using FTIR [20].

3. Results and Discussion

3.1. FTIR Spectra of Atlox, Bio-Oil, Diesel, and Methanol

FTIR analysis of the input components (Atlox, food waste liquefaction bio-oil, diesel, and methanol) was performed and the spectra are shown in Figure 2. Diesel, methanol, and Atlox 4914 are commercially available substances with distinct chemical compositions. To identify the stratification of the bio-oil in the emulsions, the chemical composition of the liquefaction bio-oil was studied. From the obtained spectra and functional groups summarized in Table 3, the liquefaction bio-oil contained C-H-(CH$_2$) (732 cm^{-1}), C-O (1034 cm^{-1}), C-O (1258 cm^{-1}), C-H-(CH$_3$) (1414 cm^{-1}), CH$_3$ (2918 cm^{-1}), and polymeric O-H (3320 cm^{-1}) functional groups.

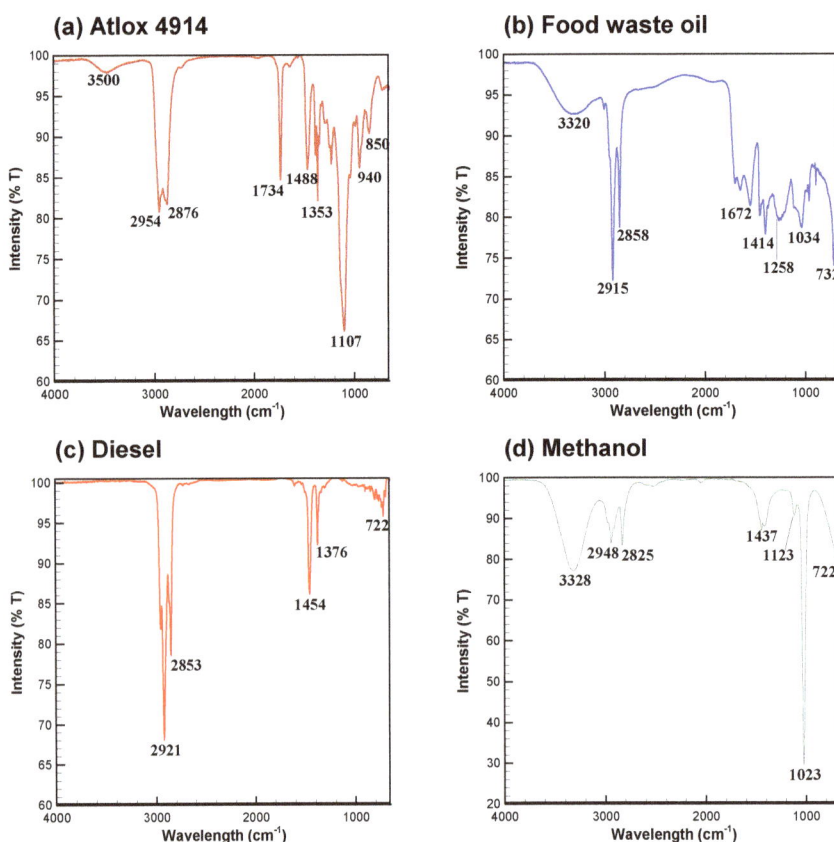

Figure 2. FTIR spectra of (**a**) Atlox, (**b**) bio-oil, (**c**) diesel, and (**d**) methanol.

Table 3. Main functional groups of emulsifiers, co-surfactants and liquefaction bio-oil.

Wave Number Range (cm^{-1})	Functional Group	Mode of Vibration	Reference
3550–3450	Dimetric O-H	Stretching	[21]
3400–3230	Polymetric O-H	Stretching	[21]
2949–2915	Aliphatic C-H	Stretching	[21,22]
2924–2858	Aliphatic C-H	Stretching	[23]
2835–2815	CH_3-O-CH_3 (ethers)	Stretching	[21]
1740–1732	C=O (ester)	Stretching	[23]
1651–1640	C=C (aromatic)	Skeletal	[24]
1652–1579	C-C (ring)	Stretching	[21]
1486–1446	C-H- (CH_3)/(CH_2)	Symmetric Asymmetric	[24]
1461–1377	C-H- (CH_3)	Symmetric	[24]
1290–1211	C-O	Stretching	[21]
1230–1140	C-O (phenol)	Stretching	[21]
1239–1100	C-O-C (ester)	Stretching	[22,23]
1090–1049	C-O	Stretching	[21]
1043–1006	C-O	Stretching	[24]
970–960	C=C-H	Bending	[21]
730–710	C-H-(CH_2)	Bending	[21]

3.2. Bio-Oil Solubility and Emulsion Stability at B/E Ratio = 1

By mixing various volumetric ratios of bio-oil/emulsifier and observing the emulsion's stability, Figure 3 examines the solubility of the bio-oil in the emulsions over time. In all cases, the B/E ratio was 1 but there were different amounts of bio-oil and emulsifier in the various samples in the emulsions. In all instances, there was an increase in the amount of bio-oil in the sediment over time. The solubility in all cases at 1 h duration was as high as at 70 wt %, except for the case of 1 g of bio-oil with 55.70 wt %. The separation of bio-oil after 1 day was significant in the case of 1 g bio-oil and only 16.61 wt % of it remained dissolved into the emulsion. With an increase in storage time, the sediment of the cases involving 1, 2, and 3 g of bio-oil became pronounced. However, the solubility of the cases involving 4, 5, and 6 g of bio-oil with B/E = 1 remained above 55 wt % on the 7th day, implying that the solubility of the emulsion improved greatly when the amounts of surfactant and bio-oil were increased. On the 7th day, the maximum solubility in the case of 6 g bio-oil with B/E = 1 was 66.48 wt %.

The FTIR spectra at three regions (i.e., top, middle, and bottom liquid levels in the tube) in the emulsion under 2 g bio-oil with B/E = 1 for 1 h to 7 days are displayed in Figure 4. The FTIR spectra appear to almost overlap and coincide with each other, confirming homogeneity and successful emulsification [12].

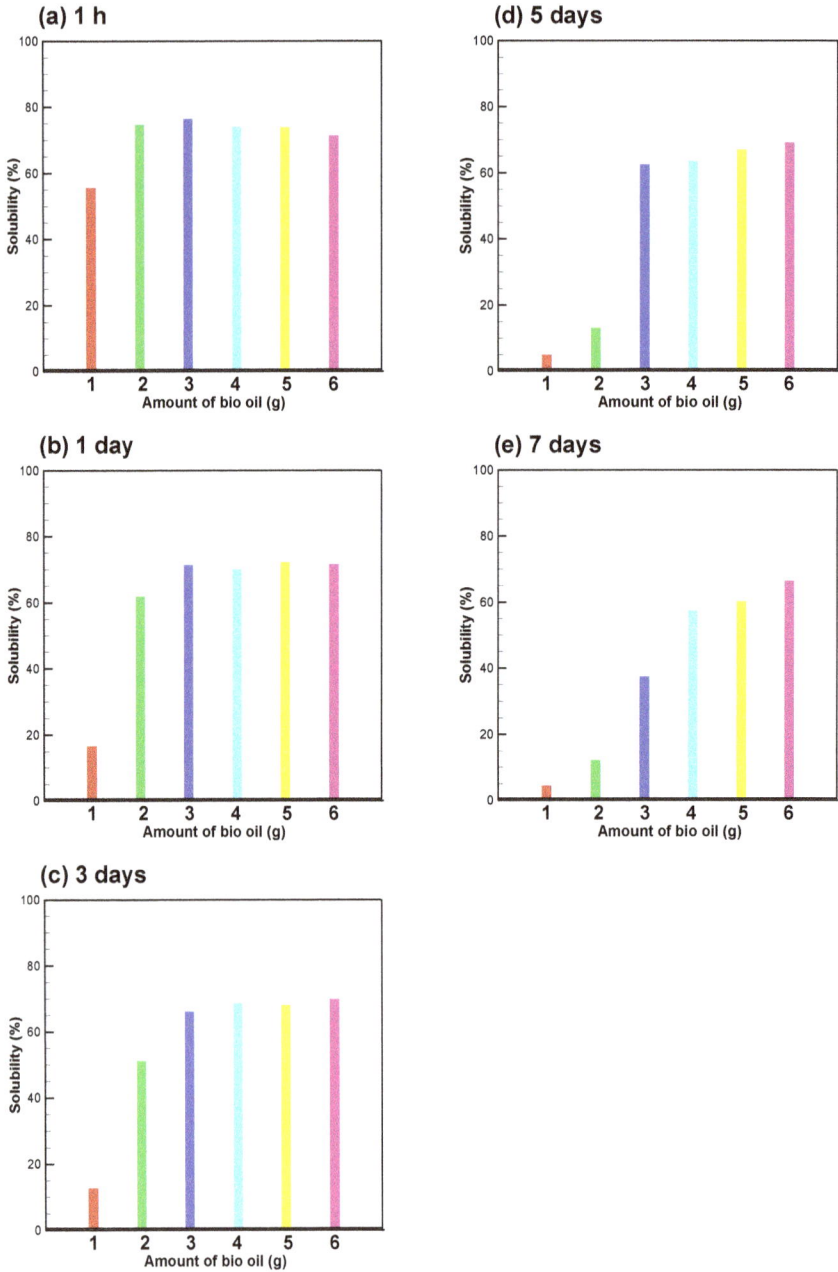

Figure 3. The solubility of the emulsions with different bio-oil amounts (B/E = 1) (**a**) 1 h, (**b**) 1 day, (**c**) 3 days, (**d**) 5 days, and (**e**) 7 days.

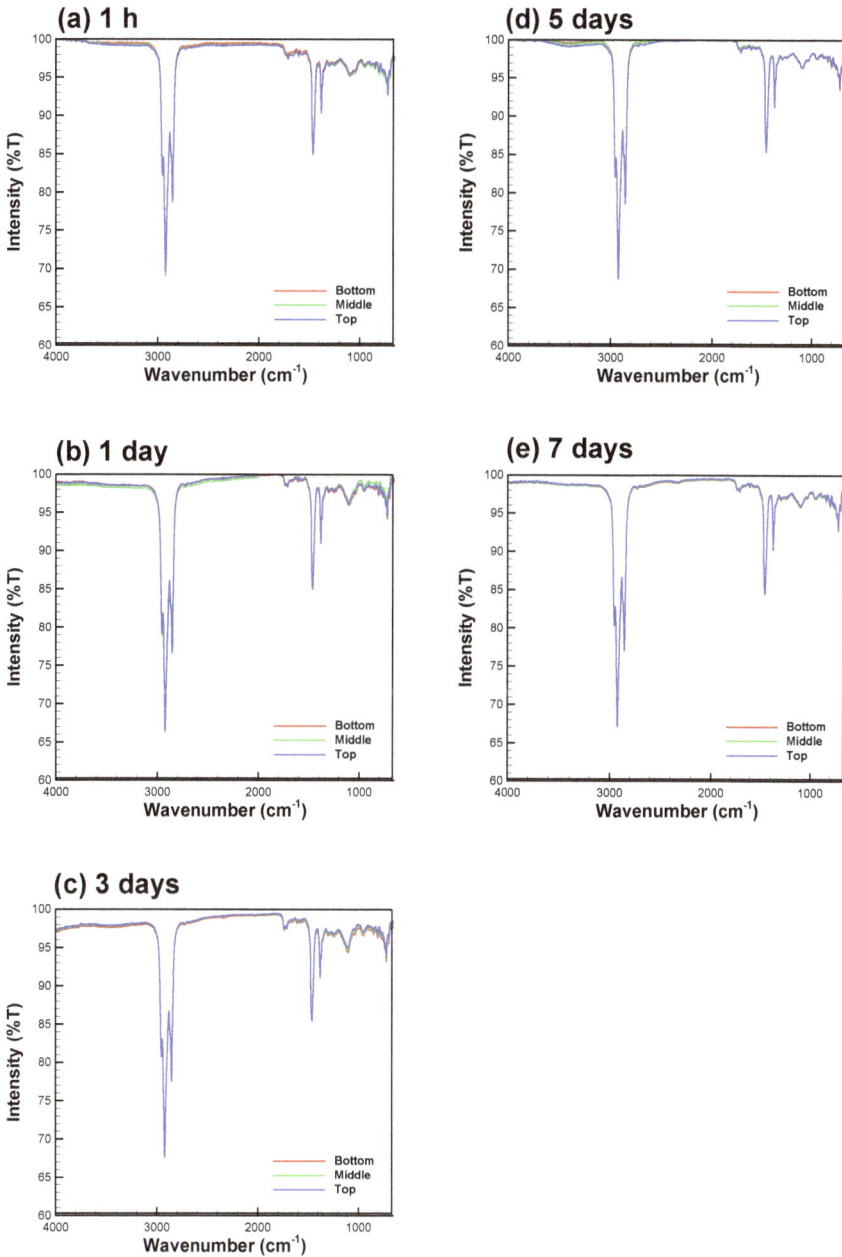

Figure 4. Fourier transform infrared spectroscopy (FTIR) spectra of emulsion (2 g bio-oil, B/E = 1) at three regions for (**a**) 1 h, (**b**) 1 day, (**c**) 3 days, (**d**) 5 days, and (**e**) 7 days.

3.3. Bio-Oil Solubility at B/E Ratios between 0.556 and 1

The separation of the emulsion in the cases of a lower amount of bio-oil with B/E = 1 was more pronounced compared to those with a higher amount of bio-oil. To study the solubility of the bio-oil, the solubility at the B/E ratios of 0.556–1 was investigated. The surfactant's concentration was one of the most important factors influencing the solubility of emulsions. The bio-oil and diesel contents were fixed at 2 g and 20 g, respectively. An increase in the amount of the emulsifier or a decrease in the B/E ratio caused more bio-oil to dissolve into the emulsion (Figure 5). The solubility at B/E = 1 on the 7th day was only 12.02 wt %. When the B/E ratio decreased to 0.556, the solubility increased to 57.81 wt %, implying that a higher emulsifier concentration rendered a higher solubility of the bio-oil into the emulsion.

The FTIR spectra at three regions (i.e., top, middle, and bottom liquid levels in the tube) in the emulsion with 2 g bio-oil and B/E = 1 for 7 days are displayed in Figure 4e, showing the stability of the emulsion. The FTIR spectra almost overlap and coincide with each other, confirming homogeneity and higher stability of the emulsion.

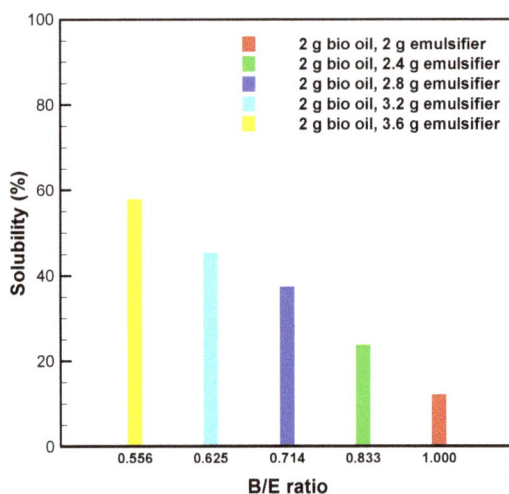

Figure 5. The solubility of the emulsions with different B/E ratios (2 g bio-oil) after 7 days.

3.4. Temperature Effect

The fluidity of the bio-oil was low at room temperature. To increase the fluidity of the emulsion during storage, the emulsion was heated at 100 °C in the oven for 7 days. In general, both the viscosity and interfacial tension decreased with increasing temperature, making the emulsification easier [15]. This resulted in an increased degree of solubility at 100 °C compared to that at room temperature. Compared to the storage of the emulsion at room temperature, the solubility in the case of 2 g bio-oil with B/E = 1 at 100 °C promoted the solubility to 59.86 wt % from 12.02 wt % at room temperature (Figure 6). This clearly indicates that the effect of the bio-oil temperature at B/E = 1 played an important role in improving its solubility. However, the impact of the storage temperature in the case of B/E = 0.556 was not of the same magnitude; the solubility being merely intensified from 57.81 wt % at room temperature to 63.45 wt % at 100 °C, accounting for an increase of 6.64 wt %. Overall, the temperature effect on the solubility of bio-oil increased when the B/E ratio increased.

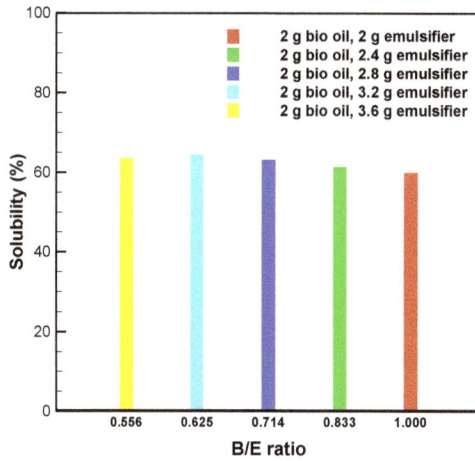

Figure 6. The solubility of bio-oil at different B/E ratios (2 g of bio-oil, fixed amount) at 100 °C for 7 days.

The FTIR spectra at three regions (i.e., top, middle, and bottom liquid levels in the tube) in the emulsion at the storage temperature of 100 °C with 2 g bio-oil and B/E = 1 for 7 days are displayed in Figure 7. The FTIR spectra also almost overlap and coincide with each other, showing the homogeneity and high stability of the emulsion again.

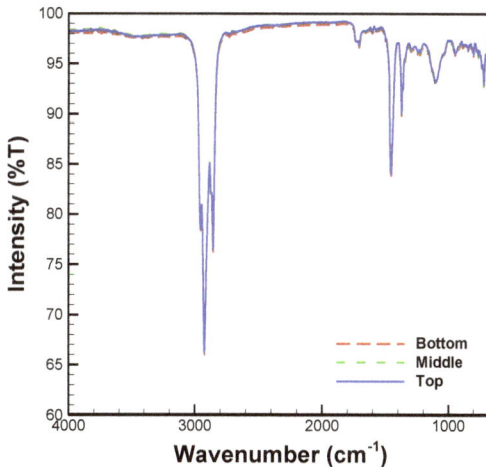

Figure 7. FTIR spectra of the emulsion with 2 g bio-oil and B/E = 1 under 100 °C at 7 days.

3.5. Co-Surfactant (Methanol) Effect

In general, the amount of the emulsifier must be increased to improve the bio-oil usage in the emulsion. The co-surfactants are surface-active compounds that can lower the surface tension of bio-oil or the dispersed phase, making them suitable for emulsification [25]. In this study, methanol was used as the co-surfactant, resulting from its influence on increasing the emulsion's stability [13]. In each experimental run, 1 g of methanol was loaded into the emulsions. The bio-oil solubility in the case of 2 g bio-oil with B/E = 1 at a duration of 7 days increased to 46.66 wt % (Figure 8), as opposed to 12.02 wt % without the addition of methanol. This evidence suggests that methanol can significantly facilitate the emulsification performance. The impact of the co-surfactant in the case of B/E = 0.556 was not pronounced. The solubility with the co-surfactant (70.96 wt %) was higher than that without methanol (57.81 wt %) by 13.15 wt %. The co-surfactants effect on the solubility of the bio-oil increased when the B/E ratio increased.

The FTIR spectra at the three regions (i.e., top, middle, and bottom liquid levels in the tube) in the emulsion after adding 1 g methanol with 2 g bio-oil and B/E = 1 for 7 days are displayed in Figure 9. The FTIR spectra also show the high homogeneity and stability of the emulsion. By comparing the influences of the two factors (storage temperature and co-surfactant) on the solubility of the bio-oil from B/E = 0.556 to 1 (Figures 6 and 8), when the B/E ratio was between 0.714 and 1, the impact of the storage temperature on the solubility was beyond the co-surfactant. On the other hand, when the B/E ratio ranged from 0.556 to 0.625, a converse behavior was observed.

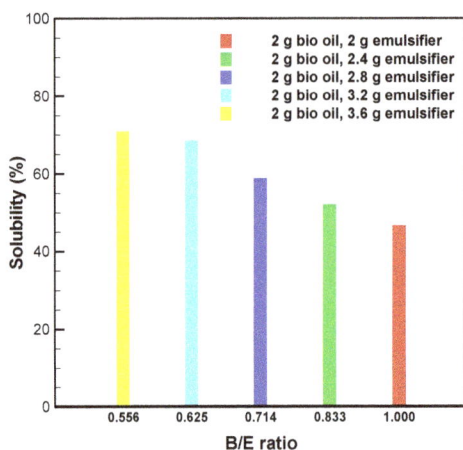

Figure 8. The solubility of bio-oil at different B/E ratios with 1 g methanol.

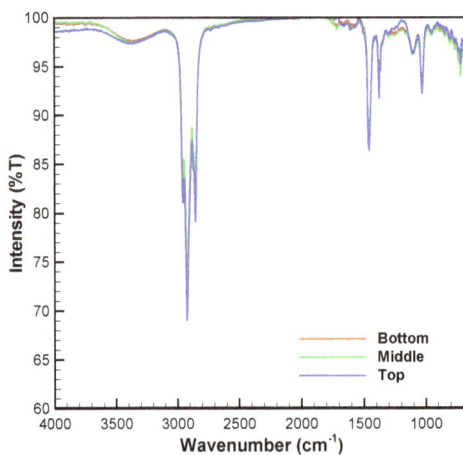

Figure 9. FTIR spectra of the emulsion with 2 g bio-oil, B/E = 1, and 1 g methanol for 7 days.

4. Conclusions

The simultaneous emulsification-extraction process of liquefaction bio-oil and diesel with the aid of a commercial emulsifier (Atlox 4914) and co-surfactant (methanol) have been explored where 20 g of diesel was used. When the bio-oil amount increased from 1 g to 6 g with the fixed B/E ratio of 1, its solubility increased from 55.70 wt % to 71.35 wt %. However, the solubility with 1 g bio-oil (B/E = 1) decreased from 55.70 wt % to 4.41 wt % after the 7 days storage and decreased from 71.35 wt % to 66.48 wt % with 6 g bio-oil (B/E = 1). This suggests that performing emulsification at a larger scale may facilitate the solubility. When a higher amount of emulsifier was used with the B/E ratio decreasing from 1 to 0.556, the bio-oil solubility also increased from 12.02 wt % to 57.81 wt % after a 7-day duration. To reduce the viscosity of the emulsions, the emulsions with B/E ratios ranging from 0.556 to 1 were stored in an oven at 100 °C for 7 days. The storage temperature of 100 °C had more impact on the solubility at higher B/E ratios, especially for the case of B/E = 1, which enhanced the solubility by a factor of 47.84% as compared to room temperature (25 °C). Moreover, the addition of a co-surfactant (methanol) at B/E = 1 at room temperature intensified the bio-oil solubility from 12.02 wt % (without addition) to 46.66 wt %. When the liquefaction bio-oil was extracted and emulsified in the emulsions, their higher stability after 7-day storage was achieved from the FTIR analysis. The results of the study indicate that hydrothermal liquefaction can be applied to waste-to-fuel technology and the liquefaction bio-oil from food waste has a high potential to replace the diesel fuel used in industrial furnaces and boilers.

Author Contributions: D.L, M.Š. and V.Č. did experiments, analyzed data and wrote the paper. Y.-Y.L. and C.-W.W. helped with the experiments and writing the paper. W.-H.C. provided this research idea and organize this study, developed the experiment, analyzed data, and provided facilities and instruments for research.

Funding: The authors gratefully acknowledge the financial support (MOST 106-2923-E-006-002-MY3) of the Ministry of Science and Technology, Taiwan, R.O.C., and the financial support (TUCENOT LO1404) of the research center for low-carbon energy technologies.

Conflicts of Interest: The authors declare no conflicts of interest

References

1. Kavitha, S.; Banu, J.R.; Priya, A.A.; Uan, D.K.; Yeom, I.T. Liquefaction of food waste and its impacts on anaerobic biodegradability, energy ratio and economic feasibility. *Appl. Energy* **2017**, *208*, 228–238. [CrossRef]
2. Bach, Q.V.; Chen, W. A comprehensive study on pyrolysis kinetics of microalgal biomass. *Energy Convers. Manag.* **2017**, *131*, 109–116. [CrossRef]
3. Chen, W.; Chen, C.J.; Hung, C.I.; Shen, C.H.; Hsu, H.W. A comparison of gasification phenomena among raw biomass, torrefied biomass and coal in an entrained-flow reactor. *Appl. Energy* **2013**, *112*, 421–430. [CrossRef]
4. Karthikeyan, O.P.; Wong, S.M.J.W.C. Bio-refining of food waste for fuel and value products. In Proceedings of the 4th International conference on Energy and Environment Research (ICEER 2017), Porto, Portugal, 17–20 July 2017; pp. 14–21.
5. Lin, C.K.; Pfaltzgraff, L.A.; Mubofu, E.B.; Abderrahim, S. Food waste as a valuable resource for the production of chemicals, materials and fuels. Current situation and global perspective. *Energy Environ. Sci.* **2013**, *6*, 426–464. [CrossRef]
6. Chen, W.H.; Lin, B.J.; Huang, M.Y.; Chang, J.S. Thermochemical conversion of microalgal biomass into biofuels: A review. *Bioresour. Technol.* **2015**, *184*, 314–327. [CrossRef] [PubMed]
7. Dimitriadis, A.; Bezergianni, S. Hydrothermal liquefaction of various biomass and waste feedstocks for biocrude production: A state of the art review. *Renew. Sustain. Energy Rev.* **2017**, *68*, 113–125. [CrossRef]
8. Chen, Z.; Long, J. Organosolv liquefaction of sugarcane bagasse catalyzed by acidic ionic liquids. *Bioresour. Technol.* **2016**, *214*, 16–23. [CrossRef] [PubMed]
9. Saber, M.; Golzary, A.; Hosseinpour, M.; Takahashi, F.; Yoshikawa, K. Catalytic hydrothermal liquefaction of microalgae using nanocatalyst. *Appl. Energy* **2016**, *183*, 566–576. [CrossRef]
10. Gollakota, A.R.K.; Kishore, N.; Gu, S. A review on hydrothermal liquefaction of biomass. *Renew. Sustain. Energy Rev.* **2018**, *81*, 1378–1392. [CrossRef]

11. Jena, U.; Das, K.C.; Kastner, J.R. Effect of operating conditions of thermochemical liquefaction on biocrude production from Spirulina platensis. *Bioresour. Technol.* **2011**, *102*, 6221–6229. [CrossRef] [PubMed]

12. Lin, B.-J.; Chen, W.-H.; Budzianowski, W.M.; Hsieh, C.-T.; Lin, P.-H. Emulsification analysis of bio-oil and diesel under various combinations of emulsifiers. *Appl. Energy* **2016**, *178*, 746–757. [CrossRef]

13. De Luna, M.D.G.; Cruz, L.A.D.; Chen, W.-H.; Lin, B.-J.; Hsieh, T.-H. Improving the stability of diesel emulsions with high pyrolysis bio-oil content by alcohol co-surfactants and high shear mixing strategies. *Energy* **2017**, *141*, 1416–1428. [CrossRef]

14. Ikura, M.; Stanciulescu, M.; Hogan, E. Emulsification of pyrolysis derived bio-oil in diesel fuel. *Biomass Bioenergy* **2003**, *24*, 221–232. [CrossRef]

15. Chiaramonti, D.; Bonini, M.; Fratini, E.; Tondi, G.; Gartner, K.; Bridgwater, A.V.; Grimm, H.P.; Soldaini, I.; Webster, A.; Baglioni, P. Development of emulsions from biomass pyrolysis liquid and diesel and their use in engines—Part 1: Emulsion production. *Biomass Bioenergy* **2003**, *25*, 85–99. [CrossRef]

16. Guo, Z.; Wang, S.; Wang, X. Stability mechanism investigation of emulsion fuels from biomass pyrolysis oil and diesel. *Energy* **2014**, *66*, 250–255. [CrossRef]

17. Leng, L.; Yuan, X.; Chen, X.; Huang, H.; Wang, H.; Li, H.; Zhu, R.; Li, S.; Zeng, G. Characterization of liquefaction bio-oil from sewage sludge and its solubilization in diesel microemulsion. *Energy* **2015**, *82*, 218–228. [CrossRef]

18. Houghton, R.A.; Hall, F.; Goetz, S.J. Importance of biomass in the global carbon cycle. *J. Geophys. Res. Biogeosci.* **2009**, *114*, G00E03. [CrossRef]

19. Jiang, X.; Ellis, N. Upgrading Bio-oil through Emulsification with Biodiesel: Mixture Production. *Energy Fuels* **2010**, *24*, 1358–1364. [CrossRef]

20. Qian, Y.; Zuo, C.; Tan, J.; He, J. Structural analysis of bio-oils from sub-and supercritical water liquefaction of woody biomass. *Energy* **2007**, *32*, 196–202. [CrossRef]

21. Mistry, B.D. *A Handbook of Spectroscopic Data Chemistry*; Oxford Book Company: Oxford, UK, 2009.

22. Assanvo, E.F.; Gogoi, P.; Dolui, S.K.; Baruah, S.D. Synthesis, characterization, and performance characteristics of alkyd resins based on Ricinodendron heudelotii oil and their blending with epoxy resins. *Ind. Crops Prod.* **2015**, *65* (Suppl. C), 293–302. [CrossRef]

23. Bora, M.M.; Gogoi, P.; Deka, D.C.; Kakati, D.K. Synthesis and characterization of yellow oleander (Thevetia peruviana) seed oil-based alkyd resin. *Ind. Crops Prod.* **2014**, *52* (Suppl. C), 721–728. [CrossRef]

24. Benavente, V.; Fullana, A. Torrefaction of olive mill waste. *Biomass Bioenergy* **2015**, *73*, 186–194. [CrossRef]

25. 1Sun, M.; Wang, Y.; Firoozabadi, A. Effectiveness of Alcohol Cosurfactants in Hydrate Antiagglomeration. *Energy Fuels* **2012**, *26*, 5626–5632.

![energies logo] *energies*

MDPI

Article

Variation in the Distribution of Hydrogen Producers from the Clostridiales Order in Biogas Reactors Depending on Different Input Substrates

Martin Černý [1], Monika Vítězová [1], Tomáš Vítěz [2], Milan Bartoš [1] and Ivan Kushkevych [1,*]

[1] Department of Experimental Biology, Faculty of Science, Masaryk University, Kamenice 753/5, 62500 Brno, Czech Republic; cernyarchaea@mail.muni.cz (M.Č.); vitezova@sci.muni.cz (M.V.); bartos.milan@atlas.cz (M.B.)

[2] Department of Agricultural, Food, and Environmental Engineering, Faculty of AgriSciences, Mendel University, Brno, Zemědelska 1, 61300 Brno, Czech Republic; tomas.vitez@mendelu.cz

* Correspondence: ivan.kushkevych@gmail.com or kushkevych@mail.muni.cz; Tel.: +420-549-49-5315

Received: 17 September 2018; Accepted: 10 October 2018; Published: 23 November 2018

Abstract: With growing demand for clean and cheap energy resources, biogas production is emerging as an ideal solution, as it provides relatively cheap and clean energy, while also tackling the problematic production of excessive organic waste from crops and animal agriculture. Behind this process stands a variety of anaerobic microorganisms, which turn organic substrates into valuable biogas. The biogas itself is a mixture of gases, produced mostly as metabolic byproducts of the microorganisms, such as methane, hydrogen, or carbon dioxide. Hydrogen itself figures as a potent bio-fuel, however in many bioreactors it serves as the main substrate of methanogenesis, thus potentially limiting biogas yield. With help of modern sequencing techniques, we tried to evaluate the composition in eight bioreactors using different input materials, showing shifts in the microbial consortia depending on the substrate itself. In this paper, we provide insight on the occurrence of potentially harmful microorganisms such as *Clostridium novyi* and *Clostridium septicum*, as well as key genera in hydrogen production, such as *Clostridium stercorarium*, *Mobilitalea* sp., *Herbinix* sp., *Herbivorax* sp., and *Acetivibrio* sp.

Keywords: biogas; *Clostridiales*; hydrogen-producing bacteria; bioreactors; anaerobic fermentation; anaerobic digestion; microbial community composition

1. Introduction

Biogas production appears to be a brilliant solution to tackle common energetics, waste management, and environmental pollution problems. Production of biogas provides us with clean energy in a form of gas mixture, utilizing common organic wastes as an input substrate and thus helping to maintain the biomass cycle without the formation of unnecessary byproducts [1,2]. The elegance of biogas production lies in the complex microbial consortia capable of anaerobic fermentation, which ultimately leads to the final step, methanogenesis. Methanogenesis itself is a dominant metabolic pathway typical for the group of microorganisms referred to as methanogenic archaea. These microorganisms rely on the end-products of bacterial fermentation, such as short-chain fatty acids (SCFA), carbon dioxide, hydrogen, and even methyl–amines and methanol [3–6]. However, relatively little is known about the microbial consortia responsible for the production of methanogenesis precursors. The composition of the bioreactor ecosystem has a direct impact on the overall fitness of biogas production [7,8]. Unbalance in the microbial consortia may lead to the collapse of fermentative processes or contamination of biogas by undesired products, such as H_2S [9].

In order to keep biogas production sustainable for a long period of time, with high yields of gas, one must be able to identify key groups of microorganisms that help to maintain steady conditions and precursors for biogas production [3,10]. Many anaerobic microorganisms were identified as potential hydrogen producers, figuring in many natural and anthropogenic processes. This knowledge then helps not only to ensure methanogenesis occurs in bioreactors but is also exploited in terms of biohydrogen production [11–13]. Hydrogen itself is one of the many final products of bacterial fermentation and also serves as a proton that is used in various microbial enzymatic pathways. In connection with SCFAs, it is also associated with interspecies hydrogen-electron transfer, which further serves as a connection among fermenting bacterial taxa [14–16]. In anaerobic digesters and bioreactors, the microbial community is greatly dependent on the inoculum and the origin of used substrate. These factors have a substantial effect on the behaviour of the microbial community and yields of the biogas plant [7,8]. With crop waste and animal manure being used as an input substrate, a great portion of the microbial community is connected to soil and gastrointestinal tract consortia. Understanding the diversity of primary hydrogen producers can foretell if a correlation exists between production, environment, and the microbial community, and if this has some effect on biogas yield. Bearing this knowledge in mind, the *Clostridiales* order was chosen as the desired microbial group. Many species belonging to this clade are considered potent hydrogen producers. These microorganisms create a substantial portion of the Gram-positive microbial community in mesophilic bioreactors [17,18]. Results of this study illustrate the variability in the composition of the *Clostridiales* order depending on the different input substrates.

2. Materials and Methods

Eight biogas plants throughout the Czech Republic were chosen based on known input material with information about the substrate provided as the ratio of substrate components (*w*/*w* %). Names of the plants, as well as information about substrate components, are summarized in Table 1. Fermenters used in these biogas plants had the operational volume of 2500–3500 m^3. Sampling was carried out anoxically and material was directly transferred from the fermenter into sterile sampling vessels. Afterwards, the samples were stored in a thermo-isolating box and were transported for immediate analysis to our laboratory.

Table 1. Characteristics of evaluated anaerobic digesters (Reproduced with permission of Kushkevych [6]).

Geographic Location	Main Substrate	Input Ratio	Temperature (°C)	pH	ORP (mV)
Modřice	primary sludge, biological sludge	50:50	34	7	−3.1
Bratčice	maize silage, whole crop silage, poultry litter	63:31:6	43	8.3	−75
Pánov	maize silage, poultry litter	92:8	49	8	−58
Úvalno	maize silage, sugar beet pulp, whole crop silage, cattle manure	44:44:6:6	48	7.69	−38.5
Horní Benešov	maize silage, sugar beet pulp, whole crop silage, cattle manure, grass silage	29:39:12:15:5	49	7.85	−47.4
Rusín	maize silage, sugar beet pulp	70:30	48	7.63	−34.7
Loděnice	maize silage, sugar beet pulp	75:25	44	7.65	−36
Čejč	pig manure, maize silage	75:25	–	–	–

Prior to DNA analysis, chemical and physical parameters were determined. The pH, redox potential, and temperature were measured, as well as total solids content, volatile solids content, and biogas composition. Measurements were conducted for each anaerobic digester of a biogas plant with data summarized in Table 1. For pH and redox, potential measurement was used pH/Cond meter 3320 (WTW GmbH, Dinslaken, Germany) in accordance with standard procedures [19]. The sample temperature was assessed by high accuracy PT100 RTD thermometer HH804U (OMEGA Engineering,

Stamford, CT 06907-0047, USA). Total solids (TS) content was determined by drying at 105 ± 5 °C, using EcoCELL 111 (BMT Medical Technology Ltd., Brno, Czech Republic) according to Czech Standard Method [20]. After drying, samples were cooled in a desiccator, until they reached constant weight and the value could be determined. Volatile solids content (VS) was assessed by combustion of the samples in a muffle furnace LMH 11/12 (LAC, Ltd., Rajhrad, Czech Republic) at 550 ± 5 °C according to Czech Standard Method [21].

According to our previous studies, we used the QIAamp Fast DNA Stool Mini Kit (QIAGEN GmbH, Hilden, Germany). This was appropriate for a fast and easy purification process of total DNA from fresh or frozen samples and was used for DNA extraction from samples of anaerobic fermenters in previous studies [9]. DNA extractions were conducted according to manufacturer protocol, only with minor adjustments, which we describe below. We extracted 100 mg of each sample, which we mixed with 1.4 mL of ASL buffer (QIAGEN GmbH, Hilden, Germany) and incubated at 95 °C for 10 min. After centrifugation, we added InhibitEX tablets to the supernatant, according to protocol, to remove impurities and possible PCR inhibitors. After another centrifugation, 200 µL of supernatant was mixed with 15 µL of proteinase K solution and 200 µL of buffer AL (QIAGEN GmbH, Hilden, Germany). Then, the solution was incubated at 70 °C for 10 min, cooled down and 200 µL of ethanol (96–100%) was added to the mixture. Following the procedure, the supernatant was centrifuged through the QIAamp kit column and was washed twice by buffers AW1 and AW2 (QIAGEN GmbH, Hilden, Germany). Finally, the DNA elution was conducted by 200 µL of elution buffer.

For amplification of the desired product, V3 and V4 variable regions of the 16S rRNA gene were targeted by universal primers [22]. Primers were marked for sample identification by molecular barcoding. According to our previous research, the Maxima™ Probe qPCR Master Mix (Thermo Fisher Scientific, Waltham, MA, USA) was used for PCR reaction. Initial denaturation was conducted at 95 °C for 10 min, followed by 30 cycles of incubation at 94 °C for 30 s, 60 °C for 30 s, and 72 °C for 120 s, with a final extension step at 72 °C with a 2 min duration. Visualization of PCR products was performed using 1.5% agarose gel. After the reaction, DNA was extracted from the gel using the QIAquick Gel Extraction Kit (Qiagen GmbH, Hilden, Germany). DNA was quantified using the Quant-iTPicoGreen dsDNA Assay (Thermo Fisher Scientific, Waltham, MA, USA) and equimolar amounts of PCR products were pooled together. Paired-end amplicons were sequenced via Illumina Mi-Seq platform. Data analysis of 16S rRNA sequences was carried out using QIIME data analysis package [23].

According to base quality score distributions, average base content per reading and the guanosine-cytosine pairs (GC) distribution in the reads, quality filtering on raw sequences was conducted. Chimeras and reads that did not cluster with other sequences were removed. The obtained sequences with quality scores higher than 20 were shortened to the same length of 350 bp and classified with RDP Seqmatch with an operational taxonomic unit (OTU) discrimination level set to 97%. The relative abundance of taxonomic groups was calculated for microorganisms detected in this study. Sequences were compared using the BLAST feature of the National Center for Biotechnology Information (NCBI) [24]. The sequences were uploaded to Geneious (Geneious 7.1.9) for comparative genomic analyses [25]. Alignments of sequences were performed in Geneious 7.1.9 using Muscle (Clustal W) with the BLOSUM cost matrix and clustering was performed by the neighbor-joining method [26].

Sequences of selected microorganisms were then deposited in GenBank under accession numbers: MH045949, MH045950, MH045951, MH045952, MH045953, MH045954, MH045955, MH045956, MH045957, MH045958, MH045959, MH045960, MH045961, MH045962, MH045963, MH045964, MH045965, MH045966, MH045967, MH045968, MH045969, MH045970, MH045971, MH045972, MH045973, MH045974, MH045975, MH045976, MH045977. Origin7.0 software (www.origin-lab.com) was used for further processing data and analysis of the obtained results.

3. Results

3.1. Commenting on the Control

After the analysis of raw data, the most abundant genera were investigated thoroughly. First, the overall abundance of the Clostridiales order was compared against the total microbial background (Figure 1A). The mean ratio was 5.84% of the whole microbial background, with a minimum of 0.64% in the case of the Modřice bioreactor, which serves as a control, and a maximum of 9.51% in the case of the Čejč bioreactor. This discrepancy is even more visible on the graph visualizing the ratios of detected OTUs in the manner of the whole set of raw data (Figure 1B). Variation in the structure of the microbial community of the Modřice digester is probably a consequence of a significant difference in input substrate. The input material, in this case, consists of biological and waste sludge, which has a slightly less alkaline pH and a higher redox potential compared to other reactors, with crop silage or manure being used as the main input substrate. Thus, the explanation probably lies in the different substrate, its properties, and microbial load [27,28]. Biological sludge is a complex microbial ecosystem, comprised of many bacterial, archaeal, and eukaryotic taxa such as bacteria *Acinetobacter* sp., *Flavobacterium* sp., *Cloacibacterium* sp., *Desulfovibrio* sp., *Desulfomicrobium* sp. or archaea such as *Methanobrevibacter* sp. and *Methanosarcina* sp. [8,29–32]. Microorganisms coming from this environment may favour these conditions, leading to a significant shift in the *Clostridiales* or total bacterial ratio due to different conditions (Figure 1A).

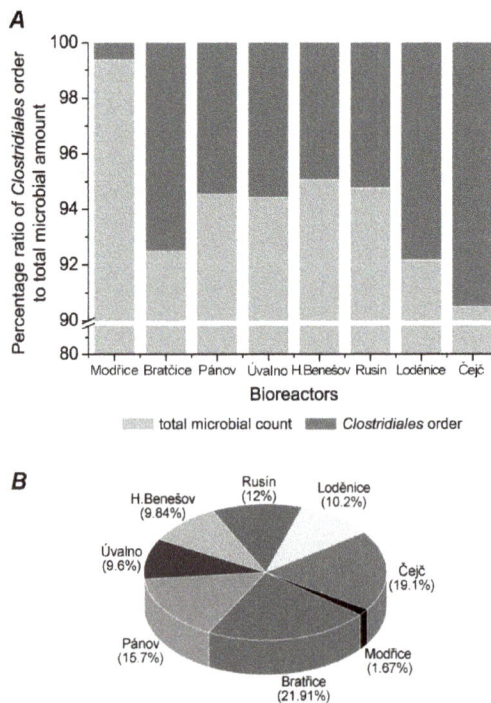

Figure 1. Ratio of Clostridiales representative composition: ratio between *Clostridiales* order and total microbial amount (**A**); pie-graph indicating portions of each biogas plant to *Clostridiales* order ratio (**B**).

3.2. Comparison of the Clostridiales Order Communities

Comparing the rest of the bioreactors, it seems for the Figure 1A that there is a shift from the *Clostridiales* order microorganisms towards other bacterial species. The lower ratios detected in Pánov, Horní Benešov, Úvalno, and Rusín (Figure 1B) may be connected to the optimal growth temperature and temperature threshold for many mesophilic microorganisms. Many *Clostridiales* order organisms manifest poor growth above 44 °C [33]. Judging by this, it is probable that higher temperatures may lead to changes in the taxa, moving from *Clostridiales* candidates to more thermophilic organisms from different taxonomic groups. Analysis of the most abundant genera revealed domination of the *Clostridiales* group by the genus *Clostridium*. The mean value for the *Clostridium* genus is 58.98%. In addition to Figure 1, Figure 2 shows that in the case of extreme results, the microbial consortium in the Čejč bioreactor was dominated by this genus with the ratio of 89.44%. In the case of the Čejč bioreactor, the *Clostridium* genus was dominated by the *Clostridium novyi* species, belonging to cluster I, which comprised 65.63% of all OTUs retrieved from this bioreactor sample, deposited in GenBank under accession number MH045969.

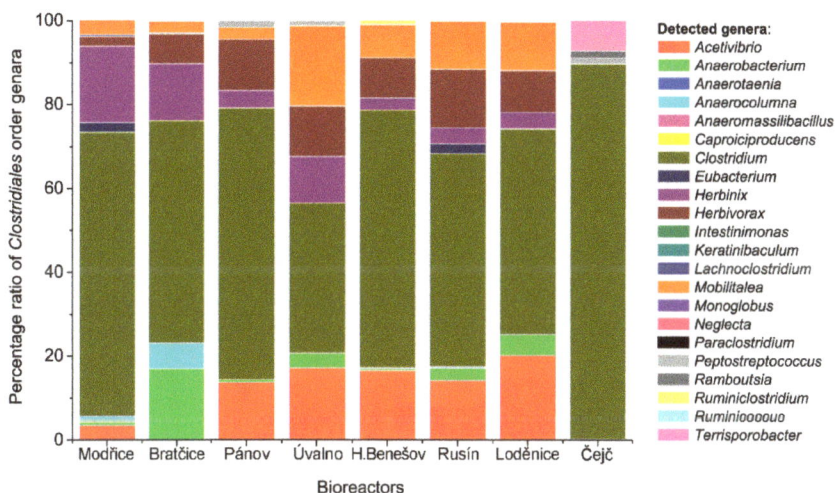

Figure 2. Visualization of the genera composition in each biogas plant and mean of its abundance to total microbial count.

Analysis by BLAST gives 99% similarity to the sequence with GenBank accession number LC193834.1 (Figure 3). *C. novyi* is characterized as 0.6–1.4 × 1.6–17 µm long gram-positive rods, which usually occur singly or in pairs with good gas production. Among the common substrates belong carbohydrates such as raffinose, glucose, melibiose, and cellobiose, which in turn produce products such as acetate, propionate, isobutyrate, butyrate, isovalerate, and more importantly hydrogen, and CO_2. *C. novyi* is a sporulating rod with sub-terminally to terminally localized spores, which may cause swelling of the cell. Optimal growth temperatures range from 30–45 °C, with an alkaline pH threshold of 8.5. However, *C. novyi* is regarded as a soil-dwelling microorganism and a known pathogen of animals and even humans. It is the first time that *C. novyi* was detected in such a high ratio compared to other data. *C. novyi* was formerly detected in bioreactors only in low abundances as described by Fröschle et al., 2015. Its potential to pose a threat to biogas plant operators remains unclear [33–36]. Another notable species is *Clostridium septicum* from cluster I and *Clostridium stercorarium* belonging to cluster II. As in the case of *C. novyi*, *C. septicum* may also be considered a human pathogen. The optimal temperature is 44 °C. However, its growth is inhibited by temperatures above 46 °C. It forms straight to curved cells that are motile. This species creates oval spores, located sub-terminally within the cell, which may swell if present. It is considered a strong gas producer [33].

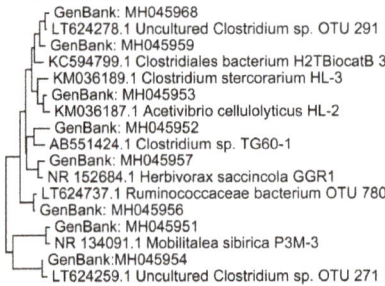

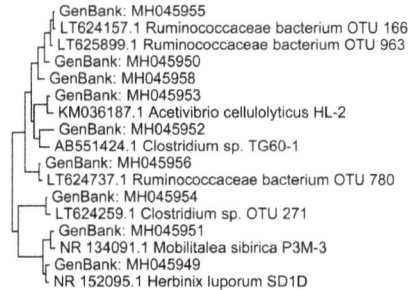

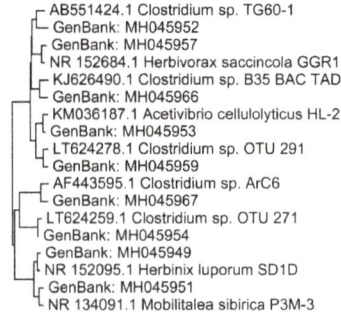

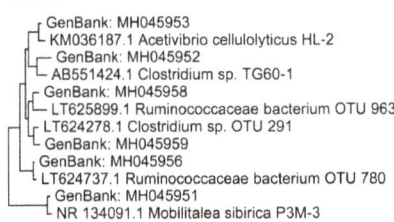

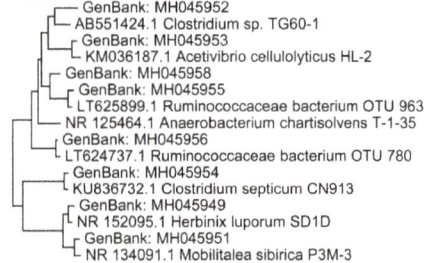

Figure 3. Phylogenetic relationship of detected species in terms of their source.

C. stercorarium is a representative of thermophilic clostridia. Its temperature optimum lies around 65 °C and under given conditions forms relatively small rods 0.3–0.4 × 2–4 μm and produces oval terminal spores. This species is able to ferment cellulose, cellobiose, melibiose, raffinose, fructose, and sucrose in turn for the production of acetate, lactate, ethanol, hydrogen, and CO_2. A possible explanation for the occurrence of pathogenic clostridia may be their connection with the gastrointestinal tract of farm animals and soil. In the case of *C. stercorarium*, it naturally occurs in compost, thus probably promoting anaerobic digestion of silage in the bioreactor [18]. Other notable species detected in this study are *C. clariflavum*, *C. propionicum*, *C. neopropionicum*, and *C. sulfidigenes* (Figure 3). The importance of these species lies mainly in the formation of hydrogen and SCFAs, which are connected with methanogenesis and linked to interspecies hydrogen transfer. Their metabolism ensures the thermodynamic stability of anaerobic fermentation, providing a favorable environment for deeper and fine digestion of the given substrate [33].

Apart from the genus *Clostridium*, minor clades such as *Acetivibrio* sp. (10.64%), *Herbinix* sp. (8.26%), *Herbivorax* sp. (8.38%), and *Mobilitalea* sp. (8.40%) were detected in the bioreactors, comprising 10–20% of all retrieved OTUs (Figures 2 and 3). *Acetivibrio cellulolyticus* was the sole member of this genus detected in this study, which was similar from 98% to GenBank sequence NR_025917.1 and was deposited under accession number MH045953 (Figure 3). Its morphology consists of slightly curved gram-negative rods with a single flagellum having dimensions of 0.5–0.8 × 4–10 μm. Major substrates consisted of cellulose, cellobiose, and salicin, producing acetate, hydrogen, and CO_2 in return [36]. Similar to *Acetivibrio*, the genus *Herbinix* is represented by the sole species *H. luporum* (NR_152095.1), with a sequence similarity of 99% (MH045949). This thermophilic cellulose-degrading bacterium can be found in bioreactors, where it is connected with plant biomass degradation. *H. luporum* represents a novel genus in the family *Lachnospiraceae*. It is a non-motile rod, growing to 2.0–6.0 × 0.5 μm, with an optimal growth temperature of 55 °C and main fermentation products such as ethanol, acetate, butyrate, and hydrogen [37]. *Herbivorax saccinicola* is a key player in plant biomass remineralization during anaerobic digestion processes and a member of the family *Ruminococcaceae*. These non-motile long and thin rods with dimensions of 5–10 × 0.2 μm grow between 45–65 °C. However, the growth of this microorganism was observed even in lower temperatures (40–44 °C), one may deduce that temperature specificity is not so strict. The main fermentation products are acetate, ethanol, and hydrogen [38]. This organism (NR_152684.1) was similar to the sequence deposited in GenBank under the accession number MH045957 (Figure 3). *Mobilitalea sibirica* is the last of the moderately abundant members of curved rod-shaped anaerobic bacteria with the main products of fermentation being ethanol, acetate, hydrogen, and CO_2 [39]. Our sequence (MH045951) possessed 99% similarity with type strain P3M-3 sequence under accession number NR_134091.1 (Figure 3). Although many genera were detected in this study, the remaining clades created only a negligible minority which abundance ranged mostly in order of tenths of percentages compared to other taxa (Figure 2). Thus, we didn´t evaluated their occurrence in a deeper context.

4. Discussion

As for methanogenesis and methanogenic archaea, understanding fine arrangements in microbial community relationships are key to maintaining steady and profitable biogas production. As it may seem from some studies, the amount of hydrogen may be misjudged as a smaller portion of gas production. However, it is essential for CO_2 to be reduced into methane by methanogenic archaea and naturally turned into biogas. The smaller portion of produced hydrogen also serves as a mediator in interspecies hydrogen transfer, resulting in a high level of well-being of the anaerobic microbial community [3,4,6]. Our data suggests that the composition of anaerobic fermenters depends on the input substrate and may vary greatly because of the given condition. The impact of these findings remains to be evaluated, also whether this knowledge can be exploited to the artificially potent microenvironment, as well as to design inoculum, and knowing possible impacts of using different input substrates. The stable and diverse microbiota of the bioreactor may prevent colonization and succession of undesired microorganisms such as sulfate reducing bacteria. Their metabolic products may damage not only the microbial consortia by strong competition for hydrogen, originally utilized by methanogens, but may also lead to corrosion of the bioreactor itself [40–42]. Additionally, this study confirmed the presence of potentially harmful microorganism such as *C. septicum* and *C. novyi*, which are connected to myonecrosis and other dangerous illnesses, this raises questions about aseptic and sanitation protocols when operating bioreactor processes. Their presence agrees with previous findings, however, we detected *C. novyi* in a much higher ratio, comprising 65.63% of the whole *Clostridiales* microbiota in the bioreactor for the first time [35]. Apart from the genus *Clostridium*, families *Lachnospiraceae* and *Ruminococcaceae*, also have their place in the anaerobic consortia. These microorganisms have similar physiological and biochemical properties as the *Clostridium* genus, thus also significantly contributing to the methane production process. On the contrary, the input substrate has a direct impact on the microbial community, as well as the well-being of the reactor. It is not only modulating accessibility to essential nutrients, but also bears an additional microbial load with which native microbes must compete for resources.

Author Contributions: M.Č. and I.K. conceived and contributed to the writing of the paper. Together with, M.V., T.V., and M.B. who further designed and performed the experiments and measurements described in this paper. M.Č. together with T.V. analyzed the data and optimized process parameters. T.V. performed analysis of the data and validation of the results.

Funding: This research received no external funding.

Conflicts of Interest: The authors declare no conflicts of interest.

References

1. Hobson, P.N. Biogas production from agricultural wastes. *Experientia* **1982**, *38*, 206–209. [CrossRef]
2. Papurello, D.; Silvestri, S.; Tomasi, L.; Belcari, I.; Biasioli, F.; Santarelli, M. Biowaste for SOFCs. *Energy Procedia* **2016**, *101*, 424–431. [CrossRef]
3. Oppermann, R.A.; Nelson, W.O.; Brown, R.E. In vivo studies of methanogenesis in the bovine rumen: Dissimilation of acetate. *J. Gen. Microbiol.* **1961**, *25*, 103–111. [CrossRef] [PubMed]
4. Miller, T.L.; Wolin, M.J. *Methanosphaera stadtmaniae* gen. Nov., sp. nov.: A species that forms methane by reducing methanol with hydrogen. *Arch. Microbiol.* **1985**, *141*, 116–122. [CrossRef] [PubMed]
5. Papurello, D.; Soukoulis, C.; Schuhfried, E.; Cappellin, L.; Gasperi, F.; Silvestri, S.; Santarelli, M.; Biasioli, F. Monitoring of volatile compound emissions during dry anaerobic digestion of the Organic Fraction of Municipal Solid Waste by Proton Transfer Reaction Time-of-Flight Mass Spectrometry. *Bioresour. Technol.* **2012**, *126*, 254–265. [CrossRef] [PubMed]
6. Dridi, B.; Fardeau, M.-L.; Ollivier, B.; Raoult, D.; Drancourt, M. *Methanomassiliicoccus luminyensis* gen. nov., sp. nov., a methanogenic archaeon isolated from human faeces. *Int. J. Syst. Evol. Microbiol.* **2012**, *62*, 1902–1907. [CrossRef] [PubMed]
7. Kushkevych, I.; Vítězová, M.; Vítěz, T.; Kováč, J.; Kaucká, P.; Jesionek, W.; Bartoš, M.; Barton, L. A new combination of substrates: Biogas production and diversity of the methanogenic microorganisms. *Open Life Sci.* **2018**, *13*, 119–128. [CrossRef]
8. Kushkevych, I.; Vítězová, M.; Vítěz, T.; Bartoš, M. Production of biogas: Relationship between methanogenic and sulfate-reducing microorganisms. *Open Life Sci.* **2017**, *12*, 82–91. [CrossRef]
9. Kushkevych, I.; Kováč, J.; Vítězová, M.; Vítěz, T.; Bartoš, M. The diversity of sulfate-reducing bacteria in the seven bioreactors. *Arch. Microbiol.* **2018**. [CrossRef] [PubMed]
10. Nath, K.; Das, D. Improvement of fermentative hydrogen production: Various approaches. *Appl. Microbiol. Biotechnol.* **2004**, *65*, 520–529. [CrossRef] [PubMed]
11. Hawkes, F.R.; Dinsdale, R.; Hawkes, D.L.; Hussy, I. Sustainable fermentative hydrogen production: Challenges for process optimisation. *Int. J. Hydrogen Energy.* **2002**, *27*, 1339–1347. [CrossRef]
12. Levin, D.B.; Pitt, L.; Love, M. Biohydrogen production: Prospects and limitations to practical application. *Int. J. Hydrogen Energy.* **2004**, *29*, 173–185. [CrossRef]

13. Nanqi, R.; Wanqian, G.; Bingfeng, L.; Guangli, C.; Jie, D. Biological hydrogen production by dark fermentation: Challenges and prospects towards scaled-up production. *Curr. Opin. Biotechnol.* **2011**, *22*, 365–370. [CrossRef]

14. Baek, G.; Kim, J.; Kim, J.; Lee, C. Role and potential of direct interspecies electron transfer in anaerobic digestion. *Energies* **2018**, *11*, 107. [CrossRef]

15. Ozturk, S.S.; Palsson, B.O.; Thiele, J.H. Control of interspecies electron transfer flow during anaerobic digestion: Dynamic diffusion reaction models for hydrogen gas transfer in microbial flocs. *Biotechnol. Bioeng.* **1989**, *33*, 745–757. [CrossRef] [PubMed]

16. Cord-Ruwisch, R.; Seitz, H.-J.; Conrad, R. The capacity of hydrogenotrophic anaerobic bacteria to compete for traces of hydrogen depends on the redox potential of the terminal electron acceptor. *Arch. Microbiol.* **1988**, *149*, 350–357. [CrossRef]

17. Kong, X.; Yu, S.; Fang, W.; Liu, J.; Li, H. Enhancing syntrophic associations among *Clostridium butyricum*, *Syntrophomonas* and two types of methanogen by zero valent iron in an anaerobic assay with high organic loading. *Bioresour. Technol.* **2018**, *257*, 181–191. [CrossRef] [PubMed]

18. Fang, H.H.P.; Zhang, T.; Liu, H. Microbial diversity of a mesophilic hydrogen-producing sludge. *Appl. Microbiol. Biotechnol.* **2002**, *58*, 112–118. [CrossRef] [PubMed]

19. *CSN EN 12176, Characterization of Sludge—Determination of pH-Value*; Czech Standards Institute: Prague, Czech Republic, 1999.

20. *CSN EN 14346, Characterization of Waste—Calculation of Dry Matter by Determination of Dry Residue or Water Content*; Czech Standards Institute: Prague, Czech Republic, 2007.

21. *CSN EN 15169, Characterization of Waste—Determination of Loss on Ignition in Waste, Sludge and Sediments*; Czech Standards Institute: Prague, Czech Republic, 2007.

22. Nossa, C.W.; Oberdorf, W.E.; Yang, L.; Aas, J.A.; Paster, B.J.; DeSantis, T.Z.; Brodie, E.L.; Malamud, D.; Poles, M.A.; Pei, Z. Design of 16S rRNA gene primers for 454 pyrosequencing of the human foregut microbiome. *World. J. Gastroenterol.* **2010**, *16*, 4135–4144. [CrossRef] [PubMed]

23. Caporaso, J.G.; Kuczynski, J.; Stombaugh, J.; Bittinger, K.; Bushman, F.D.; Costello, E.K.; Fierer, N.; Peña, A.G.; Goodrich, J.K.; Gordon, J.I.; et al. QIIME allows analysis of high-throughput community sequencing data. *Nat. Methods* **2010**, *7*, 335–336. [CrossRef] [PubMed]

24. Altschul, S.F.; Gish, W.; Mille, W.; Myers, E.W.; Lipman, D.J. Basic local alignment search tool. *J. Mol. Biol.* **1990**, *215*, 403–410. [CrossRef]

25. Kearse, M.; Moir, R.; Wilson, A.; Stones-Havas, S.; Cheung, M.; Sturrock, S. Geneious Basic: An integrated and extendable desktop software platform for the organization and analysis of sequence data. *Bioinformatics* **2012**, *28*, 1647–1649. [CrossRef] [PubMed]

26. Larkin, M.A.; Blackshields, G.; Brown, N.P.; Chenna, R.; McGettigan, P.A.; McWilliam, H. Clustal W and Clustal X version 2.0. *Bioinformatics* **2007**, *23*, 2947–2948. [CrossRef] [PubMed]

27. Venkiteshwaran, K.; Bocher, B.; Maki, J.; Zitomer, D. Relating anaerobic digestion microbial community and process function. *Microbiol. Insights* **2015**, *8*, 37–44. [CrossRef] [PubMed]

28. Moestedt, J.; Nilsson Påledal, S.; Schnürer, A. The effect of substrate and operational parameters on the abundance of sulphate-reducing bacteria in industrial anaerobic biogas digesters. *Bioresour. Technol.* **2013**, *132*, 327–332. [CrossRef] [PubMed]

29. Shchegolkova, N.M.; Krasnov, G.S.; Belova, A.A.; Dmitriev, A.A.; Kharitonov, S.L.; Klimina, K.M.; Melnikova, N.V.; Kudryavtseva, A.V. Microbial community structure of activated sludge in treatment plants with different wastewater compositions. *Front. Microbiol.* **2016**, *7*, 90. [CrossRef] [PubMed]

30. Kushkevych, I.V. Activity and kinetic properties of phosphotransacetylase from intestinal sulfate-reducing bacteria. *Acta Biochim. Polonica* **2015**, *62*, 103–108. [CrossRef]

31. Kushkevych, I.; Kollar, P.; Suchy, P.; Parak, K.; Pauk, K.; Imramovsky, A. Activity of selected salicylamides against intestinal sulfate-reducing bacteria. *Neuroendocrinol. Lett.* **2015**, *36*, 106–113. [PubMed]

32. Kushkevych, I.; Vítězová, M.; Kos, J.; Kollár, P.; Jampílek, J. Effect of selected 8-hydroxyquinoline-2-carboxanilides on viability and sulfate metabolism of *Desulfovibrio piger*. *J. Appl. Biomed.* **2018**, *16*, 241–246. [CrossRef]

33. Rainey, F.A. Clostridiales. In *Bergey's Manual of Systematics of Archaea and Bacteria*; John Wiley & Sons, Inc.: Hoboken, NJ, USA, 2015; pp. 1–5. [CrossRef]

34. Kim, M.D.; Song, M.; Jo, M.; Shin, S.G.; Khim, J.H.; Hwang, S. Growth condition and bacterial community for maximum hydrolysis of suspended organic materials in anaerobic digestion of food waste-recycling wastewater. *Appl. Microbiol. Biotechnol.* **2010**, *85*, 1611–1618. [CrossRef] [PubMed]

35. Fröschle, B.; Messelhäusser, U.; Höller, C.; Lebuhn, M. Fate of *Clostridium botulinum* and incidence of pathogenic clostridia in biogas processes. *J. Appl. Microbiol.* **2015**, *119*, 936–947. [CrossRef] [PubMed]

36. Patel, G.B.; Khan, A.W.; Agnew, B.J.; Colvin, J.R. Isolation and characterization of an anaerobic, cellulolytic microorganism, *Acetivibrio cellulolyticus* gen. nov., sp. nov. *Int. J. Syst. Bacteriol.* **1980**, *30*, 179–185. [CrossRef]

37. Koeck, D.E.; Hahnke, S.; Zverlov, V.V. *Herbinix luporum* sp. nov., a thermophilic cellulose-degrading bacterium isolated from thermophilic biogas reactor. *Int. J. Syst. Evol. Microbiol.* **2016**, *66*, 4132–4137. [CrossRef] [PubMed]

38. Koeck, D.E.; Mechelke, M.; Zverlov, V.V.; Liebl, W.; Schwarz, W.H. *Herbivorax saccincola* gen. nov., sp. nov., a cellulolytic, anaerobic, thermophilic bacterium isolated via *in sacco* enrichments from a lab-scale biogas reactor. *Int. J. Syst. Evol. Microbiol.* **2016**, *66*, 4458–4463. [CrossRef] [PubMed]

39. Podosokorskaya, O.A.; Bonch-Osmolovskaya, E.A.; Beskorovaynyy, A.V.; Toshchakov, S.V.; Kolganova, T.V.; Kublanov, I.V. *Mobilitalea sibirica* gen. nov., sp. nov., a halotolerant polysaccharide-degrading bacterium. *Int. J. Syst. Evol. Microbiol.* **2014**, *64*, 2657–2661. [CrossRef] [PubMed]

40. Kushkevych, I.V. Kinetic Properties of Pyruvate Ferredoxin Oxidoreductase of Intestinal Sulfate-Reducing Bacteria *Desulfovibrio piger* Vib-7 and *Desulfomicrobium* sp. Rod-9. *Polish J. Microbiol.* **2015**, *64*, 107–114.

41. Kushkevych, I.; Fafula, R.; Parak, T.; Bartos, M. Activity of Na$^+$/K$^+$-activated Mg^{2+}-dependent ATP hydrolase in the cell-free extracts of the sulfate-reducing bacteria *Desulfovibrio piger* Vib-7 and *Desulfomicrobium* sp. Rod-9. *Acta Vet Brno* **2015**, *84*, 3–12. [CrossRef]

42. Kushkevych, I.; Vítězová, M.; Fedrová, M.; Vochyanová, Z.; Paráková, L.; Hošek, J. Kinetic properties of growth of intestinal sulphate-reducing bacteria isolated from healthy mice and mice with ulcerative colitis. *Acta Vet Brno* **2017**, *86*, 405–411. [CrossRef]

MDPI

St. Alban-Anlage 66

4052 Basel

Switzerland

Tel. +41 61 683 77 34

Fax +41 61 302 89 18

www.mdpi.com

Energies Editorial Office

E-mail: energies@mdpi.com

www.mdpi.com/journal/energies

www.ingramcontent.com/pod-product-compliance
Lightning Source LLC
Chambersburg PA
CBHW051705210326
41597CB00032B/5372